THE VERTICAL BUILDING STRUCTURE

THE VERTICAL BUILDING STRUCTURE

Wolfgang Schueller

Professor of Architecture
Virginia Polytechnic Institute
and State University

VNR VAN NOSTRAND REINHOLD
New York

For information about our audio products, write us at:
Newbridge Book Clubs, 3000 Cindel Drive, Delran, NJ 08370

To my wife, Ria

Copyright © 1990 by Van Nostrand Reinhold

Library of Congress Catalog Card Number 89-27498
ISBN 0-442-23910-6

Printed in the United States of America

Van Nostrand Reinhold
115 Fifth Avenue
New York, New York 10003

Chapman & Hall
2-6 Boundary Row
London SE1 8HN, England

Thomas Nelson Australia
102 Dodds Street
South Melbourne, Victoria 3205, Australia

Nelson Canada
1120 Birchmount Road
Scarborough, Ontario M1K 5G4, Canada

16 15 14 13 12 11 10 9 8 7 6 5 4 3 2

Library of Congress Cataloging-in-Publication Data

Schueller, Wolfgang, 1934–
   The vertical building structure / Wolfgang Schueller.
      p.   cm.
   Includes bibliographic references.
   ISBN 0-442-23910-6
   1. Tall buildings—Design and construction.   2. Structural design.
I. Title
TH845.S374   1990
721—dc20   89-27498
                                                            CIP

CONTENTS

PREFACE

A wealth of vertical building structure types are presented in an ordered manner and comparative fashion, according to form and structural behavior. They encompass types from the massive building block to the slender tower, from structures above or below ground or in deep water to structures in outer space. They range from simple symmetrical to complex asymmetrical forms, from boxes to terraced buildings, from low-rise to high-rise buildings, from ordinary bearing wall, skeleton, and core constructions to bridge buildings, cellular clusters, suspension buildings, tubes, superframes, and the new breed of compound hybrid skyscraper forms.

In order to develop an understanding for the building structure as a system that supports and, as a pattern, orders space, it must be seen in the context of the building's anatomy in addition to other design determinants. Therefore, the buildings are also studied from a geometrical, aesthetic, historical, functional, environmental, and construction point of view. Through this approach, the various design specialists can learn to appreciate each other's concerns, and thus develop a basis for successful team work. At the same time, the treatment of structures is broadened and enriched by relating it to the traditionally separate fields of construction, structural analysis and design, materials, history, geometry, and graphics. Nearly equal emphasis has been placed on the descriptive, analytical, and graphical investigations of the topics. The visual portion of the book is essential to the education of the building designer; the graphical analyses of hundreds of images attempt to support this point.

Only elementary mathematics is employed to break down some of the complex structural problems, and to develop a feeling for structural behavior through approximate structural design methods, as well as the capability for checking computer output with hand calculations.

Although the computer is a necessary tool, it may also become institutionalized to the point where its results may not be doubted. The complete reliance of the designer on answers from purchased software may result in a loss of feeling for the overall behavior of complete structures, and thus the designer's capability for independent checking. It can only be hoped that the approach used in this book will aid the designer in further developing a sense of structural behavior and knowledge about structures, as well as to bridge the gap between the various fields of building design and construction, thus helping to reduce the risk of building failures. This book attempts to overcome the current tendency of architects and engineers to be isolated from each other. It establishes a position for structures in architecture which constitutes a complementary role to the structural engineer's traditional view.

This book grew out of a need to make the field of structures a part of the domain of building, and accessible to building designers in general, rather than relying on separate literature in the fields of structural and mechanical engineering, architecture, and construction. Structural engineering books usually place their emphasis on solving given problems where the physical reality often is lost in mathematical abstractions, so that the field of structures becomes nearly inaccessible to other building professionals. Similarly, architects use broad conceptual interpretations, developed during the 1960s, that result in generalizations only partially relevant today, and that do not respond to the explosion of knowledge which has taken place since then.

The writing of this book was also motivated by the current methods of teaching the field of structures in colleges. Graduating civil engineers are familiar and comfortable with computers—they know how to input information and to get solutions, but they do not understand the overall structural behavior of the building, and thus may use an abstract model that imperfectly simulates the reality of its structure. Young engineers have not learned to draw, or how a building is physically constructed, so naturally, they tend to overanalyze and underdetail the structure. On the other hand, architectural students in the design studios are primarily concerned with presentation and appearance, independent of the building as an organism and how it is made. They are familiarized with technology in lecture courses that are separate from design. These courses introduce information about basic concepts only, and do not keep up with the explosion of technical knowledge that has occurred in recent years.

Furthermore, there is a lack of communication, on the academic level, between structure, construction, and building design. In courses on structures, the structural element, together with its loading, is usually already isolated from its context, so that the primary emphasis can be placed on the various analytical methods of solution. In contrast, this book emphasizes the setting up of problems based on actual buildings— loads, member shapes, and boundary conditions are derived from the layout of a structure, from construction methods, and from detailing and appearance requirements. The student may have to resolve a complex continuous system into several basic ones, so that the structural elements can be evaluated and proportioned quickly. By deriving the mathematical and behavioral concepts of the field of mechanics from an actual building, the student can better understand the purpose and logic of structure, and be helped to perceive how it functions and affects the building form. But most importantly, this experience teaches students to see mathematics as an efficient means of communication rather than an end in itself. Through this approach, the student will develop a sense for the behavior of structures, and thus will gain confidence in dealing with the concept of structure.

The descriptions of the numerous building cases should help students to expand their ability to communicate about structure, as well as to relate abstract principles to physical reality; they should understand that structure does not just happen on a mathematical level. The building cases were selected solely to exemplify structural concepts and to develop a feeling for structure and form, rather than to support specific architectural styles or structural acrobatics. Design concepts are investigated in this book visually, through a morphological approach in which specific characteristics in existing buildings are studied, abstracted, and presented as simple line drawings or diagrams, in order to discover the logic of form. The presentation of actual buildings will help designers to explore the three-dimensional order of structure, as well as to visualize the force flow and spatial interaction of structural elements. Architects will be able to perceive the order of structure, and learn to control the interplay of material and nonmaterial spaces during the design stage. Similarly, structural engineers will be able to translate the structural behavior of building elements into abstract images with which they are already familiar.

Not all of the illustrations are of analytical nature or precisely explained in accordance with engineering tradition. Some of the drawings are not simple and clearly defined, on the contrary, they are complex and presented in such a manner as to deliberately be open to interpretation. They tell a story, forcing the reader to participate in the design through the drawing; there is no intention to define the limits (i.e., to solve the problem) by explaining exactly what is portrayed in the figures. These types of presentations stand by themselves, only connected with the respective text in its overall spirit. They project beyond the scope of the written descriptions, to transfer a feeling to the designer that there is much more than the limits set by the statements. This approach relates to visual thinking, a form of art, rather than to graphical analysis— the illustrations are supposed to initiate curiosity with respect to the sources of building design, in addition to providing aesthetic pleasure.

One of the most important purposes of this book is to interpret the structural behavior of the elements within various building forms by using a minimum of mathematics, while estimating the preliminary member sizes with reasonable accuracy. The mathematics is deliberately kept at a basic level so that the primary emphasis on behavioral aspects is not hidden behind complex analytical processes. Throughout the book, however, the reasons for any of the simplifications are explained—the process of design is never reduced to merely plugging data into ready-made formulas. The discipline of thought, as established by the engineering sciences over a long period of time, is an important aspect of the designer's education.

Though there are many handbooks available for the sizing of beams and columns, it is the intention here to develop some feeling for structural behavior, and some self-confidence in being able to quickly proportion members, and thus being in full control of the design without having to depend completely on ready-made solutions, as provided by handbooks and computer software. Simple equations have been developed that make the quick estimation of member proportions possible. This is especially important at various times during design and construction. Examples of their use would be when a sense for the order of magnitude of forces and stresses must be developed at the early design stage, when different structure schemes are compared and evaluated, when member sizes must be known for aesthetic purposes during the developmental stage of architectural design, when a fast judgment of sizes is needed for checking drawings and computer outputs, or for the designer to sense on the construction site when member sizes don't seem right, or finally, when first trial sections may be needed for the solution of indeterminate structures.

The student of structures will never truly comprehend the complexity of structural behavior just by reading descriptive material and/or listening to fascinating lectures. He or she must actually solve problems in order to really learn the subject matter and to discover what is not understood. To support this goal, emphasis is placed on analytical exercises; more than 200 problems are given, nearly all of which are solved in the *Exercise Manual*, which is available from the author (Prof. W. Schueller, College of Architecture and Urban Studies, Virginia Tech, Blacksburg, VA 24061; Tel. 703-231-7735). Eventually, the student will have developed that certain feeling which brings the building structure alive, because suddenly he perceives himself as being the structure, and thus experiencing the pain of stress concentrations and distortions of his bones.

It should be noted here that the many problems provided at the ends of the chapters serve not only as practice exercises designed to promote better comprehension of the material, but also, in a number of instances, to further develop and clarify concepts or even, in a few cases, to introduce new ideas not covered in the text. Occasionally, reference is made to problems that go beyond the information given in the text. The reader wishing to study the material in greater depth, whether student or practicing professional, will find the available *Exercise Manual* especially useful.

This text does not replace books on the *precise* analysis and design of structural components or on construction. It is assumed that the reader has taken elementary courses in the fields of statics, as well as in the design of steel and reinforced concrete. Some of these important basic topics are, however, reviewed in Chapter 4.

Basic concepts of design are introduced in Chapter 1 to provide the reader with a general understanding of structures as a part of buildings. Important historical events, architectural considerations, the effect of building context, functional organizers, mechanical and electrical systems, and safety considerations are all included. In Chapter 2, general structure principles, as applied to the building as a whole rather than to its components, such as beams and columns, are presented, thereby laying the foundation for the following chapters. Basic principles of construction and foundation systems are also discussed. The complexity of load actions such as that due to gravity, dynamic forces, wind, earthquakes, temperature, and other hidden loads (including energy dissipation systems), structural integrity, and redundancy are treated in Chapter 3. Chapter 4 deals with approximate design methods for reinforced concrete, steel, and prestressed concrete members. The distribution of lateral forces to the resisting vertical structures, including torsion, is studied in Chapter 5. Also presented are floor structure systems, which are seen as diaphragms and gravity distribution systems. Various types of reinforced concrete slabs are approximately designed. In Chapter 6, wall systems are covered, including curtain walls, partition walls, and bearing walls. Approximate design methods for masonry and reinforced masonry are introduced—several masonry wall buildings are designed. A wealth of building structure systems are presented in Chapter 7, in an organized, comprehensive fashion. The topics include masonry, reinforced concrete, steel, composite, and mixed construction; the structures range from small buildings to skyscrapers and megastructures. Numerous typical building structures are designed approximately. Finally, in Chapter 8, some other large structures, such as towers, offshore architecture, underground, and outer space structures are introduced.

This book can be used not only as a text for courses in building structures and design engineering, but also as a reference for design studios and classes in construction. The book will be extremely helpful to the young engineer or architect who is faced with the reality of a building for the first time. The comparative presentation of the many building cases, often given in an historical context, together with the references, should be an asset to the practicing architectural and structural designers during the preliminary design stage.

The subjects in this book are obviously dealt with only on an introductory level. However, it is hoped that the process presented here will not only familiarize the designer with the wealth of vertical building structures, but will also develop critical thinking, initiate enough curiosity for further studies, and strengthen the creative response to the design and construction of buildings. This book should provide another bridge for the communication and understanding of the various professionals involved with the making of buildings.

ACKNOWLEDGMENTS

My sincere gratitude to the dedicated group of students of the Department of Architecture at Virginia Polytechnic Institute and State University, who have, under my guidance, diligently developed in my studio many of the illustrations in this book. The students who have been involved more substantially over a period of several years, and to whom I am deeply indebted are: Mark Bittle, Louis Conway, Mario Cortes, Lisa Davis, Luciano Galletta, William Galloway, Jennifer Gillespie, Gerry Gutierrez, Mark Jones, Charles Matta, Mark Pruitt, Eka Rohardjo, George Russel, Andrea Varner, Mitzi Vernon, Robert Wright, as well as Bryan Berthold and Ron Weston of Syracuse University.

Also greatly appreciated is the help of my graduate assistants at Virginia Tech: Mario Cortes and Eka Rohardjo did a superb job in drawing several of the illustrations. I thank David Johnson, Thomas Kostelecky, John Moench, Peter Paoli, and Daniel Shear for their commitment in preparing many of the graphical presentations. I am grateful to Glen Oakley, P.E., for his thorough proofreading of my manuscript and the valuable suggestions he has offered. Finally, my thanks go to Kurt Hedrick for editing the *Exercise Manual* to this book. The positive attitude of all of my assistants has been of stimulating support to me.

I also wish to thank those other students and assistants whose names I have not mentioned, but who have also been involved in the preparation of drawings and proofreading, and who have supported me, through their critical and constructive thinking, in the writing of this book.

My sincere appreciation to Brenda Lester Hale for her patience and cooperation over many years in preparing the manuscript using *GML script* word processing on Virginia Tech's mainframe.

This book would not have been possible without the contributions of the many architects and engineers whose design of buildings, or whose mathematical interpretation of structural behavior, has provided a basis for this text. These individuals are too numerous to identify here, but they are given credit in the references and the list of illustrations.

I must pay a special tribute to the eminent engineer and educator Lynn S. Beedle, Director of the *Council on Tall Buildings and Urban Habitat*; the council was born at Lehigh University under his direction in 1969. The numerous publications of this organization have been most helpful in writing this book.

Finally, I wish to thank the publisher's editorial and production staff for their sincere support.

1 GENERAL CONCEPTS OF BUILDING DESIGN

The design of a building evolves out of a complex interactive process. The many form determinants range from the effect of the environmental context, be it cultural or physical, to the building organism itself, which must function properly. To develop a clear understanding of the *building as structure*, which is the primary concern of this book, some of the form determinants for buildings, in general, are briefly introduced in this chapter. They are:

- The evolution of the high-rise building, considering technology and style
- Architectural design theories, as well as geometry and morphology
- The environmental context
- The building function, including activities and supportive mechanical systems, as well as building safety

THE DEVELOPMENT OF THE VERTICAL BUILDING STRUCTURE

Before the nineteenth century, tall buildings were terraced temple mountains, pyramids, amphitheaters, fortresses, city halls, temples, mosques, cathedrals, towers of various types, and so on. Their design was generally motivated by political or religious reasons—these structures were symbols of power and faith. Other multistory buildings were three- to four-story residential urban housing, possibly in conjunction with commercial use at their base. Tall buildings were mostly masonry structures, although the Romans had built with concrete as early as the second century B.C. and brought its use to a high level of sophistication as demonstrated by the famous Pantheon in Rome (c. 123 A.D.). Concrete was usually not exposed on the outside of buildings, but rather, a brick veneer was used as a permanent formwork.

In the ancient urban centers of Babylon, Athens, Byzantium (i.e., Constantinople), etc., four-story apartment buildings constructed from mud-brick with timber floors were quite common. The Romans built ten-story tenement buildings, mainly of wood, although the building height was later, under Emperor Augustus, limited to 70 ft to reduce the risk of fire. Often, concrete was substituted for the timber floors by using floor vaulting and concrete stairs. In the following centuries, through the Middle Ages, the common materials for wall construction in Europe were stone, brick, timber, and timber framing with masonry infill. Other structural elements used in combination with these materials were arches, vaulting, and the post-beam concept. Floors were supported by timber beams or by relatively deep and heavy masonry vaults. It was not until the nineteenth century that, first, the lighter shallow vaults with hollow bricks or clay tiles, and later, reinforced concrete slabs, were introduced as floor structures.

The traditional method of high-rise construction is still found today in such places as southern Saudi Arabia, and North and South Yemen. Here, for centuries, the Arabs have been building cities of towers, where houses reach up to 100 ft in height, or eight stories. In many regions, these tall tower houses are made entirely from unbaked sun-dried clay brick.

In the following sections, the development of modern high-rise construction is discussed. First, iron and steel are introduced, as well as the other technical components necessary for modern tall buildings, which eventually gave rise to the skyscrapers of the 1920s and 30s. Later, reinforced concrete will be considered.

Technical Considerations

It was not until the turn of the nineteenth century that the metal skeleton began to slowly replace heavy masonry construction in multistory buildings. The traditional tall masonry buildings were massive gravity structures, where the walls were perceived to act independently—their action was not seen as a part of the entire three-dimensional building form. Since loads increase with the building volume and height, a limit to height was rapidly reached, because the resisting surface area at the base could not become reasonably larger, owing to the fact that brick masonry could only support about 200 psi, in contrast to a steel column with a supporting capacity of 16,000 psi. Thick, massive walls were the result; their heavy weight and the inflexibility in plan layout made masonry construction inefficient for multistory applications. This becomes apparent in the last great bearing-wall brick structure of that era: the original north half of the 16-story Monadnock Building [1891] in Chicago (Fig. 1.1e). In this building, based on cage construction, the walls had to be more than 6 ft thick at the base, as derived from empirical rules of design. The designers braced this long narrow building with masonry shear walls—the interior skeleton consists of cast iron columns and steel beams. The solid brickwork occupies nearly one-fifth of the building area at the lowest story!

The lack of strength and the inflexibility of plan layout of masonry bearing-wall construction clearly point toward the metal frame as a necessary component for the evolution of the skyscraper. Although iron and steel have been known for thousands of years, it was not until the eighteenth century, in England, that the iron industry developed methods for the mass production of, initially, cast iron, and then wrought iron, members. It took the metallurgists of the early years a long time to develop volume methods of separating iron from its ore, of how to get a high enough temperature for melting large quantities of iron, and of how to get the metal into its most useful condition of strength and hardness. Finally, in 1855, the Englishman Henry Bessemer introduced the first large-volume process for producing steel.

The iron frame is often treated as a symbol of the Industrial Revolution, which was accompanied by urbanization and rapid population growth. During this period, new building types were born, ranging from railway stations, factories, large warehouses, market halls, and exhibition enclosures to mass-produced housing. During this age of invention and industrialization, not only were mass-production methods and new approaches to construction developed, but also new structure systems and building forms. Large spans required the invention of new structure types or the further development of existing ones. This is demonstrated by the arches used for railroad stations; the girders, trusses, and suspension spans for bridges; and the single-story long-span portal or gable frames for exhibition halls. Multistory skeleton construction must surely have been influenced by the balloon wood frame used for residential buildings, invented by Deodat Taylor in Chicago [1833]. The advances in technology would not have been possible without the growth of science—the structural theory developed during the eighteenth and nineteenth centuries, mainly in France, where the first technical university was founded in Paris in 1747.

As convincing landmarks for the expansion of structure concepts and the inventive minds of engineers, with respect to immense enclosures, large spans of roofs and bridges, and great heights during the nineteenth century, the following structures may be noted:

- Paxton's Crystal Palace of 1851 in London
- Roebling's 1595-ft Brooklyn Bridge of 1883 in New York
- Eiffel's 984-ft high Tower of 1889 in Paris

In comparison to the evolution of long-span roof and bridge structures or single-story frames, which were all designed by engineers, the development of skyscrapers, for which architects were responsible, took much more time, because of the involvement of many other factors. Not only support structures, but also fireproofing methods and elevators, as well as mechanical, electrical, and plumbing systems, had to be developed. In addition to these technical considerations, architects were very much concerned with aesthetics, particularly related to the preservation of the past and what image a skyscraper should have within an urban context.

The beginning of multistory construction goes back to the English mills and warehouses around the turn of the nineteenth century, when full interior cast-iron framing, often mixed with masonry, was used to support the floors. Later, cast-iron columns were also placed along the building perimeter, but were hidden behind self-supporting stone or brick facade walls. During the mid-1840s in the United States, wrought-iron beams began to replace the low-tensile-capacity and brittle cast-iron beams. The typical multistory frame construction that began to slowly evolve from this consisted of an interior skeleton of cast-iron columns and wrought-iron beams carrying the weight of the floors, while the exterior masonry-bearing walls and piers provided lateral support to the building. This so-called *cage construction* represents the first step towards true *skeleton construction*, or the birth of the skyscraper.

The transition from the exterior load-bearing wall to the frame is reflected by cast-iron facades such as found in Liverpool, Glasgow, and the warehouses on the river front in St. Louis. Most influential is the work of the design inventors Daniel Badger and James Bogardus, who established foundries in New York in the 1840s. They manufactured two- to six-story iron skeleton buildings, including prefabricated iron facades which could easily be shipped to any location and quickly assembled on-site.

In the six-story Harper and Brothers Printing House in New York (built in 1854, demolished in 1920), by John B. Coliers and James Bogardus in Venetian Renaissance style, the slender iron facade framework, together with the glass, expressed the lightness and strength of the material. The interior girders were cast in the form of shallow arches and tied with wrought-iron rods; these built-up girders spanned between interior cast-iron columns and exterior bearing walls, they supported wrought-iron I-beams, which in turn carried the masonry brick arches upon which the floors rested. This structure is often considered the first fire-resistant building in New York.

The Saint-Ouen Dock Warehouse in France (built in 1865, demolished later), by Préfontaine and Fontaine, surely represented one of the more important steps towards the skyscraper. This, the first fireproof, all-skeleton multistory structure, consisted of cast-iron columns and wrought-iron girders that supported the shallow hollow-brick vaults resting on their lower flanges and topped with concrete. The facades clearly exposed the iron skeleton with brick infill. The Menier Chocolate Factory at Noisiel-sur-Marne near Paris [1872] by Jules Saulnier, together with the engineer A. Moisant, went even further, by employing lateral bracing of the framework and cantilever action. The four-story building, with no interior stiffening walls, rests on four massive piers. It is supported by parallel long facade walls consisting of deep cantilever box girders at the bottom, which in turn carry the wrought-iron facade frames, the network of diagonal wind bracing, and the hollow bricks used as filling. In a way, this exposed braced framework acts like a huge lattice cantilever truss, as found in bridge construction.

The technical evolution of the skyscraper in the second half of the nineteenth century was only made possible by the earlier independent development of its essential components. These included the following:

- *Structure:* all-skeleton metal construction, and the ability to provide lateral stability
- *Curtain wall:* separation of building support structure from its enclosure wall—enclosure becomes cladding
- *Safety:* fireproofing
- *Elevators*
- *Mechanical systems and sanitation:* plumbing, central heating, artificial lighting, ventilation

In parts of the United States, reliable plumbing systems, electric lighting systems, steam heating systems, and steam-operated power-driven fans for ventilation were available in the 1860s; air conditioners were introduced in the 1920s.

With the arrival of passenger elevators, the height of a building was no longer limited to five stories, the distance people would reasonably climb. Often this date is associated with Elisha Otis's presentation of his famous "safety elevator" at the 1854 New York exhibition. There, he convincingly demonstrated his invention by cutting the cables supporting the elevator platform but remaining safely suspended in midair. His first safety passenger elevator, driven by steam, was installed in the Haughwout Building, a New York City department store designed by Daniel Badger and John P. Gaynore in 1857, and is considered to have been the first in the United States The construction of this building is also often taken as the beginning of the *preskyscraper period.*

By the late 1870s, the steam passenger elevator had been replaced by the much faster hydraulic elevator, and finally, in the early 1890s, Otis had basically developed the modern electric elevator. In 1889, not only was Gustave Eiffel's 984-ft Eiffel Tower a sensation, but also the double-decked hydraulic lifts in the inlined legs, one of which was able to reach the Tower's top.

After 1857, further development toward the skyscraper entered an important new phase in 1870, with the Equitable Life Assurance Building in New York by Gilman and Kendall, together with the engineer G. B. Post. It is considered an early version of the skyscraper because it included all of the technical factors necessary, with exception of height and an all-frame construction (it still used the cage construction principle). The building, in the ornate French Mansardic style, had only five stories, but rose to a height of 130 ft, and for the first time a passenger elevator was used in an office building. This building is often considered to introduce the so-called *elevator building* or *proto-skyscraper*.

The failure of iron members in buildings in the disastrous Chicago fire of 1871 led to many new innovations in fire protection as a necessary requirement for the further development of the skyscraper. Builders began to cover iron beams and columns with tiles, and to use hollow tile floors; in other words, they protected the metal skeleton with more adequate heat-resistive insulation.

Finally, in 1885, the birth of the true skyscraper occurred with construction of the ten-story Home Life Insurance Company Building (Fig. 1.1b) in Chicago, designed by William Le Baron Jenney (the building was demolished in 1931). There were no bearing walls, the entire building weight being carried solely by the metal skeleton, consisting of round cast-iron columns filled with cement mortar, with wrought-iron I-beams for the first six floors, and steel beams for the remaining floors. The typical beams were on 5-ft centers, and supported the flat tile floor arches. The masonry facade, that is the piers and spandrels, were hung from the frame like a curtain—they were carried on shelf angles fastened to the spandrel beams.

The architects Burnham and Root developed the concept of the vertical shear wall in the 20-story Masonic Temple Building, in Chicago [1892]. Because of the height of the building, the tallest at its time, the facade frames were laterally braced with diagonals, thereby inventing the vertical truss principle.

The Evolution of the Skyscraper

The early development of the skyscraper occurred in Chicago, from about 1880 to 1900, where block- and slab-like building forms reached merely 20 stories. Then the soaring towers of New York introduced the true skyscraper, the symbol of American cities. Several of the important buildings in the following section are shown in Figure 1.1, emphasizing essential visual characteristics.

The First Skyscraper Period

With the construction of the Home Life Insurance Company Building, the first skyscraper period was introduced—it took place primarily in Chicago, and is associated directly with the *Chicago School*. Although this first phase of skyscraper evolution was very much involved with the development of new technology, architects also had to struggle to find an image, or new form language, expressing this large evolving building type in an urban context. This was not a simple undertaking. On the one hand, engineers were being creatively inventive, in response to the spirit of the broadening vistas of science and technology: they efficiently constructed bridges, railway stations, exhibition halls, and so on; a world of thought most clearly reflected, possibly, by the Crystal Palace. On the other hand, architects (according to their tradition) were inseparably bound to historical styles. They needed to find a way out of the dilemma of having to preserve the formal values of the past while, at the same time, having to respond to the completely new conditions of an explosive urban environment—that

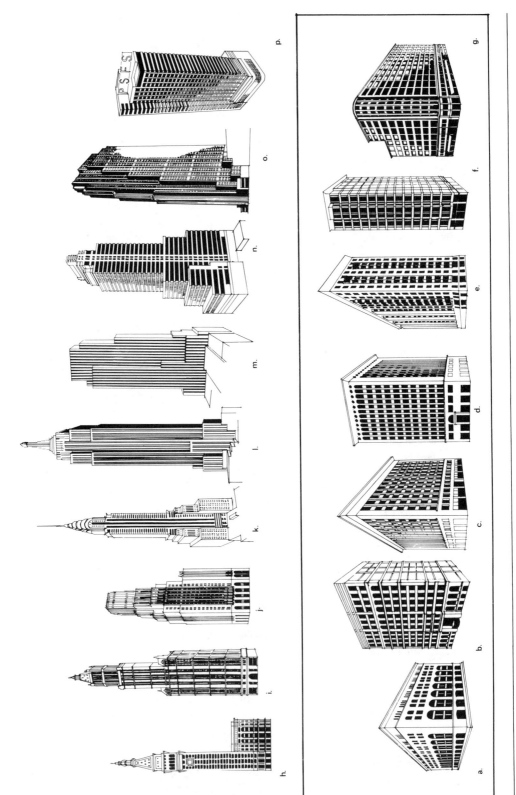

Figure 1.1. *The evolution of the skyscraper.*

is, the pioneering spirit of fast-growing Chicago, as well as new methods of building construction.

The Renaissance Palazzo style was a compositional feature prior to the first skyscraper phase. But as buildings grew taller, the palazzo organization became unsatisfactory and a new compositional solution had to be found. In Chicago, the first skyscraper period introduced flat roofs for its block- and slab-like skyscrapers, replacing the mansardic ones. In their search for an attractive compositional solution, designers experimented by varying the grouping of stories clothed in a variety of revival styles (e.g., Romanesque, classical, Queen Anne), often based on mathematical (geometric) progressions. They piled palazzi on top of each other, or they stretched the middle portion of the palazzo, thereby generating giant orders (e.g., romanesque arches and columns stretched vertically). The following models of design developed from each other: 1) An emphasis on horizontal division, by stacking palazzi on top of each other; 2) An extended palazzo-type or tripartite system of composition, using the metaphor of a column (i.e., base, shaft, capital), by stretching the middle zone; and 3) The *Chicago frame* or *Commercial Style* where the facade composition is not imposed but articulates the support structure and the nature of the skin.

The Home Life Insurance Company Building (see fig. 1.1b) represents the first model. The building facade did not express the skeleton-skin concept, but rather that of the traditional load-bearing masonry piers—it is organized in horizontal layers in a vaguely Romanesque revival style. The significance of the building lies in the technology, rather than in its aesthetics.

A masterpiece of architecture, with respect to the first model, is the Marshall Field Warehouse (Fig. 1.1a) in Chicago, by Henry Hobson Richardson [1887]. The building is modelled after the Romanesque precedent, but ordered in a manner similar to a Renaissance palazzo. The perfect proportion of the horizontal layers, together with the vertical progression of the windows from the heavy base to the top exposes a relaxed balanced consistency and simplicity, as well as a visual clarity. The nature, strength, and massiveness of the bearing walls are clearly expressed by the arches, which at the same time cause an upward thrust, thereby making the building appear taller.

Louis Sullivan, the great master of the Chicago School, felt deeply indebted to Richardson. Rather than its style, it was the visual order of the Marshall Field Warehouse, reflecting simplicity, austerity, and perfect composition, where no element can be removed without destroying the whole, which had a large influence on Sullivan and helped him to formulate the skyscraper shape. Sullivan was convinced that technology represented a meaningful component of design, although he believed it not sufficient by itself to assure good architecture. To him the ornamental system was inseparable from the building itself. He tried to generate some overall unity through a visual union of function and emotion. Sullivan kept on searching for this new architecture—he was not impressed by the architect ideologists, who were preoccupied with reviving historic styles. He believed that a well-balanced composition of the ideal skyscraper must express the *nature of the construction*, the *idea of height*, and the *spirit of industrial society*. Sullivan tested his ideas about the interaction of the art of composition, the nature of the construction, and the spirit of time in his two masterpieces, the Wainwright Building and the Guaranty Building, by employing the extended palazzo-type order.

In the Wainwright Building (Fig. 1.1c) in St. Louis [1891], Sullivan, together with his partner Dankmar Adler, used a fully developed steel frame, which is revealed along the exterior only by the wide column spacing at the base. Above, only every other column is structural—this false-column approach naturally disguises the nature of the steel frame. The height of the skyscraper is articulated by the verticality of the shaft, and its upward thrust is contained by the heavy horizontal cap in the form of a cornice; the subtle ornaments are to give a more organic quality to the facade.

Sullivan's greatest achievement is most probably the Guaranty Building (Fig. 1.1d) in Buffalo [1895]. Here, he does not as strongly express the tripartite subdivision, as derived from abstract stylistic considerations. It seems that the large load-bearing base structure carries the upward striving tower shaft, somehow predating the building on pilotis of the international style. The cornice not only terminates the thrust of the soaring shaft, but also seems to grow naturally out of the vertical elements.

The 16-story Monadnock Building (see fig. 1.1e) in Chicago [1891], designed by John Wellborn Root of Burnham and Root, is not only famous as the last great bearing-wall structure or proto-skyscraper, but also for its quality of design. It frankly expresses structure and function with no ornament or reference to historical styles. The extended-palazzo model is only vaguely indicated. The taper of the load-bearing brick wall toward the top, in response to diminishing weight, enhances the impression of strength, and the vertical upward thrust is terminated by the cornice. Although there is a powerful expression of vertical support and height, at the same time there is a plastic quality of skin predating the minimalism of about thirty years later in Europe—the visual difference to a skeleton building may not be at all obvious.

While the Wainwright, Guaranty, and Monadnock buildings were experimenting with a new architectural language that included the values of the past, the Tacoma Building in Chicago [1889], by Holabird and Roche, may have been the first to break away and introduce the so-called *Commercial Style*. It was composed of horizontal layers defined by the floor structures, rather than built on the palazzo model. It was the first building to express skin and an open facade, to allow natural light to penetrate the interior. The vertical upward movement caused by the outward projecting bays was balanced by the top story. This 13-story building (demolished in 1929) was the first to use riveted connections rather than bolts; also, its floating rafts, consisting of I-beams and reinforced concrete for the foundations, were revolutionary. The lightness and transparency of the facade is even further articulated in the 15-story Reliance Building (Fig. 1.1f) in Chicago [1894], designed by C. B. Atwood of Burnham and Company. The horizontal window bands are visually stressed, and are almost completely glass, largely concealing the load-bearing frame; the facade framing is faced with terra cotta. The transparency and pure proportions of the facade structure lead directly to the glass towers of the 1950s.

However, William Le Baron Jenney had already (in 1879) introduced, with the First Leiter Building in Chicago (demolished in 1972), a bold statement of the Chicago School philosophy; this building is often considered to be the first notable representation of the movement. Jenney abolished the exterior wall by letting the window openings reach from floor to ceiling and by articulating the slender character of the piers. Cage construction was used, where the exterior masonry piers, together with the wrought-iron spandrel girders, were kept separate from the interior iron skeleton. Although this seven-story structure was not a skyscraper, it was most influential from an aesthetic point of view, and already suggested the Chicago frame of later years.

Finally, a high point (but also a conclusion) of the Commercial Style at the end of the first skyscraper period was reached with Sullivan's last masterpiece in Chicago:

the Carson Pirie Scott Department Store (Fig. 1.1g) of 1904. The building exposes the static, neutral nature of the frame, clad with white terra cotta. The spans and heights of the skeleton bays were determined by functional requirements alone, and not by abstract rules of regularity and symmetry. The building is an impressive statement of modern architecture and underscores Sullivan's famous declaration that "form follows function."

The Second Skyscraper Period

When, in 1895, the 21-story American Surety Building (which is also believed to be the first free-standing tower building) topped the neogothic 20-story Masonic Temple Building, the race for height moved from Chicago to New York. Up to roughly the turn of this century, it was the state of technology that set limits to the growth of buildings—the many tall structures built in the short time before World War I prove that technical components were no longer the primary determinants in restricting building height.

Already, the 47-story Singer Building [1908], by Ernest Flagg, reached the soaring height of 614 ft. One year later, the 50-story Metropolitan Life Insurance Tower, by Napoleon LeBrun and Sons, topped it at 675 ft. Finally, the 57-story Woolworth Building [1913], by Cass Gilbert, at 792 ft in height, was to remain the world's tallest building for seventeen years before the 77-story Chrysler Building was completed. Only one year later, in 1931, the golden age of the American skyscraper was crowned with the 102-story Empire State Building, at the incredible height of 1250 ft.

The *second skyscraper period*, or the tower phase of skyscraper construction, occurred in New York. It can be organized according to tower shapes and styles as follows:

First phase: Freestanding towers and towers with base
   Classical model (taken from all periods)
   Gothic model
Second phase: Setback towers (c. 1925 to World War II)
   Art Deco

The concern of the Chicago School in expressing the function of the building was not shared by New York designers. They strongly believed in a separate ornamental system within the order of past styles (as represented primarily by the Classic and Gothic periods), attached to the steel frame and disguising it. While Sullivan tried to make ornament and building inseparable, the designers of the New York School composed the ornament through skilled academic exercises directly from eclectic forms, by concentrating on the overall image of the building. When buildings continued to grow taller, the palazzo model became outmoded, and designers were forced to search for new historical precedents. Tall towers had replaced the relatively low building blocks of the Chicago period. They thrust into the sky above other buildings and became symbols defining the skyline and the silhouette of the American city: the true skyscraper was born, connecting the sky to the earth. But the symbolic role of the skyscraper was much broader—it was also supposed to be beautiful and reflect an image of the power of American corporations.

The beginning of the second skyscraper period is marked by the Singer [1908] and Metropolitan Life Insurance (Fig. 1.1h) [1909] buildings, designed in the classical fashion. Here, for the first time, the tower model (as derived from historical precedents) was introduced in response to the skyscraper as a symbol to reach into the sky. While in the Singer Building the tower is laterally attached to the primary building mass, in the Metropolitan Life Building it grows out of the mass at the base and is designed after the medieval campanile of San Marco in Venice, using a classical treatment of the facade.

The Gothic style of the cathedrals was successfully introduced as a more appropriate model for skyscraper design with the Woolworth Building (Fig. 1.1i) in 1913, often called the Cathedral of Commerce. As the world's tallest building of its time, it was praised as the first true skyscraper, ideally expressing verticality and upward thrust, as well as the power of commerce in American life. The tower was not primarily just a symbol, as with previous cases, but provided a reasonable amount of floor area. The building consists of three parts: the massive 27-story base, the 30-story tower, and the spire. The central piers on the front facade rise from the bottom to the crown, and tie all of the parts together as a unit. The thick piers around the steel columns, together with the thinner piers forming the window mullions, strongly accentuate the uninterrupted verticality. Although the Gothic facade structure is decorative and does not articulate the order of the building support structure, one can easily imagine the extensive bracing that this tall tower building required, which included, besides rigid frames, knee braces, K-bracing, and diagonal bracing in the floor planes.

It became apparent how strong the image and influence of the Woolworth Building was when the Chicago Tribune competition of 1922 was won by Raymond Hood's Gothic tower solution. This 36-story building, completed in 1925, is often considered to represent the conclusion of the eclectic, or first, phase of the second skyscraper period.

The international Chicago Tribune Tower competition represented a most important architectural event. Not only did it identify the various positions in architecture, as related to high-rise buildings in an urban context, but it also stimulated the further development of skyscraper design. The entry by Walter Gropius and Adolf Meyer expressed the spirit of technology, or structure and function—in a way, an advanced stage of Sullivan's Carson Pirie Scott Department Store. This modernist view had already been taken further by Mies van der Rohe in his glass skyscraper projects of 1920. In contrast, Eliel Saarinen's entry (Fig. 1.1j), rather than being revolutionary and confrontational, worked within the spirit of American architecture and opened up new aesthetic possibilities. Saarinen did not use the exaggerated vertical organization of structural elements attached to the tower form, as in the Gothic style Woolworth Building, but rather, he let the massing of the building itself convey the upward movement, as opposed to the style. The stepped-back building, similar to a mountain, seems to grow out of the ground organically; the central tower shaft tends to thrust out of its enclosure mass or its thicker, solid base. Although the ornamental treatment is entirely vertical, and suppresses the horizontal floor structure, it is not Gothic—it is personal and artistic, expressing a more plastic character reminiscent of Art Nouveau, or is often described as Nordic Romanticism. For the contemporary critics, Saarinen's design was considered more successful than Hood's first prize, and it has been most influential on a great number of subsequent building designs.

Saarinen's design seems to have somehow predicted the effect of the revision of the New York building code in 1916, which attempted to control the impact of gigantic buildings upon the urban environment by defining the maximum building envelope for each lot, so as to protect the light and ventilation of adjacent sites. The new building form that the code produced was the *setback tower*. Hugh Ferris's dramatic and mystic drawings of 1922 envisioned the impact of the zoning laws, by letting buildings be sculptures which rose like mountains, thereby establishing the skyscraper as the basis for urban order, and also reflecting the optimistic vision of the 1920s of skyscraper cities as the symbol of the American spirit. Ferris's artwork, together with Saarinen's personal interpretation, were used as models or solutions for the setback skyscraper. This resulted in the arrangement and composition of the masses as a symbol becoming the most important design element. Upon the image of the building mass as a whole was superimposed a romantic utopian notion of the urban metropolis, as well as (most importantly) the modernist view of technology in the Machine Age. Art Deco evolved out of this spirit by integrating art, architecture, and a personal philosophy, by reflecting a sort of picturesque romanticism. The Art Deco style in skyscrapers forms a decorative envelope or facade architecture expressing the modernist spirit fused with an awareness of history—this includes the reflection or manipulation of other styles, although it resents revivalist dressing as dishonest and artistic fraud. The Art Deco skyscraper was primarily treated as an aesthetic structure with its clothing playing the primary role—the dressing developed its own composition and organic qualities independent of the interior organs of the building. The most prominent architectural events in the Art Deco skyscrapers occurred in the spires, pinnacles and crowns, as well as in the entrances and lobbies, where decoration was celebrated.

The Art Deco style is highlighted by William Van Alen's Chrysler Building in New York (Fig. 1.1k), and in particular in its stainless steel spire [1930]. Its facade architecture uses modern form language with a composition reminding us of industrial design concepts, perhaps symbolizing the building as a machine; a dressing that seems to reflect the flamboyant jazz age of Manhattan. The composition of the setback massing (consisting of a large base, enclosing slabs, tower, and spire, together with the conventional vertical piers and recessed spandrels at the center portion of the tower, but horizontally banded around the corners) truly built vertically up into the sky, proving that it was the tallest building in the world. The Chrysler Building is surely an advertisement, and represents one of the most powerful skyscraper images in America.

In contrast, the Empire State Building (Fig. 1.1l) by Shreve, Lamb, and Harmon [1931], uses purer forms. Soaring verticality, as achieved by the setback idiom and the continuous piers with recessed Art Deco aluminum spandrels, is clearly dominant. The upward thrust culminates in the spectacular top as a transition into the sky, celebrating the achievement of height as the tallest building in the world for many years. The building, however, does not externally express the complexity and effort of the organism in making this incredible height of 1250 ft possible, a feature that the Modern Movement is very much concerned with.

Toward the end of the 1920s, the more austere, linear geometric patterns of the modernists had become predominant. They expressed function, efficiency, and economy as ornament to glorify materialism and business, or the building as advertisement, thereby creating the basis for the great streamlined modernist skyscrapers of the early 1930s. Raymond M. Hood was the most inventive and brilliant representative of this period, due to his sense for composition, together with a clear understanding of pragmatic requirements.

In the Daily News Building (Fig. 1.1m) in New York [1930], Hood represents a modern abstraction of Hugh Ferris's romantic mountain-like skyscraper. This flat slab tower is stretched upward without interruption by the successive setbacks, and terminates in no crown. The facade is treated as a uniform series of continuous vertical heavy masonry strips, bare of any special architectural event, and completely subdues horizontality by recessing the spandrels. Hood used false columns between the real ones, just to make the skyscraper stand tall.

Quite in contrast is the McGraw-Hill Building (Fig. 1.1n) in New York [1931]. It is less pure in expression and not as massive. Here, Hood makes the building feel light by strongly supporting the horizontal dimension. The two facades, expressing the steel frame layout together with the horizontality, seem to have been inspired by the European modern movement, although these features are somewhat subdued by the Art Deco composition and the image of the tower and its crown.

Hood was largely responsible for the conceptual design of the 70-story RCA Building (Fig. 1.1o) of Rockefeller Center in Manhattan [1933], possibly his greatest achievement. Here, he molded the vertical slab masterfully with thin setbacks to transcend its character by resolving the flat mass into a series of floating slender slabs. The stepping is based not only on aesthetic, but also functional, considerations, as it occurs at floor levels where elevator shaftways terminate. The verticality of the facade composition is expressed by the uniform arrangement of the piers, where the 27-ft spacing of the load-bearing columns along the long faces is clearly articulated by the wider piers alternating with two narrower ones. The upward thrust is somehow softened by the horizontality of the cast aluminum spandrel panels, which are nearly flush with the piers.

The first pure skyscraper that rejected the shaped-tower configuration, and which clearly exposed its load-bearing columns and let them carry the floor levels, was the Philadelphia Saving Fund Society (PSFS) Building [1932] (Fig. 1.1p), by Howe and Lescaze. The architects produced a building style of distinction with a radically different look in this 39-story steel structure, which was built in Philadelphia rather than New York; it was also only the second skyscraper in the United States to be completely air-conditioned. The building has a T-shaped configuration in plan, where the narrow tower web is positioned asymmetrically on its base and attached to the elevator spine (flange) at the end. The design clearly grew out of the European modern movement: its massing discloses functional considerations and expresses the different internal functions of the vertical core and horizontal office layers, as well as the nature of the support structure, including the forceful cantilevering at the narrow face of the tower slab. It does not use the wall as enclosure and is not concerned with the romantic notion of image or expression of meaning. Architectural composition is not imposed upon the exterior as clothing, but evolves out of the interior building organism; it uses a most revolutionary interpretation of "modern" thinking, a Cubist vision supporting a balanced asymmetry. The PSFS Building marks the end of the second period of the American skyscraper. Although ahead of its time in the early 1930s, it foreshadows the high-rise buildings of the post-World War II period in the International Style.

The Third and Fourth Skyscraper Periods

The third skyscraper period is determined by modernism; in a way, it continues where the Commercial Style of the first skyscraper period in Chicago left off. High-rise buildings (as they were called then) had flat tops and were built all over the world—

they were built not just in steel, but also in reinforced concrete and masonry. In contrast to the emphasis on external dressing and the meaning of the building or historical styles, and by addressing the emotions more closely, as had been the case in the preceding period, emphasis had now clearly shifted to reason, using an approach derived from functional and technological facts: construction had become one of the prime motivations of the form-giving process. The organic qualities of the building were expressed, such as the bones and the transparent skin—a spirit reflected in Figure 1.2. Designers were innovative, inventive, and experimented with materials and building techniques to celebrate technology and function—technology was raised to the level of art. During this period, the engineering development of tall buildings was phenomenal. Efficient high-rise structure systems were developed that, by the early 1970s, resulted in the world's currently tallest buildings: first, the World Trade Center Towers in New York, and then the Sears Tower in Chicago; both of these topped the Empire State Building (see fig. 8.1).

The fourth and current skyscraper period occurs in conjunction with Post-Modernism and Late-Modernism, in the 1970s. Especially for Post-Modernism, Art Deco again became a strong source of inspiration and brought back the early skyscraper image. In contrast, structural exhibitionism (a form of late Modernism) expresses the space age spirit, reminding us of robotics and advanced space technology. The discussion of the last two skyscraper periods occurs throughout this book, and will therefore not be dealt with any further in the present context.

The Development of the Reinforced Concrete High-rise Building

All of the tall skyscrapers of the second period were constructed with steel skeletons. Reinforced concrete was used only for lower buildings, and was treated merely as a substitute for steel. It was not until the 1950s that reinforced concrete, as a material and structure, established its own identity in high-rise construction.

It is apparent that, in comparison to the rapid growth of steel in the construction of tall buildings, the progress in reinforced concrete was slow. Important milestones for the development of reinforced concrete were Joseph Aspdin's invention of Portland Cement in 1824, in England (an artificial cement made from a mixture of limestone and clay), as well as the patents on reinforced concrete construction by the early pioneers Joseph-Louis Lambot, François Coignet, and Joseph Monier in France, just after the middle of the nineteenth century. Other renowned contributors who should be mentioned were the British William B. Wilkinson and Thaddeus Hyatt, the German G. A. Wayss, the Frenchman François Hennebique, and the Americans William E. Ward and Ernest Leslie Ransome.

The Ward House in Port Chester, New York, which the mechanical engineer William E. Ward designed as his own residence in 1876, is often considered to be the first true all-reinforced concrete building in the United States; its floor construction was based on Coignet's patent. In the following years, to the turn of the century, there was a boom in concrete architecture and the science of reinforced concrete likewise developed rapidly. Prior to World War I, tall steel buildings used reinforced concrete primarily for foundations and occasionally for floor slabs.

Ernest Leslie Ransome is known not only for founding a factory for the manufacture of precast concrete blocks in Chicago in 1872, but also as being responsible for the derivation of the design principles for multistory reinforced concrete frame construction

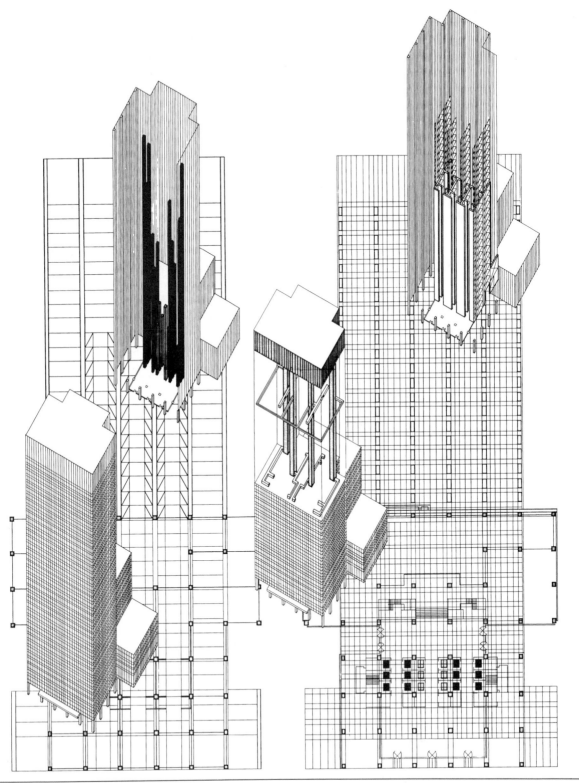

Figure 1.2. *The modern spirit.*

in the U.S. at around the turn of this century. François Hennebique did similar frame studies in France at about the same time. In 1902, the German engineer Emil Mörsch presented a final theory of reinforced concrete in his famous book *Der Eisenbetonbau*, which remained the standard method of design for many decades, until the ultimate strength design method replaced it about twenty years ago.

Finally, in 1903, the world's first reinforced concrete skyscraper was completed: the daring 16-story, 210-ft high Ingalls Building, by Elzner & Anderson in Cincinnati, designed under the Ransome system. At the same time, Auguste Perret and the engineer François Hennebique were the first to employ the reinforced concrete skeleton concept in high-rise construction in Europe. In the eight-story rue Franklin Apartment Building in Paris [1904], Perret exposed the concrete skeleton to unify structure and architecture. Besides Auguste Perret, Anatole de Baudot and Tony Garnier (all of France) were crusaders for concrete, trying to make it a material in its own right.

In the following decades, American designers concentrated primarily on steel and neglected concrete construction, with a few notable exceptions. Frank Lloyd Wright built the Unity Temple [1906] in Oak Park, Illinois, in reinforced concrete, and Albert Kahn designed his reinforced concrete factories for Packard, Ford, and Hudson in Detroit. In contrast, the European designers of the modern movement turned away from steel after World War I and got fully involved in the aesthetic use of concrete, as a material, for its own sake. Some of the brilliant pioneers of this period who explored the potential of concrete were the engineering designers Eugène Freyssinnet, Robert Maillart, Eduardo Torroja, and Pier Luigi Nervi, as well as the architects Le Corbusier, Dominikus Böhm, Erich Mendelsohn, and many others.

It was only after World War II that reinforced concrete buildings that exceeded 23 stories were constructed in the United States, although some taller concrete buildings had already been built in Argentina and Brazil. Before this time, concrete construction in high-rise buildings had been treated merely as a substitute for steel, by exactly imitating steel frames. This practice resulted in heavy members and especially large column sizes for the lower floors of tall buildings. It took some time for reinforced concrete to develop its own identity and new concepts. An important step toward establishing this special character was flat plate construction, together with shear walls, for the 13-story apartment buildings of the Clinton Hill Housing Project in Brooklyn during the mid-1940s, pioneered by Joseph DiStasio. These new design concepts of flat slab, shear wall, and later the load-bearing facade grid wall, began to slowly challenge the one-way slab on beams, braced frame, and curtain wall typical for steel-frame structures. With the Johnson Wax Tower [1950] at Racine, Wisconsin (see fig. 8.3), Frank Lloyd Wright became one of the first designers to break away from the traditional skeleton concept in high-rise construction. Although the 15-story building cannot be considered advanced in terms of height, Wright innovatively integrated the nature of concrete into the building design. He used the tree concept, in his urge toward the organic, by letting the mushroom-type floor slabs cantilever from the central core, which is rooted deep in the ground. With the 16-story Price Tower [1956] in Bartlesville, Oklahoma (see fig. 7.3j), Wright freely used the plastic quality of concrete and helped to even further identify the potential of the material. Finally, in 1962, the 60-story Marina Towers in Chicago (see fig. 5.7e) by Bertrand Goldberg, reached 588 ft. These may be considered the first true reinforced concrete skyscrapers expressing the character of the material. The race for height in concrete then began in 1968, with the 70-story, 645-ft high Lake Point Tower apartment building in Chicago, which utilized the flat plate–shear wall principle (see fig. 5.12d). The world's tallest concrete building is currently the 75-story 946-ft high 311 South Wacker Drive office tower in Chicago (see fig. 7.28b) which is to be completed in 1990; there are only 11 buildings taller.

Further development of the materials steel and reinforced concrete, as well as the recurrence of masonry, together with the introduction of precast concrete during the 1960s, and the use, later, of mixed and composite construction, including the evolution of compound free-form tower buildings, will be discussed in Chapters 6 and 7.

GENERAL ARCHITECTURAL DESIGN CONSIDERATIONS

Unfortunately, architecture in our culture cannot be simply defined by an all-encompassing universal theory or by formulas, as in the sciences, because architecture is also an art as well as a science; it translates abstract ideas into form. One side of architecture is objective and intellectual, while the other side is subjective, irrational, and emotional. The architecture, as an object, may be measured. However the architecture as an experience depends on the rather unpredictable response of the human being, out of his needs and views, which cannot be easily quantified or may be immeasurable. Further, external effects influence this experience of architecture; architecture depends on the situation (that is, values and attitudes), but at the same time, derives its order from culture, which requires an understanding of the unity of the past and the present, or in other words, the balance between tradition and innovation.

Early architectural theories by the Roman Vitruvius, and later in the Renaissance by Alberti and Palladio, can only be considered general in the context of their time and culture; in earlier periods, styles lasted for a prolonged length of time, reflecting the fact that architectural concepts were unified. In contrast, there are a multitude of architectural theories, expressed in a diversity of styles, in the twentieth century, partially because of the democratic structure of society in many countries and also because new building materials and construction methods make any form possible. For example, contemporary theories may be based primarily on functionalism, science, constructivism, nature, economy, behavioralism, philosophy (e.g., symbolism), and on art history. These individualistic and pluralistic theories of architecture may contradict each other, thereby reflecting the fact that life is dominated by conflicts or an apparent disorder (i.e., entropy); it is this disorder, however, which is the source of the continuous search for truth, which in turn is order. But all of these theories more or less include Vitruvius's basic principles of: *commodity* or functional serviceability, *firmness* or strength, safety, and durability of materials to assure permanence, *economy*, and *delight* or appealing appearance. The classic concept of beauty, as expressed by a well-balanced compositional equilibrium and visual integrity of the whole, has been transmitted through the ages from Vitruvius, Palladio, and Ledoux to Mies van der Rohe and Le Corbusier.

In this text, with the intention of building and developing a sense of constructibility, only the fundamental principles necessary for designing a building are investigated; hence, only a part of the ordered, rational, measurable dimension in architecture, and not the intuitive unmeasurable one, will be described. It is not the purpose here to define the aesthetic qualities of architecture, or the various styles. Architectural styles are seen only as a geometrical composition, as derived from formal rule systems or grammars of form. Any style can be developed from a geometrical point of view, although the significant dimensions of meaning, or its symbolic contents as the outcome of ideas and the necessary source of the design, will be ignored. It is, therefore, irrelevant if a compositional system is determined by either aesthetic criteria, as derived from idealistic formal orders, or as a response to structure and function, and the spirit of time. The building is seen in this context as a composite organism controlled by specific form determinants. These basic design organizers attempt to explain architectural space in terms of how it is defined, how it functions, how it is made, and its compositional characteristics. The primary form determinants are:

> Formal space:
>> the building as a formal object
>> the building as an idea
>> the building as an assembly of spaces
> Functional space:
>> the building as an organism
>> the building as activity
>> the building as a functioning system
>> the building as enclosure or climatic control
> Material space:
>> the building as material
>> the building as a support structure
>> the building as an assembly of components
>> the building as a construction process

Any of these form-giving determinants may have a predominant position in the design. Underground, hydro-, and aerospace environments are clearly controlled by function and material, for example. Other significant dimensions, for instance architecture as a "place," or as exciting, or its symbolism as a carrier of expression and images as derived from its context, these essential and complex dimensions of design are beyond the scope of this discussion, as previously mentioned.

The definitions of the *formal, functional,* and *material spaces* are approached only from the point of view of actually making architecture that is architecture as a built, three-dimensional reality. They are described by the following basic design organizers, which are introduced in the first three chapters:

> Geometry
> Environmental context
> Function
> Mechanical systems
> Structure
> Construction
> Safety considerations

The design concepts are investigated in this book visually, through a morphological approach in which specific characteristics of existing buildings are studied, abstracted, and presented as simple line drawings or diagrams, in order to discover the logic of form. That is, for purposes of understanding, the building form may be dissected and described by force diagrams representing the *physical*, *philosophical*, and *psychological* design determinants. Morphology is a branch of biology that deals with the form and structure of animals and plants. It is also a branch of linguistics that is concerned with the internal structure and form of words, which, together with syntax, is a basic division of grammar. The morphological technique is used here to display the logic of the building or detail form as reflected by geometrical and other organizational features. The concern here is for how to build, which is a syntactic dimension similar to structuralism, where meaning is revealed through the process of establishing order and relationships, rather than through image. An attempt is not made to search for the unity, essence, or totality in a building, nor to comprehend its full complexity or contradiction. It is the meaning through syntax, rather than through image and semantics, that is of primary concern in this analysis. This approach, however, is not treated simply as a classification of ideas similar to a pattern book, but as a creative process and powerful tool for learning.

It is also not the purpose of this investigation to present processes of architectural design or a model of how to design; typology is seen solely as a functional device, and not as a low-level theory of architecture. The design process may consist, on the one hand, of assembling and integrating the various design requirements by using some kind of method (e.g., *analysis-synthesis model*). On the other hand, it may consist of the more personal, subjective approach, that of dealing initially with the whole (a gestalt or parti) by conceptualizing the diversity of requirements through the use of images, metaphors, analogies, signs, symbols, models, and allegories, and then proceeding through successive design cycles by integrating the other form determinants (e.g., *conjecture-analysis model*). But, whatever the design method, the fundamental design determinants must be understood.

Traditionally, building technology and the building sciences have been taught by presenting various parts, for example, the site, foundations, floors, walls, roofs, materials, doors and windows, environmental control, mechanical systems, function, finish work, etc., as separated from the building as a unit. By concentrating on knowledge and on the information about the various parts, technology tends to be isolated from the whole that is also the concern of design process. The morphological approach attempts to supplement this learning experience by deriving the parts from the whole, so that the physical and organizational components can be discovered as design fundamentals, as the interacting parts of a system, and thus as an integral element of the total design process. In addition, by using appropriate building cases as models for learning about technology, current and possibly innovative construction methods can be discussed. The reader will develop self-confidence and a sense for the physical reality of the building.

It may be concluded that the strength of the morphological technique lies in the discovery of the design fundamentals and specific geometric forms as properties of the whole building. Through this deductive approach to a typological study, form models or prototypes are developed, and a vast number of variables are organized to establish a basis for visual and behavioral ordering principles. But it must be emphasized that this does not mean that the organizational diagrams will automatically lead to good design, and it must be kept in mind that the whole is usually more than simply the sum of its parts.

GEOMETRICAL ORGANIZERS

Geometry is the language of form, it is the study of spatial order; it does not distinguish between architects, engineers, artists, scientists, and others. Drawing is a media for expressing geometry; it not only provides the constructive basis for making buildings, but also a means for the formulation and articulation of architecture in general. The drawing can blend the design and the constructing of a building into a single process. But it must be emphasized that geometry not only provides a quantitative measure of Euclidean space, but also unveils an awareness of pattern and relationship, as well as a sense of scale and proportion. As early as the time of ancient Greece, Plato considered geometry and number a minimal form of expression, and therefore an ideal philosophical language.

The following typological studies, collecting various kinds of building shapes, plans, sections, and elevations, are seen as part of a theoretical tool or method for generating contrast, and hence a sense for formal order. This approach serves as a process of thought to clarify interrelationships, and it aids in the perception of formal configurations, as generated through permutations and combinations of geometrical elements. But the approach should not be treated as just a classification device or a catalog for product design. The method is limited solely to the study of individual and isolated geometric properties. Obviously, the richness of architectural form and the significant dimension of the whole cannot be defined simply by the abstracted, oversimplified line drawings.

The complexity of building forms can only be suggested in Figure 1.3, as described by plan, side, front, and spatial views, by realizing that this exploration has been based solely on a repetitive rectangular spatial network, reflecting the organic quality of a high-density urban architecture. The formal variations include single and cluster houses, free-standing and merging buildings, terraced and inverted stepped buildings, open and closed shapes, and so on. In contrast, the richness and limitless formal potential of individual buildings is suggested, for instance, by the study of unconventional architecture (see fig. 7.41). The purpose of the following study is to conquer the spatial geometries of high-rise buildings, to define and reconstruct the given form, and to expose geometric ordering principles.

The basic elements of geometry, or its *elementary dimensions*, are the point, the line, the surface, and volume. Lines may be straight or curved, continuous or broken. They may be organized according to their position, direction, or relationship to other elements. A similar approach can also be used for planes, where the various types of lines define the boundaries and hence the shape. The basic surface forms in architecture have been, in the past, the circle and ellipse, square and rectangle, and the triangle. The surface may be a grid or solid. Volumes may be formed by solids or by intersecting planes and lines.

These elementary dimensions, in turn, combine in infinite variations to form patterns and compositions. Some of the more familiar basic concepts for evaluating the arrangement of geometry, or its *composed dimension*, in order to establish visual order and clarity (but which are not presented in any specific sequence), are: modularity, form, size, scale, proportion, boundaries, regularity, degree of symmetry (axes), center of gravity, unity, chaos, dominance, polarity, articulation, gradation, contrast, monotony, austerity, radiation, tension, equilibrium, balance, accent, concentration,

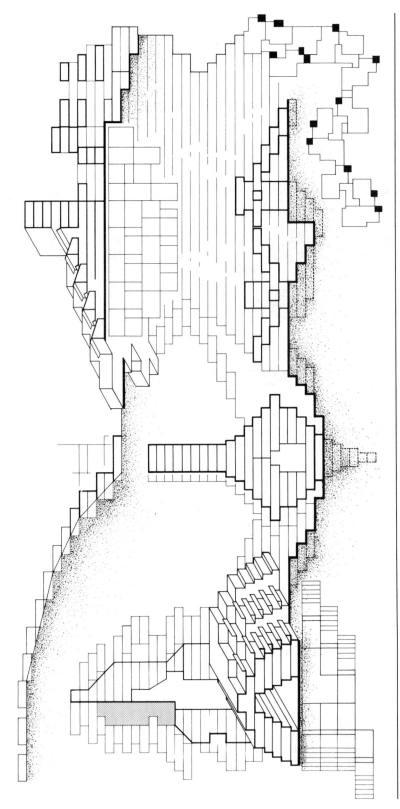

Figure 1.3. *From the single house to the urban building.*

punctuation, legibility, identity, diversity, hierarchy, harmony, rhythm, repetition, simplicity, reflection, density, sparsity, texture, color, massing, similarity, figure-ground (solid-void), territory, linking, transition, zoning, separation, partitioning, compactness, integrity, degree of openness, continuity, movement, progression, growth, directionality, disturbance, redundance, and so on.

The evaluation of the geometric patterns, as related to the *performance* or *goal-oriented dimension* of the building, is discussed in the following sections on building context, environmental control, structure, and the construction process. The highest level of evaluation of the geometric composition describes an overall concept of the whole, a dimension of meaning that reflects the unity of the *compositional dimension*, which may be functional, social, symbolic, and so on. In this context, the dominant, minimum but essential, geometric ordering idea expressing form represents the *gestalt* of a building or *parti* of a plan. This main idea of the whole design should be recognized without effort and should be caught at a glance, clearly representing an order of priority. Naturally, to draw images and to do an analysis of a building that expresses the complexity of an architecture that is deeply rooted in culture requires an extensive theoretical investigation.

In this section, on the analysis of basic geometrical dimensions, high-rise building shapes are first identified in terms of wire-frame models, and then the general principles of the geometrical growth of the building are investigated. The inner geometric organization of the building is studied through its plans and sections, and then through the interplay of the horizontal and vertical building planes; finally, the building appearance (as expressed by enclosure) is discussed. This approach corresponds to the basic methods of design, using the fundamental tools of visual communication to define ordinary objects: plans, sections, elevations, and spatial views, together with conceptual sketches, diagrams, and working models to check the design decisions.

The high-rise building shapes in Figure 1.4 range from boxy, rigid pure shapes to compound hybrid forms with rich facade profile modulations. The traditional buildings of the modern architecture period have generally been prisms, as based on the square, rectangle, cruciform, pinwheel, and other planar rectangular forms and their combinations. In contrast, the high-rise buildings of the postmodern era seem to have complete freedom of form-giving. The building masses may be broken up vertically and horizontally into interacting blocks, to reduce the scale of the building and make it appear as a bundle of smaller towers joined together. The stepped-back design may provide sky gardens and terraces for the occupants, or it may just be in response to zoning regulations.

On the one hand, the massing may express the life and behavior of the interior building organism, or, on the other hand, it may just present a unifying formal symbol, for instance, expressing the mountain-like skyscraper or the classical column analogy, where the building is divided into base, shaft, and capital. Spires and ornamental roof tops, such as a stepped top with a pyramidal cap, are contrasted by flat roofs. Outdoor extensions of lobbies may define a base; shafts may be altered by setbacks reflecting plan changes at certain levels.

From the point of view of overall proportions, building blocks can be organized as:

- horizontal slabs
- massive blocks
- vertical slabs
- towers

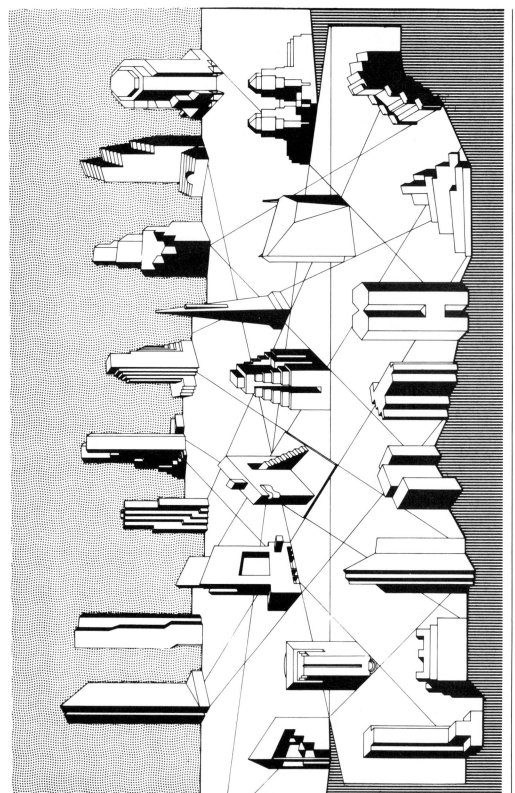

Figure 1.4. *Building shapes.*

The basic building forms may also be derived from solid geometry, as:

- spheres
- prisms/cylinders (right/oblique, regular/irregular)
- pyramids, cones
- truncated prisms and pyramids
- antiprisms
- prismatoids
- other polyhedra

In the past, pyramids, prisms, cylinders, cones, and spheres had been considered to be the only means of organizing architectural space.

Other building shapes are of the letter type, possibly with dimensional variations. They were quite popular in the United States before World War II, as free-standing blocks defined by the street fabric. Typical forms are:

$$I, L, T, +, F, |, Z, H, U, \theta, E, \square, Y, X, \Delta, O.$$

The formation of building shapes can also be accomplished through a process of growth such as the addition of lines, planes, lattices, and volumes, or by the packing of polyhedra and the formation of spatial networks. There are many other processes that can generate space geometry, such as:

- Addition, duplication, combination, stacking, mounting;
- Subtraction, reduction, erosion (holes, openings, cavities, voids, slots), truncation, terracing, cutouts;
- Reflection, convexity versus concavity;
- Division, partitioning, zoning, layering, concentration;
- Transformation, translation, rotation (twisting);
- Deformations, distortions, bending, extensions, tapering, stepping;
- Connection, linkage, penetration, intersection, interlocking, overlapping, merging, coupling, clamping, embracing.

The list could go on. It is apparent from Figure 1.4 that, from a geometrical point of view, there seems to be no limit to the formal manipulation of building shapes.

Before proceeding from the overall building shape to the internal geometrical order (as studied in the plan, section, and spatial views), some general ordering principles have been briefly introduced in Figure 1.5 as being derived from the special but typical cubic network. One should keep in mind that the chosen geometry tends to express equilibrium of space in static terms, in contrast to the curvilinear geometry reflecting controlled movement and balance of the dynamics of space, as is the case for the flamboyant forms generated by strict geometries in the Baroque era.

The spatial organization is further subdivided by an invisible grid of basic modules (i.e., fundamental units), which is usually taken as M = 100 mm ($\simeq$ 4 in.). The multimodules, as needed for the planning and structural grids, are multiples of the basic module as, for example, the typical ones of 3M = 300 mm ($\simeq$ 12 in.), or 12M = 1.20 m ($\simeq$ 4 ft). This spatial lattice reflects the modular coordination—it imposes dimensional restraints. The location, dimensions, and relationship of columns, walls, cellular units, and space are controlled by the three-dimensional network. It guarantees the proper fit of prefabricated elements and allows for the necessary tolerances.

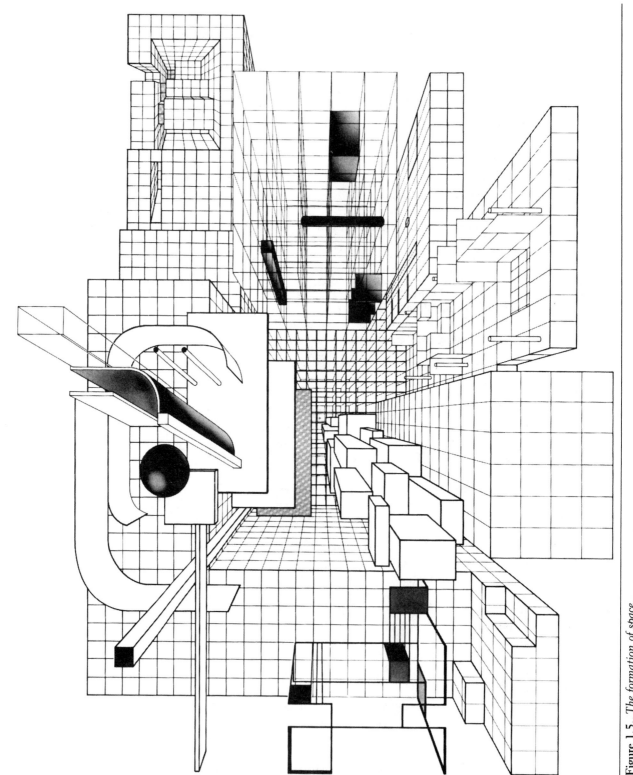

Figure 1.5. *The formation of space.*

The investigation of spaces and solids in Figure 1.5 is concerned not only with expressing the restraint of the gridwork, but also attempts to show that it is the source for flexibility, adaptability to change, and that repetition does not necessarily result in boredom, but can result in rich diversity. The formation of space is generated from the grid matrix through various operations, with no space left over. Spaces may grow, in the positive sense, by the addition of volumes, by spatial partitioning, or through a deductive process resulting in an inverted space with the inside being the outside— or this architecture by subtraction may represent underground construction. The continuous spatial flow of lines, planes, and volumes expresses that there is no limit to the manipulation of space.

The building plan is one of the most important components of the spatial organization. It ties the horizontal dimensions into a unit, and, as a vertical succession of horizontal layers, forms space. Geometry holds the building plan together and allows for the partitioning and zoning of space. The conventional plans are made up of rectangular, square, cruciform, pinwheel, letter shapes, and other linked figures usually composed of rectangles. The odd-shaped high-rise structures, in contrast, have irregular plans that may change with height. For example, the plan may be trapezoidal at the base and triangular at the top. In other words, floors are not necessarily repetitive.

The complexity of new plan forms is convincingly expressed in Figure 1.6. They range from closed regular polygonal forms, possibly with truncated or rounded corners and sawtooth faces, to irregular open eroded forms; or they range from multisided folded forms to curvilinear ones. The plan morphology may be analyzed by identifying the following geometrical ordering principles:

- *The degree of complexity of plan shape*: the degree of regularity.
- *Plan outline/contour/profile*: continuous versus interrupted straight line or curve, open versus closed, the number of sides, perimeter length, degree of regularity.
- *Plan formation*: single shapes, possibly truncated, reduced, and/or distorted with different angles, multishapes linked or overlapped, or partially merged but rotated.
- *Plan grid* (organizational network): modularity, regular versus semiregular versus demiregular equipartitions versus irregular planar lattices: the relationship of the plan grid to the building contour and to the plan formation (e.g., closed-path patterns extracted from the grid).
- *Plan composition*: arrangement of points/columns, lines/walls, and cells/cores/tubes; degree of interaction; center of gravity, eccentricity; exterior (perimeter) versus interior symmetry.

There are many other criteria for establishing a visual order or clarity. The various plan forms in Figure 1.6 can only suggest a process as to how they were generated and organized by circles, polygons and lattices, or how the solids are arranged and located, forming uniform regular patterns or open discontinuous ones. Plan forms, as related to functional organization and structure layout, are studied throughout this text.

It should be emphasized that the intention here is not to glorify the plan as a primary determinant of design by designing the building from the plan upward, but to recognize the plan as an essential tool to explain the building as a spatial organism. In the early days of the modern movement, the dynamic *open plan* (plan libre) was an important concept that opposed the static *closed plan*, where space is contained. In the open plan, the interior spaces flow into each other, are integrated, and allow the inside to

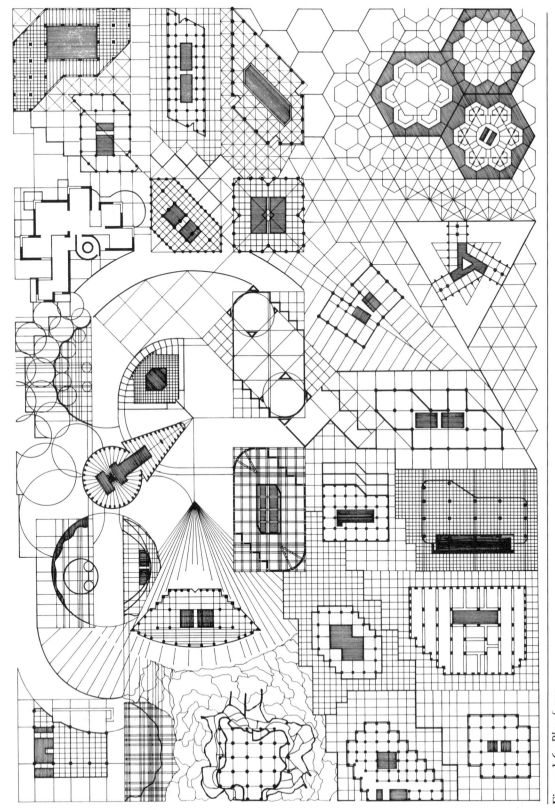

Figure 1.6. *Plan forms.*

penetrate the outside and vice versa; somehow the interior and exterior melt into each other.

What the plan does for the horizontal dimension, the section does for the vertical dimension, keeping in mind that a horizontal section also forms the elevation of the plan. The section translates or ties the plan into space, and is thus essential for comprehending the continuity of the vertical building geometry; it expresses variations in volume; it is a space-maker and should not be treated lightly, as is often done.

The various cases in Figure 1.7 attempt to illustrate some of the richness of formal expression. The section not only reveals the configuration of the building contours, the massing, the interaction of the horizontal layering, the interior enclosures, and the degree of vertical continuity, but also identifies the stacking of blocks, the cavities for atriums, and the relationship of the vertical traffic flow to the horizontal sky lobbies and various other zones. The section describes the formal manipulation of the building, the breakdown into blocks; it shows setbacks and voids, reflecting the degree of asymmetry of the form. Further evaluation of the section geometry can be done by using a formal vocabulary similar to that for the plan analysis.

Rather than simply perceiving a building in its true dimensions (that is, as plan, section, and elevation), it must also be understood as a spatial interactive organism, as indicated by the building study in Figure 1.8. In the three-dimensional view, the building can be seen as a sequence of linked and interacting spaces, either in terms of material spaces or as movement systems where services and energy are supplied along vertical cores to the horizontal activity spaces. Although this view is essential and necessary for an appreciation of spatial qualities, it does not provide the true, undistorted view of the dimensional order, as the section, elevation, and especially the plan, do. That is the reason why working drawings are generally presented as plans and sections.

The building elevation deals with the vertical dimension, in a manner similar to the section—it displays a composition or character—but it does not necessarily identify the interior order of the building that is its anatomy. Since one of its primary tasks is that of an enclosure, which must protect the inside environment from the outside, the building envelope may be completely separated from the spatial building organism and just express the independent structure of the curtain—often its appearance only reflects an overall building image.

In Figure 1.9, various building elevations have been investigated; naturally, there is no limit to the formal expression. While in high-rise buildings, the interior space is usually controlled by functional and structural considerations, it is the envelope which is given the freedom to present personal architectural values. The facade may express the strength of the building support structure by exposing the inside bearing elements, or a curtain may cover the structure. The envelope may express the tactile quality of the material and the tectonic quality of support, thereby articulating surfaces as buildable ideas; alternately, the nature of the material may be subdued in the architecture of fantasy, or the curtain may be dematerialized to occur as a skin. The skin may be scaleless, faceless, and lacking detail, or it may be decorative and articulated by contrasting solids, voids, and grids. The visual effect of the curtain may express weight or it may be weightless, it may be minimal or ornamental, solid or transparent, continuous or stepped. It may symbolize a corporate image or remind us of local traditions or of past styles, be it neo-Classical, neo-Gothic, or Art Deco. The envelope may form the primary building structure, as with tubular systems, where the perimeter

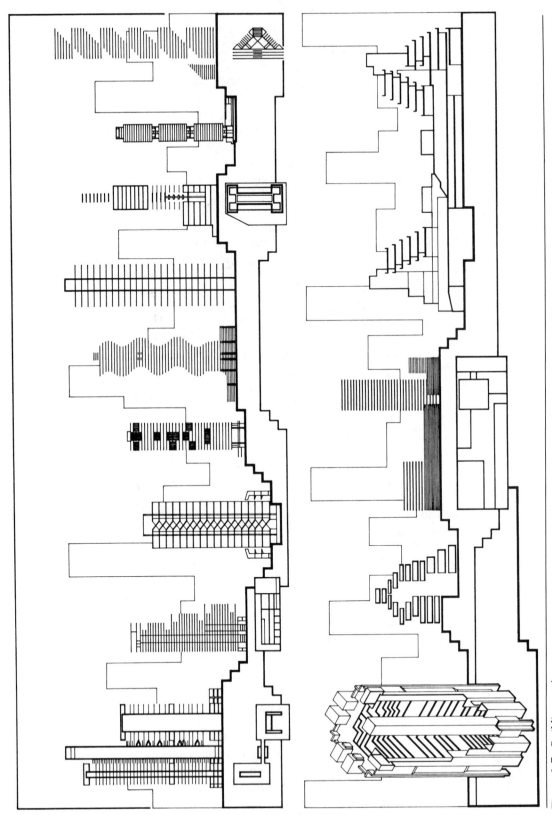

Figure 1.7. *Building sections.*

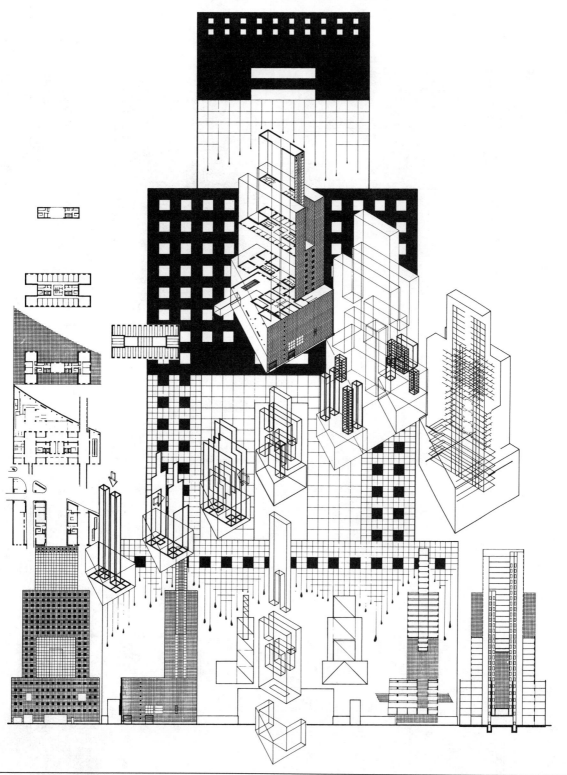

Figure 1.8. *Spatial order.*

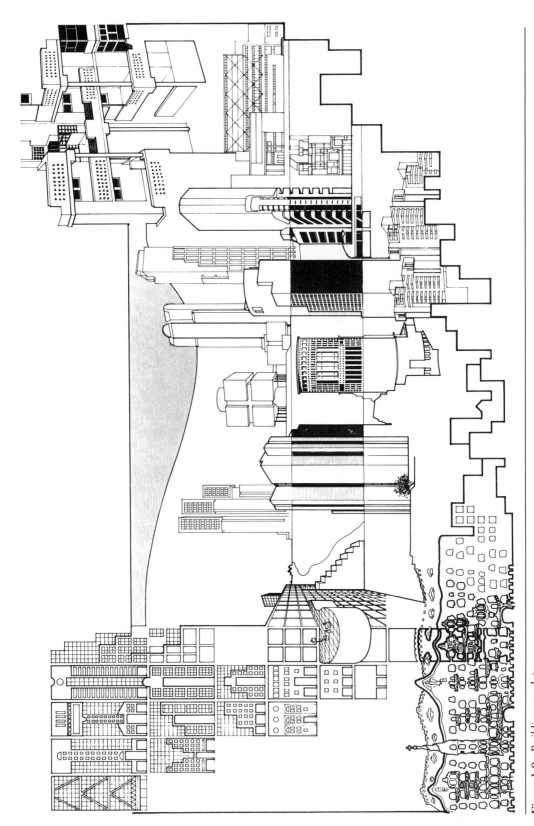

Figure 1.9. *Building envelopes.*

walls may be trussed; or holes may be punched into the vertical surfaces for fenestration. It may, however, only have the appearance of a bearing wall, as expressed by undulating heavy stone facades—in reality, these may just be thin veneers hanging onto a concealed steel frame in the back. One should also not forget that decoration is not necessarily only applied, but can be an integral part of the structure or assembly of parts. Rather than letting the facade appear as a closed and purely decorative curtain wall, a sense of life and identity may be given by allowing the cellular zoning of the building organism to expose itself to the outside. In one case in Figure 1.9, stainless steel capsules containing the toilet installation are hung outside, clear of the building.

The chosen examples in Figure 1.9 can only attempt to identify a portion of the infinity of formal expression and visual features of the exterior. A further study should clarify the degree of enclosure or openness, the quality and texture of the material, the geometrical composition, the play between structure and skin, between solid and void, the repetitive versus the unique, and the many other features.

ENVIRONMENTAL CONTEXT

The building's form, as influenced by its context, may be broadly approached from a cultural and physical point of view. The wide scope of culture ranges from political, economic, and social aspects to historical and aesthetic ones. Included in this group are the natural amenities of views and the significance of the site, as well as the legal restrictions of zoning ordinances, codes (city, regional, state, and national), and building design specifications (e.g., AISC, ACI, and AISI for structures), which attempt to protect the user and public at large (i.e., public health, safety, and general welfare). The building design is obviously also influenced by the type of client. There are distinct differences in the expectations of the speculative developer, the corporation, the institution, and the government.

The physical context includes the climate (orientation of the site, sun angles and intensities, maximum precipitation, prevailing winds, etc.), topography and geology (vegetation, unique land forms, contour intervals, potential flooding, ground-water table, drainage, bearing capacity of ground, etc.), existing urban fabric (land use, services, etc.), accessibility (access to pedestrian and vehicular traffic, parking), capacity and location of existing utilities (sewer, water, electricity, gas, etc.), and ecology (effect of the building upon its environment).

On the tight sites of high-density urban areas, the building shape is controlled very much by the lot size and zoning regulations, while the orientation and shapes of high-rise buildings on larger, more open land present many more options, as influenced by the climate, sun, topography, view and, some of the other criteria previously mentioned.

From an economic point of view, the location of the building on the site may be dependent on the contour lines, which indicate where to excavate and fill so that the least amount of earth has to be moved a minimum distance. The site geology, that is, the bearing capacity of the soil, may also determine the building location and plan shape.

The form of buildings and building groups, as influenced by context, is a rather complex issue. In this section, only some basic concepts related to orientation, scale, layout, economy of form and size, and zoning regulations are briefly investigated.

In energy-conscious design, the effect of the climate is a basic consideration. Its interaction with geological and topographical features of the site, however, is ignored in the following discussion. Ideally, the orientation, shape, and size of a building should be such as to efficiently distribute the daylight to the interior spaces, to allow natural ventilation, if needed, to efficiently absorb the winter sun but keep out the summer sun, and to provide protection from the cold north winds. For building groups, the distance between the buildings should not only ensure solar access and thermal comfort in winter, but also, in areas of strong winds, the buildings should be arranged as to break the wind and to provide shelter but still allow some access to breezes for natural ventilation (see fig. 3.6). Deciduous trees should screen the sun in summer and provide shade, as well as break the wind, while allowing the sun to enter in winter.

In the United States, the steep noon summer sun angle is about 40° higher in the sky than the low noon winter sun angle. The large sun path in summer has the following effects on the three critical facades:

- East elevation: the intensity due to solar radiation is as high as for the western side, but the air temperature is lower.
- South elevation: the solar radiation is less intense because of the higher sun angle; here the sun can be controlled, for instance, by horizontal overhangs.
- West elevation: this is the hottest side, and the sun can be controlled by a vertical sawtooth design, for instance.

In contrast, the short low sun path in winter causes most of the sun's impact to be on the south side. The sun can penetrate into the building by using a maximum of glazing on the south elevation. In cool climates, interior dark-colored wall and floor surfaces should be exposed to absorb solar radiation and thus provide a high interior thermal mass, which stores heat during the day and releases it overnight. While sunlight is needed for the natural lighting of buildings, it also heats them, which is an advantage for cold climates (where heating controls the building design), but a disadvantage for hot climates (where the primary goal is cooling). In severe dry and hot, as well as cold, climates, buildings should provide the least contact surface for a given floor area, such as provided by cubes and round cylinders, in order to have less heat gain and heat loss. A low surface-to-volume ratio allows the volume to efficiently store heat (high thermal capacity), since the surface area controls the rate of heat gain or loss. But by reducing the exterior surface, daylighting is also reduced, thus possibly requiring electric lighting, which in turn is a major heat source. The designer must deal with these opposing criteria and optimize the solution.

In general, buildings with their long axis in east-west direction and spaced in the north-south direction have been found most efficient from an energy-conscious design point of view. The degree of elongation of the building (i.e., exterior surface area-to-volume ratio), however, depends on the climate.

- In *cool climates*, a minimum of surface should be exposed to control heat loss, thus making more compact shapes with a low surface-to-volume ratio efficient. But solar gain in winter is of primary concern, together with daylighting, thus requiring a maximum of glazing along the south face to allow the winter sun to enter, together with a high interior thermal mass of dark surface color; the summer sun obviously must be controlled. A minimum of windows should be selected along all of the other elevations.

- In *hot, dry climates* with cool nights, ventilation and exposure to the sun should be minimized, and the external thermal mass maximized, by using a minimum amount of windows and light exterior colors, thus making compact building shapes efficient; that is to say, massive buildings with a low surface-to-volume ratio are advantageous. The massive adobe construction of the building envelope in indigenous architecture clearly expresses this type of energy control. One must keep in mind, however, that a taller slab-type building may be more beneficial because the sun intensity on the horizontal roof is higher than on the vertical wall, which can be more easily shaded.
- In *hot, humid climates*, the sun must be controlled on the critical east and west elevations with a minimum of wall and glass area. In this instance, long, thin slab-type buildings are most efficient, although the shallow depth may be uneconomical. Because of the low daily temperature swing, the thermal mass must be minimized for quick cooling purposes, and cross-ventilation maximized, requiring many large openings and short distances across the building. Reflective, light-colored, insulated exterior surfaces, together with shading devices for openings, should be used.
- In *temperate climates* solar gain in winter together with daylighting may be of primary concern. In this context, slab-type buildings may be most effective, since they utilize a maximum exposure to the winter sun and a minimum exposure to the east and west summer sun, together with a maximum cross-ventilation.

It should, however, be emphasized that also the building *scale* must be considered. For example, for a cube with edge length L, the surface area is proportional to the square of the length ($A_s \propto L^2$), while the volume is proportional to the cube of the length ($V \propto L^3$). Should the member length be increased by 10 times, then the surface will have increased 10^2, or 100 times, while the volume will have increased 10^3, or 1000 times. Hence, the property of the *surface-to-volume* ratio, which is investigated in the bottom part of Figure 1.10, becomes very important in building design. The surface-to-volume ratio for a single block unit is $A_s/V = 5/1 = 5$, but for a large cube, consisting of 125 single block units with a volume of $V = 5(5)5 = 125$ ft^3, the ratio becomes $A_s/V = 5(25)/125 = 1$. Hence, it is quite clear that, as the building grows in volume, the effect of the perimeter surface and the static passive energy considerations decrease, while the importance of the dynamic HVAC systems may increase. In other words, *skin-load dominated buildings* (such as residential and light commercial construction) change, with increasing scale, to *internal-load dominated buildings*, where only the perimeter zone is still skin-load controlled. This effect of scale is clearly expressed in nature, where mammals and birds require a food supply related to their surface area rather than to their weight, since the heat produced is a function of the body volume, and the heat loss is dependent on the skin surface. A grown man may need about one percent of his body weight of food each day, a mouse requires about 25 to 50 percent, but a shrew (the tiniest mammal) must eat almost continuously to stay alive, which is particularly true in winter. Hence, large species do not need as much food, because of lower heat loss and larger heat gain due to their small surface-to-volume ratio. Obviously, there must be a limit to size in this respect because of cooling requirements in summer.

The surface-to-volume ratio of a building is not only important from an energy point of view and that of continuing maintenance costs, but also with respect to the initial costs of construction. It is known from mathematics that the circular cylinder

Figure 1.10. *Basic considerations with respect to building arrangement.*

encloses a maximum volume with the least perimeter surface and no corners, assuming no change in plan. In Figure 1.10 (above the bottom cases), various rectangular, mostly single-story, building clusters with the same roof area are investigated for a constant enclosed volume. They range from a single compact building unit of square shape with the least perimeter and only four corners, to sixteen separate buildings with the largest perimeter area, that is, a four times larger perimeter and the number of corners is increased sixteen-fold.

Here, the external wall area efficiency of the rectangular shapes can be expressed by dividing the constant volume, V, by the envelope area, A. Hence, for the single compact unit, the efficiency is

$$(V/A)100\% = (16/16)\ 100 = 100\%$$

When this unit is elongated to the proportion of 2:8, the perimeter efficiency decreases to

$$(V/A)100\% = (16/20)100 = 80\%$$

Expressed in terms of the exterior wall area-to-volume ratio,

$$A/V = 2(8 + 2)/16 = 20/16 = 1.25$$

Hence, this shape requires 25 percent more envelope area. When the compact unit is broken up into 16 separate basic units, the perimeter efficiency is only

$$(V/A)100\% = (16/(16 \times 4))100 = 25\%$$

Expressed in terms of the wall surface-to-volume ratio, the units require $A/V = 16(4)/16 = 4$ times more envelope area. In addition, the number of corners is increased 16-fold, together with the corresponding increase of costs. When these individual units are stacked vertically to form a tower, the large perimeter surface-to-volume ratio remains the same!

In the discussion above, the roof area was considered constant because the buildings were all of the same height, with the exception of the tower case. But, when the basic spatial units are assembled vertically, the proportion of roof area to total surface area changes. The following conclusions can be drawn by comparing the four building types at the bottom of Figure 1.10, while keeping the enclosed volume constant:

- The surface area of the *cube* consists of 80 percent envelope area and 20 percent roof area. In this case, the ratio of the total surface area, A_s, to the volume is $A_s/V = 80/64 = 1.25$.
- For the single-story *horizontal slab building*, the roof takes about 62 percent of the total surface area and the remaining 38 percent is for the exterior wall area. In this instance, the ratio of $A_s/V = (40 + 64)/64 = 1.63$, or, when related to the cube as the most efficient shape is $1.63/1.25 = 1.3$; in other words, the horizontal slab needs 30 percent more surface area.
- For the *vertical slab-like tower* building, only about 14 percent of the total surface area is used by the roof, the remaining 86 percent is for the envelope area. Under these conditions, the surface-to-volume ratio is high, and equal to $A_s/V = (96 + 16)/64 = 1.75$, or $1.75/1.25 = 1.4$. Hence, the vertical building needs 40 percent more surface area, as compared to the cube.
- For the two-story *court building*, the central atrium is covered by a glass dome, hence only the exterior surface has to be considered. In this case, the roof takes

about 43 percent of the total surface area, and the remaining 57 percent is for the exterior wall area. The ratio is $A_s/V = (48 + 36)/64 = 1.31$, or, when related to the cube, is $1.31/1.25 = 1.05$; in other words, the atrium building needs only 5 percent more surface area, as compared to the cube. However, when the critical exterior wall-to-volume ratio is taken as the basis of evaluation, the atrium building is far more efficient than the cube, especially when the effect of the maximum daylight exposure is also included.

It is apparent that the cube provides the most efficient surface-to-volume ratio for a given volume. This ratio increases for a low-rise building, because the proportion of the roof surface increases. It also increases with an increase of building height, because the proportion of the wall area gets larger. It is obvious that the flat building is controlled by the roof surface, and the tower by the wall surface. When, however, the critical exterior wall area is the basis for evaluation, the atrium building is more efficient than the compact building block.

The use of the total surface area is deceiving, since the nature of the roof surface is quite different from that of the facade surface. The solar intensity on a vertical plane is less than that on a horizontal one; aside from this, the sun on a wall may be more easily controlled by shading than by insulation, as for the roof, and, in addition, the wall offers the benefit of daylighting. The effect of height by stacking the cubic units vertically rather than adding them horizontally, decreases the roof area, but also increases the structural load action, the effect of air infiltration, and heat loss; further, the construction process becomes more costly at a certain building height. One may conclude that, for a constant building volume, the costs per square foot tend to decrease with an increase of plan size, but increase with an increase in building height. The effect of composite plan forms and hybrid building shapes obviously increases the construction costs.

While at the bottom of Figure 1.10, surface-to-volume ratios were investigated, at the upper portion along the perimeter typical layout patterns of building units are shown, as based on a constant density, where the total building volume and the site area are not changed. For example, with respect to housing, the typical arrangements include

- Isolated, individual (one-family) nuclear forms
- Blocks of attached units of the row house type, possibly forming courtyard or other types
- Low-rise chain or net-like multifamily building forms
- Isolated high-rise slab and tower forms

Naturally, a combination of these types may form a continuous urban architecture (see fig. 1.3) of the cluster and terraced type, integrating low-, medium-, and high-rise elements.

Although each of the four groups represent the same population density (number of persons per acre) or the same number of dwellings per acre (DUA), the land utilization (that is, the proportion of public to private areas, or the open space ratio, OSR) and the circulation length for each of the layouts is quite different. In the center portion of the top diagram in Figure 1.10, the use of the land (according to various densities) by keeping the building height constant, is illustrated (see also Martin and March, 1972). The top and bottom represent 90 and 50 percent land coverages by using either concentrated, nuclear forms, as at the top, or inverted court forms, as at the bottom. The 10 percent land coverage at the center demonstrates the progression from the concentrated to the dispersed form.

Densities depend on economic considerations, the availability of land, social and cultural needs, and so on; they are regulated by zoning ordinances. They range from high-density downtown urban areas to low-density rural areas. Some typical population densities for the United States (with that of Hong Kong added for contrast) are

- Single family detached housing, eight dwelling units per acre   8 DUA
- Two-story row houses   20 DUA
- Four-story walk-up apartments   40 DUA
- Combined low-/high-rise housing   90 DUA
- Multistory slab-type apartment buildings   90 DUA
- Tower apartment buildings   120 DUA
- Typical downtown urban high-rise   200 DUA
- Maximum for New York   425 DUA
- Maximum for Hong Kong   1000 DUA

To have a rough first idea of the number of people per acre, an average of 3 people per dwelling unit and 1.25 parking spaces per unit may be assumed for the United States.

The size, shape, location, and function of a building in an urban context is affected very much by zoning ordinances, which not only control the population density and the use of buildings (e.g., residential, commercial, manufacturing, mixed), but also the location, height, and bulk of buildings. The space between buildings and setbacks from property lines and streets are specified, thus possibly limiting the shape of the building plan and the location on the lot. These minimum distances are increased with the height of the building to allow access to light, air, and sun at the street level (Fig. 1.11) and adjoining sites. In the first part of this century, up to about World War II, the massive letterlike building blocks were not formed just by the rectangular grid patterns of American cities but also in response to natural light and air ventilation, limiting the plan depth to about 50 ft. After that, the development of artificial lighting and air conditioning made isolated tower construction possible.

From a conceptual point of view, the zoning ordinances are based on a combination of the following rules, as illustrated in Figure 1.11, keeping in mind that there is usually a minimum setback zone at the base.

- *Tower coverage zoning*. A minimum amount of open space at ground level is ensured by restricting the building coverage of the lot to a specified maximum area, not to exceed, for example, 40 percent.
- *FAR zoning*. The amount of floor area is limited, as based on a multiple of the site area. The floor-to-area ratio (FAR) represents the amount of floor space as compared with the lot size. Hence, a FAR is the number by which the designer multiplies the site area to obtain the allowable building area. For example, for a FAR of 16:1, a building without setback can reach 16 stories when it covers the entire site, or it can be a 32-story building using 50 percent of the site, or a 64-story building when it covers only a quarter of the site.

    For dense urban areas the typical floor area ratios are FARs of 15 to 25. Higher FARs may be achieved by transferring area from an adjacent site's air rights (i.e., *zoning lot merger*) or by *incentive zoning*, where height and bulk limits may be waived in exchange for special amenities such as plazas, interior atriums, shopping malls, through-block shopping arcades (see fig. 1.15), transit connections, urban parks, housing, and so on.

- *Envelope zoning.* When the site is fully occupied, the building shape is controlled by an envelope that is made up of the so-called *sky exposure planes*, as related to the street width(s), which allow light to get to the street level and reduce the shading of adjacent buildings. The guideline envelope is partially or fully filled out by the building structure. Usually, a maximum height of the building base along the lot boundary is given, above which the building must be set back in order not to penetrate the sky exposure planes. Typical rise-to-setback ratios are 2.5:1 to 3:1, which correspond roughly to an average inclination of 70° for the exposure plane. This approach yields a stepped-back or ziggurat type of building configuration. Rather than using this *regular setback approach* by occupying the entire lot, in the *alternate setback approach* the building is placed further back on the lot, away from the streets and property lines, to allow for more open space at ground level. For this condition, a steeper rise-to-setback is allowed, which results, in turn, in a higher building. The upper portion of the building (that is, the tower part) can, possibly, pierce the sky exposure planes, if its floor area at any level does not exceed, for example, 40 percent of the lot area. Rather than using the set-back approach, a *straight tower* can be built that penetrates the sky exposure planes, but is only permitted to occupy, as an example, 25 to 40 percent of the site, together with a minimum setback zone at the base to allow for plaza space, reflecting the typical trend of the 1960s, where the tower became an image of prestige. In this case, the height of the tower may be controlled by FARs.

Example 1.1

For an urban site of 0.25 acres, the zoning ordinances allow 300 DUA and a FAR of 15 to 1, when the building covers no more than 55 percent of the total site. For this rough, preliminary investigation, a simple rectangular nonstepped high-rise tower with only two-bedroom apartments for upper income users is assumed, together with 1.25 cars/unit.

Lot area: 134.8×81.6 ft^2 = 11,000 ft^2 = 0.25 acres

Max. number of units: 0.25 acres @ 300 = 75 units

Max. total gross floor area: 15 FAR (11,000) = 165,000 ft^2

Max. floor area per typ. story: 0.55(11,000) = 6050 ft^2

Max. building height: 165,000/6050 = 27 stories

Assuming, conservatively, 15 percent of the gross floor area of a typical story as nonrentable space, and ignoring the effect of the different conditions at the building base, including parking for this rough approximation, yields a maximum rentable area of: 0.85(6050) = 5142 ft^2/floor. This is equivalent to four 2BR units bundled around a central core area, where each unit has a maximum of 5142/4 = 1285 ft^2, which is above the approximately 1200 ft^2 intended for use.

Assuming a building depth of about 60 ft to satisfy the typical minimum setbacks required at the base, results in a maximum building length of 6050/60 = 101 ft, which is satisfactory and fits the lot size. Hence, the number of stories above the building base, using a total of 75 units is: 75/4 = 19 stories.

The high urban density for the given condition clearly does not allow on-grade parking and requires a multilevel garage, that is, considerable space at the building base above and/or below grade. The approximate required gross garage floor area estimating 350 ft²/car, is: $1.25(75)350 = 32,813$ ft². Should the entire parking be provided under the building, without a base extension, then the number of stories required are: $32,813/(60 \times 101) \simeq 6$!

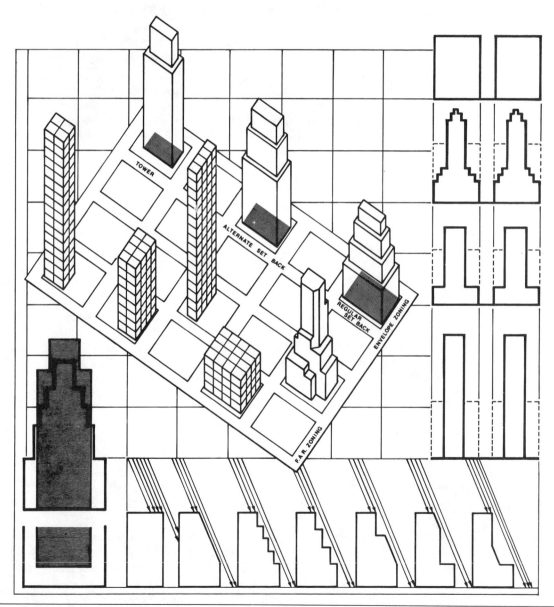

Figure 1.11. *The effect of zoning ordinances on building form.*

While the zoning ordinances just described tend to fix the building envelope, newer regulations tend to be more of the *performance type*. They are based on the quality of daylight and other needs related to the immediate building context, including the openness of the streets. Design controls for big urban sites, using an overall large-scale development rather than individual parcels, include permitted guideline envelopes with a required base structure that must hold the building line, setback lines and planes, circulation routes, requirements for open spaces, and so on.

It is quite apparent from this discussion that the shape of high-rise buildings in cities will be very much affected by the zoning ordinances. The incentive zoning regulations provide additional public amenities and give more freedom to architectural design, but may also threaten the basic needs for sunlight and openness, as well as the quality of context, because of the increase in density and building height. In any case, zoning ordinances represent an important part of the grammar of the architectural design language.

In contrast to zoning ordinances, which address external effects, codes are concerned with health and safety as related to the type of occupancy and type of construction, and are thus more interested in the method of building. Among the many code considerations are: fire protection, exit regulations and travel distances, construction and material standards, minimum requirements as related to light and air ventilation, and so on. Energy conservation codes may offer solar access protection (solar fence) and control the building orientation, vegetation, and building expansion. Sometimes codes are performance rather than specification oriented in order to not just provide "what-to-do" guidelines that tend to result in stereotyping designs. Some of the effects of codes and other regulations, and standards upon the building shape, are treated in other parts of this book.

FUNCTIONAL ORGANIZERS

Architectural space is not only static and fixed, as characterized by the material space, but (as enclosure space) it is alive and dynamic with people and life-supporting energy moving horizontally as well as vertically, from one location to another. There is circulatory space for service and movement, space that is served, as well as the critical space that connects. The visual study in Figure 1.12 helps to develop some feeling for the power of flow systems. It is intriguing to compare the crude traffic systems of a building, or even of the large-scale systems of a city, to the complex organism of the human body. The blood and nerve distribution systems, together with the fueling of the body's machinery, seem to relate directly to the life-supporting traffic network and the energy supply of piping water and steam, ducting air, and the distribution of electric power along wires in a building. This biological metaphor should provide a powerful tool for conceptualizing the basic generic principles of a functioning building environment.

The organs of the respiratory, digestive, and urinary systems fuel the body with oxygen and food and remove wastes along various branching conduits; but this process slows down during sleep, as does city life during the night. Under the heart's pumping action, blood moves away from the heart along the complex network of the arteries, and it moves back toward the heart in the veins. The intricate nerve system branches out from the spinal cord to embrace the entire body, as illustrated in Figure 1.12,

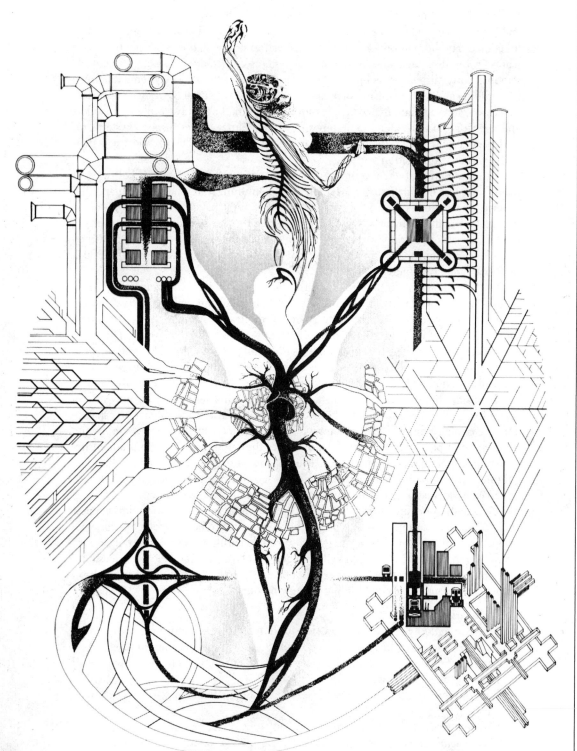

Figure 1.12. *Movement systems.*

and coordinates the body's activities as ordered by the brain; it operates the impulses and initiates the body's muscular activity, as well as regulating mental and physical functions. The human organs may be perceived as power stations, sewage treatment plants, waste disposal plants, water supply systems and reservoirs, factories, railway stations, ports, airports, machines, and so on. Some of the traffic systems that keep the building and city organisms alive range from highways, tunnels, bridges, railroads, waterways, subways, automated people movers, pedestrian walkways, building corridors and elevators, to the water supply pipes, utility pipelines, sewer disposal systems for wastewater, solid waste, and hazardous waste, airhandling ductwork, electrical conduits, and communication wiring.

Forms are born out of movement, as the meanders reflect the natural pathways of rivers, similar to the oscillating movement of the swimming water snake. The branching patterns of the fluid networks remind us of the natural patterns of drainage due to flowing water, the paths of a lightning bolt as electric energy is discharged, or of the fluid and electrical distribution systems in a car. When the regularity of flow is disturbed at connection points, a complex transition may be necessary as, for example, at a multilevel highway interchange.

The circulation distribution in a tower building, that is, its branching patterns, also remind us of a tree, where the trunk and the branches represent the life-supporting dynamic flow structure that distribute energy to the leaves or living units, and where the member sizes decrease in proportion to the decrease in flow intensity. The trunk may be visualized as the core of a high-rise building, its nucleus and central spine. Another simile would be to a vertical dead end street, with all of its services. It is the main distribution system and contains the elevators, stairs, utility ductwork, and so on. It ties the building to the urban fabric at the ground level, and possibly also at upper levels. It provides access to the pedestrian traffic, it links to the parking spaces, and occasionally directly to the subway, and it connects to the energy supply and waste-disposal pipe lines. The circulation zones the building volume and establishes a hierarchy among the various spaces. From a functional point of view, the building volume can be grouped into: *activity zones*, or served spaces; the *circulation network*, or service spaces; and *access*, or connective spaces.

The activity zones are clearly defined by the closed spaces of certain building types, such as housing, dormitories, and hotels, but they may also be open office landscapes or open-plan flexible layouts, possibly with movable partitions, as may be the case for exhibition spaces. Often, composite floor plans with both open and closed spaces offer a dynamic exchange between privacy and openness. The type of activity depends on the use or occupancy of the building. It may be organized as

- *Residential:*
  Low-rise buildings (walkups): low-density (e.g., detached housing), high-density (e.g., semidetached housing such as duplex and triplex, and row housing).
  Mid-rise buildings (say four to six stories with or without elevators): party-wall housing, block housing.
  High-rise buildings: block housing, terrace housing, slab blocks, towers, clusters, etc.
  Combinations of the above: urban housing.
- *Commercial:* offices, stores, shopping centers, hotels, restaurants, etc.
- *Industrial:* manufacturing, warehouses, etc.

- *Institutional*: schools, hospitals, prisons, churches, museums, government buildings, etc.
- *Special*: towers, sports complexes, bridges, airports, offshore structures, parking facilities, etc.
- *Mixed occupancy* (multiuse): Urban high-rise buildings have been combining living, working, and servicing activities. For example, a mixed use 70-story skyscraper may consist of: 3 underground levels of basement parking and service areas, 8 floors of retail, 20 floors of office space, 2 mechanical levels, 18 floors of hotel rooms, 2 mechanical levels, 20 floors of condominium apartments, and the mechanical penthouse.
- *Megastructure*: The large scale urban building, as realized by some of the new towns in Europe, such as around Paris, include functions for living, working, education, and recreation.

The following space allocations per activity or person may be used in the United States to estimate dwelling unit sizes, elevator loads, and other circulation requirements:

- For open office space with, for example, an average of 75 percent utilization and 25 percent traffic, 110 ft^2 per employee may be assumed, where roughly 5 to 10 percent of the gross area is for the mechanical space.
- For commercial buildings, 60 ft^2 per person may be assumed.
- For residential buildings, the average area per person is about 300 ft^2, with a minimum of 150 ft^2; 2 people per bedroom may be assumed, or an average of 3 people per dwelling unit, but 4 people for low-rent housing. In other countries, the average area per person may be by far less, for instance, 35 ft^2 for a squatter settlement in Hong Kong.
- For hotels, 1.5 persons per room may be assumed.
- Typical corridor widths are in the range of 5 to 7 ft; but for schools and hospitals, they are from 8 to 12 ft.
- Typical floor-to-floor heights for office buildings are in the range of 11 to 14 ft, where heights of 13 ft are now essential to accommodate the wiring needed for high-tech communications. For apartment buildings, this height is (on average) 9.5 ft for frame construction, but can be about one foot less for flat plate construction, where the slab may provide the ready ceiling surface. For floor beam construction, suspended ceilings may be required, though there may be no need for mechanical space, with the exception of in the corridors. The ceiling height is usually in the range of about 8 to 8.75 ft.

The typical vertical movement systems in high-rise buildings can be the passenger and freight elevators, paternosters, escalators, ramps, pneumatic message tubes, dumbwaiters (miniature elevators), stairs, plumbing stacks, exhaust shafts, air ducts, water pipes, electrical conductors, refuse shafts, etc. Elevator cabs are either pulled by traction or pushed by hydraulics. In the *traction elevator* system, the car is raised and lowered in guide rails by cables or ropes passing over the grooved sheave of the hoisting machine, usually located above the shaft in the penthouse, and are fastened to a counterweight which balances the car weight. A *hydraulic elevator* is supported by a hydraulic cylinder that pushes the car up. Since it is self-supporting, it does not impose vertical loads on the building structure, thus allowing a lighter shaft construction and smaller foundations. There is no machinery penthouse required for heavy overhead hoisting equipment, but it needs a much larger motor than an equivalent traction machine because it is not counterweighted. Whereas traction elevators travel rapidly and are used for high-rise buildings, hydraulic elevators are of low speed and limited to low- and mid-rise buildings only.

The vertical and horizontal distribution of people along the cores and branches to the respective activity zones, which consist of either closed cellular aggregates or open layers, can only be suggested in Figure 1.13. Some of the variables that influence this flow are:

- building shape
- plan form (simple, closed versus open, polygonal forms)
- plan depth (shallow versus deep versus variable depth)
- number, location, and orientation of core(s) (single versus multicore, exterior versus interior location, eccentric versus concentric)
- arrangement of, and access to, activity units (e.g., ring-like, radial, linear, grouped, polygonal)

The horizontal flow is in response to the plan organization. It may represent a clearly ordered circulation system of the radial or linear type, as for residential buildings consisting of closed cellular living units, or it may represent a relatively undefined free movement, as for open office space. Naturally, there are an infinite variety of combinations possible in response to the space use. The study of the interaction of the various building forms and plan types with their respective circulation patterns in Figure 1.13, can only develop a feeling for plan organization, coordination of space, use of space, and an order of priority.

In high-rise office and mixed-use buildings, elevator shafts, exhaust shafts, emergency stairs, and toilets are usually grouped together in the core area to keep the unusable building volume to a minimum; they cover about 20 to 25 percent of the floor area. The central location of the core allows short distribution paths to the perimeter zones. The core zone is serviced by the circulation corridors to the elevators, stairs, and toilets via a perimeter corridor encircling the core (i.e., split elevator core); the elevator shafts are arranged side-by-side and/or opposite but separated by the lobby (Fig. 1.14. See also fig. 5.4). The core is the spine of the building body, which ties it together vertically and horizontally and connects it to external energy sources.

In tall buildings, the efficient vertical distribution of people becomes a complex traffic problem; it is apparent that more people cannot just mean more or larger elevators. The skyscraper is divided by nonstop, high-speed shuttle elevators at the sky lobbies into vertical zones, similar to express stops for mass transit systems, which differentiate between express and local runs, thereby reducing shaftway space (i.e., core size) and movement time.

The selection of the number of elevators is a complex problem. It depends on the number of stories, the elevator capacity and speed (say 700 fpm (feet per minute) for office buildings, but 2000 fpm as an absolute maximum for human comfort), building size (population per story), building use (diversified versus single tenant occupancy, high morning and evening peaks, tenants versus visitors, etc.), and maximum waiting time (say 30 sec for office buildings, and 90 sec for residential buildings). A rule of thumb for preliminary estimation purposes is that there should be one elevator per 25,000 ft^2 of rentable office area. For typical apartment buildings, 12 to 40 stories

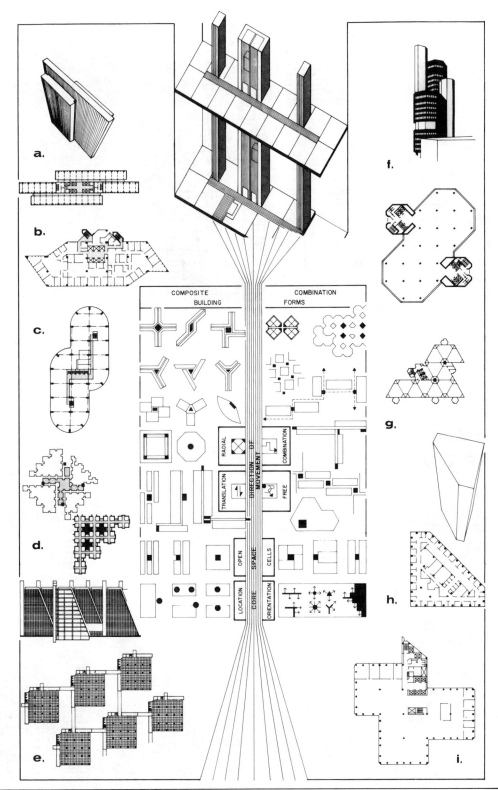

Figure 1.13. *Plan forms, circulation, and core location.*

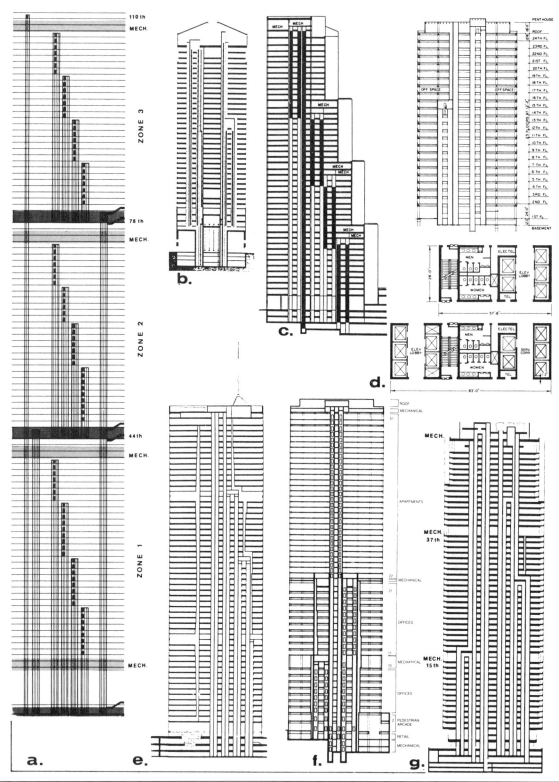

Figure 1.14. *Examples of elevator shaft systems and mechanical floors.*

high, an average of three regular-speed or two high-speed elevators, and one service elevator, may be assumed; for up to about 25 stories, one of the two elevators may also serve as a freight elevator. The elevator system consists of the penthouse machine rooms (elevator machinery supports the car and counterweights), basement pits, and shaftways. From a traffic flow point of view, there are three basic types of elevator systems:

- Local elevators
- Express elevators, which act as local elevators over a certain floor range.
- Shuttle elevators, which directly connect the ground level with sky lobbies nonstop.

The elevator cars can be of the single-deck or multiple-deck type (i.e., double-deck, triple-deck, and quadruple-deck). For example, the double deckers consist of two normal cabs on top of each other, where the bottom car serves only odd-numbered floors, and the top one serves the even-numbered floors.

The elevators in the 110-story, twin tower, World Trade Center in New York (Fig. 1.14a) move 50,000 workers and about 80,000 visitors daily. Each tower is zoned by the sky lobbies into three sections. Nonstop shuttle elevators run directly to the sky lobbies, where the passengers transfer to the local express elevators servicing different levels within the sky lobby zone. There are 23 shuttle elevators, each holding 55 passengers, operating at speeds of 1600 fpm, and 72 local express elevators (plus four freight elevators) in each tower.

In buildings where banks of multiple elevators handle complex and heavy traffic, a new generation of smart elevators considerably reduces the travelling time, and thus serves many more people. Elevators are seen as a whole system rather than just individual cars. These "thinking" elevators are programmed to sense how many people are riding by measuring the weight of the cars. They can determine how many people are waiting, they can adjust to changing demands, and they may even anticipate riding patterns.

Various features of vertical flow patterns are convincingly illustrated by the cases in Figure 1.14. The subdivision of the World Trade Center tower into three vertical zones is visually accentuated. Other examples show that the core area decreases towards the top because elevators can be eliminated at certain levels. The sudden change in the shaftway pattern of case "f" demonstrates that, when the building use alters from offices to residential, fewer elevators are used for the apartment zone. The effect of the stepped building configuration in case "c" upon the elevator shaft layout is self-explanatory.

From a functional point of view, atria, including passages, may be seen as a connective space, as access systems, and as climatic and functional buffer zones; their precedents are the colonnades, piazzas, and the famous nineteenth century gallerias in Brussels, Milan, Paris, and Rome. Atria may provide landscaped plaza space and extend the outdoor urban context into the building, they may open up and connect underground pedestrian spaces through craters, or even provide semipublic space, possibly together with vegetation and fountains at upper building levels. They open the interior space to natural light and exterior views, shelter against the external environment, and modulate the interior climate through solar gain and venting by creating a microclimate that helps to conserve energy in the entire building. From a climatic rather than spatial, communal point of view, the atrium may be compared to the lungs, which are part of the body's fueling system. They represent its respiratory unit, through which oxygen is taken in and delivered to the blood and carbon dioxide let out. The shape (i.e., proportions), orientation, and the ratio of the atrium volume to the building volume should be properly designed so that mechanical systems do not have to be used, and maximum advantage can be taken of passive energy sources— that is, daylighting, cooling, heating, and ventilation, together with plants and water for thermal comfort. The atrium can be perceived as a massive ventilation chimney (see fig. 3.6, center, stack effect) drawing the exhaust air from the surrounding activity spaces to replace it with the fresh air from the outside, thereby venting and cooling the atrium and occupied spaces. The atrium also acts as a thermal buffer, or transition zone, to shield the occupied portion of the building from the exterior environment. Under these conditions, the temperature of the atrium air mass should be between that of the outside and inside. In summer, heat is controlled through shading of the atrium walls and roof devices, and since the cold night air is stored by the building mass, it can absorb heat during the day. In winter, the sun has direct access, and the heat can be stored by the walls and floors.

The infinite potential of the formal application of the atrium principle can only be suggested in Figure 1.15. Atria may be dividers or connectors, they may form a corridor or galleria between or through buildings or arcades along buildings, they may be enclosed in buildings or give access to buildings, they may form horizontal connective spines at ground level (possibly a galleria network), or they may be stacked vertically. From a formal point of view, they are either one-sided (attached to the building), two-sided (parallel open-ended or corner units), three-sided (partially enclosed), or four-sided (fully enclosed). In high-rise buildings, the atria may be sky lobbies, or multistory atria varying in height and stacked to create buildings within a building. A ground-floor atrium lobby may be extended to the full building height; for example a 520-ft, 48-story atrium soars in the 52-story Atlanta Marriott Marquis [1985].

Rather than providing a given floor space by using the typical high-rise tower or slab-type buildings, which cover only part of a larger site, the same building volume could be arranged around the periphery of the site to yield a more compact court building of much less height (see fig. 1.10, bottom). The larger floor area per story results in a reduced building weight per square foot. Further, the shorter building is less vulnerable to lateral force action, and requires less construction time and fewer elevators; in addition, the exterior surface-to-floor area ratio is much smaller than for an ordinary high-rise tower. Since the atrium allows a maximum exposure to daylight, and since the building has much less exterior surface area, the mechanical system's design load requirements can be proportionately reduced. The proper orientation of the atria is similar to that of the facade and rooms, as has been discussed in the section in this chapter on "Environmental Context." The atrium not only acts as a light well, but also as a buffer zone that (in cold and temperate climates, for example) shelters against the cold winds in winter, and traps and stores the warm air for heating purposes and also uses it as insulation to reduce heat loss; naturally, heat gain must be controlled during summer.

A further investigation of the functional organization of residential-use buildings has been addressed in Figure 1.16. These structures lend themselves to a fairly easy typological study, since the assembling of repetitive dwelling units around circulation

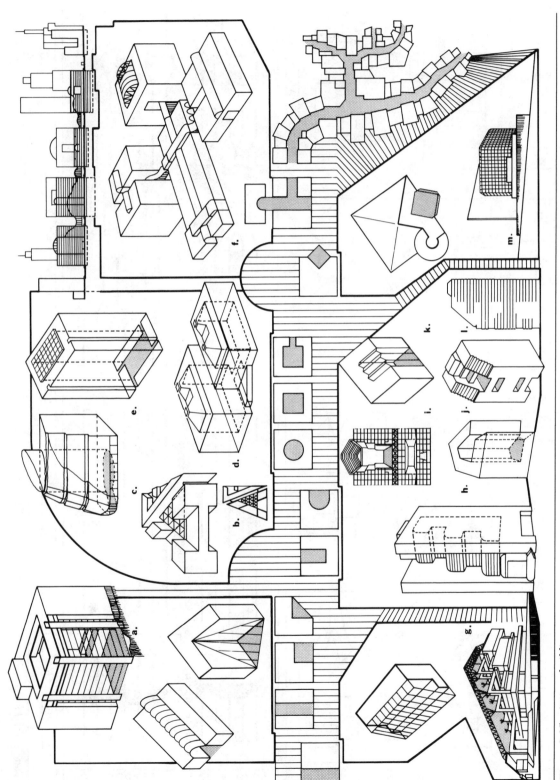

Figure 1.15. *Atrium buildings.*

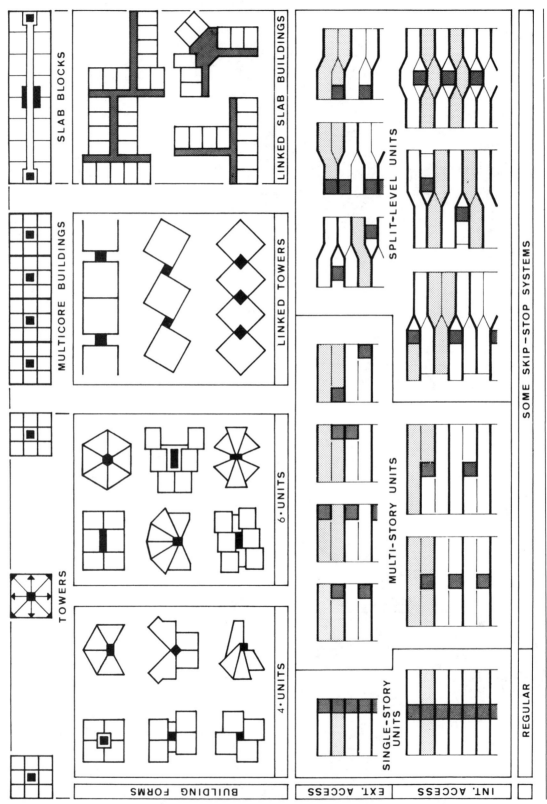

Figure 1.16. *Functional organization of residential buildings.*

systems is common to most cultures. The basic building types, from a functional point of view, are

- corridor buildings
- tower or point block buildings
- multicore buildings
- terraced buildings

From an economic point of view, the corridor and core area (as nonrental space) should be kept to a minimum, similar to the minimum path networks in nature. This nonrental space should be not larger than 10 to 15 percent of the gross floor area, to be economical. The core is centrally located in the building with respect to the apartment units to minimize travel distances, assuming that the access at the ground level does not require otherwise. Codes usually require two independent exits per floor, and they limit the travel distance from the most remote location to an exit, thus influencing the length of the building. This maximum distance, for high-rise unsprinklered residential construction, is usually between 75 to 100 ft from the dwelling unit entrance to the emergency stairs, but is increased by 50 percent for buildings with automatic sprinklers. Thus, the typical location of the primary elevator core is central and that of the secondary stairwell cores are at the ends of long-corridor slab-like buildings, where common U-type stairs with intermediate landings are used, but are part of the central service core for tower construction where scissor stairs are employed. In this case, two straight-run stairs span directly from floor to floor without an intermediate landing, but cross each other in the opposite direction. They are completely separated by a fire wall without having any access to each other.

Corridor buildings are of the slab type, where the apartment units are arranged along the corridor spine which, in turn, is served by the main elevator and secondary stair cores. The term slab-type building is somewhat deceiving, since the building shape is not necessarily a vertical flat slab, nor do the dwelling units have to be aligned in a linear fashion along the corridor. They can be arranged in curvilinear, staggered, or any other irregular fashion. The placement of the corridors is not necessarily constant; they may have different positions at the various floor levels. This location of the corridors may be the basis for other building type organizations (Fig. 1.16):

- Single-loaded corridor buildings (every floor, every second floor, etc.)
- Double-loaded corridor buildings (every floor, every second, third, or fourth floor)
- Double-loaded split level buildings
- Buildings with corridors in alternating positions

Should codes require multiple fire stairs with exits from each floor, some of the corridor systems above may not be permissible.

The central location of corridors does not permit daylight to enter; it may be perceived purely as a distribution system and as a means for reinforcing human isolation where lack of security would otherwise make tenants vulnerable to crime, particularly in subsidized public housing developments. In contrast, the open single-loaded corridors along the outside, with sunlight and view, give a more personal quality to the space and initiate contact with neighbors; they also act as climatic buffer zones against the exterior environment, although they are also more expensive to build.

For the condition where the apartments are laid out not as horizontal single-story units, but are treated as stacking of two- or three-story cells, corridors do not have to occur at every floor level, and hence elevators do not have to stop either. The dwelling units can be entered from above or below, aside from using interior stairs within the units.

In tower construction, the units are bundled in radial, rather than linear, fashion around the central core along a short peripheral corridor. The building shape is directly related to the number and configuration of the activity units, as illustrated in Figure 1.16 by the nearly square or round, tripod, cross, fan-shaped, or pinwheel and ring-shaped buildings. When towers are placed directly adjacent to, or linked with, each other, multicore buildings are formed. This multiple vertical access solution is also typical for walk-up multifamily housing.

For preliminary design purposes, the following typical apartment sizes may be assumed for the United States, realizing, however, that these values change with culture and income:

Efficiency	450–600 ft^2	(at least 16 ft wide is typical)
1BR	650–900 ft^2	($\approx$30 ft wide)
2BR	950–1200 ft^2	($\approx$45 ft wide)
3BR	1250–1600 ft^2	($\approx$55 ft wide)

Example 1.2

Assume that the program for the building tower in Example 1.1 calls for 20 percent each efficiency apartments, 1BR, and 3BR with 550, 900, and 1500 ft^2 respectively, and 40 percent 2BR with 1200 ft^2. According to the zoning ordinances, the site can accommodate 75 dwelling units.

The total net rentable floor area is equal to:

Units	Square Footage	Total Square Footage
15 efficiency	550	8,250
15 1BR	900	13,500
30 2BR	1200	36,000
15 3BR	1500	22,500
75 units total	1070 average	80,250 total

Adding 15 percent of the gross floor area or, conservatively, 20 percent of the net floor area to typical tower stories, to take into account corridors, stairs, elevator shafts, walls, etc., yields: $0.2(80,250) = 16,050$ ft^2 for a total gross tower floor square footage of 96,300 ft^2.

Adding roughly 6 percent of the gross tower area for public and mechanical spaces at ground level yields: $0.06(96,300) = 5,778$ ft^2 for a total gross floor area, excluding parking, of 102,078 ft^2. Adding the total parking floor area (see Example 1.1) of 32,813 ft^2 a total gross floor area, including parking, of 134,891 ft^2.

This total building floor area can now be multiplied by an average dollar cost per square foot to give a rough idea of the construction cost. In this case, the structure may represent nearly a third of the total costs, and the costs for the mechanical systems may come close to this value.

Using a typical apartment depth of 25 to 30 ft for a double-loaded corridor building, in order to have effective daylight, yields the approximate apartment widths given in the list of typical apartment sizes. Therefore, the number of dwelling units that fit into the given building length can be identified. Note that the effect of the support structure layout has been ignored. The typical building depth, with a standard 5-ft central corridor, is in the range of 55 to 65 ft. For an ordinary tall office building with an optimal dimension of 45 ft, from the glass line to the central core, the building depth may be in the range of 120 to 150 ft, thus being twice as deep as a typical apartment building.

The typical dwelling unit is divided into an exterior zone, containing the habitable space (i.e., living and sleeping) and an interior zone (i.e., kitchen, bathroom, entry hall) which is mechanically ventilated and artificially lit. Narrow apartment units deeper than 25 ft are usually more economical than shallow wide ones, because they are using less exterior wall area, which is expensive. However, it must be kept in mind that the plan depth significantly influences the performance requirements of the windows, with respect to daylighting. The apartment unit types may be organized according to following criteria:

- Orientation: *Single-orientation units* that are unilaterally lighted (e.g., double-loaded corridors, terrace housing), or *double-orientation units* that are bilaterally lighted (e.g., corner-type units, open-ended units).
- Form: rectangular (various proportions), angular, etc.
- Single-level units versus multi-level units.

Dwelling units are composed of various rooms (as shown in fig. 1.18). However, their arrangement, use, and need for exterior exposure are not further studied here. For minimum room sizes, the reader may refer to the HUD guidelines or Time Saver Standards.

To give some more life to the diagrammatic study in Figure 1.16, various building cases have been investigated in Figure 1.17. The richness of the spatial interaction of the vertical and horizontal circulation with the cellular units is clearly expressed. The cellular growth of an urban housing environment from the room unit to the urban megastructure is studied conceptually in Figure 1.18. First the basic modularity is defined, which coordinates the *room units*, and then the room units are assembled into *dwelling units*. These dwelling units, in turn, may be gathered into *social groups*. The social groups form the house or *building unit*. The clustering of the houses results in the *housing block*. Finally, the blocks connect to form the *urban megastructure*; hence building and town merge into each other. Though the arranging and connecting of the units can be done in many different ways, the concept of the cell as a basic building block of growth remains constant and a powerful tool for the design of large-scale built environments. Herman Hertzberger, the renowned Dutch architect, does not see this structuralistic design approach solely as a convenient method of the organization of adding standardized functions, but as a means of breaking down the monotony of the closed building block and its isolation from the urban context. The building is treated as multiple living cells or a conglomerate of autonomous parts that are clustered together to form a whole, or a building that behaves like a small town. In this context, the corridors are not just circulation, but are streets—they are communal.

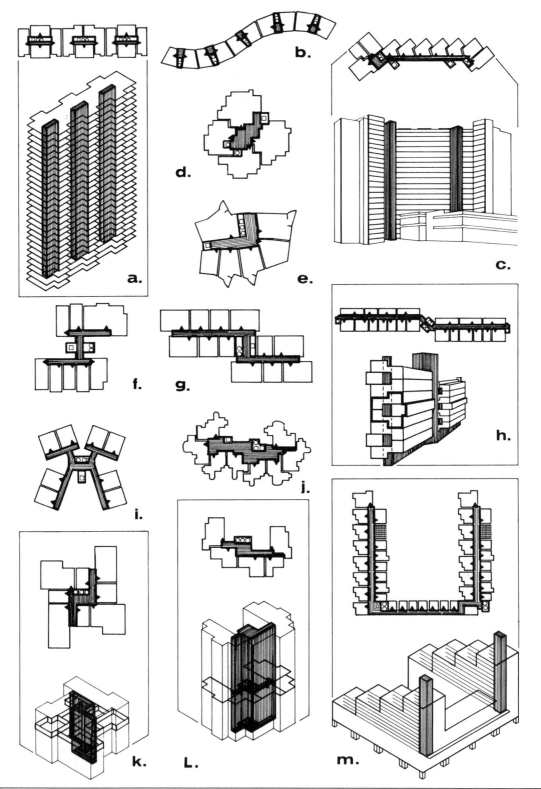

Figure 1.17. *Circulation systems for apartment buildings.*

Terraced buildings remind us of the Mesopotamian ziggurats, the early Egyptian step pyramids, the terraced pyramids in pre-Columbian Central America, the Andean and Mediterranean hill towns, and (in a way) of the stepped tower blocks in North American cities. Terraced buildings are stacked in a stair-like fashion by sitting either directly on hills, possibly cutting into the hillsides (see fig. 1.3), or they form man-made hills in the plains. They can be linear blocks that are terraced on one or both

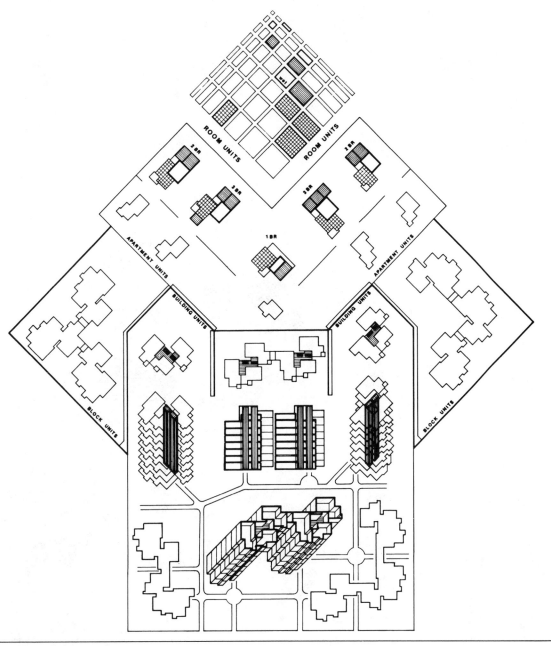

Figure 1.18. *Cellular growth.*

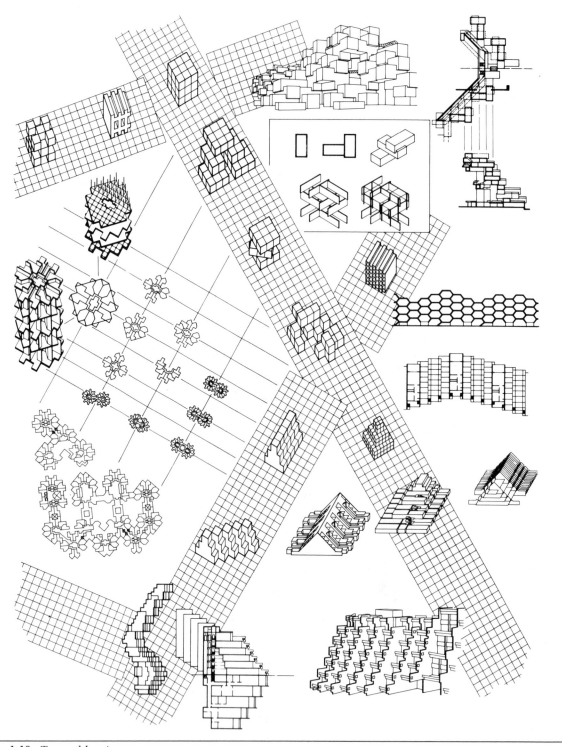

Figure 1.19. *Terraced housing.*

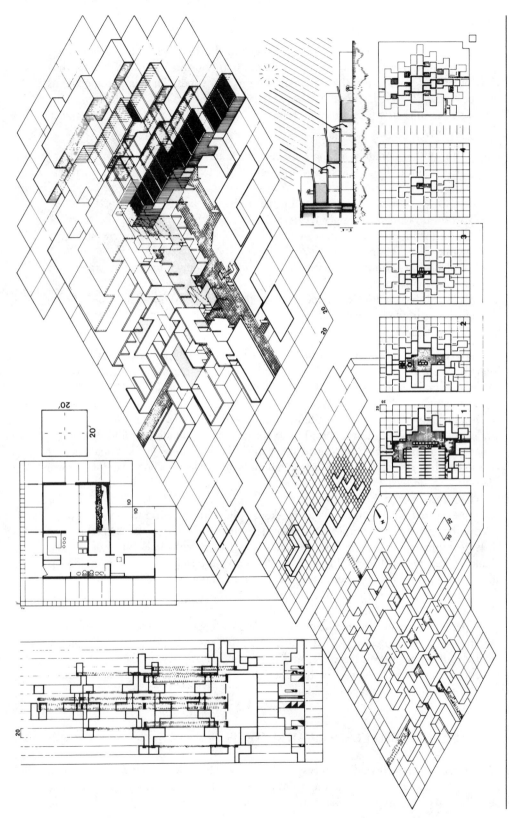

Figure 1.20. *A terraced building.*

sides, or they may be the connecting element at the bases of high-rise buildings, providing the transition to low-rise units. Rather than forming pyramidal shapes, they may be built as the inverse, that is, crater-like inverted cones. The plan arrangements can be linear, staggered, or curvilinear, or they can be closed polygonal forms, as the various cases in Figure 1.19 illustrate. Some of the intentions for the design of terraced housing are:

- To break the monotony of the high-rise block and to provide an urban building with a community atmosphere—in other words, to establish a more human environment
- To provide more outdoor privacy for the individual living units
- To provide gardens at upper levels, since the roof of one unit may serve as the terrace for the next unit above, which is set back; to possibly create a biological environment (biostructure)
- To receive a maximum exposure to sun and allow the light to efficiently penetrate into the living units
- To establish a controlled microclimate in the interior pyramid-like void that contains the circulation, with access to the dwelling units and protection against the exterior climate

Terraced buildings can be constructed by the conventional methods of bearing walls and skeletons, or by using cellular construction (as is further discussed in fig. 6.18). A famous example of this type of construction is the Habitat 67 in Montreal (Fig. 1.19, top right), where load-bearing precast concrete boxes were stacked 12 stories high and bound together by post-tensioning. Each box is 38 1/2 ft long, 17 1/2 ft wide, and 10 ft high, and weighs 80 to 95 tons. The vertical elevator and stair cores, together with the horizontal elevated streets (i.e., beams), provide additional support to the building assembly. The spatial prefab units do not have to be rectangular boxes; they can be derived from cellular agglomerates in nature, or the close packing of polyhedra can be the basis for the shape of the units and the space definition. For example, Alfred Neumann's and Zvi Hecker's polyhedric architecture is well known. Hecker based the housing project at Ramot, Jerusalem [1973], on a pentagonal pattern extended into a space-packing arrangement of regular dodecahedral units. For terraced buildings on hills with low slope angles, the cut-and-cover construction method is commonly used. This approach, however, becomes impracticable for steep banks with slopes exceeding 30 degrees, where pole- or cantilever-supported structures follow the contours.

Some of the complex geometrical organization and coordination of a terraced building has been conceptually studied in Figure 1.20. The cellular growth, the nature and layout of the support structure, and the functional space with the interaction of dwelling units and corridor space have been investigated to develop an appreciation for the restraints in this type of building.

MECHANICAL AND ELECTRICAL SYSTEMS

As the corridor network allows access to the many activity units in a building, so do the energy supply lines provide the necessary service to the same units. The tree-like distribution of the circulation space and the primary lattice of ducts, pipes, and conduits are usually conveniently grouped together, since both systems serve the same occupancy zones, similar to the streets, underground utilities, and the public works infrastructure of a city.

Rather than hiding the energy supply systems above the ceiling and in vertical shafts, they can be exposed along corridors and service areas to be easily reached, and thus become an architectural design feature. The large-volume air ductwork may be placed along the exterior of the building in order not to take up an extensive amount of interior space. As a necessary supportive element of large-scale structures, the outside circulation routes express the logic of a building and its servicing function; they display an exciting visual feature and a powerful part of the total architecture, as exemplified by the Pompidou Center [1977] in Paris (see fig. 7.41), the University Medical Center [1979] in Aachen, W. Germany (see fig. 7.6), and the Lloyd's of London [1986] (see fig. 7.10). In the Lloyd's of London, the lavatory modules are exposed and give the appearance of plug-in units. Four of the six satellite service towers are crowned with three-story plant rooms, from where the air ducts run vertically downwards and branch off horizontally to feed into the floor voids. The complex health care machine of the University Clinic of Aachen articulates along the outside in an organic fashion, an intricate web of the circulatory systems of the ventilation ducts, the pipes for water and waste for lavatories and laboratories, and the gases needed in anaesthesia and intensive care. For the Richards Memorial Laboratories [1961] in Philadelphia (see fig. 7.3), Louis Kahn exposes the exhaust, air intake, and circulation shafts as large exterior towers clustered around the laboratories, and he uses them as an essential ordering element to clearly articulate the purpose of the architecture. In contrast, for the Jonas E. Salk Institute [1965], La Jolla, CA, Kahn distributed the mechanical service systems horizontally rather than vertically, by employing the interstitial structural concept (see fig. 7.21). Structural and mechanical systems can also be integrated by using the structural members as ducts. For example, the exterior hollow columns and spandrel girders may serve as supply and return ducts for the mechanical system.

The importance of the mechanical and electrical systems in the overall design process becomes apparent when one realizes that they represent between 25 to 45 percent of the total building cost. Further, in contrast to the structural system, which constitutes a significant portion of the initial construction cost, the mechanical and electrical systems, in addition very much affect the operating costs.

It has been shown in the section in this chapter on "Environmental Context" that the effect of the perimeter surface decreases with the decrease of the surface-to-volume ratio of a building. Hence, with an increase of building scale, that is, for buildings with deep plans and massive volumes, skin-load dominated buildings such as residential low-rise construction change to internal-load dominated buildings, where the interior can no longer feel the influence of the perimeter; in this case, only the perimeter zone is still skin-load dominated (Fig. 1.21a).

The determination of the heating and cooling loads, together with energy consumption, is a complex process. It is dependent on many factors, such as the massing of the building, the type of occupancy, the orientation and nature of the building envelope (which controls the heat exchange between the exterior and interior environments), lighting, ventilation requirements, and the selected HVAC system, including the delivery system and the hours of operation. A building is zoned into spaces that have similar air quality requirements; these zones may be served by centralized

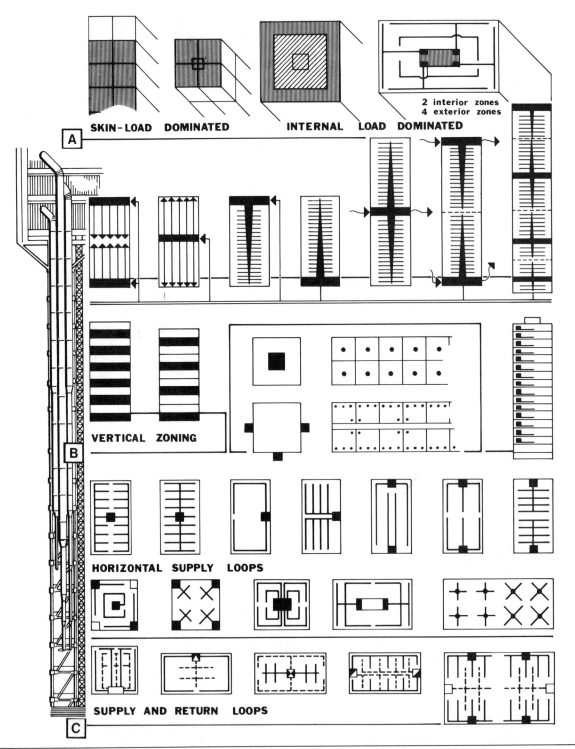

SKIN-LOAD DOMINATED

INTERNAL LOAD DOMINATED

2 interior zones
4 exterior zones

A

B

C

VERTICAL ZONING

HORIZONTAL SUPPLY LOOPS

SUPPLY AND RETURN LOOPS

Figure 1.21. *The distribution of mechanical systems.*

or decentralized HVAC systems. The typical mechanical and electrical systems in high-rise buildings are:

- HVAC (heating, ventilation, and air conditioning)
- Domestic cold and hot water systems
- Plumbing systems (storm drainage and sanitary drainage)
- Fire protection and security (see the section in this chapter on "Fire Safety")
- Electrical distribution, including communications
- Lighting
- Transportation (e.g., elevators—see the section in this chapter on "Functional Organizers")

The mechanical systems basically consist of the equipment in the mechanical rooms and the distribution trees that deliver fuel to the terminal units. Similar to the human body, which needs food to function, the mechanical systems require fuel consisting of air, water, heat, light, and energy to move it; the energy is provided as electricity, gas, oil, coal, or solar power. This fuel is distributed to the various building parts along a duct, piping, and wiring network, which reflects a tree-like distribution, where the vertical trunks are large enough to nourish all of its horizontal branches. This distribution is similar to the circulation patterns of blood through the arteries and veins. The vertical trunks represent the chimneys, flues for combustion gases, plumbing shafts, mechanical shafts, waste disposal chutes, and fresh air and exhaust air shafts.

The complex electrical wiring network distributed all over a building is similar to the nerve system that branches out from the spinal chord to embrace the entire body (see fig. 1.12). In this instance, the transmission of electricity can be visualized as being similar to the water distribution system. The electrons flow through wire conductors within a conduit along a continuous path (i.e., a closed loop) or a circuit, due to the separate charges on the opposing terminals. This potential difference across the ends of the conductor is a force causing the electrical current in the wires, analogous to the flow of water in pipes from a higher pressure reservoir to a lower one; the force that moves the current is the voltage, similar to the pressure that causes the flow of water. As the water flow can be measured in gallons per minute, the electron flow or current is measured in amperes (charge per unit time). The resistance in the circuits to the flow (measured in ohms) relates directly to the skin-friction drag in the pipes that slows down the flow of water. The available power, or the rate at which the energy can be used, is expressed in watts, and the total energy consumption is given in kilowatt hours (kWh).

The flow of the distribution systems is dependent on the function of the building. In buildings with fixed cellular subdivisions, such as for apartments, dormitories, and hotels, a decentralized branching occurs, where each activity unit must provide adjustable heat service and water supply including plumbing stacks. In contrast, in open-office landscapes of high-rise buildings or in the skyscrapers of the mixed-use type, a much more centralized branching of the mechanical services is the norm.

HVAC Systems

The need for heating and cooling in apartment buildings depends on the geographic area and the type of occupancy (i.e., luxury housing versus low-cost public housing); in certain climates no heating and cooling are required. For typical apartment buildings in moderate climates, daylighting and natural ventilation through open windows are used, together with central heating and possibly individual cooling (e.g., through-the-wall self-contained air conditioning units). Central heating is achieved by circulating hot water or steam, while central cooling is done with chilled water, usually through the same pipes that carry hot water in winter. Interior kitchens and bathrooms (i.e., wet units) need mechanical ventilation through exhaust shafts. The air that is exhausted, is replaced either directly from outdoors or by makeup air fed into the corridors. Mechanical ventilation is needed for interior corridors, if no windows are present. The wet units are stacked vertically on top of each other, with plumbing walls sandwiched between them. By backing up the plumbing facilities, the water and drainage lines can efficiently serve a maximum number of fixtures. The mechanical equipment is usually located in the basement (e.g., boilers, chillers) and on the roof (e.g., air-handling equipment, cooling towers). The pipes for heating are distributed vertically along the perimeter walls directly to the various rooms, quite in contrast to the centralized vertical approach with horizontal runs for large-scale open spaces. Whereas in apartment buildings dropped ceilings are rarely needed, they are common in hotels along access corridors.

High-rise, multiuse, or office buildings in the United States generally have central air conditioning; they mostly have fixed windows so that the thermal comfort and the air quality of the interior environment must be controlled mechanically. The factors to be considered are: air temperature, relative humidity, air movement, air cleanliness and odor, and radiant temperatures of surfaces. Most high-rise buildings in the United States have a cooling problem; their interior zone is heat-dominated, due to electric lighting, electronic office equipment, computers, and people, who act like radiators set at 98.6°F. This heat build-up in the central core area needs cooling and ventilation all year round. The perimeter zone, however, requires cooling or heating, depending on the transmission load of the building skin; it is skin-load dominated and influenced by the outdoor climate, the orientation, nature of the skin (degree of insulation), and the level of occupancy (internal heat gain).

A multiuse building is not only divided horizontally, but also vertically, into primary *thermal zones*, according to their respective use. It makes sense that each zone may be served by a different HVAC system, some zones possibly by central systems and others by local systems. For the stable internal zone, cooling is required throughout the year, while heating and cooling are needed for the perimeter zone. For example, during the winter months mechanical ventilation may redistribute the heat from the interior zone to the perimeter zone, while during summer it flushes out the heat to reduce the cooling load.

The selection of a system depends (among other criteria) on the building size, the placement of the zones, the thermal loads, and on the comfort determinants in response to space use. The typical centralized equipment that generates heat or cooling and provides circulation consists of: furnaces (warm air), boilers (hot water, steam), a chilled water plant (i.e., refrigeration equipment producing cold water for cooling) together with cooling towers (or evaporative condensers), which handle the heat rejected from the chillers and discharge it to the outside, pumps (control of water flow), motors, and air-handling equipment (supply and return fans for air circulation). The heating equipment converts the electrical or natural fuels (gas, oil, or coal) to heat; it either heats a fluid (air, water), or it generates a heat-carrying medium from water (steam). The various media for moving heat are air, water, steam, and (in a way) electricity, although it does not require any other medium (though it is costly to heat in this way). In other words, heat can be delivered to a space as warm air, hot water, steam, or electricity, while cooling is delivered by cooled air or chilled water. The HVAC systems are classified according to the heat-carrying medium as

- All-air
- All-water
- Air-water
- Steam
- Combination of the above

The heat and cooling is disseminated at the terminals, for example, by air diffusers, radiators, convectors, electric radiant baseboard heaters, or fan coil units. Some of the more common terminal devices are:

- *Fin-tube radiation systems* (e.g., baseboard units):
  Steam or hot water is circulated within the tubing, and the heat transfer from the surface to air is achieved by natural convection. This system may be used as perimeter heating, together with a ducted air-conditioning system for cooling.
- *Unit heater with fan* (e.g., through-the-wall units):
  These can obtain the heat from steam, hot water, and electricity.
- *Fan coil units* (e.g., wall or ceiling units):
  These consist of a heating and/or cooling coil, fan, and filter. They may be connected to a piped water distribution system to heat or cool the air; the fan blows the air over the coil. Also, in this case, the air ventilation must be provided by mechanical or natural means.
- *Induction units* (e.g., floor-mounted cabinet, or overhead outlet):
  These are similar to fan coil units, but without a fan. Hot or chilled water is piped to the unit and air is ducted under high pressure, passing through a coil for reheating or cooling.
- *Radiant heating or cooling devices* (e.g., infrared heaters, radiant ceiling or wall panels, embedded piping):
  These use either air, water, steam, or electricity.

The size of air-handling equipment ranges from a simple air-conditioning unit for a room or window unit, to through-the-wall heat pumps, factory-assembled unitary systems with ductwork, to a central system consisting of several air-handling units, as is typical for hospitals and multistory buildings. In the *all-air* system, the temperature, humidity, cleanliness, odor, and distribution of the air is controlled. Fans move the air for heating or cooling along ducts to the various building spaces, first vertically along shafts that may be located at the two opposite ends of the core (Fig. 1.21, top), and then horizontally along ducts at each floor level to the interior and perimeter zones. The exterior and interior zones, in turn, are further subdivided into secondary zones, which must be reached by the branches of the supply loop ducts. The ducts are terminated by outlets (e.g., ceiling diffusers of various shapes, floor or wall grilles and registers, perforated ceiling panels). The return air flows through the ceiling plenum either directly to the open return shafts, or is first collected in horizontal return ducts (Fig. 1.21c); the return inlets are located near the floor. The ducts (round, rectangular) are classified according to the air pressure and velocity within the ducts. A high-velocity, high-pressure ductwork results in a significant reduction of duct size.

The heating and cooling control can be accomplished centrally by the air-handling equipment before the conditioned air is ducted through the building to the various zones, or it can be controlled and modified locally by terminal boxes (e.g., reheat systems, mixing boxes, induction units).

Cooling basically consists of the extraction of heat from the interior space and its disposal into the atmosphere, since it may not be required for other purposes, and since it cannot be destroyed. The medium for removing the heat is a refrigerant liquid (such as freon), which is continuously recycled. In the general refrigeration cycle, a liquid must absorb energy (heat) to vaporize, and the gas vapor, in turn, must release energy (heat) to liquefy (i.e., the process of vaporization and condensation). When this principle is applied to refrigeration machines, cooling takes place as the liquid refrigerant passes through the low-pressure evaporator. At this stage, the refrigerant evaporates as it draws heat either directly from the air, as for small-capacity air-conditioning systems, or indirectly from the surrounding water, thereby producing chilled water, which is circulated through its own pipe network to the spaces to be cooled (i.e., supply and return pipes run to fan rooms or terminals). In the next cycle, the refrigerant vapor must be converted back to its original liquid state by condensing it in the condenser, and by rejecting the heat generated to a heat sink (such as a cooling tower) or to an evaporative condenser, which combines the tasks of the condenser and the cooling tower. For example, the hot water from the condensers is cooled in the cooling towers and is then circulated back to the chillers; in other words, a condenser water (supply and return) piping network connects chillers and cooling towers. This change of state from gas to liquid requires energy either from mechanical work, from some type of compressor, or it requires more heat, as through an absorption chiller.

Forced-air systems may be classified as constant-volume or variable-volume systems. These systems are of the centralized type, in contrast to fan coil and heat pump systems, which are decentralized. In the *constant-volume system*, a constant flow of conditioned air is ducted to the different zones; in this case, the variable temperature is controlled centrally, or locally by mixing cooled air with the heated air, or reheating the cooled air. The common constant-volume systems are: single duct/zone systems, terminal reheat systems, double duct systems, multizone systems, and induction systems. In the *single zone system*, often used in department stores, a single duct delivers the air from the conditioning supply directly to the required location, and a return duct brings the air back to the supply unit. When the single zone system includes a reheat coil (which uses hot water, steam, or electric energy) in the duct for the various spaces, it is called a *terminal reheat system*. This system is controlled locally by thermostats, and responds to the fluctuating conditions of laboratories and hospitals, or to the load variations in large office spaces. In the *double duct system* one duct carries the cold air and the other the warm air. The separate hot and cold air streams are extended to the point of usage, where they are combined in terminal mixing boxes to serve specific zone/room requirements. Should the air be mixed centrally at the fan system rather than locally, and be distributed with a single duct to the respective space, then the system is called a *multizone system*. These systems are often used for smaller commercial buildings. *Induction systems* provide high-pressure, high-velocity conditioned air with small ducts to the terminal induction units. The primary constant volume systems for interior zones may be supplemented by secondary perimeter systems such as reheat systems, induction systems, or fan coil unit systems.

Variable-volume systems (VAV systems) have become most popular in office buildings. In this case, a varying amount of constant temperature conditioned air is supplied through a single duct to each zone; the air is not mixed or reheated, mainly cooling is provided as is needed for the heat-dominated interior zones. Hybrid systems include the variable-volume double duct system, which uses an approach similar to the constant-volume terminal reheat system. The VAV system (for cooling) is often used together with a separate perimeter system for heating, such as fin-tube radiation systems or constant-volume systems (dual conduit system).

In the *all-water* systems, water is heated in a boiler or cooled in a chiller, and distributed along supply and return pipes to a central or terminal air-handling unit. Air is then circulated locally, rather than centrally as for air-water systems, thus eliminating ductwork. It is common for apartment and hotel buildings to distribute the water along the perimeter to the terminal room devices such as fin-tube radiation systems and unit heaters for heating, or fan coil units and radiant panels both for heating and cooling. The water circuits are organized according to following distribution systems:

- *Two-pipe systems*, with one supply and return line for either heated or chilled water.
- *Three-pipe systems*, with one hot supply pipe, one cold supply pipe, and one common return pipe.
- *Four-pipe systems*, with separate piping circuits, where one loop supplies and returns the hot water and the other the chilled water.

Because water has a much higher density than air, it also has a much higher cooling and heating capacity, thus requiring a much smaller duct size and possibly less story height; water pipes require the smallest distribution trees. Since this all-water system does not provide humidity and the control of air quality, it may have to be assisted by natural or mechanical ventilation. Water, like air, expands when heated, rising as it becomes lighter, and as it cools, falling. Thus, the temperature differences cause a continuous but weak movement in the hot water piping and tanks, but this will have to be assisted by pumps or gravity water tanks to control the flow to the various zones.

The *air-water* systems are used generally for the exterior spaces of buildings. In this instance, the air is conditioned at a central source and circulated to the terminals (e.g., air-water induction units), where the air temperature is modified by water piped from the boilers or chillers. These systems allow a reduction in the volume of duct distribution, since most of the heating or cooling is done through the water piping network, and the rest through the ducted air system.

In the *steam* heating system, as used along the perimeter of apartment buildings, steam rises in the pipes as hot vapor under its own pressure and changes back to water as it gives heat off in the terminal units, such as a fin-tube radiation system or the familiar cast-iron radiators, and then flows back in the return pipes to the boilers to be boiled again to steam. The pipes for steam heating are larger than for water, but smaller than air ducts.

In high-rise buildings with large window areas, a forced cool air system is often used together with a secondary perimeter system such as an all-water heating system, to deal with the large difference in heating and cooling demands. The ductwork for cooling is branched into space from above, while the piping is either routed vertically in the interior core area and then horizontally branched, or it is distributed directly vertically along the perimeter walls or columns to feed the terminal units underneath the windows, from below.

Plumbing Systems

The piping network of the plumbing systems carries the water, air, gas, and waste to the various building parts. In high-rise buildings, the *domestic water supply* must be pumped to the elevated storage tanks at the mechanical levels. *Pressure boosting systems* are needed for the hot and cold water systems, as well as for the water supply for fire protection, in order to provide adequate pressure to all parts of the building. Tall buildings are subdivided vertically to maintain this pressure in each zone, which may be provided by gravity water tanks, pressure tanks (closed water tanks with compressed air), or booster pump systems. When the source of pressure is at the top of the zone, it is called a *downfeed zone*, otherwise it is an *upfeed* or *combination zone*. For automatic fire sprinkler systems, with their extensive horizontal branching network (cross mains and branch lines), pumps are naturally required. For example, the sprinkler system of the Sears Tower in Chicago has seven vertical zones, where each zone is covered by pumps.

Waste disposal is moved by gravity through the drainage pipes to the sewers, and solid waste can be disposed by burning it in an incinerator. Plumbing vent-stacks may be required; the venting pipes provide fresh air circulation in the plumbing stacks to remove the gases that have been generated by the decomposition of waste. Localized exhaust systems are not only necessary for interior toilet rooms, bathrooms, and kitchens, but particularly for laboratories. To avoid the backup of foamy agents, waste stacks for upper floors of tall buildings cannot be connected to the lower floors. The storm drainage system carries water from roofs and paved areas to the storm sewer system.

Electrical Distribution

High-voltage electric power is transmitted along underground mains through primary feeders to a terminal, usually located in the basement of high-rise buildings; this service entrance equipment consists mainly of a large transformer, which reduces the power to lower voltages that can be utilized in the building. From there, the electrical power is distributed vertically along the main riser(s) in a tree-like pattern, via low-voltage cables, to secondary subterminals, which consist of interior distribution equipment. Finally, it is brought to the utilization points such as motors, machines, lighting and heating devices, outlets and so on. Electricity is needed for lighting, which consumes a large portion of the total electric energy in tall buildings, besides generating a considerable amount of heat. It is needed for mechanical, electrical, sanitary, and data-processing equipment, for elevators, communication systems, and other machines. Large, heavy battery systems (e.g., 400 lb/ft^2), or generators (i.e., engines, turbines), possibly together with fuel oil storage tanks, may be required as a source of emergency power, to ensure a continuous power supply, as is essential for hospitals and computer centers. Power demands have recently increased rapidly because of the explosion of information age technology and the development of the *intelligent building* concept. Extensive electronic equipment requires a greater power supply capacity and a more flexible underfloor electrical power distribution network to service the *electronic office*. In the modern computer building, a major portion of the electrical power is consumed by venting and cooling down machinery—in this context, equipment energy needs, by far, supersede human energy needs.

Building Organization

A typical multiuse high-rise building is divided into many different primary thermal zones; there are usually at least five zones at each floor level (see fig. 1.21a). Vertically, the building is zoned into several sections according to occupancy, such as parking, retail, offices, mechanical levels, and the apartment portion. The mechanical, electrical, and sanitary equipment is concentrated in blocks at the so-called mechanical levels. These may be located (Fig. 1.21b) at the bottom, roof-top, midheight, and/or intermediate levels, spaced at roughly 12 to 20 floors, or at points where the occupancy changes, for instance, from offices to hotels; they should be located centrally to minimize piping. The mechanical level must furnish the air requirements for the number of floors that it serves. The fresh air intake and exhaust at these levels may be visually expressed by louvers along the facade. It also should be remembered that the natural fuel burning furnaces and boilers always need an unobstructed supply of air for combustion, besides flues for the removal of the waste gases.

With an increase in the number of floors, not only the vertical duct shafts and the size of the equipment, but also the height of the mechanical floor(s) increases. Some of the typical equipment in these service spaces are chillers, boilers, pumps, fans, motors, and sanitary equipment. In addition, there will be electrical and plumbing service rooms and fuel storage at the basement level, storage tanks and hot-water tanks at the intermediate levels, as well as an elevator penthouse and cooling tower or other condenser types at the roof level. Heavy equipment should be located at the basement level on grade so that the building does not have to carry these loads.

In the conventional centralized approach, the primary energy of electricity and water is first routed along the central tree trunk (core) to the mechanical stories, realizing that the required air can enter these levels directly through grilles. From these locations, HVAC services are distributed in a tree-like fashion to the various building portions; for example, huge fans blow air through large vertical shafts and horizontal ducts—naturally, complex control systems are needed. This vertical distribution of mechanical services is not necessarily concentrated centrally along one big core, but could consist of separate shafts, depending on the building shape and occupancy distribution (e.g., laboratories); each of these smaller distribution trees services its own building zone. In the case of multiple small trunks along the building periphery, possibly integrated with the columns, only short branches within the walls are required.

The zoning of a building can be centralized or decentralized both from a horizontal and vertical point of view. In the decentralized approach, rather than concentrating the mechanical equipment at certain levels, a greater number of smaller local mechanical equipment rooms, with local air intake and exhaust, may be placed (for instance) at every floor level; these may provide space for heating, cooling, and ventilating equipment, while a cooling tower may serve the entire building. The advantage of the decentralized approach lies in the reduction of distribution trees, a stricter control of the air quality according to local needs, assurance of only a local effect due to an individual system breakdown, and better safety control, because the fans do not easily spread smoke and gases to the other floors. In the interstitial approach, the spaces provide the means for flexibility and expansion; additional branches, equipment, and controls can be provided as the floor functions change. In this case, the basic trunk lines of the mechanical systems run horizontally rather than vertically.

As has already been emphasized, there is a clear hierarchy of the duct/pipe/wiring systems, reflecting in their size decrease a decrease of loads as they approach the final point of usage. The distribution follows from the main vertical trunk to secondary and tertiary horizontal branches, and correspondingly decreases in size with distance from the central plant. The horizontal network is quite elaborate for office spaces. It may occur below the structural slab, within the slab (e.g., wires), and above the slab. While the horizontal supply and return piping for the plumbing is concentrated only at certain locations, the complex network of underfloor electrical conduits is distributed all over the floor area, as is the sprinkler piping, although not as intricately. Similarly, the supply and return ductwork for the forced air system is branched to the various zones that it must service.

There are a minimum of five thermal zones for a typical floor of a high-rise building. The interior zone is heat-dominated, it is buffered from outside conditions and usually has no net energy gain; the interior zone of the example in Figure 1.21a has been subdivided into two sections. The perimeter zone, in contrast, is directly affected by outside temperatures, losing heat in winter and gaining it in summer. It is usually divided according to orientation into east, south, west, and north zones. Single-path systems with a common duct distribution to feed all of the terminals, or dual-path systems using separate ducts for heating and cooling, with a common return, must service these different interior and perimeter building zones. Various horizontal branching systems are identified in Figure 1.21c. The typical supply branches form either *circumferential loops* within the perimeter and/or interior zones, *linear tree-like loops*, or *radial loops*. Any combination of these basic layout patterns can be used to cover the interior and perimeter zones. Should the return air be collected in horizontal ducts rather than flow directly through the ceiling plenum to the vertical return shafts, then the return loops (dashed lines) and supply loops (solid lines) can (for instance) take the form of branching pattern shown in Figure 1.21c. The horizontal branch distribution naturally depends on the number and location of the vertical supply and return shafts, ranging from a centralized approach with a single core at the building center to a multiple zone decentralized approach with many small distribution trees.

The ducts and pipes are suspended from the floor structure and run below it over the suspended ceiling, or they are directly supported by it (e.g., trusses, latticed or framed girders, beams with penetrations in the webs, castellated beams, stub-girders). In this situation, the floor beams allow the passage of the mechanical ductwork and piping perpendicular to the span so that the mechanical system can be accommodated in the same space volume as the structural system. The penetrations of the pipes are usually less than for the ducts because of their smaller sizes, but the weight of the pipes is significantly higher. The pipes and ducts, together with the electrical and communication services, can also be placed above the structural slab by providing a raised floor, as is common for computer rooms. The typical wiring distribution of the power and telephone cables occurs within the cellular deck, or as flat undercarpet cable, or below the raised floor. The fact should not be overlooked that the horizontal mechanical space for centralized air-handling systems may take up to about one-third of the overall floor-to-floor height! The typical ceiling-floor assembly height for tall office buildings is about 46 in.

Some basic considerations of efficiency for the layout of networks are:

- To use a minimum amount of space by integrating with other building components.
- To cluster closely around the vertical distribution shafts the major equipment spaces that feed them (e.g., boiler rooms, chilled water plant, cooling towers, fan rooms, water pumps, waste compactor, etc.).
- To allow for horizontal and vertical continuity by using straight lines connected with smooth curves, including easy joining of horizontal ducts with vertical ducts in the shafts.
- To use fewer branches with the shortest or smallest runs.
- To use branching systems of a balanced, symmetrical layout.
- To use ducts with a minimum frictional resistance to flow, thus minimizing central power requirements for movement.

The interior environment is automatically controlled in the new breed of smart buildings. Office automation and communications provide an optimal environment for productivity, and control systems monitor the building performance. For example, a network of sensors are alert to temperature changes and gather data about the building's interior environment. This information, in turn, is used by computers to adjust the building controls to efficiently respond to the changing conditions and manage energy consumption. The degree of automated intelligence depends on how many of the components of the mechanical systems, fire and life safety, security, elevators, telephone, data and work processing, and office automation are connected to the electronic network. However, this trend in today's buildings towards intelligence to create sophisticated microclimates requires more energy for the electronic equipment and flexibility of layout. The technology demands much more power, but also yields more heat output, which in turn increases the cooling capacity requirements, even when the heat flow is more efficiently controlled.

FIRE SAFETY

The protection of life during a fire is one of the primary and most complex design considerations. Not only the integrity of the structure, as expressed in its fire resistance, but also the escape routes for the occupants and the safety of the fire fighting teams must be considered, which includes proper access to the building on the site. The fire safety in high-rise buildings and its potential failure is much more critical than for low buildings, which are accessible from the outside with aerial ladder equipment up to ten floors in some cities. It must also be kept in mind that fires in high-rise buildings tend to spread much faster than in low-rise buildings, due to the stack effect of the temperature difference between inside and outside. In tall structures, rescue efforts must be fought from the inside; ground-based fire services cannot be provided. Only people in the lower floors can be evacuated, while all of the other occupants must be moved to safe refuge areas within the building, thus requiring the vertical and horizontal compartmentalization of a building. These compartments form fire-tight cells that consist of a continuous fire barrier membrane of wall and floor/ceiling surfaces with special tightly closed doors that form an envelope with a thermal resistance capable of containing the fire for a certain time period, and ideally surviving a burn-out of the contents without failure of the barrier.

However, this is only a theoretical model, because the walls and floors of these cells cannot be completely fire-tight. They are penetrated and weakened by service ducts, vertical shafts that act like chimneys (i.e., stack effect), open doors, cracks, and other cavities, as well as insufficient fire stops along edges and other penetrations, such as poke-through systems for the conduits through the floor slab. The flames also spread externally outside of the building through the windows from floor to floor. The size and shape of the flame plume will vary with the window shape, fuel loads, wind, room geometry, exterior finish, and so on. Even the radiant heat transfer from the flame plume may ignite combustible items on the story above. It is apparent that windows should be as small as possible to reduce fire spreading. To deflect the flames away from the facade, the windows should be taller than wide or flame barriers should be used.

The layout of the building must provide smoke-free (vented) and fire-protected horizontal and vertical enclosures as escape routes and adequate exits, as well as access paths for the fire fighters; in this context, vestibules may act as smoke barriers between shafts and floors. The minimum dimensions for the escape routes to accommodate the occupants in case of an emergency, as well as the maximum travel distance along the corridor to the exits or emergency staircases (and elevators, which may become unsafe, since the shafts act as smoke routes) is given by the codes, and has partly been introduced in the section in this chapter on "Functional Organizers." Actually, the length of the travel paths should be directly related to the combustible contents. It should also not be overlooked that the use of a fire wall may eliminate the need for a staircase. It is obvious that the early detection of fire by some alarm system is essential.

Most people get trapped by smoke and are injured or killed by toxic gases. Thus, the control of smoke infiltration to protect the egress routes is critical. Just to control the spread of smoke through a passive barrier may not be satisfactory because of leakage and the air pressure difference across the barrier; in addition, smoke zone venting or smoke control zones may have to be introduced. The venting of the smoke can be done directly through the exterior wall, by smoke shafts, or by mechanical venting using the exhaust fans of the HVAC system, keeping in mind that the mechanical system may not always work. To limit smoke movement, the mechanical fans can be used to pressurize certain zones. Examples of this are the pressurization of the vertical stairwell and elevator shafts to combat the stack effect, or the formation of a pressure sandwich by pressurizing the floors above and below the smoke zone (Fig. 1.22D).

In contrast to conventional buildings, where the floor layers resist the growth of fire, fire and smoke can easily move in atrium spaces, and thus become a complex problem in fire safety design. Large volumes of smoke must be removed not just by using automatic smoke-extraction fans and by other methods of venting the smoke through the roof glazing, but also by leaving some floor levels open to the atrium while the other floors are enclosed by a fire- and smoke-protective barrier. In addition, heat radiation through the glass may have to be considered.

In order to control the development of the fire, some understanding of its infinitely complex nature must be acquired. Fire is a combustion or a chemical reaction between a substance and the oxygen in the air, which is started by applying heat from an external source. Once the initial stage of ignition has occurred, more heat is released

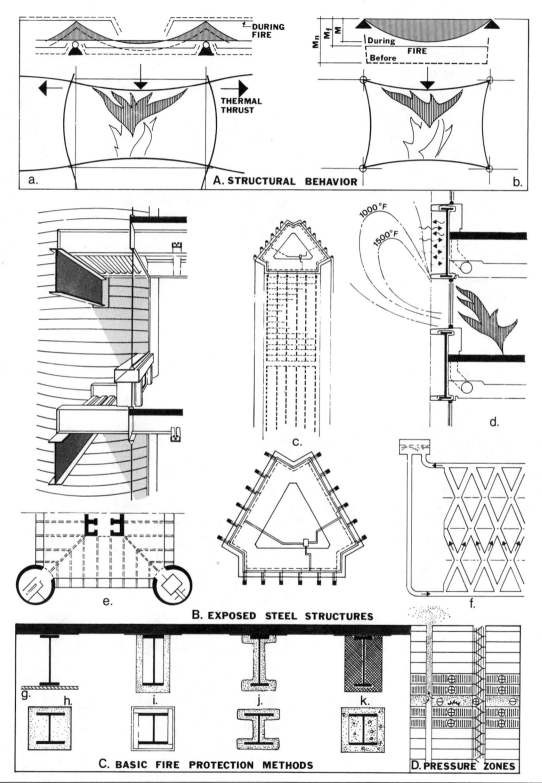

Figure 1.22. *Considerations related to fire safety.*

than needed to maintain the reaction, and the fire grows in intensity. The severity and duration of the fire depends on the fuel and the geometry, height, openness, and degree of thermal insulation of the space in which it is contained, which in turn influences the air supply and heat loss. When an adequate air supply is available, the fire is *fuel controlled*. It is dependent on the nature, amount, and arrangement of any combustible materials, while the ventilation depends on the size of the window area, doors, HVAC system, and other penetrations. However, when the air supply is restricted, the fire is *ventilation controlled*, causing the fire to spread and to search for oxygen. The most severe fire seems to occur between these two extreme conditions.

The combustion process not only creates flames and extreme heat, but also smoke together with toxic gases. As the fire grows and the flames become larger, the air supply usually decreases, resulting in an incomplete combustion, with a corresponding increase of smoke. It is the spread of smoke and toxic gases, and not the flames, which constitutes the major source of injuries and fatalities. Fires along the building perimeter spaces usually don't last as long, their high intensity being only of short duration. These spaces cool down faster because they are ventilated through the broken glass windows.

Large fires spread more rapidly by *radiation* and *convection* of heat from the flames and hot gases than by *conduction*. Combustible materials absorb heat from convection currents that are generated by heat and through radiation until they ignite. Thin materials are ignited earlier than thick ones, since less heat is needed to reach the ignition temperature. Fire spreads only slowly by conduction, where materials are in contact with each other, for example, along a wooden floor through a wall to adjacent spaces. However, in contrast to wood, steel is a good conductor. On the other hand, fire spreads by convection through vertical shafts as a result of temperature change, or heat is transferred by air flowing over hot surfaces. Fire can spread by radiation across a clear open space to nearby buildings.

A fire can be suppressed by taking its fuel away. But since different materials burn differently, special fire fighting techniques are required. Most fires in high-rise buildings can be extinguished with water to remove the heat; as the water is converted to steam, the temperatures of burning materials are reduced below their ignition points. Automatic sprinkler systems are most effective at an early stage of a fire, and they are effective in confining a fire to where it started, although it must be kept in mind that they may not always work. When water is not suitable, certain chemicals can be used that create foams, which suffocate the flames by excluding the oxygen. However, fires of high intensity, such as those due to flammable liquids and chemicals, must be handled by foaming agents, dry chemicals, carbon dioxide, or other gases—whatever is applicable. Other automatic fire suppression systems besides sprinklers are the Halon systems, which usually are employed in critical areas where potential water damage cannot be allowed. The Halon is stored in tanks as a liquid, and when released becomes a gas, creating an atmosphere not supporting combustion.

A basic consideration for fire safety and smoke control at the design stage is the restriction of the combustible content or the *fire loading*; fire loads are generally given in terms of the heat produced (in Btu/ft^2). The designer must be aware that plastic fuels cause much more intense fires than cellulosic fuels, besides a much higher production of smoke and toxic gas. The use of synthetic insulation, finishes, and carpeting must be restricted. Whereas the fixed fire loads of the exposed structural and nonstructural materials can be controlled by avoiding a highly combustible finishing, this is not the case for the movable fire loads of the furnishing materials or contents, over which the designer has no control. The duration and severity of a fire depends, among other criteria, on the type of combustible contents, which in turn is directly related to the occupancy of the enclosure! Therefore, codes require minimum times of fire resistance for the supporting structures, such as 1 1/2 hours for apartment buildings, 2 hours for office and hotel buildings, and 3 hours for industrial and mercantile structures.

A building structure must be designed with the necessary fire endurance, to guarantee stability and to guard against the spread of fire beyond the compartment of origin, so that people's lives are protected by making an escape route possible. Building codes protect structural integrity by regulating the fire resistance of the horizontal and vertical building planes that form fire-tight cells, on the basis not only of occupancy, but also of the type of construction, which is determined by the building size (height and floor area), structural materials, building location, sprinkler protection, and other criteria. In addition, insurance rates reflect the degree of fire resistance of structures.

It is apparent that the structural members must be able to withstand, for a certain time period, the stresses due to fully developed fires, together with the other load actions, without collapse. Therefore, fire resistance is rated in terms of the number of hours that the building element must resist the exposure. However, that does not mean that, for a given occupancy rating, all building components must have the same rating. In this case, the more critical columns may have the maximum rating of a 4-hour fire resistance while the floor assembly is rated at only 3 hours. In the United States, typical code requirements for tall buildings, to insure fire integrity for the main structural members, are:

Floor construction:	2–3 hours
Frames, including columns and interior bearing walls:	3–4 hours
Shaft enclosures:	2 hours
Roofs:	$1^{1}/_{2}$–2 hours

The fire-resistance rating for various structural assemblies, such as floors, ceilings, walls, columns, beams, and so on, are based on fire tests according to ASTM E119, such as those conducted by Underwriters' Laboratories, Inc., and as accepted by the building codes. A list of fire-resistance ratings is available from various agencies, such as: Underwriters' Laboratories, Inc. (UL); American Insurance Association (AIA); and the National Bureau of Standards (NBS).

For example, the minimum concrete slab thickness for a 2-hr fire resistance rating is from 3.5 to 5 in. depending on the type of aggregate. Typical for office buildings is the 2 or 3-in. steel deck with 2 1/2 in. normal-weight concrete, with spray-on fire proofing, or with 3 1/4 in. lightweight concrete (115 lb/ft^3) without fireproofing, to provide the required 2-hr fire rating. Concrete columns of 8×8 in. have a fire rating of 1.5 hours, while 12×12 in. concrete columns may have roughly a 3-hr rating. For preliminary design purposes, it may be assumed that 2 inches of fire insulation around steel I-sections will provide a 4-hr rating.

Rather than using the prescriptive code approach, structural fire engineering has developed more performance-oriented rational methods. Analytical techniques are available (rather than tests) for estimating fire-resistance ratings for steel beam and column designs, by extending the fire test data. The fire resistance of a structural member varies with its mass and shape, besides other criteria. It is apparent that a heavier and more massive section with a smaller perimeter surface exposed to the flames takes more time to heat up than a light slender section with a large exposure surface. For example, for a steel I-section without cladding, the temperature change will vary inversely with the cross-sectional area, A, but directly with the exposed perimeter, P. One may conclude that the type of fire protection must vary directly with the P/A ratio, where a small ratio, as provided by a bulky heavy section with a small heated perimeter, requires less insulation.

Structural fire engineering has been doing extensive research into developing rational methods for designing structures under gravity loads in a fire. For this design condition, the fire environment must be defined, the heat transfer to the structure predicted, and the performance of structural members understood. It is quite apparent that this is an extremely difficult undertaking. Some introductory basic concepts of analysis of thermally-induced deformations and stresses for unprotected steel elements are discussed in Chapter 3 under "Hidden Loads." However, since the thermal effects on the material properties result in a decrease of strength and stiffness, and an increase of the coefficient of expansion at high temperatures, the analysis becomes much more complex. For example, the coefficient of expansion for steel at temperatures of 100°F to 1200°F is approximately equal to $\alpha = (6.1 + 0.0019T)10^{-6}$, where T is the temperature in degrees Fahrenheit. The thermal expansion for normal-weight concretes may be estimated as roughly equal to steel up to about 1000°F, for preliminary design purposes. The modulus of elasticity for both concrete and steel reduces greatly with an increase in temperature. For steel, it decreases linearly from $E = 29000$ ksi (k/in^2) at 70°F to about 25000 ksi at 900°F, and then drops more rapidly. In contrast, the compressive strength of concrete remains relatively stable to at least 900°F, while the strength of steel at this temperature is very much reduced, and approaches a critical stage, as is discussed later.

It is extremely difficult to assess the effects of a relatively short exposure to high temperatures upon material properties, besides having to determine the temperature that the member attains, how the temperature is distributed, and how long the member is exposed. All of these factors may be impossible to determine, and hence can only be estimated. For the analysis of reinforced concrete members in Figure 1.22a and b, not only must the material properties at high temperatures be taken into account, but also the temperature distribution within the member; some typical thermal deformation patterns, with the corresponding moment diagram, are shown.

- A simply supported beam, when heated from beneath, will keep on deflecting with an increase of temperature, and the strength of the reinforced concrete will decrease. The member will fail when the steel strength reaches its stress level under the existing loading. The concrete cover over the steel controls when the critical temperature is reached.
- A continuous member also undergoes a change of stress as the temperature is increased. Here, however, the moments are redistributed; whatever is taken away from the field moment is added to the support moment.

Furthermore, the expansion of the heated structure will cause a thermal thrust upon the adjacent cooler structure, thereby causing an increase in the stresses at certain locations. While the expansion of columns and beams due to a uniform temperature increase results in an increased axial thrust, the nonuniform heat causes a bending of the members, hence also eccentric action of the axial forces, and therefore additional bending moments.

Of the major structural materials, wood, steel, concrete, and masonry, only wood is combustible, but only concrete and masonry are fire resistant. Although timber burns, heavy members will retain strength under fire longer than some unprotected metals. To achieve the necessary fire rating, the wood members should be oversized, allowing the outer layer of char to reduce further burning, but the steel connections must be hidden behind insulation. Most aluminum alloys start losing strength immediately, as the temperature is raised, and melt at 900 to 1200°F. Masonry materials have been used for a long time as fire protection for buildings. The compressive strength of masonry under high temperatures follows a similar pattern to concrete. Concrete is one of the most highly fire-resistant materials. The strength of normal-weight concrete remains relatively stable up to about 900°F, while lightweight concretes perform much better. To control the temperature in the reinforcing, the concrete cover below/above the bars must be thick enough. Usually, the minimum thickness of the concrete cover for the reinforcement, as specified by the ACI code, satisfies the requirements based on the fire-resistance rating.

In contrast to concrete, when a steel member is exposed to the high temperatures of a fire for long enough, its strength will decrease substantially in a short time. This loss of strength, together with excessive deflections and distortions that cause additional stresses in continuous structures, can lead to failure of the steel structure. At roughly 600°F, the capacity of steel rapidly decreases, and at about 1000°F is almost equal to its allowable stress; steel melts at 2400 to 2750°F. Steel does not burn and contribute fuel to feed a fire, but is an excellent heat conductor and has a low thermal capacity, so that uniform critical temperatures will be quickly achieved throughout the member early in the fire. It is apparent that structural steel members must be adequately fire protected for some anticipated fire intensity and duration, so that the average steel temperature does not exceed the critical temperature, of approximately 1000°F. This is quite significant considering that ordinary building fires may reach temperatures in the order of 1300 to 1700°F!

Fire protection of steel members in buildings can be achieved either by using a *membrane* as a fire-resistant barrier, such as a wall or a ceiling to protect the floor framing and ductwork (see fig. 5.7), or it can be achieved by direct *contact*, that is, by protecting the individual steel member with insulation to keep the heat away or to absorb the heat with special coatings. It must be emphasized that, in the membrane and compartment approach, proper fire stops along the edges (such as ceilings and walls) must be provided, similar to fire breaks between floors and exterior cladding.

Typical methods of *heat-resistive insulation* for steel members are based on solid encasement, a box-like membrane or assembly enclosure of the dry or wet type, a contour protection of the spray-on type (Fig. 1.22c), or fire-retardant paints. A steel member can be embedded directly in concrete (Fig. 1.22k), or can be encased with insulating fill surrounded by a nonbearing metal enclosure (Fig. 1.22h). A steel member can be enclosed like a box (Fig. 1.22i) by using plastering, cladding, or jackets, which can consist of one of the following typical insulating materials: cement plaster, lightweight concrete (e.g., perlite concrete, vermiculite concrete, cellular concrete), gypsum, fiber or mineral boards, or blankets. The plaster is applied to the wire mesh lath that is wrapped around the member.

The member may be cladded with boards, planks, or blocks, or occasionally also with masonry, or the member may be enclosed with mineral wool batts or blankets. A steel member can also be insulated with mineral fibers or cementitious mixtures applied with a spray gun directly to the member contours (Fig. 1.22j). Mineral fibers of the asbestos type have been drastically restricted because of health hazards! *Intu-mescent materials* (e.g., coatings or sheets) are only activated as protection under high temperatures. Usually, intumescent mastic coatings are painted or sprayed on steel. Under elevated heat exposure, the coating chemically reacts by expanding into a thick thermal barrier, thus forming an insulation blanket that protects the steel member.

Subliming materials (e.g., coatings or lightweight foam) do not insulate a steel member from the fire's heat, but absorb heat and act in a manner similar to a coolant system. High temperatures cause sublimers to turn from a solid to a gas. Since this change of state requires a large amount of heat, they are effective as a coolant in removing heat, that is, as a heat absorber.

In the structural analysis of lightly loaded steel members, the thickness of the fire insulation can be reduced, since the critical (failure) temperature is a function of the applied stress, which is low. One must not forget that tension members elongate under elevated temperatures, while the protective plaster encasement may shrink. The com-bined effect may possibly result in the exposure of the steel hanger at the top, a situation that must be prevented by extending the encasement. Attention must also be given to the fire-resistant sealants (such as putties or caulks) that are gunned, or foams that are pumped, into penetrations.

While the methods of fireproofing steel inside the building are standardized, this is not necessarily true for the exterior structure, where visual considerations become a very important criterion. Should the structure be hidden behind a facade wall, then conventional methods of fire insulation can be used. However, if the designer wants to express the structure, special techniques must be developed. In the traditional approach, the load-bearing structure is enclosed by a steel cladding; in other words, the exterior structure is insulated with fire-resistant material and then wrapped in steel covers to simulate the structural shapes, but thereby altering the appearance by dis-torting the true proportions of the frame (see fig. 3.15).

To eliminate the need for cladding, and to expose the true structure, some special techniques have been developed to keep the exposed members to an acceptable max-imum temperature level, for example, of 600°F. These methods are based on concepts of *isolation*, *separation* or *shielding*, and *cooling*.

1. *Air separation principle.* An exposed exterior steel structure does not need to be fireproofed against the flames inside the building as long as it is located sufficiently far away from the glass line or curtain wall. For example, the major exterior girders of the Knights of Columbus Building [1969] in New Haven, CT (Fig. 1.22e), which span between the corner concrete towers, did not have to be insulated, because they were isolated five feet outside, away from the glass line. It should not be overlooked, however, that columns outside a fire-resistant glass membrane, while protected from the flames, may still be exposed to radiant heat from the inside!

2. *Flame-impingement shields.* Shielding can be achieved by providing a fireproof wall barrier behind the exposed columns, together with a protective ceiling to prevent the flames from reaching the exterior structure. In the case of glass walls or windows, a fire-resistive deflector may be employed to move the flames away from the exposed structure. The layout (i.e., geometry) of the assembly may be designed such that the structural steel is kept away from the window, the primary source of heat and flames. The principle of shielding is applied to the deep

exterior spandrel girders of the One Liberty Plaza Building [1972] in New York (Fig. 1.22d). In this case, the web is exposed to the outside, but fire protected on the inside, which also serves as thermal insulation. It is shielded from flame impingement by the steel flange cladding that deflects the flame path from an interior fire, so the flames do not directly touch the web, and the web will not reach the critical temperature. The web receives the heat by radiation, but will dissipate part of it to the cooler surroundings, both by radiation and convection. Vertical flame spread along the outside of an exterior wall may also be prevented with flame barriers such as overhangs at least 3 ft wide (e.g., balconies), recessed openings, and possibly with operable panels.

3. *Liquid-filled column systems.* In this system, the exposed exterior steel columns consist of hollow sections that are filled with water and possibly with antifreeze and corrosion-inhibiting additives, to act as a heat sink. Therefore, the steel does not reach critical temperatures, as long as there is an uninterrupted water supply available to carry the heat away. As the member heats up, the heat is absorbed by the liquid while it rises by convection to storage tanks and is replaced by cooler water from below, thus causing circulating currents from cold to warm between the members and storage tank. At severe fire exposures, steam is generated, which is vented from the top of the storage tank, to avoid pressure build-up in the members and tank. The principle was first applied on a large scale to the U.S. Steel Building [1970] in Pittsburgh, PA (Fig. 1.22c, see also fig. 3.15), where the free-standing hollow, massive box-section columns, and the cantilevered stubs that connect the columns to the spandrel beams, are filled with chemically treated water. Because the columns are located 3 ft from the building face, their exposure to fire is considerably less severe. The building had to be divided into four vertical water zones to control the hydrostatic pressure in the columns. Located at the top of each of these zones, inside the triangular service core, is a vented storage tank that is connected by pipe loops to all exterior columns, at the top and bottom of each zone. An interesting application of the water cooling system is for the nine-story Bush Lane House [1976], in London, U.K. In this instance, the exterior stainless steel tubular lattice frame is water-filled, to provide one-hour fire protection, as is schematically shown in Figure 1.22f.

The Pompidou Center [1977] in Paris (see fig. 7.41) uses virtually all types of fire protection. The exterior columns are water-filled, the cantilever brackets are shielded by fire-resistant panels in the facade, the outside tension rods are 25 ft away from the windows, sprinklers are used on the external wall, and the lattice trusses and floor beams are encased to provide heat insulation.

In buildings with low fire exposure, the steel should not reach critical temperatures, as may be the case for open-deck multilevel parking garages, the roof framing of single-story industrial buildings, and horizontal-span enclosures (such as railway stations) where the steel is protected by its height above the floor. The roof structures of open, one-story buildings are usually exposed if enough escape routes are available and adequate redundancy of the structure allows local failure. However, for exhibition spaces with potentially flammable displays, the steel roof framing may still have to be partially protected, for instance, with intumescent paint and a sprinkler system.

2 CONCEPTS OF STRUCTURE AND CONSTRUCTION

In this chapter, the nature of space enclosure systems (or material space) is introduced, in contrast to the immaterial space (or zoning) of the building volume, as derived from the functioning of the space, which was one of the primary concerns of the previous chapter. The space enclosure systems consist of the support structure, the exterior envelope, the ceilings, and the partitions. Structure makes the spaces within a building possible, it gives support to the material. Whereas the structure holds the building up, the exterior envelope provides the protective shield against the outside environment, and the partitions form interior space dividers. Investigated in this chapter is not only the behavior and the purpose of the structure in place, but also how it can be constructed; other space enclosure systems are introduced in Chapter 6. In addition, the transition from the structure to the foundations, and then to the ground, is briefly discussed.

The problems at the end of this chapter are presented for the purpose of reviewing basic concepts of mechanics and the strength of materials only, and therefore the text does not provide any support work or example problems that parallel these exercises. Since many of the problems refer to towers, the reader may also want to read the first section in Chapter 8.

STRUCTURAL ORGANIZERS

The structure makes enclosure possible; it holds the building up by resisting gravity and lateral force action, so that the building does not collapse. However, the structure also acts as a spatial and dimensional organizer, similar to the skeleton in the body, with respect to the life supporting systems.

Introduction to Buildings as Support Structures

Structures range from the massive gravity block, which can be perceived as stacked one-story building slices, to the slender compact pure structure that resists lateral loads, such as a giant vertical cantilever (Fig. 2.1). This contrast is even more pronounced with the Pyramids at Gizeh in Egypt (Cheops' pyramid was originally 481 ft high), built more than 4000 years ago, which appear like mountains or seem to grow like natural extensions out of the earth. Their solid mass of gravity is inherently stable,

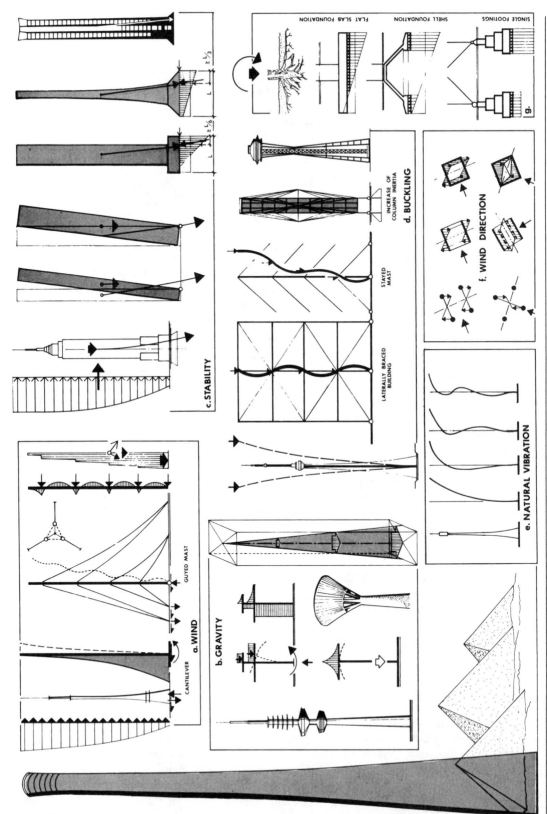

Figure 2.1. *Basic behavioral concepts for towers.*

quite opposite to the tallest chimney, of 1250-ft height, which seems to provide a bare minimum of structure, where one is afraid it may tilt over and fail as the wind pushes and twists it.

While the slender tower cantilevers out of the ground and uses all of its energy to resist the lateral forces, the flat multibay single-story building is spread out on the ground. It hardly provides any resistance to wind, and the path of the vertical gravity flow is short; in this case, the structural behavior is primarily based on the horizontal gravity flow, that is, gravity bending. The massive building block, in contrast, is controlled by axial gravity flow. Not gravity, but the lateral loads due to wind and seismic action become dominant design determinants as the building increases in height. For the typical, slender slab-type building, axial gravity, together with lateral force action, must be considered. As the slenderness of towers increases from 5:1 to 8:1 for buildings (12:1 is considered as the upper limit for buildings), to 30:1 for TV towers, the effect of wind, together with oscillations and the flexibility of the structure, become extremely critical. For the purpose of simplicity, it has been assumed in the discussion above that the building form is equal to the form of the lateral force-resisting structure, which obviously does not have to be true.

Some basic concepts of structure as a supporting system are initially introduced by concentrating on the tower structure described in Figure 2.1. The vertical, slender structures range from tapered, free-standing shafts to laterally supported guyed masts. They range from stocky, solid gravity towers to light, slender antigravity cantilevers, possibly with large heads. They may be stiff gravity towers or flexible guyed masts, where the masts not only act as compression struts, with regard to the prestress forces in the cables, but also as vertical beams to transfer the wind pressure to the spring-like cable supports. In the broader sense, the tower shafts can also be visualized as the hidden or exposed cores of more massive buildings, to which they provide lateral stability.

Basic Concepts of Structural Language

Some of the fundamental principles, as related to height, slenderness, and shape of tower structures in response to minimizing the effects of load action, will now be presented.

Slenderness

A gravity tower, which is slender enough to buckle under its own weight before its compressive capacity is reached, can be treated as a free-standing cantilever column fixed at the base and free to rotate at the top. However, one should realize that overall buckling is rarely a problem, especially not for ordinary buildings of normal height-to-width ratio.

It can be shown that a slender cantilever tower of varying cross-section buckles under its own weight (i.e., axial load p_{cr} distributed along the height) when the following critical compressive load is reached (Timoshenko and Gere, 1961).

$$P_{cr} = p_{cr}H = mEI_o/H^2, \quad \text{or} \quad p_{cr} = mEI_o/H^3 \qquad (2.1a)$$

where: m = 7.84 for tower of uniform cross-section; 5.78
for linearly tapered shafts; or 3.67 for curvilinearly tapered shafts
EI_o = stiffness of tower at the base
H = height of tower

It is apparent that a slender tower has a much lower buckling capacity when the axial force occurs at the top, as for a TV tower with a large head. For this condition, assuming a tower of uniform cross-section, A, that has an effective column length of twice its height (K = 2), the critical axial load, according to Euler's formula of elastic buckling, is

$$P_{cr} = \frac{\pi^2 EA}{(KH/r)^2} = \frac{\pi^2 EAr^2}{4H^2} = \frac{\pi^2 EI}{4H^2} = \frac{2.47EI}{H^2} \qquad (2.1b)$$

where: $r^2 = I/A$

For example, for the typical tower shape with a circular shell cross-section of an area A = πDt, the buckling stress, ignoring any safety factor, is equal to

$$F_a = \frac{P_{cr}}{A} = \frac{\pi^2 E}{(KH/r)^2} = \frac{\pi^2 ED^2}{32H^2} = 0.31E(D/H)^2 \qquad (2.2)$$

where

$$r^2 = \frac{I}{A} = \frac{\pi D^3 t/8}{\pi Dt} = \frac{D^2}{8}$$

It is interesting to derive the approximate proportions of a circular tower as based on elastic buckling. The maximum compressive stress due to the tower weight, for a shaft of uniform cross-section, cannot be more than the buckling stress as based on Equation 2.1a.

$$f_a = \frac{\gamma AH}{A} \leqslant \frac{7.84E}{H^2} \left(\frac{I}{A}\right) \simeq E(D/H)^2$$

In this instance, γ is the unit weight of the building. Rearranging the terms yields

$$\frac{H^3}{D^2} \simeq E/\gamma, \qquad \text{or} \qquad \frac{H\sqrt{H}}{D} = \frac{H^{1.5}}{D} \simeq \sqrt{E/\gamma} = \text{constant} \qquad (2.3)$$

Therefore, the tower proportions are roughly equal to the material constant (or the square root of the specific elasticity E/γ) which, for metals and timber, is approximately in the same range, while for concrete and stone it is nearly one-half of that value. If a tapered tower configuration would have been used in the derivation of the expression above, the value of the constant would have changed, but the tower form still remains: $D \propto H^{1.5}$.

From this approximation, it becomes quite clear that proportions do not simply remain constant, as expressed by H/D, but rapidly increase. This is clearly expressed in nature, where an ant's legs are much more slender than the plump legs of elephants. Hence, the proportions of cereal grass cannot simply be transferred to the large scale of a tree, for it would buckle and collapse under its own weight. Notice the decrease of the slenderness H/D from 500 for a wheat stalk, to 133 for bamboo, and finally to about 36 for a giant redwood tree.

It is interesting to compare the efficiency of nature, as expressed by the extreme slenderness of a 5-ft high wheat stalk with a stem proportion of H/D = 500, to a 1000-ft high TV tower of similar shape. The tower will bend under the weight of the conical pod and cantilever platforms at the top (Fig. 2.1b), when it sways in a manner similar to the wheat stalk under the heavy ear of grain (see fig. 8.2). Equating the known proportions of the wheat stalk to the unknown ones for the tower, yields

$$\left(\frac{H\sqrt{H}}{D}\right)_{stalk} = \left(\frac{H\sqrt{H}}{D}\right)_{tower}$$

$$500\sqrt{5} = (H/D)\sqrt{1000}$$

$$H/D \simeq 35$$

This result corresponds roughly to the proportions of about 27:1 for the Frankfurt Communication tower (see fig. 8.2), especially when the lower specific elasticity would have been taken into account. Besides, nature allows much more flexibility and lateral movement than can be allowed for man-made structures, which must be relatively stiff. It should also be realized that nature is not more efficient than man-made structures. On the contrary, a free-standing, 250-ft high steel antenna pole may reach a slenderness of 220, which is not found in nature.

It may be concluded from this discussion that the slender proportions of one scale cannot simply be transferred to a larger scale. Further, it must be realized that there is a limit to height. Galileo Galilei, as early as 1638, anticipated that no tree can be taller than 300 ft, since it would bend under its own weight as it sways and fail. Trees are the largest living organisms on Earth. California Redwood trees have reached heights of more than 360 ft while the Giant Sequoias represent the largest form of life in terms of total size (e.g., some weighing about 4300 kips), with some known to be more than 4000 years old.

The Tower of Equal Stress

As a structure's proportion decreases, the strength of the material, and not axial instability, will control the design. The resistance of the material to gravity action is briefly investigated in the following paragraphs.

For a tower under pure gravity loading, the axial stresses $f_a = N/A$ increase with height (Fig. 2.1b). The supporting cross-sectional area, A, of the columns and/or walls is directly proportional to the gravity action when a maximum allowable stress, F_a, is produced: $N \propto A$.

The fact that a considerable portion of the column/wall strength is used for the lower part of a building can be explained by letting the building be under a constant compression stress due to its own weight and a single load applied at the top of the building, which may be treated as the resultant live loading, for the sake of simplicity. To satisfy this condition of *structure of equal stress and strength*, the building tower must respond with a specific form. In other words, its shape is in response to its own weight plus a superimposed load, and is optimized by keeping the axial stresses constant and equal to the compressive strength of the material, but disregarding the effect of slenderness.

The resisting cross-sectional area, A_o, at the top (Fig. 2.2) can be found from the superimposed load, P, as based on the allowable compressive stress, F_a, or from other functional requirements.

$$A_o = P/F_a \qquad\qquad (a)$$

The required cross-sectional area, A, at any level, z, measured from the top of the tower, is

$$A = (P + W_z)/F_a \qquad\qquad (b)$$

At a distance dz, slightly below this level (Fig. 2.2), the required cross-sectional area is

$$A + dA = (P + W_z + dW)/F_a \qquad (c)$$

Subtracting Equation (b) from Equation (c), and substituting for the differential weight $dW = \gamma dV = \gamma A dz$, yields

$$dA = dW/F_a = \gamma A dz/F_a, \qquad or$$

$$\frac{dA}{A} = (\gamma/F_a)dz, \qquad or \qquad \int_o^z dA/A = (\gamma/F_a)\int_o^z dz \qquad (d)$$

Integrating the two sides of this differential equation results in a logarithmic or exponential function.

$$\ln A = (\gamma/F_a)z + C_1, \qquad or$$

$$A = e^{(\gamma/F_a)z + C_1} = e^{C_1}e^{\gamma z/F_a} = Ce^{\gamma z/F_a} \qquad (e)$$

For the special condition of $z = 0$, the cross-sectional area at the top is A_o, or

$$A_{z=0} = A_o = C \qquad (f)$$

Substituting Equation (f) into (e) gives the general equation for the cross-sectional area of a tower at any level.

$$A = A_o e^{\gamma z/F_a} \qquad (2.4)$$

For a round tube of radius r, where $A = t(2\pi r)$ and the wall thickness t is assumed constant (to simplify the calculations), the equation is equal to

$$r = r_o e^{\gamma z/F_a} \qquad (2.5)$$

Hence, the tower shape is defined by an exponential function and looks similar to the trunk of large conifers, or reminds us of the Eiffel Tower in Paris. One may conclude that, when this building tower is treated solely as a column under gravity loading, it represents a form of equal stress with no limit to height. However, the limit in height

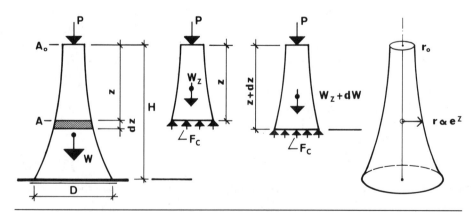

Figure 2.2. *The tower of equal stress.*

would rapidly be reached when the tower form remains constant, since gravity increases with its volume (i.e., cube), while the resisting surface area remains constant.

For example, using a 6000 psi normal-weight concrete, and ignoring the effect of slenderness, yields a maximum tower height, H, of

$$f_a = \frac{W}{A} = \frac{\gamma(HA)}{A} = \gamma H \leq F_a = 0.22 f'_c, \quad \text{or}$$

$$H_{max} = F_a/\gamma = 0.22(6)12^2/0.150 = 1267 \text{ ft} \quad (2.6)$$

The Gravity Tower versus the Cantilever Tower

Naturally, there is a limit to the height of the tower. It cannot just be treated as a gravity column and a form of equal stress and strength.

As the tower becomes more slender and increases in height, wind becomes critical. With an increase of height, the wind effect increases from the top down at a much faster rate than the gravity loads. Therefore, the cantilever action becomes more critical than the column action, or the rotation, M/S, is more important than the axial action, N/A. The section modulus, S (or the moment of inertia, I), therefore, rather than the cross-sectional area, A, controls the stress and becomes the determinant of form.

Again, it can be shown that the ideal tower shape for gravity loading is also optimal for wind loading. Hence, Equation 2.4 can be put into the more general form for the combined action of wind and gravity as

$$A = \text{const. } e^z \quad (2.7)$$

Naturally, this oversimplified approach does not include many essential design considerations such as wind torsion, temperature differences between the sun and shaded sides, flexibility of the tower allowing additional moments due to gravity, asymmetrical live loads on the elevated platforms (e.g., tower head), and particularly, dynamic considerations. The wind, with its larger periods, may come close to the natural period of vibration of a flexible tower, and thus generate resonant loading. In other words, the tower oscillations and the corresponding lateral forces may build up, depending on how much damping is present in the system.

Under certain stress conditions, the lateral load action can be ignored for the design of a building tower. The familiar interaction equation for the combined loading of gravity and lateral forces (i.e., axial action and bending), considering that the allowable stresses are increased by one-third, is

$$\frac{f_a}{F_a} + \frac{f_b}{F_b} \leq 1(1.33)$$

Letting $F_a = F_b$ gives

$$f_a + f_b \leq 1.33 F_a, \quad \text{or} \quad 0.75(f_a + f_b) \leq F_a$$

One may conclude that, for $f_a = F_a$ to be true, the bending stresses due to wind must be

$$f_b \leq f_a/3, \quad \text{or} \quad f_a/f_b \leq 3 \quad (2.8)$$

Hence, when the bending stress, f_b, due to the lateral force action, does not exceed one-third of the axial gravity stress, f_a, due to live and dead load, then the effect of wind or seismic actions can be ignored, and the structure can be treated as a gravity tower, where wind does not effect the primary member sizes.

However, should the axial loads be minimal, then the tower behaves primarily as a vertical cantilever. For this condition, the wind moment increases with the square of the height (H^2), and the lateral deflection with the height to the fourth power (H^4), thereby clearly indicating that (with an increase of height) flexibility or stiffness EI, and not strength, becomes most critical. In other words, while the moment of inertia of the tower cross section increases with the square of the tower height ($I \propto H^2$), at a certain height it changes to $I \propto H^4$.

Building Stability

When a building is loaded concentrically only with gravity, then the contact pressure at its base is uniform. However, when rotation is generated, in addition, by lateral force action, then the base contact pressure is no longer constant. For a trapezoidal contact pressure diagram no tensile stresses are present. In this instance, the compression prestresses the building, so that the tension due to lateral force rotations is suppressed, and the resultant force due to the vertical gravity and the horizontal force (see fig. 2.1c) falls within the middle third, or the so-called *kern*, of the supporting base. This resultant force is transferred through the foundations to the ground, similar to the roots of a tree (see fig. 2.1g). When the resultant falls outside the kern, as is the case for a triangular contact pressure distribution, then tensile stresses or partial uplift occurs. Naturally, when the resultant force falls outside the base, as for narrow buildings with a minimum of weight, then the building will topple, if it is not anchored to the ground. For this condition, the building base may also be widened. It is apparent that the gravity loads should be collected to the lateral force-resisting elements in a building, which should be located, in turn, where the base is broadest.

Using weight to stabilize or prestress a building or building component is not an invention of the modern age. The medieval master builders sensed the "kern" concept when they suppressed the lateral arch thrust from the roof and vaulting on the flying buttresses of the Gothic cathedrals by increasing the weight of the supporting piers through the addition of small spires.

The overturning moment must be resisted by the weight of the lateral-force resisting structure above the level of investigation and the capacity of the material or anchorage establishing continuity with the structure below. Considering, conservatively, that only the weight or minimum dead load counteracts, then according to the codes, the stabilizing, resisting moment, M_{react}, must be at least 50 percent larger than the acting moment M_{act} caused by the lateral loads, keeping in mind that some codes require higher safety factors.

$$\text{S.F.} = \frac{M_{react}}{M_{act}} \geq 1.5 \qquad (2.9)$$

This minimum requirement yields the location of the resultant force at a minimum distance of $L/6$ away from the base edges (see Problem 2.10). Critical, with respect to the stability of a skyscraper, is the relation of its height to the width at the base parallel to the wind direction, assuming that the lateral force-resisting structure is located along the building perimeter. It is apparent that a wider base spreads out the loads. The maximum aspect ratios for skyscrapers are in the range of 6 to 8, although the so-called *sliver* apartment buildings in New York, which are squeezed (for example) in a 20-ft wide space between town houses, are extremely narrow slab-type structures possibly 20 stories high or more. These slivers have reached a slenderness of above 10! The most slender structure among the world's tallest buildings is believed to be the 810-ft high, 72-story City Spire [1988] in New York. The mixed-use slim concrete tower, with a footprint width of only 80 ft has a slenderness ratio of 10:1. One should keep in mind that the lateral forces also tend to slide a building horizontally, although most codes do not require a particular safety factor against sliding.

High-rise Structure Systems

A building structure can be visualized as consisting of horizontal planes or floor framing, and the supporting vertical planes of walls and/or frames. The horizontal planes tie the vertical planes together, as to achieve a box effect and a certain degree of compactness. It is quite obvious that a slender, tall tower building must be a compact, three-dimensional closed structure, where the entire system acts as a unit. The tubular, core-interactive, and staggered truss buildings are typical examples of three-dimensional structures. On the other hand, a massive building block only needs some stiff, stabilizing elements that give lateral support to the rest of the building. In this sense, the building structure represents an open system where separate vertical planar structure systems, such as solid walls, rigid frames, and braced frames are located at various places and form stand-alone systems that provide the lateral stability.

Every building consists of the load-bearing structure and the non-load-bearing portion. The main load-bearing structure, in turn, is subdivided into the *gravity structure*, which carries only the gravity loads, and the *lateral-force resisting structure*, which supports gravity forces, but also must provide stability to the building. For the condition where the lateral bracing only resists horizontal forces but does not carry gravity loads, with the exception of its own weight, it is considered a *secondary structure*. Failure of secondary members is not as critical as that of main members, where an immediate collapse of a building portion may occur, depending on the redundancy of the structure. The non-load-bearing structural building elements include wind bracing, as well as the membranes and skins, that is the curtains, ceilings, and partitions, which cover the structure and subdivide the space.

The lateral-force resisting structure in a building tower may be concentrated entirely in the central core, for instance when an optimal view and thus a light perimeter structure is required. Conversely, rather than hiding the lateral-force resisting structure in the interior, it may be exposed and form the perimeter structure, as for tubes.

The structure represents an assembly system that consists of components and their linkages. The basic elements are lines (columns, beams), grids (floor framework, frames), surfaces (slabs, walls, plates), spatial units (cells, tubes), and any combination of the above. The interaction or degree of continuity between these elements depends on the type of linkage (hinged, semirigid, or rigid). Naturally, these basic components can be combined in an endless variety to form a building.

Before discussing some fundamental concepts of structure behavior, typical structure systems are first introduced, but purely from a geometrical point of view. The study of the building structures is approached from the organization of building plans and sections. The interaction of plan and section forms the building and is treated extensively in other parts of the book, in the study of the many cases.

Although buildings are three-dimensional, their support structures can often be treated, from a behavioral point of view, as an assembly of two-dimensional vertical planar elements in each major direction of the building. In other words, structures can usually be subdivided into a few simpler assemblies, since structural elements are rarely placed randomly in plan. The most common high-rise structure systems are identified in the block on the left side of Figure 2.3. They are shown simply as planar, two-dimensional structures, although they may act in combination with each other, and in context of the building may form spatial structures as indicated by the building plans in Figure 2.4. They range from pure structure systems, such as skeleton and wall construction, and systems requiring transfer structures, to composite systems and megastructures. As the buildings increase in height, different structure systems are needed for reasons of efficiency. The following classification of the various systems is roughly in accordance with their increase in stiffness.

- *Two-dimensional structures*
    Bearing wall structures: combinations of single walls and connected walls, cross-walls, long-walls, two-way walls, stacked boxes
    Skeleton (frame) structures: rigid frame, braced frame, truss, flat slab, Vierendeel wall beam (interspatial, bridge type)
    Connected walls and frames
    Core structures: They may be considered three-dimensional from a structural element point of view, but do not necessarily integrate the entire building shape: cantilevered slab, bridge structures (multicore), cores with outriggers on top (suspension), at the bottom, and at intermediate levels.
    Combinations of the above systems
- *Three-dimensional structures*
    Staggered wall beams
    Cores plus outriggers plus belt trusses: single, double, and multiple outrigger systems
    Tubes: Vierendeel tube, deep-spandrel tube, perforated wall/shell tube, trussed tube, tube with belt trusses and head, etc.
    Megastructure: superframe, superdiagonals
    Hybrid structures

Typical combinations of structure systems are:

Walls + core(s)
Frames + core(s)
Tube + frame(s) or wall(s)
Tube + core (tube-in-tube)
Tube + tube (bundled tubes)

Other combinations:

Vertical stacking of structures: connected towers of the bridge type
Series of superframes
Internally braced structures
Cellular structures
Stayed structures
Other mixed systems

The selection of a structure system is not a simple undertaking. Among other criteria, it depends on the overall geometry, the vertical profile, height restrictions, the slen-

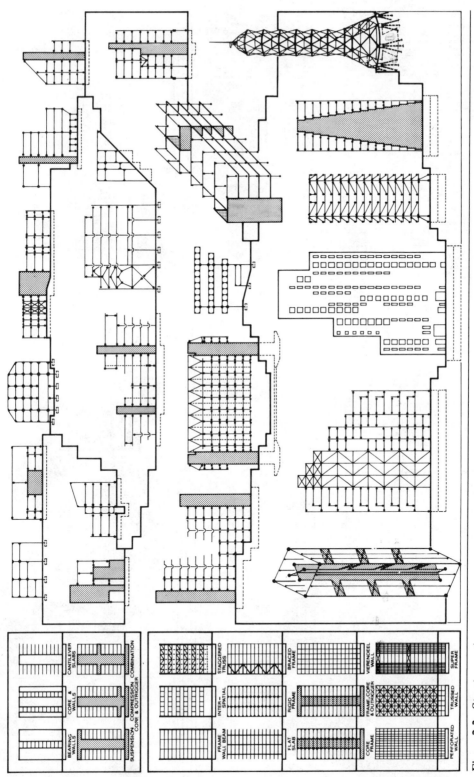

Figure 2.3. *Structure systems.*

Legend panels:

BEARING WALLS · CORE & WALLS · CANTILEVER SLABS

SUSPENSION · COMPRESSION · CORE & OUTRIGGER · COMBINATION

FRAME WALL BEAM · INTER-SPATIAL · STAGGERED TRUSS

FLAT SLAB · RIGID FRAME · BRACED FRAME

CORE FRAME · FRAME, CORE & OUTRIGGER · VIERENDEEL WALL

PERFORATED WALL · TRUSSED WALL · SUPER FRAME

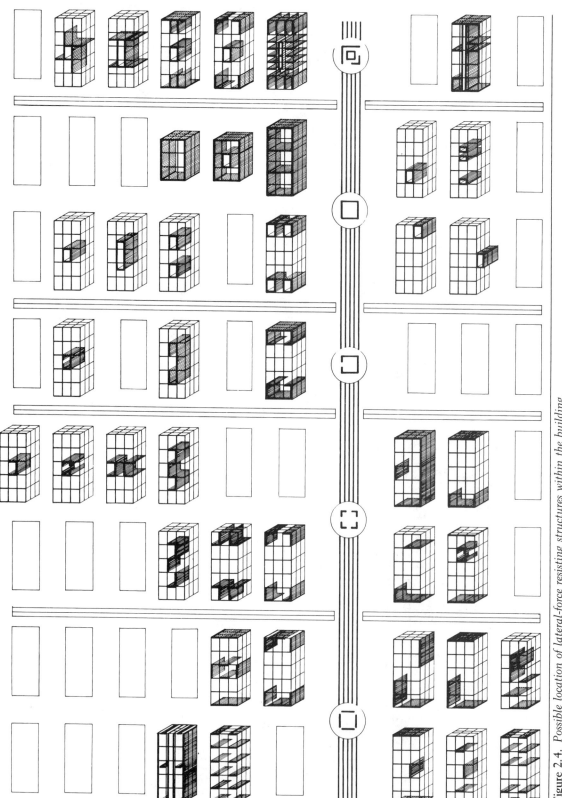

Figure 2.4. *Possible location of lateral-force resisting structures within the building.*

derness (that is the building height-to-width ratio), the plan configuration (depth-to-width ratio, degree of regularity, etc.), and is a function of strength, stiffness, and possibly ductility demands in response to loading conditions. It also depends on the building base conditions, site conditions, and on construction coordination, including preconstruction and construction time. In order not to give the impression that the pure structure systems above are imposed upon the architecture, and do not allow any flexibility in the form giving process, various building cases are presented in the right portion of Figure 2.3. They demonstrate some of the endless possible combinations of the structure systems for low-rise and mid-rise buildings, realizing that the smaller scale buildings obviously allow more freedom than large-scale towers. It is shown that the structures respond to setbacks, cavities, changing spans, varying story heights, altering bay proportions, sudden changes of stiffness, sloping site conditions, space inclinations, and so on. Most structures are treated as planar, with the exception of the central core-type buildings with diagonal outriggers connected to the corner columns or that are stabilized by a tensile network along the perimeter.

To gain a better understanding of the structure systems in Figure 2.3, they must be seen within the building space, hence their location must be known. For this reason, solid surface elements have been placed into the uniform beam-column grid of the various plans in Figure 2.4. They represent the lateral-force resisting structure systems of walls, cores, frames, tubes, or any other combination; they may form either planar or spatial assemblies. The organization of the cases is based on the shape, number, degree of continuity, location, and the arrangement of the lateral-force resisting elements. For the purpose of convenience only, a rectangular plan form has been used, and the effect of building shape and discontinuity (irregularity) has not been considered in this study.

The transformation from open to closed assemblages (i.e., from the left to the right of Figure 2.4) is the basis of the organization:

- Single-plane surfaces (⫴, ⊢)
- Bundled planes forming angles and T-sections (L, T)
- Single, open cells, such as channels and wide-flange sections (U, I, Y, X, C)
- Open multicells, such as single open cells or perforated cores (II, ⫾, ▯)
- Closed core forms such as: single shafts (○, □, △), multiple shafts (⊏⊐, ∞), and combinations (e.g., pinwheel)

The arrangement of the vertical surface elements may be in a symmetrical fashion, as indicated in the upper portion of Figure 2.4, or asymmetrically as shown in the bottom part. They may be located outside of the building, or they may form envelopes along the perimeter, or they may be hidden as cores inside the structure. The layout patterns range from a low-density structure with a minimum of three nonparallel, nonconcurrent isolated walls stabilizing the skeleton, to an all-wall, dense multicell building.

The strength and stiffness of a building is very much related to the type and arrangement of the structural elements, as is studied further in Figure 2.5. The density and interaction, or continuity, of the elements, together with the degree of symmetry, indicate the degree of compactness of the structure. Naturally, as buildings increase in height, approximately beyond 30 stories, in addition to the plan, the vertical planes become essential in describing the more spatial action of the structure.

The top portion (a) of Figure 2.5 identifies the typical structure layouts for open systems, ranging from a perimeter structure, through a concentrated interior spinal structure, to a uniform structure layout. The bottom portion (Fig. 2.5k) identifies some structure layouts for closed spatial structures, ranging from the circumferential rings of the tube-in-tube, intersecting tubes, cellular perimeter tube, and cellular or bundled tubes, to the multitubes forming a megaframe.

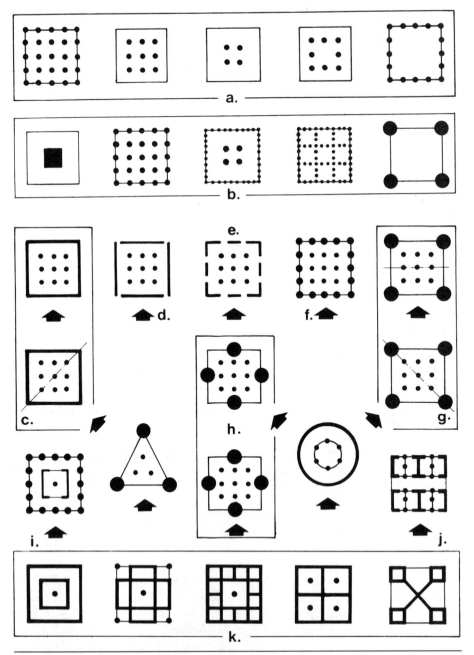

Figure 2.5. *The structure in plan.*

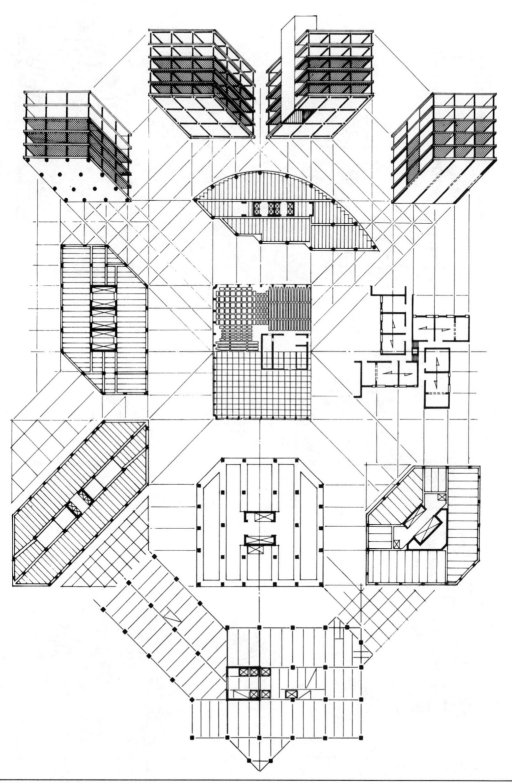

Figure 2.6. *Floor framing systems.*

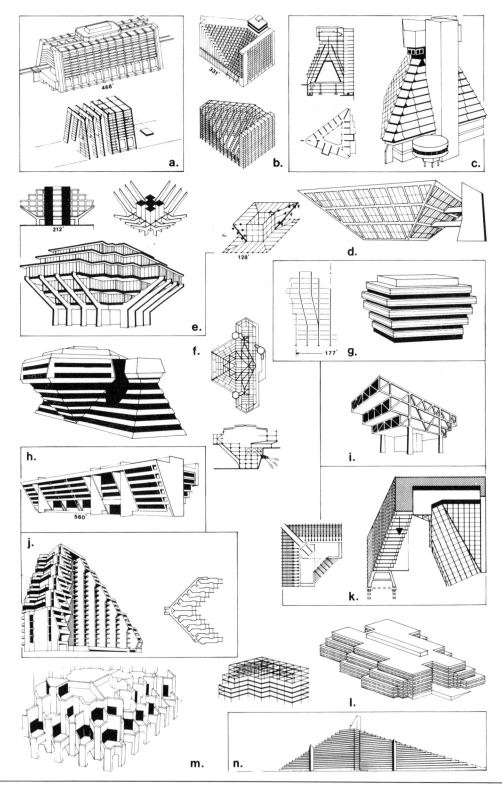

Figure 2.7. *Sloped building structures.*

In the center portion of Figure 2.5, various structure systems are investigated, taking into account the direction of the lateral force action, as is discussed later in this chapter. The transition of the tube from a solid exterior shaft (Fig. 2.5c), through a perforated or a flexible open unconnected tube (Fig. 2.5d), an open-type exterior bearing wall envelope (Fig. 2.5e), or a rigid perimeter frame (Fig. 2.5f), and finally to a bridge-type structure, where the columns are only provided in the corners (Fig. 2.5g), is investigated further in Problems 2.13 to 2.16.

Up to this point, only the vertical structure systems have been defined, while the horizontal building planes, or floor structures, which tie the vertical planes together, have been treated simply as solids. The purpose of the floor structure is not only to transfer the gravity loads to the vertical structural planes, but also to act as a huge, flat horizontal deep beam that must carry the lateral forces to the respective lateral-force resisting vertical elements. Some typical floor framing layouts are shown in Figure 2.6, and are discussed further in Chapter 5. How the floor framing patterns derive from the building shape and the layout of the vertical structure elements can only be suggested by the various cases. The density of members ranges from the wide spacing of beams which support relatively thick slabs, to the close spacing of beams supporting thinner floor slabs or the joist slab.

The effect of the building shape has been ignored in the previous discussion in order to concentrate on the principles of high-rise structure systems. To develop some intuitive appreciation for this additional dimension, sloped building structures have been studied in Figure 2.7. They range from the pyramid, the inverted pyramid, the inverted trapezoid, and the reversed stepped pyramid, to other irregular cantilever structures using modified A-frames or bearing walls as support systems.

Building Loads

The primary loads on ordinary buildings are due to the vertical action of gravity and the horizontal action of wind and earthquakes. These static or dynamic, external or internal loads may represent distributed or concentrated forces, and may act concentrically or eccentrically. Building loads are only briefly introduced here, so as to develop some initial understanding of what forces act and how the building structure must react; they are treated in more detail in Chapter 3. The following simplifying assumptions are used:

- The weight is considered uniform and a function of the floor area or the building volume.
- The wind pressure normal to the exposed surface area is treated as uniform.
- The building mass is taken as uniform and proportional to the building volume.

Concrete and masonry buildings generally weigh much more than steel buildings. While the overall average gross weight for ordinary steel buildings is in the range of 50 to 80 psf (lb/ft²) or approximated as 5 to 8 pcf (lb/ft³), non-prestressed concrete buildings may have a density of twice as much. Should the live load be included, then the overall weight ranges roughly from 10 pcf for steel office buildings to about 14 to 18 pcf for concrete office buildings and 20 pcf for concrete apartment buildings. The use of high-strength materials results in less weight. This may be advantageous when strength, rather than stiffness, controls the structural design. The dead load of the structure can further be reduced by 10 to 20 psf when using lightweight concrete for the floors.

The weight of the structure itself only constitutes a relatively small portion of the total building dead load; it may be in the range of 20 to 50 percent for frame buildings, but varies with height. A 10-story steel frame structure, for example, may weigh as little as 6 psf, in contrast to a 100-story building with about 30 psf (see Table 7.1). This effect of scale is known from nature, where animal skeletons become much bulkier with an increase of size, since the weight increases with the cube, while the supporting area only increases with the square. The bones of a mouse make up only approximately 8 percent of the total mass, in contrast to about 18 percent for the human body.

The live loads due to equipment and people are of variable character, in contrast to the material weight or dead loads. The live loads of approximately 80 psf for an office building are twice as high as for a residential building. Similarly, the live loads in public areas, including interior corridors, are at least twice as much as on living areas with 40 psf. Live loads for mechanical rooms are 150 psf and for plaza areas may be as high as 300 psf. Live loads can be reduced for members supporting an area larger than 150 ft^2, according to some codes. In multistory structures, it is improbable that every floor is fully loaded, so that the live loads on columns or walls may be reduced by up to 60 percent.

Wind and seismic loading cause horizontal force action upon a building. They are dynamic loads, but can often be treated as quasistatic lateral forces. This approach is reasonable with respect to wind action as long as the building is not of unusual shape, and as long as it is stiff enough so that it does not oscillate and give rise to accelerations, with the corresponding increase in force action. Naturally, the shape of the building as seen in plan and elevation considerably affects the lateral force resistance, remembering that the least resistance for a given wind direction is provided by the streamlined tear-drop shape. Not only is the rigidity of a building improved by sloping the exterior columns, such as the truncated pyramid of the John Hancock Center (see fig. 7.25e) in Chicago, but also the lateral force resistance is reduced, thereby resulting in a large decrease of the lateral drift.

A constant, uniform wind pressure may be assumed for the purpose of visualizing the lateral force action upon a building as a whole, realizing that the actual nonuniform pressure does generate torsion, which can, however, be treated as insignificant for symmetrical buildings, at least for preliminary design purposes. Typical pressure values for the inland United States range from 20 to 40 psf for ordinary high-rise buildings. For inclined and curvilinear surfaces, the wind pressure may be taken as perpendicular to planes projected vertically from the building, as explained at the top of Figure 2.8. It is also shown that the building shape has a substantial effect on the design. For example, the efficient round building has to resist only 60 percent of the wind load on a comparable rectangular building.

As the building increases in height and slenderness, the dynamic action of the wind becomes of major concern. The flexible building not only responds along the wind direction but suddenly also in the across-wind direction. Vortex shedding occurs when strong winds flow past a building and vortices form alternately on one and then on the other side (see fig. 3.6), causing low pressure areas, which in turn cause the largest motions in tall, flexible buildings surprisingly normal to the wind direction rather than in the wind direction. This change of pressure on the side faces occurs in a periodic

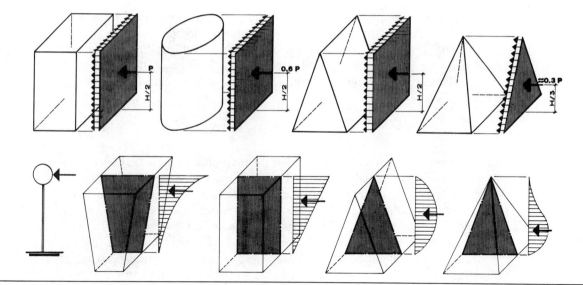

Figure 2.8. *The effect of building form upon wind and seismic load distribution.*

manner, depending on the wind speed and building form. If these fluctuations act at intervals close to the natural period of the building, it starts to resonate and loads build up drastically. For wind speeds of about 200 mph, the vortex shedding on the current skyscrapers may reach this critical stage. However, the new breed of superskyscrapers, which may reach a slenderness of up to 12 to 1, may already be excited by the lower wind speeds that we encounter. In this case, the tapering of the tower reduces vortex shedding, but mechanical damping systems may be required in addition, to control the vibrations. It has also been found that, for some building forms, more vortex shedding occurs than for others. Surprisingly, very tall round buildings are much more vulnerable than rectangular buildings, where wind turbulence acts as a damping agent. For example, the dropping of the modules for the Sears Tower (see fig. 7.33) not only reduces the wind sway by decreasing the exposed surface, but also causes turbulence, which in turn minimizes oscillations. This effect is similar to providing apertures in the upper building portion through which the wind blows and causes turbulences, as can also be generated by large spoilers; these turbulences disrupt the formation of vortices. Also, special damping devices can be used to control vibrations, as is discussed further in the section in Chapter 3 on "Dynamic Loads."

While the wind exerts external lateral forces, the ground motion due to an earthquake causes internal lateral forces, besides vertical forces, which (however) are neglected. One can visualize the building as riding on an unstable earth. As the ground abruptly accelerates in a random fashion, the building portion above the ground will be left behind, thereby activating lateral inertial forces. In other words, the inertia of the mass tends to resist the movement, similar to the experience of a person in a car which suddenly increases in speed. The time it takes for the building to respond to the base-induced acceleration due to the fluctuating seismic ground loads becomes an important characteristic of the building; the fourth dimension, that of time, is introduced as a consideration of loading.

Assuming, for this introductory discussion, that the building is rigid by ignoring the effects of flexibility, structure type, mass distribution, location, and site geology, then the lateral inertial forces, according to Newton's second law, are the product of the building mass, M, and the ground acceleration, a. In this case, the mass is equal to the building weight, W, divided by the acceleration of gravity, g, where the seismic coefficient is $C = a/g$.

$$F = M(a) = W(a/g) = WC \qquad (2.10)$$

This equation clearly expresses the fact that the magnitude of the lateral force, F, is directly related to the building weight. However, one must keep in mind that this magnitude may be less or more when the other, previously ignored factors are taken into account. It is common practice to express the magnitude of the seismic forces as a percentage of the building weight. Typical values for high-rise buildings in major seismic zones may range from about 5 percent for flexible rigid frame structures, to about 20 percent for stiff bearing-wall buildings. Because of the heavier weight of concrete buildings, the lateral seismic forces are much higher than for steel buildings. In addition, since masonry and concrete walls have less reserve capacity than a steel frame, a higher safety factor results in an even larger lateral action.

For a typical rectangular building with a uniform mass distribution, the lateral forces due to seismic ground movement may be visualized as an equivalent static, triangular load, as indicated in Figure 2.8. At the bottom of the same figure, the lateral force distribution for some other common building configurations is identified. For a uniform mass arrangement, this lateral force distribution is proportional to the shape of the building volume, thereby clearly demonstrating the pyramid as an efficient form.

It may be concluded that the building form and mass distribution, as reflected by the plan organization and vertical massing, determine the location of the resultant lateral seismic force. Further, the form of the lateral-force resisting structure, and its location within the building volume, determine the type of action of the seismic force. When the centroid of mass does not coincide with the center of resistance, twisting is generated, as may be the case at floor levels with abrupt changes of stiffness. The effect of asymmetry, as seen in section and plan, is typical for the new breed of hybrid, compound building forms currently so much in fashion.

While earthquake forces constitute internal lateral loads generated by the mass and stiffness distribution, wind causes external forces on stiff buildings that depend on the exposed facade surface area. Seismic loading is usually critical with respect to the performance of stiff low- and mid-rise structures, while wind loading generally dominates the design of tall, slender buildings. The optimum design of high-rise buildings in areas of strong earthquakes conflicts with that for wind loading. Here, seismic action calls for ductility with much redundancy, while the wind resistance requires stiffness for occupant comfort.

Force Flow

The horizontal and vertical structural building planes must disperse the external and internal forces to the ground. Some basic concepts of vertical and lateral load transmission for various structure systems are discussed in a simplified fashion in Figures 2.9 and 2.11.

Visualize a gravity load acting upon the slab and transferred, by the floor framing in bending (Fig. 2.9, top, left), to one of the vertical structural building planes, which may transmit the load axially directly to the ground. The type and pattern of force flow naturally depends on the arrangement of the vertical structural planes, as indicated at the top of Figure 2.9 for two-dimensional structures. The columns may be vertical or inclined, continuous or staggered; they may be evenly distributed, or concentrated in the center or along the periphery, to possibly form cores. The path of the force flow may be continuous along the columns or may be suddenly interrupted and transferred horizontally to another vertical line. The transmission of the loads may be short and direct, or long and indirect with a detour, such as for a suspension building. From an efficiency point of view, the vertical loads should be carried along the shortest path possible to the foundations.

When columns are inclined, gravity will cause lateral thrust, which increases as the column moves away from the vertical supporting condition. The cases at the bottom of Figure 2.9 (see also fig. 2.7) indicate that the horizontal floor beams at the top act as ties in tension when the columns lean outward, but as struts in compression at the bottom. For a symmetrical structure, the thrust due to the dead load will self-balance, but the horizontal forces due to asymmetrical live loads must still be resisted, as for an asymmetrical building where also the weight causes thrust. Hence, not only wind and earthquake, besides the centrifugal outward effect of cars upon curved bridges, but also gravity, together with the respective geometry, may cause lateral force action upon a building.

Optimum, free ground level space with a minimum of columns is often required for high-rise buildings. Examples range from grand entrances and wide lobby spaces, loading docks, and parking aisles, to open public plazas. For these conditions, the upper building mass must be linked to the ground by using a different structure system. The geometrical pattern of the building structure cannot extend to the foundation walls; it becomes discontinuous and is replaced by another structure system. For preliminary design purposes, the upper and lower structure portions can be analyzed separately. Various transition types are shown in the central part of Figure 2.9. They range from suspension buildings and lifting an entire building up on frames or stilts, to changing the column spacing to a wider pattern by using transfer systems within the framed tube grid. The latter may be accomplished, for example, by increasing the spandrel beam sizes towards the main columns, or by changing the column sizes in a tree-like fashion, thereby generating a natural arch-like gradual transition of the loads. The load transition can also be achieved by heavy transfer systems such as girders, trusses, wall beams, arches (direct or indirect action), or V- and Y-shaped tree-columns (two- and three-forked columns) to collect the columns above. These V-columns are effective in resisting wind and earthquake forces, but unfortunately respond to asymmetrical gravity loading with horizontal thrust, as has been discussed before. The reader may want to study the behavior of the various inclined column cases in Figure 2.9.

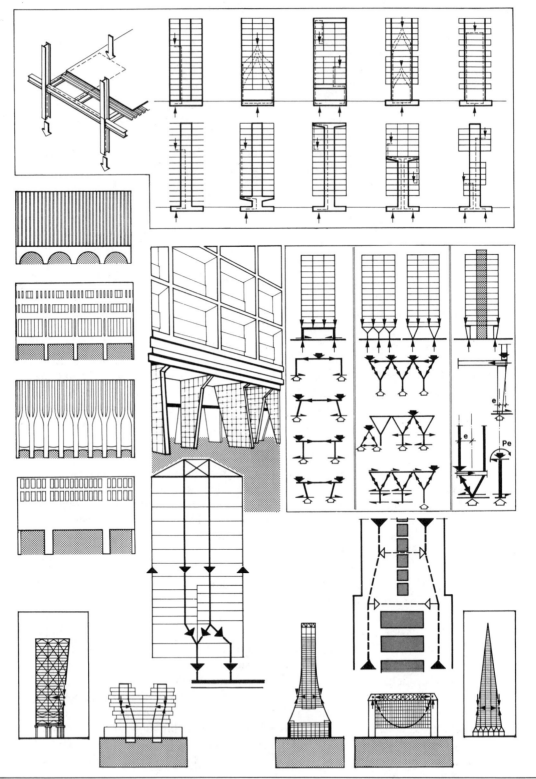

Figure 2.9. *Vertical force flow.*

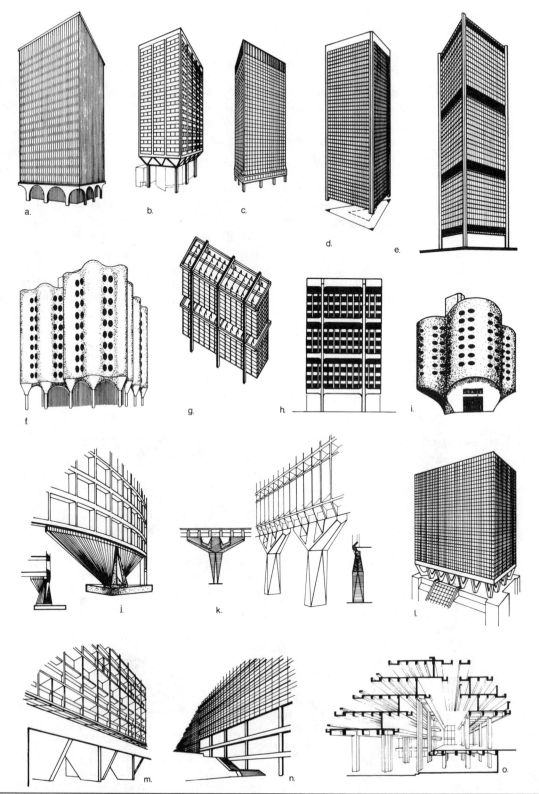

Figure 2.10. *The transition of building to base.*

The many design possibilities for providing free space at the base are studied further in Figure 2.10 by identifying the following possibilities:

- An entire facade acts as a deep Vierendeel wall beam that spans between the corner columns (as in Fig. 2.10e), or together with a transfer girder at the top (as in Fig. 2.10d).
- A steel space truss carries the entire building on 24 columns to 5 base columns and the core (in Fig. 2.10b); similarly (in Fig. 2.10o), the building can also sit on large base columns.
- From 3 rigid superframes two 4-story building blocks are suspended (in Fig. 2.10g), similar to Figure 2.10h, where the bearing frame walls are relieved at every fourth floor level by transferring the loads to 8 massive exterior columns.
- The perimeter concrete shell walls are cantilevered from the core on immense arches (Fig. 2.10i).
- Columns flare out to form half-cones and arches to support the undulating shell walls (Fig. 2.10f). Various other column types range from the sculptured tree-like columns in Figure 2.10k, and the two offset V-supports (Fig. 2.10m), to the hyperbolic section supports in Figure 2.10j. In Figure 2.10n, the post-beam structure is exposed at the base, while the 16-story building (Fig. 2.10l) is sitting on V-shaped steel stilts and a service core.

After this discussion of gravity force flow, the distribution of lateral loads is briefly introduced in Figure 2.11. The horizontal forces are transmitted along the floor planes,

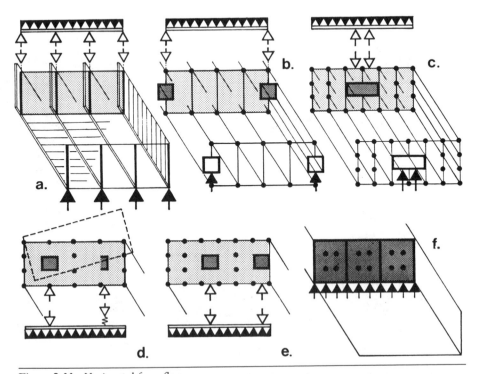

Figure 2.11. *Horizontal force flow.*

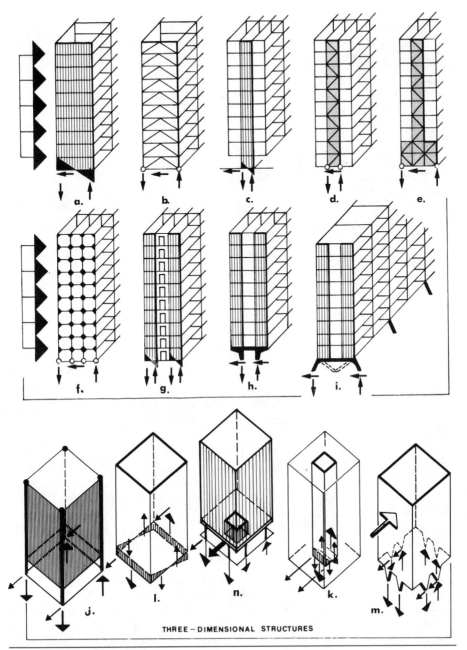

Figure 2.12. *The resistance of various structure systems to overturning.*

which act as deep, flat beams spanning between the vertical lateral-force resisting structures. The floors are treated as rigid diaphragms supported by elastic vertical shear/ bending elements. The deep beam action of the floors under uniform lateral loading is shown in Figure 2.11 for typical structure systems such as bearing wall buildings, symmetrical and asymmetrical core structures that have the same or a different stiffness, and bundled tube structures. Shear connections must be provided between the hori-

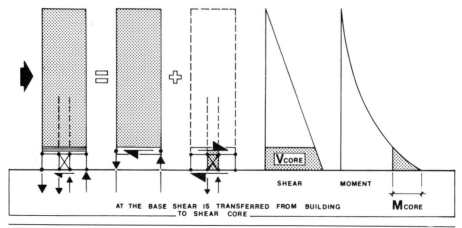

Figure 2.13. *The redirection of lateral force flow.*

zontal and vertical planes to transmit the lateral forces; this is especially important for precast, large-panel, and skeletal construction.

Should the resultant force action not pass through the center of rigidity, then twisting is generated. This torsion is efficiently resisted by a perimeter structure, as in Figure 2.11f, rather than by a concentric core (as in Fig. 2.11c). Because of the large lever arms, it is beneficial to let solid walls at least wrap around the exterior corners of a building.

Once the lateral forces are distributed to the resisting vertical structure planes, these systems must act as vertical cantilevers to carry the forces down to the ground. The various two- and three-dimensional structure systems in Figure 2.12 demonstrate how the overturning moment due to wind action is resisted at the base by the different cantilevers, consisting of solid walls or skeleton construction. In this case, the skeletons may act either as *flexural systems* when they are rigid frames (Fig. 2.12f), where the members react in bending, or they may be predominantly *axial systems* as for braced frames where the forces are effectively resisted directly in tension and compression (Fig. 2.12b, d, e).

The various structure cross sections or building plans (in fig. 2.5) indicate the many ways that lateral forces can be resisted in bending, realizing that concrete shear walls provide additional weight, helping to minimize uplift problems. For tall buildings, the cases range from the uniform rigid frame pattern (Fig. 2.5a, left) of the early skyscrapers, the single perimeter tube (Fig. 2.5c) of the 1960s and 1970s, the hybrid structure (Fig. 2.5k), to the megaframe (Fig. 2.5k, right) of the present. Structure elevations (in fig. 2.3 left), in turn, demonstrate the various possibilities of how shear can be transmitted along solid plates, various truss types, rigid frames, or framed tubes. The reader may want to study further the behavior of the vertical cantilevers in Problems 2.13 to 2.20.

Conflicts of force flow are generated when plan forms or structure systems change, possibly at locations of setbacks often found at the base, the top, or intermediate levels of buildings. For example, when a triangular plan changes to an L-shaped base, or when a perimeter structure such as a tube cannot be continued to the base, then an extensive horizontal transfer structure is necessary, not only to redirect the vertical forces, but also to act as a diaphragm to transfer the horizontal forces. For the case in Figure 2.13 (see also fig. 2.12n), the forces due to lateral loading must be redirected from the perimeter through a heavy diaphragm structure to the core. Hence, the transition from a tubular structure to an open base structure can be achieved by letting the outside columns take the overturning moment (i.e., rotation) in compression and tension only, while the entire shear is carried by the interior cores, or the so-called shear tubes. Below ground, the shear may be carried back by diaphragm action to the basement walls.

Introduction to the Behavior of Building Structures

Strength and stiffness are the primary characteristics activated when the building structure responds to the load action described in Figure 2.14; in areas of strong seismic activity, ductility also becomes an important criterion. Considerations of stability and other load-related effects have already been briefly introduced under slender towers (in fig. 2.1). In structural analysis, the real structure is replaced with an idealized model. Its response to loading is based on an analytical theory, which is derived from material properties and member behavior.

First, the reaction of the structure to lateral loading is conceptually investigated. It has been shown (in fig. 2.11) that, with respect to horizontal force action, the floor structure acts as a rigid diaphragm that is supported on elastic vertical elements. These lateral-force resisting elements in Figure 2.14 are interior cores, core plus walls (Fig. 2.14j), and a perimeter tube (Fig. 2.14m). The cores are of an open or closed type, and may be located at the centroid of the building, or their location may be eccentric. The building form determines where the resultant wind pressure acts, and the arrangement of the building masses defines the location of the resultant seismic force, which the core must resist. Since lateral stiffness requirements reduce in the upper portion of the building, the core walls can be dropped off or stepped back at the termination of the low- and mid-rise elevator banks.

The tall buildings in Figure 2.15 respond to lateral forces primarily as *flexural cantilevers* if the resisting structure consists of shear walls or braced frames. The behavior of these systems is controlled by rotation rather than shear; they have a high shear stiffness, as provided by the solid wall material or axial capacity of the diagonals, so that the shear deformations can be neglected. But tall buildings act as *shear cantilevers* when the resisting elements are rigid frames, since the shear can only be resisted by the girders and columns in bending. In this context, the effect of rotation (i.e., axial shortening and lengthening of columns) is secondary and may be ignored for preliminary design purposes. The combined action of different structure systems, such as rigid frames, together with a braced core (depending on the relative stiffness of each of the systems), may have the appearance of a flat S-curve with a shear-type frame building sitting on top of a flexural cantilever-type structure.

In the discussion above, it has been assumed that the structure systems were for tall buildings and were of the same height. It is apparent that, when the shear wall or braced frame is no longer shallow and slender, as for the extreme case of a horizontal panel in a low-rise building, they do behave like shear cantilevers and not flexural cantilevers.

As is known from basic strength of materials, the flexural resistance to lateral loads is expressed by the axial bending action M/S and shear action of roughly V/A for

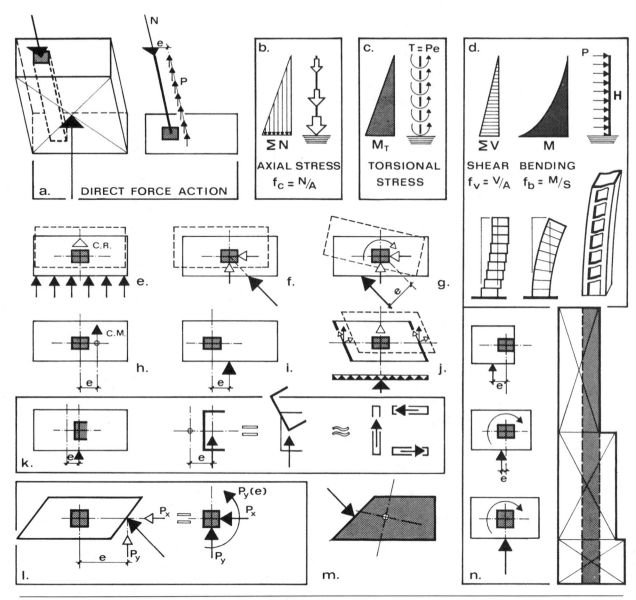

Figure 2.14. *The building response to load action—the effect of asymmetry.*

certain conditions of symmetry (Fig. 2.14d). For the given uniform lateral loading case, the shear increases linearly towards the base ($V \propto H$), while the moment grows much faster, following a second degree parabola ($M \propto H^2$). Since this special condition of simple bending due to symmetry is often not present because of asymmetry of the resisting structure and/or the eccentric action of the resultant lateral force, some general concepts of structural behavior of bending members are briefly reviewed first. Furthermore, thin-walled beam behavior, as for tubular structures, is ignored in this context.

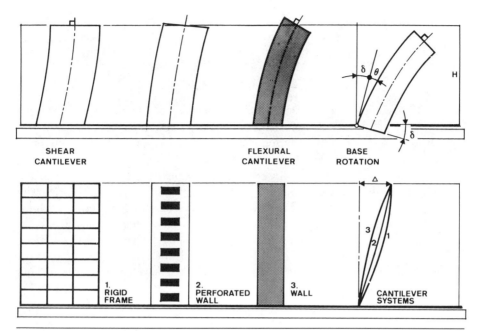

Figure 2.15. *The effect of structure type upon cantilever action.*

In general, to determine the stresses due to pure bending of an unsymmetrical section with no *axes of symmetry*, requires complex calculations. First, the *principal axes* that are always mutually perpendicular, and about which the moments of inertia are maximum and minimum, respectively, must be located; then the direction of the *neutral axis* has to be found. All of these axes, together with the *centroidal axes*, pass through the centroid of the cross section. Naturally, for this general condition, the simple bending formula $f_b = Mc/I = M/S$, which applies only to symmetrical bending, cannot be used!

In addition, the loads must act through the shear center or center of twist, which is located at the intersection of the *shear axes*, in order not to generate torsion in addition to unsymmetrical bending. Therefore, this shear center must be located; it does not coincide with the centroid of the cross section. One may conclude that, when the load is applied at the centroid, the member will twist as it bends. Lack of symmetry results in eccentric loads, unsymmetrical bending, and torsion! Fortunately, cross sections usually have a certain degree of symmetry which simplifies the understanding of the behavior and the stress calculations, remembering that an axis of symmetry is always a principal axis. The following cases describe basic cross-sectional shapes in which most lateral-force resisting building structure systems can be classified.

- *Shapes with one axis of symmetry* (T,U,A,E). Here the centroidal axes are also the principal axes.

  When the load acts parallel to one of the two principal axes, but through the shear center, simple bending occurs with no torsion: $f_b = M/S$. Should the load through the shear center not be parallel, then it can be resolved into components parallel to the principal axes, causing simple biaxial bending with different maximum stresses at the extreme corners.

$$f_b = +[(M/S)_x + (M/S)_y]_1$$

$$f_b = -[(M/S)_x + (M/S)_y]_2 \tag{2.11a}$$

When the load does not act through the shear center, but (for example) through the centroid, then torsion is generated. The approximate response of a section to torsion is discussed further in the section in Chapter 5 on "The Distribution of Lateral Forces to the Vertical Resisting Structures."

• *Doubly-symmetrical shapes with two axes of symmetry* (|, I, H, □, ○). In this case, the centroidal axes, the principal axes, and the shear axes all coincide with the axes of symmetry. They are also neutral axes when the load acts through the centroid of the cross section, which is also the shear center; in other words, the bending axis is the centroidal axis of the beam. For this condition (Fig. 2.14f), the force is resolved into components in line with the principal axes and causes simple biaxial bending, where the maximum stresses may occur at any of the four extreme core corners.

$$f_b = \pm[(M/S)_x + (M/S)_y]_{1,2} \tag{2.11b}$$

There will be torsion, together with simple bending, however, when the resultant force does not act through the centroid, as is the case for the rectangular core that is located eccentric with respect to the resultant wind pressure (Fig. 2.14i). Alternatively, there will be torsion when the center of the building mass does not coincide with the center of rigidity in earthquake areas (Fig. 2.14h); for other cases (Fig. 2.14j,l), there will be torsion together with biaxial bending.

It is apparent that closed tubular sections are much stronger and stiffer than the equivalent perforated open shafts. In the first case, the torsion is efficiently resisted in simple torsional shear (St. Venant's torsion), while in the open section (Fig. 2.14k), bending (warping) torsion occurs, in addition to the torsional shear, as is discussed further in the first section in Chapter 5. The round, closed thin-walled tube is the ideal shape for resisting twisting; in this case, only uniform torsional shear stresses are generated, although one must keep in mind that the weakening of the core by openings will disturb the uniform shear stress flow, and will also cause warping.

Open tubular sections will warp due to twisting as well as bend due to direct force action. When open shafts have at least two parallel walls (I-, E-, H-, or U-shapes, etc.), which is usually the case, then the torque will effectively be resisted by them in bending, since the torsion is resolved into a couple (Fig. 5.1); however, open shafts with no parallel walls (L-, T-, or +-sections, etc.) will be quite flexible and their behavior is extremely complex.

For symmetrical ordinary building structures (Fig. 2.14e, f, g) the effect of wind torsion due to nonuniform pressure can usually be ignored for primary design considerations, but this is not true for earthquake action where accidental torsion must be taken into account.

It is shown that torsion is generated by the asymmetrical buildings (Fig. 2.14j, l, m); eccentric and asymmetric cores (Fig. 2.14i, k), when the center of mass does not coincide with the center of rigidity; or due to the direction and location of the resultant load action (Fig. 2.14g).

The compound, hybrid building forms that are currently in fashion cause extensive eccentric loading due to lack of symmetry. The effect of asymmetry of geometry is seen in plan in Figure 1.6, in section in Figure 1.7, and in space in Figure 1.4. Asymmetry is caused by the alteration of the regular mass distribution and the abrupt or slow change of stiffness as for stepped or tapered forms; it is caused by discontinuity at the base or other levels, or by cavities, cut-outs, and so on. For example, between a triangular tower on top of a rectangular base there will be a radical change of mass and stiffness, especially if the structural systems of the two shapes are also different. The degree of symmetry of a building is a most important consideration; there should be as much symmetry as possible to avoid generating torsional action which is especially critical in earthquake regions.

As shown, torsional action may be uniform across the full building height (Fig. 2.14c), or it may be changing at points of setbacks (Fig. 2.14n); it may be resisted within the building or by the building shape (Fig. 2.14m). From a structural point of view, it may be more efficient to resist the torsion along the building perimeter, because the forces will be smaller due to their larger lever arms. The inherent torsional stiffness of the tubular building structure is well known, especially when trusses wrap around the corners to tie the building planes together to form a spatial unit.

The analysis of the simple flexural stresses in the lateral-force resisting elements can often be greatly simplified for preliminary design purposes. For example, in the typical cross-wall building in Figure 2.16f, the shear walls have, at the ends, single or double returns, and may also be coupled across the corridors. These walls can be visualized as vertical cantilever plate girders, where the wall webs are stiffened by the floors. It

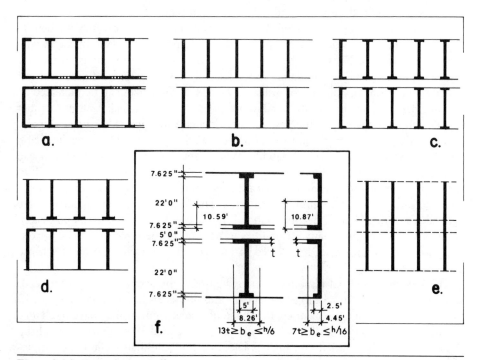

Figure 2.16. *Typical cross-walls.*

would surely be conservative to ignore the flange action in resisting rotation, and would take much less calculation time to use just the planar wall elements (Fig. 2.16b), rather than having to first determine the location of the centroidal axis and then to find the moment of inertia. Various other simplified approaches are investigated further in Problem 2.21.

The Building Response to Force Action

It has been discussed previously (in fig. 2.14) that the magnitude of rotation due to lateral force action grows at a much faster rate towards the base than the gravity load flow. The necessary reaction of the building to this force intensity distribution can be achieved in various ways, as is investigated in Figure 2.17. Typical solutions can be:

- Arched or tapered building forms
- Geometrical arrangement of members
- Increase of member sizes
- Increase of material strength
- Construction methods
- Combinations of the above

The logical formal response of the building would be a shape defined by an exponential function (Fig. 2.2, Equation 2.7). The curvilinear logarithmic profile, as reflected by the 1093-ft Tokyo Tower (Fig. 2.17d) or the buildings adjacent to it, gives the appearance of stability either by spreading the pier walls apart at the base, or by placing a flexible tower on a rigid triangular base. The efficiency of a lateral-force resisting perimeter structure is obviously related to the geometry of the shape, as expressed by the depth-to-width and height-to-width ratios.

The increase of force flow towards the base is convincingly expressed by the density of the stress trajectories and the truss analogy, in the respective cases. In the middle portion of Figure 2.17, it can be seen how a column and wall arrangement may visually respond to the accumulation of the loads by deviating from a uniform layout pattern. For example, columns or walls can be added, their sizes can be increased by stepping, tapering, or curving; the window openings can be decreased; the truss member density can be increased, or the base can be widened with inclined or tree columns. The stepping of the walls not only somewhat follows the increase of the force intensity, but also provides excellent lateral stability. Walter Netsch of SOM designed the facade structure of the 28-story Administration Building of the U. of Illinois in Chicago (Fig. 2.17a) by increasing the bay dimensions as the building goes up, as based on a uniform concrete strength and a minimum steel ratio. The facade zoning or proportions are derived from the "Golden Section" ratio. From a construction or structural system point of view, there are many possibilities to respond to the force flow. The various structure systems have been introduced in the section of this chapter on "High-rise Structure Systems."

The hidden, nonvisual response of the structure to the accumulation of the loads is usually done by the thickening of members and/or increasing the material strength. For example, at roughly 40 stories, the shape of a steel column may have to change from a heavy W-section of the jumbo type to a coverplated W-section, or a built-up shape, such as a box, cruciform, folded plate, or a multiple shaft section (e.g., triple-web). The column size may be kept constant from a visual point of view by decreasing

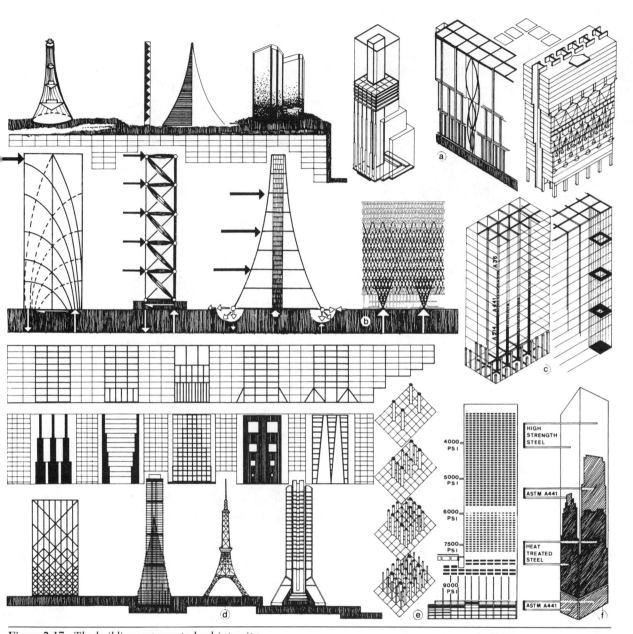

Figure 2.17. *The building response to load intensity.*

the density of the cross section, for instance, from a hollow concrete tube to a solid section at the base, and by changing the material strength.

The magnitude of the load flow along the load-bearing steel lattice walls of the IBM Building in Pittsburgh (Fig. 2.17b) is covered with various steels ranging from A36 at the top, A441 (F_y = 50 ksi) in the lower portion, to 100 ksi steel in the triangular columns at the base, where all of the building loads are concentrated. The exterior tube Vierendeel walls of the World Trade Center in New York (Fig. 2.17f) include twelve different steel grades, with yield points ranging from 42 to 65 ksi in the top half, and 100 ksi in the lower half, where the additional wind stresses become predominant. At the building base, the larger, wider spaced columns required 50 ksi steel. All interior core columns are A36 steel.

The typical material change of steel columns from the top of a tall structure downwards, as shown in Figure 2.17c, could be

- Carbon steel: ASTM A36: F_y = 36 ksi. This steel is most common.
- High-strength low-alloy steel:
    ASTM A572: F_y = 42 to 65 ksi
    Grade 50 is the typical choice for bolted and welded buildings using high-strength steel.
    ASTM A441: F_y = 40 to 50 ksi, is used occasionally.
    ASTM A588: F_y = 50 ksi for shapes. This is a corrosion resistant weathering steel used primarily for exposed conditions.
- High-yield strength quenched and tempered alloy steel:
    ASTM A514: F_y = 90 to 100 ksi
    This very strong steel is currently only available in plate form. It may be used for welded, built-up columns at the base of tall buildings.

For an elliptical, 34-story office building at 53rd St. at 3rd Ave. in New York [1985], only four massive steel columns were used in the core, to carry 50 percent of the entire gravity load and 80 percent of the wind loads. These jumbo columns had to be built up from 6 inch thick × 30 inch wide steel plates, welded together to form tubular shapes in the bottom portion of the building. The columns weigh almost five times as much as the largest rolled wide-flange section available.

For the lower portion of tall building structures, high-strength steels are usually used when strength considerations control, as (for instance) where the dead load constitutes the major portion of the total design load, while lower-strength steels, with correspondingly larger member sizes, may be required when stiffness criteria govern, as for the exterior columns of tubular buildings that must resist critical lateral forces occasionally. In summary, high-strength steels tend to be economical where strength controls the design, but standard carbon steels may be more economical when stability, deflection, elongation, or stiffness is the governing design factor.

In high-rise concrete buildings, the reduction of column sizes for the lower portion is essential. Columns of conventional-strength concrete would be quite large, severely restrict the flexibility of the floor layout, and consume a great deal of floor space. Therefore, the greatest use of high-strength concrete to date has been in high-rise buildings, to minimize column sizes at the lower levels. Currently, the 19,000 psi concrete columns for Seattle's Union Square building (see fig. 7.38i) constitute the greatest strength concrete ever used in conventional structures. In addition, the high modulus of elasticity of 7.2×10^{-6} psi is about twice as high as for conventional concretes, which was essential for the stiffness requirement, to meet the occupants' comfort criteria. It is remarkable how the useful strength of concrete has been increasing from about one-tenth that of steel in the past to nearly one-half, at present.

In contrast to conventional ready-mix concrete with a compressive strength of 3000 to 5000 psi, the quality of high-strength concrete is achieved through the following changes from standard construction practice, which include: the use of chemical admixtures, fly ashes, stronger cements, and grading of aggregates; and minimizing the water-cement ratio by using water-reducers is the most important factor. The less water in a concrete mix, the higher the strength and the lower the shrinkage and creep, but also the more unworkable the mix becomes. The excess water beyond that required for hydration reduces the strength, but the addition of a water-reducing superplasticizer makes the concrete workable. It is a liquid chemical that makes the mix fluid, and hence workable, so that less water is needed. Another additive is used in conjunction: powdered fly ash, or superfine micro silica (also known as silica fume), for high-strength concretes behave similar to cement, but do not generate as much heat when reacting with water. By filling the tiny voids between the cement grains, the silica thereby strengthens and densifies the concrete, besides giving a greater durability. Other factors necessary for increasing the concrete strength are: stronger cement; a high cement content; a stronger but smaller round aggregate, together with an optimum grading; and finally, a thorough quality-control program.

The first notable example for the application of high-strength concrete in tall building construction is the 74-story Water Tower Place (Fig. 2.17e) in Chicago [1976]. The slender tubular tower portion consists of a soft perimeter tube braced transversely by an internal perforated shear wall, which takes 65 percent of the wind loads. Lightweight concrete was used for the floors to reduce the building weight and the column loads. The column strengths range from 9000 psi at the base to 4000 psi at the top, while the column sizes change from 4-ft square with a vertical reinforcement of 8 percent at the basement level, to 16×48 in. at the twenty-fourth level, with only 0.8 percent of steel.

For engineered and inspected masonry, the same wall thickness may be used for the entire height of a building rather than progressively increasing the wall thickness. For example, for concrete block construction, the following approach could be used:

- Top portion: partially grouted, not inspected, low masonry strength 50 percent solid (e.g., 1000 psi block)
- Center portion: solid grouted, inspected, and higher strength masonry (e.g., 3500 psi)
- Bottom portion: solid grouted, inspected, higher masonry strength at least 75 percent solid, and possibly reinforced concrete masonry columns

Multistory load-bearing masonry buildings, beyond approximately 15 stories, may require a compressive strength of masonry units of 4000 to 6000 psi with the corresponding masonry strength of 2000 to 2400 psi for Type M and S mortar. In very tall buildings, compressive steel reinforcement in the masonry wall will reduce the wall area and increase vertical continuity.

The Effect of Building Height

The structural design of ordinary buildings in the 20- to 30-story range, where the gravity structure also provides the lateral resistance, is usually controlled by gravity loading. In this instance, the gravity structure can absorb the lateral forces so that the member sizes may not have to be increased. As the building increases in height, lateral force resistance becomes a dominating design consideration. For ordinary buildings up to approximately 40 stories, but occasionally up to 60 stories, in height, the lateral-force resisting structure can be contained within the building volume, and thus does not have to influence the appearance of the architecture. As the building increases further in height, the overall building form must be activated, so that the structure must be concentrated along the perimeter to efficiently resist overturning by lateral forces. However, when the tall building also becomes slender, that is, when its height exceeds about five times the minimum base dimension, then the lateral deformation of the building due to wind becomes of primary concern. Now stiffness, rather than strength, becomes a controlling design determinant—this includes the critical torsional deflections at building corners, hence the concern for torsional stiffness as well. The wind causes a dynamic response in the flexible building, especially in the across-wind direction, thereby generating twisting and oscillations. Further, secondary moments due to the large drift (P-Δ effect) must be considered, in addition to other secondary effects, such as differential shortening of vertical elements, that now become primary design determinants. It also may be necessary to collect and concentrate the gravity loads at just a few locations along the perimeter to help to overcome overturning, which is an essential consideration today since buildings have become much lighter. One must realize, nevertheless, that even for a building of lesser proportions than 5:1, when all of the lateral forces are solely resisted by a central core, the structure may still have to be considered extremely slender from a structural engineer's point of view. Also with an increase of building height, the control of critical temperature variations between the exterior and interior structural elements becomes an essential consideration; a thermal jacket is required for at least a part of the exterior building structure.

As the building increases in height and slenderness, stiffness rather than strength determines the amount of material required, keeping in mind that ductility must also be considered in critical seismic areas. The overall stiffness of a building depends on its height-to-width ratio, its form, and the structure system. The lateral deflection of tall buildings has become of primary concern to the design of tall buildings, realizing that the World Trade Center in New York sways in the wind nearly twice as much as the heavier and less efficient Empire State Building, with only 6.5 in. of sway.

The lateral sway of buildings is primarily caused by wind and earthquake. One must, however, keep in mind that an earthquake, in contrast to wind, may sway tall buildings not any more than lower ones. Whereas low buildings will vibrate primarily in their first mode, taller ones will also deflect in the higher modes, similar to the movement of a snake. The sidesway produced by unsymmetrical vertical loading usually balances itself over several stories and can be neglected for relatively stiff buildings. The maximum static deflection of a slender tower with uniform cross section under a uniformly distributed load, w, due to bending by ignoring shear deflection, is given by the familiar expression of

$$\Delta = wH^4/8EI \qquad (2.12a)$$

For a flexural cantilever of linear taper, this expression becomes

$$\Delta = wH^4/6EI_o \qquad (2.12b)$$

For a tower of curvilinear profile, the maximum sway at the top may be approximated as

$$\Delta = wH^4/4EI_o \qquad (2.12c)$$

where: EI_o = stiffness at tower base

Refer to Table A.14 for other lateral displacement cases, as based on uniform tower cross sections. Codes limit the lateral building deflection to Δ_{max} = H/500, that is, by the drift ratio R = Δ/H = 1/500 for wind. For earthquake displacement, this ratio is Δ_{max} = H/200, except H/400 for buildings with unreinforced masonry. The same drift ratios also apply to the sway of a story relative to adjacent stories, by letting H = h. This limiting deflection may be based on the strongest wind anticipated in a 100-year period for strength design criteria, but it sometimes may be based on a 50-year mean recurrence interval, for serviceability design requirements. For example, a 1000-ft high building tower should not deflect more than 1000/500 = 2 ft.

The lateral stiffness (k = 1/Δ) of a building also controls its dynamic properties, that is, its natural period. For the preliminary design of ordinary buildings, it may be considered as proportional to the building height (T $\propto$ H, see Equations 3.21 and 3.22), as based on a linear increase of material resistance (i.e., similar to a linear taper of a cantilever beam). However, for a curvilinear taper, the natural period appears to be proportional to approximately the square root of the height (T $\propto \sqrt{H}$) as has been verified for an example for trees.

The stiffness of buildings or their lateral drift must be controlled for several reasons:

- **Architectural integrity**: A flexible building may distort so much that it stresses the architectural subsystems or the curtains and partitions, as well as the mechanical systems.
- **Occupant comfort**: Excessive lateral sway, together with oscillations (i.e., acceleration) due to gusty winds, may not be acceptable for human comfort. People may be motion-sensitive and not be able to work under high wind storms, and some may even experience motion sickness. Naturally, the simple drift ratios above do not control the performance of the structure with respect to the building occupants' sensitivity to motion.
- **Structural stability**: Under lateral loads, the mass of a flexible building will shift and move the centroid of each floor, and thus the weight of the structure increases the tendency toward overturning. The resulting large lateral deflections cause the so-called P-Δ effect, where the eccentric action of gravity loads results in secondary moments due to first order displacements. The consideration of the P-Δ phenomenon is essential for the design of flexible tall buildings.

Linear shear deformations generally dominate the lateral sway of open rigid frames, whereas flexural deformations control the behavior of shear walls and braced frames. However, since most tall slender building structures of relatively uniform mass and stiffness distribution displace in a mode somewhere between the ones for pure shear and flexural cantilevers, it is reasonable to assume (for preliminary design purposes) that the structure deforms in a linear fashion, approaching a shear cantilever as shown in Figure 2.18, for rigid base conditions. From the elementary strength of materials, it is known that the moment is proportional to the curvature (i.e., the flexural stiffness

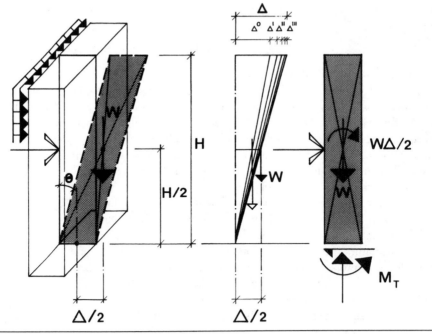

Figure 2.18. *The P-Δ effect upon the building.*

is assumed constant) or to the rate of change of the slope, which in this case is constant because of the straight line deflection.

$$\tan \theta \simeq \theta = \Delta/H = M/k, \quad \text{or}$$

$$M = k\theta, \quad \text{or} \quad k = M/\theta \quad (2.13)$$

The final moment, M_T, is caused by the initial moment, M, due to the lateral forces and by the moment due to the weight of the structure, as shown in Figure 2.18.

$$M_T = M + W(\Delta/2) \quad \text{(a)}$$

The initial additional moment due to the weight, W, is $W\Delta°/2$, which in turn causes a further deflection Δ', with a corresponding increase in moment of $W\Delta'/2$. This process continues at a decreasing rate, but may result in collapse unless the structure has sufficient stiffness to stabilize the movement. The final deflection, Δ, is equal to

$$\Delta = \Delta° + \Delta' + \Delta'' + \dots . \quad \text{(b)}$$

The ratios of the moments with respect to the deflections they generate, remain constant.

$$\frac{\Delta}{M_T} = \frac{\Delta°}{M} = \frac{\Delta'}{W\Delta°/2} = \frac{\Delta''}{W\Delta'/2} = \dots \quad \text{or:}$$

$$\Delta' = \frac{W\Delta°}{2}\left[\frac{\Delta°}{M}\right], \quad \Delta'' = \frac{W\Delta'}{2}\left[\frac{\Delta°}{M}\right] \quad \text{(c)}$$

Substituting Equation (c) into (b) yields

$$\Delta = \Delta^\circ + \frac{W\Delta^\circ}{2M}(\Delta^\circ + \Delta' + \Delta'' + \ldots) = \Delta^\circ + \frac{W\Delta^\circ}{2M}\Delta$$

$$= \Delta^\circ \left[\frac{1}{1 - \dfrac{W\Delta^\circ}{2M}} \right] \tag{d}$$

However, from Equation (c) it is known that $\Delta = (M_T/M)\Delta^\circ$, hence

$$M_T = M \left[\frac{1}{1 - \dfrac{W\Delta^\circ}{2M}} \right] \tag{e}$$

Substituting into Equation (e) $M = k\theta^\circ$ and $\Delta^\circ = H\theta^\circ$ (Equation 2.13) yields

$$M_T = M \left[\frac{1}{1 - \dfrac{WH}{2k}} \right] = M(MF) \tag{2.14a}$$

In this case, the magnification factor, MF, is similar to the moment magnifier used for slender column design. This expression can be further simplified for a rectangular building shape by letting $k = M/\theta = M/R = M/(\Delta/H) = \dfrac{wH^2/2}{\Delta/H}$, and $W = qHB = pH$.

$$MF = 1/\left[1 - \frac{WH}{2k} \right] = 1/\left[1 - \frac{p(\Delta/H)}{w} \right] = 1/\left[1 - \frac{p}{p_c} \right] \tag{2.14b}$$

Since the P-Δ effect is a consideration of stability, the working load is determined by dividing the buckling load by a safety factor or by including the load factor γ and the resistance factor ϕ (see Equation 3.45).

$$MF = 1/\left[1 - \frac{\gamma p}{\phi p_c} \right] \tag{2.14c}$$

where: q = unit building weight (pcf)

w = uniform wind pressure (psf)

B = building depth in the direction of wind loading

$R = \Delta/H$ = drift ratio

$p = qB$ = building weight per unit height

$p_c \approx w(H/\Delta)$ = building weight per unit height causing lateral elastic buckling as derived from Equations 2.1a and 2.12

γ = load factor

ϕ = resistance factor

Although the magnification factor was derived from moment action on the overall structure, it may also be applied to the shear effects for first approximation purposes. For the preliminary design of structural members, the moments, shears, and axial forces due to lateral force action are multiplied by the magnification factor.

Rigid base conditions have been assumed in Equation 2.13, realizing that foundations and soil are not infinitely stiff. Taking into account tilting of a slender building, then the rotation of the foundation δ (in fig. 2.15), which is treated as constant, causes the following total lateral drift:

$$\Delta = (\theta + \delta)H$$

Example 2.1

Investigate approximately the P-Δ effect for a building which is 860 ft high, 120 ft wide, and 180 ft long. Assume that the building weighs 10 pcf, and must resist a uniform wind pressure of 40 psf. The limiting drift ratio is H/500, and the load factor is taken as 1.4.

The height-to-width ratio is H/B = 860/120 = 7.17 > 5, which indicates that the building is quite slender and must be considered flexible.

The magnification factor, according to Equation 2.14c by ignoring ϕ, is

$$MF = 1/\left[1 - \frac{\gamma P}{P_c}\right] = 1/\left[1 - 1.4\frac{10(120)1/500}{40}\right] = 1.09$$

The wind moment due to the P-Δ effect is 6.4 percent larger, ignoring the safety factor of 1.4, but for the design of the structural members the moments and forces due to wind action should be magnified by 9 percent. Hence, the magnified wind moment about the base is

$$M_T = M(MF) = [0.040(180 \times 860)860/2]1.09 = [2662560]1.09 = 2902190 \text{ k-ft}$$

CONSTRUCTION

Designs are often only concerned with the final building performance; they may not place much emphasis on the constructibility, that is, the actual building of structures with the interacting flow of activities of the various trades. The construction of buildings involves a sophisticated process, not just in terms of fabricating, shipping, storing, and assembling the various building components, but also in organizing and scheduling this flow of labor and equipment.

The designer is often lacking the field experience to visualize how the contractor is going to build it. But he should take basic concepts and methods of construction into account during the design stage, when he lays out and dimensions the building and responds to the economy of construction, possibly including the contractor in the design team, and during the later quality control stage when he makes sure that the building is constructed according to the specifications and drawings. By communicating and working together with the fabricator and erector, the design team will be familiar with the flow of component fabrication, transportation from the shop to the site, accessibility of site, the handling of materials, the size and location of the storage space on the site, the energy supply sources, the process of assembly, including the allowable geometric tolerances and how to allow compensations, as well as the capacity and position of the erection equipment, and the availability of local materials and construction expertise. All of this should make the designer appreciate the complexity of the construction process and give him a sense of control, so that he can produce complete specifications and clear detailed drawings to facilitate the building process and thus prevent future confusions and misunderstandings. When the architect uses the result oriented *performance specifications* rather than the common *descriptive specifications* which describe in detail the method of construction in a cookbook fashion, then he can take advantage of the rapid development of new technologies in certain fields such as curtain wall design, although he must deal with elaborate documents that specify design criteria and testing procedures.

The greatest economy and efficiency of construction are achieved by using a *minimum number of operations* on site, which includes minimizing the *number of different components* which should be assembled in a *repetitive, continuous* process, *simplifying* the field connections, and minimizing the *start-stop* of any activity, that is, each trade should be in control of its own activity without interference. Hence, a proper construction sequence of the different trades is required, which (in turn) has an effect on the speed of construction. A *dimensional organization* of the building is necessary that reflects *simplicity* and *modular coordination* of standardized elements, including duplication of parts and methods of linkage, and thus includes a means of the control of accuracy, tolerances, and fit. For a regular building, this means uniform bay sizes, symmetry and alignment of columns; but even a nonregular structure of complex configuration should be the result of a simple *unit addition*. This repetition allows less piecing together, less waste and less time; that is, a smooth flow of construction, where machines and other equipment can be effectively utilized. Considering the high cost of borrowing money to finance construction, time is money. Hence, moving men and materials onto the site as rapidly as possible is a basic requirement. Traffic time on the site can be kept to a minimum by a computer system that schedules and monitors the time frame of deliveries through specific gates and for specific hoists. Rationalization and automation of the building process will eventually make construction robots feasible, as most manufacturing industries are already in the process of robotizing their production activities.

Naturally, the designer must not allow this repetition of elements and processes to take over and result in visual monotony and boredom, as experienced by so many large-scale housing blocks of the 1950s and 1960s. The components must be combined in such a way as to yield an optimum number of formal variations!

The typical construction methods that are presented in the following sections are the *conventional, industrialized,* and *special construction techniques*. The selection of a construction system includes life-cycle costs, which is to say, not just the initial construction costs, but also the future maintenance and operating costs or capital-investment and maintenance replacement costs. Usually, the speculative developer is concerned only with the short-range construction costs, while corporate and institutional clients must consider the total project development costs.

The owner's choice of contract substantially determines the designer's and contractor's roles, that is, the degree of their interaction as a team. Among the numerous construction contracts, the most common forms are: the guaranteed maximum price contract, the cost plus fixed fee contract, and the typical lump sum contract. An important factor is the time of construction. A building system may very well cost more to construct, but may also be ready at an earlier date for occupancy, hence may still be more cost-effective. In the *fast track method*, or phased construction, in contrast to the standard design/bid/build sequence, the time of construction is compressed by starting with the construction of the foundations, while the working drawings for the building are still being completed. Since the project costs are generally higher under this approach, it should only be used when a fast construction time is essential to achieve early occupancy, and it is worthwhile to take the larger risks involved. Fast tracking a project requires a well-managed team, because of overlapping of construction and design; the speed of construction now makes quality control a much more critical issue. Another design consideration, besides the length of time of construction, may be the *flexibility* or *adaptability* of a system to future change. In the choice of a structural system, the following criteria (among other aspects) must be considered: cost, construction time, level of industrialization, availability of materials and labor, degree of expertise, nature of the site, labor group requirements, construction risk, and adaptability to integration of structural, mechanical, and architectural needs. Important in the selection of a construction system is the degree of dimensional accuracy, which can only be achieved with skilled craftsmen. One also should keep in mind that the structure can be exposed or hidden behind a skin. While in exposed structures every detail must be studied, thus requiring high quality standards of work, in covered structures many of the sins can be obscured.

Whatever construction system is selected, precise measurements must be taken when the building components are assembled. Since it is impossible in the production, fabrication, and erection process to have exact dimensions, permissible tolerances must be taken into account; it is obviously necessary to maintain vertical and horizontal tolerances. Though most buildings are (to some extent) out-of-plumb, close plumbing tolerances must be required for elevator shafts and facade framing; in addition, levelling of the floors must be observed, which is critical for tall buildings.

In the control of permissible tolerances, the joint is the critical element, but it has to satisfy many other requirements. As the traditional methods of construction are being replaced more and more by automated operations, which includes the assembly of prefabricated components, an increasing importance is placed on building joints and jointing. At the junction of every building component occurs a joint (disregarding the joints within the prefinished components). The joint type is dependent on the location and position of the adjacent members. The components may be closely fitted, or a gap may be left deliberately between them (e.g., dry or wet type). This gap may have to be sealed with familiar materials such as bedding, caulking and glazing compounds, putties, mastics, or gaskets. The members at a joint may either stay clear of each other or they may be interlocked in some fashion. Similarly, on a large scale, joints may separate or partially separate entire building blocks that have different mass and stiffness characteristics, allowing certain movements; in this instance, the joints are usually formed between double members (e.g., beams and columns) or at the ends of cantilevered members. A joint may have to satisfy any of the following performance requirements (see fig. 3.14): *environmental control* (sealing against moisture, wind, dirt, sound, insects, etc.); *dimensional control* (product, and erection tolerances); *functional control* (maintenance, replacement, etc.); *movement control* (sliding, rotation, etc.); *structural control* (for load supporting and lateral connections). It is quite apparent that a joint provides a buffer zone where tolerance variations must be absorbed.

The joints may be designed as so-called *open joints* to allow for more movement and less precision in the fitting of members (i.e., larger dimensional tolerances). In *closed joints*, the gap between the components is simply closed by a weatherproofing seal of mastics or gaskets. Whatever the joint type, the width of the joint and the sealant must respond to the performance criteria; the sealants must be sufficiently elastic to permit movement between the elements.

It is not the purpose of this discussion to study and classify the seemingly endless number of joint systems that have been established by the various traditional building trades or that are presently developing from the new technologies, or to teach joint detailing, but rather, only to give some general introduction.

The speed of construction, together with an efficient quality control procedure, is a basic economic consideration; it is directly related to the amount of hoisting equipment needed and the number of lifting cycles per shift.

Erection Equipment and Scaffolding

Important considerations in building construction relate to scaffolding, shoring, forming, and hoisting. A structure during the erection process may have to be temporarily braced and shored, similar to concrete formwork that must be supported by simple or scaffold-type shoring. Scaffolding provides the working platform for labor, or it may be used for safety purposes to protect construction personnel. At present, it is most often used for existing rather than new construction because of the active renovation market. The massive aluminum scaffold for the Statue of Liberty restoration is still in the public mind. It enclosed the figure but was not permitted to touch it.

The most common material for scaffolding is steel, although aluminum and wood are also being used; bamboo is common in the Far East. Scaffolds fall into two main groups:

- *Standing scaffolds* (in place, or mobile units such as rolling tower scaffolds), which may be further subdivided into the following main types: *Frame scaffolding* is the most common type and quickest to install, where welded frames or sections are stacked on top of each other and are braced; *pipe-and-clamp scaffolding* is the most versatile system, which is clamped together piece by piece in the field; and *system scaffolding*, which is a variation of the pipe-and-clamp type.
- *Swing stages*, or suspended platforms (i.e., one-level platforms hanging from an overhead support), which are mostly used for routine maintenance and window washing.

The hoisting of materials and equipment for the erection crew and finishing trades is done primarily by cranes or derricks and hoisting towers (platform hoists, material and personnel cages, elevators). Jacks are used for vertical slip-forming, push-up and lift-slab construction. Often, pumps are employed rather than material hoists with conveyors and deck hoppers for the placement of the concrete mix; but a pump may also feed a separate placing boom(s) with a horizontal reach of 100 ft or more. For the lower floors, concrete may be pumped through rigid pipe or flexible hose on one building side, but by crane and bucket on the opposite side. Concrete has been pumped over a horizontal distance of up to 2000 ft and has been vertically moved, by a single pump at ground level, higher than 1000 ft with a volume, for instance, of 120 cu yd/hr on lower levels to about 50 cu yd/hr for the highest levels; the conventional pipeline for pumping is 4 or 5 in. in diameter. For the pumping of lightweight concrete, special attention must be given to preventing the water required for workability from being lost to absorption, since the lightweight aggregates are very porous. Therefore, the aggregate must be presoaked, so that the lightweight concrete can be pumped without higher pressure to the same heights as normal-weight concrete. Also, super-plasticizers make the concrete workable and fluid so that it can be efficiently pumped to great heights.

For special construction conditions, helicopters and helium-filled balloons, as employed for the unloading of cargo ships, may be used. The size, height, and shape, as well as the location of the building (in addition to the site layout and accessibility) determine the type of erection equipment. While material and personnel hoists move vertically, cranes and derricks provide the spatial interaction. The economy of construction is partly related to site organization, how and where the cranes are placed, and how often they are used. The flow of material from the fabricator or stock yard must be done with a minimum of movement. Often there is a choice between the number, size, and type of cranes: they can be *stationary*, *travellers*, or *climbers*. The nature of construction, such as the weight, size, and shape of building components (and their location) determines the lift capacity of the crane, with its maximum moment capacity of load times radius ($P \times R$) under certain wind conditions. The number of cranes and their reach capacity are dependent on the size of the site that must be covered (i.e., building layout, storage yard, delivery space, property line, and other clearances) and the number of crane lifts required or *hook time*. Different construction techniques must also be considered—building components may be delivered to the site or manufactured on site. For example, the assembly of a steel skeleton building story may require, for instance, 25 lifts of steel, whereas an equivalent poured-in-place concrete story may take six times as much. The selection of crane types is not simple. A high-rise, slab-like apartment block (Fig. 2.19o, r), for instance, could be served either by one travelling, rail-mounted (or truck-mounted) tower crane, by two climbing cranes/derricks, or by two stationary cranes, where the last two solutions operate in a circular pattern. In this context, the hook time is an important concept, keeping in mind that a crane has only one hook. Though stationary cranes may be less flexible, they grow with the building and take less space on the ground and ease traffic congestion at the site. The primary hoisting equipment of skyscrapers are climbing cranes or derricks; they are often seated on the interior elevator shaft.

The typical force actions upon a crane are identified in Figure 2.19m, although one must keep in mind that impact loads due to sudden slewing motions cause horizontal force action perpendicular to the boom similar to dynamic vertical loads due to hoisting. Stability against overturning may be more critical than strength, particularly for mobile cranes.

Cranes may also become part of the architectural design language in expressing the spirit of the building. In the early 1960s, Archigram (an architectural group in England) dreamt about an entire urban environment that could be programmed to be adaptable to change. In this case, the cranes became a symbol for serving and moving the plug-in units within this living dynamic urban network-stucture. Today, a small part of this dream may be visually expressed by the five permanent service cranes crowning the staircase towers of the Lloyd's of London, or by the maintenance cranes on top of four of the main structural masts of the Hongkong Bank of Hong Kong, which capture the dynamic quality of the building as an organism that must be serviced. In the following paragraphs, various types of cranes and derricks are discussed.

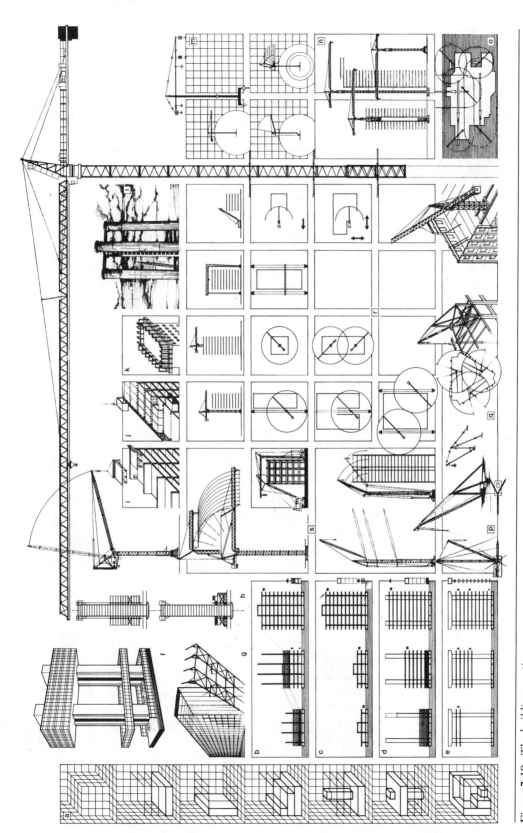

Figure 2.19. *The building erection.*

Cranes

The most common cranes for high-rise building construction are the *tower* or *jib cranes*. For less building height and smaller loads, *truck cranes* with telescopic booms and *mobile cranes* of the rubber-tired and crawler types with latticed booms, are sometimes used. *Portal* or *bridge cranes* are found in factories, but also occasionally used for the construction of large apartment blocks of the slab type. Here, only the jib cranes are briefly discussed. The jib is the long latticed boom which may be luffed (derricked, raised or lowered) and slewed (rotated, turned in a circle). Thus, a crane operator must control the slewing, luffing, and lifting. The boom is supported by the vertical lattice tower, called a mast.

The three common tower crane types are:

- *Diagonal boom* or *derricking jib cranes*: This crane type consists either of a slewing mast with counterweights at the mast base (Fig. 2.19s), or of a fixed mast with a counter jib and counterweight located at the top of the mast (Fig. 2.19s).
- *Hammerhead boom* or *horizontal jib cranes*: The mast is usually fixed, and the fixed horizontal jib with the travelling trolley and the counterweight are rotating from a mobile platform (Fig. 2.19n).
- Combinations of the above (Fig. 2.19s).

Tower cranes can be:

- Travelling on rails beside the building
- Stationary and free-standing beside the building
- Stationary and anchored to the building for larger heights
- Stationary and climbing within the building, as for skyscraper construction. In this case, derricks could be used to lift the cranes, or the cranes are jumped by employing jacks.

With an increase in building height, cranes either climb with the structure, taller new cranes must replace the original ones, or the masts of the existing cranes must be extended. There is a wide range of tower cranes available for any erection situation. Horizontal jib cranes have a much higher capacity than luffing boom cranes. The largest cranes have a reach (radius) of over 300 ft and a lift capacity of over 100 tons. One must keep in mind, however, that the crane range-weight-capacity may not be reached under certain unfavorable wind conditions.

Derricks

The use of climbing derricks is quite common for high-rise construction. They usually sit on the top level and operate above the framework. They may be located anywhere within, that is, attached to the central core or perimeter structure—or adjacent to the building. They are used for heavy lifts over a wide radius. Their typical capacity range may be up to 10 and 20 tons, though much higher capacity derricks do exist. The two common derrick types are:

- The *stiffleg derrick* consists of a boom (jib) pinned at the base to the slewing mast, with a swing of about 290° and a vertical mast, which is held in position by two diagonal stifflegs. It sits on a platform and creeps up on the structure. For example, for the U.S. Steel building (Fig. 2.19q), three climbing creeper derricks raised themselves by means of electric winches. With a 52-ton capacity at the end of 60-ft booms, the three derricks provided total coverage for the building.

- In the *guy derrick*, the stifflegs are replaced by guys, which support the vertical rotating mast (guyed mast). The mast is slightly taller than the boom when it is in vertical position, in contrast to the stiffleg derrick, where the mast is often much shorter than the jib. The derrick is easily jumped vertically by letting the boom and mast reverse position, each (in turn) serving to lift up the other. The swing of the boom is controlled by the clearance of the boom end and the guys.

Conventional Construction Methods

The assembly of building components, or the casting of concrete in place, is generally done from the *bottom up*, on a story-by-story basis (Fig. 2.19a). When larger component systems are used, like column trees (i.e., beam-column assemblies) and floor-bay panels, a stepped story approach may be convenient. For example, in the step-form technique, the concrete core may be advanced three to four floors ahead of surrounding steel framing. In suspension construction, the core is erected first, and, after the cantilevering support at the top has been positioned, the floors are usually assembled in place, but *from the top down*. In other construction processes, structures are erected bay-by-bay, sector-by-sector, as for large massive buildings, or using any combination of the above systems, if appropriate and as suggested in Figure 2.19a. Other approaches may include building down together with building up. The *up/down* construction procedure may involve the installation of caisson foundations, followed by simultaneous erection of the superstructure and excavation of the below-grade portion for the substructure. In conventional construction, nearly all structural components are custom designed for a specific building from standard rolled sections and standard formwork. In contrast, in the industrialized approach, manufacturers have mass-produced more finished component assemblies, like panels and boxes, with which the preengineered building is designed. In a way, conventional construction uses rationalized traditional methods with a certain degree of industrialization. Although it still depends primarily on conventional skilled trades, it does include mechanization of the building process and the usage of prefabricated components. One should keep in mind that the size of prefabricated subassemblies may be limited, especially in cities, by the permissible truck load envelope, which may be, for example, 10 ft wide by 14 ft high and 60 ft long. The primary erection equipment in conventional construction generally consists of cranes or derricks, and pumps for concrete work.

Steel Construction

Typical steel buildings consist of a skeleton framing, using concrete only as topping for the floor framework, for basement walls, and foundations. In contrast to cast-in-place concrete construction, where formwork is needed to construct a building, steel framework is shop fabricated, and marked members and subsystems are assembled on-site, in accordance with erection plans. Hardly any shoring and scaffolding is required, with the exception of temporary lateral bracing of frames, cross-bridging of beams, and shoring of decking for composite construction. Wet operations are kept to a minimum, saving curing time; often dry floor systems are used. Steel buildings require less field labor than concrete and the speed of construction is more rapid; they weigh less, and thus have less foundation costs, and can provide the desired ductility for high seismic areas. But one must keep in mind that steel costs more, and has to be fire protected. For instance, on the exterior facade columns, concrete may be used for protection, interior columns may be cased against fire, and hidden beams may have sprayed-on fire protection if nonrated suspended ceilings are used.

In steel construction, the floor plans must be ready at a very early stage, so that the fabricator can prepare and submit his fabrication drawings to the architects and engineers for approval, and so that he can order the material. The contractor can then schedule the sequence of construction, namely fabrication, with the necessary drawings, shipment to a storage yard or directly to the site, and the erection process. For tall buildings, the contractor must take differential column shortening into account by ordering some columns longer than others.

Steel construction is erected faster than cast-in-place concrete once the construction drawings are prepared. Further, it is more adaptable to possible changes during the construction process, or even in the future, which may be important today where it is common practice to reskin and expand buildings—steel members can be more easily reinforced than concrete. Although steel is stronger than concrete, and thus spans further with lighter members, it also lacks rigidity, especially when high-strength steels are used.

The fabrication and erection costs of the steel are more than double the material costs; the connection of members in particular constitutes a significant portion of the total costs. To reduce the expenses, standard shapes should be used in a repetitious fashion, and other time-consuming processes of assembly should be reduced to a minimum, thus requiring the field assembly to be simple. Often, it is more economical to cut, fit, join, weld (i.e., assemble), and finish structural elements in the fabrication shop under controlled conditions, and to form subsystems that can be erected faster on the site, such as two- or three-story tree-columns with two or three levels of spandrels, and floor assemblies. Usually, connections are shop welded and field bolted. In general, field connections consist of high-strength bolts, or are welded, for which welding machines are required.

Though, for the majority of buildings, the standard structural steel is A36, for portions of taller buildings, high-strength steels may be more cost-effective because they offer an improved strength-to-weight ratio. The choice between carbon steel and high-strength steel, from an economic point of view, is not always simple. Since steel prices are based on a per-pound basis, the total material cost of the lower-cost carbon steel, yielding a heavier total weight, must be compared with the higher cost of high-strength steel, resulting in less overall weight. It should also be kept in mind that not all high-strength steels are available in the thickness and shape variety of A36. Generally, when deflection (stiffness) or stability (slenderness) are the controlling design parameters, standard carbon steels may be economical, whereas high-strength steels tend to be more efficient when strength governs the design. Nevertheless, one should remember that the lightest structure does not necessarily cost the least—detailing expenses must also be considered, besides the material costs. Furthermore, the designer should be aware that, with an increase of steel strength, brittleness becomes critical; lower-strength steel has reserve strength: because of its ductile character, it avoids crack formation and sudden failure without warning. For typical conditions, beams and girders (as well as the lateral-force resisting frames) are ASTM A36, while columns carrying gravity loads are ASTM A572.

Cast-in-Place Concrete Construction

For certain conditions, the heavier concrete building, as opposed to a steel building, may be advantageous, since its weight may be used effectively as a stabilizing element against lateral wind action. But, since the dead load may be twice as much as for steel buildings, deeper and more massive foundations are required. Concrete may also permit a height saving in the building, particularly in flat plate construction, where the floor framing just consists of a slab, rather than a slab on beams, as in steel construction. Concrete members are larger than steel members, and thus provide more ridigity (besides having good damping qualities), which may be preferable for conditions where stiffness and acceleration control the design. Concrete allows easier modelling of shapes and does not require special fireproofing. In the design of columns, especially, but depending on locality, concrete may have cost advantages over steel. The present advance in concrete strength has made taller buildings (to about 1000 ft) possible, although beyond this height, the massiveness of members may cause space inefficiencies, unless a super-high-strength concrete can be developed.

In steel and precast concrete construction, time must first be spent to design the building components and prepare the necessary fabrication drawings, similar to a product design approach. In contrast, concrete construction work can start right away, since the structure is produced in place, that is, in the case of a regular building layout, where a maximum of standard forms and a minimum of custom forms can be used. In steel and precast concrete construction, most structure elements are assembled and connected to form the building assembly, while in concrete construction, first the formwork is set up and the steel reinforcing is placed (or vice versa), and then the concrete is cast, in turn, requiring a large labor force on site. This temporary formwork, together with the falsework, constitutes a structure that must provide the necessary strength and stiffness to resist the gravity loads of the material weight; the dumping of concrete; construction loads, including effects due to the movement of equipment; lateral wind forces; and the outward pressure of the wet concrete. The minimum vertical dead and live loads should be at least 100 psf and the minimum live loads 50 psf, assuming no motorized carts are used.

When fresh concrete is placed in wall or column forms, it acts temporarily as a liquid, causing lateral hydrostatic pressure. The effective lateral pressure depends on the rate of placement (R, in ft/hr) of the concrete, since the concrete has already started to harden at the bottom portion when the concreting is completed. It also depends on the concrete weight (e.g., 150 pcf); the temperature (T, in degrees Fahrenheit) of the concrete in the forms, which influences the setting time of the concrete; and the method of consolidation; for instance, when the forms are vibrated externally, the concrete is kept in a liquid state, thereby significantly increasing the lateral pressure. Stronger forms, because of larger lateral pressure, are required for lower temperatures and a faster rate of placements. For example, should normal weight concrete be placed in a 10-ft high wall form at 65°F at a rate of R = 4 ft/hr, using internal vibrators, then the semiliquid concrete exerts the following equivalent liquid pressure, p, according to the ACI Specifications for R ≤ 7 ft/hr:

$$p = 150 + \frac{9000R}{T} \le 150h \le 2000 \text{ psf} \qquad (2.15a)$$

$$= 150 + \frac{9000(4)}{65} = 704 \text{ psf} \le 150(10) = 1500 \text{ psf}$$

For the structural design of the sheathing, studs, wales, and ties, a uniform lateral pressure of 704 psf over the full height may be conservatively assumed, although the pressure near the top is less and controlled by 150h, where h (ft) is the depth below the surface of freshly placed concrete. Should the rate of placement of the concrete be greater than 10 ft/hr, then the equivalent liquid pressure can be taken as:

$$p = 150h \qquad\qquad (2.15b)$$

The formwork constitutes a large portion of the cost of a reinforced concrete structure; more than 50 percent of the total cost of a concrete frame is generally for the formwork. It is quite apparent that the design of the formwork is of utmost importance with respect to economy. The cost of formwork per unit of concrete surface formed is inversely proportional to the number of times a form can be used without any changes. The designer must take this fact into account by developing a regular plan layout, by controlling the dimensionality of the building, and by avoiding irregularities in the concrete shapes. The simplicity of its geometry as reflected by repetition from bay to bay and floor to floor allows for standard form sizes, and thus maximum reuse of forms, and allows common column- and beam-forming systems to be used instead of job-built forms, thus resulting in less waste of formwork material and less assembly time. It may be more economical not to change member sizes from floor to floor, but to vary reinforcement ratios. The increase in material costs due to member size uniformity is more than offset by the time savings from avoiding formwork changes and designing for constructibility. It is clear that the mechanization of formwork is of primary importance to the economy of construction.

Flat surfaces without offsets or irregularities, such as flat plates and core walls, are the fastest to construct; intricate shapes require custom-built forms. It is apparent that economy is achieved if the building layout allows for repetitive large-size reusable forms and construction cycling for the vertical building planes of frames and walls, as well as for the horizontal floor structure planes. It may also mean that, for formwork to be cost-effective, concrete buildings should have a minimum height.

Handset formwork, using conventional form components for columns, walls, and floors, which requires considerable labor, is replaced in larger scale high-rise building construction by various techniques of sliding, dropping, folding and slipping form-units. The most common forming systems are the flying forms, gang forms, and slip forms, together with automated form lifters. Special custom forms may be used, for instance, for stair construction and other high-quality, exposed concrete surfaces.

Flying forms are generally used for large floor sections (Fig. 2.19i,j). They may be more economical than handset systems when the initial fabricating costs can be balanced by more than, for instance, 10 to 20 repetitive reuses of the same large-size forms. They constitute platforms (i.e., table forms) supported on a shoring system (e.g., wood or metal posts, or scaffold-type frames) and may be inserted between or directly sitting on columns or walls that have already been cast (i.e., drawer-like forms), or they may be a tunnel form system for bearing wall construction, providing collapsible forms for walls and slabs. The typical flying forms are prefabricated decks composed of sheeting (e.g., plastic-coated plywood deck), joists, and possibly scaffold supports (e.g., adjustable post shoring with cross-bracing), which are hoisted in one piece from one floor level to the next by a climbing crane, in (for example) a floor per week cycle.

Gang forms (or ganged forms) are often used for facade frames and walls (Fig. 2.19k). They consist of prefabricated, repetitive modular panels that are joined into larger units that, for example, can be unfolded, as for gang-hinged forms, so that the vertical surfaces can be moved in one piece. Typical gang form systems for walls consist of sheathing (e.g., 4 × 8 ft plywood sheets) that retain the concrete, the vertical studs which support the sheathing (e.g., aluminum, steel, or wood studs), the horizontal wales (e.g., double channels) that support the studs, and ties and spreaders that hold the forms at the correct spacing under the lateral pressure of the wet concrete. For a typical bearing-wall apartment building, steel gang forms may be used for the walls and flying tables for the slabs.

In *slip forming* of cores and walls, forms are raised by electric, hydraulic, pneumatic, or manual jacks as the concrete hardens. A typical vertical rate is one to two floors per working day cycle, or 10 to 12 in. per hour, depending on the strength and setting time of the concrete. Vertical slip forming may not be cheaper than the ganging of modular panels, but it is faster. Notice that the bridges between the five slip-formed cores of the Knights of Columbus Building (Fig. 2.19l) act as spacers to hold dimensions, besides serving as links for material distribution.

The use of *self-jumping forms*, employing hydraulic lifts or all-electric jacking systems, frees the crane(s) for other usage, and thus economizes on the need for critical hook time. For example, the forms for the sides of the exterior columns and bottoms of spandrels may be hinged to swing out through the window openings to be raised along the exterior to the next level. The development of self-elevating jump forms is also a key reason for the economics of exterior composite construction.

Formwork costs may be reduced for some types of structures by using certain techniques. For instance, for long-span beam structures, the beam may be cantilevered during construction, as is practiced in bridge building, rather than being vertically supported by the falsework. For cast-in-place concrete floor construction, shoring may be eliminated to save erection time by designing the beam reinforcement as prefabricated open-web box trusses capable of supporting the construction loads. In this case, time need not be spent waiting for the concrete to gain significant strength, and the forms can be stripped for reuse earlier than in conventional construction.

Composite Steel-Concrete Construction

The composite interaction of reinforced concrete and steel, which is usually achieved by embedding the steel member in reinforced concrete or by bonding the adjacent materials to each other, as through mechanical shear connectors, is used not only for structure members such as the familiar slabs, beams, and columns, but also (on a larger scale) for entire buildings. These composite building structures are made up of a combination of composite members. Some of the typical systems are:

- composite framed tubes
- composite steel frames
- composite panel-braced frames
- composite interior core-braced systems
- hybrid composite structures

Composite construction takes advantage of the positive characteristics of both steel and reinforced concrete. The strength, light weight, speed of construction, and flexibility of interior layout of steel are combined with the low material costs (including a steel reduction of around 50 percent), stiffness, larger mass for damping, fire resistance, and moldability of concrete.

Maximum speed of construction is achieved by confining the steel and concrete trades to different building levels, so that there is a separation of several floors, often taken as at least ten levels, to ensure minimum interference. Generally, the gravity steel framing, with the corresponding small erection columns, is erected first (up to a predetermined level) and stabilized with temporary cable bracing. Then, the envelope frame is encased in concrete, as for a composite tube, or the concreting of the perimeter frames and core walls for composite frame-core structures is done with gang forms, as in conventional concrete construction.

Composite steel-concrete construction may be more economical from a materials and labor point of view, but also may take additional construction time, due to a more difficult forming process, thus resulting in interim finance charges and lost leasing time.

Other composite systems occasionally found in building construction are thin precast-prestressed concrete slabs or exterior wall/frame panels used as formwork for cast-in-place concrete, or composite masonry and frame systems, where the masonry forms the infill wall for the skeleton and provides the lateral shear resistance. In all steel, stressed-skin tubular construction, the framing of the perimeter tube provides all of the strength and some stiffness, while the composite steel cladding provides the rest of the required lateral stiffness, and does not need fire protection.

Precast Concrete Construction

In a way, precast concrete construction is similar to steel construction, in that it must concern itself with fabrication, transportation, and erection, which includes the ease of connection. Having larger prefabricated components with fewer connections is generally more economical, but is also more critical from a force resistance point of view.

The connections in precast concrete are the weak points, not only from an assembly, but also a behavioral, point of view. To achieve overall structural integrity, large scale assemblies must be tied together vertically and horizontally. In wall construction, the large panels are vertically post-tensioned together, while precast slab panel floor structures may have to be tied together by mechanical ties between the units, cast-in-place concrete ring beams (i.e., peripheral ties), and topping, so as to achieve the necessary diaphragm action, particularly in earthquake regions. The connections in precast concrete often consist of wet joints rather than dry joints, as in steel construction. Instead of casting the concrete in place, a contractor may precast the structure members either on-site to eliminate trucking, or in a precasting yard away from the site, assuming that he has the expertise to make the precise molds to achieve the necessary dimensional accuracy and quality finish of the components; in addition, a versatile prestressing bed for the production of the beams and planks may be required. On the other hand, it may be more economical for the contractor to commercially purchase available factory-made precast units, if they fit the building layout, rather than producing them himself; naturally, that depends also on his skills versus the level of performance demand.

Masonry Construction

In high-rise reinforced or unreinforced masonry bearing-wall structures, the typical clay bricks or concrete blocks are laid by hand, and thus represent a traditional construction system, although masonry panels have been prefabricated. There are fewer trades involved in the building process; the enclosing skin and interior partition walls are usually of the same material as the supporting structure. The construction process can be visualized as setting one-story structures on top of each other at a rate, for instance, of one floor each week; the masonry crews need the scaffolding only for one floor to lay the masonry. When precast floor systems or steel joist floors are used, no formwork for casting the concrete slab is required. Also, no floor beams are needed, since the slabs span directly from wall to wall, thus permitting a reduction in the floor-to-floor height, which may result in an extra floor for a 15-story building. Although mortar has been traditionally site-manufactured, it is now also available for site delivery similar to ready-mixed concrete. In addition, epoxy and other adhesive bonding systems have been introduced to facilitate erection, and to strengthen the bond capacity (i.e., tensile strength).

Cast Metal Construction

With the Pompidou Center in Paris [1977] cast metal reentered the building field as a primary structural material. In building construction, cast iron has always been considered to be merely a forerunner of steel; besides, the modern movement of architecture had no use for the decorative cast iron.

The Coalbrookdale Bridge in Scotland [1779] was the first all-iron structure ever built. In the first half of the nineteenth century, iron frames were used for railway stations, market halls, greenhouses, mills, exhibition halls, and industrial buildings, with the Crystal Palace in London [1851] representing the climax of cast iron architecture. Since then, first wrought iron and then rolled mild steel slowly replaced cast iron as a structural building material, though it remained extremely popular up to the end of the nineteenth century for numerous architectural products (e.g., railings, radiators, gates, balconies, and even entire facades) of any style, which were sold through catalogs.

The decline of cast iron from a structural point of view was primarily due to the fact that it is a compressive material, with a brittle character and a relatively low tensile and bending strength. Recently, however, the material sciences have developed cast metal alloys (cast steel, iron, and aluminum) with properties comparable to steel, that have sufficient ductility, plastic behavior, and high tensile strength to be useful. In addition, certain cast metal alloys show corrosion and heat resistance. Since the metal is cast in molds, its forms can be more continuous and rounded, especially at transition points, quite in contrast to the right-angled steel systems. Thus, a new formal potential is opened, allowing a smooth transition of force flow. This is quite apparent from the truss connections and gerberette shapes of the Pompidou Center, and the various fittings and anchors for the cables of the Olympic Stadium roof in Munich. The structure of the Pompidou Center (see fig. 7.41) consists of a mixed steel construction, where the columns, the gerberettes, the truss joints, and the wind bracing anchors are cast steel; the composite floor trusses consist of cast steel joints and steel tubes welded to them.

Mixed Construction

It is quite common to use precast concrete or cast-in-place concrete slabs in steel or masonry buildings, but in this type of mixed construction, the concrete plays a secondary role. Of interest here is the relatively new development of mixed steel-concrete construction where major components of concrete buildings, steel buildings, or composite buildings are combined. Vertically mixed systems may result from the mixed-use of buildings, where structure systems may alter with occupancies. For example, flat plate concrete construction for the upper residential portion of a building may change to a long-span steel framing for the lower office levels. Some typical examples of mixed construction are:

- Concrete core(s) and steel framing: In this case, the tall concrete core or shear walls may be slip-formed, possibly to the top, ahead of the steel erection, so as to avoid interference between the concrete and steel subcontractors. But, in order not to delay the steel erection, a rapid slip forming of the walls, as based on a 24-hour five day per week cycle, with the corresponding overtime pay, may be required. These higher costs, however, may be outweighed by the inconveniences associated with constructing the concrete core(s) simultaneously with the steelwork, for instance, in a two-level operation.
- Concrete frame on top of a steel structure: For example, a 40-story reinforced concrete frame with shorter spans is used for the upper residential section of the building, while the bottom 30-story steel structure provides the large spans required for the retail and office spaces. The closer concrete column spacing will be accommodated by transfer girders at the mechanical level. The lateral forces are resisted by the perimeter frame.
- Exterior composite framed tube plus interior simple steel framing.
- Slip-formed concrete core plus precast concrete floor framing plus precast concrete interior walls plus cast-in-place perimeter frame.
- Precast concrete panels used as formwork for cast-in-place concrete along the exterior of the building, thus providing the desired exposed surface quality. The forms may act compositely with the cast-in-place concrete.

Industrialized Construction

The application of industrialized techniques, using continuous automated processes of repetitive character, to the manufacture and assembly of buildings, has resulted in so-called *building systems* or subsystems, which are mass-produced in factories. Inherent in mass production is a reduction in time (costs), as achieved through sufficient repetition. Industrialized construction is part of the larger group of *systems building*, which includes, in addition to the hardware technology, the coordination of design, production, and site operations, financing, marketing, and management considerations. The degree of industrialization from a hardware point of view depends on how finished or unfinished the component systems are. Industrialization is very much associated with housing (single-family dwellings), school systems, and a variety of low-rise commercial and industrial buildings.

The concept of mass-producing building components and assembling them into a building unit was introduced on a large scale by Joseph Paxton for the Crystal Palace in London [1851]. By the middle of the nineteenth century, American trade catalogs were offering standardized mass-produced building components. Also, the development of the balloon frame at about the same period reflects the transition from craftsmanship to the trade approach, emphasizing the assemblage of standardized components by less skilled workmen in a relatively short time. Walter Gropius talked, as early as 1910, about the industrialization of housing through machine-produced standardized building parts. HUD's Operation Breakthrough of 1970 intended to establish a self-sustaining mechanism for rapid volume production of marketable housing at progressively lower costs for people of all income levels.

Some of the prefabrication manufacturers for the housing market, representing various degrees of industrialization are:

- Mobile home manufacturers
- Home manufacturers
- Sectionalized home manufacturers
- On-site fabricators using modular and panelized construction
- Patented building structure systems
- Fabricators of building subsystems, such as: elevator modules; interior partitioning systems; and service and utility modules, containing kitchen, bathroom, and utility centers, with all of the mechanical and electrical connections.

In high-rise construction, industrialization consists primarily of the addition of structure assemblies, mechanical systems, and architectural sytems (partitions, curtains, ceilings, etc.), rather than the finished product of the home manufacturers. Generally, they do not represent closed finished systems, but rather unfinished, more open ones; in other words, buildings delivered to the site may only be about 40 percent complete. The more common prefabricated structure systems are:

- *Skeleton systems:* the two basic building elements of column and beam are combined in various ways to form planar or spatial tree units (see fig. 7.15).
- *Panel-frame systems:* the three basic building elements of column, beam, and slab are combined to form planar or spatial column-slab or beam-slab units (see Fig. 5.5).
- *Wall-panel systems:* large-panel units (room-sized panels) are mostly used as interior bearing walls (see fig. 6.19), or finished curtain wall panels, of various materials, shape, and finish.
- *Box-systems* of the skeleton type, stressed-skin type, or monolithic shell type may be of any polyhedral shape. They are stacked to form self-supporting cellular clusters. Typical box kits are illustrated later (see fig. 6.18).
- *Mixed systems:* any of the structure systems above can be combined in an endless variety of ways, such as: Finished box units, which may be plugged into the supporting skeleton or spatial network (see fig. 2.7a); finished box units, which may be clipped on to a core (see fig. 7.41); or finished box modules, which may be used in combination with panel and skeletal systems.

The weight, size, and shape of prefabricated units are limited by shipping clearances, available storage space, especially for city sites, and the handling capacity of trucks and hoisting equipment.

Special Construction Techniques

It may be economical to construct entire building portions at the ground level, and then raise them into their final position by either lifting them in tension or pushing them up in compression. These packages can also be assembled at other building levels, so that the building is erected from the top down rather than from the bottom up. Some of the more common special construction systems are discussed in the following paragraphs.

Lift-slab Construction

First the columns are erected, and then, at ground or some intermediate building level, the steel floors (space frame or beam-joist floor framing) or flat plate concrete floors (post-tensioned or conventionally reinforced) are assembled or cast, one on top of the other, in stacks of up to 20 slabs, but separated by bond breakers. The slabs are then pulled up singly or in groups by hydraulic jacks, which are mounted atop the columns. At each floor level, the bottom slab is separated from the stack and connected to the columns.

Push-up Construction

This system may be seen as a reverse construction process, where the top floor is built first and the bottom floor last. The roof story is constructed at ground level, with the columns and/or walls supported by jacks. It is then pushed up one story, so that the next level can be built underneath. Then the two-story block is jacked up another level so that the next lower floor can be assembled; this process is continued until the bottom floor is constructed. On the completion of each step, the jacks are replaced by spacer blocks. The height of the building is dependent on the capacity of the jacks. Interior finishing can be done while the building is raised. In 1962, under the Dutch Jackblock system, a 17-story apartment building was built in England, using the push-up method for the first time.

Suspension Erection Systems

In this method, the central concrete core(s) is slip-formed first, to the full height. Steel-framed floor decks are then assembled on the ground, or concrete slabs are cast and singly, or in clusters, hoisted by hydraulic jacks into place, starting with the top floor assembly and working downward (Fig. 2.19d). The floors are then either attached to suspension cables or they are cantilevered from the core(s). In this approach, the top floors are finished first, and the exterior cladding and interior finish can be applied even before the floors are in their final position. The 20-story apartment building for the Russian U.N. officials [1974] in New York (SOM, Fig. 2.19f) and the BMW Tower [1972] in Munich, W. Germany (Karl Schwanzer, Fig. 2.19h) are examples of this type of construction method. Rather than assembling the floors at the ground and lifting them up, they can also be assembled at the top or intermediate levels and then let down (Fig. 2.19e).

Other Construction Methods

There are many other special construction methods, such as the balanced cantilever erection method, used in bridge construction, or multistory steel-framed bents assembled on the ground and pivoted upright (Fig. 2.19g), if enough space is available, similar to the tilt-up concrete wall system applied to low-rise buildings. The floor framing panels, which were also assembled on the ground, can then be raised and put into place. Post-tensioning of concrete is used for segmental construction and other prefabricated building systems, to horizontally and vertically tie the components together so that the building acts as a unit.

FOUNDATION SYSTEMS

Foundations are necessary as transition structures from the building to the ground, since buildings are not usually founded directly on hard rock. The bearing capacity of the soil is generally much lower than that of the structural materials, hence transfer structures with deep roots, or flat roots with a wide base, similar to base plates for steel columns, are required at the junction where columns and walls meet the soil. Foundations are either of the shallow type, which may consist of individual spread footings and continuous mats, or they are deep foundations of piles, piers, and caissons. Building columns and walls are either supported by individual or combined footings directly bearing on the ground or by piles. Foundations on firm soils form a natural extension of the bearing elements of the superstructure, while foundations on soft soils may form a large mat to spread the loads over the entire footprint of a building, because the weak soil cannot support pressure concentrations.

Ordinary high-rise building foundations consist of either a collection of individual rectangular and strip footings, or a large mat combining all of the single footings. In seismic areas, the individual spread and pile foundations have to be linked and tied together by bracing struts, so that the entire building foundation can act as a unit in sharing the load resistance. In crowded urban areas, the deep basements for tall buildings, particularly adjacent to other heavy buildings, cause excavation problems. In this case, special foundations, such as sheet piling, slurry walls, bracing of walls, or walls with tie-backs, along with underpinning of adjacent buildings and subways, are required. Pumps may be needed for conditions where the water table is high, which (in turn) may cause settlement problems with adjoining buildings.

The building foundations distribute the loads due to gravity and lateral force action to the ground. The resulting forces to be transferred are horizontal and vertical, depending on the structure systems (as indicated for various cases in fig. 2.12). The vertical forces are generated by gravity (e.g., building weight) and overturning. When a building is loaded only concentrically, then the contact pressure at its base may be assumed to be uniform, for preliminary design purposes. But, when overturning due to lateral force action, and less so due to asymmetrical gravity action, is generated in addition, then the overall base pressure is no longer constant. The type of force transition to the ground depends on the type of load-bearing element. While the foundations for a *gravity structure* of relatively symmetrical layout may cause a primarily uniform pressure distribution, the foundations for a *lateral-force resisting structure* may generate a nonuniform pressure due to rotation, thereby concentrating large forces on certain areas. The foundations of lateral-force resisting structure elements, such as walls, react in a nonuniform fashion (e.g., see fig. 2.12a, c, g, k, l), while the frame columns may respond in a uniform manner (see other cases in fig. 2.12). The eccentric loading of footings should be kept to a minimum, otherwise, the maximum contact pressure occurs only along the foundation edges, and the capacity of the soil is thus not efficiently used, besides causing nonuniform settlements. Considerations of nonuniform pressure distribution and stability have already been briefly discussed (see fig. 2.1c). Occasionally it may be necessary to employ under spread footings, which transfer compression forces, tension piles or piers in order to resist uplift forces.

Horizontal forces (i.e., base shear), as from wind and unbalanced earth pressures, can be transmitted to the soil by one or more of the following methods:

- Shear resistance at the base, such as provided by friction between the footings and soil, as well as by side friction
- Passive soil pressure, as provided by deep basements, piers, etc.
- Shear and bending resistance of piles or piers
- Axial resistance of battered piles
- Any combination of the above.

For large lateral loads, isolated footings and pile foundations should be braced by struts so that the foundations are tied together and the force resistance is more evenly shared.

The foundation systems of the past are not that different from the ones of today, from a conceptual point of view. Before reinforced concrete was introduced, around the turn of this century, shallow foundations were constructed of stone and masonry, possibly placed on timber grillages when the loads were high. For example, in Chicago in the early 1880s, piers were supported on stepped masonry pyramids, and slightly later columns on steel grillage foundations. These steel grillages consisted of two layers of railroad rails placed at right angles and embedded in concrete. Even thousands of years ago, the Babylonians (and later the Romans) used thick rafts on soft soils to carry their heavy buildings. Similarly, some of the medieval cathedrals are supported by reversed vaulting, which connects the foundation walls to form mat-like structures. Piled foundations go back to the lake dwellings of prehistoric times. The Roman Vitruvius, in the first century B.C., described the details for timber pile foundations. Venice and Amsterdam could not exist without pile construction. In 2000 B.C., the Egyptians were already using timber and stone caissons for shaft construction. Some of the early high-rise buildings in New York were founded on timber piles. In the early 1890s, the pneumatic caisson technique, as known from bridge pier construction, was used in New York for the first time for buildings, so that piers or columns could be carried down to bedrock. Until the beginning of this century, foundation design was based on empirical knowledge or experience as developed over a long time, rather than on a rational approach of analysis, as derived from an understanding of the behavior of the substructure soil under loading. The introduction of soil mechanics as a scientific discipline is closely associated with the Austrian engineer Karl Terzaghi and the publication of his famous book on the topic in 1925.

In order to select an appropriate foundation system, the subsurface conditions of a building site must be investigated, which includes (among other criteria) the nature of the soil, the soil stratification and the thickness of the layers, and any existing underground structures. Soil engineers must determine the strength and settlement characteristics, as well as other criteria pertinent to the design of foundations. Often, only confirmatory soil explorations are required, especially in urban areas, where the local geology and groundwater conditions are known and subsurface maps are available, and the performance of adjacent buildings can be evaluated. In more detailed investigations, borings are usually used to determine the soil profile, along with the location of the water table. The spacing and depth of borings depend on the complexity of the site and the type of building. The testing of the soil samples obtained from the borings yields important material properties, such as shear strength, compressibility, and permeability, in addition to the necessary basic soil properties of density, porosity (void ratio), water content, grain size distribution, relative consistency (soft to hard), and so on. The grain sizes range from boulders, gravel, and sand, to clay and silt. The subsurface material is broadly divided into *rock* and *soil* (see Table 2.1), that is, cemented and loose materials. Soils, in turn, may be divided into *cohesive* and *non-cohesive materials*, excluding organic deposits such as top soil.

Cohesive soils such as silts, clays, and clay mixtures contain a large proportion of fine particles. These expand and shrink with a change of water content; they are compressible and may creep under constant load action, and hence are prone to long-term settlements. They may have a low shear strength, which is derived primarily from cohesion (tension) and may lose part of it upon wetting and other disturbances. In contrast, granular soils such as sand, gravel and granular soil mixtures, are cohesionless, so that their shear strength must depend on the internal friction between grains; these are prone to immediate settlement. They do not exhibit elastic properties, and thus do not rebound when the load is taken away, as many other soils do.

Critical, with respect to foundation design, are:

- The bearing capacity of the soil
- The control of excessive settlement
- The control of differential settlement between the various vertical support elements.

The ultimate bearing capacity, that is, the shear failure of the soil, may only be controlling for soils such as rocks and certain clays. The tilting of the Transcona concrete grain elevator, Winnipeg, Canada, in 1913 at about 30 degrees due to uneven loading, is a rare example of a complete ground rupture. Ground failure due to slope instability is treated as a special condition. In this case, a landslide may be initiated by a heavy structure, together with rainfall or seismic action, which make the building slide down the hill. An earthquake may cause saturated sands to liquefy so that soil flows out from under foundations, and thereby may cause a building to rotate. For most soils (e.g., soft clays, sands, and mixtures), excessive settlement must be controlled, but may not necessarily be intolerable. For example, the Monadnock Building [1891] in Chicago has settled almost two feet with virtually no damage. Originally, the National Theater in Mexico City had sunk as much as ten feet, due to the pumping of water from deep wells, but then was pushed up again by the weight of high-rise structures built nearby.

One of the primary reasons for building failures is due to large differential ground settlements; hence, not the soil but the building fails. It is the stiffness of the superstructure and its tolerance for vertical movement which, in turn, influences the performance of the foundations, that is critical. Naturally, the smaller building structure

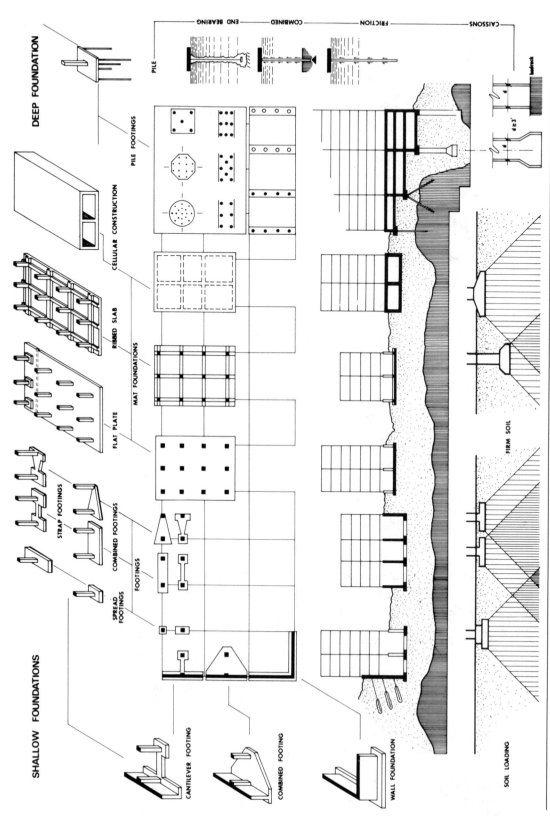

Figure 2.20. Foundation systems. Reproduced with permission from Horizontal-Span Building Structures, Wolfgang Schueller, copyright © 1983 by John Wiley & Sons, Inc.

can be designed as a statistically determinate system to allow large movements rather than to resist them with continuity and thereby to develop significant stresses. Various examples of differential settlements are investigated later (see fig. 3.14, under "Hidden Loads"). They range from footings of different sizes and elevations, different foundation types with varying ratios of live and dead loads, to complex nonuniform subgrade conditions. A spectacular example is the 179-ft Leaning Tower of Pisa [1350] (see fig. 8.2), which is currently about 18 ft out of plumb, due to consolidation of the clayey soil under a layer of sand, and which continues to tilt. It must also be kept in mind that the heavy weight of a new skyscraper may cause settlement problems on existing adjacent buildings. In the following discussion, the various shallow and deep foundation systems are briefly introduced as organized in Figure 2.20.

Shallow Foundations

As the name suggests, shallow foundations transfer loads in bearing close to the surface. They either form individual *spread footings* or *mat foundations*, which combine the individual footings to support an entire building or part of it. The two systems may also act in combination with each other, for example, where a service core is seated on a large mat while the columns are founded on pad footings.

The base size of column and wall footings, or of mat foundations, depends on the allowable bearing capacity of the soil, q_a. The allowable bearing capacity for shallow foundations supported on cohesive soils, such as stiff clays, depends on their shear strength, as determined by dividing the ultimate bearing capacity by a safety factor, for example, of 3. In this instance, the ultimate bearing capacity is derived from an average contact pressure that causes the soil mass to fail in shear. However, for cohesionless soils such as sands and gravel, and cohesive soils of soft clay, the allowable bearing capacity is derived from the control of settlement, which is much lower than that obtained from the ultimate bearing capacity. The settlement for noncohesive soils, such as loose sands, is primarily of an immediate nature, because they are permeable, and hence the water is squeezed out quickly; it consists of the contact settlement directly under the foundation displacing the soil laterally and changing the soil profile, and of compression settlement. In contrast, compressible clays have a very low permeability, so that a substantial part of the final settlement is due to long-term consolidation movements; in other words, settlement as a result of soil volume reduction due to the extrusion of water from the voids is a very slow process.

For the evaluation of the bearing capacity of the soil, the reader should refer to literature on soil mechanics. Building codes give allowable bearing capacities for various rocks and soils, which may be used for preliminary estimation purposes. Typical values are shown in Table 2.1. They range from 3 ksf for medium-density soft clay, 8 ksf for loose gravel or compact coarse sand, and 50 ksf for compacted shale, to 200 ksf for hard rock.

The distribution of the contact pressure between the soil and bottom face of the foundation is an important consideration for the structural design of the foundation and for determining allowable bearing pressures. This distribution is rather complex, because a uniform force action does not generate a uniform response in the soil. The response of the soil to loading is not only dependent on the type of force action, but also on the stiffness of the foundation and that of the ground as well. For example, a rigid foundation concentrically loaded is supported at points that have deformed the

Table 2.1. Typical Average Allowable Bearing Capacities
of Various Soils

Soil Type	Bearing Capacity (ksf)
Silt (sandy or clayey silt)	1
Clay (soft), soft broken shale	3
Clay (stiff), wet sand, sand-clay mixture	4
Sand (fine, dry), sand-gravel mixture	6
Sand (compact coarse, dry), loose gravel, hard dry clay	8
Gravel, gravel-sand mixture	12
Gravel-sand mixture well cemented	16
Soft rock (broken bedrock, compaction shale), hardpan	20
Sedimentary rock (hard shale, sandstone, limestone, siltstone)	50
Medium hard rock (slate, schist)	80
Hard rock (basalt, diorite, dolomite, gneiss, granite)	200

least. These points of maximum pressure occur at the center for cohesionless sandy soils, but along the footing's outer edges for cohesive clayey soils. Nearly the opposite is true for flexible foundations deforming in a bowl shape. The flexible foundation on clay causes maximum pressure at the center, while on sand, the pressure distribution is more uniform, with slightly larger values at the ends. Because of the nonuniform character of the soil, and since foundations are generally neither rigid nor flexible, it is the common practice to treat the contact pressure as uniform.

Occasionally, contact pressure may not control the design of a foundation. Even by causing the largest contact pressure, a soft strata below may still only be able to support a fraction of this amount. Similarly, when footings are very closely spaced, or adjacent footings are located at different levels, overlapping soil stresses are created, which may govern the design (Fig. 2.20, bottom). In this context, the problems which may be generated by new foundations influencing nearby older ones should also not be forgotten. Rarely are the subsoils uniform; they may consist of irregular deposits and of alternating layers of varying stiffness (i.e., soft and stiff soils) and thickness. Also for this condition, it is necessary to evaluate the vertical stress distribution for determining settlement. For a homogeneous elastic soil, this may be done through the bulb of pressure (i.e., lines of equal vertical stress) concept, which shows the volume of influence of a foundation, and thus the corresponding potential settlement. It indicates that the maximum vertical pressure at the contact surface decreases with an increase of depth, while the horizontal pressure increases (see fig. 6.9, bottom left). The spread of the load may be crudely approximated at an angle of 30 degrees to the vertical, or at a slope of 2 to 1, thus yielding a pressure under the footing of width B and length, L, at level, z, as caused by the load $W = wLB$, of

$$q = \frac{W}{(B + z)(L + z)} \qquad (2.16a)$$

The pressure at a depth of twice the width ($z = 2B$) under a square footing following the outline of a truncated pyramid is only about 10 percent of the contact pressure.

$$q = W/(B + z)^2 = W/(3B)^2 \simeq 0.1w \qquad (2.16b)$$

It may be concluded that soil investigations for ordinary building site conditions may only have to be made to a depth of up to 1.5 to 2 times the width of the largest footing, assuming that they are not spaced too close together.

Climatic effects make it necessary to place shallow foundations at a depth below that of frost penetration and seasonal moisture change. In semiarid regions, the so-called expansive soils are found, which absorb rain water and swell in the rainy season, and dry and shrink in the dry season. As the soil expands it may create forces of up to 30 ksf and higher, and movements of more than six inches with a corresponding uplift pressure; hence perimeter footings should be founded on soil at a depth with constant moisture. Similar to the seasonal up and down movements due to the *shrink-swell* cycles are the *freeze-thaw* cycles in cold climates. It is apparent that footings should not be founded on frozen soil unless it is of a permanent nature. Interior footings that are not affected by frost may be placed higher than the frost line. However, some designers use as a minimum depth two feet below ground to be sure of encountering undisturbed soil.

In order to control settlements and tilting, foundations on compressible soils should only be concentrically loaded; in other words, the column and wall bases should not be fixed to the footings, and lateral shear forces should not be transferred in bending. On the other hand, footings on highly compacted soils may be loaded eccentrically. Single footings should only be used on soils of low compressibility, because the independent displacements of the foundations may cause significant stresses in the superstructure. Columns may be joined by continuous footings to control vertical differential movements between them. Mat foundations are most effective in reducing differential movements on compressible soils.

Spread Footings

Spread footings are divided into *isolated footings* (e.g., column footings), *strip footings* (e.g., for walls and rows of columns), *combined footings*, and *strap footings*. In conventional foundation design, bearing walls are supported on reinforced concrete spread footings of the strip type, as shown in Figure 2.21g; footings for transverse partition walls are reinforced thickened areas of the grade slab. When the spread footings and basement are below the ground water table, the basement slab must be designed to resist hydrostatic uplift pressure. Naturally, should the building stand on landfill, the footings would have to be supported on piles or piers.

Combined footings carry two or more columns or walls that are either so close to each other that their individual footings would overlap, or where a column or wall is too close to the property line and would cause a large rotation on a single footing due to eccentric action. The designer must determine the shape of the combined footing, be it rectangular or trapezoidal, so that only uniform pressure is generated. In other words, to avoid rotation and unequal soil pressure, as well as unequal settlements, the centroid of the bearing area of the combined footing should coincide with the resultant of the loads acting on the footing. Should the distance between an eccentrically loaded perimeter column and an interior column be large, and the bearing

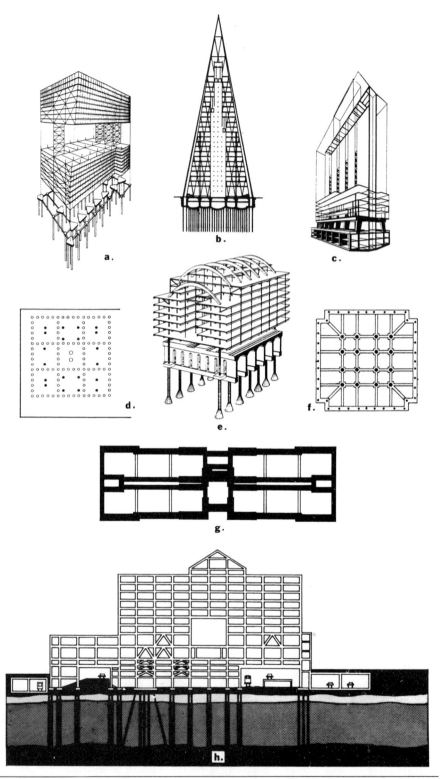

Figure 2.21. *Case study of building foundations.*

capacity of the soil be high, then the single footings could be linked by a strap beam rather than by a slab. For this so-called *strap* or *cantilever footing*, the individual footings should be proportioned as to generate only uniform soil pressure.

Individual footings are usually constructed in reinforced concrete, but for lighter loads they are also constructed as plain concrete footings, possibly of the pyramidal or stepped type. Occasionally, steel grillage foundations encased in concrete are necessary to spread heavy loads from steel columns to a wide base; in the past, timber grillages have been used to support masonry footings. A contemporary example of grillage foundations are the World Trade Center Towers (see figs. 2.17f and 5.3j) which rest 70 ft below ground, directly on bedrock, with a permissible bearing capacity of 80 ksf. The concrete footings carrying the columns were placed in the bedrock to support the grillages, a massive two-layer framework of steel beams; a typical grillage unit is about 80 ft^2 and approximately 5 ft high.

Mat or Raft Foundation

A mat foundation is basically one large continuous footing upon which the building rests. In this case, the total gross bearing pressure at the mat-soil interface cannot exceed the allowable bearing strength of the soil. The system is used when the soil bearing capacity is low, and it may prove to be more economical when more than about one-half of the plan area of a building is required for single footings; it also provides a uniform excavation depth. Mat foundations may be useful when individual footings touch, when a concentration of high soil pressure must be distributed over the entire building area, when small soft soil areas must be bridged, when compressible strata are located at a shallow depth, so that settlement will be minimized, when differential settlement of variable soils must be minimized, since individual footings would create unequal settlement, or when hydrostatic pressure of groundwater and possible uplift must be controlled.

The common mat or raft foundations (see fig. 2.20) are:

- flat plate/slab mats
- ribbed mat (two-way slab on beams) with ribs above or below slab
- cellular raft (slabs plus basement walls)

The mat foundation is designed as an inverted floor structure, but settlements must also be taken into account where the loading or soil pressure distribution depends on the layout of the columns or walls, the magnitude and type of loads, and on the stiffness of the foundation and soil. For preliminary design purposes, an average or uniform soil pressure distribution may be assumed, and the effect of differential settlements may be ignored, because of the rigidity of the mat. In other words, the contact pressure is distributed in a straight line, where the centroid of the soil pressure coincides with the line of action of the resultant vertical loads. This condition actually only applies for a rigid mat, such as a cellular one on cohesionless soil, or a uniformly loaded flexible mat on an elastic compressible soil.

Examples of high-rise buildings on mat foundations include the 569-ft high Rainier National Bank Tower (see fig. 7.5e) in Seattle, resting on a 106-ft square, 12-ft thick mat, as based on an allowable soil pressure of 16 ksf; and the 841-ft high U.S. Steel Building (see figs. 5.7f, 7.4e, and 7.30f) in Pittsburgh, supported on a continuous concrete mat up to 12 ft thick, which bears on hard, shaley sandstone. The 360-ft high BASF Administration Building [1961] in Ludwigshafen, W. Germany (Fig. 2.21c), sits on a 2.63 ft thick arched concrete mat stabilized by seven cross-walls and six long walls to form a stiff cellular raft, which is founded on firm soil over a soft stratum.

Floating or Compensated Mat Foundations

When tall buildings are founded on weak soils of very thick deposits of soft materials, such as compressible clay, the bearing capacity of the soil is extremely low and controlled by settlement criteria. To control this movement due to heave, as well as compressive forces and long-term consolidation of the soil stratum, the fully compensated mat foundation is often used. In this case, so much soil is excavated that the weight of the soil removed plus any uplift from water pressure is replaced by the combined gross loading of the substructure and the superstructure. In other words, the pressure at the base of the excavated soil will not change; the pressure of the displaced soil will be equal to the pressure caused by the building, thus theoretically resulting in no settlement. The structure seems to float on soil like a ship in water, as caused by a buoyant force equal to the weight of the soil displaced by the basement volume, balancing the weight of the floating structure according to the Archimedes principle.

Naturally, this is only a theoretical model, since some settlement of the mat will occur due to the change of live loads and the water table, nonhomogeneity of the ground, and due to recompression of excavation heave with subsequent settlement, realizing that, as a result of the removal of the soil overburden, pressure heave of the bottom of the excavation has occurred.

For tall buildings, this floating foundation concept may require several basements, which may not take just the form of a box or pedestal, but rather a stepped rootlike extension into the ground. These deep substructures, in turn, generate extremely large loads, clearly indicating that the building should be a lightweight structure. Hence, it may be more economical to use a partially compensated mat foundation rather than a fully compensated one, when some settlement due to the net bearing pressure of building weight minus the weight of the soil excavated at the mat-soil interface is tolerable.

Because the floating mat is constructed deep into the ground, groundwater may have to be considered, especially the high water table in spring, thus requiring a watertight box-type foundation. In this instance, the buoyancy effect and the lateral pressure must be considered in the design.

Most of the high-rise buildings in Houston use the floating mat on thick deposits of compressible clay as foundations, since the bedrock is several thousand feet below the ground surface. For example, the 1049-ft high Texas Commerce Tower (see figs. 5.7b and 7.33a) rests on a partially compensated mat foundation 9.9 ft thick over most of its area and 12.75 ft thick at the elevator pits. The mat is located 63 ft below street level, 20 ft below the water table, and 15 ft below sea level.

The 714-ft high One Shell Plaza (see fig. 5.4h) is supported on an 8.25-ft thick raft that extends 20 ft beyond all of the building sides. Because of the use of lightweight concrete, the mat had to be placed at a depth of only 60 ft below grade; for normal-weight concrete the mat would have to be located at about 85 ft. The 57-story Republic Bank Center (see fig. 7.33) is supported on a fully compensated, 8-ft thick mat foundation. In this instance, the design of the mat is complicated because of asymmetrical gravity loading due to the varying tower heights causing the mat not only to dish, but also to rotate.

Compensated Mat Foundations with Friction Piles

Should the thickness of the mat required to resist differential settlements become too large, friction piles may then have to be used, together with a thinner slab, which can be visualized as a giant pile cap. In addition, when the buoyancy effect (the hydrostatic uplift of the groundwater) is too large, the mat may also have to be supported by friction piles.

For example, the 417-ft high, A-shaped Nonoalco Tower [1964] in Mexico City (Fig. 2.21b) required a huge watertight cellular raft. The 22-ft deep concrete basement is divided into compartments, each with an inverted thin-shell arch bottom. The raft is stabilized by 233 friction piles driven 56 ft into the subsoil in order to allow the building to float like the hull of a ship.

Example 2.2

A 550-ft high steel building resting on thick deposits of compressible soil is investigated by using the floating mat foundation concept. The building is assumed to weigh 10 pcf, and the density of the soil is taken as 120 pcf.

For a fully compensated mat foundation the depth, h, of the substructure is required, so that the weight of the soil removed is equal to the weight of the building

$$0.010(550 + h) = 0.120h, \quad \text{or} \quad h = 50 \text{ ft}$$

Hence, the applied pressure at the base of the mat is

$$q = 0.010(600) = 0.120(50) = 6 \text{ ksf}$$

It is assumed that this height of the substructure is uneconomical, so that only two basement levels of the 36-ft total height are considered, and a partially compensated mat foundation is used. In this case, the soil pressure at the base hardly changes, but the weight of the soil removed is less than the weight of the building. Therefore, settlement must be considered for the net applied pressure of the building weight relieved by the weight of soil excavated.

$$q_{net} = 0.010(550 + 36) - 0.120(36) = 1.54 \text{ ksf}$$

However, should settlement-related problems not be acceptable, and friction piles unable to be used, then the building height must be reduced to

$$0.010(H + 36) = 0.120(36), \quad \text{or} \quad H = 396 \text{ ft}$$

A further increase in building height beyond 396 ft requires a thicker mat to control settlement, or the mat must extend beyond the building outline.

Deep Foundations

Deep foundations are used when adequate soil capacity is not available close to the surface and loads must be transferred to firm layers substantially below the ground surface. When settlement is a primary problem, then a pile length must be selected to minimize differential settlement.

The common deep foundation systems for buildings are *piles* and *piers* (caisson piles). While the small-diameter slender piles are normally driven into the ground, the large diameter piers are placed by first excavating a hole; this distinction, however, may not always be that clear. Other deep foundation systems occasionally used are *slurry walls* (i.e., a method of construction for earth retaining walls) and *caisson foundations*, which are generally used for the construction of bridge piers and abutments. A caisson is a massive, cellular hollow box structure that is sunk into position, and also provides the bracing for the excavation. The three major types are the *box caisson* or *floating caisson* (open at top and closed at bottom), the *open caisson* (open at top and bottom), and the *pneumatic caisson* (closed at top, open at bottom, and filled with compressed air to prevent water from entering the working chamber) as may be used for constructing an underground garage. In the following paragraphs, the most common deep foundations for buildings are briefly discussed.

Pile Foundations

Piles are of small cross section with diameters ranging from 6 to about 24 inches, and are of a slenderness ratio larger than 10; their typical load capacities range from 20 to about 400 kips. Piles are usually driven by hammers. They are made of treated timber, steel, cast-in-place concrete, precast, prestressed concrete, or composite material; they are manufactured in various shapes of solid or hollow configuration. The bearing capacity of a pile depends on the strength of the pile and the supporting strength of the soil. The estimation of the bearing capacity of piles is quite complex; it is determined by static analysis, dynamic analysis, or pile load tests. In static analysis, which is often used for preliminary design purposes, the pile-bearing capacity depends on the soil-to-pile connection (i.e., soil properties and pile geometry), that is, the sum of end-bearing resistance and the skin friction, as well as (possibly) on group effect. For long slender piles, the pile-tip resistance becomes insignificant, so that they act mainly as *friction piles*, although in clayey soils the resistance is primarily provided by adhesion. In dynamic analysis, so-called pile formulas have been developed where the pile capacity is directly related to the resistance offered to driving with a hammer (e.g., Engineering News formula). The estimation of the pile length is not always easy. When *point-bearing piles* are supported directly on rock-like material, the pile length is known. For this condition, the pile is treated as a short column braced by the soil, hence assuming it is not surrounded by soft mud; its size is dependent on the load-bearing capacity of the base material and on the strength of the pile. Occasionally, it may be the case that the bedrock is stronger than the concrete. For the opposite condition, where there is no firm soil available at a reasonable depth, friction piles must be used. In this case, the length of the piles depends on the pile size and the skin friction along the pile, which is derived, in turn, primarily from the shear strength of the soil or the adhesion on the pile face. Often, piles pass through a soft soil layer where they are supported by skin friction but then must be extended several feet into firm soil to act as *end-bearing friction piles*. The evaluation of skin friction through layered soil systems, and hence the determination of pile length, is extremely difficult. Where settlement is a serious problem, the pile must be long enough to withstand differential settlement. Short piles may be driven in granular soils to compact the soil close to the ground surface; they are called *compaction piles*.

Piles are generally used in groups, such as at least three piles to support a major column. A concrete cap is always necessary to distribute the loads from the super-structure to the piles. Pile-cap footings are designed like spread footings but for con-centrated pile loads. Pile clusters may be of any arrangement below column, wall, or combined footings; usually, the location of the resultant pile load coincides with the resultant applied load. The piles should be spaced far enough from each other, so that the load-bearing capacity of each individual pile is not reduced, otherwise group behavior must be taken into account, which not only results in less soil resistance but also larger settlements.

Where pile groups are subject to lateral forces, and the lateral resistance of the vertical piles in bending (as partially supported cantilevers) is exceeded, then inclined or batter piles must be employed. When, in addition, an overturning moment is applied, then some of the piles may have to act in tension to resist uplift.

Since throughout New Orleans compressible clay interspersed with sand layers ex-ists to about 150 ft, and consolidated clays below that level, tall buildings must stand on deep piles. For example, the composite tube of One Shell Square (see fig. 7.38e) is founded on piles of 210-ft lengths and spaced 7.5 ft apart, in order to control settlement. An 8.25-ft thick reinforced concrete mat was required to redistribute loads uniformly over the 558 octagonal prestressed, precast concrete piles, each 18 in. in diameter and of an unusually high load capacity of 280 tons. Because the higher column loads in the core area would have caused a greater settlement at the center of the structure, the core was built up higher than the edges so that when the total load was applied, the mat was in level.

Pier Foundations

Piers or drilled piers are large-diameter piles with a slenderness of less than 10, placed vertically. Their load capacity can be several thousand kips, depending on their size and soil conditions. They are sometimes called caissons, but should not be confused with caissons sunk into position. For the construction of piers or caisson piles, a hole is usually drilled by machines, or circular steel shells are driven into the ground and the soil inside is excavated. Piers can be of any cross section, but are at least 3 ft in diameter, to allow for inspection. They are large enough so that a single pier can replace a group of piles. Piers can penetrate dense soil, which piles may not. Piers may be *belled* or *straight shafted*, they usually are supported by end bearing, and occasionally may be supported, in addition, by skin friction. They are classified ac-cording to material as concrete, concrete in steel pipe, and concrete plus steel core piers.

The size of a typical concrete pier is determined from the soil capacity in bearing and sometimes in skin friction, while the pier itself is designed as a compression member. It is assumed that it does not have to resist lateral forces, which are usually absorbed by shear resistance at the building base and passive earth pressure on basement walls. Occasionally, piers must resist uplift forces, as may be the case in core columns and corner columns of trussed tubes. Here belled piers act in tension, or piers are post-tensionsed and anchored in bedrock.

In Chicago, pier foundations are often needed to transfer the heavy loads of the skyscrapers more than 100 ft down to the bedrock. However, a larger number of the shorter belled caissons may be used, as for the 74-story Water Tower Place (see fig. 2.17e), to go down only about 80 ft to bear on the hardpan located above the bedrock. The AMOCO Building (see figs. 1.2, 2.21f, and 5.4k) is founded on 40 perimeter caissons that vary from 5 to 6.75 ft in diameter and support a transfer girder that carries the 64 chevron columns of the tube. Every one of the 16 core columns is supported on a caisson varying from 8.75 to 10.25 ft in diameter. Similarly, each column of the John Hancock Center (see fig. 7.25e) is supported by a caisson reaching down to the bedrock at an average depth of 140 ft. The caissons of both buildings are tied together with a concrete grade beam grid, or so-called egg crate, where the beams are connected at the top by a deep concrete slab. This pad is anchored and embedded in the clay to transfer the wind shear to the soil, so that the caissons carry only vertical loads and no lateral forces.

The Sears Tower (see figs 2.21d, 5.4m, and 7.33) rests on 114 bedrock caissons. Every tower column has its own bedrock caisson; they are 6 ft in diameter and go to a depth of 100 ft, with some being anchored as much as 9 ft into the bedrock. In the center module, two 10-ft diameter bedrock caissons support two columns rising the full height of the building. Twenty belled hardpan caissons, which support the tower's three basements (of conventional steel skeleton construction), go to a depth of 60 ft. The caissons are connected by a 5-ft thick floor mat at the lowest basement, to transfer the wind shear to the clay. The plaza and levels below are carried by 59 hardpan caissons, while its perimeter is supported by a 30-in. thick slurry wall, which goes to a depth of 58 ft. The relocated sewer lines are supported by 47 of the 240 caissons used for the building.

Another example is the 24-story, 349-ft high Georgia Power Tower [1980] in Atlanta (Fig. 2.21a). It is supported by 68 caisson piles, where the deepest goes down 90 ft, and the diameters range from 2.5 to 6 ft; piers supporting columns in potential uplift are drilled 8 to 10 ft into the bedrock. The installation of clusters of three caisson piles topped with concrete caps proved more economical than installing a single large one; they also provided more skin friction surface area to resist uplift for the same end bearing capacity. Around the tower perimeter, the piers are tied together with grade beams.

Constraints on urban sites are often not visible because they are underground. Subway tunnels, commuter tunnels, ventilation shafts, old foundations, adjacent building footings, sewers, water mains, gas and steam lines, electric and telegraph conduits, and other utilities, all located at different depths (see fig. 8.4, center), may require complex foundation layouts. To avoid extensive excavation and stressing of a tunnel, for example, a transfer grade beam structure may be necessary, possibly a cantilever system, to carry the building columns to the belled caisson piles passing through the myriad underground lines. In addition, tunnels and building foundations may have to be underpinned with piles to be carried deeper because of adjacent excavation work or because they have to carry higher loads. For Boston's Copley Place (see fig. 2.21h), 14 and 16 in. square prestressed precast concrete piles were squeezed between turnpike and railroad structures underneath and driven as deep as 120 ft. Another example is the Embankment Place [1989] in London, U.K. (Fig. 2.21e), constructed as an independent structure over Charing Cross Station. The main building columns that carry the arches at the top, from which the floors are suspended, rest at the base on nearly 100-ft long piles that penetrate through the station's brick vaults, which (in turn) support the railway tracks and platforms.

Choosing a Foundation System

The selection of a foundation system depends on the type of superstructure, that is, the location, magnitude, and kind of forces to be transmitted to the ground, and also on the subsurface conditions, the bearing capacity and settlement characteristics of the soil, as well as the groundwater conditions. Usually, there are several possible economical solutions for a given situation. For example, when forces are concentrated locally, such as for a lateral-force resisting central core, it may be more economical to use piles or piers directly under the core rather than a thick mat to redistribute the loads over the footprint of the building. For complex subsurface conditions, such as nonuniform soils across the building site or other underground interferences, it may be necessary to adjust the layout of the superstructure.

In this context, only larger, heavy buildings are considered, where load resistance is critical. The selection of foundations for small buildings is usually based on local practice. These buildings are light and cause hardly any bearing pressure, so that the design of their foundations depends less on loads and more often on resistance to movements in the soil.

A building can be founded on soil or, directly or indirectly, on rock. While New York City sits directly on rock, the tall buildings in Chicago are carried by caisson piles roughly 100 feet down to bedrock. Naturally, the ideal situation exists when the bedrock is located near the surface and seismic action is nonexistent, so that shallow foundations can be supported directly and settlements are usually not a problem. However, when this bedrock is at great depth, then it depends on the nature of the overlying soil and the magnitude of the loads to determine whether the building should be founded directly on soil or indirectly on rock. When buildings are founded on soil rather than rock, the selection of the foundation type depends on the bearing capacity and settlement characteristics, as well as the necessary compatibility with the super-structure. For a thick, firm stratum, individual shallow foundations may be satisfactory, or mat foundations (for tall buildings). However, for the other extreme, where there is only a thick stratum of weak, soft soil present, such as in Houston, tall buildings cannot be founded just on mat foundations because of excessive settlement problems, unless it is placed in a deep excavation to behave like a floating mat. Softer soils may have to be stabilized by friction piles, which carry the mat, as is the case in New Orleans. For the condition where weak soil is overlying firm soil at a reasonable depth, piers or end-bearing piles may be used, although it may be more economical to use spread footings and stay closer to the surface, than to go deeper and take advantage of higher bearing values, assuming differential settlement is tolerable. On the other hand, when firm soil is over a soft stratum, mat foundations may be required for heavy loads, possibly together with piles, to control settlements, while individual shallow footings may be satisfactory for light flexible buildings. Hybrid foundations are required, for example, when the bedrock underlying a site is sharply sloped and has an overlay of loose sand and a wedge of clay. For this situation, a foundation mat may be used beneath a portion of the building, while the remainder is founded on caisson piles drilled into the bedrock.

Particular attention must be given to building foundations on sloping ground (see fig. 3.14, part 5). For a firm soil and a stable slope, step footings transverse to the slope or trenched footings parallel to the slope with transverse grade beams may be used. For sloping soft ground or semistable slopes, slope stability must obviously be considered. In this case, the building may be supported by piles (poles) anchored into a firm layer and cantilevered above the ground. There are several methods for the prevention of a potential landslide, which include: soil stabilization (e.g., chemical grouting, compaction, reinforced earth), retaining structures, tiebacks, surface and subsurface drainage, and flattening the slope, possibly with a series of terraces.

In the following example, it is demonstrated that, for constant soil conditions with an increase in building height (i.e, loading increase), the individual footings for a low-rise building must be replaced by the combined and strip footings for mid-rise buildings, and change to the mat foundation for tall buildings. As the building height increases further, either pile foundations are used in addition, or floating mats with sufficient excavation (Example 2.2), or drilled piers to transfer the loads to stronger soil layers, assuming that the basement cannot be enlarged to provide a larger mat.

Example 2.3

A 120-ft square building is supported by 36 columns arranged on a 24-ft square grid. The overall average gross dead load of this concrete structure is assumed to be 150 psf, with a 50 psf live load, that is, an equivalent average floor load of 200 psf or an average weight of 18 pcf of building volume. Neglected in this rough, preliminary investigation is the increase of the building unit weight with height, and the overturning effect due to the lateral forces which, however, controls only for tall narrow buildings.

For a bearing capacity of $q_a = 4$ ksf for stiff clay (Table 2.1), the proportion of the floor area at the base used for shallow foundations, or the proportion of the average load intensity to the soil capacity, is a function of the number of floors, n, or the height of the building, H, measured from the foundations. It is roughly equal to

$$0.018H/q_a, \quad \text{or } 0.2n/q_a$$

1). For a 50-ft high building, the proportion of the floor area used for the foundation is

$$0.018H/q_a = 100(0.018)50/4 = 23\%$$

Since the average load intensity is smaller than approximately one-quarter of the soil capacity, individual column footings can be used. A typical interior footing size is approximately

$$A = P/q_a = 0.018(24)^2 50/4 = 129.6 \text{ ft}^2$$

Try an 11 ft, 5 in. square footing, $A = 130.34 \text{ ft}^2$.

2). For a 100-ft high building, the proportion of the floor area used for the foundation is

$$0.018H/q_a = 100(0.018)100/4 = 45\%$$

Since this proportion is larger than 25 percent but less than 50 percent, parallel long strip footings should be investigated. Hence, the required footing width, a, for the 120-ft building length should be approximately equal to

$$A = P/q_a, \quad \text{or} \quad 120(a) = 0.018(120 \times 24)100/4, \quad \text{or} \quad a = 10.8 \text{ ft}$$

Try a footing width of 10 ft, 10 in.

3). For a 180-ft high building, the proportion of the floor area used for the foundation, or the average load intensity of the allowable soil capacity, is

$$0.018H/q_a = 100(0.018)180/4 = 81\%$$

Usually, a mat or grid foundation should be investigated when the load intensity is larger than roughly one-half of the soil capacity. In this case, it is clearly indicated that a mat foundation is required.

By treating the mat foundation as an inverted flat plate carrying a soil reaction of $0.018(180) = 3.24$ ksf, the approximate maximum moment can be determined according to Equation 5.30. Then the slab thickness can be estimated from the allowable bending stress (Equation 4.3), or it can be derived from punching shear requirements, where the minimum steel for flexure (Equation 5.19) is usually satisfactory.

4). For a 360-ft high building, the average load intensity at the base is $0.018(360) = 6.48$ ksf, which is clearly above the bearing capacity of 4 ksf. Hence, either the building must obtain a wider basement base, or deep foundations are required.

4a). Determine the approximate size of a square mat for a basement acting as a pedestal for the building, to transfer the loads.

$$A = P/q_a = 0.018(120)^2 360/4 = 93312/4 = 23328 \text{ ft}^2$$

Try a 153-ft square basement mat with $A = 23409$ ft^2, if the underground space is available, keeping in mind that the additional basement weight must still be considered.

Check whether the mat foundation is alright for a uniform wind pressure of 33 psf.

The overturning moment due to the wind is

$$M = 0.033(360 \times 120)360/2 = 256608 \text{ ft-k}$$

This rotation causes maximum bending stresses along the edges of the mat of

$$\pm f_b = M/S = 256608/(153^3/6) = 0.43 \text{ ksf} \leqslant f_C/3$$

Hence the combined stresses due to gravity and rotation are

$$f_a = 93312/153^2 + 0.43 = 3.96 + 0.43 = 4.42 \text{ ksf} \leqslant 1.33(4) = 5.32 \text{ ksf}$$

Notice that, as has been shown in Equation 2.8 when the bending stresses due to lateral force action do not exceed one-third of the axial stresses due to gravity action, then the effect of the lateral load can be ignored and the building can be treated as a gravity structure.

4b). Assume that each column is supported by one pier, rather than by a cluster of piles, to carry the loads down to the bedrock in end bearing.

A typical interior column carries the following load

$$p = 0.018(24)^2 360 = 3733 \text{ k}$$

The corresponding diameter of the caisson pile for a bearing capacity of the bedrock of 80 ksf, which is less than the compressive capacity of the concrete pier of $0.25 f'_c = 0.25(3) = 0.75$ ksi $= 108$ ksf, is

$$A = P/q_a, \quad \text{or} \quad \pi d^2/4 = 3733/80, \quad \text{or} \quad d = 7.71 \text{ ft}$$

Rather than using a straight pier, the pier could also be belled out at the bottom to achieve the required diameter of 7 ft., 9 in. assuming that the strength of the concrete is increased to satisfy the shaft design.

4c). Assume a pile foundation, tied together by a concrete mat, must hold up the building weight of

$$P = 0.018(120)^2 360 = 93312 \text{ k}$$

If 288 piles are to be used, then each one must carry a load of

$$93312/288 = 324 \text{ k}$$

Settlement investigations have shown that friction piles must be taken to a depth of at least 110 ft below ground level. Determine the size of a square precast concrete pile to be driven into the homogeneous stiff clay, which will provide an average allowable skin friction of 500 psf. The pile spacing is large enough so that the effect of group behavior can be ignored. For these long piles, the bearing capacity is primarily based on skin friction, so that end bearing may be ignored for this preliminary investigation.

$$500(4 \times d)110 = 324000, \quad \text{or} \quad d = 1.47 \text{ ft}$$

Try 18-in. square, reinforced concrete piles.

PROBLEMS

2.1. Determine the maximum height of a conical reinforced concrete tower, as based on its own weight and a 5000 psi concrete strength. Use an allowable concrete strength of $0.25f'_c$ and reason whether the result makes sense.

2.2. A free-standing tubular antenna pole is 100 ft high and tapers from a maximum 16-in. bottom diameter to 3 in. at the top. Assume a 30 psf wind load on the projected area of the cylindrical surface. In addition, allow for 5 ft^2 of projected area at the top to accommodate antennas, lights, and so on. Determine the approximate wall thickness of the steel pole by considering only bending due to wind and ignoring the effect of axial action, as well as the secondary effects due to pole deflection. Use A595 steel ($F_y = 55$ ksi), and the diameter/thickness ratio shall not exceed $3300/F_y$.

2.3. A 225-ft high mast is guyed at a level of 150 ft from the base by four cables spaced at 90 degrees and inclined under 45 degrees in elevation parallel to the base line. The mast must resist a wind pressure of 50 plf (lb/ft) by acting as a vertical cantilever beam pinned at the base.

   a. Determine the prestress force in the guys and the preliminary cable size using $F_u = 250$ ksi. Assume that each of the cables resists an equal portion of the wind load. Ignore the flexibility and sag of the cables and the lateral displacement of the support for this fast approximation. Increase the cable prestress force, which must be at least equal to the compressive wind force, by 20 percent.

   b. Determine the critical location, together with the forces for the mast, so that the mast can be designed. Assume the weight of the latticed tower as 30 plf. Ignore the effect of change in wind direction as a design consideration.

   c. Determine the increase in the cable tensile stress due to a temperature drop of 80°F. Compare the result with the primary stresses.

2.4. Find the approximate maximum deflection of a 50-ft high tubular aluminum flagpole. The pole is one foot in diameter with a wall thickness of 1/2 in. and must resist a wind pressure of 20 psf, as well as a single 200 lbs load, due to the flag at the top.

2.5. To the top of a 30-ft high tubular steel pole (A36), a 3 × 5 ft sign is fastened, with the 5 ft dimension cantilevering. The panel weighs 30 lbs and must resist a wind pressure of 20 psf. The tubular section is 6 in. in diameter, has a wall thickness of 3/8 in. and weighs about 20 lbs/ft. Determine the approximate shear and bending stresses, and check to see whether the pole is satisfactory.

2.6. Investigate the behavior of the Washington Monument. Treat the solid masonry obelisk as a hollow truncated pyramid with a pyramidal cap; it tapers from a 55-ft square base to a 34 ft., 6 in. square at the 500 ft level, with an average hollow core size of 29-ft square. The weight of the stone is 0.150 k/ft^3, and the weight of stairs, elevators, and tower cap may be taken as roughly 3 percent of the total wall weight. Use a uniform wind pressure of 50 psf, which is assumed to act on a straight tower of an average width. The allowable compressive masonry stresses are assumed to be 750 psi. Check the safety against overturning, the stresses at the base, and draw your conclusions about the behavior of this type of tower.

2.7. Study the behavior of the 984-ft Eiffel Tower in Paris under gravity and wind loading. Assume the four arched corner columns, which lean against each other to form the tower, to be of parabolic shape for this preliminary investigation. At the ground level, the four huge corner box columns are leaning inward at an angle of 54° and are spread apart to form a 328-ft square base. Assume a total weight of 11,000 tons, which includes a 15 percent live load. It is quite difficult to determine the wind pressure distribution in this lattice tower with large openings in the bottom portion. Since the pressure increases with height but the form decreases, a uniform pressure of 35 psf may be assumed, as based on a straight solid tower with a width of 75 ft, which is derived from the tower dimension at approximately midheight. Use an allowable compressive stress of 15 ksi for the wrought iron. Investigate the compressive stresses in the columns at the base, assuming 800 in.^{2} resisting material, and check overturning.

2.8. A tapered, four sided, self-supporting trussed microwave tower must resist a wind load of 35 psf. The solidity ratio is assumed as 30 percent, that is, the ratio of the exposed steel work to envelope area; further, the projected area of the windward face is increased by 75 percent to take into account the pressure upon the back face. The free-standing tower is 500 ft high and has a constant batter; the truncated pyramid forms a 10 ft square at the top and a 60 ft square at the base. Determine the critical forces upon the legs, which form the chords of the trussed tower faces; these forces can then be resolved into the member forces. Consider diagonal wind direction as well as wind normal to one face.

2.9. A 100-ft high brick chimney tapers from a diameter of 10 ft at the bottom to 4 ft at the top; the wall thickness is assumed to vary linearly from 15 to 7.5 in. at the top. Determine the critical stresses at the base due to a wind pressure of 23 psf on the projected area of the cylindrical surface and due to the weight of the brick, which is equal to 120 pcf.

2.10. Assume the brick chimney in Problem 2.9 to be supported by a 10-ft square concrete spread footing. Is the chimney stable against overturning for this condition?

2.11. Assume the brick chimney of Problem 2.9 to be a four-sided truncated pyramid with a square cross section, but otherwise the same dimensions. Find the critical stresses.

2.12. A cylindrical steel chimney, with no taper, is 120 ft high, has a diameter of 10 ft and a steel shell thickness at the base of 7/16 in. which reduces to one-half this value at the top; the shotcrete lining is 2 in. thick and weighs 110 pcf. Determine the critical stresses at the base, assuming a wind pressure of 35 psf.

2.13. A braced tubular building, 80-ft square from outside to outside (see figs. 2.5c, 2.12l), is 40 stores high and has an average floor height of 12 ft. It must resist an average uniform wind pressure of 32 psf. Use a hypothetical wall thickness of 12 in. at the base.

  a. Find the bending and shear stresses for the wind action perpendicular to the facade. Also determine the moment capacity provided by the flanges.

  b. Find the bending stresses due to wind action in the diagonal direction for the wind on the vertical projection. Compare these stresses to the ones in the perpendicular direction.

2.14. Assume the lateral-force resisting exterior tube of Problem 2.13 to be flexible or equivalent to an open, unconnected tube (see fig. 2.5d). Determine the stresses as caused by wind. Show that, for this case, the walls perpendicular to the wind action can be neglected. Compare the open tube with the closed one.

2.15. Assume the closed tube of Problem 2.13 not to be braced, and to be weakened by window openings, and therefore to be a perforated tube, where the shear lag causes axial stresses in the four corners to be 50 percent larger, compared to the linear stress distribution. Show the stress flow and find the magnitude of the stresses.

2.16. Study the transition of the tube in Problem 2.13 from a solid exterior shaft through an open system, a perforated tube, an open-type exterior bearing wall envelope with 40 percent window openings, a rigid frame, and finally to a system where columns are only provided in the corners (see figs. 2.5c to g). Discuss the behavior of these different systems to lateral force action.

2.17. Compare the column systems for the cases of Figures 2.5g and h. Use the same wind pressure as for Problem 2.13a.

2.18. Consider the building of Problem 2.13 to be laterally stabilized, first by an interior 40-ft square open core (see fig. 2.5i), and then by 35-ft long cross walls, as indicated in Figure 2.5j. Determine the approximate stresses in the 12-in. wide walls at the base of the building, as caused by the wind pressure.

2.19. Replace the tubular building of Problem 2.13 by the triangular legged tower and by the round tube with 30 percent window openings (see fig. 2.5). Use a wind reduction of 40 percent for the circular building. Determine the approximate stresses/forces at the base for the buildings to be investigated.

2.20. Study the various lateral-force resisting structure systems in Figure 2.12, which could stabilize a 100-ft high, 10-story building that has a width of 60 ft with 3 equal bays of 20 ft. Determine the magnitude of the axial forces or maximum stresses at the building base due to an average uniform wind pressure of 24 psf for Figures 2.12a,d,e, and g, and for 2.12i with a base width of 90 ft. Assume the lateral-force resisting structure systems to be spaced 20 ft apart.

2.21. Investigate the response of a typical 8-in. cross-wall for an apartment building (see fig. 2.16f) to lateral load action, by visualizing it to behave as a vertical cantilever. Use the following assumptions:

   a. Neglect flange capacity for this fast check (see fig. 2.16b).
   b. Assume symmetrical I-sections with 5-ft wide flanges for a typical interior cross-wall, as a simplified method of investigation (see fig. 2.16c).
   c. Only use the corridor flanges for a typical interior wall, considering the facade walls as not load-bearing (see fig. 2.16d).
   d. Assume continuous walls across the corridor by ignoring the wall openings (see fig. 2.16e).
   e. Investigate the true layout with the effective flange width, b_e, as given in Figure 2.16f, for both the exterior and interior walls.

Determine the moment of inertias for the various conditions to evaluate the results.

3 LOADS AND FORCE FLOW

A structure must be strong enough to resist the many types of physical forces imposed upon it. The magnitude and direction of these forces vary with the material, type of structural system, purpose of the building, and the locality. The most obvious loads are due to gravity action, as caused by the self-weight of the building, snow, and occupancy. Lateral forces are exerted upon the structure by wind and earthquakes, as well as by earth and hydrostatic pressure. The lateral forces tend to slide and rotate the building block, and the wind attempts to lift up the roof; gravity, in contrast, will counteract and stabilize the structure. While weight and lateral pressure induce a direct force action, movement or deformation (and not applied loads) generate an indirect action. There must also be a distinction between the forces acting on the overall building and those acting locally upon the individual framing elements. Loads may be distributed as point, line, or surface loads.

Among the many examples of lateral-force action are the large lateral pressures and impact forces generated by a crane. A travelling elevator causes pumping action, particularly in a single-elevator shaft, together with pressures from the wind entering through the vent shafts; the shaft walls must be designed to resist these forces. Similarly, the walls of the stair shaft for a tall building must resist the lateral air pressure due to pressurization in case of fire, especially close to the fan location. A fire in an interior bay of a multistory building will cause an expansion of the concrete floor, with corresponding horizontal forces. This thermal thrust must be resisted by the cooler adjacent frames and/or shear walls; it acts similar to an eccentric external prestressing force that increases the moment capacity. Other examples of lateral-force action include possibly excessive stresses in columns, resulting from the internal hydrostatic pressure of liquid used for fire proofing due to the height of the columns. Should a large-panel structure be designed to avoid a progressive collapse, then the building must withstand an internal blast pressure of 5 psi.

Loads may also be distinguished according to their variability with respect to location and time. They may be permanent, such as the dead load due the structure itself, or they may be variable, as is the case for live loads (such as occupancy loads or wind). Variable loads may be of short duration (such as due to people) or of long duration (such as caused by movable partition walls and furnishings). While these loads are fixed in place, car loads, in contrast, are free in location. The duration of the live load is also of importance for deflection considerations (e.g., creep). Live loads may

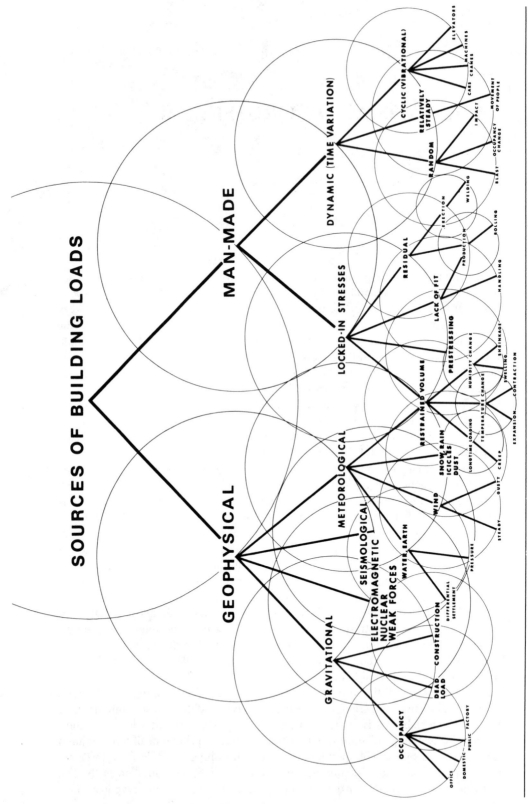

Figure 3.1. Sources of building loads. Reproduced with permission from High-Rise Building Structures, 2nd ed., Wolfgang Schueller, copyright ©1986 by Robert E. Krieger Publ. Co.

be static, or they may be dynamic, as when they cause vibration of the structure. While the ever-changing occupancy loads are generally static, since they do not change rapidly, gusty winds (depending upon the stiffness and mass of the building) may have to be considered dynamic. The dynamic loads may be cyclic, as due to vibrations caused by a machine, or random, such as the impact loads due to a collision; the dynamic sea wave action is both regular and random. But there are not only the constant or repetitive load actions, but also accidental loads, as due to explosions.

Vibrational loads may be transmitted to the occupied portions from the mechanical equipment rooms, which may be located at any level of a high-rise building. Machinery such as fans, pumps, chillers, etc. must be isolated by pad materials (e.g., neoprene, cork, fiberglass), steel spring isolation material, or floating concrete bases. In this case, the isolators must resonate at a much lower frequency than that of the machinery, requiring large deflections of the isolators to eliminate potential vibration problems.

Forces may be induced deliberately, as in prestressing, or involuntarily, such as residual stresses due to the production and fabrication process. They also may be locked into members when the material is prevented from responding to changes in temperature and humidity, and when the material cannot creep and displace, as caused by constant loading or support movement, thus causing reactive loading. Some loads are time-independent (such as the dead load), while others are time-dependent (for instance, the shrinkage of concrete occurs at a decreasing rate in its early stage of hardening). Most loads, whether geophysical or man-made (Fig. 3.1) are extremely complex, and care must be taken by the designer to properly predict their action.

The following discussion of the various loads should only be considered introductory, as needed for preliminary design purposes. For a precise description of loads, the appropriate state or local construction codes, as well as one of the model codes listed below, should be consulted:

- The Standard Building Code (SBC)
- The BOCA National Building Code
- The Uniform Building Code (UBC)

The model codes become law when adopted by the state or city. In the absence of any governing code, the best reference for building loads is the American National Standards Institute—*Minimum Design Loads for Buildings and Other Structures* (ANSI A58). The designer must always keep in mind that the information given in codes is only for minimum loading and may be inadequate for special loading conditions.

DEAD AND LIVE LOADS

The loads caused by the weight of the building, called dead loads, and its occupancy, called live loads, represent gravity loads. In contrast to dead loads, which are static (since they do not change), live loads are dynamic and not constant, although they are usually treated as static because they are slowly applied. Dead loads include the weight of the load-bearing structure, ceilings, flooring, partitions, curtains, storage tanks, mechanical and electrical distribution systems, and so on. It may be important to consider the portion of the dead load that is superimposed in a manner similar to the portion of the live load that is sustained. The gravity loads that are not part of the dead load must be considered under live loads. In this century, especially since the 1960s, buildings have become much lighter, so that the effect of live load in comparison to dead load has become much more significant.

Some approximate material weights are given in Tables 3.1 and 3.2 to develop some feeling for loads. For preliminary estimation purposes, typical weights for floor and wall systems have been selected in Table 3.2. They are averaged and given in terms of pounds per square foot (psf) of their own projected area. The weights of suspended ceilings together with the mechanical and electrical loadings usually range between 2 and 10 psf. In office buildings, where partition locations are subject to change, a uniformly distributed partition dead load of 20 psf should be used. Glass curtain walls weigh roughly 12 psf, in contrast to the much heavier precast or masonry facades of 40 to 80 psf and more.

Although it appears to be a simple matter to determine the weights of materials or the dead load of a structure, it may be impossible at an early design stage to accurately predict the weight of materials not yet selected. Specific nonstructural materials to be chosen include prefabricated facade panels, light fixtures, ceiling systems, pipes, ducts, electrical lines, and other interior components. The weight of stiffening elements and joinery systems for steel structures is estimated only on a percentage basis. The unit weights of materials given by producers or codes are not always consistent with those of the manufactured product. The nominal sizes of building elements differ from the actual sizes; the formwork for poured-in-place concrete may have inaccuracies of 1/2 inch. It must also be kept in mind, in order to accommodate the future computerization of offices, that heavy duty floor loads have to be planned with high electrical capacity and provisions for raised floors and an uninterruptable power supply. These few examples indicate that, in absence of precise information, dead loads cannot be accurately predicted, and may be in error by 15 to 20 percent or more.

Concrete and masonry buildings are heavier than steel buildings. The overall average gross weight for steel buildings is roughly in the range of 50 to 80 psf, while nonprestressed concrete buildings weigh approximately twice as much. When the live load is included in the overall weight, the concrete building may weigh only 30 to 40

Table 3.1. Typical Approximate Material Weights

Material	Weight (pcf)
Steel	490
Reinforced Concrete (normal weight)	150
Aluminum	170
Granite	165
Marble	170
Brick	120
Earth, wet loose sand and gravel	125
Water	62.4
Lumber	35

Table 3.2. Typical Approximate Design Dead Loads

Building Components	Dead Load (psf)
Ceilings	
Plaster on tile or concrete	5
Suspended metal lath and	
gypsum plaster	10
cement plaster	15
Acoustical fiber tile on rock lath and channel ceiling	5
Floors	
Reinforced concrete slab per inch of thickness	
normal weight	12.5
lightweight	9
Plywood per inch of thickness	3
3/4-in. tile on 1/2-in. mortar bed	16
Linoleum or asphalt tile	1
Wood joist floor (no plaster)	6
3-in. concrete slab on steel deck and open web steel joists	60
8-in. precast planks plus 1 1/2-in. topping	94
Metal deck (7 psf) + concrete floor (38 psf) + ceiling (7 psf) + mech. and elect. (6 psf) + floor beams (15 psf) + fire proofing (2 psf)	75
Partitions	
Masonry, 4 in. thick	
Hollow concrete block (light)	20
Clay brick	40
Gypsum tile	12.5
Plaster on one side of wall	5
2-in. solid plaster on metal lath and studs	20
2 × 4 studs, 1/2-in. gypsum drywall on two sides	8
Movable metal partitions	5–10
Walls	
12-in. hollow concrete block	
heavy	80
light	55
12-in. clay brick	120
12-in. reinforced concrete	150
4-in. brick plus 8-in. clay tile backing	75
Glass wall, large plate, heavy mullions	10–15
Window (glass, frame, and sash)	8

percent more than the steel building. The weight of the structure itself depends on the height, slenderness, loading conditions and efficiency of the structure system (see fig. 7.1 and Table 7.1). Especially important for the weight of tall structures are the vertical building planes rather than the horizontal planes. It is apparent that the structure weight constitutes only a relatively small portion of the total dead load, and may be in the range of 20 to 50 percent. Weight reduction is an important design

Table 3.3. Typical Uniform Occupancy Live Loads

Occupancy or Use	Live Load (psf)
Residential (private dwellings, apartments, and hotel guest rooms), private rooms and wards in hospitals, classrooms in schools, dressing rooms in theaters, etc.	40
Office buildings (no computer office use), private passenger car garage, etc.	50
Laboratories in hospitals, reading rooms in libraries, balconies not exceeding 100 ft^2 for one- or two-family residences, orchestra floor in theater, etc.	60
Office building corridors (above first floor), court rooms, etc.	80
Assembly areas, exterior balconies, terraces, public corridors, dance halls, restaurants, gymnasiums, lobbies, public stairs, repair garages, office buildings (office computer use), public dining rooms, public garages, etc.	100
Wholesale stores, computer floor (load must be verified), etc.	125
Armories and drillrooms, stage floors in theaters, stack rooms in libraries, mechanical rooms (transformer rooms, elevator machine room, fan room, but weight of actual equipment should be used when greater), etc.	150
Manufacturing, storage warehouses, etc.	125–150
Boiler room (but use weight of actual equipment when greater)	300

criterion, since it will result in significant savings in materials, freight, foundations and erection. On the other hand, it may be beneficial and necessary, when it must act as a stabilizing agent for slender buildings—that is, when the building layout takes advantage of gravity resistance in counteracting uplift forces. It should also not be forgotten that the building mass acts as a damping agent.

Gravity loads that are not part of the dead loads must be considered under live loads. Live loads are not permanent—they are variable and unpredictable. Live loads not only change over time but also depend on location and building type. Floor live loads caused by the contents or objects are often called *occupancy loads*, while roof live loads are snow, rain, and ice loads. Floor live loads include the weights of people, furniture, books, filing cabinets, fixtures, and other semipermanent loads that were not considered under dead loads. Codes provide values for live loads (i.e., for the sustained portion based on regular use and the variable portion due to unusual events), mostly in terms of equivalent uniform loads distributed over the floor area, as given in Table 3.3.

The equivalent floor loads have evolved empirically from experience and not from systematic surveys of loading. For example, code provisions for office floor live loads vary widely from 50 to 100 psf. A survey taken on the actual occupancy load in various office buildings showed a maximum load of only 40 psf. Similarly, a load survey on apartments noted that the maximum load intensity measured in a 10-year period was about 26 psf, hence quite a bit less than the usual code value of 40 psf. When the live load acts on smaller areas, it may have to be considered as a concentrated load. Concentrated loads given in codes indicate possible single-load action at critical locations such as stair treads, accessible ceilings, parking garages (e.g., a jack for changing a tire), and other vulnerable areas that are subject to high concentrated stresses. Stairway and balcony railings must be designed for a simultaneous vertical and horizontal thrust of 50 plf applied at the top of the railing.

From the values in Table 3.3 it is apparent that public areas such as corridors must carry more live load than living or working areas, or that office buildings weigh more than apartment buildings by ignoring the difference in dead load. The live load for office buildings may have to be further increased to say 100 psf, particularly adjacent to interior cores, to keep up with future developments in making buildings more intelligent. For spaces supporting computer equipment and backup batteries, the live load may be as high as 400 psf.

Although it may appear that some of the floor live loads are too conservative, there is always the unpredictable element to consider. For example, the live loads for exterior and interior (with movable seats) balconies of 100 psf are high, because the consequences of failure can be severe. The minimum regulated safety factors are warranted by such uncontrollable, extraordinary situations as people crowding because of ceremonies, parties, and fire drills, or the overloading of parts of a building due to a change in occupancy that will exert more load on a specific area.

It seems efficient to reach a high live load/dead load ratio according to strength and stiffness criteria, however, a floor system (e.g., open-web steel joists) may experience vibration problems that can be prevented by adding more weight or using other methods of damping.

It is improbable that, in multistory structures, every floor simultaneously carries the full live load; in general, the larger the area or the number of floors, the smaller the potential load intensity. Building codes take these conditions into account by allowing the use of floor live load reduction factors. The following reductions in live loads for the design of columns, piers, walls, foundations, trusses, beams, and two-way slabs are permitted. There is no reduction for one-way slabs, and according to ANSI, for two-way slabs only if the panel area is larger than 400 ft^2. According to the BOCA Basic Building Code/1981 a design live load of 100 psf or less acting on any member that supports at least 150 ft^2 or more, may be reduced at the rate of 0.08 percent per square foot of area supported by the members, except that a reduction shall not be made for garages, open parking structures, roofs or for areas to be occupied as places for public assembly. However, the reduction is not to exceed 60 percent or R, as determined by the following formula:

$$R = 23(1 + (D/L)) \hspace{3cm} (3.1)$$

where: R = reduction in percent
        D = dead load per square foot of area supported by the member
        L = design live load per square foot of area supported by the member

For live loads exceeding 100 psf a reduction cannot be made, except that the design live loads on columns may be reduced 20%.

Although a bearing wall may be investigated on a linear foot basis, the live load reduction, as related to the tributary area, depends on the degree of continuity of the wall. A continuous wall in high-rise construction supports a substantial floor area, so that the live load reduction may be taken only as a function of the D/L ratio. A bearing wall is treated as discontinuous in length due to the entrance and window openings, so that each length of wall between the openings can be considered as a member.

With the 1984 and following issues, the live load reduction regulations of BOCA have changed and adopted the ANSI version, although several codes still use an approach similar to that presented above. According to the ANSI A58.1–1982, members having an influence area of 400 ft² or more may be designed for a reduced live load determined by the following equation:

$$L = L_o(0.25 + 15/\sqrt{A_i}) \tag{3.2}$$

where: L = reduced design live load (psf)
L_o = unreduced design live load (psf)
A_i = influence area (ft²) taken as four times the tributary area for a column, two times the tributary area for a beam, and equal to the panel area for a two-way slab.

The reduced design live load, however, cannot be less than 50 percent of the unreduced live load L_o for members supporting one floor and not less than 40 percent of L_o for members supporting more than one floor. Furthermore live loads of 100 psf or less cannot be reduced for general-use parking structures, one-way slabs, roofs, or for areas for public assembly. Live loads that exceed 100 psf and live loads in garages for passenger cars only, acting on members supporting more than one floor, may be reduced 20 percent; no reduction is allowed otherwise.

Codes do not take into account that live load action on a building element is reduced because of the ability of the continuous structure to redistribute loading as it deforms. On the other hand, one may argue that the load capacity of many of our building assemblies is reduced, since they are subject to fatigue brought about by years of combating wind loads, vibrations, temperature changes, settlements, and the continuous change of environmental forces, as well as occupancy. However, compact buildings made from concrete and masonry materials have the advantage of gaining strength with age, therefore increasing their load-carrying capacity.

The roof live loads are usually smaller than the floor live loads, that is not considering any mechanical equipment loads. The distribution of the maximum ground snow loads in the United States, as recorded by the U.S. Weather Bureau, is identified in the codes by snow load maps. They indicate that the snow loads range from 80 psf in the Northeast to 5 psf in the South. Special values must be used for the mountain regions, where certain localities record snow loads of up to 300 psf. One inch of snow weighs about 0.5 to 1 psf depending on the moisture content.

Since the effect of the roof loading due to snow is less critical for the overall design of multistory buildings, it will not be discussed any further in detail. The following roof loads may be used for preliminary design purposes

South of latitude 37°	20 psf
Between latitudes 37° and 45°	30 psf
North of latitude 45°	40 psf

Except at high altitudes, the loads may be taken as 10 psf greater than as given above. The pitch and the shape of the roof are ignored. The snow load is considered on the horizontal projection of the roof surface, rather than on the inclined surface. Usually, for a pitch larger than 20°, the snow load can be reduced. Excessive accumulation of snow in the vicinity of obstructions such as penthouses, signs, parapets, adjacent buildings, etc. must be considered by the designer. Roofs used for roof gardens or assembly purposes shall be designed for 100 psf. A minimum roof live load of 20 psf is usually required for felt roofs, to allow for construction loads (i.e., workers and materials).

Water loads may become important if a flat longspan roof is not properly drained, if it does not have sufficient slope, if the drains clog, or if it is too flexible. Rain, with a weight of 5.20 psf/in. of depth, will collect and form standing pools. That is, water from heavy rain storms, rain on snow, or snow meltwater may accumulate and con-centrate as ponds, and may cause a flexible roof structure to deflect, thereby attracting more water, causing a deeper pool. This process continues until either equilibrium is reached or collapse occurs. Ponded rain water on flat roofs is controlled by having sufficient slope; a proper drain, which should not become blocked; and by providing a sufficiently stiff roof structure to avoid ponding failure.

One may also have to consider in the design heavy loads of icicles which may form on protruding roof elements, and one may have to consider the formation of ice surfaces, which in turn attract wind forces.

Distribution of Gravity Loads: Load Flow

At the initial design stage, the dead weight of the structural element must be estimated. The best procedure is to follow the force flow, starting at the roof level. The live loads are supported by decking, which carries them (together with its own weight) to the beams or joists, from where these loads and the additional beam weights are transferred to the main beams (trusses, girders, etc.), which then transmit the accumulated loads to the columns or walls. At the different design stages, the estimated loads and member sizes should be checked, and adjustments should be made, so that errors do not accumulate. In order to facilitate the estimate of the dead weight for the preliminary design of the structural elements, approximate weights per square foot of various construction materials are given in Table 3.2.

Various horizontal load distribution systems, in addition to some vertical flow sys-tems, are shown in Figure 3.2. Various one-way and two-way slab systems span directly to the bearing wall structures of Figure 3.2a–e. In the one-way systems (Fig. 3.2a, b, e), the loads are assumed to flow only in one direction to the perpendicular walls; the effect of the wall intersections at the corners and the effect of the parallel walls is ignored. A two-way slab distributes the loads in a rather complex fashion. The walls support triangular and trapezoidal tributary floor areas (see the subsection in Chapter 5 on "Two-way Slabs"), where continuous boundaries attract more load than hinged ones. The dividing lines for slabs with the same boundaries (i.e., support conditions) are assumed at 45°; but where a restrained or continuous slab boundary meets a simply supported condition, the fixed condition will attract more load so that the dividing line may be taken at 60° (Fig. 3.2d). In reality, boundary conditions are neither free to rotate nor fixed, and it has also been found that the stresses due to vertical loading in high-rise bearing walls are approximately uniform due to arch action, where the stresses are redistributed to lightly loaded wall portions as the relative deflection in-creases under the more heavily loaded portions. Thus, the difference of boundary

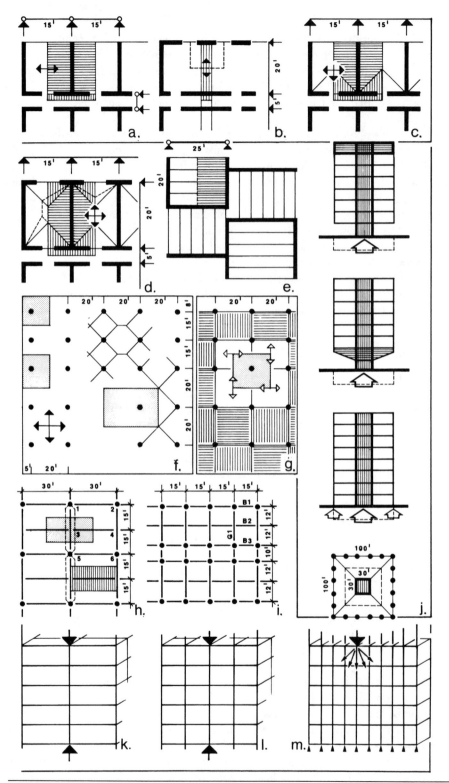

Figure 3.2. *Gravity load flow for various structure systems.*

conditions for preliminary design purposes may be ignored, and 45° dividing lines may be assumed (see fig. 3.2d). In addition, since the centroid of the loads acting on the tributary floor area does not generally coincide with the centroid of resistance of the wall unit, such as for a typical asymmetrical wide flange section, rotation is generated. Again, this effect of the vertical load distribution is ignored because of arch action, and because any rotation is assumed to be transferred by the slabs to the shear walls. One may conclude that, for tall bearing-wall buildings of ten stories and more, the axial gravity loads are spread evenly over the total cross-sectional wall area, composed of web and flange(s); some related concepts are discussed in Problem 3.2.

The distribution of the floor loads along one-way slabs to the beams, and from the beams to the girders, is shown in Figure 3.2h and i, and further discussed in Problems 3.3 to 3.6. The fact that some of the floor loads are carried directly to the girders is conservatively ignored. The load flow of a two-way slab to the columns (Fig. 3.2f and g) is described in Problem 3.1.

From a gravity load point of view, corner columns carry only about one-quarter of the load of an interior column and one-half of the load of a facade column for a regular building layout and light curtain construction. Naturally, with respect to lateral loading, the corner columns may act as chords and carry much heavier loads. The size of the corner column may also be larger than required structurally for construction reasons, or aesthetic reasons for exposed columns, where a visual sense of expressing the transition around the corner may need solidity.

While the vertical load flow is along linear paths in traditional skeleton construction (Fig. 3.2k and l), in tubular construction, with closely spaced columns, the gravity loads tend to spread out, as in a bearing wall (Fig. 3.2m).

Various types of vertical load distribution systems for an eight-story core structure (Fig. 3.2j) are investigated in Problem 3.7. In this case, the gravity loads along the periphery are either brought up to the top and from there to the core, brought down to an outrigger at the base cantilevering from the core, or they are separated from the core loads and flow directly to the ground.

Live Load Placement

The effect of live load arrangement is a necessary consideration of structural design, which is especially critical for continuous structures, as is apparent from Figure 3.3, where various loading patterns are investigated for the vertical and horizontal structural building planes.

In statically determinate structures, member rotations are not transferred, considering only primary effects. In this instance, a beam deflects only locally—its shear and moment diagrams are affected solely by the loads that the beam directly supports. In statically indeterminate structures, however, member rotations are transferred to other members because of the continuity of the joints; hence, the member stresses not only depend on the loads, but also on the stiffness of the boundaries. It is quite apparent then, that for the individual design of members, the critical arrangement of the live loads that create the maximum shear, bending moments, and axial forces must be considered.

In column and wall design, the loading patterns producing the maximum end moments in the column, and those causing the maximum axial loads, do not occur simultaneously. Using checkerboard loading with full dead load on all spans, but live

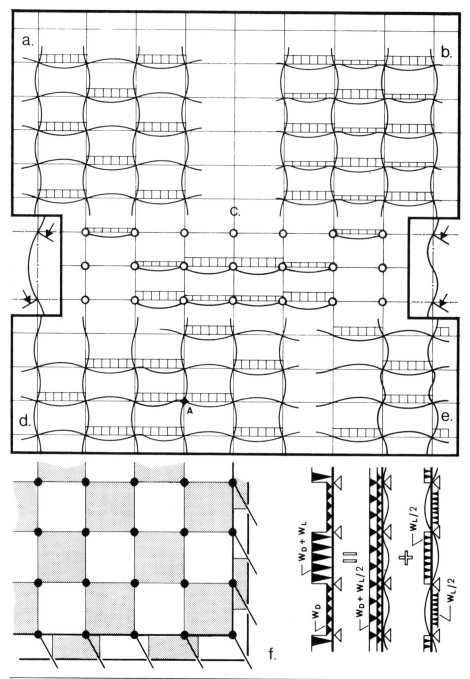

Figure 3.3. *Live load placement for maximum effects.*

load only on alternate spans (Fig. 3.3b), causes maximum end moment action, as reflected by the single curvature bending (Fig. 3.3, center left), but at the same time reduces the total axial load. Uniform loading, in contrast, will generate maximum axial loads, but no moments in interior columns for a symmetrical frame layout and the less critical double curvature bending in the exterior columns (Fig. 3.3, center right).

For the condition where the ratio of the live load to dead load is high, as for office buildings in steel, full gravity loading may initially be assumed for the investigation of the lower floor columns (or walls), while column bending due to checkerboard loading may be critical for the upper floors. However, it must be kept in mind that, when uplift or overturning is checked, the lateral loads act together only with the minimum dead loads and not with the combined dead and live loading.

For the frame beam or slab design, the checkerboard patterns of Figure 3.3b and f generate the maximum and minimum field moments, while the maximum support moment at point A, for example, is caused by the live load arrangement on adjacent spans of Figure 3.3d.

The controlling load positions for the maximum and minimum field moments in two-way slabs are given in Figure 3.3f. In this case, one can obtain the critical field moments by simply superimposing the uniform load case of $w_D + w_L/2$ and the loading condition of $w_L/2$ acting in alternating directions, which is equivalent in behavior to simply supported slab conditions. Hence the approximate maximum field moment, ignoring conservatively the portion of bending resisted in torsional shear and considering the two-way action of the square slab, is equal to

$$M_{max} \simeq \frac{w_D + w_L/2}{2} \left(\frac{L^2}{24}\right) + \left(\frac{w_L}{2}\right) \frac{1}{2} \left(\frac{L^2}{8}\right)$$

$$\simeq w_D L^2/48 + w_L L^2/24 \simeq (w_L + w_D/2)L^2/24 \tag{3.3}$$

Live load placement must also be considered for the design of a building core or other large scale elements. For example, a nonuniform distribution of the live loads may cause the core to bend, though the axial loads are reduced. One may conclude that, for the sizing of each member in a structure, an influence area with its respective loading must be established as to achieve critical design conditions.

CONSTRUCTION LOADS

Though a structure is generally designed for the gravity and lateral load action on the finished building, some of its members and bays may be subject to larger loads during the erection process. Every contractor has developed a construction procedure proven economical to him. For instance, equipment and material may be stockpiled on a small area of the structure, which is especially true for city sites, where little ground storage space is available and material is placed on the top floor. For this condition, shoring is required to distribute the weight, or the members have to be designed for these critical, concentrated loads. Further, certain building portions, like cores or frame bays, may have to support the dead and live loads of derricks or cranes, which includes the vertical and horizontal components of the derrick guy forces, and the gravity, uplift, and wind forces from a climbing tower crane.

Construction loads are also generated due to prevented volume change, as may be the case during winter construction, when the upper floor of a multistory building is cast under controlled temperature conditions, while the rest of the building may be exposed to freezing temperatures. Stresses are generated when the controlled temperature is stopped and a temperature drop occurs, resulting in differential movement. A major problem in concrete construction results when the contractor fails to allow sufficient curing time before the removal of shoring and formwork. Concrete increases in strength with time, particularly at the early stage of hardening; but since time is money to the contractor, he may remove the forms before the concrete has reached its minimum design strength. It must be remembered that the weight of the concrete, equipment, formwork, and workers on the upper floors must be supported by the lower floors, and cannot exceed the live load for which they were designed. Usually two stories of shores and one of reshores are required for high-rise structures for a rate of construction of one story per week. But for buildings designed for light live loads, such as apartment buildings, or for conditions of faster rates of construction, more floors have to be shored.

Construction loads must be considered for beams designed to act compositely with the concrete slab when no temporary shoring is used during the construction process. In this case, the beams have to be checked with respect to carrying construction loads in noncomposite action. Further, the lifting of prefab concrete components or stone slabs with cranes may generate much higher stresses during installation and handling than when the member is in place, and may very well control the design of the member. For instance, a solid wall panel, when in place, may just have to carry axial loads, but when it is lifted horizontally it will behave similar to a flat plate, bending under its own weight; also critical is when the panel is being lifted from a flat to a vertical position. Naturally, its behavior depends on the specific rigging situation; the number of lifting points and their location must be given. Other types of stresses due to handling may be caused by accidental impact, vibration during transportation, and force fitting on site.

DYNAMIC LOADS

In contrast to static loads, which are stationary or change slowly and cause a static *deflection*, dynamic loads vary more rapidly, or occur abruptly, and generate *vibrations*, thus introducing another dimension—that of time. The dynamic properties of the building which are activated are the mass, stiffness, and damping, in contrast to the static property of the building, which is only stiffness. Dynamic loads are not always cyclic—they can be sudden impact loads causing shock waves of short duration, such as blast loads due to explosions, sonic booms, or when moving loads like cars, trains, cranes, elevators, etc. suddenly change their speed. They cause, in addition to more periodic vibrations, an impact upon the supporting structure that results in longitudinal forces in the direction of movement, or a centrifugal effect upon curved structures in a radial direction. Vibrational loads may come from within the building or from the outside. Internal sources are elevators, escalators, oscillating machinery, mechanical equipment, cranes, etc. Outside sources are the wind, earthquakes, noise, blasting, driving piles, and traffic systems (e.g., streets, railways, subways, and bridges).

The building dead loads are stationary and fixed in magnitude, direction, and location, hence are static loads causing permanent deflections, but when they are set in motion they will generate dynamic loads due to inertia forces. Live loads, in contrast, are movable. They may be considered static, if they are applied slowly, as in the case of occupancy loads (though the natural period of the supporting structure may have to be considered, as is discussed later), but they are dynamic when they are applied abruptly or change rapidly and cause vibrations in the building. Vibrations due to people walking may be generated in the floors in the vertical direction, while wind-induced vibrations are primarily in the horizontal direction, as are seismic oscillations, though earthquakes may also cause significant vertical motions. Dynamic action causes larger loads than a comparative static one: the shorter the period of action, the greater the load increase.

But it is not just a question of the rate of application or of how fast a load fluctuates, but also one of how the building responds. The respective dynamic property of the structure is measured by the natural period of the building, that is, the time it takes for a building to freely swing back and forth to complete a full cycle of vibration without any external excitation. It is known from physics that, whenever a system is acted on by a periodic series of impulses having a vibrational period nearly equal to the natural period of the system itself, the oscillations in the system will gradually build up until it starts to resonate, and can lead to failure if undamped. For example, when soldiers cross a bridge, they must break step so that they do not march in rhythm with the bridge's natural frequency and cause the bridge to collapse. These vibrational forces are called *resonant loads*; they are quite different from the dynamic impact forces that produce large immediate effects.

The fundamental natural periods of typical buildings range from 0.1 sec for a single-story building, 0.6 sec for a 10-story bearing-wall brick building, 1 sec for a more flexible 10-story rigid frame building, 2 sec for a 20-story rigid frame building, 7 sec for the 59-story Citicorp building (see fig. 7.36) in New York, 7.6 sec for the 109-story Sears Tower in Chicago, to 10 sec for the 110-story World Trade Center in New York. The natural periods of long-span bridges are in the same range as for tall slender buildings.

While earthquakes apply sudden, violent, almost random forces with short periods, usually only fractions of a second, the wind appears smooth in contrast: it grows to its strongest pressure and then decreases in a period of several seconds. Wind turbulence contains a wide range of periods, so that the structure is only excited by a small portion of it; in other words, it is extremely variable in size and frequency. The critical, most intense oscillations with the largest amplitudes occur typically at long periods of about 15 to 20 seconds and more, while short gusts with low amplitudes may occur more frequently, for instance, at time intervals of less than 2 seconds. This only refers to in-wind oscillations, that is, vibrations parallel to the wind direction, and not to the usually more serious cross-wind oscillation due to vortex shedding, which is discussed in the next section, under "Wind Loads." In contrast, with respect to earthquakes, the natural periods of vibration of the site roughly vary from about 0.1 seconds for firm stiff soils to 5 seconds for soft flexible soils, with the range of 0.5 to 1.0 seconds being dominant in the critical regions of the United States. Comparing the natural periods of buildings with the ones of the exciting sources, it becomes apparent that the state of partial resonance may be approached for stiff buildings under seismic action, and for flexible superskyscrapers with their larger natural periods, possibly under smaller wind deflections. In addition, tall slender buildings with their large natural periods are sensitive to short wind gusts; for example, a wind gust of 2 seconds may have to be treated as a dynamic load for a tall, flexible building but as a static load for a low-rise building. In seismic design, the fundamental site period, as related to the fundamental natural elastic period of the building, is taken care of by the site-structure resonance factor S (see Equation 3.23).

It may be concluded that, when the period of the source is much longer (i.e., the frequency is much lower) than the period of the building, as is the case of stiff buildings under wind loading, the load can be treated as static. But when the period of the source is much shorter (i.e., the frequency is much higher) than the natural period of the building, as for seismic action of the waves caused by an explosion, and as is experienced by people as vibrations, then the load must be considered dynamic with a corresponding increase in stresses. However, one must keep in mind that dynamic conditions may occur even when the loads vary slowly but with a period close, for example, to the one of a flexible, tall structure with long natural periods, thus causing resonant loading. Since the natural period of the structure is such a significant parameter, it is briefly discussed, so as to develop some feeling for dynamic force action.

For preliminary design purposes in the elastic range, a building structure sometimes can be reduced to a single-degree-of-freedom elastic system, where the entire building mass is lumped at the top and assumed to be supported by a massless spring. This inverted pendulum allows only one type of motion (Fig. 3.4). This case, of an equivalent one-degree system, is reasonable for one-story buildings, where the mass of the columns (springs) is so small in comparison to the lumped mass at the roof level that it can be neglected; even an offshore structure may be visualized as a giant inverted pendulum. The shafts of water towers, observation towers, and TV towers are not massless; in these cases, the mass is (to some degree) more continuously distributed, thus representing a higher degree of freedom system. In multi-degree-of-freedom systems, such as high-rise buildings, the mass and stiffness may be assumed to be uniformly distributed as a rough first approximation, and then only the fundamental mode of vibration may be considered as critical. The foundations of a building are assumed to stay fixed and not to rotate and translate (i.e., fixed base).

Under repeated ordinary cyclic loading, the material must stay elastic. Occasionally, however, as for example under severe seismic action, it may be stressed inelastically. A linearly elastic member subjected to an alternating load returns to its original position, as expressed by the straight line of the force-deformation diagram in Figure 3.4c. In contrast, large cyclic loads due to a severe earthquake may cause substantial plastic deformations, as indicated for a ductile steel structure in Figure 3.4d. There will be a permanent deformation in tension during the loading; the unloading branch is assumed to have the same slope as the initial loading branch. When loaded in the opposite direction in compression and then unloaded, a reverse curve is generated. This loading cycle forms a closed stress-strain loop called a *hysteresis loop*, where the areas inside the loop represent the energy dissipated per cycle by heat, friction and so on. For an elastic member, the hysteretic stress-strain relationship is merely a straight line, where the energy is stored and then released to drive it in the opposite direction, similar to a spring. Only a first-cycle idealized loop for a ductile structure is shown in Figure 3.4d. Most important is the behavior of various materials, as well as that of the member, connection, and structure types under repeated loading. When subse-

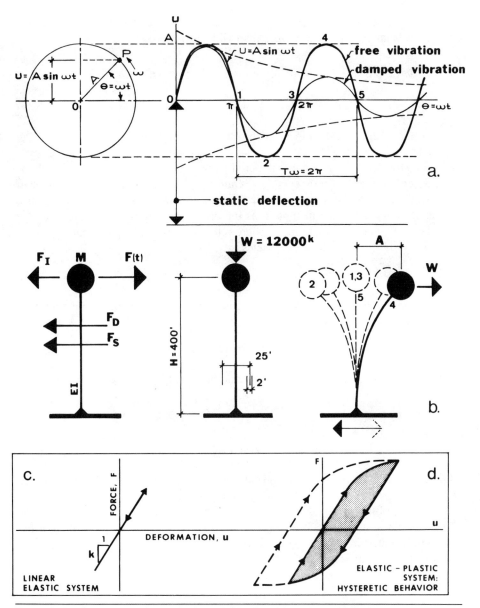

Figure 3.4. *The response of a single-degree-of-freedom system to cyclic loading.*

quent loops follow the first cycle plot closely, the hysteresis loops are stable and show no strength degradation, in other words, the structure has a large energy dissipation capacity (e.g., rigid steel frame). In contrast, degrading hysteresis curves indicate that the capacity and stiffness of the materials and structure systems deteriorate with load reversals.

In the following, an idealized one-degree elastic system, shown in Figure 3.4b, is investigated, where excitation is caused either by a direct external dynamic load, such as wind, or by an equivalent force, like a seismic ground acceleration. This dynamic load $F(t)$ is balanced by the following internal resisting actions:

- The *elastic resisting force*, F_s (spring force), resists the deformation. It is equal to

$$F_s = ku \tag{a}$$

  Where k is the lateral stiffness of the structure (see Equation 5.1) and u is the lateral displacement. The elastic force is proportional to the relative displacement.
- The *inertia force*, F_i, is equal to the product of the mass, m, and the acceleration, a, according to Newton's second law of motion. Mass is the quantity of matter in a body, and a measure of its inertia. Galileo had introduced the tendency of a body to resist a change in motion as inertia. Expressing the acceleration as the time rate of change of motion, v, (where velocity, v, equals the time rate of change of the position, u, of the mass), yields

$$F_i = ma = m(dv/dt) = m(d^2u/dt^2) \tag{b}$$

- The *damping force*, F_d, resists velocity (i.e., is proportional to the relative velocity) where c is a damping coefficient

$$F_d = cv = c(du/dt) \tag{c}$$

Therefore, the dynamic equilibrium conditions in Figure 3.4b can be expressed as

$$F_i + F_d + F_s = F(t)$$

$$m(d^2u/dt^2) + c(du/dt) + ku = F(t) \tag{3.4}$$

Of interest here are the basic natural dynamic properties of the structure as reflected by a free vibration (that is, the natural period of the building), which is in response to an initial disturbance in the absence of external forces, as for a forced vibration $(F(t) = 0)$. The structure is considered conservatively as an undamped system; it has been found that, for ordinary conditions, particularly for loads of short duration, damping does not appreciably affect the natural period of vibration; besides, only the maximum dynamic response is of interest. Hence, the equation of motion, Equation 3.4 can be reduced to

$$F_i + F_s = 0$$

$$d^2u/dt^2 + (k/m)u = 0 \tag{3.5}$$

This differential equation represents a simple harmonic vibration, which can be expressed in sine and cosine functions. Most conveniently, this sinusoidal oscillation is described in terms of the projection of a point, P, moving at a constant angular velocity or circular frequency, ω, around the circumference of a circle (Fig. 3.4a); the angular displacement, θ, of the point at any time, t, is ωt. Since any point on the circle represents the coordinates of the vibrational displacement, the lateral deflection of the mass can be expressed as a function of the velocity and the maximum amplitude, A.

$$u = A \sin \omega t \tag{3.6}$$

The second derivative, with respect to time, yields the acceleration, a, as experienced by the mass during the periodic motion.

$$d^2u/dt^2 = a = -A\omega^2 \sin \omega t \tag{3.7}$$

By substituting Equations 3.6 and 3.7 into Equation 3.5, $\omega^2 = k/m$ can be obtained, or the natural circular frequency of undamped vibration is

$$\omega = \sqrt{k/m} \quad \text{(rad/sec)} \tag{3.8}$$

In this case, the distance of one full wave length of harmonic motion is $\omega t = 2\pi$; or the natural period of vibration, T, to complete one full cycle is

$$T = 2\pi/\omega = 2\pi\sqrt{m/k} = 2\pi\sqrt{W/kg} \quad \text{(sec)} \tag{3.9}$$

The mass, m, has been expressed in terms of the weight of the structure, W, and the acceleration due to gravity, $g = 32.2$ ft/sec^2, as $m = W/g$.

The vibration can also be described with respect to time, that is, the number of cycles of vibration per second (cps), often also called hertz (Hz) as the natural cyclic frequency of vibration, f, which is the reciprocal of the natural period, T.

$$f = 1/T \quad \text{(cps, or Hz)} \tag{3.10}$$

The equations clearly express that the natural period of the building is dependent on its mass and stiffness. A building has a shorter natural period or higher frequency if it is either lighter or stiffer.

The maximum acceleration, a, can be found from Equation 3.7 at maximum displacement when $\sin \omega t = -1$, as

$$a = A\omega^2 = A(k/m) = A(2\pi f)^2 = A(2\pi/T)^2 \tag{3.11}$$

Humans are very sensitive to acceleration and changes of acceleration, which are the principal factors associated with the perception of motion of tall slender buildings. Often, a peak horizontal acceleration of below 0.005g to 0.015g (i.e., 0.05 to 0.15 m/sec^2) for a 10-yr return period is taken for the comfort of the occupants (see Example 3.1); 0.02g is considered disturbing and 0.05g painful. It is apparent from Equation 3.11 that it is not a simple task to reduce the acceleration, since it requires either more mass, which may not be feasible, or the building frequency must be decreased (i.e., a longer period), or special damping must be provided. Should the vibrational deflection, A, be decreased, then the building must be stiffened. However, by adding stiffness, the period will shorten and cause an increase in acceleration. In other words, an increase in stiffness results in a corresponding increase of the square of the frequency. These opposing phenomena, of reducing building sway, but at the same time increasing the frequency, hardly effect the acceleration. Further, one must realize that increasing the mass of the structure at the same time results in an increase of stiffness, and also in a larger accumulation of lateral inertial forces. Here, the higher natural damping of a concrete building is beneficial for a reduction of perception levels.

Example 3.1

The Empire State Building in New York deflects 6.5 in. under an 80 mph wind, and then vibrates 7.2 in. about the deflected position. Hence the maximum deflection is

$$\Delta_{max} = 6.5 + 7.2/2 = 10.1 \text{ in.} \leq H/500 = 1248(12)/500 \approx 30 \text{ in.}$$

The vibration amplitude, or peak displacement, is: $A = 0.5(7.2) = 3.6$ in.

The natural frequency was found to be: $f = 0.12$ Hz,
or, the natural period is: $T = 1/f = 1/0.12 = 8.33$ sec,
or, the natural circular frequency is: $\omega = 2\pi f = 0.24\pi$.

The maximum acceleration, expressed as a fraction of g, where the acceleration due to gravity is $g = 32.2$ ft/sec^2, is equal to

$$a = A\omega^2 = 3.6(0.24\pi)^2 \tag{3.11}$$

$$= 2.047 \text{ in./sec}^2 = 0.171 \text{ ft/sec}^2 = 0.052 \text{ m/sec}^2$$

$$= 0.171/32.2 = 0.0053g$$

Therefore, the behavior of the Empire State Building under wind can be considered to be hardly perceptible by the occupants.

Example 3.2

Determine the approximate natural period of a concrete TV tower with a large head at 400 ft from the base. Assume a constant 25-ft circular shaft with 2-ft walls, as indicated in Figure 3.4b, and a modulus of elasticity for the concrete of $E_c = 4000$ ksi. The entire mass is lumped together at the head, representing a weight of 12,000 k; the shaft is treated as massless.

The approximate moment of inertia of the thin-walled circular shaft is

$$I = \pi R^3 t = \pi(12.5)^3 2 = 12272 \text{ ft}^4$$

The lateral stiffness of a vertical cantilever is a function of the lateral static deflection, Δ, due to a single unit load $P = 1$, at the top and $EI = $ constant, as based on Equation 5.1.

$$k = 1/\Delta = 3EI/H^3$$

Substituting this spring constant into Equation 3.9 yields the approximate natural period of the tower.

$$T_1 = 2\pi\sqrt{W/kg} = 2\pi\sqrt{WH^3/(3EIg)} \tag{3.12}$$

$$= 2\pi\sqrt{12000(400)^3/[3(4000)12^2(12272)32.2]}$$

$$= 6.66 \text{ sec,} \qquad \text{or } f = 1/T = 0.15 \text{ Hz}$$

The maximum inertial force for a given peak displacement (i.e., vibration amplitude), u = A, can be obtained from Equation 3.11, as

$$F_1 = ma = mA\omega^2 = mA(2\pi/T)^2$$

Should it take 2 seconds for a peak gust to develop (i.e., a quarter of the assumed sinusoidal wave), then the corresponding period may be taken as $T = 4t = 4(2) = 8$ sec. The relative small difference between the natural period of the tower and the wind period indicates that the resonant loading may have to be considered, though vibrations in the tower will not be felt by people. Rather than immediately increasing the static loading for impact and repetition, the designer may consider decreasing the natural period of the building to another value, such as $T = 5$ sec, which results in following mass-to-stiffness ratio

$$m/k = (T/2\pi)^2 = (5/2\pi)^2 = 0.63$$

One may conclude that the tower design not only depends on the steady static wind pressure, but also on the vibrational component, and thus also the stiffness and mass of the building; the effect of damping is discussed later.

The simple, single-degree of freedom systems have been briefly introduced to develop some feeling for dynamic analysis, and because an actual structure may occasionally be represented by a one-degree system that consists of an equivalent mass with a massless spring. It is apparent that high-rise buildings are multiple-degree-of-freedom systems with many possible patterns of lateral deformation. They represent distributed mass systems and not massless shafts with mass concentration solely at the top. A continuous-mass structure is an n-degree-of-freedom system which, however, can be converted into an equivalent lumped-mass system. In other words, a building can be visualized as a lumped-mass multiple-degree system, where the masses are concentrated at the floor levels so that the weight of columns and walls is considered negligible in comparison to that of the floors. Hence, each mode of deformation of a lumped-mass system can be visualized as the superimposition of the independent actions of many one-degree systems. The n-story building has n degrees of freedom with n independent types of motions and n natural periods, thus requiring n equations for solution. Its dynamic response results in a complex series of vibrations, which involves tedious computer calculations. But there are methods available for a faster resolution of the critical first modes of vibration, particularly the dominant fundamental (first) mode, with the longest period. It must be kept in mind, however, that the disturbing impulses may not be close to harmonic vibrations, but quite random and erratic in nature!

Some approximate formulas for estimating the natural period of buildings are given below. The natural period of a flexible building of uniform mass and stiffness distribution that deflects in a relative linear fashion (as shown in fig. 2.18) is derived in Problem 3.8c. It can be roughly approximated as

$$T = 2\pi H\sqrt{W/(3gk)} = 2\pi H\sqrt{WR/(3gM)} \tag{3.13}$$

In this case, the structure stiffness is $k = M/\phi = M/R$, according to Equation 2.13, as based on the limiting drift ratio such as $R = \Delta/H = 1/500$ for wind.

For the vertical single-degree-of-freedom cantilever beam (see fig. 3.4b), the ratio of W/k in Equation 3.9 can be interpreted as the maximum static deflection, Δ, at the top of a building, caused by W, as initiated by the seismic ground acceleration, a:

$$T = 2\pi\sqrt{W/kg} = 2\pi\sqrt{\Delta/a}$$

However, the ground acceleration can be expressed as a percentage of the acceleration of gravity, g or a = C(g). In this case, the seismic coefficient, C, together with some other factors discussed in the section in this chapter on "Seismic Loads" (see also Problem 3.22, C_1 = ZICS), is called C_1 and is substituted into the equation above, together with g = 32.2(12) in./sec^2:

$$T = 0.32\sqrt{\Delta/C_1}$$

Since this period is based on a circular period of vibration of a single-degree-of-freedom system, Edward Teal proposed the following expression for a building with multiple-degrees-of-freedom (Teal, 1975):

$$T_1 = 0.25\sqrt{\Delta/C_1} \quad \text{(sec)} \tag{3.14}$$

Often, two-thirds of the allowable drift is used as an initial approximation for this equation:

$$\Delta = 0.67(0.005H) = 0.0033H$$

While Equation 3.13 applies to flexible buildings of uniform mass and stiffness distribution, with a constant straight line deflection, the response of a multistory building to vibrations in general is extremely complex. The Uniform Building Code gives the following empirical formulas for the preliminary approximating of the natural period of a building in the fundamental mode:

- For rigid frames with N stories

$$T_1 = 0.1N \quad \text{(see Equation 3.22)}$$

- For other building structures

$$T_1 = 0.05H/\sqrt{D} \quad \text{(see Equation 3.21)}$$

The Teal equation (3.14) gives a more reasonable period estimate for short, rigid frame buildings, and for tall and narrow braced frames. Equation 3.22 may also be used for first estimation purposes for other tall flexible buildings such as braced frames as well as tubes.

It is also important to point to lightweight long-span floor structures that may be susceptible to dynamic excitement, such as that due to people walking or dancing, causing a typical period of 0.25 to 0.5 seconds. It is apparent that when the floor system has a natural period of 0.2 seconds or more, resonant loading may occur, together with motion objectionable to humans. The natural period of vibration of the floor can be estimated by treating the uniformly loaded structure as an equivalent one-degree system and ignoring its mass. Hence, according to Equation 3.9, letting W/k = Δ (see Example 3.2), where the maximum static deflection Δ is caused by the dead and live loads present when the floor is vibrating:

$$T = 2\pi\sqrt{W/kg} = 6.3\sqrt{\Delta/g} = 0.32\sqrt{\Delta} \quad \text{(sec)} \tag{3.15}$$

Actually, this factor of 0.32 depends on the floor type and varies from 0.32 to 0.25; often, a factor of 0.29 is used.

For the condition where the vibration of the structure is small, it is common practice to simply increase the static loads, so as to take into account the stress increase due to acceleration, while a dynamic analysis must be used for large vibrations. When members are subject to impact loads, it is the usual practice (for typical conditions) to increase the live loads that induce the impact by impact factors to cover the dynamic effects, otherwise dynamic analyses are performed to compute the maximum forces. If not otherwise specified, codes typically indicate that the increase shall be as follows:

For supports of elevators	100 percent
For cab-operated travelling crane support girders and their connections	25 percent
For pendant-operated travelling crane support girders and their connections	10 percent
For supports of light machinery, shaft or motor driven, not less than	20 percent
For supports of reciprocating machinery or power-driven units, not less than	50 percent
For hangers supporting floors and balconies	33 percent

To provide for the effect of a moving crane, lateral force action is given in codes as a percentage of the weight of the crane and loads to be lifted. In areas receiving heavy impact and dynamic loads, where structures are subject to vibrations causing alternate compression and tension, members may experience fatigue; codes do take into account the resulting reduction in strength due to stress reversal!

Large dynamic forces must be controlled, not just by determining their magnitude so that the building can be designed accordingly, but by changing the period of the source (e.g., driving motor), by isolating the source of the excitation, by damping, and by controlling the mass-to-stiffness ratio. The stiffening of skyscrapers to control oscillations is a most important design consideration. The amount of damping that will prevent all vibrations (i.e., critical damping) is proportional to the product of the stiffness and mass. Hence, for a given structure, dynamic response can be reduced by:

- an increase in stiffness
- an increase in mass (for purposes of damping)
- damping

However, since stiffening a structure results in an increase of material, as does the increase of mass, other methods of damping must be explored to possibly reduce costs. Besides the *natural damping* of structural materials, *artificial damping* may be employed to control dynamic excitation, similar to shock absorbers on our cars.

Natural damping consists of internal damping and other natural ways of reducing vibrations. *Internal damping* is a property of every building and is activated as the building deforms; it results in the decay of the amplitude of vibration (Fig. 3.4a). It primarily consists of three main types, solid (hysteresis), viscous, and frictional damping, which resist movement. *Hysteresis damping* reflects the energy absorbed by internal friction of the material in the stressed state forming a closed loop under cyclic loading.

While *viscous damping* is provided by fluids such as air, *frictional damping* arises from friction between the moving adjacent parts. The extent of damping depends upon the method of construction, and is on the order of 2–15 percent of the critical damping; the amount of damping necessary to stop an oscillation before it finishes one cycle is called the critical damping. In continuous monolithic structures, such as welded steel frames and cast-in-place concrete frames, vibrations are dissipated mainly by internal friction of the materials; this hysteresis damping is relatively small in the elastic range. In this case, cracked conventionally reinforced concrete has a higher damping capacity than either crackfree post-tensioned concrete or steel. It has been found that tall concrete skyscrapers tend to have roughly 30 percent more damping than steel structures! In contrast, in constructions with many joints and large contact surfaces, for instance, in panel and composite structures, and for buildings with many additional nonstructural fill-in components (partitions, exterior walls, windows, etc.), resistance to acceleration is provided more effectively by frictional forces. It is apparent that highly redundant structure systems and mixed construction methods are preferred.

Besides the internal material damping other *natural methods of damping* include the following special design considerations:

- The plastic behavior of materials
- The isolation of the building
- The form and texture of the building

Plastic behavior is similar to damping since it interferes with the elastic vibrations. The plasticity of steel and of ductile reinforced concrete construction allows absorption of energy in the inelastic range by dissipating part of the energy through friction, heat, cracks, and so on. Ductile detailing makes possible controlled energy dissipation in selected members under earthquake action, for instance, at the base of a building (i.e., soft story, Fig. 3.5a), or by eccentric truss connections, off-column plastic hinges in reinforced concrete frames, slotted wall construction, and coupled walls, where the coupling beams dissipate most of the energy by yielding.

In the *shock absorbing soft story concept*, the columns at the first floor, at top and bottom, may yield under seismic action before shock absorbers will come into action. The late, eminent structural engineer Fazlur Khan of SOM and Mark Fintel of the Portland Cement Association developed this concept in 1969, which permits the control of damage caused by an earthquake. In this construction technique, the first story is allowed to deform with the earthquake while the upper portion of the building remains almost unaffected, and thus stays in the elastic range. The concept was first applied in the St. Joseph Hospital, Tacoma, Washington in 1973 (Fig. 3.5a). The concrete columns surrounding the stair and elevator shafts of the core area at the mechanical floor level are diagonally x-braced with post-tensioning tendons. After the columns yield under earthquake motion, the energy is effectively absorbed by the diagonals, which transmit the shear forces to the shear walls below. Should the structure come to rest with a small off-center set, the tendons can be used to jack the building back into plumb.

To control vibrations transmitted from the ground to the building, for example, due to railroad traffic directly below the structure, or due to earthquakes, the structure may be isolated from the ground to allow it to float as a rigid body, as for medium-rise buildings in the five- to ten-story range—their behavior is similar to the vibration

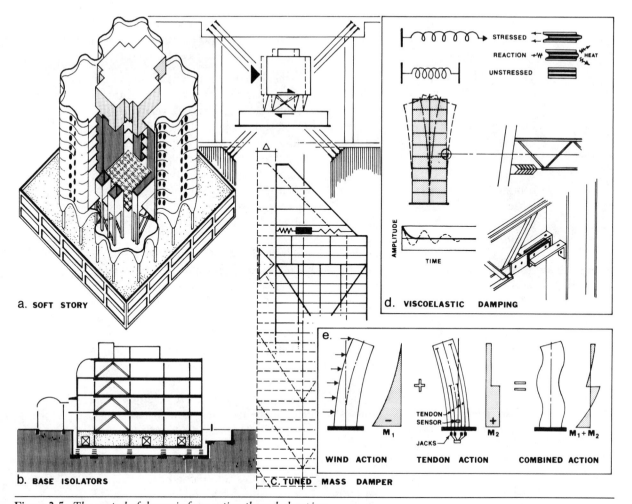

Figure 3.5. *The control of dynamic force action through damping.*

isolation mounts for mechanical equipment. While in the fixed-base structure ground motions are sent upward, possibly amplifying the ground accelerations on the upper floors, buildings on isolators hardly deform, because the movement occurs at the isolators. The following criteria are to be considered for *base isolation* systems:

- *Flexibility:* Flexible, soft support must allow, for example, lateral movement due to seismic action (e.g., elastomeric bearings, rollers, sliding supports, etc.)
- *Energy dissipation:* Dampers must control the movement between ground and structure by using hysteretic, friction, or hydraulic devices. They change the building's natural period of vibration so that it is different from that of the ground, and amplification cannot occur. While the stiff building moves slowly back and forth, the energy of the rapid seismic ground accelerations is mostly dissipated in the isolators.

- *Rigidity*: A minimum initial lateral stiffness is required for low service loads, such as wind and minor earthquakes. Special elastomers and mechanical devices have been developed that provide high initial rigidity and soften as strain increases.

The Foothills Communities Law and Justice Center [1986] in San Bernardino County, CA (Fig. 3.5b) is the first U.S. building to stand on an isolation system consisting of ninety-eight laminated sandwiches of rubber and steel bearing pods, weighing about 1100 pounds each. The base isolators are located at the bottom of the building under each steel column; they have a high vertical stiffness, as they must support the building weight, but they have a very low horizontal stiffness, in order to absorb horizontal seismic ground movement. Since the building acts like a rigid body, the cyclic motion is primarily concentrated at the isolators, which shift the building's period into the longer range away from the frequency of the earthquake. The base isolation system is an important concept that allows the building to float, rather than fight the motion of earthquakes when anchored to the ground.

For very tall buildings, natural damping can be achieved, with respect to high wind speeds that cause vortex shedding, through *aerodynamic damping*. In this instance, the shape of the building, the texture of the facade, chamfered corners, notches, cutouts, openings, cantilevering portions, etc. cause wind turbulence, which in turn acts as a natural means for damping oscillations. The notched corner, triangular shape of the U.S. Steel Building (see fig. 7.4e) in Pittsburgh was selected after an extensive aerodynamic study of various building shapes. Other design methods to achieve damping involve the manipulation of building masses and how to connect them, including the use of cable systems.

In addition to the natural damping of a building, artificial damping systems (similar to door dampers and shock absorbers in cars) may be required, particularly for tall, slender skyscrapers, to effectively offset wind gusts. The *artificial damping systems* may be classified as active and passive.

Passive dampers are usually based on Coulomb friction, viscous friction, or shock absorbers of the piston type. In Coulomb friction joints, the natural dry friction of members in contact opposes the sliding. The friction force is constant and depends only on the coefficient of friction and the weight of the body. More common is viscous damping, where the friction is based on shearing a fluid film between the sliding surfaces, as is discussed in the following section.

Examples of passive damping are to be found in New York's World Trade Center [1972] and Seattle's Columbia Seafirst Center [1985]. In the World Trade Center, viscoelastic dampers were attached to the bottom chord ends of the open web steel joists and to the adjacent exterior columns (Fig. 3.5d). As the name suggests, viscoelastic material is both elastic (returns to its original position like a rubber band) and viscous (tends to flow under pressure like a liquid). Viscoelastic material resists forces in shear; it does not store energy like a spring, but converts it into heat that is diffused to the surrounding environment. Therefore, after the forces are released, the material does not snap back and forth, as a spring does, but slowly returns to its unstressed position. The building does not oscillate with damping; instead, gust winds emit heat inside the building because of the response of the dampers (A.R., 1971).

In the 76-story Columbia Center tower, viscoelastic dampers (consisting of steel sandwich plates joined by a rubberized plastic material) are added to some diagonal bracing of the rigid frame core. As the bracing members move back and forth, the viscoelastic material goes through shear deformation, thereby dissipating energy and reducing the sway of the building in large storms.

There are several *active damping systems* currently being investigated, but only the *powered passive tuned mass dampers* have actually been used, such as for Boston's John Hancock Tower [1977] and New York's Citicorp Center [1978].

In the Citicorp Building, a 410-ton mass of concrete, which is connected to the structure by spring damping mechanisms in the two major horizontal directions, is located at the top of the building. The mass is tuned so that it can move in a special period equal to the building's natural period of about 7 seconds. When the building sways, the mass remains still, since it is free-floating on a nearly frictionless film of oil, but will compress and pull the springs (Fig. 3.5c). The springs, in turn, pull and push the building back to the center. In other words, the tuned mass tends to oscillate with the same period as the building, but in the opposite direction, thereby effectively counteracting and damping the building motion.

Some other active damping systems that are currently being researched are:

- *Active mass impact dampers:* These are similar to the tuned mass dampers, but the sliding mass is not connected to the supporting structure.
- *Active tendon control system:* As early as the 1960s, Eugene Freyssinet of France and Lev Zetlin of the United States proposed to control lateral deflection of a building by introducing stressed tendons within the structure to generate an opposing deformation (Fig. 3.5e). This eliminates the need to achieve the lateral stiffness of a building by additional material, most of which may be used only once in 100 years to absorb maximum wind velocities. Cables near the exterior facade are attached to jacks at their base. A sensor unit measures the wind velocity and direction. This information is transmitted to a control unit at the base, which causes the jacks to tension the cables. This off-center tensioning induces a bending moment opposing the wind moment. Therefore, the moments are neutralized and the lateral deformation is greatly reduced. The amount of tension in the cables and in the building side being stressed vary according to the magnitude and direction of the wind pressure. The concept is analogous to the tension felt in the muscles of an outstretched arm when the hand accepts a heavy object and the arm attempts to maintain its position. The active tendon control can also be applied to diagonal bracing systems.
- *Pulse control systems:* To reduce the response of the building to vibrations, jets of compressed air from tanks at one or more levels are released and cause short duration counteracting lateral forces (pulses).
- *Air release systems:* Since the positive pressure on the windward building face generates suction on the leeward side, the pressure difference could be overcome through the discharge of air on the leeward side. Thus, not only is the magnitude of the lateral action reduced, but there is also a diminution of the oscillations that form after a change in wind velocity allows air to rush into the negative pressure region and push the building in the opposite direction.
- *Aerodynamic appendages:* Wing-shaped movable appendages of various sizes are attached near the top of a building, similar to wing flaps on airplanes, and are controlled by sensors. They can effectively reduce the response of the building to wind.

WIND LOADS

Wind forces are among the most violent sources of destruction in nature. It is extremely difficult to predict their complex behavior, since they are not constant and static like dead load, but are dynamic and fluctuate in an unpredictable manner, not only in magnitude but also in direction. Wind behavior is influenced by the topography (open, wooded, rolling, hilly, urban, vegetation, roughness, etc.), the building type (shape, size, height, texture, flexibility, degree of tightness, openness, etc.), and the nature of the airflow (air density, direction, velocity, degree of steadiness, etc.).

Tornadoes are the most devastating winds. In the United States, they appear mostly between the Rocky and Appalachian Mountains, with a tangential wind velocity as high as 500 mph within the tornado tunnel. The probability of a particular building being hit by a tornado is extremely small, therefore a structure is usually not designed for tornadoes. Buildings designed for the 200-mph range should generally be safe 95 percent of the time and this should also take care of severe hurricanes. While tornadoes only act over a relatively small area, of about 3 square miles, hurricanes affect many thousands of square miles, and thus are much more destructive. Hurricanes occur in the United States primarily along the Atlantic coastal zones. They cause damage not only because of the high wind velocity but also because of the waves (erosion) and rainfall (flooding). The maximum winds appear close to the eye of the hurricane. Typical design values for strong hurricanes are 150 mph, and 90 mph for average ones. The radius of curvature of the rotating winds of hurricanes is so large that their path may be considered straight. Similar in behavior to the hurricanes in the Atlantic and South Pacific are the typhoons in the Western Pacific and the tropical cyclones in the Indian Ocean, Arabian Sea, and offshore Australia.

Early skyscrapers were not vulnerable to the lateral drift and oscillations caused by wind. Heavy stone facades with small openings, closely spaced columns, massive built-up frame members, and heavy partition walls made the buildings extremely stiff and provided high damping; the buildings had a great deal of mass. In addition, the facade texture and many setbacks generated effective aerodynamic damping. In contrast, today's functional glass-walled or cladded tall buildings use much stronger materials; they have long-span beams allowing large open spaces and have light movable partition walls. They are optimized lightweight structures that are much more flexible, and thus more vulnerable, to lateral sway and to vibrations excited by turbulent winds. On tall, flexible buildings, the wind not only causes a response along the wind direction but also in the across-wind direction, besides generating torsion.

The lateral wind loading normal to a building face, among other criteria discussed later, is directly related to the wind velocity, which consists of the constant mean velocity (steady component) and the varying gust velocity (dynamic component). Therefore, a building deflects along the direction of the wind due to the mean wind pressure, and vibrates from this position due to gust buffeting, which can be larger than the static sway! However, it must be emphasized that wind velocities are nonuniform in time; they are full of gusts.

As air moves along the earth, it is retarded close to its surface due to frictional drag. The wind near the ground behaves in an erratic manner; though the wind velocity will generally be less in cities, the gustiness may be much greater. In the boundary layer, where the buildings are located, the mean wind velocity increases with height, with the rate of increase being a function of the ground roughness. As shown in the top portion of Figure 3.6, the greater the interference by surrounding objects (i.e., trees, land forms, buildings), the higher the altitude at which the maximum and steady velocity occurs. When the air stream passes the building, its behavior is drastically altered. It is deflected and then rejoins the original flow pattern, exerting pressure on the front side and suction along all of the other sides, as illustrated by the plan views of the circular shaft and the rectangular building in Figure 3.6. The building must resist the total wind force that is the sum of the pressures. The degree of disturbance of the original flow pattern depends, among other criteria, on the building shape and the nature of the surface; a sawtooth face may cause significant wind drag. It is apparent that, for a given wind direction, the streamlined tear-drop shape (as exemplified by aircraft wings and the body profile of the dolphin) provides the least resistance to the air flow.

A typical rectangular tower (Fig. 3.6, top left) causes a descending wind movement along the front side, thereby producing a vortex at the ground, which can be an annoyance for pedestrians. As the wind is deflected around the vertical and horizontal corners, a series of eddies and vortices, or regions of accelerated flow, are generated. Though eddies are similar to vortices, they are only slow-moving circular air currents, creating little perceivable building motion. The vortices adjacent to the building are high-velocity air currents that produce circular updrafts and suction streams. They form alternately on each building side, and create low pressure areas when they break away, thereby pulling the building in this direction and producing cross-wind oscillations (see also the section in Chapter 2 on "Building Loads"). As a result, the largest movements of slender flexible towers under strong winds are mostly not in the direction of the wind but perpendicular to it; resonant amplification is not usually generated by the gustiness of the wind, but rather by vortex shedding. The rate at which the building may shed vortices, that is, rhythmic sideway movements, depends (besides other criteria) on the building slenderness, the damping qualities, and the wind velocity, as well as direction. At high wind speeds, the building's natural period may coincide with the period of the vortices and cause dangerous resonant loading conditions. The triangular U.S. Steel Building in Pittsburgh (Fig. 3.6, right) has cutout corners to create turbulence, which provide a natural means of damping. One of the newest techniques proposed for reducing vortex shedding on very tall buildings is to place openings near the top that penetrate completely through, to generate turbulences, which reduces the perception of the critical vibrational motion. This concept was applied for the first time to the 56-story Texas Commerce Tower [1987] in Dallas.

In addition to the vortex-flow, the through-flow is critical. The Earth Sciences Building at MIT (Fig. 3.6, left) originally used an open entrance passage through the building, which generated a wind tunnel, with the corresponding strong winds at the ground level, and causing a maximum pressure at the center of the windward face, rather than at the top, as assumed by codes. Wind tunnel tests also demonstrated that the maximum pressure for the building without an opening would have occurred at the base due to the channel effect formed by the adjacent buildings. The Venturi effect is produced when an air mass is funneled through the narrow spaces of two tall buildings, resulting in high-speed winds. Adjacent buildings should be placed so as to protect each other and reduce the wind loads and the wind at the pedestrian level. Also, the building should be oriented on the site to allow a reduction of wind pressure and possibly a cut of the solar load. The shape of an irregular building should be closer to the streamlined cylinder than the rectangle, so that a reduction of the code

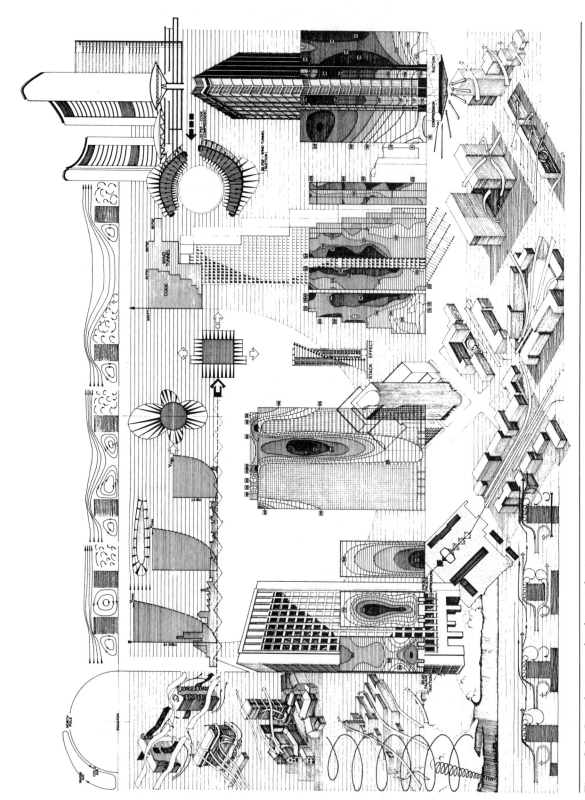

Figure 3.6. *Considerations related to wind action.*

wind pressure is allowed. The discussion above shows that the wind pressure does not necessarily increase with height, but is dependent on the nature of the building and the surrounding topography. The reader may want to study the examples in Figure 3.6 illustrating the complex interaction of air flow and building masses, as influenced by their arrangement.

The performance of a building may have to be investigated for wind action from different directions. Wind flow on more than one building face causes double flexure and possibly torsion (see fig. 2.14). For a typical rectangular building, the primary wind direction can be separated into the two components—perpendicular to the faces and those causing twisting. Direct wind action perpendicular to a building face is usually greatest and controls the design of ordinary buildings, since much of the airflow striking a smooth building surface at an angle other than 90° will dissipate naturally, though a facade with much frictional resistance (such as a sawtooth face) may result in significant wind drag. For unusual curvilinear shapes, it is extremely difficult to predict the most critical angle of attack; in wind tunnel tests, many probes must be used to analyze numerous angles at velocities possibly up to that of a hurricane.

Though codes are adequate for the design of ordinary buildings, they are not for wind-sensitive buildings, that is, tall buildings of unusual shape, slender buildings of a height-to-width ratio of approximately more than five, buildings over 400 ft in height, lightweight flexible buildings, and skyscrapers within a particular city-scape. For these buildings, wind tunnel testing or computer modeling must be done to produce the wind pressure distribution and response of the building. Besides the static behavior (pressure loading), other tests produce the dynamic response which, however, requires a special building model to replicate the building's mass, flexibility, and natural frequency. It should also be mentioned here that Professor Davenport of the University of Western Ontario must be given credit as one of the pre-eminent contributors to the development of wind engineering with his extensive wind tunnel testing of buildings since the 1960s.

As recently as the early 1960s, designers of Toronto's City Hall (Fig. 3.6, top right) became aware of the unpredictable nature of wind action due to the unconventional wing-shaped towers and their close position, although efficient wind resistance was the source for the shapes of the cantilevering vertical cylindrical shells stiffened by their slabs. Wind tunnel tests indicated design loads nearly four times higher than specified by the code; notice the unusual pressure distribution with the extremely high suction. In addition to the large pressures, the critical wind direction had to be determined that would create the highest torsional moments. The contours for the various cases of irregular building shapes in Figure 3.6, measured in psf, convincingly illustrate the complex wind pressure distribution across the faces, clearly demonstrating the deviation from code assumptions.

Although codes are generally sufficient for wind forces on ordinary main load-bearing structures, this may not be true for local loading on components, especially cladding. In 1974, strong winds smashed more than 50 windows in the Sears Tower and Standard Oil Co. Building in Chicago. The failure of the glass facade of Boston's John Hancock Tower [1973] cannot be forgotten. Wind tunnel tests indicated pressure values in some places on the facade up to three times higher than required by the local building code. In addition, the larger loads occurred closer to the base, rather than the top, as was assumed according to the code. Accordingly, the thickness of the glass increased from the bottom to the top of the building instead of in the opposite direction. The wind pressure diagrams from testing should be the basis for the choice of curtain-wall design loads. Of special concern are sharp corner pressures that vary with the direction of the wind, and which may be twice as high as the typical center wall loading, thus requiring much thicker glass along the edges. The typical code approach for lateral loading along the wind direction is briefly discussed in the following paragraphs.

According to Daniel Bernoulli (1700–1782), for a steady streamline air flow of velocity V, the local dynamic or velocity pressure, q, upon a rigid body is related to the square of its velocity and equal to

$$q = \frac{1}{2}\rho V^2 \tag{a}$$

Where the air mass density ρ is the air weight divided by the acceleration of gravity $g = 32.2$ ft/sec^2. For the special condition of a standard atmosphere with a temperature of 15°C (59°F) at sea level, the air weighs 0.0765 lb/ft^3, which yields the following value corresponding to a wind speed in mph (see Equation A2, in ANSI A58.1–1982). Thus, Equation (a) can be expressed for V (in mph), as

$$q = 0.00256V^2 \text{ (psf)} \tag{b}$$

It should be kept in mind that the air mass density varies with locality, altitude, time, and weather; a different numerical coefficient may be derived if respective climatic data are known. The basic wind speed, V (mph), for the United States is recorded by weather stations and given as the annual fastest speed 30 ft above ground, based on a 1-, 25-, 50-, and 100-year mean recurrence interval. Ordinarily, wind velocities with a mean recurrence period of 50 years should be used as given in Figure 3.7, if there is no high degree of risk to life and property in case of failure, thus allowing some damage for a stronger wind occurring once in 100 years. It should be noticed in Figure 3.7 that the wind speeds range from a high in Miami to less in New York, to a further reduction in Chicago, and finally to a low in Los Angeles. For example, the wind loads in Miami may be twice that of Houston and four times that of Chicago.

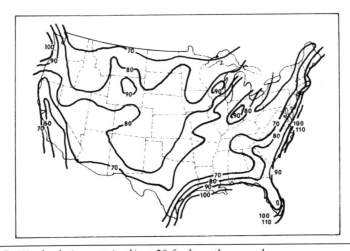

Figure 3.7. *Wind velocity map (mph) at 30 ft above the ground.*

According to the American National Standards Institute: *Minimum Design Loads for Buildings and Other Structures* (ANSI A58.1–1982), the dynamic pressure is converted to the velocity pressure q_z (psf) at height z, taking into account the effect of exposure conditions.

$$q_z = 0.00256K_z(IV)^2 \qquad \text{(c)}$$

For buildings of 15 ft or higher, the velocity pressure exposure coefficient, K_z, is defined as

$$K_z = 2.58(z/z_g)^{2/\alpha} \qquad \text{(d)}$$

This equation expresses the increase in wind velocity with height according to an exponential velocity profile. ANSI A58.1 identifies four exposure categories for the main wind-force resisting structure systems, reflecting the effect of surface roughness (topography, vegetation, and other man-made objects) upon the wind speed. They are:

- Exposure A (large city centers, $\alpha = 3.0$, $z_g = 1500$ ft)
- Exposure B (urban and suburban areas, wooded areas, $\alpha = 4.5$, $z_g = 1200$ ft)
- Exposure C (flat, open country, slightly rolling with minimal obstructions, $\alpha = 7.0$, $z_g = 900$ ft)
- Exposure D (flat, unobstructed coastal areas, $\alpha = 10.0$, $z_g = 700$ ft)

Where z_g is the gradient height or the elevation at which ground friction no longer effects the wind speed, and where α is the power law coefficient.

For the design examples in this text, the severe "Exposure C" condition is conservatively assumed. Hence, Equation (d) gives

$$K_z = 2.58(z/900)^{2/7} \simeq (z/30)^{2/7} \qquad \text{(e)}$$

The importance factor I in Equation (c) adjusts the wind speed to annual probabilities of being exceeded, reflecting the difference in probability along coastlines and inland regions; but it also depends on the nature of occupancy of the building—it ranges from 0.95 to 1.11. In this context, buildings of Category 1 (Equation 3.18) that are located at least 100 miles inland from the hurricane oceanline are assumed with I = 1. Hence Equations (c) and (e) can be combined and simplified to the following form

$$q_z = 0.00256V^2(z/30)^{2/7} \qquad \text{(f)}$$

This equation is given by the Standard Building Code (1985 ed.) for a 100-year wind. One must keep in mind, however, that for flat, unobstructed coastal areas, the velocity pressure has to be increased. For suburban and urban areas it can be decreased, though at the centers of large cities the effect of adjacent buildings will always cause a much higher gustiness and may also generate channeling effects, resulting in an increased pressure.

Up to this point, only the local pressure upon a rigid, nonmovable large body has been considered. In reality, the air flow does not come to a complete halt, but only changes as it passes the building object and deviates from its original path in response to the structure. Until now, the nature of the building had no effect upon the wind pressure distribution; this obviously cannot be true. As the wind strikes a structure, its shape, size, texture, openness, and flexibility have an influence upon the magnitude of the pressure distribution. A building body can be visualized as immersed in flowing water, where the change of the originally even flow pattern results in the velocity pressure changes. The building is subjected to *normal* and *transverse forces*. The transverse action is caused by side and lift forces, such as the familiar lift action on aerofoil shapes. The normal forces are also called *drag forces* because the body is dragged in the direction of the flow; they are caused on buildings primarily through direct normal pressure, that is, positive pressure on the windward face and simultaneously negative pressure (suction) on the leeward face, but also through surface friction tangential to the building's surface. The magnitude of the drag forces is defined for buildings by the external pressure coefficients, C_p, and by the shape or force factors, C_f, for other structures like lattice frameworks, open structures, chimneys, trussed towers, tower guys, etc. The external pressure coefficients for the main force-resisting building structure represent average loads by relating to the projected building face area normal to the wind, while the local external pressure coefficients for loads on the elements of structure and on the nonstructural components, such as cladding, refer to the force action per square foot. Negative pressure coefficients indicate suction, while the positive ones represent pressure. These factors have been determined in tests for various building types, profiles, and wind directions, and are available in various codes and ANSI A58.1.

The pressure coefficient for the windward wall of a flat-roofed rectangular building is 0.8 regardless of the building proportions, while the suction value on the rear side wall, depending on the plan dimensions, is between -0.2 and -0.5. Most codes give a factor of $0.8 + 0.5 = 1.3$ for typical rectangular prismatic buildings, which reflect the resultant effect of pressure and suction as required for the overall building design. Similarly, the resultant pressure coefficient for hexagonal or octagonal prisms is 1.0, while for round or elliptical tubes it may be assumed to be 0.8. The conversion from the velocity pressure, Equation (f), into an equivalent average static wind pressure, p_z, for the design of the main lateral-force resisting ordinary structure, where gust is not critical, is given by

$$p_z = q_z G_h C_p \simeq q_z C_p \qquad (3.16)$$

Tall, flexible buildings are sensitive to wind gusts that last only about one to two seconds. Therefore, it is not the maximum mean wind speed of 70 to 120 mph with an average period of 35 to 60s that controls the design, but rather the gust speeds. In Equation 3.16, the gust response factor, G_h, accounts for the additional loads due to wind turbulence and dynamic amplification for flexible buildings, but does not include effects of crosswind deflection, vortex shedding, flutter, galloping, etc. The expression for the gust response factor is quite complex, and is beyond the scope of this introduction; the reader may refer to ANSI A58.1 for further clarification. The factor relates to the dynamic properties of the wind and the building, that is, its fundamental period in the direction parallel to the wind and damping properties; it depends on the building size, dimensions, and shape, as well as its surface roughness and exposure conditions. Naturally, "Exposure A" (in large city centers) causes more wind turbulence than the assumed "Exposure C," and thus requires larger gust response factors. According to the codes, a building must be considered sensitive to dynamic effects when it is slender and has a height-to-width ratio greater than five, when it is taller than 400 ft, or when it is prone to wind-excited oscillations due to other reasons. For the preliminary investigation of the ordinary, relatively stiff buildings not of unusual shape or exposure, the effects of turbulence in urban areas may be neglected, particularly since the larger wind pressure values for open country have been used.

While ANSI A58.1 uses a rationalized approach towards wind load analysis and recognizes the dynamic interactions of wind and structure, most of the codes do not. A typical approach of the codes is to obtain the wind pressure by multiplying the velocity pressure q_z by the respective pressure coefficient C_p. For example, for ordinary rectangular prismatic buildings using a mean pressure coefficient of $C_p = 1.3$ according to Equations (f) and 3.16, the effective design pressure is

$$p_z = q_z C_p = 0.00256 C_p V^2 (z/30)^{2/7}$$
$$= 0.00333 V^2 (z/30)^{2/7}$$

(3.17)

This equation is plotted in Figure 3.8 for the maximum wind velocities of 70 mph and 80 mph, which is typical for most of the United States' inland areas. For instance, for the range from 201 to 300 ft, using an average height of 250 ft, the wind pressure for a 70 mph wind is

$$P_{250} = 0.00333(70)^2 (250/30)^{2/7} = 29.90 \text{ psf}$$

Although the typical stepped wind pressure distribution normal to the vertical surfaces along the height of the building is considered critical, wind actions from other directions may have to be considered. For example, the Canadian Code recommends an application of 75 percent of the wind simultaneously to each building face, which

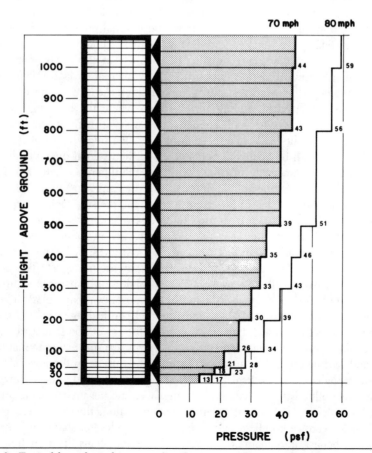

Figure 3.8. *Typical lateral wind pressure distribution according to codes.*

approximately corresponds to 71 percent when resolving the diagonal wind force into components along the two main building axes (i.e., $(P/\sqrt{2})\ 100\%$).

For structures which are hexagonal or octagonal in plan, the projected area (see fig. 2.8) should be used and the values in Figure 3.8 multiplied by 0.8; similarly, for structures that are round or elliptical in plan, the projected area is used and the values multiplied by 0.6. The building shape is an essential consideration; a rectangular building must resist nearly twice as much load as a round building!

Local wall pressures at overhangs, around corners, etc. may be considerably higher than the average resultant wind pressure used for the overall building design, and they are not uniform at a given height, as is apparent from the more concentric patterns of the wind pressure contours in Figure 3.6. This aspect is important for the design of the secondary members, such as cladding and glazing, which transfer the wind loads to the primary wind-force resisting system. In this case, the gust factors must take into account the local effects of turbulence, as well as resonant loading. The period of the cladding is usually below 0.2s, and thus much shorter than gust fluctuations, so that the cladding can be designed as based on equivalent static loads. In addition to the external wind pressures, it must be considered that buildings are not airtight; walls have some permeability and windows may be open. This air movement through the building itself, as well as air movement from heating and air-conditioning systems, cause internal pressures that must also be considered for the curtain and other secondary member design. The internal wind pressure is determined, as is the external one, by multiplying the velocity pressure by the internal pressure coefficient, which is given in codes for various conditions; under these conditions, the gust factor is neglected. The internal wind pressure may be visualized as having a balloon-like effect, with a uniform pressure acting outwardly on all internal surfaces, when the wind enters primarily on the windward side; but openings on the leeward or side walls result in an inward-acting pressure. For the condition where the building walls on the windward and leeward sides each contain more than 30 percent window and door openings, each of the walls may be considered open-sided separately, so that the internal pressure can be treated as being equal to the external one. In tall buildings with fixed glazing, especially in colder climates, the difference between the outside wind pressure on the facade and the interior pressure causes the air to move from the area of high pressure to that of lower pressure, as the pressure tends to equalize; this condition creates air infiltration on the windward side and exfiltration on the leeward side. Pressure differences, however, are not only caused by the wind effects, but also by the temperature differences that cause wind themselves, or by the so-called *stack or chimney effect* (see fig. 3.6, center), as well as by the *fan effect*. The amount of pressure due to thermal differences, when it is colder outside than inside, is proportional to the height of the "stack"; hence, this warm air buoyancy is particularly a behavioral phenomenon of high-rise buildings, and extremely critical for very tall skyscrapers, which tend to act like flues, drawing wind from the lower floors upward. As for the chimney draft, during winter hot air rises as it warms, causing exfiltration at the upper floors and infiltration of cold air drawn in from the outside in the lower floors. This generates a high-velocity vertical air flow and critical increased positive outward pressures at the top levels in the leeward building face, and negative pressure (or suction) on the lower floors. The route of the air movement is along the path of least resistance, especially the vertical building shafts (i.e., stairwells, elevator shafts, and mechanical shafts) that behave in a manner similar to ventilation shafts. The air flow must be controlled by sealing the facade walls and all openings in the shafts, together with horizontal separation (airlocks) for very tall buildings, and by controlling the building entrance, perhaps with a double entry. During summer, air conditioning causes similar pressures, but the flow directions are reversed. The outward pressure due to the fan effect may be assumed uniformly over the envelope, as for a balloon. In pressurized tall buildings, the stack effect due to the vertical air circulation and the air conditioning may cause an increase of as much as 5 psf within the building; the true evaluation of the internal pressure is, at this time, still in an early developmental stage. For the preliminary design of curtains and exterior panels under typical loading conditions, the wind pressure values of Figure 3.8 may be used.

Human tolerance to wind action, both inside and outside buildings, has become an increasingly important factor in the design of high-rise buildings. Excessive lateral sway and oscillations, which a building's structural system may be able to withstand, must still be reduced to the acceptable limits for human comfort; besides building strength, the building stiffness and damping must be considered. Some inhabitants in existing buildings have experienced motion sickness caused by sway; people feel the movement and sense the twisting of the building. Humans have been found to be particularly sensitive to acceleration, rather than displacement or velocity. In some restaurants atop tall buildings, wines are not clear when served because wind action has caused the sediment to become stirred up. At times, minor damage to furniture and equipment has occurred; strange creaking sounds from shaking elevator shafts, or noises from elevators bumping against the sides of their shafts, and air leakage around windows have been noticed; the unpleasant whistling of the wind around the sides of the building itself has been heard. In several buildings in the 40- to 50-story range in New York City, excessive lateral sway, including torsional motion and floor tilt (aside from noise) have made it impossible for people to work at their desks; employees are regularly excused from work during high wind storms. Strange occurrences observed outside high-rise buildings also cause discomfort and annoyance to both inhabitants and neighbors. Changes in the local wind character, such as vortex currents formed in the wake of buildings, have torn wash from clotheslines, damaged gardens, wrenched opened doors off automobiles, and scattered debris through the air. Some building occupants find it impossible to use balconies except on totally calm days, because of constantly turbulent winds on the building face. Worse yet, windows can be smashed or sucked from buildings, causing serious injury or death to people walking below. The list of examples can go on and on. What is important, however, is the need to recognize that a concern for human tolerance and the activities to be performed in and around the building must be a major factor in the design of today's high-rise buildings.

SEISMIC LOADS

Earthquakes are among the most awesome of natural forces; the Xian quake of 1556 in China killed 830,000 people, by far the most destructive quake ever. Earthquakes occur suddenly, without warning, and within 10–20 seconds can turn cities into wastelands, make islands disappear, alter the flow of rivers, and give birth to new land and lakes. Building designers in the United States often automatically relate earthquakes to California and Alaska; they do not realize that the earthquakes of Charleston,

South Carolina (1886) and of New Madrid, Missouri, near Memphis (1811, 1812) were of nearly the same magnitude as the most severe one in San Francisco (1906), and that more than a third of the U.S. population lives in areas of high to moderate seismic risk. Not only the large western cities of Los Angeles and San Francisco, but also metropolitan areas of Buffalo, Providence, Boston, Charleston, Memphis, St. Louis, Salt Lake City, Seattle, and Anchorage are regions prone to major earthquakes. Severe earthquakes have occurred in many other parts of the world, such as around the rim of the Pacific Ocean (e.g., Chile, Peru, Nicaragua, Guatemala, Mexico, California, Japan, New Zealand), along the Mediterranean Sea (e.g., Morocco, Greece, Italy, Yugoslavia, Romania, Turkey), around the Himalayan Mountains (e.g., China, India) and in Iran. Building designers must be acquainted with the effects of quakes, so they can make buildings earthquake-resistant and safeguard life.

According to the theory of plate tectonics, the earth's crust consists of separate plates that float upon the earth's molten interior. Each of the plates moves (creeps) a few inches every year. At points of convergence, the plates want to slide past each other, as (for instance) along the San Andreas fault, where the Pacific plate and the North American plate move in the same northwestward direction but at different rates. If the plates are prevented from doing so at certain locations, by friction or by being locked into each other, elastic strain is stored and accumulates until the forces can no longer be resisted by the material. When the capacity is exceeded, sudden rupture and slippage occur, which may cause the upper crust of the earth to fracture and form a fault. This abrupt release of strain energy results in complex vibrations propagating in high speeds from the source (focus or hypocenter) in all directions through the earth and along its surface, reaching a given building at different time periods with different velocities, from different directions (Fig. 3.9). The more significant seismic wave types are the faster longitudinal *P-waves* (primary waves, compressional waves, push-pull waves) which compress the earth in front and move building foundations back and forth in the direction of travel. The later and slower transversal *S-waves* (shear or secondary waves) oscillate in a plane perpendicular to the direction of propagation and tend to move the building foundations up and down and side to side at right angles to the P-waves. The P- and S-waves are body waves, travelling through the interior of the earth, keeping in mind that only compressional waves are possible through gases and liquids. In contrast, the *Q-waves* (Love waves), with no vertical movement, and the *R-waves* (Rayleigh waves), with both vertical and horizontal movement, are relatively slow-moving surface waves that propagate along the earth's crust.

Since the high-frequency components of the seismic waves, in contrast to the low frequency, larger-period waves, tend to weaken rapidly as they propagate away from the source, short-period low-rise and massive buildings tend to be excited more near the epicenter, while long-period tall buildings are activated more by the low-frequency waves that are transmitted over larger distances. Naturally, the nature of the regional geology that is the transmission path to the site influences the behavior of the seismic waves. Usually, the most destructive part of an earthquake is over in 10 to 20 seconds.

Earthquakes are broadly classified as *shallow-focus earthquakes*, where the focus is located near the epicenter, thus causing more intense local effects, and *deep-focus earthquakes*, effecting much larger areas. Most western earthquakes are generated along the plate boundaries near the surface. However, the potential earthquakes east of the Rocky Mountains are of the midplate type, originating deep in the earth. They do not occur along recent active fault lines; in this case, the weight of the overlying rock,

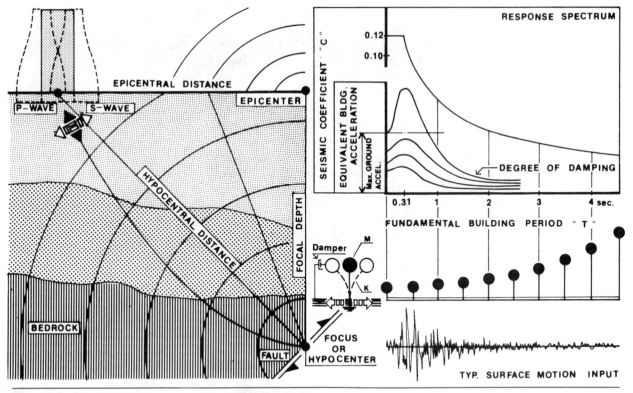

Figure 3.9. *Seismic action.*

sand, and clay squeezes the understructure of the earth. For example, the area in southeastern Missouri that represents a critical seismic zone around New Madrid is located over an ancient subterranean fracture called the New Madrid fault zone. Eastern earthquakes are less frequent, but, because of the more ancient geology with fewer faults to slow the travel of the waves relatively quickly, shock waves travel much further than in the West, and hence can damage a much wider area, thus making them potentially far more destructive. Recently it has been discovered that in California not only the familiar *surface faults* exist but also a network of *hidden faults* four to ten miles beneath the surface. These hidden faults do not break the earth's crust but form folds and eventually hills causing earthquakes during this process.

A major percentage of earthquakes are of tectonic origin, but a second cause may be due to volcanic eruptions or subterranean movements of magma. Besides the natural tectonic and volcanic earthquakes, there may be artificially induced ones, such as due to underground nuclear explosions and large water reservoirs. The primary effects of earthquakes causing possible building damage are:

- Ground rupture in the fault zone.
- Ground failure due to landslides, where ground displaces horizontally and/or vertically without ground rupture; mud slides, avalanches, ground settlement; shaking of ground resulting in loss of bearing capacity due to liquefaction, where saturated granular soil is transformed from a solid state into a liquefied state due to porewater pressure, causing the soil to behave like quicksand.

- Tsunamis, which are large sea waves generated by the sudden displacement of land at the ocean bottom.
- Ground shaking: its effect is influenced by the magnitude of the energy released, the location of the focus, its duration, the geology of the site, and building characteristics.

Among the secondary and often more critical effects of earthquakes are fire, disease, looting, explosion, flooding, and disruption of economic and social life.

The remaining discussion of earthquake-resistant building design will be concerned only with ground shaking as the source of damage, since the chances for ground rupture under a building are extremely small. The structural design for earthquake forces, because of their random character, cannot be considered as an exact engineering science; some engineers even claim it to be more of an art. The design is based on assumptions rather than on precise data; the prediction of the ground motion, together with the response of the building riding on the earth along three directions may be impossible to determine accurately.

A network of seismic stations is distributed across the United States in regions having a past history of earthquakes to record and to attempt to predict earthquake motions. These stations are equipped with instruments such as seismographs, strong-motion accelerographs, and tiltmeters. This monitoring grid can locate the focus of an earthquake, that is, its location deep in the earth's crust where the rupture originated, and the epicenter, which lies on the earth's surface directly above the focus. The network also records the point of highest wave intensity, which is generally centered around the fault, and thus can be far away from the epicenter. The magnitude of the earthquake is calculated from the seismogram, which is the response of the seismograph to the motion of the ground, recording a zigzag line reflecting the varying amplitude of the vibrations. The Chinese not only use seismological instruments for the prediction of earthquakes, but also pay close attention to the oddities of animal behavior.

Earthquakes are classified either according to the magnitude of the energy they release or according to their intensity, that is, destructiveness. The Richter scale, invented by Charles F. Richter of the California Institute of Technology in 1935, is a measure of the energy released at the focus. It ranges from 3 to 9, based on a logarithmic scale, where each unit increase reflects an increase of about 32 times more energy. Comparing the energy release of the Hiroshima atom bomb (about 6.4 on the Richter scale) with that of the 1964 Alaska earthquake of 8.4, shows that about 1000 (i.e., 32 × 32) times more energy was freed by the earthquake. The largest known earthquake is 8.9 (Colombia, 1906); the 1906 San Francisco earthquake measured 8.25, and the 1812 New Madrid one was 8.2. Earthquakes above 6 on the Richter Scale are considered severe, while the ones between 4 and 5 are considered moderate; a magnitude of 8 or higher is classified as a great earthquake. In one of the most devastating quakes in recent history, nearly a quarter of a million people died in the 1976 Tangshan quake, of Richter magnitude 7.8, in China. The Richter scale does not concern itself with the effect of the earthquake. An earthquake of Richter magnitude 6 may be far more destructive when it hits a densely populated region directly, than one of magnitude 8, with its focus far away from inhabited areas. Nearly five times more people were killed in the Armenian earthquake of 1988, which registered 6.9 on the Richter scale, as compared with Mexico City's 8.1 quake of 1985, where 10,000 people died.

The Modified Mercalli Intensity Scale (MMI), as initially developed by Guiseppe Mercalli (1850–1914) for use in Italy, and then modified by H. O. Wood and F. Neumann in 1931 for the conditions in the United States, is of a subjective nature. It describes the degree of damage based on 12 intensity divisions. The Seismic Risk Map (see fig. 3.10) is used by the Uniform Building Code and correlated with the MMI Scale. Buildings should be able to resist minor earthquakes without damage, moderate earthquakes without structural damage, but with some nonstructural damage, and major earthquakes without collapse, but with some structural, as well as non-structural, damage.

For the sake of simplicity, the complex, random ground vibrations can be visualized as known horizontal movements travelling back and forth. Ignored are the vertical vibrational displacements, since the building is already designed for gravity in this direction. As the earth abruptly accelerates horizontally in one direction, taking the building foundation along, but leaving the portion of the building above the ground behind, it causes lateral inertia forces to act. This phenomenon is similar to that experienced by a person travelling in a car that suddenly increases in speed, or the vertical inertia force experienced when an elevator abruptly rises. Assuming initially that the building and its foundation were rigid, by ignoring the properties of the structure, as well as the particular character of the exciting motion, which includes the effect of the site geology, then the acceleration of the building would be equal to the ground acceleration. The lateral inertial forces would then, according to Newton's second law, be the product of the building mass, M, and the ground acceleration, a, where the mass, in turn, is equal to the building weight, W, divided by the acceleration of gravity, $g = 32 \text{ ft/s}^2$.

$$F = Ma = W(a/g) = WC \qquad (2.10)$$

Letting $a/g = C$ (the seismic coefficient), results in the ground acceleration expressed as a percentage of g: $a = C(g)$. For example, a ground acceleration of 0.2g is equal to $0.2(32) = 6.4 \text{ ft/sec}^2$, or the lateral inertia force is equal to 20 percent of the weight of the rigid building. Ground shaking accelerations vary from 0.2g to 0.3g for a moderate earthquake, and greater than 0.4g for a severe earthquake.

To obtain the magnitude of the inertial forces, the building has been assumed to be rigid and to rock and heave similar to a ship at sea, which is obviously not true. The inertia of the building mass resists ground movement and causes the building to deform, and hence activates the building stiffness; the random shaking of the base results in a series of complex oscillations of the building. However, it has been found that the elastic response of multistory buildings is largely determined by the first or fundamental mode of vibration, so that the building can be treated, for preliminary design purposes, as a single-degree-of-freedom (SDF) system or cantilever pendulum. To evaluate the effect of the building stiffness upon the magnitude of the lateral inertia force, or building acceleration, several simple regular buildings of decreasing stiffness are considered as inverted pendulums (see fig. 3.9, right), visualizing an effective mass for each case. The increase of the pendulums in height indicates the increase of flexibility, and also the increase in the natural period, T, of the building, where the natural period represents the time it takes for the structure to swing back and forth and to go through one full cycle of free motion. The pendulums selected show the typical fundamental natural periods of 0.1 to 0.4 seconds for low-rise buildings, and roughly 1 to 3 seconds for ordinary high-rise buildings. An average damping value, as provided by the natural damping, is included. The structure is assumed to behave linearly, although in reality some members may be stressed beyond their elastic limit, resulting in a nonlinear response with permanent deformations.

The pendulums are attached to a movable base and are shaken back and forth under a simulated ground acceleration, which is based on the maximum one experienced as obtained from an actual or simulated earthquake accelerogram. The record most often used in the United States is the one for the EL Centro earthquake of 1940, with a maximum acceleration of about 0.3g. If the maximum response (i.e., acceleration, absolute displacement, or lateral inertia force) of each of the buildings is plotted, a so-called *response spectrum* is formed for the given seismic movement, keeping in mind that every site actually has its own response spectrum. The one shown in Figure 3.9 represents the plot of the seismic coefficient, C, as a function of the fundamental period of the building, T, according to the UBC formula 12-2 (1985 edition) or Equation 3.19. The following conclusions may be drawn from the selected response spectrum.

- A rigid (infinitely stiff) building will move together with the ground without deflecting (T = O), so that its acceleration is equal to the one of the ground.
- A relatively stiff building with a natural period of about 0.3 seconds reaches a lateral acceleration much larger than the one of the ground. This is the state where the soil period coincides with the one of the building, thus theoretically resulting in resonance (infinitely large forces) if there were no damping present. In other words, the ground motion is amplified as it moves through the building, reaching an acceleration on the upper floors much higher than that at the ground.
- A flexible building with a natural period, for example, larger than 1.4 seconds shows an acceleration of less than the ground. This clearly indicates that a tall, flexible structure may be advantageous on sites where ground motion is of a short period. However, it must be kept in mind that these most intense waves occur during the first few seconds of the earthquake and that the later long-period waves may come close to the natural period of the building, and thus be much more critical.

It may be concluded, according to UBC, that flexible buildings with long periods of vibration attract less lateral force than stiff buildings with short periods.

The amount of damping in buildings depends on the degree to which the building will dissipate motion as, for instance, through conversion to heat due to friction, which is a function of the number of building components and how they are connected, as well as the types of materials and the finish. Different damping produces different response spectra where the shape of the curves, however, remains the same. The higher damping will yield less acceleration, that is, a flatter curve (see fig. 3.9).

The assumption in the response spectrum that the behavior of a multistory building can simply be represented by its fundamental period as a single-degree-of-freedom system is oversimplistic, especially for flexible structures. Visualize a high-rise structure to be modeled as lumped masses at each floor level along a vertical column (e.g., fig. 3.11), where a typical lumped mass consists of the floor and the one-story high columns, walls, and partitions. When this lumped-mass, multiple-degree system is excited by ground vibrations, many different types of motion, with their corresponding deflected shapes, are possible. There will be many degrees of freedom (i.e., patterns of deformation), where a natural period is associated with each mode.

However, for stiff multistory buildings, it has been found that the first or fundamental period is the dominant one, and that even for more flexible buildings it contributes the largest influence.

The principle of the response spectrum is used by the major building codes. Its shape is approximated by the Uniform Building Code (1985 edition), as mentioned before, and defined mathematically by the seismic coefficient, C. In Figure 3.9, the seismic coefficient, C, has been multiplied by the soil factor, S, to take the possible increase of the foundation effect into account.

In addition, material behavior must be considered, for severe earthquakes. When some members (e.g., girders, beams) go through elasto-plastic deformations under repeated cyclic loading by alternating from the tensile to the compressive range, the stress-strain diagram for the material transverses a hysteresis loop. In this case, the load-bearing capacity of ductile steel is not reduced, in contrast to more brittle materials, where it diminishes rapidly (i.e., a degrading hysteresis curve) and leads to failure.

The code further modifies Equation 2.10 by considering, in addition to the building weight and the dynamic characteristics of the building as expressed in the fundamental period, the seismic risk zone, the type of structure, the geology of the site, and the importance of the building, keeping in mind that dynamic analysis is required, unless the building is of regular shape and reasonably uniform stiffness. Hence, the minimum lateral force for which a regular structure has to be designed is given in terms of the equivalent static base shear, as follows:

$$V = ZIKCSW \qquad (3.18)$$

where: V = the total lateral force or shear at the base

Z = seismic probability zone factor (see seismic risk map, Fig. 3.10)

= 0   in zone 0 (no damage)

= 3/16 in zone 1 (minor damage)

= 3/8   in zone 2 (moderate damage)

= 3/4   in zone 3 (major damage)

= 1   in zone 4 (major damage, close to major fault systems)

I = occupancy importance factor

= 1.5 for essential facilities, such as hospitals and fire and police stations

= 1.25 for any building where the primary occupancy is for assembly use for more than 300 persons in one room

= 1.00 for all other cases

K = building type factor (Table 3.4)

C = seismic coefficient (Equation 3.19)

S = soil–structure interaction factor as determined according to Equation 3.23 or in accordance with the following values:

= 1.0 for rocklike formations, or stiff soil conditions overlaying rock at a depth of less than 200 ft

= 1.2 for deep cohesionless or stiff clay soil conditions overlaying rock at a depth greater than 200 ft

= 1.5 for soft to medium-stiff clays and sands 30 ft or more deep, or if the soil profile is not known

W = the total dead load and appropriate portions of the live load

The seismic coefficient (Fig. 3.9) is taken as:

$$C = \frac{1}{15\sqrt{T}} \leqslant 0.12, \quad \text{or} \quad T \geqslant 0.3 \text{ s} \tag{3.19}$$

$$CS \leqslant 0.14 \tag{3.20}$$

In Equation 3.19, the fundamental natural period of vibration, T, of the building is in seconds in the direction under consideration. The value of T can be computed or measured by sophisticated methods; at the stage of preliminary design, however, T cannot be known. On the basis of measurements made on many existing structures, an approximation of T has been formulated:

$$T = \frac{0.05 h_n}{\sqrt{D}} \tag{3.21}$$

where: h_n = building height (ft) above base

   D = dimension of the building (ft) in a direction parallel to the applied force

In the case of a moment-resisting space frame that resists 100 percent of lateral forces, and is not connected to more rigid elements, the code permits 0.1s per floor.

$$T = 0.10N \tag{3.22}$$

where: N = the total number of stories above grade in the main portion
   of the structure

A fast approximation of the base shear can be conservatively obtained when the natural period of the building and the site conditions are not known, by letting CS = 0.14.

$$V = ZIKCSW = 0.14ZIKW \tag{3.18a}$$

The influence of the building structure, with respect to lateral force action, is taken into account by the framing factor (ductility factor, horizontal force factor, building type factor), K. It ranges from 0.67 to 1.33 for buildings, based on the past performance of the structure type, evaluated empirically, and reflecting the differences in the ductility or energy-dissipation capacity, as well as the degree of structural redundancy.

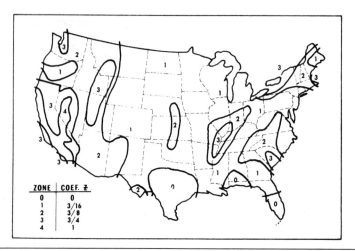

ZONE	COEF. Z
0	0
1	3/16
2	3/8
3	3/4
4	1

Figure 3.10. *Seismic zone map.*

A low factor is assigned to the ductile, continuous rigid frame, where members and their joineries have the required strength and deformation capacity. Ductility is the ability of members within the structure to go through a number of inelastic cycles of deformation without a significant loss of strength, thus clearly establishing the high redundancy of the structure and its capacity for carrying overloads. A factor nearly twice as high is given to bearing-wall and all concentrically braced-frame structures, which can be visualized as stiff boxes responding in shorter periods and thus causing larger lateral forces. They also lack ductile behavior because of their inability to deform much. Brittle materials, such as masonry and concrete, will form cracks and lose their load-carrying capacity rather than redistribute stresses, should they not be properly reinforced; they are incapable of deforming much beyond the elastic range and lack reserve strength. Therefore, structures that lack ductility must have more strength, and they must remain within the elastic range of the materials, allowing no inelastic deformations.

In buildings where the lateral-force resistance is provided by shear walls or braced frames, together with rigid ductile frames capable of resisting at least 25 percent of the lateral force, a frame factor of $K = 0.8$ is assigned (Table 3.4). However, one should not overlook the fact that rigid steel frames can also fail in a brittle fashion if the connections are not properly designed. The trade-off between ductility and stiffness for tall buildings is critical. Seismic design calls for a ductile structure with much redundancy, while wind design requires stiffness for occupant comfort and to control damage to nonstructural components.

The soil or site factor, S, takes into account the site-structure quasi-resonance effects, which includes the effect of the travel path of the earth vibrations from the source to the site. It is equal to

$$S = 1.0 + \frac{T}{T_s} - 0.5\left[\frac{T}{T_s}\right]^2, \qquad \text{for } \frac{T}{T_s} \leq 1.0 \qquad (3.23a)$$

$$S = 1.2 + 0.6\left[\frac{T}{T_s}\right] - 0.3\left[\frac{T}{T_s}\right]^2, \qquad \text{for } \frac{T}{T_s} \geq 1.0 \qquad (3.23b)$$

where: T_s = site period(s) as established from geotechnical data but, $0.5 \leq T_s \leq 2.5$

$T \geq 0.3$ s

As the site period and structure period approach each other, they also approach a quasi-resonance, with a corresponding increase of seismic forces. Therefore, the maximum amplification is obtained from the above equations when

$$T/T_s = 1, \text{ then } S = 1.5$$

The dynamic characteristics of a site are determined by the depth and type of the various layers of soil or rock over the underlying bedrock. Sites with shallow, dense, stiff deposits tend to have short fundamental periods, but which are larger than the ones of the underlying bedrock, particularly when the incoming motions are of low intensity. However, sites with deep flexible soil layers tend to have long fundamental periods, where the bedrock accelerations are amplified for low intensity motions but reduced for high-intensity motions, since the deep soft soil cannot vibrate rapidly. The typical fundamental periods of vibration of a site vary from about 0.1 seconds for stiff sites to 5 seconds for a flexible soft ground, with the range of 0.25 to 1.0 second being more common.

Table 3.4. Horizontal Force Factor (K) for Buildings

Type of Building Structure	K-Value
All systems not otherwise classified	1.00
Building with a box system: lateral forces resisted by shear walls or braced frames.	1.33
Buildings with a dual bracing system of ductile moment-resisting space frame plus shear walls or braced frames meeting these criteria: <ul><li>The frame and shear walls shall resist lateral forces in accordance with their relative rigidities.</li><li>Shear walls acting alone shall be able to resist total lateral force.</li><li>The frame acting alone shall be able to resist 25% of total lateral force.</li></ul>	0.80
Ductile moment-resisting space-frame, capable of resisting total lateral force.	0.67

The Mexico City earthquake of 1985, with a Richter magnitude of 8.1 (where close to 10,000 people were killed), had a lateral ground acceleration of 0.18g, which is not extreme for California conditions. But the severity of the damage was partially due to a double resonance phenomenon of earthquake-ground and ground-building, as well as the long duration of the earthquake, of a period of 80 to 110 seconds. In this instance, the ground motion period of about 2 seconds coincided with the natural period of mid-rise buildings.

The occupancy importance factor, I, takes into account that not only must life safety be guaranteed, but also that damage control must be considered for certain situations. Facilities that provide services that are essential after an earthquake, such as hospitals, fire stations, power plants, as well as police and communication stations, must be protected.

It may be concluded, according to Equation 3.18a, that the maximum lateral inertia forces for ordinary, but relatively stiff buildings in a highly seismic region range from approximately 9 percent of the building weight (V = 0.09W), for a more flexible rigid frame structure, to 19 percent for a stiff shear-wall structure. Notice that, in this context, the height of the building has no direct effect on the magnitude of the lateral forces, in contrast to the design for wind action. On the other hand, for a flexible 30-story rigid frame structure, the lateral inertial force is only about 4 percent of the building weight. For preliminary estimation purposes, the magnitude of the lateral seismic forces in major damage seismic zones may be assumed to be in the range from about 5 percent, for tall flexible frame structures, to 20 percent, for stiff bearing-wall buildings.

Lateral Distribution of Base Shear

The formula V = ZIKCSW does not indicate how the shear force is distributed throughout the height of the structure. The shear force, at any level, depends on how the structure deforms, that is, on the mass at that level and the amplitude of oscillation, which may be assumed to vary linearly with the height of the building. Earthquake forces deflect a structure into certain shapes, known as the natural modes of vibration.

Only the most important first three modes are shown in Figure 3.11, but it must be realized that a high-rise building actually is a multiple-degree-of-freedom system with many possible patterns of deformation. For example, a 30-story rigid frame building has a first period of vibration of roughly $T_1 = 0.1N = 0.1(30) = 3$ s and, according to the linear approximation of the first three modes of vibration in Figure 3.11, has the following second and third periods of approximately $T_2 = T_1/3 = 3/3 = 1$ s, and $T_3 = T_1/5 = 3/5 = 0.6$ s. In any case, the first or fundamental mode, as exemplified by the pendulum, has still been found to contribute the largest influence, especially for stiff or short-period buildings responding more abruptly. Flexible long-period buildings respond in slower, longer, and more complex movements; in this case, the higher modes of vibration indicate the whiplash effect, which is taken into account by the concentrated load F_t at the top of the building (Fig. 3.11). Related to the shape of each mode is a certain distribution of lateral forces. As long as there is no large inelastic deformation, instantaneous lateral forces are found by superposition of the forces resulting from each vibration mode; sometimes the forces add and some-times they cancel one another. The resulting maximum shear envelope (Fig. 3.11) can be visualized as being generated by a triangular load. The Uniform Building Code uses a triangular lateral load configuration for a building with uniform mass distribution along its height, that is, for a building structure of regular rectangular shape with equal floor weights and heights and no vertical and horizontal irregularities in geometry, stiffness, strength and mass. For the effect of the mass arrangement of other building shapes on the lateral force distribution refer to Figure 2.8.

Recognizing the whiplash effect of flexible buildings, the code places part of the total lateral base shear, V, as a concentrated load, F_t, at the top of the structure, while the balance $(V - F_t)$ is to be distributed in a triangular fashion over the entire building height. The top load F_t is only present if the fundamental period of vibration $T > 0.7$ seconds, but does not have to be larger than $T = 3.57$ seconds. It is equal to

$$F_t = 0.07TV \leq 0.25V \qquad (3.24)$$

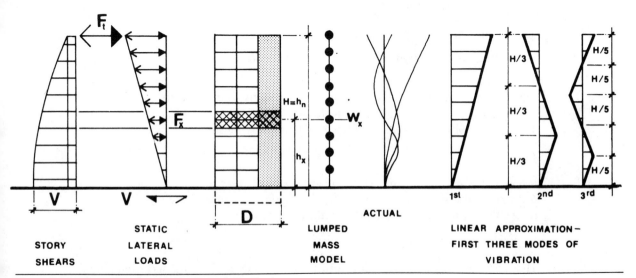

Figure 3.11. *Equivalent lateral seismic load distribution.*

According to UBC, the balance of $(V - F_t)$ is distributed over the entire building height, generally as concentrated loads at the floor levels, in a triangular fashion, as discussed above.

If the weight, w, is taken to be constant for every floor level, then the force, F, is proportional to the height, h (see fig. 3.11),

$$\frac{F_1}{h_1} = \frac{F_2}{h_2} = \frac{F_3}{h_3} = \ldots = \frac{F_x}{h_x}$$

Horizontal equilibrium yields

$$V - F_t = F_1 + F_2 + \ldots + F_n$$

However, $F_1 = h_1(F_x/h_x)$, $F_2 = h_2(F_x/h_x)$, and so forth, so that

$$V - F_t = \frac{F_x}{h_x}(h_1 + h_2 + \ldots + h_n)$$

Solving for F_x yields

$$F_x = \frac{(V - F_t)h_x}{h_1 + h_2 + \ldots + h_n} = \frac{(V - F_t)h_x}{\sum\limits_{i=1}^{n} h_i} \tag{3.25a}$$

Taking into account that the weight at the floor levels may not be constant, then the magnitude of the distributed forces, F_x, is given by

$$F_x = \frac{(V - F_t)w_x h_x}{\sum\limits_{i=1}^{n} w_i h_i} \tag{3.25b}$$

where: w_i, w_x = that portion of W which is located at or is assigned to level i or x, respectively

h_i, h_x = the height (in feet) above the base to level i, n, or x, respectively. The level n is the uppermost level in the main portion of the structure.

Some Additional Code Requirements

Structures with Irregular Shapes or Framing Systems

The equivalent static lateral force method is only applicable to regular building shapes with no vertical and horizontal irregularities in geometry, stiffness, strength and mass. For irregular shapes, the distribution of lateral forces in these structures must be determined, considering the dynamic characteristics of the structure.

Drift

Lateral deflections or drift of a story, relative to its adjacent stories, shall not exceed 0.005 times the story height (0.0025 for unreinforced masonry buildings), unless it can be demonstrated that greater drift can be tolerated.

Accidental Torsion

The shear-resisting elements shall be capable of not only resisting the shear resulting from the horizontal torsion due to the eccentricity between the center of rigidity and the center of mass, but also should resist a torsional moment assumed to be equivalent to the story shear, acting with an eccentricity, e, of not less than 5 percent of the maximum building dimension, D_{max}, at that level due to mass displacement each way.

$$e = 0.05D_{max} \tag{3.26}$$

Overturning

At any floor level, a building shall be designed to resist overturning caused by wind or earthquake, whichever governs. Moments can be calculated from the triangular load distribution formula for preliminary design purposes. The overturning moment at the base of a building is

$$M = F_t h_n + \sum_{i=1}^{n} F_i h_i \tag{3.27}$$

Lateral Force on Elements of Structures and Nonstructural Components

The failure of appendages in buildings, such as parapets, veneers, fixtures, and appliances, represents a serious problem to the safety of people.

The Uniform Building Code (1985 edition) further specifies that parts or portions of structures and their anchorages shall be designed for lateral forces, F_p, proportional to their weight and function, in accordance with the following formula.

$$F_p = ZIC_p W_p \tag{3.28}$$

where: F_p = lateral force on the part being considered
    Z = zonal probability factor
    W_p = weight of the part being considered
    C_p = coefficient as determined from Table 3.5
    I = occupancy importance factor

Table 3.5. Horizontal Force Factor (C_p) for Elements of Structures and Nonstructural Components

Part of Structure	Value of C_p
Interior and exterior bearing and nonbearing walls	0.30 (force normal to surface)
Cantilever parapet walls	0.80 (force normal to surface)
Ornamentations, appendages	0.80 (force in any direction)
Connections for prefabricated structural elements other than walls	0.30 (force in any direction)
Other parts	

Buildings Taller than 160 Feet in Seismic Zones 3 and 4

Such buildings shall have a ductile moment-resisting space frame capable of resisting at least 25 percent of the required seismic force of the total structure. All buildings designed with a K-factor of 0.67 or 0.80 shall have a ductile moment-resisting space frame of structural steel or cast-in-place reinforced concrete. Reinforced concrete frames must meet certain special code requirements (see the section in Chapter 7 on "Skeleton Buildings").

Other Design Requirements

- All portions of structures shall be designed and constructed to act as a unit in resisting horizontal forces, unless separated structurally by a distance sufficient to avoid contact under deflection from seismic or wind forces.
- In seismic zones 2, 3, and 4 (see fig. 3.10), masonry or concrete elements resisting seismic forces must be reinforced.
- Concrete or masonry walls shall be anchored to all floors and roofs that provide lateral support for the wall.
- Individual pile caps and caissons of every building shall be connected by ties, each of which can carry, by tension and compression, a minimum horizontal force equal to 10 percent of the larger pile cap or caisson loading.

Conclusions

The following conclusions can be drawn from the base shear formula: Heavy buildings attract more force than light ones. An increase of stiffness, for the sake of reduction of deflections and vibrations, also results in larger inertia forces for sites with shallow dense deposits that efficiently transmit short periods of vibration. On the other hand, a tall flexible building on a deep layer of flexible soft soil may experience higher lateral forces than a stiff building, since the soil cannot vibrate rapidly and transfer high frequency waves from the bedrock beneath to the building. A rigid ductile frame may be preferable to a braced-frame or shear-wall structure, but it is quite flexible and must endure large deflections in severe earthquakes, which may cause extensive non-structural damage of partitions, curtain panels, ceilings, etc. Stiff buildings, in contrast, resist deflection (and thus nonstructural damage), particularly in moderate earthquakes, but do require more structural material because of a lack of ductility. The medium-stiff building may be the best solution, providing the safety of the ductile frame and the rigidity of the shear walls.

The mass and stiffness of a building should be selected to yield a period different from the one of the ground motion, in order to prevent the building response from being amplified. The mass and stiffness of a building should be uniformly distributed, and any discontinuities of structural parts should be avoided; structural continuity is essential in case individual members fail. To prevent torsion, the centroid of the mass should not deviate from the centroid of rigidity (see the section in Chapter 5 on "The Distribution of Lateral Forces to the Vertical Resisting Structures")—when an unsymmetrical building bends, it will twist. Ideally, a building should have two axes of symmetry, so that only accidental torsion may be present. It is apparent that irregular and odd building shapes, or the lack of symmetry of the structure, will cause major torsional eccentricities and will amplify deflections at building ends furthest away from the center of rigidity. Abrupt changes of structure, with a sudden change of stiffness, causes stress concentrations, resulting in potential failure. For example, these stress concentrations occur at the intersection of different building parts; it may be better to separate buildings at junction points, so they can behave as individual building blocks. Further, a building with several supports has a higher redundancy than one that is supported only by a single core, which is more vulnerable to failure. Since earthquake motion causes rapid stress reversals in structural members, the structure must be able to withstand the effects of fatigue. Large permanent deformations can be controlled and concentrated at a story level by using the soft-story concept (see fig. 3.21). Also, the use of special damping devices like shock absorbers and mass dampers may be considered.

Seismic design is still based largely on trial and error. Researchers are constantly trying to find more accurate ways of predicting, and new ways of responding to, earthquakes. The equivalent static lateral force approach used by the codes can obviously only be considered as a rough first approximation and may give reasonable results for simple regular buildings. The coefficients Z, I, and K are not based on a quantitative evaluation but on a qualitative judgement, and only the coefficients C and S attempt, vaguely, to include dynamic considerations which, however, are based on records of earthquake measurements taken far away from the location of high-intensity waves; data for actual ground motions for building sites are not known. The framing factor for box structures does not take into account all of the variations of shear walls and their combinations, and neither does it take into account different construction materials. The frequency of building vibration is affected by nonstructural elements, such as nonbearing partitions that increase a building's stiffness. The response spectrum neglects the duration of the force action, and the seismic coefficient has largely underestimated the spectrum values of several earthquakes. The structure is assumed to behave elastically under dynamic loading; under strong earthquakes, however, a building will deform partially inelastically, dissipating part of the seismic force action. In consequence, buildings are often observed to resist forces far greater than their rated capacity on the basis of elastic analysis. The reserve strength provided by inelastic deformation should allow the designer to lower the seismic coefficient. However, because of the unpredictable nature of earthquakes, it is essential to provide redundancy, that is, a backup system, in case failure should occur.

The following earthquake example problems explain the quasi-static approach of the codes.

Example 3.3

The seven-story building in Figure 3.12 is located in seismic zone 2. The lateral forces are resisted by rigid frames along the perimeter of the building; the interior columns carry only gravity loads. The characteristic site period has been determined as $T_s = 0.8s$. The following loading conditions are used for this preliminary investigation:

Floor weight, including girders, columns, and spray-on fire proofing: 85 psf
Curtain wall, including column and spandrel covers: 15 psf
Uniform wind load for a wind velocity of 70 mph: 21 psf

The total building weight is

$$W = 7[0.085(90 \times 75) + 0.015(90 + 75)2(11.5)] = 4414.73 \text{ k}$$

The fundamental period for a rigid frame is

$$T = 0.1N = 0.1(7) = 0.7s$$

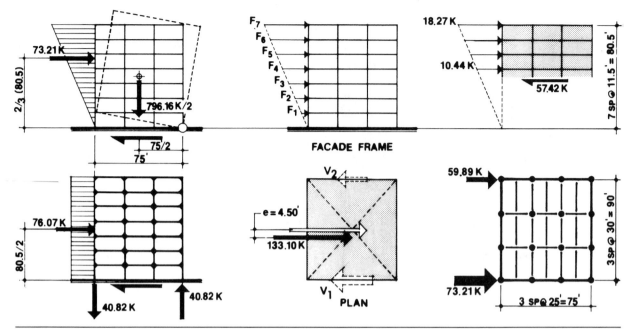

Figure 3.12. *The response of the building structure in Example 3.3 to lateral force action.*

The ratio of the building-to-site period is

$$T/T_s = 0.7/0.8 = 0.88 \leqslant 1.0$$

The soil factor is

$$S = 1.0 + (T/T_s) - 0.5(T/T_s)^2 = 1.0 + 0.88 - 0.5(0.88)^2 \simeq 1.5 \quad (3.23a)$$

The seismic coefficient is

$$C = \frac{1}{15\sqrt{T}} = \frac{1}{15\sqrt{0.7}} = 0.0797 < 0.12 \quad (3.19)$$

$$CS = 0.0797(1.5) = 0.12 < 0.14 \quad (3.20)$$

Therefore, the base shear is

$$V = ZIKCSW = (3/8)(1)0.67(0.12)W = 0.03W \quad (3.18)$$
$$= 0.03(4414.73) = 133.10 \text{ k}$$

There is no whiplash effect, since $T \leqslant 0.7s$

$$F_t = 0$$

The base shear is to act with a minimum eccentricity of 5 percent of the maximum building dimension.

$$e = 0.05D_{max} = 0.05(90) = 4.5 \text{ ft} \quad (3.26)$$

Each of the facade frames carries the following portion of the total shear along the cross direction: First, taking moments about V_2 in Figure 3.12, yields

$$V_1(90) - 133.10(45 + 4.5) = 0, \qquad V_1 = 73.21 \text{ k}$$

Horizontal equilibrium yields

$$V_2 = 133.10 - 73.21 = 59.89 \text{ k}$$

The distribution of the base shear along the building height (Equation 3.25a) is

$$F_x = V \frac{h_x}{\sum\limits_{i=1}^{n} h_i} = 73.21 h_x/322 = 0.227 h_x$$

where: $\sum\limits_{i=1}^{n} h_i = 11.5(1 + 2 + \ldots + 7) = 322 \text{ ft}$ or,

$$= h[n(1 + n)/2] = 11.5[7(1 + 7)/2] = 322 \text{ ft}$$

The lateral forces at the seventh and fourth floor levels are

$$F_7 = 0.227(7 \times 11.5) = 18.27 \text{ k}, \qquad F_4 = 0.227(4 \times 11.5) = 10.44 \text{ k}$$

The seismic shear at the fourth floor level is

$$V_4 = \left[\frac{18.27 + 10.44}{2} \right] 4 = 57.42 \text{ k}$$

The total wind pressure that each frame carries is

$$V_w = 0.021(90 \times 80.5)/2 = 76.07 \text{ k} > 73.21 \text{ k}$$

The wind shear is slightly larger than the seismic shear.
The wind moment about the base is

$$M_w = 76.07(80.5/2) = 3061.82 \text{ ft-k}$$

The seismic moment about the same level is

$$M_s = 73.21(80.5)2/3 = 3928.94 \text{ ft-k} > M_w \qquad (3.27)$$

Hence the seismic moment controls.
The weight of the lateral-force resisting facade frame is

$$W_f = 7[0.085(75 \times 15) + 0.015(75 + 2(15))11.5] = 796.16 \text{ k}$$

The safety factor against overturning is

$$\text{S.F} = M_{res}/M_{act} = \frac{796.16(75/2)}{3928.94} = 7.60 > 1.5 \text{ OK} \qquad (2.9)$$

Example 3.4

Check the stability of a typical pair of 25 ft interior cross walls for a ten-story reinforced masonry building using a wind velocity of 80 mph and seismic zone 2; for the building layout refer to Figure 3.13b. All walls are assumed to be 8 in. lightweight concrete block, plus finish weighing 80 psf. Assume a 100 psf floor dead load and 80 psf roof dead load. The typical story height is 10 ft. For this preliminary check, ignore wall openings and accidental torsion (e_{min}).

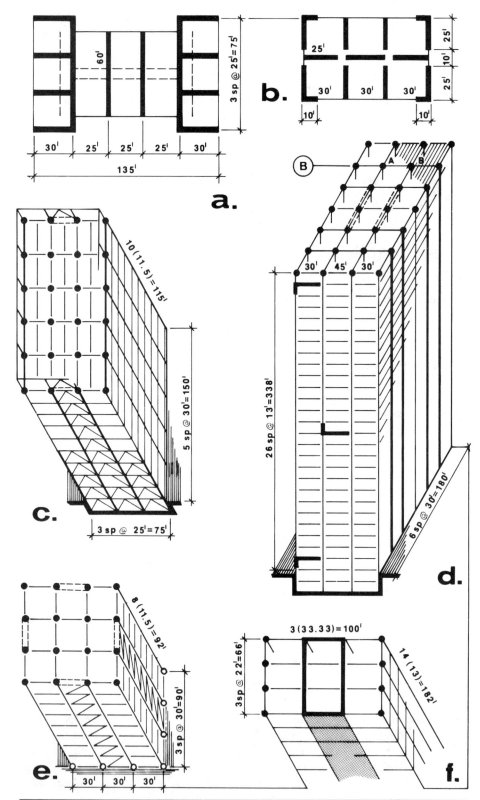

Figure 3.13. *Investigation of the response of various building structures to lateral force action.*

First the building weight is found.

The total wall weight is $\qquad$ $10[(25 + 10)4 + 25(6) + 30]0.080(10) = 2560\text{ k}$

The roof dead load is: $\qquad$ $(90 \times 60)\,0.080 \qquad = 432\text{ k}$

The floor dead load is: $\qquad$ $9(90 \times 60)\,0.100 \qquad = 4860\text{ k}$

The total building weight is: $\qquad = 7852\text{ k}$

The fundamental period is

$$T = \frac{0.05h_n}{\sqrt{D}} = \frac{0.05(100)}{\sqrt{60}} = 0.646s \leqslant 0.7s \qquad (3.21)$$

The seismic coefficient is

$$C = \frac{1}{15\sqrt{T}} = \frac{1}{15\sqrt{0.646}} = 0.083 < 0.12 \qquad (3.19)$$

$$CS = 0.083(1.5) = 0.125 < 0.14 \qquad (3.20)$$

The framing factor for shear wall structures is K = 1.33.
Hence, the total base shear is

$$V = ZIKCSW = 0.375(1)1.33(0.125)W = 0.062W \qquad (3.18)$$
$$= 0.062(7852) = 486.82\text{ k}$$

The total wind shear is based on a conservative uniform wind pressure of 28 psf, as obtained from Figure 3.8:

$$V_w = 0.028(90 \times 100) = 252\text{ k} < 486.82\text{ k}$$

There is no whiplash force, since $T \leqslant 0.7s$

$$F_t = 0$$

For this preliminary investigation, all cross walls have been assumed to be of the same shape and thickness (i.e., the same stiffness), and the beneficial effects of the wall flanges and wall interaction across the corridor have been neglected. Also, the minimum eccentricity requirement has been ignored.

Therefore, each of the eight cross walls resists an equal shear of

$$V = 486.82/8 = 60.85\text{ k}$$

The weight of one interior wall is

$$W_w = 30 \times 30(0.080 + 0.100(9)) + 25 \times 100(0.080) = 882 + 200 = 1082\text{ k}$$

The safety factor against overturning is

$$S.F = \frac{M_{res}}{M_{act}} = \frac{1082(25/2)}{60.85(100)2/3} = 3.33 < 1.50 \qquad OK \qquad (2.9)$$

HIDDEN LOADS

Entire buildings, parts of buildings, individual building components, and the materials, all of them move more or less in response to the direct force action of gravity and lateral loads, or indirectly due to earth settlement or changes in material movement. Their response, that is, the degree of movement, depends on the stiffness of the building structure and its members, including the flexibility of the connections. Whatever the source is for the displacement and volume/shape change, if it should be held back from freely moving, additional forces, sometimes called locked-in forces or reactive loading, are induced in the structure element and the adjacent members, preventing the free deformation.

Each structural element in a building may bend and deform axially under force action. The horizontal floor framing deflects vertically under gravity loading, while the vertical structural building planes sway laterally under wind and seismic action. The recent development of the increased strength of the major structural materials has resulted in reduced member sizes and a decrease of rigidity, hence an increase of member deflection. The dead loading causes a permanently deformed state to the structural members if they are not cambered, as in the case of floors. All materials more or less expand or contract as a result of changes in temperature. Some materials, such as brick and timber, swell and shrink with variations in moisture content. Other materials, like concrete and concrete masonry, go through the chemical process of drying shrinkage caused by air-drying during the early months of construction. The same materials also creep and shorten under sustained loading, which is especially true for prestressed concrete.

In general, a high degree of continuity should be avoided to prevent stress concentrations, and the exposure of elements should be reduced to a minimum by using sufficient insulation. For example, to control the cracking of low-tensile capacity material like masonry, *movement joints* should be provided at critical locations. These movement joints include **soft joints** (e.g., expansion joints, isolation joints), which may be used where independent parts of a structure are cast or placed directly against each other, and where either of them may interfere with the freedom of movement of the other. They include **control joints** (e.g., contraction joints), which are made by weakening the section so that eventually controlled cracking along the joint results due to contraction, and **shrinkage strips,** which are temporary joints that are left open for a certain time to allow the early drying shrinkage for concrete. *Construction joints* may be required when, for instance, the process of concreting is interrupted so that the concrete pour is no longer continuous.

First, the movement of entire building blocks is briefly investigated in Figure 3.14. It is quite clear that stiffer units deflect less than flexible ones, where each employ different structure systems. The same situation of differential movement occurs between building sections of dissimilar mass distribution, slenderness, and load distribution, as visually expressed by the various geometrical structural patterns (Fig. 3.14, parts 1d–g), using the same material. The differential vertical, horizontal, and rotational movements between the building units should be controlled by joints so that the building blocks are separated and can behave independently; one should keep in mind that each unit has its own structure, otherwise adjacent units must support each other, which then requires special *structural joints* (Fig. 3.14, part 2d).

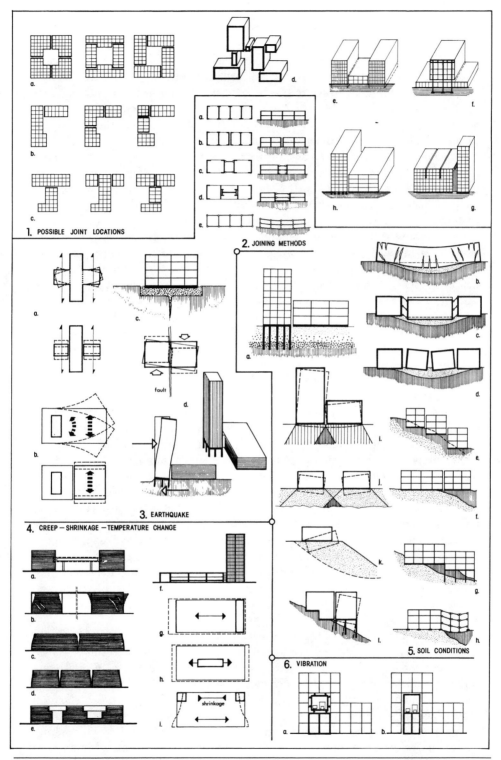

1. POSSIBLE JOINT LOCATIONS

2. JOINING METHODS

3. EARTHQUAKE

4. CREEP — SHRINKAGE — TEMPERATURE CHANGE

5. SOIL CONDITIONS

6. VIBRATION

Figure 3.14. *Building joints. Reproduced with permission from* Horizontal-Span Building Structures, *Wolfgang Schueller, copyright © 1983 by John Wiley & Sons, Inc.*

Similar reasoning can be applied from a visual point of view, where the nonalignment of the structural grids requires some type of rigid or flexible transition system. Some of the possible joining methods are identified (Fig. 3.14, parts 2a–e). They range from a continuous structure (Fig. 3.14, part 2a), the structure with only one joint at midlength (Fig. 3.14, part 2b), to the structure with a soft, hinged bay that acts as a large-scale joint (Fig. 3.14, parts 2c,d), and finally, the flexible hinged structure capable of adjusting to any differential motion (Fig. 3.14, part 2e).

Vibrational movements, such as those caused by oscillating machinery, escalators, cars, and other equipment should be isolated, either by separating the structure that supports the source (Fig. 3.14, part 6b), or by damping the movement with special joints so that the vibration is not transmitted to other building parts (Fig. 3.14, part 6a). *Seismic joints* (Fig. 3.14, part 3) control lateral building sway, as initiated by ground movement, by separating building masses so that they can behave independently. For instance, the shallow wings of a building with an irregular plan shape (Fig. 3.14, part 3a) may tend to move in different directions, while the deep (and hence stiffer) central portion may hardly displace. These torsional, as well as translational, movements will cause large stresses at the junction of the wings to the central portion. In general, buildings with irregular configurations due to different mass or stiffness distribution should be separated as indicated in Figure 3.14, parts 1a–g.

Torsion is also generated in the regular building with an eccentric core (Fig. 3.14, part 3b), since the center of the mass (i.e., the location of resultant seismic force) does not coincide with the center of rigidity. The clearance gap between the tall flexible building and the much stiffer adjacent low building (Fig. 3.14, part 3d) must be wide enough so that they will not touch each other during an earthquake. For the extraordinary condition where a building is constructed across a fault, its rupture may be prevented by tying the building together with a mat foundation and letting it float on a sand cushion (Fig 3.14, part 3c). This cushion is similar to a joint in that it allows the structure to rotate, and possibly slide, while under the differential movement along the fault line.

While the portion of a building below ground is kept at a constant temperature, the portion above is exposed to the climate, and expands and contracts with temperature changes. When this differential movement between the lower and upper parts is prevented, stresses are set up. This may be critical for composite building forms (Fig. 3.14, parts 1b–d), where it may be advantageous to separate the wings to prevent stress concentrations at the junction lines. As a result of higher temperature variations in the upper building portion, intermediate expansion joints may have to be placed there (Fig. 3.14, part 1g).

The examples in Figure 3.14, part 5 show how joints control settlement movements so that potential stresses are reduced to a minimum. Earth settlements can be caused by soil consolidation, shear failure of soil, change of groundwater table (e.g., sinkhole), shrinkage and swelling (e.g., clay), collapse (failure of sewer, mine, cave, etc.), expansion (ground freezing), landslides (unstable slopes (Fig. 3.14, part 5k)), chemical attacks, and by many other factors. A long flat building with uniformly distributed loads sitting on compressible soils deflects in a concave manner (Fig. 3.14, part 5b), while it may respond in a convex form if the primary support structures and foundations are placed eccentrically. It is advantageous to subdivide the long building, including foundations, into small blocks separated by expansion joints or hinged bays (Fig. 3.14, parts 5b and c). Differential settlements are also generated when the same building block sits on different soils (Fig. 3.14, part 5f), on sloping sites (Fig. 3.14, parts 5e,g), or is unevenly loaded (Fig. 3.14, part 5j), and (in addition) uses different foundation types. On sloping sites, the tendency of the upper soil layers to slide downward, as initiated by the building weight, may be avoided by anchoring the outer building unit with deep foundations to the lower soil strata (Fig. 3.14, part 5l). Tall, slender buildings tend to tilt under differential settlement (Fig. 3.14, part 5i), as (for example) when the ground below is unevenly loaded by an adjacent building. The effect of settlements can be reduced by replacing the indeterminate stiffer continuous structure with the determinate, more flexible, hinged structure (Fig. 3.14, part 5h).

It may be concluded that, whatever the main purpose of an expansion joint is, be it for the control of temperature and moisture, creep and shrinkage, settlements, seismic or dynamic loading, or any other reason, the following criteria should be considered:

- Long buildings should be subdivided into units, realizing that the joints are spaced close for stiffer than for flexible buildings.
- The spacing of expansion joints for open structures should be closer, since they are subject to greater temperature changes.
- Buildings with compound or irregular plan shapes should be separated into units, as indicated in Figure 3.14, parts 1a–c.
- Buildings consisting of different blocks of high- and low-rise sections, each with its own structural system, should be separated at their junction.

The maximum spacing of expansion joints through the entire building mass, keeping in mind that they are subject to difference in opinion, is recommended as 200 to 300 ft for steel and concrete buildings, and 300 to 400 ft for bearing-wall masonry buildings. Buildings have been built that have exceeded these limits by far and have performed well, including large-scale concrete buildings without any joints.

In the following, the effect of climate on the behavior of the exterior structure of wall and roof is briefly investigated. While the inner layer of the exterior structure faces the interior temperature, and a portion of the exterior climatic conditions, depending on the degree of insulation, the outer layer of cladding must respond to the extreme outside temperature and humidity fluctuations, among other criteria. The differential movement between dissimilar materials, such as steel and reinforced concrete, a concrete frame enclosing an aluminum window frame, in turn enclosing the glass, are investigated in some of the Problems.

The early high-rise buildings of this century were less susceptible to temperature changes, since the structure was well-protected behind heavy stone or masonry cladding, which usually had only small window openings. These exterior walls were quite capable of carrying considerable loads and also acted somewhat as a composite part of the steel frame. The hidden steel structure was neither vulnerable to weather changes nor susceptible to shrinkage and creep, and concrete frames were rarely higher than 20 stories. In addition, the building skeleton was composed of much larger members than today, and the building was braced internally by masonry partition walls, resulting in quite stiff buildings.

Today, structural framing is less protected and may even be fully exposed. The curtain has replaced the wall; it consists of lightweight panels or a flexible skin that will move independently of the supporting back-up structure. This trend toward the reduction of building stiffness started in the late 1950s. It was not only initiated by the thin suspended facade membranes, but also by smaller frame sections based on stronger materials and a more precise understanding of structural behavior that came about with the development of computers. Further, the height of concrete buildings has jumped drastically, resulting in column shortening due to creep and drying shrinkage, primary effects that can no longer be ignored.

Interior members of buildings are faced with a relatively constant temperature of approximately 70°F, therefore not causing any change in length, while exterior members are exposed to weather variations ranging from the coldest temperatures in winter to the hottest in summer. In addition, it is not only the ambient air temperature that influences the temperature of the exposed surface, but also wind, solar radiation, and condensation. The average material temperature depends not only on the thermal resistance of the material covering a column, but also on the duration of the exterior peak temperature. This change in temperature causes movement in the facade structure that is partially resisted by the inside structure, but which cannot follow this movement, since it is under the constant state of interior temperature, T_i, and the two structures are rigidly connected to each other. The difference in temperature, ΔT, between the inside and the average material temperature, T_m, causes stresses that become critical for buildings taller than about 20 stories.

$$\Delta T = T_i - T_m \tag{3.29a}$$

In this case, temperature drop, $T_i > T_m$, is assumed, since it usually controls the design. Naturally, when the outside wall is not restrained by the inside structure, and is allowed to move independently, stresses do not occur in the main structure. Curtain walls should, however, be able to accommodate movements induced by their own mean temperatures, and the following movements should be considered for their design:

- relative movement between cladding and building structure
- movement between the curtain wall components
- relative movement between curtain wall components.

The evaluation of the average material temperature is quite complex and beyond the scope of this discussion. However, some basic concepts can be developed by using the steel column as an example (Fig. 3.15, top). The degree of column exposure depends on the position of the column as related to the curtain wall, and the thermal resistance of the material covering the column. The columns may be located inside or outside the building, or along the wall line; thus they may be fully, semi-, face-, or unexposed. The column may be bare, covered only with sheet metal, for example, but with an air space (for low-rise buildings up to ten stories), covered with fireproofing and metal, insulated on the cold side, or the column may be part of a composite skin construction (Fig. 3.15, center). Inside insulation results in greater temperature variations in the column. For taller buildings, the insulated column may also need gravity-type vertical air circulation or forced mechanical ventilation to control the column temperature (Fig. 3.15, right).

In the preceding discussion, the effect of fireproofing and insulation of partially exposed columns is conservatively neglected, as is the fact that the surface exposed to the sun will have a temperature due to solar radiation well above the air temperature; the member is assumed to be relatively thin (for example < 12 in.), so that the time lag of penetration of the ambient temperature does not have to be taken into account. In the following, the average member temperature for various exposure conditions (Fig. 3.15)

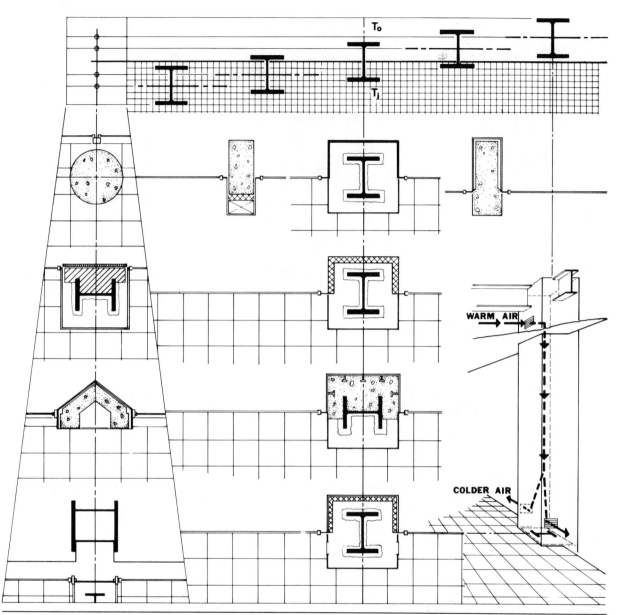

Figure 3.15. *Column exposures.*

is found, as based on simplifying assumptions, so that a feeling for thermal mechanics can be developed. As an example, an indoor temperature of $T_i = 70°F$ and a critical outdoor temperature of $T_o = -20°F$ are used for northern climates. The actual design temperatures for members less than 12-in. thick can be obtained from maps prepared for lowest winter and highest summer mean daily temperatures with frequencies of recurrence of once in 40 years. It should also be kept in mind that, for office buildings, the weekday and weekend indoor temperatures may not be the same.

Where a column, wall, or roof is *partially exposed*, the change of temperature across the section causes axial deformations due to the average material temperature, and bowing due to the temperature gradient. The effect of thermal bowing is disregarded here, but will be discussed later (see fig. 3.18). The average material temperature, T_m, is determined by assuming the face temperatures as equal to the ambient air temperature and by using a linear temperature gradient between the two faces: $T_m = (T_i + T_o)/2$. Then the temperature difference is

$$\Delta T = T_i - T_m = T_m - T_o = (T_i - T_o)/2, \text{ when } T_i > T_o$$
$$\Delta T = T_m - T_i = T_o - T_m = (T_o - T_i)/2, \text{ when } T_o > T_i$$

(3.29b)

Assuming minus outdoor temperatures, $-T_o$, causes the following temperature difference for the conditions given above

$$\Delta T = (T_i - T_o)/2 = (T_i - (-T_o))/2 = (T_i + T_o)/2$$

$$= (70 - (-20))/2 = (70 + 20)/2 = 45°F$$

Notice that, for the given conditions, both the average steel column temperature and the temperature differential are equal to each other. In the sunbelt, say with mean low winter temperatures of $T_o = 20°F$, and $T_i = 70°F$, the temperature differential is only $\Delta T = (70 - 20)/2 = 25°F$.

Where a column, wall, or roof structure is *fully exposed*, or where a high percentage of the column area is outside the glass line, or where the insulation is on the inside, the average material temperature may be assumed to be equal to the ambient air temperature, $T_m = T_o$. Therefore, there is a constant temperature change across the section, which causes only axial deformations and no bending—the effects of solar radiation and insulation are ignored for this preliminary investigation. When the member is restrained from movement by the interior structure, the temperature difference is:

$$\Delta T = T_i - T_m = T_i - T_o, \text{ when } T_i > T_o$$
$$\Delta T = T_m - T_i = T_o - T_i, \text{ when } T_o > T_i$$

(3.29c)

For the given example, the average steel column temperature is $T_m = T_o = -20°F$ and the temperature difference is $\Delta T = 70 - (-20) = 90°F$.

For the condition where sufficient insulation is placed on the outside of the building structure, temperature changes may be ignored with respect to its design. However, they must be taken into account for the fully exposed cladding in front, which acts independently of the interior building environment. In this instance, the average temperature is $T_m = T_o$, and the member movement to be considered is due to the temperature difference between the original temperature, T_{or}, when it was built and the final temperature, T_f

$$\Delta T = T_{or} - T_f$$

(3.29d)

The final temperature is the critical highest temperature in summer, T_{max}, or the lowest temperature in winter, T_{min}. The original temperature may be taken as the mean temperature during the normal construction season in the locality of the building.

Rather than taking the difference, as based on maximum rise or drop from the mean, as the design temperature change, it is obviously conservative to use the difference between the extreme high and low temperatures, as is recommended by some material specifications. This approach is based on the assumption that the member was built at T_{max} or T_{min}, so that the temperature difference represents the critical temperature range for a given locality.

$$\pm \Delta T = T_{or} - T_f = T_{max} - T_{min} \qquad (3.30)$$

The temperature range for the northeastern part of the United States may be conservatively estimated as

$$\pm \Delta T = 100 - (-20) = 120°F$$

However, one must keep in mind that the temperature range for dark facades, and especially for metal curtain walls, may be substantially higher than that of the air. For example, dark-colored metal facades may reach a temperature as high as 175°F in summer, while in winter they may be 10 degrees colder than the outside temperature, thus having an extreme temperature range of 215°F.

The change in member length, ΔL, is proportional to temperature changes, ΔT, and expressed by the coefficient of linear thermal expansion, α. It is equal to

$$\Delta L = \varepsilon_t L = \alpha L \Delta T \qquad (3.31)$$

In this case, L is equal to the original member length (in.), and the thermal strain $\varepsilon_t = \Delta L/L = \alpha \Delta T$. Notice that the displacement of the linear element increases directly with its length.

The average coefficients of expansion for some materials are approximately:

- $\alpha = 5.5 \times 10^{-6}$ in./in. per degrees Fahrenheit, for normal-weight concrete and granite
- $\alpha = 6.5 \times 10^{-6}$ in./in. per degrees Fahrenheit, for steel and sandstone
- $\alpha = 12.8 \times 10^{-6}$ in./in. per degrees Fahrenheit, for aluminum
- $\alpha = 17.5 \times 10^{-6}$ in./in. per degrees Fahrenheit, for glass/polyester panel
- $\alpha = 5 \times 10^{-6}$ in./in. per degrees Fahrenheit, for window glass
- $\alpha = 4.3 \times 10^{-6}$ in./in. per degrees Fahrenheit, for lightweight concrete masonry, and limestone
- $\alpha = 3.6 \times 10^{-6}$ in./in. per degrees Fahrenheit, for clay brick masonry in the longitudinal direction. The movement occurring in the vertical direction is about 50 percent greater.

In addition to the effect of temperature, clay bricks swell considerably because of absorption of moisture; an average *coefficient of moisture* of 2.0×10^{-4} in./in. is recommended. Clay brick also has a larger modulus of elasticity than concrete masonry, but for concrete block walls, the initial drying shrinkage and creep is often more critical. Sixty percent of the shrinkage occurs by an age of about 90 days; it is sometimes treated as an equivalent temperature drop of about 33°F for estimation purposes, or a *shrinkage coefficient* or roughly 2.0×10^{-4} in./in. is used. For lightweight concrete, the shrinkage factor may be taken as 50 percent higher. Creep is a function of stress, time, and the concrete properties and is similar to shrinkage in that after one-half year, 70 percent of the inelastic strains have taken place; it is on the order of 5 to 10×10^{-4} in./in.

The axial stresses due to prevention of movement can be easily derived. Visualize a restrained beam allowed to move freely (Fig. 3.17, top left). Now the force, P, or stress, f_a, can be found that is necessary to bring the member back to its original restrained position.

$$\pm \Delta L = PL/AE = f_a L/E = \epsilon_t L = \alpha L \Delta T,$$

or

$$\pm f_a = \epsilon_t E = \alpha E \Delta T \qquad (3.32a)$$

It must be remembered that these stresses are only present when restraint does not allow freedom of movement. Notice that the thermal stresses are independent of member dimensions and member span!

As an example, an exterior clay masonry wall is investigated, and is conservatively treated as fully exposed. The long masonry wall (Fig. 3.14, parts 4c, d) tends to contract due to thermal and moisture changes, as well as movement of the roof, but is restrained from doing so by the cross walls and, at its base, by the frictional resistance along the contact surface with the foundation, which does not have to face temperature and humidity fluctuations. Tensile stresses will gradually build up in the wall, and may cause cracks to relieve the stresses. To absorb some of the horizontal movement, *vertical expansion* joints must be provided to reduce the wall length and thus to keep the stresses within their allowable limits.

Example 3.5

Determine the tensile stress of a 100-ft long restrained clay masonry wall that is exposed to a temperature drop of

$$\Delta T = T_{max} - T_{min} = 100°F. \text{ Use } \alpha = 3.6 \times 10^{-6} \text{ in./in./°F and}$$
$$E = 2,400,000 \text{ psi.}$$

The tensile stress is

$$f_t = \alpha E \Delta T \qquad (3.32a)$$
$$= 3.6(10)^{-6} (2,400,000)100 = 864 \text{ psi} > F_t = 72 \text{ psi}$$

The tensile stresses are far beyond the material capacity. Rather than using horizontal steel reinforcing or bond beams, vertical movement joints are provided to allow horizontal movement, and thus to relieve the horizontal tensile stresses.

Determine the spacing of the joint, considering a temperature increase of $\Delta T = 100°F$ and swelling due to moisture using a coefficient of $\epsilon_s = 2.0 \times 10^{-4}$ in./in. Neglect the restraining effects of cross-walls, slabs, vertical reinforcing steel, and foundations. Assuming a 1-in. wide joint with an elastic sealant that has a maximum allowable strain of 50 percent, yields a joint spacing, L, of

$$\Delta L = (\epsilon_s + \epsilon_t)L = [0.0002 + 0.0000036(T_{max} - T_{min})]L \qquad (3.33)$$
$$1/2 = [0.0002 + 0.0000036(100)]L$$
$$L = 892.86 \text{ in.} = 74.41 \text{ ft.}$$

For elastic joint sealants with a compressibility of only 25 percent, wider joints or a closer spacing of joints is required, otherwise sealant failure will occur. In general, expansion joint spacing for brick masonry walls should not exceed 50 to 100 ft, depending on the joint width, but should be within 30 ft of corners and wall intersections, and along columns or pilasters.

While vertical joints are typical for long walls and are necessary to accommodate horizontal movement in high-rise buildings, horizontal relief joints are particularly important for masonry, natural stone cladding, and precast concrete panels to control vertical movement (Fig. 3.16). Horizontal expansion joints should be provided immediately beneath the shelf angles that support the cladding to prevent the stacking of panels, which may result in bowing (buckling), spalling, or even failure of the veneer or anchorage system. Usually, the cladding is not directly supported by foundations, but by the building structure, and since it does not act compositely, but is separated from the structure, it moves independently from the back-up system. While a concrete frame shortens due to elastic deformations, creep, and drying shrinkage, and a steel frame may have significantly larger elastic deformations, the veneer expands vertically due to extreme summer temperatures, and possibly due to moisture (see Problem 3.32). When the joint width is not sufficient to allow for the combined effect of frame shortening, causing a reduction of floor height, and the expansion of the facing, then loads are transferred from the frame to the cladding, which may fail or force panels to pop out. It is quite apparent that a masonry wall absorbs water and

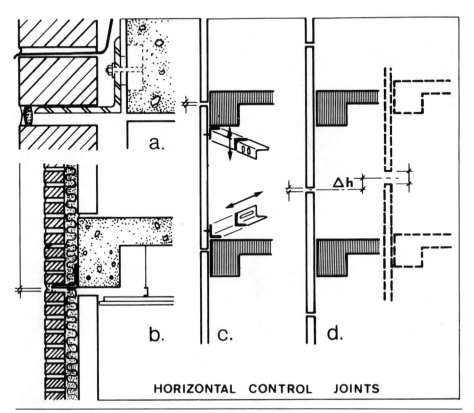

HORIZONTAL CONTROL JOINTS

Figure 3.16. *The importance of horizontal relief joints for curtain walls.*

lacks flexibility and ductility, and thus is much more vulnerable to failure than a metal curtain wall. Often overlooked is the fact that the coefficient of expansion for marble is nearly twice that for granite, therefore, granite joints would have to be of less width. Some of the plastic materials have a coefficient in the order of 5 to 10 times that of conventional masonry, thus requiring special attention for proper joint design.

Glass must not only resist wind loads, but must also float within the frame opening so that mechanical and thermal stresses are kept to a minimum. The thermally induced stresses depend upon, aside from the primary critical temperature differences, the effect of orientation, the glass type, size, and shape, the shading conditions, the type of framing system, the location of heating registers and other criteria. The variations in thermal loading over a glass pane may induce significant stresses.

The effect of temperature differentials is quite pronounced for roof slabs, which are subject to large movement, especially for concrete slabs on masonry walls (Fig. 3.17c). The hot sun on black roofing produces a membrane temperature far higher than the ambient air temperature. The quality of the insulation is critical, but will not prevent the bowing and horizontal expansion of the concrete slab, which (in turn) causes stress

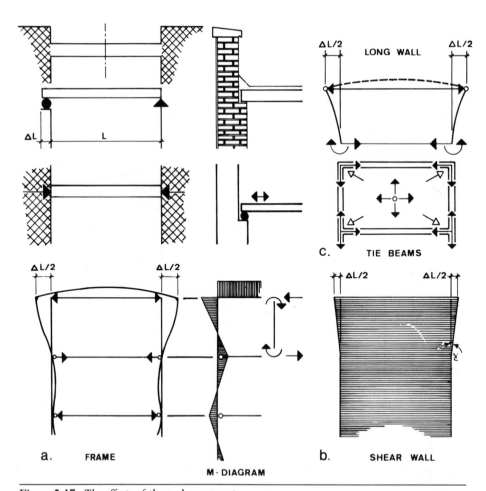

Figure 3.17. *The effects of thermal movement.*

in the masonry facade walls if slip planes, for instance with pads of roofing felt and soft end joints between the slab and wall, are not provided. If this free movement is prevented, lateral deflection is imposed upon the vertical structure (Fig. 3.17a,b,c). This is critical for the long wall masonry structure, with its low modulus of elasticity (see Problem 3.23). To reduce the temperature movement, longitudinal joints along the center of the slab and parallel to the long walls can be provided for expansion control. Also, horizontal tie beams can restrain and equalize the slab movement by not allowing the walls to be stressed. The lateral expansion of the roof resisted by the frame as indicated in Figure 3.17a, shows the rapid decrease of the bending from the top down. Similarly, the poured-in-place concrete slab shrinks, and thus attempts to induce inward bending of the walls if slippage is not allowed. In other words, the slabs above ground tend to shorten, particularly in flexible structures, where the vertical elements provide less resistance, while the basement does not follow, and thus resists this movement.

Should the lateral forces on a building not be resisted in a uniform manner, but rather by concentrated stiff core structures, then the location of the core within the building is an important consideration (Fig. 3.14, parts 4g,h,l). Uniform lateral displacement occurs for the condition where the core is symmetrically arranged at the center of the building (Fig. 3.14, part 4h), while only one-directional movement will appear when the core is located eccentrically at the outer end (Fig. 3.14, part g). This lateral deflection may cause large stresses in the facade masonry curtain, if not taken into account and properly controlled.

When the restrained exterior column or wall is only partially exposed, it is subject to a temperature difference between the outside face and inside face that cannot be ignored, as has been discussed. Similarly, for exposed concrete decks (such as open parking ramps), the surface temperature in summer may be 50°F above the ambient temperature, or in winter the surface may be well above freezing, while the underside of the slab is well below freezing, thus causing a temperature differential across the section. In this instance, the face with the lower temperature shortens, while the warm face expands, resulting in bowing of the section. For this condition axial stresses are generated when the member is restrained from changing in length as well as bending stresses when bowing is prevented; the average temperature change in the member causes axial forces, while the remainder of the temperature change induces the moment (Fig. 3.18). According to Equation 3.29b, by assuming a linear temperature gradient between the opposite faces, and by ignoring the effect of insulation for $T_i > T_o$, the change in member length at the centroidal axis is

$$\Delta L = \alpha L \Delta T = \alpha L (T_i - T_o)/2 \tag{3.34}$$

The bowing due to the temperature gradient can be derived as follows for the differential temperature of $\Delta T = T_i - T_o$ at the outer face, which is twice as much as at the centroidal axis. The constant angle change from Figure 3.18, for a symmetrical section of depth d, is: $\tan d\theta \simeq d\theta = dx/R = (\alpha \Delta T dx/2)/d/2$, or

$$d\theta/dx = 1/R = \alpha \Delta T/d \tag{a}$$

From elementary strength of materials, it is known that the curvature is directly proportional to the moment.

$$1/R = M/EI \tag{b}$$

From Equations (a) and (b), the temperature-induced rotational deformation, or the fixed end moments required to prevent the member from bowing, are derived as

$$M = \alpha \Delta T E I / d \qquad (3.35)$$

Notice that the thermal moment is independent of the span, L, but is a function of the member stiffness EI. While in the determinate beam strains and bending deformations exist without thermal stresses, in the indeterminate restrained member significant bending stresses are generated by the gradient effect but without any thermal strains. The corresponding axial deformations at the centroidal axis of the bent member if it were not restrained, are given by Equation 3.34. For symmetrical sections, when bowing but not linear movement is prevented, the thermal bending stresses at the faces, by substituting Equation 3.35, are

$$\pm f_b = \frac{Md/2}{I} = \alpha E \Delta T/2 \qquad (3.32b)$$

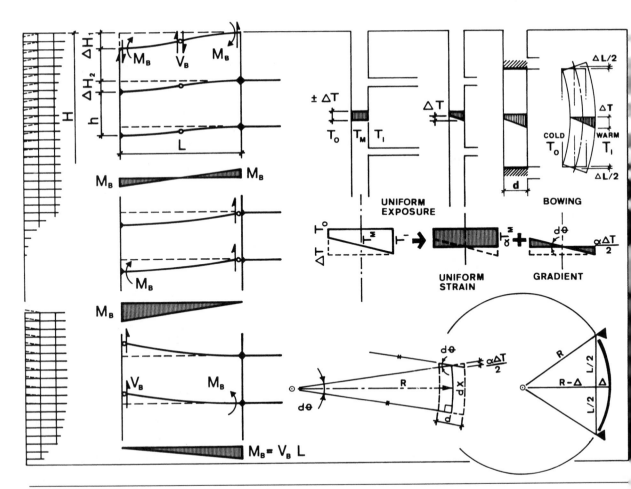

Figure 3.18. *Bending due to differential movements.*

When the same member is restrained from linear movement to conform to the average temperature T_m by ignoring bowing, then the resulting axial stresses due to $T_i - T_m = (T_i - T_o)/2 = \Delta T/2$, are

$$f_a = \alpha E \Delta T/2 \qquad (3.32c)$$

The superposition of the thermal bending and uniform axial stresses constitute the final normal stresses when the member ends are fully restrained from linear and rotational movement. In other words for a symmetrical section, the bending stresses and axial stresses each constitute one-half of the normal stresses at the critical face. For example, tensile stresses are generated for the condition of $T_i > T_o$. They are zero at the interior face and are maximum at the cold exterior face with a magnitude of

$$f_t = f_b + f_a = \alpha E \Delta T \qquad (3.32d)$$

Continuous members, whether short or long, are equally vulnerable to temperature variations. However, one must keep in mind that in taller buildings there is more resisting beam stiffness available to prevent movement in columns than in lower buildings.

The maximum beam deflection can be found from the geometry of the deflected beam as a circular arch (Fig. 3.18), since for a constant bending moment the curvature is constant.

$R^2 = (L/2)^2 + (R - \Delta)^2$, or $2R\Delta - \Delta^2 = (L/2)^2$, where the term Δ^2 is so small that it can be ignored.

$$\Delta \simeq L^2/8R \qquad (c)$$

Substituting Equation (a) into (c) yields the maximum deflection.

$$\Delta_{max} = \alpha \Delta T L^2/8d \qquad (3.36a)$$

Notice that the member deflection is independent of the member stiffness, but does depend on the span and member depth. The effect of the thermal bowing of wall panels is usually not significant with the exception of corners, where the panels tend to separate and damage the joint sealants, if they are not restrained by connectors from doing so. More critical is the upward bow of roof members that are not properly insulated, and of floor/roof members of nonheated and open structures, such as parking decks. When these members are prevented from freely rotating at their ends, stresses are developed at those locations.

Also, the approximate maximum lateral movement of the exposed core structure and the tubular tower in Figure 3.19a and b, due to uneven temperature exposure, can be estimated from Equation 3.36a, by treating the buildings as vertical cantilevers and letting $L = 2H$.

$$\Delta_{max} = \alpha \Delta T H^2/2d \qquad (3.36b)$$

For the thermal behavior of structural members under fire loading, the reader may refer to the brief discussion in Figure 1.22.

For tall buildings, the shortening of the facade frames is quite critical. Visualize the exterior columns able to freely move up and down with respect to the inside core structure by assuming the exterior bay beams to be hinged to the columns, allowing

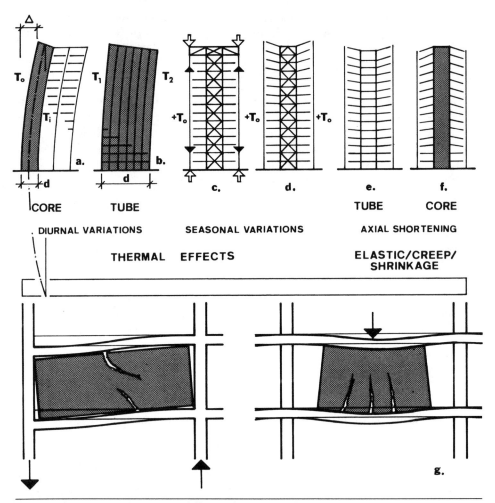

Figure 3.19. *Typical effects due to differential movements.*

this freedom of motion by simply tilting. In reality, however, the slabs, together with the beams (which may be moment-connected to the columns), do resist part of the free movement (see fig. 3.18, left). The degree of restraint depends on the stiffness and span of the exterior bay floor structure and on the relative stiffnesses of the outside and inside columns of the exterior bay. Naturally, if all of the columns should be on the outside of the building, with no vertical support on the inside, then seasonal temperature differences are not a problem! The vertical movement of the exterior structure is not even, since the south side walls are subject to a larger temperature differential than the shaded building parts; also, since the four corner columns have different exposure conditions, there will be a differential movement between the adjacent exterior columns. The differential movement between the exterior and interior structure (see fig. 3.18, left) primarily causes axial forces in the columns and bending of the floor structure.

The free movement reaches a maximum at the top of the building, and thus causes maximum vertical wracking in the upper portion, quite in contrast to the critical wind-induced wracking in the lower portion. Not only must the beam sizes be increased, but also additional slab reinforcement at the perimeter of tall concrete buildings must be added in the upper floors to accommodate the higher stresses due to the increased vertical movement of the exterior frame. The large frame distortions in the upper floors due to the thermal movement may cause stressing and cracking of the partitions; particularly vulnerable are plaster walls, stiff ceilings, and masonry cladding. The maximum thermal column loads occur at the first floor level, where the restraint against movement is the largest; hence, with an increase of height and floor stiffnesses, the column restraint and the corresponding stresses increase. The axial forces in the exterior columns reach maximum compression when expansion is prevented by in-finitely stiff floor beams or by an outrigger truss, as in Figure 3.19c. In winter, the axial loads in the exterior columns are reduced but increased in the interior columns; the opposite condition occurs in summer.

The late Fazlur R. Khan, of SOM, and Anthony F. Nasetta proposed in 1970 to limit thermal movement of exterior columns relative to interior columns/walls to a maximum of $\Delta H = 3/4$ in. in the upper floors. They assumed (for concrete structures) a typical span of the exterior bay floor framing of L = 37.5 ft, which corresponds to a center beam deflection of $1/2(L/300) = L/600$, which will not cause excessive distress to the partitions, and should be added to the gravity load or other load deflections.

When nonstructural systems (like partitions or cladding) are placed beneath beams, enough clearance space must be chosen so that, during future live load deflections, the beams do not touch these components and impose loads upon them that may cause the material to crack (Fig. 3.19g). A partition, in a manner similar to glass, must float in the opening. Similar reasoning may be applied to the rather stiff plaster ceilings, which may form cracks if the beams they are directly attached to are too flexible. Care must be taken in handling the large deflections of cantilevers, as well as the considerable extensions of cable structures. Special staircases may have to be installed that compensate for the building's motion. In addition, the flexibility of the overall building structure must be controlled so that the distortions under lateral loads do not become large enough to stress the non-load-bearing systems.

Example 3.6

The columns of a 50-story, 650-ft high steel building are partially exposed; the span of the exterior bay beams is 36 ft. Because the structure is located in the northeastern part of the United States, the effect of the coldest winter temperature of $T_o = -20°F$, which causes the largest differential movement, is investigated. An inside temperature of $T_i = 70°F$ is assumed.

The average column temperature is estimated as

$$T_m = (T_i + T_o)/2 = (70 + (-20))/2 = 25°F$$

Therefore, the corresponding difference in temperature between the inside and average material temperature is

$$\Delta T = T_i - T_m = (T_i - T_o)/2 = (70 - (-20))/2 = 70 - 25 = 45°F \quad \textbf{(3.29b)}$$

Allowing free movement of the column with no restraint from floor framing and treating it as an equivalent column with an average temperature change of 45°F over its full height where the free movement at each story is equal because of the assumed constant column cross section and equal story height, yields a maximum contraction at the roof level according to Equation 3.31 (letting $\Delta L = \Delta H$), of

$$\Delta H = \alpha H \Delta T = 6.5(10)^{-6} (650 \times 12)45 = (0.0456 \text{ in./story})50 = 2.28 \text{ in.}$$

Assuming a tolerable movement of only 0.75 in., a final movement of $2.28 - 0.75 = 1.53$ in., or $(1.53/2.28) 100 = 67$ percent of the free movement must be resisted by the structure.

In general, the following separate thermal effects must be considered: length changes of columns producing frame moments (i.e., moments in beams and columns)—axial forces in columns due to load transfer from exterior to interior columns—column stresses due to restrained bowing.

For rough preliminary estimation purposes, the thermal bending stresses in the critical floor beam at roof level and the approximate axial thermal column stresses at the first floor level generated by resisting the actual movement of 67 percent of the free movement, can be estimated as follows (see fig. 3.18, left and top).

The approximate beam stresses due to the support settlement can be obtained by assuming the beam to have rigid boundaries (i.e., the actual stiffness of the columns is ignored) and by visualizing it to consist of two horizontal beam cantilevers each of $I_B/2$ (see fig. 3.18, top left).

$$\Delta H/2 = \frac{V_B(L_B/2)^3}{3EI_B}, \qquad \text{or } \Delta H = \frac{V_B L_B^3}{12EI_B}$$

Therefore, the constant beam shear, by letting the beam stiffness $k_B = I_B/L_B$, is

$$V_B = \Delta H(12EI_B/L_B^3) = \Delta H(12Ek_B/L_B^2) \tag{3.37a}$$

This shear causes the following beam moment (see also Table A.14)

$$M_B = V_B(L_B/2) = \Delta H(6EI_B/L_B^2) = \Delta H(6Ek_B/L_B) \tag{3.37b}$$

Hence the bending stress, for f_b, for a symmetrical section of depth, d, is

$$f_b = M_B d/2I_B = \Delta H(3Ed/L_B^2)$$

Approximating the beam depth as $d = L/24$, yields

$$f_b = E(\Delta H)/8L_B \tag{3.38}$$

where: L (in.), f_b(ksi), E (ksi), ΔH (in.)

For this case the beam stresses at the top floor are

$$f_b = \frac{29000(1.53)}{8(36)12} = 12.84 \text{ ksi}$$

This is $(12.84/24)100 = 53.5$ percent of the allowable bending stress, using A36 steel keeping in mind that rotation of the joints was ignored and conservatively rigid boundaries were assumed. Notice also that the temperature induced moments in the exterior top-story columns are quite significant, as is easily derived from joint equilibrium.

The thermal axial forces in the exterior and interior columns generated by frame wracking consist of the thermal beam shears that accumulate from the top down to the base. Whereas in winter the axial loads in the exterior columns are reduced, they are increased in the interior columns; the opposite situation occurs in summer. The critical thermal axial stresses in the columns at the building base are equal to the sum of the beam shears from above divided by the resisting cross-sectional area of the column, A_c.

$$f_a = \Sigma(V_B)_n/A_c$$

Since it is time consuming during the preliminary design stage to determine the stiffness, k_B, of each of the floors in order to find the beam shears, V_B, the following conservative approximation may be used for estimating the axial column stresses. The stiffer the floor structures, the smaller the final column movements, hence the greater the thermal axial stresses in the columns. Hence based on the assumption of rigid floors and treating the exterior columns in this example as a single 650-ft equivalent column where the contraction of 1.53 in. is prevented, the following axial tensile stresses at the bottom story of the building are generated where the resistance to movement is greatest (notice that in this context ΔT is already $(T_i - T_o)/2$):

$$f_a = \alpha E \Delta T = 6.5 \times 10^{-6} (29000)45(0.67) = 5.68 \text{ ksi} \qquad \textbf{(3.32a)}$$

The temperature loads determined in this example must be superimposed upon the other loads such as gravity and wind in order to find the critical stresses. In addition, the stresses caused by bowing must be investigated.

It is shown in Problem 3.26 that when the columns of Example 3.6 are fully exposed, the thermal column and beam stresses become quite large. In general, large thermal or other vertical movements in tall buildings can be controlled by any of the following approaches.

- Less exposure of structure.
- Story-high outrigger steel trusses or post-tensioned concrete wall beams connected to interior shear walls (e.g., head truss on top of the building). In addition, belt trusses or collar walls can equalize differential movements (see fig. 3.19c and fig. 7.30).
- Creation of a thermal break by subdividing the building into smaller secondary structures.
- A deep girder system rigidly connected to the columns in the upper floors.
- Using a suitable construction process (see the discussion following).
- Forced mechanical or gravity-type ventilation of the columns, so that the material temperature is close to the interior temperature $T_i \simeq T_m$.
- Relief of stresses by using simply supported beams in the upper floors. The shear connections should allow sliding and rotation only under larger forces, since they still must give lateral support to the exterior columns under normal loading conditions. When only one hinge is provided in the beam, it should be located at the interior column/wall when winter conditions control the design, so that the negative moments due to the differential movement and gravity counteract each other, while the field moments only slightly increase. Correspondingly, when summer conditions control, the beams should be hinged to the exterior columns (see fig. 3.18, left).

The contrast between the thermal/creep/shrinkage behavior of a concrete tube structure, where the exterior framing is exposed, and the same structure, but with an insulated panel cladding system, is quite apparent. The decision as to which of the systems to use is not that simple, since each approach has its advantages.

Differential linear deformations of the vertical structure, or *vertical wracking*, in tall buildings occur not just because of temperature fluctuations, but also because of the immediate elastic axial deformations, as well as the time-dependent shortening due to creep and shrinkage for concrete, as is the case (for example) between a peripheral lateral-force resisting structure and the interior gravity structure. Elastic shortening, as expressed by the familiar expression $\Delta L = \Sigma PL/AE = \Sigma f_a L/E$, refers to the immediate axial deformations due to the column loads, and can readily be determined from the stress levels—it is predominant for strong steel columns. For example, a typical 12-ft high steel column that is stressed in compression to 16 ksi, shortens by $\Delta L = 16(12)12/29000 = 0.08$ in./story, which accumulates to 8 in. for a 100-story building, or to one-half of this value for a 50-story building.

But, for concrete columns, in addition to the elastic deformations, shortening due to creep and shrinkage must be added. The shortening due to creep when the concrete is stressed in compression is influenced, among other criteria, by the stress level, the concrete strength, the percentage of longitudinal steel, and the age of the concrete at loading. The shortening due to shrinkage, on the other hand, is caused by the evaporation of the free water not needed for the hydration of the cement; in other words, as the concrete dries, it shrinks. Up to roughly 65 percent of the shortening due to creep and shrinkage takes place during the first weeks, while the remaining portion happens over many years. The total shortening of concrete and steel columns may be in the same range. However, it must be kept in mind that their components of deformation do not occur at the same time.

Since all columns and/or walls in a tall building cannot be stressed equally (P/A is not constant) because of the variable nature of the live loads, they will not shorten by the same amount. For example, in tubular steel buildings, the interior columns are usually designed of high-strength steel, and thus require smaller sections to resist the gravity loading—they also support larger tributary areas. These are subject to much more axial shortening than the perimeter columns. The perimeter columns may be more closely spaced and support smaller tributary areas—they are designed as beam columns that must resist the lateral forces, but much smaller gravity loads. Their sizes are based on high wind forces that occur only rarely and then just for short periods, and (in addition) because of stiffness requirements, carbon steel (A36) may be used. Hence, there is an imbalance in the gravity stress level that causes the interior columns to shorten much more than the exterior ones, resulting in the so-called *dish effect* (Fig. 3.19e). This effect may be worse for concrete or composite columns where, in addition, shrinkage and creep must be considered. However, it must be kept in mind that the elastic shortening of steel columns because of their higher strength is usually substantially larger than that for concrete columns.

Opposite conditions exist where a central concrete or braced core resists the lateral forces, and the steel frame along the periphery with widely spaced columns carries only axial forces. Here, under gravity action, the central core walls shorten much less than the exterior steel columns (Fig. 3.19f).

To compensate for this differential movement, not only must the construction sequence be taken into account, but also overlength columns for the upper building portion may have to be used during erection in order to maintain floor levels. Therefore, the nonwind columns at the top floor levels must be longer than the wind columns, so that the columns come out to the same length in the finished state. The corrective column lengths can be obtained by lumping together, for instance, the 0.08 in./story of the previous example over 10 floors, adding 0.8 in. every tenth floor, rather than increasing the column lengths at each floor level. For cast-in-place concrete construction, the correction for differential movement is much more difficult because of the time-dependent creep and shrinkage. It is made during construction by adjusting the level of the form-work to the expected shortening at some time after the initial occupancy.

The control of differential column shortening is most important with respect to the performance of the curtain wall, the vertical mechanical pipes, and floor leveling. The bending of the floor structure (slab tilt), which may be especially critical for flat slabs, is not only caused by differential movement between the exterior perimeter structure and the interior core, but also between interior columns, due to live load arrangement or due to the difference in stiffness between adjacent columns and walls; this may, in turn, result in a redistribution of loading, where the stiffer and less deformed element will attract more load. Although differential vertical movements are most critical in high-rise buildings, exposed spandrels will cause lateral thermal strains and stress the floor slabs, should they be monolithically connected, as discussed in Problem 3.31.

Differential vertical shortening not only happens between individual building components, but also with entire building portions. For example, to prevent differential shortening due to a substantial height difference between the two integrated concrete towers of the 65-story Rialto Towers Building [1985] in Melbourne, Australia, partial vertical prestressing was induced in the shorter tower to generate simulated gravity loads to match the nonelastic deformations.

WATER AND EARTH PRESSURE LOADS

Structures below ground, including buildings with floating foundations resting on compressible soil, are subject to loads that differ from those encountered above grade. The substructure of a building must support lateral pressures caused by earth, earthquakes, ground water, expansive thrust from frozen earth, and surcharge loads on the ground surface. The magnitude and distribution of lateral earth pressures perpendicular to the substructure walls are highly indeterminate. They not only depend on the nature of the soil, but also on the flexibility of the resisting structure. The lateral pressure ranges from an active stage, where the earth is allowed to move laterally (as is true for the more flexible cantilever retaining walls), to a state of rest for absolutely rigid structures (as is usually applicable to the basement wall condition, assuming that the opposite sides balance each other), to the passive stage, where the wall moves towards the earth. In this context, the minimum pressure is caused by the active state and the maximum one by the passive state, with the earth at rest causing a pressure between these two extremes.

For preliminary design purposes, the lateral earth pressure may be treated as an equivalent liquid pressure due to some percentage of the soil weight. For cohesionless soils, sands, and gravels, the minimum active pressure values range from 30 pcf for dry granular soils and a horizontal ground surface to twice as much for wet soils approaching the weight of water. In contrast, cohesive soils (such as clays and silts)

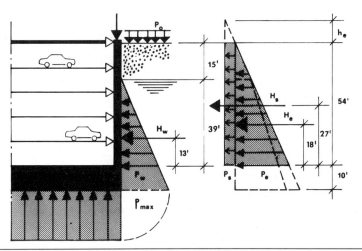

Figure 3.20. *Lateral earth and water pressure on building substructure.*

swell when they absorb water, thereby causing large lateral forces; in this case, the equivalent liquid pressure may be as high as their unit weight of approximately 120 pcf, generating a pressure four times larger than for dry granular soils. It is apparent that the wall backfill should consist of cohesionless material and not saturated soils, so that proper drainage can be provided. It may be concluded that cohesionless soil may be treated as a kind of fluid weighing twice as much as water and exerting about half the pressure.

The equivalent, minimum lateral liquid pressure on the wall in Figure 3.20, for a dry granular soil, is roughly equal to the equivalent fluid unit weight of 30 pcf multiplied by the distance from the top of the soil to the depth in question, thus causing a simple linear pressure variation as a function of height.

$$p_e = 30h \qquad (3.39)$$

Therefore, the maximum lateral pressure at the intersection of wall and foundation is

$$p_e = 30(54) = 1620 \text{ psf/ft of wall}$$

Alternatively, the total resultant earth pressure for the given conditions in Figure 3.20, is

$$H_e = 1.62(54/2) = 43.74 \text{ k/ft of wall}$$

In cases where groundwater is present, a lateral pressure is caused by the unit weight of water of 62.4 pcf.

$$p_w = wh_i = 62.4h_i \qquad (3.40)$$

Due to the hydrostatic pressure, the weight of the soil below the water level is reduced, as the buoyant weight of the soil is equal to its normal dry weight reduced by the weight of the water it has displaced. It is conservative for approximation purposes to neglect the buoyant weight of the soil by considering the dry weight. Also, as a first approach, an equivalent fluid pressure of 80 pcf is often used for the combined action of the noncohesive soil and water, or a 20 percent reduction of the water pressure may be taken for the combined action.

$$p_w = 0.8(62.4)h_i \simeq 50h_i \qquad (3.41)$$

Therefore, the maximum water pressure, as based on a groundwater table 15 ft below grade, is approximately

$$p_w \simeq 50(39) = 1950 \text{ psf/ft of wall}$$

Alternatively, the total lateral water pressure is

$$H_w = 1.95(39/2) = 38.03 \text{ k/ft of wall}$$

The maximum lateral water pressure at the base of the foundation is equal to the buoyancy pressure attempting to lift the building. In the early stages of construction, the upward lift is of major concern. The basement floor slab has to be designed for the uplift force.

$$p_{max} = wh_i = 62.4(49) = 3057.6 \text{ psf}$$

Additional pressure that must be resisted by the wall may be caused by loads acting on top of the backfill due to people, cars, slabs, and buildings. These loads are called surcharge, and may be approximated by uniform strip, line, or point loads. The lateral pressure, p_s, due to a uniform surcharge, p_o, may be approximated by transforming it into an equivalent height, h_e, of earth backfill. The equivalent height of the soil is assumed to be equal to the surcharge, p_o, divided by the weight of the soil, w.

The equivalent lateral pressure due to the surcharge from fixed or moving loads depends on the type of soil. It may be treated as constant along the wall and may be approximated, for noncohesive dry soil conditions and a unit weight of soil equal to $w = 100$ pcf, as

$$p_s = 0.30p_o \qquad (3.42)$$

For a vehicular surcharge load of $p_o = 150$ psf, the uniform lateral pressure is

$$p_s = 0.30(150) = 45 \text{ psf/ft of wall}$$

Alternatively, the total lateral pressure due to the surcharge load is

$$H_s = 0.045(54) = 2.43 \text{ k/ft of wall}$$

LOAD COMBINATIONS

Many of the loads just discussed may act simultaneously and should be combined if they are superimposable. However, these loads are maximized where the probability of their combined action is less than their separate action, hence the separate design loads can be reduced when they are combined. Sometimes, it is unreasonable to let loads act together; for instance, the probability of a wind with a 50-year recurrence interval occurring at the same time as a major earthquake, where most of the damage is done in a period of say 30 to 60 seconds, is extremely small. Hence building codes do not require the design of structures for simultaneous action, but for that of either wind or seismic action. Similarly, for most roof structures, the probability of maximum snow acting together with maximum wind is small; besides, 70 mph winds will blow at least part of the snow off the roof.

In the allowable stress design, as is used in this book for steel and masonry design, the allowable stresses only contain a single safety factor, which takes the effect of overloads, variations in the material properties, residual stresses, and other uncertainties into account. Here, for some of the load combinations, *load combination factors* are used to reduce the combined effect. A similar approach is used in plastic design of steel.

In contrast, in the strength design for reinforced concrete and the load and resistance factor design (LRFD) for steel, the uncertainty of the load is taken into account by *load factors*, while the uncertainty of the material resistance is reflected by the *resistance factors*; the design is based on the actual strength of a member, rather than on an arbitrarily calculated stress. For example, larger load factors are assigned to the live loads than to the dead loads, thus identifying the more unpredictable character of the live loads. The AISC LRFD approach is more rational than the ACI Strength Design, since it is based on probabilistic methods. The following service loads are commonly encountered:

D = dead load due to the weight of the structural elements and the permanent features on the structure (e.g., built-in partitions)
L = live load due to occupancy, movable equipment, and movable partitions
L_r = roof live load
W = wind load
S = snow load
E = earthquake load
R = load due to initial rainwater or ice, exclusive of ponding contribution
T = self-straining forces due to temperature changes, shrinkage, moisture changes, creep, differential settlement, etc.
H = Loads due to the weight and lateral pressure of soil and water in soil
F = loads due to fluids with well-defined pressures and maximum heights
P = loads, forces, and effects due to ponding

As based on allowable stress design, the loads shall be considered to act in the following combinations:

$$D$$

$$D + L + (L_r \text{ or } S \text{ or } R)$$

$$D + (W \text{ or } E)$$

$$[D + L + (L_r \text{ or } S \text{ or } R) + (W \text{ or } E)]0.75$$

$$[D + L + (L_r \text{ or } S \text{ or } R) + T]0.75$$

$$[D + (W \text{ or } E) + T]0.75$$

$$[D + (L_r \text{ or } S \text{ or } R) + (W \text{ or } E) + T]0.66 \qquad (3.43)$$

When the structural effects of F, H, P, or T are significant, they must be considered in the design. To take conditions like that into account, codes allow an increase of allowable stresses or, equivalently, a reduction of loads for certain load combinations. For example, where dead load D acts together with live load L and wind load W, the allowable stresses can be increased by 33 percent or the loads reduced by 25 percent. This is equivalent to multiplying the loads by a load combination probability factor of 0.75. The load combinations above, however, are not all-inclusive. Some material specifications (e.g., masonry, timber) take into account the increased material strength under short-term loads due to wind, earthquake, or blast, and permit the allowable stresses to be increased by a factor of 1.33, or the loads to be reduced by 25 percent. This consideration is applied in this text to masonry walls, when their design is controlled by tensile stresses due to slab action (i.e., curtain wall action): 0.75(D + (W or E)).

When continuous steel members or frames are proportioned on the basis of plastic design, then their strength should be capable of supporting the following factored loads:

$$1.7(D + L) \tag{3.44}$$

$$1.3[D + L + (W \text{ or } E)]$$

In the *elastic* or *working stress theory*, members are designed as based on stresses derived from elastic analysis, with the corresponding material behavior maintained within the elastic range. The service loads cause stresses that must be kept below the allowable ones. The permissible stresses are specified by codes as a fraction of the yield stress as, for example, for steel beams with $F_b = 0.66F_y$, or as a fraction of the crushing strength of the material, such as $F_b = 0.45f_c'$ for reinforced concrete beams, or as a fraction of the buckling load for slender columns. In the *plastic design* of steel, the process is rather similar, only here the ultimate forces and moments at failure due to factored (i.e., ultimate) loading are determined, which cause the total or partial collapse of a structure. The plastic theory takes into account the ability of a ductile continuous structure to form plastic hinges that rotate at constant moment, thereby permitting redistribution of moments from highly-stressed portions to less-stressed members. However, in ordinary reinforced concrete design, due to lack of member ductility, only a partial redistribution of moments, as based on an empirical percentage, is allowed. The factor of safety for both the elastic and plastic theories is determined subjectively, it is applied to all load situations, that is, to the reasonably predictable dead load and the highly variable live loads.

In contrast, the *strength design* method recognizes the uncertainty in loading between the load types and the uncertainty of resistance. Here, the combined effect of the loads is not to exceed the structural resistance to particular failure modes. In other words, the design resistance ϕR_n must be equal to or larger than the required resistance $\Sigma \gamma_i Q_i$, which is computed by structural analysis as based on several load combinations.

Design Resistance $\geq$ Effect of Design Loads, or Required Resistance

$$\phi R_n \geq \sum \gamma_i Q_i \tag{3.45}$$

The term resistance includes both strength limit states and serviceability limit states. The load factors γ and the resistance factor ϕ reflect the degree of uncertainty—that is, inaccuracies in the theory, variations in the material properties and member dimensions, as well as uncertainties in the determination of the loads. Probabilistic methods have been used for the selection of the factors by the AISC LRFD Specification. The resistance factor ϕ is equal to or less than 1.0, because there is always the chance for the actual resistance to be less than the nominal value R_n. The load factors γ are equal to or larger than 1.0, reflecting the fact that the actual load effects may deviate from the nominal mean loads. The factored resistance R_n depends, among other criteria, on the material and limit state, the mode and consequence of failure, and possible errors in the model used for the analysis.

The AISC LRFD Specification for Structural Steel Buildings permits the use of both elastic and plastic structural analyses. It lists the following load combinations with the corresponding load factors, as based on the ANSI Specs:

$$
\begin{align}
&(1) \quad 1.4D \\
&(2) \quad 1.2D + 1.6L + 0.5(L_r \text{ or } S \text{ or } R) \\
&(3) \quad 1.2D + 1.6(L_r \text{ or } S \text{ or } R) + (0.5L \text{ or } 0.8W) \\
&(4) \quad 1.2D + 1.3W + 0.5L + 0.5(L_r \text{ or } S \text{ or } R) \\
&(5) \quad 1.2D + 1.5E + (0.5L \text{ or } 0.2S) \\
&(6) \quad 0.9D - (1.3W \text{ or } 1.5E) \tag{3.46}
\end{align}
$$

Exception: The load factor on L in combinations (3), (4), and (5) shall equal 1.0 for garages, areas occupied as places of public assembly, and all areas where the live load is greater than 100 psf.

When the structural effects of F, H, P, or T are significant, they shall be considered in designs as the following factored loads: 1.3F, 1.6H, 1.2P, and 1.2T.

The more common resistance factors range from $\phi = 0.9$ for tension yielding and for beams in bending and shear, $\phi = 0.85$ for columns, and $\phi = 0.75$ for tension fracture, to $\phi = 0.6$ for bearing on A307 bolts.

The ACI Building Code for Reinforced Concrete lists the following load combinations with the corresponding load factors, γ,

$$
\begin{align}
&1.4D + 1.7L \\
&1.4(D + T) \\
&0.75 (1.4D + 1.7L + 1.7(W \text{ or } 1.1E)) \\
&0.75 (1.4D + 1.4T + 1.7L) \\
&0.9D + 1.3(W \text{ or } 1.1E) \tag{3.47}
\end{align}
$$

The resistance or strength reduction factors, ϕ, for several situations are:

Axial tension, flexure, tension plus flexure	0.90
Shear and torsion	0.85
Flexure in plain concrete	0.65
Bearing in concrete	0.70
Columns with ties	0.70
Columns with spirals	0.75

$$\tag{3.48}$$

ABNORMAL LOADS AND STRUCTURAL SAFETY

Because buildings are made by humans, there always will be failures, be it a catastrophic collapse, a near-collapse, or simply defects of materials, which continue to deteriorate. This failure may be local or may involve the entire building; it may be sudden or occur over time through environmental exposure, incompatibility of dissimilar materials, aging, lack of maintainability, or settlement. Structural failures may happen during construction or when the building is finished. They may be due to design errors or simply ignorance; unanticipated building use; poor construction practices; lack of quality control; inadequate communication between design professionals, builder, and project manager; incomplete or incorrect working drawings; or a lack of field supervision. Failures may also be due to abnormal loads such as explosions, fire, flooding, tornadoes, and vehicle collision. The risk of failure is greater when the building lacks structural redundancy, that is, a back-up system with reserve strength.

Especially since the late 1970s and early 1980s, the media has been reporting many building failures, including the "sick building syndrome" of air-conditioned buildings with fixed windows, and has questioned the competence of designers. This state of bad quality workmanship, malfunctioning buildings and systems, indoor air pollution, and so on, is reflected in rising liability insurance premiums or the unavailability of insurance, and the increase in court cases. There has also been much concern about the durability or life expectancy of buildings constructed in the last 25 years or so because of changing building methods, although it may be reasonable to design a 20-year obsolescence into special building types, as long as control of the process exists. While the failures of the past were usually due to experimentation with new construction techniques and record-setting innovations, the current failures are often due to a fragmentation of responsibility that separates the designer, fabricator, contractor, manager, and inspector. One of the frightening developments to come out of the recent concern for safety at any cost, and the reduction of any risk of failure through regulations, however, has been the stifling of one of the most exciting aspects of building design, that of experimentation, adaptability, innovation, and invention.

It is the large-scale structure failure and threat to life that remind the designer, contractor, and manager to keep searching for means of protection to guarantee structural safety. As early as c.1760 B.C., Hammurabi, the king of Babylon, included in his famous law code a rather severe penalty for defective construction. The Romans also had building regulations emphasizing safety.

There have been many classic building failures in the past from which designers have learned. Among them is the first collapse of the dome of the Hagia Sophia in Istanbul, due to an earthquake in 557 A.D., only twenty-one years after it had been completed, because of insufficient buttressing. The Gothic cathedral at Beauvais, France, collapsed twice. It was supposed to reach the ideal expression of lightness and delicacy of structure, and an unprecedented vault height of more than 157 ft. First, the main vaults of the choir collapsed in 1284, just after it had been finished, probably due to excessive loading. Then, in 1573, the stone tower of an incredible height of 490 ft collapsed because of an incorrect layout of the support structure. In more recent times, the 1940 collapse of the too-flexible Tacoma Narrows Suspension Bridge, Washington, has become a classic example of the lack of proper aerodynamic design. The progressive collapse, in 1968, of one corner of the 22-story Ronan Point Apartment

Building (see fig. 6.19) in London, England, due to a gas explosion on the eighteenth floor, caused a reappraisal of the structural integrity of large-panel building construction, and industrialized systems for tall buildings in general. Several long-span roofs failed in the late 1970s, among them the space frame collapse of the Hartford Civic Center Coliseum in 1978, due to a design error and inadequate inspection, among other reasons. Proper procedures were not followed, causing the collapse of a jack-up formwork system on a power plant cooling tower in West Virginia in 1978, killing 51 workers.

The eroding infrastructure of cities has become of major concern. Bridges have been falling at an alarming rate in the 1980s. More than 40 percent of all U.S. bridges have become structurally deficient or functionally obsolete. Bridges in particular wear out with age because of exposure to the weather, impact loading, and stress reversals. Not only ordinary concrete and steel bridges, but also most of the cable-stayed bridges in the world show premature cable deterioration due to corrosion. Bridges must be inspected and repaired regularly, which often is not the case.

Still fresh in memory is the disaster of the Kansas City Hyatt Regency sky walkways collapse of 1981, due to a design error; 113 persons were killed and 186 were injured. One of the worst construction accidents in recent years was the collapse of the highrise L'Ambiance Plaza Apartment Building at Bridgeport, CT, in 1987, using the liftslab construction technique. In this case, the failure of one of the critical support details for the lift slab triggered a chain reaction and progressive collapse that killed 28 workers and injured many others.

In the building field, the word failure is often used quite loosely to represent a wide range of imperfections. Failure does not only mean a building collapse, but also component defects that may possibly result in the breaking down of members. For example, some reinforced concrete buildings, especially parking garages, built more than 20 years ago show spalling, cracking, and rusting of reinforcing steel, causing progressive deterioration. Most critical are deficiencies at joints or connections. For example, the connection details of prestressed concrete members often do not take into account the continuing motion (i.e., creep) caused by the constant compressive prestress force. When this movement is prevented, the buildup of stresses at the joints may result in subsequent cracking. Another problem is the quality of materials and products, made by foreign manufacturers (and passing through several middlemen), which may be substandard and/or counterfeited.

Recently, there have been troubles with exterior facade walls, adding to the already prevalent problems of the cracking of partition walls and traditional roofing failures. The corrosion of attaching devices for cladding and water intrusion are constant problems. Only 15 years after completion, the entire skin of the 1136-ft high AMOCO building in Chicago had to be replaced with 2-in. granite because of bowing of the original $1\frac{1}{4}$-in. marble panels. There are numerous cases of broken glass sucked out of its framing and blown by turbulent winds against other glass on its way down, possibly injuring people. Often, this is due to a lack of rigidity in the building structure because of the development of high-strength materials and the trend towards optimization, as well as noncomposite action of the nonstructural materials; buildings have become lighter, more flexible, and vulnerable to dynamic excitement. A noted example is the 60-story John Hancock Tower [1973] in Boston. This mirror-glazed, 790-ft high slender steel structure, a 117×290 ft rhomboid in plan, has won national honors for its aesthetic qualities, but at the same time has earned a worldwide reputation for its technical flaws. Early during its construction, the foundation walls caved in, causing broken utility lines and settling of adjacent buildings. An investigation into the performance of the building started after a January 1973 storm caused extensive glass breakage, and eventually about one-third of the tower's glass had to be replaced with plywood. It was found that the excessive movement of the building, in the short direction and in torsion, caused accelerations that would have discomforted the inhabitants; tuned mass dampers had to be installed at opposite ends of the 58th floor. Further, it was found that the building was much too flexible along its long axis, so that, surprisingly, the P-Δ effect, together with resonant loading, could have caused the building to topple endwise. As a result, diagonal core bracing had to be added to the steel frame to double the stiffness in the long direction. It seems that there was no relationship between the building movement and the breakage of the large double-glazed reflective units, which appears to have been caused by fatigue failure of the seal between the two pieces of glass at the edges, due to cyclic wind loading and thermal stressing, since the joint lacked flexibility. One of the beneficial aspects of the technical problems of the John Hancock Tower was the further development of the science of wind tunnel testing.

Building failures are thoroughly studied and recorded by experts to learn from mistakes and to prevent similar events in the future. The discipline of *forensic engineering* has gained considerable recognition in recent years; it has developed procedures of failure investigation and evaluation. Concern over the short lifespan of newer buildings, as reflected by numerous defects, and also over the decay of older buildings, as well as the need for the evaluation of energy efficiency and the trend towards building restoration and preservation, has made necessary the development of *building diagnostics* as a new field of specialization. Similar to preventive medicine, it assesses and judges how well the building performs.

In this context, the concern is primarily the prevention of structural failure during construction and after completion of the building. These failures may be due to uncertainties of loading conditions (e.g., overloading, or accidental and low-probability loading), or uncertainties in the strengths of the materials, but probably most often due to human errors. Structural failures attributable to human error can include any of the following: poor design, inadequate layout of support structure, sloppy workmanship, substandard materials, faulty shop drawings and documentation conflicts, lack of quality control during the design and construction stages, etc.

While ordinary loading conditions have been discussed in the previous sections, *abnormal* or *accidental loads*, which are generally not considered in the design because of their low probability of occurrence, and because they are assumed to be resisted by the reserve load capacity, are often the cause of building failure. Examples are:

- Explosions, external or internal blast loads (e.g., gas service system, bombings)
- Collisions, impact loads (e.g., wind-blown debris, crane, vehicle, aircraft)
- Sonic boom
- Tornado
- Flooding
- Fire
- Vandalism and terrorism

The safety factor in structural design can only very crudely deal with all of the uncertainties just mentioned. It evaluates the performance of the building primarily with respect to safety, but it also reflects economy and serviceability. As has been stated in the section of this chapter on "Load Combinations," the safety factor, with respect to public safety, is measured in working-stress design as the ratio of steel yield or concrete crushing strength to allowable stress, and in load-factor design it is measured as the ratio of the factored load to the design strength. While in elastic and plastic design, a simple safety factor applies to all load situations and other uncertainties, in the strength design for reinforced concrete and the LRFD for steel, the uncertainties of the various loading conditions and material resistance are taken into account. This limit design concept reflects the rapid development toward quality assurance. In this instance, the quality is controlled by functional and more critical strength requirements, but the expertise, competence, and reputation of the design and construction team is not taken into account, nor is peer review considered. *Limit states of strength* refer to public safety, and are based on the maximum load-carrying capacity, derived from the actual collapse of the whole or a part of the structure, taken as for brittle structures (fracture) or hinged structures (instability), or may be derived from the conceptual failure, such as formation of plastic hinges and plastic mechanisms in ductile structures, allowing extensive plastic deformations without collapse. *Limit states of serviceability* refer to functional performance under normal service conditions, and are concerned with the performance of the structure with respect to the control of member deflection, crack widths in concrete, building drift, accelerations, and leakage of cladding and roofing.

The design of beams, slabs, columns, walls, and connections is generally based on the individual element, taking into account different modes of local failure (e.g., shear, flexure, instability, deflection), hence the safety factor is a function of the limit states of strength and serviceability of each building element, and is independent of the building system as a whole. The effect of the *importance* of a member as part of the entire structure is not taken into account by the resistance factors. For example, a column at the base of a skyscraper is treated like one at the top or one of a two-story building, and the *uniqueness* of a diagonal column for a perimeter tube, as contrasted with an interior gravity column, is not recognized. Further, the amount of *redundancy* provided by a structure system against total collapse, such as the failure of a column in a tubular frame versus one in a braced hinged frame which may cause a progressive collapse, is ignored.

Redundancy is a most important consideration for the prevention of building failures; it is a basic concept of design. If something should be happening to the structure it should be able to defend itself; it should tolerate overloading and provide multiple independent load paths, and it should at least allow for adequate warning of impending disaster. Failure of a single element should not result in a chain reaction of failures and the progressive collapse of a larger building portion. For example, if one cable of a stayed bridge should snap, a progressive collapse could be triggered that would be sudden and without warning. The designer should anticipate failure at various critical locations. In a way, it is similar to the demolition of a building by detonating explosives at key positions so that the critical supporting members are blasted away, thereby causing a progressive collapse of the building under its own weight.

The element as a nonredundant system should be designed for a lower allowable stress range, but as a redundant member showing signs of distress, such as sagging and cracking, the structural layout will be capable of redistributing the forces, and, as a back-up system, will allow load flow along multiple, alternate paths. The degree of redundancy of a structure must be checked by anticipating failure mechanisms at various locations, so that the number of potential modes of failure can be established, and the consequences of failure can be evaluated. The degree of static indeterminancy of a structure is equal to its redundancy. It is apparent that a typical high-rise building has a much larger number of potential failure mechanisms than an ordinary low-rise building. A tubular structure has a higher redundancy than a nonbraced, flat plate structure, which is especially vulnerable to shear failure around a column. It must be emphasized that redundancy does not mean overdesign and a lack of economy, but is, rather, a function of the type of structure system.

Continuous ductile structures have much more redundancy than brittle structures of the same layout. Although masonry buildings are made of brittle material, they are quite rigid, due to their dense cellular arrangement of walls and heavy weight, and usually never fail in sudden fracture, as may have been the case in cast-iron frame structures of the nineteenth century. The familiar failures of medieval masonry structures in cathedrals were generally caused by a reaction to lateral thrust and foundation settlements, and hence were problems of form and much less of material strength. Because of the lack of ductility, brittle structures must be designed for near-elastic response, with more strength built in to resist unpredictable force action, such as severe earthquakes, thereby requiring much larger safety factors.

However, it must be kept in mind that, under extraordinary conditions, a ductile steel structure may suddenly fail in a mode quite different than expected. These sudden failures can occur either due to brittle failure or fatigue. Brittle behavior may be generated under certain unusual situations such as those that result from the fabrication process (e.g., residual tensile stresses), low temperature, high stress concentrations, and other conditions. An example of this type of failure is the brittle fracture of welded ships during World War II. Fatigue is associated with repeated cyclic loading causing sudden failure at a stress considerably below the ultimate tensile strength. Fatigue failures are most likely to occur in tension members of bridges and aircraft, or in machine parts and supporting structures for cranes and vibrating equipment—that is, in structures with a large number of load fluctuations and a wide range of stress reversals. In conventional buildings, however, repeated maximum service loads are generally not a design factor because they are below 20,000 cycles, which is approximately equivalent to two applications every day for 25 years. Therefore, no consideration needs to be given to fatigue loading in ordinary buildings, which includes wind and earthquake loads, because they occur too infrequently.

The material characteristics of ductility and brittleness may not be involved at all when the members lack continuity and are hinge connected. In this case, failure occurs in the joints representing the weakest link in the chain, as in precast concrete construction. A familiar example is the potential lack of structural integrity in large-panel concrete buildings, where local failure at the joints, which often depend only on simple gravity and friction action (i.e., friction connection), may cause a progressive collapse, if alternate paths of load flow are not provided and redundant systems created (see fig. 6.9). It is clear that, in seismic regions, continuity and ductility are necessary to provide sufficient reserve strength. A structure must act as a unit and not as an assembly of individual components. Because of the brittle nature of concrete frames, special ductile concrete frame construction (see fig. 7.20) is required to absorb severe earthquakes without causing a progressive collapse after a local failure. Similarly, reinforced masonry was introduced on a larger scale in the 1930s, with the reinforcing

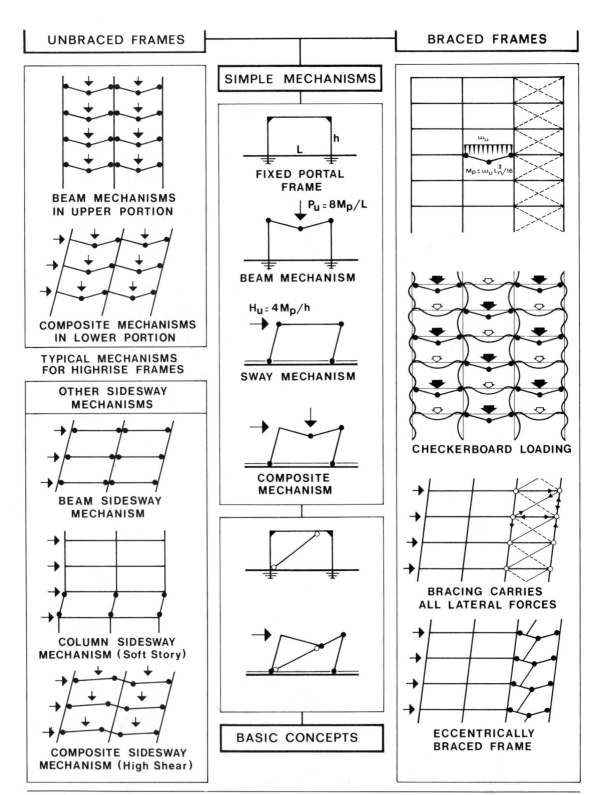

UNBRACED FRAMES

BEAM MECHANISMS
IN UPPER PORTION

COMPOSITE MECHANISMS
IN LOWER PORTION

TYPICAL MECHANISMS
FOR HIGHRISE FRAMES

OTHER SIDESWAY
MECHANISMS

BEAM SIDESWAY
MECHANISM

COLUMN SIDESWAY
MECHANISM (Soft Story)

COMPOSITE SIDESWAY
MECHANISM (High Shear)

SIMPLE MECHANISMS

L   h

FIXED PORTAL
FRAME

$P_u = 8M_p/L$

BEAM MECHANISM

$H_u = 4M_p/h$

SWAY MECHANISM

COMPOSITE
MECHANISM

BASIC CONCEPTS

BRACED FRAMES

ω_u

$M_p = \omega_u L_n^2/16$

CHECKERBOARD LOADING

BRACING CARRIES
ALL LATERAL FORCES

ECCENTRICALLY
BRACED FRAME

Figure 3.21. *Common collapse models for ductile frame structures.*

providing a better resistance to earthquakes by tying the masonry building together to form an integral three-dimensional structure.

Typical failure modes for shear walls are shown in Figure 6.12, and numerous common collapse models for ductile or semiductile frame structures are shown in Figure 3.21, where the various loading patterns and load intensities mobilize a number of potential load paths, depending on the member strengths, stiffnesses, and arrangement. It is not a simple task to determine, for a highly indeterminate or redundant structure, the case that produces total collapse with the smallest load. It becomes even more difficult when collapse of the connecting floor planes is taken into account (e.g., yield line theory for slabs). Gravity loading may form just local failures, while combined loading may produce complete collapse mechanisms. First, localized failures due to flexural yielding occur at points of high stress, usually at the joints, by forming plastic hinges that rotate under a constant moment and allow an increase of load action in continuous (i.e., indeterminate) structures, thus permitting a redistribution of moments. This is quite in contrast to a brittle hinge, where fracture has occurred and the resistance to rotation is lost. The complete collapse of a redundant structure happens when enough hinges have formed to make the structure a mechanism. Since structural design damage is permitted under severe earthquakes, it is apparent that a frame's stability due to hinge formation must be evaluated, so as to prevent building collapse. The reader may want to refer to the respective literature on plastic analysis and design to truly visualize the behavior of a frame structure in the post-elastic range.

In braced frames, continuous beams can be immediately designed by assuming the formation of three plastic hinges after checking local buckling. Columns fail also by the formation of three hinges, or hinges in the columns and adjacent beams. Concentrically braced, simple frames are stiff laterally, but provide only moderate ductility. Therefore, in regions with severe earthquakes, bracing must be used (together with rigid frames) to achieve the necessary ductility after the braces have buckled. On the other hand, eccentrically braced frames are stiff at moderate seismic loads and ductile at severe loads. In this case, the axial forces in the braces are transferred through shear in the beams, forming a plastic shear hinge for short weak links (see fig. 7.24).

In contrast to braced frames, the plastic hinge formation in unbraced moment-resisting frames depends primarily on beam stiffnesses. For the condition where beams are very stiff, the building sway is primarily due to column bending, thus causing *column sidesway mechanisms*, but where the beams are very flexible, sway is primarily due to bending of beams causing *beam sidesway mechanisms*. It is reasonable to assume that, at the upper portion of a multistory building, gravity controls and three hinges form in beams. However, at the base, where the combined action of gravity and lateral force action is predominant, a *composite sidesway mechanism* develops, with plastic hinges possibly occurring at the ends of the beams.

Rigid frames are capable of large deformation, and therefore are quite ductile. But because they are very flexible they also cause excessive nonstructural damage during a severe earthquake. To prevent this from happening, the story drifts must be reduced by increasing the beam sizes beyond that required for strength.

In the following discussion, some basic mechanisms are briefly described.

- *Beam Mechanisms:* Typical floor beams form a local collapse mechanism under uniform gravity loading, with plastic hinges at the ends and midspan. Therefore, the total ultimate moment, $M_u = w_u L^2/8$, is resisted equally by the plastic moments, M_p, at the hinges.

$$M_u = w_u L^2/8 = 2M_p \quad \text{or,} \quad M_p = w_u L^2/16 = 1.7wL^2/16 \leqslant Z_x F_y \quad (3.49)$$

Therefore, the plastic section modulus, Z, can be obtained from

$$Z_x = M_p/F_y \quad\quad\quad (3.50)$$

Notice that the plastic moment capacity is a function only of load and span, and is not affected by the adjacent member sizes, as is true in the elastic approach.

The plastic reserve strength after initial yielding is measured by the shape factor, f, which has a mean value of 1.12 for W-sections.

$$f = M_p/M_y = Z/S \simeq 1.12 \quad\quad\quad (3.51)$$

Substituting Equation (3.51) into Equation (3.49), to replace the plastic section modulus by the elastic section modulus, yields

$$M_{max} = wL^2/16 \leqslant 1.12S_x F_y/1.7 = 0.66F_y S_x = F_b S_x \quad (3.52)$$

However, according to the elastic theory,

$$M_{max} \simeq wL^2/12 \leqslant F_b S_x = 0.66F_y S_x$$

It may be concluded that, based on plastic design, a continuous beam can be designed for $wL^2/16$, instead of $wL^2/12$ as in elastic design, thereby saving about 25 percent of the girder weight, assuming that stability and deflection requirements are met.

Several modes of failure are formed under lateral loading and gravity.

- *Beam sidesway mechanisms* under lateral loading are formed first, when the columns are designed for a larger moment capacity than the beams. This *strong column-weak beam* concept is the preferred potential collapse mechanism for earthquake design, and recommended by most codes.
- *Column sidesway mechanisms* are formed when column hinging occurs first, and a story mechanism develops, causing lateral instability. This mechanism should be avoided for high-rise buildings, although it should be anticipated at the first story level because of the load reversal of earthquake action. The so-called *soft-story design concept* used to be popular and provided very ductile columns, usually at the base, to confine inelasticity to this level and to absorb the seismic forces like a shock absorber, thereby protecting the rest of the building and keeping it elastic.

  The concept of the column sidesway mechanism can be explained with the simple fixed portal frame in the center part of Figure 3.21. In this case, an ultimate lateral load, H_u, causes a moment which is balanced by each of the two columns equally.

$$M_u = (H_u h)/2$$

This moment, in turn, is resisted equally by the plastic moments, M_p, at the top and bottom of the column, assuming that the moment capacity of the beams is greater than that of the columns.

$$(H_u h)/2 = 2M_p, \quad \text{or} \quad M_p = H_u h/4 \quad\quad (3.53)$$

The column will be designed for the combined loading case of gravity and lateral force action.

- *Composite sidesway mechanisms* are generated both by lateral and gravity loading, as indicated in Figure 3.21 for several situations. In this context, the columns may fail for stiff beam conditions, due to the development of a mechanism under axial loads and bending, magnified by second-order effects. However, in general, the strong column-weak beam concept should be maintained throughout the structure. In other words, the sum of the plastic moment capacity of a column above and below a joint should be greater than the sum of the plastic moment capacities of the beams on either side of the joint. This condition can be achieved by selecting a high-strength steel for the columns and A36 steel for the beams.

PROBLEMS

3.1   For the 15-story, flat slab office building, determine the column loads as indicated in Figure 3.2 f and g, at the first floor level and then at the thirteenth floor level. Assume a dead load of 140 psf, and a live load of 80 psf; the curtain weighs 20 psf of wall area. Consider the roof load as 75 percent of the floor load, and assume a typical story height of 12 ft.

3.2   Determine the total floor load that a typical interior wall of the various structure systems, using one-way and two-way slabs (in fig. 3.2 a to e), must support at each floor level. Only for the long-wall structure (Fig. 3.2b), investigate the exterior wall by first assuming a continuous solid wall, and then a perforated wall consisting of 5-ft wide vertical strips (i.e., only 50 percent of facade wall is available for bearing). Use a floor load of 100 psf.

3.3   The typical floor framing of a high-rise office building consists of a 2 1/2-in. concrete slab on cellular steel decking that spans 15 ft between the beams (see fig. 3.2h). Assume the dead weight of the floor as 60 psf, which includes 15 psf for the ceiling, and assume a live load of 80 psf. Set up the loading for beams 3-4 and 1-5. Design the beams for simply supported conditions by using $F_y = 50$ ksi. What steel would you select?

3.4   Assume the floor structure of Problem 3.3 to be typical for a 20-story building with an average story height of 12 ft. The roof dead load may be considered to be equal to the typical floor dead load, but the live load is 30 psf. Determine the load of columns including fireproofing, connection material, etc. as constant and equivalent to 12 psf of floor area. Do a rough estimate of the column size (A36) at the second floor level, considering the building laterally braced.

3.5   Determine the distribution of the gravity loads for a ten-story apartment building that has an average story height of 10 ft and a typical floor framing as shown in Figure 3.2i. Assume the same floor structure as in Problem 3.3, with the same dead load; the live load for the apartment area is 40 psf and for the corridor 60 psf. Determine the sizes for the simply supported beams B1, B2, and B3, as well as for the hinged girder G1. Further, find the load on the second floor level for a typical interior column and estimate the column size, using the same assumptions as in Problem 3.4. Take A36 steel.

3.6   Determine the loading for girder G1 in Problem 3.5, ignoring the filler beam B2 and assuming a uniform load. Is this a reasonable first assumption for determining the girder size? Compare the moments.

3.7   Determine the core loads at the third-floor level, just due to the floor and roof loading of 150 psf, for the three eight-story core structures of Figure 3.2j. Also, show the axial force flow in the cores.

3.8   a. Determine the approximate natural period and frequency of the pole in Problem 2.5. For this preliminary approach place the entire weight concentrically at the top.
      b. A vertical 9-ft long cable has a cross-sectional area of 0.459 in.^{2} and supports a weight of 10 k. What is the natural period of the system? Use a modulus of elasticity of E = 24,000 ksi.
      c. Determine the approximate period of oscillation of the structure in Example 2.1 by using Equation 3.13.

3.9   Determine the critical loading for a 25-story rigid frame building in the short direction. The rigid frames are spaced 25 ft apart in the cross direction, and 20 ft in the long direction. The plan dimension of the building is 175 × 100 ft, and the structure is 25 × 12 = 300 ft high. Assume seismic zone 4 and a critical wind velocity of 70 mph. Use an average dead load of 195 psf. Ignore the minimum eccentricity. What are the seismic forces at the twenty-fifth and twenty-fourth floor levels? Check the building stability.

3.10 Determine the largest torsion at the tenth floor level for the building in Problem 3.9.

3.11 A ten-story rigid frame concrete building has an average dead weight of 185 psf. The frames are spaced 30 ft apart along the short direction, and 25 ft along the long direction; the building is 10(12) = 120 ft high. Determine the total lateral shear on the 200 × 90 ft structure, which is located in an area of high seismicity and is not an essential facility. Check the building stability and compare the seismic action with an average wind pressure of 30 psf. Also, find the seismic shear at the third floor level.

3.12 Check the stability of a typical interior cross wall for the eight-story apartment building in Figure 3.13a. The structure is located in seismic zone 4 and must withstand a wind velocity of 80 mph. The layout of the building is simplified so as to have symmetrical conditions, and an equal thickness is assumed for the partially grouted 10-in. hollow concrete block walls, with a weight of 80 psf. Use a roof dead load of 70 psf and a floor dead load of 90 psf. The typical story height is 9 ft. For this quick check ignore the wall flanges and the minimum eccentricity of the earthquake force.

3.13 For the ten-story laterally braced hinged-frame building in Figure 3.13c, determine the lateral pressure distribution due to a wind action of 80 mph against the long facade, and find the lateral forces at every floor level. Then figure the magnitude of the overturning moment.

3.14 Investigate the building in Problem 3.13 by assuming an average uniform wind pressure. Find the overturning moment, and compare the result with the precise solution.

3.15 Determine the lateral force distribution caused by seismic action (zone 3) for the ten-story frame building in Figure 3.13c. The steel frame building obtains its lateral rigidity by rigid frame action along the facade in the long direction of the building and by K-bracing along the facade in short direction of the building. Assume the following typical floor weights:

3-in lightweight concrete fill (110 pcf):	40 psf
3-in. metal deck:	3 psf
Ceiling and mechanical ducts:	5 psf
Partitions:	20 psf
Assume floor framing, columns, and spray-on fireproofing as equivalent to:	<u>15 psf</u>
Total floor weight is:	83 psf
Curtain wall, including columns and spandrel covers, is:	15 psf

For this approximate approach assume, conservatively, a roof weight equal to floor weight, and consider the first floor slab to be slightly above ground. Show load distribution for both building directions, and determine the overturning moment in the short direction.

3.16 Compare the wind and seismic action of Problems 3.13 and 3.15 and determine which one controls the design. What are the safety factors against overturning by considering the facade K-bracing as given (i.e., bracing the full width of the building at the base) and then assuming only the central bay to be braced. What conclusions do you draw?

3.17 A 26-story office building (see fig. 3.13d) is laterally stabilized by four rigid-frame bents across the short direction, which (for this preliminary investigation) are assumed to have the same stiffness, and X-braced frames in the longitudinal direction along the service core as indicated by the dotted lines. The typical story height is 13 ft. Assume the loading to be as shown in the following list.

- Dead load: Consider roof dead load equal to the floor load.

The floor system consists of open-web joists at 2-ft on center with a 2 $\frac{1}{2}$-in. concrete slab on steel decking:	30 psf
suspended fire-rated ceiling:	7 psf
mechanical and miscellaneous:	<u>3 psf</u>
	40 psf
average load for columns, bracing, connections, etc.:	<u>12 psf</u>
	52 psf
facade wall loads; metal cladding:	25 psf

- Live load:

roof	30 psf
office area, including partitions, neglecting the 100 psf for public corridor in this approximate approach:	80 psf

- Wind load: neglect gradual increase in wind loads according to codes and assume constant wind pressure for full height: 30 psf

    a. Determine the loads for a typical open-web steel joist; consider live load reduction. Show the beam with loading.

    b. Determine the loads for floor girder AB; consider live load reduction. Show the beam with loading.

    c. Determine the loads for column 'A' at first floor level; consider live load reduction.

    d. Determine the seismic forces at roof and tenth floor level for earthquake action against the long facade (i.e., parallel to rigid frames) for zone 2. What is the safety factor against overturning? Consider wind and seismic action. Which of the two is more critical?

3.18 Investigate a 14-story commercial building, located in San Francisco, across its critical short direction (Fig. 3.13f). A central core system carries 75 percent of the lateral load, while the rest of the loads are equally shared by the moment-resisting facade frames. Assume a typical average equivalent floor dead load of 130 psf and an average wind pressure of 25 psf for this preliminary investigation. The average floor height is 13 ft. Ignore e_{min}, that is, the accidental torsion.

    a. Determine the seismic shear each lateral force resisting system must carry.

    b. Determine the seismic force at the seventh floor level.

    c. Determine the total wind shear and rotation. Compare the results with the seismic action.

    d. Determine the safety factor against overturning for the controlling case by treating the building as a single block for first approximation purposes rather than checking each of the resisting structure systems.

3.19 The eight-story building structure shown in Figure 3.13e, is located in the seismic probability zone 2 (moderate damage). Assume the weight of the floor structure and the vertical framing to be equivalent to 100 psf of floor area, and the weight of the facade structure to be 15 psf of facade area. Consider the roof weight to be equal to a typical floor weight; ignore the accidental torsion.

    a. The braced frames at the central bay of each facade are designed to resist the lateral forces. Determine the total base shear and the force action at the fourth floor level, as well as the total wind force for an average pressure of 24 psf, and compare it to the seismic force, but do not compare rotations.

    b. How is the seismic force distributed to the braced frames? Give the numerical value for the force each wall carries.

    c. According to the code requirements, is the building safe against overturning? If it should turn out not to be safe, what corrections do you propose?

3.20 Check the stability conditions of the building in Problem 3.16 at the base of the fourth floor level, where the truss changes from the full width to only the central bay.

3.21 The weight of an elevator is 1200 lb, and that of the people 800 lb at rest. If the elevator accelerates 12 ft/s², what is the weight during the ascension? What is the tension in the elevator cable? Give the percentage increase in loading due to the acceleration. The acceleration due to gravity is 32.2 ft/s².

3.22 The period of a rigid frame building is estimated as T = 0.1N, which is reasonable for very tall buildings. For lower rigid frame buildings, the following expression gives more reliable estimates:

$$T = 0.25\sqrt{\Delta/C_1}$$

where: $C_1 = ZICS$,   and $\Delta \leq 0.005II$ (in.)

Check the period of Example 3.3 and compare the results.

3.23 A longwall masonry structure consists of 22-ft wide outside bays with a 5-ft wide corridor along the center. Investigate the free movement of the roof concrete slab by treating it as 1-ft wide parallel strips and considering the insulation on top of the slab as ineffective.

   a. How much does the slab contract in its early life, using an average shrinkage factor of 0.0005 in./in. and a coefficient of expansion of $\alpha = 5.5 \times 10^{-6}$ in./in. per degree Fahrenheit assuming an outside temperature of $-25°F$ and an inside temperature of 75°F. Ignore bending of the slab.

   b. Determine the forces which the slab exerts upon the facade walls if it is anchored to them. Assume a 4-in. slab of 4000 psi concrete.

   c. Investigate the expansion of the slab under a summer temperature of 110°F and an interior temperature of 75°F by providing slip planes but not vertical expansion joints between slab and wall. Roughly determine the bending stresses in the 7 5/8 in. thick, 10-ft high masonry wall for this loading condition.

   d. Determine the required joint width if slip planes are provided and the insulation is placed at the underside of the slab rather than on top, so that the slab is not restrained from movement by the inside structure.

3.24 What would happen in reinforced concrete if the coefficients of thermal expansion for steel and concrete were not approximately the same?

3.25 Determine the stresses in a concrete wall of 4000 psi at a temperature of 110°F. The wall is 120 ft long at 30°F with 1 1/2-in. expansion joints at the ends that allow a reduction in width of a 1/4 in., and is prevented from expanding by adjacent building blocks. Neglect the interaction of the wall and foundation and any other cross walls.

3.26 Treat the column in Example 3.6 as exposed, without any insulation. Determine the approximate critical stresses in the roof beams and first floor columns.

3.27 Investigate the column in Example 3.6 for expansion due to the highest summer temperature of 120°F. First treat the columns as exposed and then as partially exposed.

3.28 A 20-ft steel beam, W18 × 35 of A36 steel, is exposed to an outdoor temperature of $-5°F$ and an indoor temperature of 75°F.

   a. Determine the maximum thermal deflection of this partially exposed, simply supported beam.

   b. Consider the beam restrained, and determine the bending moment caused by it. Further, find the critical thermal stresses as caused by prevention of linear and rotational movement.

3.29 A 6-in. concrete slab of 60-ft length rests on a subgrade, with an average coefficient of friction of $\mu = 1.5$. The tension in the slab will result from the friction force on the ground that resists part of the shortening due to temperature drop and drying shrinkage. Notice that, for this condition, the slab is not fully restrained from moving and depends on the degree of frictional resistance. Any permanent live loading is ignored in this investigation.

   a. Determine the amount of reinforcing using $F_t = 0.66\ F_y \simeq 40$ ksi for Grade 60 steel.

   b. Determine the maximum spacing of the joints, using no steel reinforcing, so that the slab does not develop any cracks. Use 3000 psi concrete with a tensile capacity of $F_t = 7.5\sqrt{f_c'}$.

3.30 Determine the width of the joints for a slab with 40-ft joint intervals, assuming a temperature differential of $\Delta T = 100°F$ and ignoring the frictional resistance of the ground.

3.31 A 12×24 in. concrete spandrel beam of a floor slab is located outside the glass-line of a building. The beam is subject to an average temperature of $-20°F$ while the interior floor slab is exposed to $+70°F$. Since beam and slab are rigidly connected, the beam cannot reduce in length. Determine the amount of steel reinforcing for $F_t = 24$ ksi of Grade 60 steel to resist the tensile stresses in the beam.

3.32 Give a rough estimate of the width of the horizontal expansion joints placed at 12-ft intervals under the shelf angles that support sandstone cladding for a high-rise building. Assume, conservatively, shortening of the concrete columns of 0.0009 in./in. due to elastic strain, creep, and drying shrinkage. Use a coefficient of expansion for sandstone of $\alpha = 6 \times 10^{-6}$ in./in. per degree Fahrenheit to determine the expansion of the cladding material due to extreme summer temperatures. Use $\Delta T = 100°F$.

3.33 Estimate the cumulative shortening of the load-bearing concrete wall of a ten-story structure. The walls consist of precast 9-ft high normal-weight concrete panels. Consider an outside temperature of $T_o = -20°F$, and an inside temperature of $T_i = 70°F$. Use the following typical factors for ordinary conditions to determine the approximate volume movement: creep factor = 0.00012 in./in., shrinkage factor = 0.00020 in./in., and a coefficient of expansion of $\alpha = 5.5 \times 10^{-6}$.

3.34 An aluminum frame of 6061 alloy for a window is 15 ft long and holds a piece of plate glass 14.99 ft long when the temperature is 60°F. In other words, the glass floats in the opening. At what temperature will the aluminum and glass be at the same length?

$$\alpha_{al} = 13 \times 10^{-6} \text{ in./in./°F}, \qquad \alpha_{gl} = 5 \times 10^{-6} \text{in./in./°F}.$$

3.35 See Problem 4.22.

3.36 At the top level of a parking garage, a 30-in. deep concrete beam, spanning 25 ft, supports a joist floor. A thermal upward bow is caused by a temperature difference of $\Delta T = 35°F$. Determine the thermal-induced moments if the beam is restrained from rotation at the ends, and find the tensile force that is developed. Use $E_c = 4300$ ksi, $I = 50000$ in.^{4}, and $\alpha = 6 \times 10^{-6}$ in./in./°F.

4 APPROXIMATE DESIGN OF BEAMS AND COLUMNS

The process of proportioning structural members according to the respective code requirements is often quite complex and time consuming. In order to be able to concentrate on the building structure as a whole, simple processes and formulas for the preliminary sizing of elements have been developed. Furthermore, to facilitate the understanding of the structural design of the various buildings discussed in other parts of this book, some of the basic principles of concrete and steel design are briefly reviewed or introduced in this chapter. For the discussion of masonry and wall design refer to Chapter 6.

Though there are many handbooks available for the sizing of beams and columns, it is the intention here to develop some feeling for structural behavior and some self-confidence for being able to quickly proportion members, and thus be in full control of the design without having to depend completely on ready-made solutions as provided by handbooks and computer software. Simple equations have been developed that make the quick estimate of member proportions possible. This is especially important at various times during design and construction. For example, when a sense for the order of magnitude of forces and stresses must be developed at the early design stage when different structural layout systems are compared and evaluated, when member sizes must be known for aesthetic purposes during the developmental stage of architectural design, when a fast judgement for size is needed for checking drawings and computer outputs, or for the designer to sense, on the construction site, when member sizes don't seem right, or finally, when first trial sections may be needed for the solution of indeterminate structures.

The building design specifications of the American Institute of Steel Construction (AISC) and the American Concrete Institute (ACI) have been the basis for structural design; they have been incorporated into most building codes in the United States. For the preliminary design of steel members, the traditional allowable stress approach has been used (with which the reader is familiar), while, for the sizing of reinforced concrete elements, the strength method has been employed. In this case, the design strength must be at least equal to the required strength, as derived from the loading conditions. The required strengths are the service load effects (w, p, V, M, and T), increased by the respective load factors, as shown in Chapter 3 in the section on "Load Combinations" (see Equation 3.47). The design strength is equal to the nominal strength reduced by the capacity reduction factor, ϕ:

$$\text{Design Strength} \geq \text{Required Strength} \qquad (3.45)$$

Bending strength: $\phi M_n \geq M_u (= 1.4\,M_D + 1.7\,M_L, \text{for example})$

Axial strength: $\phi P_n \geq P_u$

Shear strength: $\phi V_n \geq V_u$

Torsional moment strength: $\phi T_n \geq T_u$

Where the strength reduction factors, ϕ, are given in Chapter 3, in the section on "Load Combinations" (see Equation 3.48).

The minimum clear cover to the reinforcement for cast-in-place concrete, as protection against corrosion by weather or loss of strength from fire exposure, is:

Concrete cast against and exposed to earth:	3 in.
Concrete exposed to weather:	
#6 through #18 bars	2 in.
#5 bar and smaller	$1\,\tfrac{1}{2}$ in.
Concrete not exposed to weather:	
slabs, walls, joists: #11 bar and smaller	$\tfrac{3}{4}$ in.
slabs, walls, joists: #14 bar and larger	$1\,\tfrac{1}{2}$ in.
beams, columns: primary reinforcement, ties, stirrups, spirals	$1\,\tfrac{1}{2}$ in.

CONCRETE BEAMS

In cast-in-place concrete construction, the beams form an integral part of the floor framing systems; this is discussed in more detail in Chapter 5 in the sections on "Floor Framing Systems" and "Reinforced Concrete Slab Systems." With respect to gravity loading, they constitute T-sections (or L-sections, for the spandrel beams) in relation to the positive bending along the midspan region, but only rectangular sections for negative bending close to the supports (Fig. 4.1). Simple rectangular sections or inverted T-sections are also typical for precast concrete construction, where the slab may rest on the beams without any continuous interaction. Since the axial forces in frame girders are relatively small, they are ignored for preliminary design purposes.

The preliminary sizing of flexural members depends on stiffness and strength considerations. Usually, stiffness controls the design of flexible elements such as slabs, joists, shallow beams, and long-span beams. For this condition, the minimum member depth can be determined from flexibility considerations. The ratio of deflection to span Δ/L of a beam can be expressed in terms of its span-to-depth ratio, L/t, multiplied by a constant, C.

$$\Delta/L = C(L/t)$$

In order to avoid the complex deflection calculations for reinforced concrete, limiting L/t ratios are given. Hence, the following approximate minimum member thicknesses with t (in.) and L (ft) for various cases, (see Problem 4.1), are:

Cantilevers:	$L(12)/t = 8$, or	$t_{min} = 1.5L$
Simple-span beams:		$t = L/1.33 = 3L/4$
Continuous-span beams:		$t = L/1.5 = 2L/3 \qquad (4.1)$

These values are based on normal-weight concrete and Grade 60 steel. For Grade 40 reinforcement, the beam depth may be reduced by 20 percent, but it must be increased by 20 or 10 percent for 90 pcf or 120 pcf lightweight concretes, respectively.

The size of ordinary beams (i.e., not short- or long-span beams, or beams with large concentrated loads near the support) that do not need compression reinforcement is controlled by flexural compression. Shear is rarely critical, with the exception of members with unreinforced webs (e.g., joists, foundations), and of punching shear around columns, in flat plates and slabs. The following expressions for the sizing of members and for finding the flexural reinforcement have been derived from the bending strength of the beam, while the shear reinforcement has been obtained from its shear strength (see Problem 4.1). The size of concrete beams is dependent on the amount of reinforcement used in the section, as measured by the steel ratio $\rho = A_s/b_w d$ at the critical support location, where the continuous beam acts only as a rectangular member. In other words, the size of ordinary floor beams is controlled by the flexural compressive strength at the maximum moment location, where the resisting compressive cross-sectional area is the smallest. In this context, the equation for selecting the concrete beam depth is based on the typical condition of Grade 60 steel, together with an average reinforcement ratio of $\rho = \rho_{max}/2 \simeq 1\%$, so that the bars can be easily placed and also provide reasonable deflection control. For this condition, the coefficient of resistance is: $R_u = \phi R_n = M_u/b_w d^2 \simeq 0.52$ ksi. This expression is now changed in form as given below, but using mixed units—it is applicable to beams with common concrete strengths of 3000 to 6000 psi. The equation can also be represented in terms of the service moment, M, by assuming an average load factor of 1.5, as is derived in Equation 4.9.

$$bd^2 = 23M_u \simeq 35M \qquad (4.2)$$

where:
M_u = ultimate bending moment (ft-k) $\simeq 1.5M$

b = beam web width (in.)

d = effective depth of beam (in.)

Notice that the expression is in mixed units. From this equation, the beam depth can easily be found by assuming a typical beam width.

This equation also brings to mind the approximate maximum concrete stress, f_c, in the working stress approach for the balanced condition of a rectangular section, as if it would have been derived from an uncracked section. Using the same material strength as before results in:

$$f_c \simeq M/S = M/(bd^2/6) \leqslant 0.45f_c', \qquad \text{or} \qquad bd^2 = 40M \qquad (4.3)$$

If Equation 4.2, however, is also used for Grade 40 steel, a minimum of approximately 1.6 percent of reinforcement is required, with 1.3 percent for 50 ksi steel, rather than the assumed 1 percent for 60 ksi steel. Assuming a continuous beam with the critical support moment due to gravity loading to control its design, Equation 4.2 can further be simplified (see Problem 4.1) to

$$d = 1.45l_n \sqrt{w_u/b_w} \qquad (4.4)$$

where:
w_u = uniform ultimate load (k/ft)

l_n = clear beam span (ft)

$$d = \text{effective beam depth (in.)}$$
$$b_w = \text{beam stem width (in.)}$$

For preliminary design purposes, the following simplified expression is derived from Equation 4.4 for typical beam widths of b = 10 to 16 in., with at least 1 percent of Grade 60 steel, but using the span length, L, from center-to-center of supports.

$$d = 0.4L\sqrt{w_u} \tag{4.5}$$

It must be kept in mind when selecting the beam proportions that wide, shallow beams may be more economical from an overall construction point of view than the narrow, deep beams that the designer might be more used to. For this condition, as for slabs under normal loads, the minimum member thickness is determined mostly by deflection limitation rather than strength, besides having to consider fire-resistance requirements.

When the cross section is known, the moment reinforcement can be found as shown below. Rotational equilibrium, as shown in Figure 4.1, necessitates the balance of the acting moment M_u and the steel strength A_sf_y, which is then reduced by the capacity reduction factor, $\phi = 0.9$.

$$M_u \leq \phi M_n = \phi Tz = \phi A_sf_yz \tag{4.6}$$

From this equation, the approximate moment reinforcement for rectangular beams, joists, and T-beams can be found by using an average internal lever arm length of z = 0.9d.

$$A_s = M_u/(0.8f_yd) \geq \rho_{min}b_wd = 0.2(b_wd)/f_y \tag{4.7a}$$

where:

A_s = moment reinforcement (in.^{2})

M_u = ultimate moment (in. k)

f_y = yield stress of steel (ksi)

d = effective beam depth (in.)

This equation gives reasonable results for typical beam steel ratios of 1 percent to about 1.6 percent. For this range, it is quite insensitive to the various common concrete strengths and the variation of the steel ratio, as is apparent from the closely bundled and nearly straight lines of the strength curves for a given steel (see Problem 4.1). For fast approximation purposes, however, this equation may even be used for its entire permitted range, as long as the steel ratio is less than ρ_{max}, which is given in Table A.4 for various steel concrete combinations. Attention must be given to 3000 psi concrete together with 60 ksi steel, where the equation becomes less precise beyond a steel ratio of about 1.1 percent.

Equation 4.6 is also used for the T-beam behavior of the composite beam-slab at midspan (Fig. 4.1), where the flexural capacity of the concrete is so large that the stress block depth, a, usually lies within the flange (slab), so that the T-section can be treated as a wide, shallow rectangular beam section of width, b_e, for preliminary design purposes; here only the minimum steel ratio ρ_{min} must be checked.

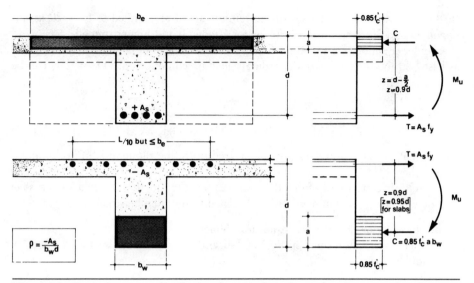

Figure 4.1. *Equivalent stress distribution for typical singly reinforced concrete floor beams at ultimate load.*

For ordinary solid concrete slabs, which have a much lower steel ratio than beams, often $z = 0.95d$ is taken as a first approximation, which results in the following required moment reinforcing:

$$A_s = M_u/(0.85f_yd) \geq A_{smin} \qquad (4.7b)$$

Generally, Grade 60 bars are used—Grade 40 requires 50 percent more steel. Substituting Grade 60 steel in Equation 4.6 yields the following simplified expression for beam design which, however, is also often used for slab design, when fast estimates are needed; the equation is in mixed units.

$$A_s = M_u/4d \geq b_wd/300 \qquad (4.8)$$

where: M_u = ultimate moment (ft-k)

A_s = moment reinforcement (in.^{2})

d = effective beam depth (in.)

Notice that the minimum reinforcement for this condition is $\rho = 0.33\%$. The maximum reinforcement ratio, ρ_{max}, which ensures that the beam is underreinforced and fails in tension, does not have to be checked if the moment reinforcement equation is only applied to beams with reinforcement ratios of up to about 1.6 percent; however, one must watch out for the combination of 3000 psi concrete and Grade 60 steel with $\rho_{max} = 1.61\%$.

According to the working stress method, Equation 4.7a can be expressed by using $z = jd \simeq 7d/8$ for first approximation purposes, as

$$A_s = \frac{M}{f_sz} = \frac{M}{f_sjd} \simeq \frac{M}{0.875df_s} \qquad (4.7c)$$

In this case, the tensile stresses in the reinforcement are limited to $f_s = 20$ ksi for Grade 40 and 50 steel, and $f_s = 24$ ksi for Grade 60 and higher strength steel.

Another equation often used for the approximate design of bending reinforcement, as based on the service loads and ordinary loading conditions of D/L = 2/1, can be derived as follows:

$$M_u \simeq 1.4(2M/3) + 1.7(M/3) = 1.5M$$

$$A_s = M_u/(0.8f_yd) = 1.5M/[0.8(60)d] = M/32d \qquad (4.9)$$

Notice that this equation does not have mixed units; the moment is in (in.k), and d is in inches.

The effective depth, d, from the compression face to the centroid of the steel reinforcement can be approximated and related to the member thickness, t, as

Beams (interior, for exterior exposure subtract 0.5 in.):
    single layer (always for top steel of T-beam):   d = t − 2.50
    double layer: d = t − 3.50
Joists: d = t − 1.25
Slabs:
    one-way slabs: d = t − 1.00
    two-way slabs (center of upper layer):   d = t − 1.50

A simplified approach to shear design is discussed next. Although the size of ordinary beams is not usually controlled by shear, shear may become critical for short-span beams that carry heavy loads and beams with unreinforced webs, such as joists.

The shear capacity of reinforced concrete consists of the strength of the concrete, ϕV_c, and of the shear reinforcement, ϕV_s. It obviously must be at least as large as the ultimate shear force, V_u, as in the case of $V_u = 1.4V_D + 1.7V_L$ at the point of investigation (Fig. 4.2).

$$V_u \leq \phi V_n = \phi V_c + \phi V_s \qquad (4.10)$$

A conservative definition of the shear strength of the concrete is

$$\phi V_c = \phi(2\sqrt{f_c'})b_wd \qquad (4.11)$$

Letting the capacity reduction factor for shear equal $\phi = 0.85$, and the typical concrete type be $f_c' = 4000$ psi, yields $\phi V_c = 0.85(2\sqrt{4000})b_wd/1000$, or

$$\phi V_c = 0.11b_wd \qquad (4.12)$$

where: b_w = web width of beam (in.)

                           d = effective beam depth (in.)

                         ϕV_c = shear strength of concrete (k)

It should be kept in mind that, for beam column design, the shear capacity of the concrete is very much increased by the axial compressive stress! The shear reinforcement is located in the beam web and is provided by the inclined (bent-up) portion of the longitudinal steel, which is approximately in line with the diagonal tension close to the support, and/or is provided by the U-shaped vertical stirrups. In this case, only the latter are considered to resist the shear; the Howe-truss analog in Figure 4.2

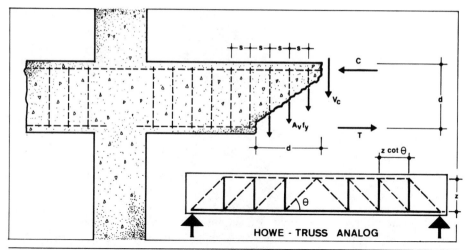

Figure 4.2. *Shear force resistance of vertical stirrups.*

convincingly reflects the idealized tension and compression action of the concrete beam. The strength of the shear reinforcement, for relatively shallow beams with transverse failure through the entire section, is independent of the concrete member dimensions and the concrete strength. The capacity of the stirrups along the section in Figure 4.2 is

$$\phi V_s = \phi(A_v f_y)n = \phi(A_v f_y)d/s \qquad (4.13)$$

Where $n(A_v f_y)$ is the entire tensile capacity of n number of stirrups along the diagonal crack. The number of stirrups, n, can be expressed in terms of the stirrup spacing, s, as $n = d/s$ by conservatively assuming the horizontal projection of the diagonal crack as equal to the effective depth, d. The stirrup spacing is thus conveniently represented as a function of the effective beam depth, d. Using the usual maximum stirrup spacing of $s = d/2$, smaller spacing intervals can be continued for standard conditions of d/3 and for, under most conditions, the closest spacing of d/4, considering that the stirrups should not be closer than about 3 inches. The stirrup spacing should be made in increments of not less than 1/2 in.

$$s = d/n, \; d/2, \; d/3, \; \text{and} \; d/4$$

For typical conditions of $f_y = 60$ ksi, $\phi = 0.85$, and #3 stirrups, with two legs of $A_v = 2(0.1) = 0.22$ in.^{2}, the following stirrup capacity is achieved,

$$\phi V_s = \phi A_v f_y n = 0.85(0.22)60n = 11n \qquad (4.14)$$

where: $n = 2, 3, \text{and} 4$

For wide beams use multiple stirrups, for example, four legs when $24 < b_w \leq 47$ in.
   The following general design process may be used:

• There is no shear reinforcement required for beams if

$$V_u \leq \phi V_c/2 = \phi\sqrt{f_c'}b_w d \qquad (4.15)$$

or, expressed in terms of the nominal ultimate shear stress for 4000 psi concrete:

$$v_u = V_u/b_w d \leq \phi\sqrt{f_c'} = 54 \text{ psi}$$

- There is a minimum shear reinforcement required for beams (but not for shallow beams, slabs, footings, and joist construction), if

$$\phi V_c/2 < V_u \leq \phi V_c \qquad (4.16)$$

or, for 4000 psi concrete, when:

$$v_u = V_u/b_w d \leq 107 \text{ psi}$$

- There is shear reinforcement required, if

$$V_u > \phi V_c \qquad (4.17)$$

When the ultimate shear force, V_{ux}, at any point x along the beam, exceeds the shear capacity of the concrete, shear reinforcement must resist the excess shear $V_u - \phi V_c$.

According to Equations 4.10 and 4.13, the required shear capacity of the stirrups ϕV_s and the corresponding stirrup spacing, are

$$V_u - \phi V_c \leq \phi V_s, \qquad \text{or} \qquad s \leq A_v f_y d/(V_u/\phi - V_c) \qquad (4.18a)$$

For instance, for the special condition of $\phi V_s = 11n$,

$$n \geq (V_u - \phi V_c)/11 = d/s \geq 2, \qquad \text{or} \qquad s \leq 11d/(V_u - \phi V_c) \leq d/2 \qquad (4.18b)$$

Not only should the maximum spacing, or the minimum stirrup reinforcement, be $s \leq d/2$, but the web reinforcement should also be able to transfer a shear stress of $\phi 50 = 42$ psi.

$$s \leq A_v f_y/50 b_w \qquad (4.19)$$

This expression does not control ordinary beam sizes of 4000 psi concrete. However, the following restriction for $s_{max} = d/2$ is required:

$$V_u - \phi V_c \leq \phi 4\sqrt{f_c'} \, b_w d, \qquad \text{but:} \qquad \phi V_c = \phi 2\sqrt{f_c'} b_w d, \qquad \text{or}$$

$$V_u \leq \phi 6\sqrt{f_c'} \, b_w d$$

Let $f_c' = 4000$ psi and $\phi = 0.85$, then: $V_u \leq 0.32 b_w d$, or $v_u = V_u/b_w d \leq 322$ psi.

Therefore, the minimum beam depth, d, required for typical conditions is approximately

$$d_{min} \simeq 3V_u/b_w \qquad (4.20)$$

It should be kept in mind, however, that the code does allow smaller beam sections for a maximum stirrup spacing of d/4.

Additional shear is generated by torsion, which may have to be covered with closed stirrups; the corresponding additional bending is resisted by longitudinal bars around the beam perimeter. Torsion may be critical for spandrel beams in monolithic floor structures, where torsion reinforcement is required. The twisting of interior beams due to an asymmetrical arrangement of live loads is generally not critical. Torsion is not further treated here, but the designer should always keep in mind the importance of attempting to reduce member twisting through special detailing, so as to minimize torsional effects upon structural components, or by sizing the member so that torsion can be neglected.

With respect to the various approximate beam sizes, it may be concluded that:

- Any beam proportion is possible, as long as the depth is at least equal to the one required for deflection control (Equation 4.1)
- The shear strength is directly related to the cross-sectional area of beams (Equation 4.11), and thus does not influence the beam proportion, but only its size; it only becomes critical for short-span beams under heavy loads or beams with unreinforced webs. The moment capacity is, however, primarily affected by the square of the depth (Equation 4.2), making narrow deep beams more efficient from a local material point of view, as reflected by the lower steel ratios. The usual depth-to-width ratio for shorter spans is 1.5 to 2, while for larger spans the ratio may be 2.5 to 3 or larger. However, it must be kept in mind that, from an overall point of view, it may still be more economical to use wide, shallow beams rather than narrow, deep beams. For instance, in pan joist construction, the supporting beams often have the same depth as the joists, in order to reduce formwork costs rather than material costs, and to reduce the overall building height. Furthermore, by changing the width of beams, only the bottom forms are affected, but not the side forms and shores. Shallow, wide beams are first checked with respect to depth according to deflection control, and then the width is found from flexural requirements. The beam widths are usually multiples of 2 or 3 inches. Often, constant beam sizes are used for one building story by only changing the reinforcement according to the span and load variations.
- The beam proportions are also influenced by the placement of the flexural reinforcement. In narrow beams, several layers of longitudinal steel may be required. Refer to Table A.3 for the minimum beam widths for various reinforcement bar combinations. Also, the shear reinforcement has an effect upon the beam sizes; small beams, for instance, may need very close stirrup spacing. Additionally, it may be advantageous to make beams at least 2 inches wider than narrow columns, so that the bars in the beam corners can pass unobstructed.

Example 4.1

A six-story concrete frame office building consists of 30 × 34 ft bays, with the floor framing layout as shown in Figure 4.3. For the description of the loading conditions and the design of the one-way concrete slab, refer to Example 5.5. In this case, the typical continuous interior beams and girders will be investigated. Use $f'_c = 4000$ psi and $f_y = 60000$ psi.

Part I:

The interior column sizes at the base of the six-story building are roughly estimated by using Equation 4.59a, which is derived later in this chapter.

$$A_g = nA/10 = 6(34 \times 30)/10 = 612 \text{ in.}^2$$

Assume 25 × 25 in. columns, $A_g = 625$ in.^{2}

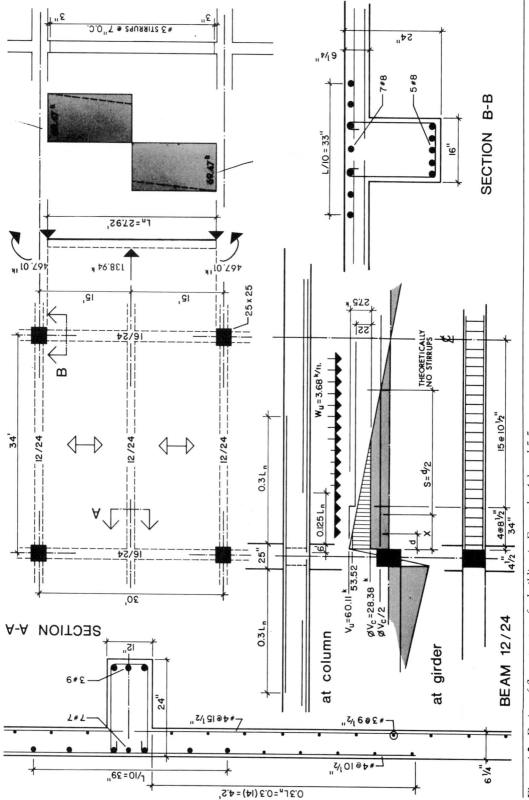

Figure 4.3. *Design of floor structure for building in Examples 4.1 and 5.5.*

Part II: Beam design.

A beam width of $b_w = 12$ in. is assumed. The clear span, for a girder width of 16 in., is

$$l_n = 34 - 16/12 = 32.67 \text{ ft}$$

Since $A = 15 \times 34 = 510 \text{ ft}^2 \geqslant 150 \text{ ft}^2$, the live load reduction is:

$$R_1 = 0.08\%A = 0.08(15 \times 34) = 40.8\%$$

$$R_2 = 23(1 + D/L) = 23(1 + 103/80) = 52.61\% < 60\% \qquad (3.1)$$

Use the least reduction of 40.8 percent. As a first trial for the beam weight, its depth may be estimated as

$$t = L/1.5 = 34/1.5 = 22.67 \text{ in.} \qquad (4.1)$$

In this case, for reasons of construction and by using increments of two inches, the floor framing is kept to a constant depth of $t = 24$ in.

The beam must support the following loads (see Example 5.5 for slab dead load):

Slab loads: $(1.4(103) + 1.7(80)0.59)15/1000$	$= 3.37$ k/ft
Stem weight: $1.4[(24 - 6.25)12/12^2](1)0.150$	$= 0.31$ k/ft
Total load:	$= 3.68$ k/ft

The moments at the supports and midspan (according to fig. 5.8a) are

$$- M_u = w_u l_n^2/11 = 3.68(32.67)^2/11 = 357.07 \text{ ft-k}$$

$$+ M_u = w_u l_n^2/16 = 3.68(32.67)^2/16 = 245.49 \text{ ft-k}$$

The required moment reinforcement at the top of the support is

$$- A_s = M_u/4d = 357.07/4(21.5) = 4.15 \text{ in.}^2 \qquad (4.8)$$

Try 7 #7, $A_s = 4.20$ in.^{2}

Where the flanges of T-beam construction are in tension, part of the flexural reinforcement is to be distributed over the effective flange width, assumed as $L/10 = 32.67(12)/10 = 39.20$ in. Place 3 #7 inside the steel cage and 2 #7 on each side in the slab (Fig. 4.3).

The corresponding steel ratio is not much larger than 1.6 percent, so that this beam section should be all right from a flexural point of view.

$$\rho = A_s/b_w d = 4.2/12(21.5) = 1.63\%$$

The required bottom reinforcement at midspan is

$$+ A_s = M_u/4d = 245.49/4(21.5) = 2.86 \text{ in.}^2 \qquad (4.8)$$

Try 3 #9, $A_s = 3.00$ in.^{2}, and according to Table A.3: $b_{req} = 9.80 \leqslant 12$ in.

Because of the high compression block resistance of the flanges in the field, only the minimum amount of reinforcing, or ρ_{min}, is checked; according to the ACI Code, just the web width is to be used.

$$A_{smin} = b_w d/300 = 12(21.5)/300 = 0.86 \text{ in.}^2 \leqslant 3.00 \text{ in.}^2 \qquad (4.8)$$

Notice that, for a simply supported T-section, ρ_{max} almost never presents a problem, since, the large resisting flange width lowers the compressive stresses.

Next, the shear will be investigated; #3 stirrups are used with $f_y = 60$ ksi. The maximum beam shears at the girder faces are

$$R = 3.68(32.67/2) = 60.11 \text{ k}$$

The maximum shear acts at d-distance adjacent to the face of the girder.

$$V_{umax} = 60.11 - 3.68(21.5/12) = 53.52 \text{ k}$$

Check whether the beam depth is satisfactory.

$$d_{min} = 3V_u/b_w = 3(53.52)/12 = 13.38 \leqslant 21.5 \text{ in. Depth is satisfactory.} \quad (4.20)$$

The concrete strength is

$$\phi V_c = 0.11b_wd = 0.11(12)21.5 = 28.38 \text{ k} \quad (4.12)$$

The stirrup strength is

$$\phi V_s = 11n \quad (4.14)$$

Establish a pattern of stirrup spacing of say d/2, d/2.5, and d/3 with their corresponding capacities.

$$s = d/2, \quad \text{or} \quad \phi V_s = 11(2) = 22 \text{ k}$$

$$s = d/2.5, \quad \text{or} \quad \phi V_s = 11(2.5) = 27.5 \text{ k}$$

$$s = d/3, \quad \text{or} \quad \phi V_s = 11(3) = 33 \text{ k}$$

The closest stirrup spacing occurs at the support, where the maximum excess shear is.

$$V_u - \phi V_c = 53.52 - 28.38 = 25.14 < 27.5\text{k}$$

The stirrup spacing of d/2.5, with a capacity of 27.5 k, covers the excess shear of 25.14 k.

The location, x, must be found, where the excess shear is equal to 22 k, that is, where the spacing d/2.5 can be changed to $s_{max} = d/2$. From the similarity of triangles in Figure 4.3 it may be obtained that

$$V_u/(l_n/2) = (\phi V_s + \phi V_c)/[(l_n/2) - x], \text{ or}$$

$$x = [1 - (\phi V_s + \phi V_c)/V_u]l_n/2$$

$$60.11/32.67/2 = [22 + 28.38]/[(32.67/2) - x], \text{ or } x = 2.64 \text{ ft}$$

Use the following stirrup spacing:

$$s = d/2.5 = 21.5/2.5 = 8.60 \text{ in., say #3 @ 8 1/2 in. o.c.}$$

$$s = d/2 = 21.5/2 = 10.75 \text{ in., say #3 @ 10 1/2 in. o.c.}$$

The stirrup layout is shown in Figure 4.3, but it should be kept in mind that many other groupings are possible. The intention here was solely to present a fast approximation.

Theoretically, there are no stirrups required beyond $\phi V_c/2$ close to midspan. However, under a partial live load on one-half of the beam span, worse shear conditions will be generated along the center beam portion, probably requiring a stirrup spacing of $s_{max} = d/2$; in addition, practical considerations of construction will require stirrups for the steel cage.

Part III: Girder design.

Assume the girder proportions as $b_w/t = 16/24$. Since the girder span is less than the beam span, but the depth is the same, deflection will not be a consideration. The live load reduction is assumed not to change from the one for the beam.

The clear span of the girder is

$$l_n = 30 - 25/12 = 27.92 \text{ ft}$$

The concentrated load reactions from the beams are:

$$P_u = 3.68(32.67) = 120.23\text{k}$$

Girder weight: $1.4(24(16)/12^2)(1)0.150$	$= 0.56$ k/ft
Additional live load: $1.7(0.080 \times 0.59)16/12$	$= 0.11$ k/ft
Total uniform girder load:	$= 0.67$ k/ft

The single load is transformed into an equivalent uniform load by equating the support moments for the two loading cases

$$M_s = Pl_n/8 = wl_n^2/11,$$

$$w_{eq} = 11P/8l_n = 11(120.23)/8(27.92) = 5.92 \text{ k/ft}$$

The total uniform load is

$$w = 0.67 + 5.92 = 6.59 \text{ k/ft}$$

The critical moments and the corresponding flexural steel are at the:

Support:

$$- M_u = w_u l_n^2/11 = 6.59(27.92)^2/11 = 467.01 \text{ ft-k}$$

$$- A_s = M_u/4d = 467.01/4(21.5) = 5.43 \text{ in.}^2$$

Try 7 #8, $A_s = 5.53$ in.^{2}, distributed across $27.92(12)/10 = 33.5$ in.

The corresponding steel ratio is close to 1.6 percent, so that the girder proportions can be considered satisfactory from a flexural point of view.

$$\rho = A_s/b_w d = 5.53/16(21.5) = 1.61\%$$

Midspan:

$$+ M_u = w_u l_n^2/16 = 6.59(27.92)^2/16 = 321.07 \text{ ft-k}$$

$$+ A_s = M_u/4d = 321.07/4(21.5) = 3.73 \text{ in.}^2$$

Try 5 #8, $A_s = 3.95$ in.^{2}, $b_{req} = 13.3$ in. $\leqslant 16$ in.

Check:

$$A_{smin} = b_w d/300 = 16(21.5)/300 = 1.15 \text{ in.}^2 < 3.95 \text{ in.}^2$$

Next, the shear is investigated using #3 stirrups, with $f_y = 60$ ksi. Because the uniform load is much smaller than the concentrated load, and also to reduce the calculations, it is conservatively assumed that it is replaced by its resultant at midspan (Fig. 4.3).

The maximum beam shears at the column faces, or at d-distance away, are

$$V_{umax} = (120.23 + 0.67(27.92))/2 = 138.94/2 = 69.47 \text{ k}$$

First, check whether the given beam depth is satisfactory, with respect to using $s_{max} = d/2$.

$$d_{min} = 3V_u/b_w = 3(69.47)/16 = 13.03 \text{ in.} < 21.5 \text{ in. Depth is satisfactory.}$$

Since the shear is nearly constant, the stirrup spacing will be constant!

$$V_u \le \phi V_c + \phi V_s$$

$$V_u - \phi V_c \le \phi V_s$$

$$V_u - 0.11 \, b_w d \le \phi A_v f_y n$$

$$69.47 - 0.11 (16 \times 21.5) \le 11 \text{ n}, \quad 31.63 \le 11 \text{ n} = 11d/s, \text{ or}$$

$$s = 11(21.5/31.63) = 7.48 \text{ in.}$$

Use #3 stirrups @ 7 in. o.c.

CONCRETE WALL BEAMS

There are many familiar examples of deep concrete beams. Included in this category are such elements as shear walls, transfer girders for columns, foundation walls supported by single footings, pile caps, and floor diaphragms. The behavior of a concrete beam changes drastically as its height, h, increases for a given span, L. It acts as a line element or shallow beam when the beam height is much smaller than the span, approximately for a depth-to-span ratio of $h/L_n < 0.2$ and uniform loading at the top—for deeper sections, special shear requirements must be considered. Deep beam behavior with respect to flexure is introduced as the h/L_n ratio increases. At approximately $h/L_n > 0.4$ for continuous spans, and $h/L_n > 0.8$ for simple spans, loads can be transferred to the supports, primarily in direct arch action with the corresponding thrust. A deep beam acts more like a surface element or plate where the strain distribution is no longer linear and the shear deformations cannot be ignored, as in the common shallow beams. The elastic flexural stress distribution is likewise no longer linear, as is illustrated in Figures 4.4a and 6.9. Here, for the positive moment range, the compressive stresses in the concrete are usually not critical. They are small in comparison to the magnitude of the tensile stresses that are concentrated at the bottom, as is clearly demonstrated by the principal tensile stress trajectories at midspan.

When ordinary beams are loaded on the top edge, it has been shown in Equation 4.7a that the internal lever arm can be estimated as $z = 0.9d$ for rectangular sections. As the depth-to-span ratio increases, however, the lever arm, z, decreases for the given rectangular shape. For preliminary estimation purposes of wall beams, it may be taken as $z = 0.5L$ or $z = 0.6h$, whichever is smaller. For example, for a simply supported wall beam that carries a uniform load on top and has a span equal to its height,

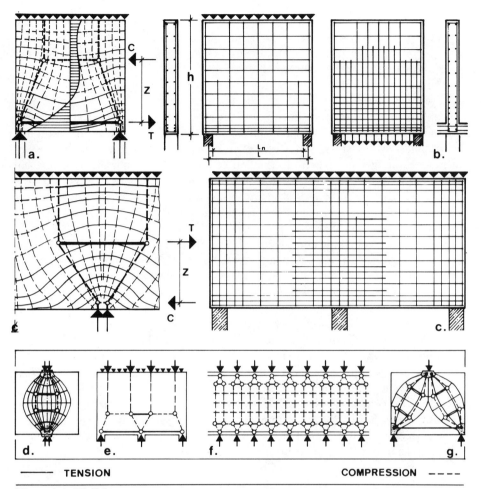

TENSION ————   COMPRESSION ————

Figure 4.4. *Wall beams.*

$h/L = 1$, the approximate flexural reinforcement at the bottom can be determined as follows:

$$M_u \leq \phi M_n = \phi A_s f_y z \qquad (4.6)$$

$$w_u L^2/8 = 0.9\, A_s f_y(0.5L), \qquad \text{let } W_u = w_u L$$

$$A_s = 0.28\, W_u/f_y \geq A_{smin} = 0.2bd/fy \qquad (4.21)$$

The lower portion of a simply supported wall beam can be visualized to act as a compression arch with a tie rod at the base (see fig. 6.1b), or as a wall sitting on a transfer girder at the wall base, which acts primarily as a tie. The tension bars that resist the field moment must be concentrated along the bottom of the beam (Fig. 4.4a), as is clearly demonstrated by the stress diagram. When the *simply supported beam* is loaded from the top, the tensile stress trajectories shown in Figures 4.4a and 6.9 are relatively flat, indicating that the horizontal bars are the primary ones resisting the tension. However, when the simple beam is loaded at the bottom, or when deep intersecting beams are carried, then the tensile trajectories are much steeper, so that the vertical reinforcement becomes more important. Since the stress trajectories look similar to the ones for shallow beams, these types of beams may be designed as shallow ones for preliminary design purposes.

The tensile trajectories for a *continuous beam* with uniform loads at the top are also relatively flat, so that the primary reinforcement is horizontal, with additional horizontal bars above the supports (Fig. 4.4c). The stress diagram at the support in Figure 6.9 indicates that the tension zone continues nearly to the bottom of the beam and is distributed over a large depth. The section may thus be treated as fully in tension, thereby requiring a more even distribution of the horizontal support reinforcement across the beam height. The actual distribution depends on the h/L_n ratio, or the corresponding shape of the stress diagram. When $h/L_n > 1$, no support reinforcement is required in the top portion of the beam, where the tensile stresses are small or nonexistent (see fig. 6.9).

For other support and loading conditions, the funicular tied-arch mechanism, the truss analogy, or the simpler strut-and-tie model, as used in Figure 4.4, may be an effective way for visualizing and representing stress trajectories, which includes the analysis of the force flow in wall beams. Similar to using the parabolic arch model for uniform loads, the A-frame analog can be applied for single load action (Fig. 4.4g).

The initial size of a deep girder that, for instance, is used as a transfer wall beam, may be estimated by ignoring the effect of the h/L_n ratio, and limiting the shear stresses according to Equation 4.20. It is apparent that the shear capacity of the concrete, V_c, for deep beams must be considerably higher than for normal beams due to the internal tied-arch action; it is obviously conservative to use Equation 4.11 for finding the shear strength of the concrete. The predominance of the inclined stress trajectories for many cases in Figure 6.9 indicates the importance of shear, although the concrete shear stresses are usually not critical in wall beams under ordinary loading conditions. Horizontal and vertical reinforcement must be provided along the faces throughout a wall beam, as is discussed further in Chapter 6, in the section on "Reinforced Concrete Walls."

STEEL BEAMS

Typical steel beams for floor framing in high-rise building construction are the common W-sections, open web steel joists, trusses, castellated beams, stub girders, plate girders, and tapered and haunched-taper beams.

For the design of open-web steel joists, the reader should refer to the loading tables of the respective manufacturers, or the specifications of the Steel Joist Institute (SJI). Steel joists have the appearance of shallow trusses, mostly of the Warren-type configuration, and are designed as simply supported uniformly loaded beams. Two series are used for floor construction: 1) K-series (standard), range from 8 to 30 in. in depth, and have a span range of 8 to 60 ft; 2) LH-series (long-span steel joists), range in depth from 18 to 48 in. and have a span range from 25 to 96 ft.

The typical joist spacing varies from two to four feet to provide efficient use of the corrugated steel deck. The joists are stabilized by either diagonal or horizontal bridging.

For the preliminary estimate of the joist depth, use a depth-to-span ratio of 1/24, which is about one-half of the typical value for the wider spaced long-span trusses.

$$t = L/2 \qquad (4.22)$$

In this section, only the structural design of the common rolled W-shapes is briefly discussed. They are produced in a very wide range of sizes; currently, the largest beam is a W36 × 848, with a section modulus of $S_x = 3170$ in.[3]. For long spans, for instance beyond 100 ft, deeper and lighter beam sections should be considered: plate girders (single-web or double-web) or trusses (that is, built-up members) can be employed. It was already mentioned in the section on concrete beams that the axial forces in frame beams are rather small and can be neglected for preliminary design purposes.

The design of beams is generally controlled by bending. The familiar flexural stress for symmetrical rolled beam sections is

$$f_b = M/S \leq F_b, \qquad \text{or } S \geq M/F_b \qquad (4.23)$$

The section modulus necessary to resist the given moment can easily be looked up. The allowable bending stress, F_b, is dependent upon the lateral support of the compression flange, in addition to the section properties. The section properties refer to slenderness considerations of the compression flange (F_y') and the web (F_y''') when an axial load is present. The yield stresses in the parentheses are hypothetical values, above which the flange and web are noncompact.

Most floor beams are fully laterally braced by the concrete slab (i.e., the unbraced length $L_b = 0$) and nearly all A36 beams have $F_y \leq F_y'$, so that they can be considered compact; in other words, the sections are capable of developing their plastic moment capacity before any buckling occurs. For these conditions, the allowable bending stress for strong axis bending of doubly-symmetrical members is

$$F_b = 0.66F_y \qquad (4.24a)$$

For example, for A36 steel $F_b = 0.66(36) = 23.76$ ksi, which is usually taken as 24 ksi. For weak-axis bending of doubly-symmetrical members, including solid rectangular and round bars, the allowable bending stress is

$$F_b = 0.75F_y \qquad (4.24b)$$

When members bend about their minor axis, they seldom need to be braced, because of their superior lateral stiffness; the same reasoning applies to the bending of box sections (e.g., tubing).

At times, the compression flange may not be braced, as is typical for building columns that may have an unbraced length, L_b, larger than the theoretical value of L_c, but which should be kept at less than, L_u, for this preliminary design approach ($L_c < L_b \leq L_u$). Further, should the beam-column web be noncompact ($F_y > F_y'''$), or the flange be noncompact ($F_y > F_y'$), then the allowable bending stresses for those conditions may be taken as

$$F_b = 0.60F_y \qquad (4.25)$$

For example, for A36 steel: $F_b = 0.6(36) \simeq 22$ ksi.

The design parameters of F_y', F_y''', L_c, and L_u are found in the S_x-Selection Table and the W-Shapes Properties Table of the AISC Manual of Steel Construction (1980) and in Table A.11 of this book.

Web crippling at the junction of the flange and web (i.e., at the toe of the fillet) at points of stress concentrations due to concentrated loads should be checked. The shear stresses for rolled and fabricated shapes $f_v = V/A_w = V/(dt_w) \leq 0.4F_y$ rarely control the preliminary design of the beam, as long as the web is not weakened by holes, or large loads are not acting adjacent to the support.

Deflection limitation is generally based on the live load and equal to L/360 for beams supporting a plaster ceiling, which should not be fractured. The Commentary on the AISC Specs proposes, as deflection limits for fully stressed floor beams and girders, $L(F_y/800)$, or L/22 for A36 steel as a minimum beam depth. Thus, a 30-ft beam should have a depth of at least $30(12)/22 = 16.36$ in., for deflection not to control. Where the floor framing is subject to vibrations with no sources of damping available, a beam depth of at least L/20 is suggested. Long-span beams are usually cambered for dead load deflection.

The actual deflection for beams carrying a uniform load can be conveniently expressed, first in terms of the simple-span moment $M = wL^2/8$, and then by letting M in (k-ft), L in (ft), I in (in.^{4}), and E = 29,000 ksi, so that the following deflection, Δ_s, for a simple-span beam is obtained

$$\Delta_s = 5wL^4/(384EI) = 5ML^2/(48EI) \quad \text{or}$$

$$\Delta_s = ML^2/(161I) \tag{4.26}$$

This expression can further be simplified by letting $f_b = (M/t/2)/I$ as

$$\Delta_s \simeq f_b(L^2/t)/1000 \tag{4.27}$$

where: f_b in (ksi), L in (ft), and t in (in.)

Now, for full loading conditions, let $f_b = F_b = 24$ ksi, so that one obtains

$$\Delta_s = 0.0248L^2/t \tag{4.28}$$

where: L = beam span (ft)

t = beam depth (in.)

Δ_s = beam deflection due to full loading (in.)

For preliminary design purposes, this expression can also be used for floor framing where the beams carry the concentrated loads of the supporting filler beams. Notice that the deflection is primarily dependent on the square of the span and the beam depth for a maximum bending stress F_b; otherwise, the deflection increases with an increase in bending stress. Hence, beams of the same span and depth have the same deflection when they are stressed to their allowable limit under full loading conditions.

The corresponding deflection for a fixed beam is $\Delta = 0.2\Delta_s$, and the maximum deflection for a continuous beam may be assumed to be 25 percent of that of the simple beam:

$$\Delta = 0.25\Delta_s$$

To quickly estimate the size of a typical simply supported floor beam (A36) under uniform load, the following rules of thumb (with mixed units) are often found in practice (see Problem 4.5):

- The section modulus, S_x, of a W-section is roughly equal to the product of beam weight (lb/ft), w, and its nominal depth (in.), t, divided by 10.

$$S_x = wt/10 \qquad or, \qquad I_x = S_x(t/2) = wt^2/20 \qquad (4.29)$$

For example, for a W16 × 36: S = 36(16)/10 = 57.6 in.^{3}; the actual S is 56.5 in.^{3}.

- The nominal depth (in.), t, of a W-section is approximately equal to one-half of its span (ft), L.

$$t = L/2 \qquad (4.30)$$

For the primary beams supporting filler beams, the nominal depth is often assumed to be:

$$t = L/1.5$$

- The beam section weight (lb/ft), w, is roughly 1.25 times the total load (k), W, that the beam must support.

$$w = 1.25W \qquad (4.31)$$

The maximum deflection (in.), Δ, is one-tenth of the nominal beam depth (in.), t.

$$\Delta_{max} = t/10 \qquad (4.32)$$

Example 4.2

Simply supported floor beams are spaced 8 ft apart and span 20 ft; each one carries a uniform load of 1 k/ft. Estimate the beam size, using A36 steel.

The nominal beam depth is: t = L/2 = 20/2 = 10 in.
The required beam weight is: w = 1.25(W) = 1.25(1 × 20) = 25 lb/ft
The maximum deflection is: Δ = t/10 = 10/10 = 1 in.

The hypothetical section is a W10 × 25, which does not exist. Try the section closest to it, i.e., W10 × 26.

COMPOSITE BEAMS

Composite beam action may be achieved by either encasing the steel beam in concrete, where the natural bond between steel and concrete is sufficient to provide resistance to horizontal shear, and which is usually done for purposes of fireproofing, or by connecting the steel beam to the concrete slab with shear connectors. The composite action of steel beams and concrete slabs, with the aid of shear connectors, is a common construction practice in high-rise buildings today, for spans larger than 25 to 30 ft; it usually increases the ultimate strength and stiffness of the member by more than 50 percent. The bonding between the interface of the two materials is generally achieved by stud or channel connectors, which are welded to the flanges and resist the horizontal shear, thus prohibiting slippage; limited slippage is allowed in partial composite action, which may be more economical. In noncomposite floor systems, the slab and beam

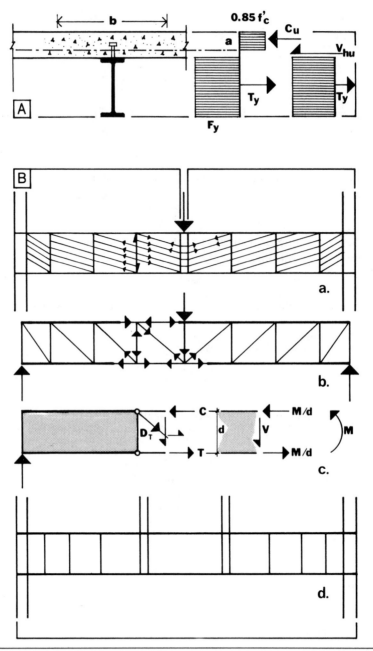

Figure 4.5. *Composite floor beams and plate girders.*

act independently in opposite directions. In composite construction, however, the steel beam, together with a portion of the slab, forms a T-beam, as in monolithic concrete construction (see fig. 4.1). When corrugated steel deck spans perpendicular to the supporting beam, only the concrete above the deck is considered to be structurally effective (in other words, the concrete below the top of the steel deck is ignored), while for the parallel condition, the full depth of the slab is effective. For continuous beams there will be composite action along the positive moment region, but noncomposite action along the negative moment region, if no special reinforcement is provided.

For the preliminary design of composite beams with shear connectors, a rough rule of thumb is that the capacity of the steel beam in composite action is increased by one-third, that is, the steel beam alone can be designed for 75 *percent of the moment.* Often, the steel beam depth is estimated as 80 percent of the noncomposite section. The designer must keep in mind, however, that during the construction stage the beam alone, if it is not shored, must be able to resist the floor dead load and construction loads in noncomposite action.

The number of connectors needed to resist the shear at the interface of the slab and steel beam can be estimated as follows. It is assumed, for preliminary design purposes, that the neutral axis (N.A.) falls within the concrete slab of the composite section, so that the concrete slab is adequate in resisting the total compressive force at ultimate load. Hence, for full composite action, the shear connectors must be able to transfer the tensile capacity of the steel beam $T_y = A_sF_y$ (see fig. 4.5A); in other words, T_y is balanced by the shear load capacity of the connectors $V_{hu} = 2V_h$ by using a load factor of 2.

$$T_y = A_sF_y = 2V_h$$

Therefore, the total horizontal shear to be resisted between the points of maximum positive moment and zero moment is

$$V_h = A_sF_y/2 \qquad (4.33)$$

where: A_s = area of the steel beam cross section

The total number of connectors, n, for a simply supported beam, for example, is equal to the shear, V_h, divided by the allowable shear load, q, for one connector (as given in the Manual of Steel Construction), and multiplied by 2 for symmetrical loading conditions.

$$n = 2(V_h/q) = A_sF_y/q$$

The connectors may be evenly spaced when no concentrated loads are present. For the condition of partial composite action, fewer shear connectors are required than for fully composite behavior.

STEEL PLATE GIRDERS

Plate girders are built-up single-web I-shaped beams, or double- or multi-web box girders. They are composed of heavy flanges and relatively thin web plates that are stiffened by vertical and possibly horizontal plates; special bearing stiffeners are required, under concentrated loads. Today, the plate assembly is generally welded together, as shown in Figure 4.5d.

For heavy loading conditions, the plate girders may provide a larger moment of inertia than that available for rolled beams. For ordinary loading, they may be more economical than rolled beams, for spans larger than approximately 70 ft. However, even at spans as short as 35 ft they may become competitive with rolled beams or built-up rolled beams, since the designer has the freedom to proportion the cross section of the girder.

The thin web and the corresponding buckling considerations make the design of plate girders quite complex; their behavior is somewhat between rolled beams and trusses. After the web of a plate girder has buckled under shear action and has become ineffective, which actually represents buckling along the compression diagonal of the web plate, the tension diagonal in the opposite direction must resist the entire shear (Fig. 4.5a). This tension field in each web panel must be stabilized by the flanges and the vertical stiffeners, which act as compression struts. Therefore, the behavior of the plate girder is similar to that of a Pratt truss (Fig. 4.5b), where the girder flanges constitute the chords and where the webs may be designed as diagonal tension-field members, together with the vertical stiffeners as compression members. In this context, the truss analogy will only be used for the design of the flanges, which are assumed to resist the entire moment, while the web carries all of the shear. It is assumed that instability in the flange and web, together with the stiffeners, does not develop prior to yield (Fig. 4.5c), otherwise instability criteria such as vertical and lateral buckling of the compression flange and lateral buckling of the web may require a reduction of the allowable stresses.

The following approach may be used to proportion an I-shaped welded plate girder (A36) for first-trial purposes.

- Assume a typical plate girder depth, d, of 1/10 of its span if no other considerations, such as headroom or aesthetic reasons, are given.
- Determine the flange area, A_f, by assuming the entire moment to be carried by the flanges, using the full depth of the section as an internal lever arm (flange-area method, Fig. 4.5c).

$$f_c = P/A_f = (M/d)/A_f \leq F_b = 0.6F_y \qquad \text{or}$$

$$A_f = bt_f = (M/d)/F_b \qquad (4.34)$$

where: $F_b = 0.6(36) = 22$ ksi, assuming proper vertical and lateral support of the compression flange

Check the compression flange thickness so that it is adequate against local buckling according to the permissible width-thickness ratio (AISC, Sec. 1.9.1.2).

$$(b/2)/t_f \leq 16, \qquad \text{or} \qquad t_f \geq b/32 \qquad (4.35)$$

- Determine the web thickness, as based on having no reduction in flange stress. Therefore, the corresponding web thickness, according to the permissible depth-thickness ratio (AISC, Sec. 1.10.6), should be

$$t_w \geq h/162 \qquad (4.36)$$

but never less than one-half of this value, nor less than 1/4 in. Check the shear stresses in the web:

$$f_v = V/A_w \leq F_v = 0.4F_y \qquad (4.37)$$

where: $A_w = t_w h , \qquad h = d - 2t_f$

Now the trial girder section can be checked by using the familiar moment of inertia method; the size, as well as the location of the stiffeners, must also be determined.

PRESTRESSED CONCRETE BEAMS

The principle of prestressing, as applied to high-rise buildings, is only briefly introduced here. It has been used for some time in segmental construction, where precast elements are assembled on site and tied together through post-tensioning to form a rigid, monolithic whole. It has also been widely employed in the precast concrete industry, for mass-producing standardized pretensioned floor elements such as hollow-core slabs and double-tee sections. Only recently has unbonded post-tensioning of cast-in-place floor framing systems for high-rise buildings, including garages, developed on a larger scale, although the post-tensioning of slab-on-grade construction (i.e., floating slabs) has been in use for some time. The span ranges of various concrete floor framing systems (see the section in Chapter 5 on "Reinforced Concrete Slab Systems") are limited by economic considerations. With an increase of span, floor structures become too heavy, thereby leaving only a small portion of their strength available for added service loads; in addition, long-term deflections due to creep become critical. By prestressing conventional concrete floor framing systems, not only can their depth be substantially reduced, but also their span range may be increased by roughly 30 to 40 percent.

The two mechanical prestressing methods are pretensioning and post-tensioning; *pretensioning* is usually associated with precast concrete. In this case, tendons are anchored to molds or abutments outside the forms and are tensioned *before the concrete is cast*. When the concrete has reached sufficient strength, the tendons are released or cut in the long-line production process, but prevented from returning to their original length by the concrete to which they are bonded, thus placing the surrounding concrete in compression.

Post-tensioning of tendons is done by hydraulic jacks *after the concrete has hardened* and has reached about 75 percent of its design strength, by inserting the wires, strands, or bars in the metal/plastic conduits that were placed in the forms before the concrete was poured. The concrete element generally carries the compression in bearing, through special anchoring devices at the end faces, rather than through bond, as in pretensioning, although the tendons may also be bonded to the concrete by grouting within the conduit.

The purpose of prestressing concrete lies in improving the behavior of the composite material, and to use less of it because of the high strength of the individual components. Here, an external prestress force is applied that bends the member in a direction opposite to the bending resulting from loading. In nonprestressed reinforced concrete, the section has to substantially crack before the steel reinforcement can fully act, while in the prestressed design approach, the composite interaction of concrete and steel is very much improved by prestretching the high-strength steel, so that the section is uncracked under service loads and the full cross-sectional concrete area is available for resistance. The locked-in, constant compressive stresses efficiently resist the tensile stresses due to external loading. The result yields shallower and stiffer sections, with a corresponding lower story height and a better control of deflection, which is especially important for long spans. However, one must also keep in mind the higher cost of prestress steels and anchorages, the high installation costs, and the necessary high quality control, together with a more sophisticated construction process, which makes post-tensioning primarily economical only for larger spans.

Since a prestressed concrete member shortens due to shrinkage and creep, part of the prestress force is lost—this loss may be in the range of approximately 40 ksi. As it is common practice to limit prestress losses to approximately 20 percent of the initial prestress force, a 200-ksi steel is required for a prestress loss of 40 ksi. Hence, it is apparent that high-strength steels must be used for post-tensioning—they may be wires, strands, or alloy bars. There are two grades of popular post-tensioning strands, with strengths of 250 and 270 ksi and a modulus of elasticity of 27,500 ksi (see Table A.5). For the geometrical properties of alloy bars of Grade 145 and 160, refer to the corresponding bars for nonprestressed reinforcement (Table A.1).

Typical span-to-depth (L/t) ratios for prestressed floor framing systems are roughly 45 for slabs, 32 for wide beam bands, 30 for joists, and 24 for beams. The depth of prestressed concrete members varies between approximately 60 and 85 percent of that of equivalent nonprestressed members. For the design of prestressed, continuous floor beams in typical high-rise buildings, the following rules of thumb may be used, where the span, L, is in feet and the beam thickness, t, in inches.

Wide band beams:	$t = L/2.5$ to $L/3$
Joists:	$t = L/2$ to $L/2.5$
Beams:	$t = L/1.7$ to $L/2$
Girders:	$t = L/1.3$ to $L/2$

For preliminary design purposes, the lower range values above may also be applied to simple span conditions.

The area of prestress steel required to resist an effective prestress force, P, in a concrete member under an external moment, M, can be roughly approximated according to Equation 4.7c (but only for quick first-trial purposes) as

$$A_s = P/f_s = (M/z)/f_s, \quad \text{where: } z \simeq 0.8t, \quad f_s = 0.56f_{pu}$$

The required cross-sectional area of the concrete member can be estimated from

$$A_c = P/f_c = (M/z)/f_c, \quad \text{where: } z \simeq 0.8t$$

In this case, the precompression stress, f_c, may be taken as approximately 350 psi for slabs, and twice as high for beams.

In this context, it is assumed that the concrete is *fully prestressed* so that the entire cross section is effective in resisting the external forces. It may sometimes be more economical, however, to allow some cracking so that the beam is only *partially prestressed*, where the tendons are stressed below the usual level, or where the member contains a significant amount of nonprestressed steel. Since the section is not cracked in the fully prestressed design, it can be treated as any other homogeneous composite member, where the allowable stress approach can be used, as based on elastic behavior:

$$f = P/A \pm M/S$$

Two critical loading stages must be investigated: 1) The initial stage, where the beam bends upward, as caused by the prestress force and possibly only counteracted by its own weight; 2) The final stage, where the beam deflects downward as governed by the full gravity condition.

Generally, the final loading stage controls the magnitude of the prestress force. However, the critical maximum tensile and compressive stresses in the concrete section may very well occur during the initial loading stage. Since emphasis here is placed on the preliminary design of common post-tensioned floor framing members, only the typical uniform loading is considered to act upon beams and slabs, although a loading case due to beam action on a girder with a straight-line tendon profile is shown in Figure 4.6d. The uniform load causes a parabolic moment diagram, remembering that a flat parabolic draped tendon or a compressed concrete arch represent the funicular shape of uniform loading, and thus respond in pure axial action. When a parabolic tendon profile is selected (Fig. 4.6a, b) and the cable is tensioned, then a uniform upward load is generated, which partly balances the downward gravity load. In the *load-balancing concept* of design, as developed by T. Y. Lin in the early 1960s, *the cable profile corresponds to the arrangement of the applied loads, and mirrors the moment diagram of the gravity loads;* the tensioned tendons generate equivalent forces acting opposite to the externally applied ones and produce a moment diagram exactly opposite to the one caused by a part of the external loads (Lin and Burns, 1981). The concept of two-dimensional load balancing by post-tensioning a flat slab (Fig. 4.6a) is rather similar to the one-directional beam approach, as is discussed further in the section in Chapter 5 on "Post-tensioned Slabs."

The designer must determine the portion of the applied load to be balanced by the prestress force, which is not a simple decision. Often, it is convenient, as a first cycle, to balance the dead load for the condition where the live loads are in the range of the dead loads; other designers assume an average compression stress for the initial estimate of the balanced load. These precompression stresses are in the range of 175 to 400 psi for slabs and 600 to 800 psi for beams, so as to avoid excessive creep.

In this context, it is assumed that the balanced load, w_p, is equal to the dead load, w_D, so that the dead load deflection is equal to the camber due to prestressing, thereby resulting, theoretically, in no deflection. Since the loads balance each other and the section is under constant compression (Fig. 4.6 b, c), the moment caused by the dead load at midspan must be equal to the prestress moment, $P(e)$. Treating the tendons as suspended cables similar to Figure 4.6c, it can be obtained from statics that the resisting moment, $P(e)$, as provided by the cable force, must balance the rotation due to the external loads, $w_p = w_D$.

$$P(e) = w_p L^2/8 = w_D L^2/8$$

Hence, for the general condition, the load, w_p, to be balanced by the prestress force, P, is

$$w_p = 8Pe/L^2 \tag{4.38}$$

Or, the required prestress force, P, to balance the dead load, w_D, is,

$$P = w_p L^2/8e = w_D L^2/8e \tag{4.39}$$

At this stage, the parabolic tendon is tensioned with such a magnitude compressing the concrete beam as to cause a uniform upward force exactly equal to the uniform dead load. The result is a zero net load and a constant compression along the beam, as long as the prestress tendon is anchored at the centroidal axis of a simply supported beam (Fig. 4.6b), or above the centroidal axis for the continuous beam portion (Fig. 4.6c). For the preliminary design of floor beams, the flange action of the slabs may be ignored, and the T-beams may simply be treated as rectangular sections where the pressure line coincides with the centroidal axis, so that there are only axial stresses, and no bending due to the prestress force, P.

$$f_a = P/A \tag{4.40}$$

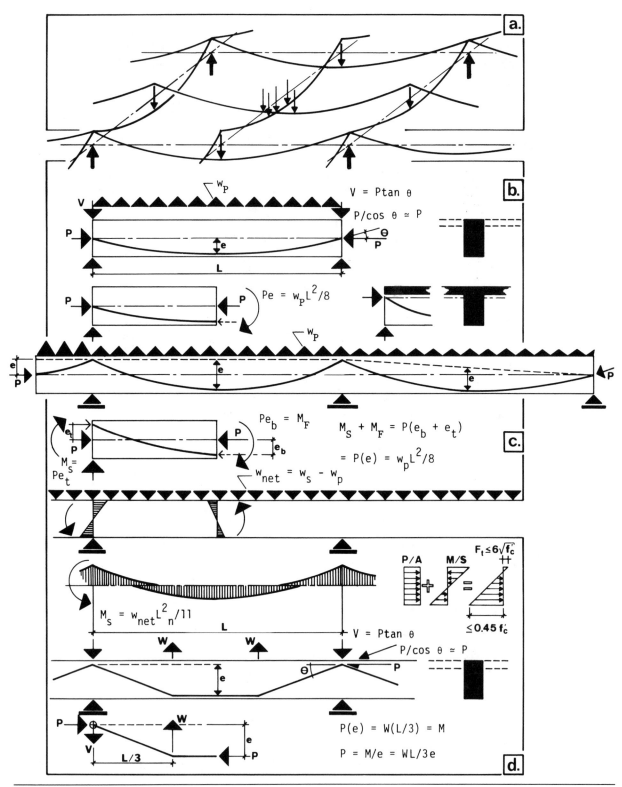

Figure 4.6. *The load balancing method in prestressed concrete.*

For a continuous beam, the cables are not just hanging over the supports with sharp breaks in curvature, as has been assumed, but form a reverse transitional smooth curvature, with points of inflection close to the support. However, it has been shown that the load-balancing method is sufficiently accurate, without taking into account the effect of the reverse tendon curvature.

Although the final loading stage controls the magnitude of the prestress force, the initial stress state must also be considered for the tendon design, since there will be a loss of the initial prestress force, P_i, due to creep, drying shrinkage, elastic shortening, slippage of steel, etc. It is beyond the scope of this discussion to deal with the complex issue of prestress loss. As a first approximation, often 15 percent to 25 percent is taken; in this case, an average of 20 percent reduction of the initial prestress force is assumed.

$$P = 0.8P_i$$

The allowable stresses, f_{pi}, in the post-tensioning tendons immediately after tendon anchorage or prestress transfer are 70 percent of the tensile strength f_{pu}

$$f_{pi} = 0.70f_{pu}$$

Therefore, the required prestress tendon area is

$$A_s = P_i/f_{pi} = (P/0.8)/0.7f_{pu} = P/0.56f_{pu} \tag{4.41}$$

The analysis of prestressed beams under the loads that are not balanced can now be done as for nonprestressed beams. The unbalanced portion of the loads, that is, the live loads for this case, will cause the following flexural stresses.

$$f_b = + M_L/S$$

The maximum combined compressive stresses due to prestressing and live load bending, after all losses, should be less than the allowable stresses.

$$f_c = f_b + f_a = M_L/S + P/A \leq 0.45f_c' \tag{4.42}$$

The maximum tensile stress in the concrete is

$$f_t = f_b - f_a = M_L/S - P/A \leq 6\sqrt{f_c'} \tag{4.43}$$

The initial stress state in the concrete, after prestress transfer and before prestress losses, is not checked in this preliminary design; the reader may refer to the ACI Code to obtain the respective allowable compressive and tensile stresses for the concrete.

Example 4.3

A continuous one-way slab in an office building spans 25 ft from center to center beams, and has clear spans of 23 ft. The slab must support a live load of 80 psf. Estimate the required post-tensioning strands for a typical interior bay, realizing that the end spans require more prestressing. Because live and dead loads are nearly equal to each other, base your design on an assumption of zero deflection under full dead load. Use $f_c' = 4000$ psi and Grade 270 strands.

Estimate the slab thickness (see the section in Chapter 5 on "Post-tensioned Slabs"):

$$t \simeq L/4 = 25/4 = 6.25 \text{ in.}, \qquad \text{try a 6 1/2 in. slab}$$

The slab weighs $(150/12)\, 6.5 = 81.25$ psf, and adding the floor finish yields a final dead load of 85 psf.

For a parabolic tendon layout, the maximum cable drape is

$$e = t - 2d' = 6.5 - 2(1) = 4.50 \text{ in.}$$

The required prestress force to balance the dead load is

$$P = w_D L^2/8e = 0.085(25)^2/[8(4.5)/12] = 17.71 \text{ k/ft of slab} \tag{4.39}$$

This causes an average precompression stress of

$$f_a = P/A = 17710/6.5(12) = 227 \text{ psi} \tag{4.40}$$

This value is within the typical range of 175 to 400 psi.

Now the flexural stresses due to the live loads must be checked. The critical support moment (according to fig. 5.8a) is

$$-M_s = w_L l_n^2/11 = 0.080(23)^2/11 = 3.85 \text{ ft-k/ft}$$

The corresponding bending stresses are

$$f_b = +M/S = 3850(12)/[12(6.5)^2/6] = 546 \text{ psi}$$

The combined compressive stresses are

$$f_c = f_b + f_a = 546 + 227 = 773 \text{ psi} \leqslant 0.45 f_c' = 0.45(4000) = 1800 \text{ psi} \tag{4.42}$$

The compressive stresses, after all prestress losses, are low as compared to the allowable ones. The tensile stresses are

$$f_t = f_b - f_a = 546 - 227 = 319 \text{ psi} \leqslant 6\sqrt{f_c'} = 6\sqrt{4000} = 380 \text{ psi} \tag{4.43}$$

Since the tensile stresses are all right, the prestress tendons can be designed. The required strand area, according to Equation 4.41 is

$$A_s = P/0.56 f_{pu} = 17.71/0.56(270) = 0.117 \text{ in.}^2/\text{ft}$$

This yields a steel weight per square foot of slab of

$$w_s = A_s \gamma = A_s(490/12^2) = A_s(3.4 \text{ lb/in.}^2/\text{ft}) = 0.117(3.4) = 0.398 \text{ lb/ft}^2$$

This value is typical for one-way slabs. For a strand with 0.5 in. diameter, the spacing is

$$12/0.117 = s/0.153, \quad \text{or} \quad s = 15.69 \text{ in.}$$

Use 0.5 in. diameter Grade 270 strands spaced at 15.5 in. on center. However, some minimum bonded, nonprestressed reinforcement must be added, according to the ACI Code. In the final precise design, the prestress member must be checked not only as an uncracked section for the elastic conditions just investigated, but also for the initial design stage. Also, the complex stress distribution at the anchorage points must be considered. Additionally, the member must be checked for deflection due to the live loads, remembering that the dead loads do not cause any deflection, for deflection due to creep (which is critical in prestressed members), and it must be checked as a cracked section for the condition of ultimate strength for flexure and shear. The shear reinforcement is determined as for ordinary reinforced concrete beams, but also taking into account the axial stresses due to post-tensioning.

In Example 4.3, the live load and dead load were nearly equal to each other, and the balancing of the dead load gave a reasonable result. For many loading conditions, however, the live load is much less than one-half of the total load. In this case, balancing all of the dead load may be uneconomical and require too much prestress force (see Example 5.10), and designers often assume, as a first trial, 80 percent of the dead load to be balanced by the equivalent prestress load, w_p.

$$w_p = 0.8w_D \qquad (4.44a)$$

Should the live load be high in comparison to the dead load, then a part of the live should also be balanced. For this condition, one may assume, as a first trial

$$w_p = w_D + w_L/2 \qquad (4.44b)$$

However, it must be kept in mind that the upward camber may be excessive when the live load and superimposed dead load are not acting—hence only the portion of the live load that occurs frequently should be balanced. As a rule of thumb, one may assume that dead and live loads are approximately equal for the design of slabs in office buildings; for slabs in apartment buildings and for beam design, however, the dead load will usually exceed the live load.

For the general condition where dead and live loads are not of approximately the same magnitude, a somewhat time-consuming trial and error procedure is required to determine the magnitude of the balancing load that represents an economical solution. For this condition, the following iterative process may be used, as based on the tensile concrete capacity, by starting with an initial trial prestress load w_p.

The flexural stresses are only caused by the loads that are not balanced, that is, the service loads minus the equivalent prestress loads.

$$w_{net} = w_s - w_p$$

The critical moment for a typical continuous interior span may be taken as

$$M = w_{net}(l_n)^2/11$$

The magnitude of the prestress force, P, is estimated from the critical tensile stresses due to the full service loading.

$$f_t = f_b - f_a \le 6\sqrt{f_c'}$$

$$f_a \ge f_b - 6\sqrt{f_c'}, \quad \text{or} \quad P/A \ge M/S - 6\sqrt{f_c'}, \quad \text{or}$$

$$P \ge (A/S)M - 6A\sqrt{f_c'} \qquad (4.45)$$

After the prestress force has been found, the equivalent uniform prestress load (according to Equation 4.38) is,

$$w_p = 8Pe/L^2 \qquad (4.38)$$

This equivalent load should be equal to the load assumed at the start of the design. If this is not the case, a new value must be assumed and the process continued until the values have converged; this convergence is usually quite rapid (see Problem 4.31).

COLUMNS

From a structural point of view, the size and shape of a column depend on both the magnitude and type of force action, as well as its slenderness. A *column* that primarily carries axial loads, and that is laterally braced by the floors, should have equal moments of inertia in both principal directions. A *beam column*, on the other hand, will require a larger dimension about its bending axis, and may possibly be braced between the floors against buckling in the other direction of less stiffness.

The reader should be quite familiar with the fact that the degree of column slenderness is measured by KL/r and that, with increasing slenderness, the strength of axially loaded columns decreases. According to their mode of failure, columns can be roughly grouped as:

- Short, *stocky columns* that fail in crushing and/or yielding (i.e., *material failure*), when the slenderness is approximately Kl/r ≤ 40.
- Long, *slender columns* that fail in buckling (i.e., *stability failure*). Examples for steel columns are:
  long, slender columns that fail in elastic buckling at approximately Kl/r ≥ 100, intermediate long columns, covering the range between short and long columns, that fail due to inelastic buckling.

Typical building columns can generally be considered short and in the low range of intermediate long columns. Cross-bracing for steel frames and the web members of trusses are examples of long column behavior, as described by Euler's buckling formula.

In addition to axial action, most columns must also resist bending. In simple hinged frames, this is caused by the eccentric beam reactions along the column faces inducing relatively small local moments. In rigid frames, moments are directly transferred to the columns, because the beams are continuously connected; the joints are rigid or semirigid, depending on the degree of continuity. In this context, bending is caused not only by the direct effect of the end moments induced by lateral loads and gravity, but possibly also by the secondary effect of the axial loads, due to eccentric action along a laterally deflected slender column (i.e., the P-Δ effect).

Although the interaction between the axial load, P, and the bending moment, M, cannot always be considered as a simple addition or percentage process, this approach is reasonable for predominate bending members, where axial loads are small, or for short columns of low slenderness (for example, Kl/r ≤ 40), where the interaction between P and M can be treated linearly, and secondary effects can be ignored. For columns with higher slenderness, however, the interaction between P and M causes a new situation: the interaction line becomes a curve due to the secondary P-Δ bending effects. The straight line equation must, therefore, be modified to take these secondary effects due to the lateral displacement of the column into account. In practice, this is done by magnifying the moment consisting of the nonsway moment due to gravity and the sway moment due to the lateral loads. These secondary effects are especially critical for tall flexible buildings with large lateral loads, as well as for bridge-type frames with only few columns and relatively flexible long-span girders.

In ordinary multistory frame construction, the effect of the K-value upon the structural design of the columns is relatively small, since the columns are generally in the low-slenderness range and can be treated as short. The magnitude of the column slenderness depends directly on the type of frame employed—that is, on its lateral stiffness. For braced frames, the columns are assumed not to sidesway, so that typical columns are usually in the low-slenderness range and there is not much reduction of column strength. For this condition, single curvature is the worst possible bending configuration (see fig. 3.3), and the actual column length is safely taken as equal to the effective length (K = 1). Although this assumption may be true for hinged frames, it is conservative for braced rigid frames. In laterally braced building skeletons with a symmetrical column layout, moments can be ignored for the preliminary design of interior columns; for the preliminary design of exterior columns, reverse curvature may be assumed, and therefore moment magnification can be neglected. Should frame columns be braced only about their weak axes and free to sway about their strong axes, it may be assumed (for preliminary design purposes) that the weak axis controls the design of the columns (K_y = 1.0).

The reader may also refer to rigid frame–shear wall interaction in the section in Chapter 7 on "Frame–Shear Wall Interaction," where the shear wall resists most lateral forces in the lower building portion, while the frame does so in the upper stories. Because of this response, it is assumed (as a first approximation) that the top two stories are unbraced, but with a magnification decreased by 50 percent.

In nonbraced buildings, such as rigid frame and flat slab construction, sidesway must be resisted by the beams/slabs and columns, although no sidesway would occur if geometrically symmetrical frames were loaded only symmetrically, a situation which does not, however, exist. The columns buckle into double curvature, but will sway laterally, resulting in an effective column length equal to or larger than the actual one (K ≥ 1). The strength of the column is very sensitive to the bending stiffness of the girders at its top and bottom; it depends on the ratio of the column stiffness $(I/L)_c$ to that of the girders $(I/L)_g$. The relative sizes of the columns and girders at each floor level control the amount of rotation, and thus govern the effective column length. As the girder sizes increase (or composite floor action is introduced) or the column sizes decrease because of higher strength, less rotation occurs and the K-factor is reduced. For instance, to reduce lateral building drift and the corresponding K-values, A36 girders and A572 Grade 50 columns may be chosen. Under the worst conditions, the effective column height may extend over several stories, as may be the case at the base of a tall flat plate structure with flexible slab beams and stiff columns (e.g., K > 3). At the other extreme, for instance, a rigid deep girder floor system swaying laterally will restrain a flexible column from rotation, and thus cause a theoretical stiffness factor of K = 1, or a minimum recommended design value of K = 1.2.

In ordinary tall buildings, the columns may be much stiffer than the beams, particularly in the intermediate story range; in this case, the K-factors may be larger than in the bottom portion of the building. In the upper part of a multistory frame, the girder sizes basically do not change because gravity loading still controls the design, whereas the column sizes increase from the top down, resulting in an increase of K-values. For tall slender buildings, lateral force action will be quite critical, especially for the design of the girders in the bottom portion of the rigid frame; their stiffness controls the lateral sway of the building, thus resulting in a possible decrease of K-values.

For ordinary, nonbraced multistory building blocks with a height/width ratio of roughly less than two, and normal floor heights, the lateral load effects are of secondary nature because of the number of columns available to resist lateral movement and the relatively small magnitude of the lateral forces; in this instance, only the exterior columns are vulnerable to the unbalanced gravity moments. However, even for typical multistory buildings, for instance, up to about 15 stories with ordinary lateral loading conditions, the design of an interior column is primarily governed by gravity loading. The effect of the moment, that is the combined load action, may only yield an increase of stress by approximately 10 percent. In conclusion, it may be said that ordinary nonbraced multistory buildings will drift laterally within reasonable limits; the columns can thus be considered relatively short, where the effect of the K-values upon the design is not too critical and the P-Δ effect may be ignored for preliminary design purposes.

Steel Columns

Typical column shapes in ordinary multistory frame construction are W10, W12, and W14 sections; occasionally pipes and structural tubing are found. The W14 series are widely used; they provide an extensive range of sections with some having wider flanges to balance the radii of gyration about both major axes. They furnish areas ranging from 6.49 in.^{2} (W 14 $\times$ 22) to the jumbo sizes of 215 in.^{2} (W 14 $\times$ 730). It must be kept in mind, however, that the deeper W-beam sections, rather than the standard W14 ones, may result in savings for peripheral rigid frame systems where shear wracking (i.e., bending) is a controlling design criterion.

Built-up columns may be needed for large loading conditions, such as occur at the base of tall buildings, and/or for long unbraced columns to increase their buckling capacity (i.e., moment of inertia), as may be the case for the first two floors of a structure. Built-up columns may also be required to provide stiffness for drift control. They basically consist of multiple members such as plates, angles, channels, and W-sections. They may form typical shapes, such as

- Coverplated W-sections
- Cruciform columns built, for instance, from two or four W-sections.
- Box columns (open or closed, with or without interior webs) forming single-, double-, or triple-shaft columns. The multiple shaft columns may be connected by lacing, batten, or perforated or solid cover plates.

The arrangement of elements and the shape of these built-up columns is not just a function of construction (connection) and aesthetic conditions, but also must provide the largest possible radii of gyration in response to local and overall buckling conditions.

Multistory columns increase in increments of size and strength with an increase of loads from the top of the building down to the bottom, for example, from A36 steel to A572 and its various grades. The column sections are usually fabricated in two- and three-story heights.

In laterally braced building skeletons with a symmetrical column layout, moments can be ignored for the preliminary design of interior columns and the effective length factor can safely be assumed to be K = 1.0. Typical columns are usually in the low-slenderness range, where there is not much reduction of column strength. The allowable axial stress, F_a, corresponding to the slenderness, Kl/r, can easily be obtained from the respective tables in the AISC Manual of Steel Construction.

For *short columns*, approximately below a slenderness of 30, there is only a decrease of the allowable compressive stress of $0.6F_y$ by approximately 10 percent.

For *long columns* with a slenderness of more than about 120 but less than 200,

Euler's formula can be used, as based on elastic buckling for the design of main members.

$$F_a = 12\pi^2 E/[23(Kl/r)^2] \simeq 149000/(Kl/r)^2 \quad \text{(ksi)} \tag{4.46}$$

For the design of secondary members, such as for diagonal bracing with a slenderness of more than 120 but less than 200 a more liberal expression for the allowable stresses is given by the AISC Specification.

The AISC allowable stress formula for *intermediate length main and secondary columns*, as based on inelastic buckling, can be simplified (for preliminary design purposes) for A36 steel by replacing the curve with a straight line between Kl/r = 30 and 120 (see Problem 4.7).

$$F_a \simeq 23 - 0.1Kl/r, \quad \text{for } 30 \leqslant Kl/r \leqslant 120 \tag{4.47}$$

where: F_a = allowable axial compressive stress (ksi)

K = effective length factor

l = actual unbraced length of member (in.)

r = governing radius of gyration (in.)

For example, for Kl/r = 44, $F_a \simeq 23 - 0.1(44) = 18.60$ ksi, which is close to the true value of 18.86 ksi.

For preliminary design purposes, it may be convenient to express the required column size in terms of its weight, which is equal to

$$w_c = A_s\gamma = A_s(490/12^2) = A_s(3.4 \text{ lb/in.}^2/\text{ft}), \quad \text{or} \quad A_s = 0.294w_c,$$

using a unit steel weight of $\gamma = 490$ pcf. Letting $f_a = P/A_s = F_a$, and substituting the respective values for F_a and A_s into the equation, yields the following approximation:

$$w_c \simeq P/(7 - 0.03Kl/r) \tag{4.48}$$

where: w_c = column weight (lb/ft)

P = axial load (k)

The typical column slenderness for laterally braced high-rise buildings is rather low. For instance, for an unbraced column height of l = 11 ft, the slenderness for typical sections are as follows:

W12 - section with r = 3 in.: Kl/r = 1(11)12/3 = 44

W14 - section with r = 4 in.: Kl/r = 1(11)12/4 = 33

Notice there is only about 5 percent difference in the allowable stresses for the two cases. Due to the slenderness, $0.6F_y$ is decreased only by about 14 percent, indicating the short column character of the typical building column in a braced structure.

Example 4.4

Estimate the size of a steel column at the first floor level of a 15-story braced frame building. The interior column has an unbraced length of 11 ft and supports a 24 × 24 ft bay. Assume an 80 psf dead load and a reduced live load of 50 psf. Investigate a W12-section using r = 3 in., and a W14-section with r = 4 in.; use A36 steel.

The total axial load is

$$P = 15(24 \times 24)(0.080 + 0.050) = 1123.20 \text{ k}$$

The required column weight is

$$w_c \simeq P/(7 - 0.03Kl/r) \tag{4.48}$$

W12: $w_c \simeq 1123.2/[7 - 0.03(1(11)12/3)] = 197.75$ lb/ft, try W12 $\times$ 210

W14: $w_c \simeq 1123.2/[7 - 0.03(1(11)12/4)] = 186.89$ lb/ft, try W14 $\times$ 193

Equation 4.48 can be further simplified for a typical condition of an interior column in a multistory braced building with a dead load of 80 psf and a reduced live load of 50 psf, yielding a total load of $w = 130$ psf, by conservatively assuming a slenderness of 50 as based on W12 or W14 column sections.

The axial load is equal to the load of one floor, wA, multiplied by n floors.

$$P = wAn = 0.130nA, \text{ but } w_c \simeq 0.130nA/[7 - 0.03(50)], \qquad \text{or}$$
$$w_c \simeq nA/40 \tag{4.49}$$

where: A = floor area of one story that the column supports (ft^2)

n = number of floors that the column supports

It should be kept in mind that the results for this equation may be somewhat conservative because of the assumed large slenderness, so that the next lower section may be investigated. The column weight in this equation could also be expressed as column area, as obtained from $A_s = 0.294w_c$.

$$A_s \simeq nA/136 \tag{4.50}$$

Example 4.5

Estimate the column sizes for the conditions in Example 4.4 using Equation 4.49.

$$w_c \simeq nA/40 = 15(24 \times 24)/40 = 216 \text{ lb/ft, try W12} \times 210 \text{ or W14} \times 211$$

In rigid frame buildings, particularly when they are not laterally braced, or for the exterior columns of buildings that are braced, the preliminary design of columns may also have to include moments, in addition to axial loads.

In the introduction to this section, the effect of sidesway upon column design in nonlaterally braced buildings was briefly described. Moments are magnified due to the P-Δ effect, as caused by the larger slenderness of the intermediate-long columns. It has been found, however, that for low slenderness ratios of roughly $1/r \leqslant 50$, the K-values in ordinary unbraced multistory buildings can be considered to be nearly equal to unity, because of the adequate frame stiffness generally available. However, even by ignoring these findings and assuming a high slenderness of 59, as for a typical unbraced column length of 11 ft in the middle or lower portion of tall buildings, a W14-section with $r = 4$ in. and $K_y = 1.8$, often used for a first trial (Hooper, 1967) causes a decrease of only about 11 percent of the column strength as compared to a laterally braced column with $K = 1.0$ and Kl/r = 33. It may therefore be concluded that the exact evaluation of the boundary rotation of the column at the initial design stage may be of importance only for intermediate length columns, though it may be of considerable importance for slender columns. In addition, it should be kept in mind that actual story heights (i.e., no K-factors) were used in the design of columns prior to 1961, though it must not be forgotten that buildings have become more flexible, particularly due to the change to lighter curtains and movable partitions.

In this discussion, beam-to-column connections are considered rigid; for semirigid connections the ability of the frame to resist sidesway is obviously reduced, since the beams have a lower effective stiffness.

Based on the above discussion, it will be assumed for the preliminary design of beam-columns that the lateral building drift is kept within reasonable limits so that the columns will be within a low slenderness range where the P-Δ effect can be ignored and the moments do not have to be amplified. It may be concluded that the column capacity can be represented by a straight-line interaction of P_a and M_a, or their respective stresses, for this low slenderness range. The familiar simple interaction equation for bending about only one axis is

$$f_a/F_a + f_b/F_b \leq 1 \tag{4.51}$$

where:

$$f_a = P/A_s = \text{axial stress}$$

$$F_a = P_a/A_s = \text{allowable axial stress}$$

$$f_b = M/S = \text{bending stress}$$

$$F_b = M_a/S = \text{allowable bending stress}$$

It is conservatively assumed that $F_b = F_a$ for preliminary design purposes.

$$P/A_s + M/S \leq P_a/A_s$$

Let $B = A_s/S$ = bending factor, which transforms the moment into an equivalent axial force P'.

$$P + P' = P + BM \leq P_a$$

Since the assumptions made were conservative, the equation tends to overestimate the column size. Therefore, rather than selecting the section required by this calculation, the next smaller one is chosen as a first trial section. In general, considering biaxial bending, the equivalent axial force, P_{eq}, is

$$P_{eq} = P + P'_x + P'_y = P + B_x M_x + B_y M_y \tag{4.52}$$

Now, the column tables in the AISC Manual can be conveniently used to select a section as based on $(Kl)_y$ or $(Kl)_x/(r_x/r_y)$, whichever controls. As a first-trial use: $B_x = 0.18$ (for W14), $B_x = 0.21$ (for W12), $B_x = 0.26$ (for W10), $B_y = 3B_x$, and $r_x/r_y = 1.7$. When more than one-half of the column capacity is used in bending, however, the column tables may not provide an economical solution, since the bending is more efficiently resisted by deeper W-sections rather than the typical column members in the tables.

Example 4.6

Determine the preliminary size of a W14 column (A36) that is 12 ft long and is not braced about its strong axis, but is braced about its weak axis, and hence does not sway in this direction. The column carries an axial load of 500 k and a moment of $M_x = 200$ ft-k.

For preliminary estimation purposes, it may be assumed that the braced weak axis controls the design (see Problem 4.8). As a first trial use $K_y = 1.0$ and $B_x = 0.18$.

$$P_{eq} = P + P' = P + B_xM_x$$

$$= 500 + 0.18(200)12 = 932 \text{ k} \qquad (4.52)$$

From the Column Load Tables of the AISC Manual for $Kl = 1(12)$, the following section is obtained.

$$\text{Try W14} \times 159, P_a = 911 \text{ k}, B_x = 0.184 \approx 0.18.$$

Example 4.7

Find the approximate size of the W14 column (A36) in the previous example, but now for an unbraced building. As a first trial assume $K_y = 1.8$, which may be quite conservative, and $Kl = 1.8(12) = 21.6$, about 22, together with $P_{eq} = 932$ k. From the AISC Column Tables,

$$\text{Try W14} \times 176, \qquad P_a = 874 \text{ k.}$$

Should the AISC Steel Manual not be available, approximate the bending factor for W-sections as: $B_x = 2.5/t$ for heavy sections, as for taller buildings; $B_x = 2.6/t$ for mid-rise structures; and $B_x = 2.7/t$ for lower buildings (see Problem 4.9) using the nominal depth, t; further, assume $B_y = 3B_x$. For the special condition where the moment is quite large in comparison to the axial load, a deep beam section may be more economical than the typical W14 and W12 column sections; for this situation estimate $B_x = 3/t$.

Substituting this information into the beam column equation yields, for heavy loads:

$$P + P' = P + B_xM_x = P + 2.5M_x/t = A_sF_a, \qquad \text{or} \qquad (4.53)$$
$$A_s = (P + B_xM_x)/F_a = (P + 2.5M_x/t)/F_a$$

The column area, A_s, can easily be estimated by assuming for a W14 (t = 14 in., $r_y = 4$ in.), W12 (t = 12 in., $r_y = 3$ in.), W10 (t = 10 in., $r_y = 2.6$ in.), and $r_x = 1.7r_y$. In addition, with respect to this equation when bending action is substantial, do not select the section required, but rather try the next lower one, because of the conservative assumptions made.

Example 4.8

Determine the preliminary column size for the case in Example 4.6, by applying Equation 4.53, assuming $K_x = 1.8$ and $K_y = 1.0$. Check to see which axis controls.

$$(Kl/r)_y = 1(12)12/4 = 36$$

$$(Kl/r)_x = 1.8(12)12/1.7(4) = 38.12 \geqslant 36$$

Therefore, the strong axis barely controls the design. Using the straight line column formula approximation yields

$$F_a \approx 23 - 0.1(Kl/r) = 23 - 0.1(38.12) = 19.19 \text{ ksi}$$

Thus, the required minimum cross-sectional area for the column is

$$A_s = (P + 2.5M_x/t)/F_a \tag{4.53}$$
$$= [500 + 2.5(200(12)/14)]/19.19 = 48.39 \text{ in.}^2$$

Try W14 × 159, A_s = 46.7 in.^{2}, r_x = 6.38 in., r_y = 4 in.

Notice that it would have made hardly any difference for the column design if the known effective length of the y-axis rather than the estimated length of the x-axis would have been used for this preliminary investigation, as was already pointed out in Example 4.6.

Concrete Columns

Most concrete columns in ordinary high-rise buildings are rectangular or round. Occasionally they are of polygonal, T-, or L-shaped, as for corner columns, or they form flat wall columns.

The typical building column contains at least 1 percent, but not more than 8 percent, vertical reinforcement ($0.01A_g \leq A_{st} \leq 0.08A_g$), which is held in position by lateral ties or spirals to form a stiff steel cage. The practical limit for fitting the bars, if they are not bundled together, is about 5 to 6 percent of the column area, A_g; when more than 4 percent reinforcement is used, however, the beam-column bar clearances at the floor level should be checked. Concrete columns may not just be tied or spiral columns, but interact with other materials in composite action. The composite column may be either a concrete-filled steel pipe or tube (e.g., jumbo column scheme), or it may be a reinforced concrete column with an encased steel shape, as is typical for high-rise composite frame construction.

Because of the inherent continuity of cast-in-place reinforced concrete construction, concrete columns must always be treated as beam-columns. They will be capable of attaining *material failure* in ordinary braced frames, while *stability failure* may occur in unbraced frames, or for very slender braced columns. Hence, one may distinguish between two groups of columns from a behavioral point of view:

- The *short columns*, where slenderness is ignored.
- The slender or *long columns*, where the effects of slenderness must be considered. In this case the axial load capacity of a column may be significantly reduced due to the moments resulting from lateral deflection of the column (P-Δ effect).

Columns of laterally braced buildings can generally be treated as short; in this case, an effective clear height of $Kl_u = 1.0l_u$ can safely be assumed. The slenderness effects for the design of braced frames can be ignored in ordinary monolithic concrete construction if the column thickness, t, is larger than 1/14th of the clear column length, l_u, measured between the floor slabs, or the top of the floor slab and the bottom of the beam or column capital (see Problem 4.13). In general, for columns of rectangular cross section:

$$t \geq l_u/14 \text{ for typical columns above the first floor level} \tag{4.54a}$$
$$t \geq l_u/10 \text{ for first floor columns with zero end restraint}$$

Thus, for typical net column heights of l_u = 8 to 12 ft, the corresponding minimum column dimensions are in the range of 8 to 10 in.; designers often consider 8 in. as a minimum column dimension, which is usually also the minimum as based on fire resistance. Surveys have shown that, for more than 90 percent of columns in braced frames, buckling can be ignored and the columns can be treated as short!

Selecting column proportions in frames not braced against sidesway such that slenderness criteria do not control may be quite unreasonable, because of the large column sizes thus required. According to the ACI Code, unbraced columns can be considered short if $Kl_u/r \leq 22$. Using a minimum effective length factor of $K = 1.2$, which is based on a nearly equal column to beam stiffness or nearly rigid beam action, results in the following expression for columns of rectangular cross section for which slenderness considerations could be ignored under certain conditions:

$$t \geq l_u/6 \qquad\qquad (4.54b)$$

This limiting condition represents roughly the changeover from short to long columns, and may be particularly applicable to ordinary low-rise buildings and the upper floors of high-rise buildings. Notice that, for slenderness effects to be neglected, the column for the unbraced frame would have to be about twice as thick as the one for the braced frame, which obviously does not make sense, although it is clear that a laterally braced column may be much more slender than the equivalent unbraced column, since it has to resist much less lateral force. Surveys have shown that 40 percent of columns in unbraced frames are actually short columns! However, it must be kept in mind that, with the development of high-strength concretes, much smaller cross sections are possible, which will (in turn) result in more slender members that are vulnerable to stability problems and secondary loading effects.

For purposes of the preliminary design of columns, it is assumed that, in ordinary braced buildings, slenderness can be ignored. The same is assumed for unbraced buildings, but here the column sizes are selected as based on a low percentage of steel, so that more reinforcement can be added later in the final design stage, when slenderness will have to be taken into account.

The columns are treated as short in the following discussion so that material failure may be assumed. In cast-in-place reinforced construction, a minimum moment for every column must be considered because of the inherent continuity and rigidity of the monolithic connections—in other words, a column must always be treated as a beam-column. Its behavior and mode of failure depend on the relative magnitude of the ultimate forces P_u and M_u, or the distance, e, that is, the statically equivalent representation of the force P_u acting eccentric with respect to the centroidal axes of a cross section where the eccentricity, e, is measured by $e = M_u/P_u$.

Most concrete columns of ordinary buildings fail in compression, as surveys have shown; that is, the eccentricity, e, of the axial force is less than the eccentricity, e_b, of the balanced loading case, at which compression and tension failure occur simultaneously. When $e > e_b$, that is, when the moment is very large in comparison to the axial force, P_u lies well outside the cross section and the column fails in tension.

In the compression range, it holds true that the larger the axial load, the smaller the moment the column can sustain. In the tension zone, the opposite is true: with an increase of axial action, larger moments can be carried because the axial load is prestressing the column and suppressing the moment-induced tension as shown in Figure 4.7.

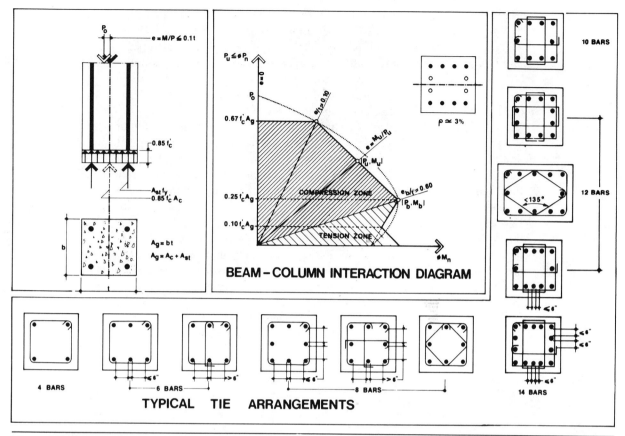

Figure 4.7. *Reinforced concrete columns.*

In conclusion, the following three short column types can be distinguished, as based on loading:

- Columns that primarily carry axial forces and only small moments; they are predominantly axial members, which fail in compression. This condition is typical for interior columns of laterally braced buildings with regular bay layouts.
- Columns that must resist axial forces and bending, with each possibly using nearly equal portions of the column capacity; these are truly beam-columns, which fail mostly in compression. This condition is typical for flexible unbraced buildings.
- Columns that carry large bending moments but small axial forces. They are predominantly bending members (beams), where failure is initiated in tension.

For the preliminary design of concrete columns, only the first two types (which are the most common in regular buildings) will be investigated in this context, that is, only compression failure is considered; only normal-weight concrete will be used.

For the condition where the column moments are small, the ACI Code (318-71) used to require, for tied columns, a minimum eccentricity of 10 percent of the overall depth of the section.

$$e \geq 0.1t \qquad (4.55)$$

This case is rather typical for the interior columns of ordinary laterally braced buildings with regular bays (i.e., where beam gravity moments balance each other). In this case, the columns do not resist much lateral force, with the exception of the upper floors, where however, the combined action of gravity and wind/earthquake is not critical. Nearly the same column sizes as derived for the interior may also be used for the exterior columns, which carry less axial load but larger moments, due to the girder rotation as caused by the unbalanced gravity loads.

This case can also be used as a first approximation for the upper floors of nonbraced buildings or for massive building blocks where (as has been discussed) lateral force action does not control the design, but slenderness must still be considered. For this condition, the column section may, for instance, be selected as based on the minimum vertical reinforcement of 1 percent of the gross area of the column ($\rho_{min} = A_{st}/A_g = 0.01$). The effects of slenderness will then be covered by additional steel in the final design stage.

The compressive capacity $\phi P_n = P_o$ of a short concrete column under only concentric loads, consists of the sum of the concrete strength $\phi 0.85f'_c A_c$ and the steel strength $\phi f_y A_{st}$, as shown in Figure 4.7. This expression is further reduced for tied columns by a factor of 0.8, to take into account the nature of continuity in cast-in-place concrete structures; in this instance, the moment action of $M_{umin} = P_u e_{min}$ is roughly equal to the minimum eccentricity concept of $e_{min} = 0.1t$ of previous codes. Therefore, the compressive strength of a short, tied rectangular column is

$$P_u \leq \phi P_n$$

$$\leq \phi 0.80[0.85f'_c(A_g - A_{st}) + f_y A_{st}] \tag{4.56a}$$

$$\leq \phi 0.80 A_g[0.85f'_c(1 - \rho_g) + \rho_g f_y] \tag{4.56b}$$

$$\leq \phi 0.80 A_g[0.85f'_c + \rho_g(f_y - 0.85f'_c)] \tag{4.56c}$$

where: P_u = factored axial load (k)

A_g = area of column cross section (in.^{2}) = cross-sectional area of concrete A_c plus area of longitudinal reinforcement A_{st} (i.e., $A_g = A_c + A_{st}$)

f'_c = compressive strength of concrete (ksi)

$\rho_g = A_{st}/A_g$ = column reinforcement ratio

f_y = yield strength of longitudinal reinforcement (ksi)

From this equation, the required cross-sectional area of the column can easily be derived. For the typical conditions of Grade 60 steel, realizing that the effect of the concrete strength in decreasing the steel strength ($f_y - 0.85f'_c$) is minimal (see Equation 4.56c), and using a capacity reduction factor for tied columns of $\phi = 0.7$, the following simple approximation can be used for the first sizing of short concrete columns:

$$A_g = P_u/(0.5f'_c + 0.3\rho_g), \qquad \text{for } e = M_u/P_u \leq 0.1t \tag{4.57}$$

where: ρ_g = column reinforcement ratio (%)

For instance, for a minimum steel ratio of $\rho_g = 1\%$ and 4000 psi concrete, $A_g = .44P_u$, but for a higher steel ratio of $\rho_g = 4\%$, $A_g = 0.31P_u$. This clearly shows that, by increasing the amount of reinforcement by four times, the column cross section can be reduced by about 30 percent. For spiral columns, the capacity is 14 percent larger than for tied columns, or alternately the cross-sectional area of a column can be reduced by 12 percent, if spiral steel is used instead of ties—however, the cost of spiral steel is about twice that of tie steel. Another, even simpler expression often used in practice can be derived from Equation 4.56c, for $\rho_g \simeq 1\%$, by dividing the axial load by an average stress.

$$A_g = P_u/0.55f_c' \qquad (4.58a)$$

The axial capacity of a round tied column can be approximated as roughly 80 percent of the capacity of a square tied column with the sides equal to the diameter and with the same number of vertical bars. In other words, the diameter of the round column should be about 13 percent larger than the sides of the square column to achieve the same strength. On the other hand, the axial capacity of a round spiral column is about 90 percent the capacity of a square tied column using the same main reinforcement. Regular polygonal columns can be approximately designed as based on a circular section enclosed within the column boundaries. Other column shapes such as L-shaped columns can be treated as intersecting rectangular columns.

As based on the *working stress approach*, an expression similar to Equation 4.58a can be derived for 1 percent steel and an allowable compressive stress of $0.22f_c'$.

$$A_g = P/0.25f_c' = 4P/f_c' \qquad (4.58b)$$

Equation 4.58a can be further simplified for the following typical loading conditions of 100 psf dead load and a reduced live load of 50 psf:

$$P_u = nA[1.4(0.100) + 1.7(0.050)] = 0.225nA$$

Substituting this expression into Equation 4.58a by letting $f_c' = 4$ ksi yields approximately

$$A_g = nA/10 \qquad (4.59a)$$

The column area A_g (in.^{2}) is equal to the total floor area nA (ft^2) that the column supports, divided by 10, where the floor area is equal to the supported area of one floor, A, multiplied by n stories.

For apartment buildings with smaller loads, the following expression is often used for the estimation of concrete column sizes:

$$A_g = nA/12 \qquad (4.59b)$$

For the preliminary estimation of laterally braced columns in office buildings, as based on one percent steel, for 8000 psi concrete, use a required cross-sectional area of $A_g = nA/20$, and for 6000 psi concrete $A_g = nA/15$. It should be kept in mind that these equations are only applicable to interior columns of laterally braced buildings, with regular column layouts satisfying the above loading conditions.

Similar to Equation 4.56, the compressive strength of a *composite column* is shared by the steel section, the longitudinal reinforcement, and the concrete. Therefore, the compression strength of a stocky composite column, where the design yield strength of the steel core is limited to 52 ksi, can be expressed for the following conditions, considering

- a concrete filled pipe or tube:

$$P_u \leq A_s F_y + A_{st} F_{yr} + 0.85 f'_c A_c \qquad (4.60a)$$

- a concrete-encased structural steel section:

$$P_u \leq A_s F_y + 0.7(A_{st} F_{yr} + 0.85 f'_c A_c), \qquad (4.60b)$$

where shear transfer between the concrete and steel must be considered.

Example 4.9

A 15-story laterally braced concrete frame building is organized on 20 × 20 ft bays, approximately satisfying the discussed loading conditions. Estimate the column sizes assuming that the same column continues for three stories.

An interior column supports a typical floor area of $A = 20(20) = 400$ ft^2, or the floor area of three stories is 1200 ft^2.

13th to 15th floor:	$A_g = 1200/10 = 120$ in.^{2}	try 12 × 12 in.
10th to 12th floor:	$A_g = 2(120) = 240$ in.^{2}	try 16 × 16 in.
7th to 9th floor:	$A_g = 3(120) = 360$ in.^{2}	try 20 × 20 in.
4th to 6th floor:	$A_g = 4(120) = 480$ in.^{2}	try 22 × 22 in.
1st to 3rd floor:	$A_g = 5(120) = 600$ in.^{2}	try 25 × 25 in.

At the building base, roughly $A_g/A = [(25/12)^2/400]100 \simeq 1.1\%$ of the floor area is taken up by the columns!

In general, as based on formwork costs, it is more economical not to change column sizes from floor to floor, but rather to alter only the material strength and the percentage of reinforcing steel. In multistory buildings, the largest column size is determined at the base of a building for a high-strength concrete of, for instance, 10,000 psi and a reasonable maximum percentage of Grade 60 steel. As the column progresses upward, the percentage of reinforcing bars decreases to a minimum of one percent. Then the 10,000 psi column is replaced by a lower strength concrete of, for example, 8000 psi. This process is continued until, after several floors, the column size can be substantially reduced. However, consideration should then be given to changing only one dimension of the column. It is often economical to change column sizes in increments of at least 30 percent to 50 percent or more.

One should keep in mind that slab openings for pipe chases at the columns may require larger column sizes as to provide an adequate slab capacity for flat plates.

Example 4.10

Investigate an interior column from Example 4.9 at the first floor level. Use a steel ratio of about 3 percent, rather than the minimum amount of 1 percent, as in the previous example, together with $f'_c = 4000$ psi and $f_y = 60$ ksi. The clear height of the column is 10 ft.

The ultimate floor load is

$$w_u = 1.4D + 1.7L = 1.4(0.100) + 1.7(0.050) = 0.225 \text{ ksf}$$

The column load is equal to the floor load multiplied by the floor area that the column must support.

$$P_u = 0.225(20 \times 20)15 = 1350 \text{ k}$$

The required column area is

$$A_g = P_u/(0.5f'_c + 0.3\rho_g) \tag{4.57}$$
$$= 1350/[0.5(4) + 0.3(3)] = 465.52 \text{ in.}^2$$

Try a 22×22 in. column, $A_g = 484$ in.^{2}.

The column can be treated as short, and the effect of slenderness ignored, since

$$l_u/t = 10(12)/22 = 5.46 < 10$$

Determine the true ρ_g ratio from Equation 4.57.

$$484 = 1350/[0.5(4) + 0.3\rho_g], \text{ which yields } \rho_g = 2.63\%$$

The required vertical steel is

$$A_{st} = \rho_g A_g = 0.0263(484) = 12.73 \text{ in.}^2$$

Use 14 #9, $A_{st} = 14$ in.^{2} by placing 5 bars on two faces and the rest along the other two faces.

According to the ACI Code, all nonprestressed bars shall be enclosed by lateral ties at least #3 in size for longitudinal bars #10 and smaller, and at least #4 in size for larger bar sizes and bundled bars. The vertical spacing of the ties is the least of the following values:

16 vertical bar diameters:	$16(1.128) = 18.05$ in.
48 tie diameters:	$48(0.375) = 18$ in.
least column dimension:	22 in.

Use #3 ties @ 18 in.

Furthermore, according to the ACI Code, ties shall be arranged such that every corner and alternate longitudinal bars shall have lateral support provided by the corner of a tie, with an included angle of not more than 135 degrees, and no bar shall be farther than 6 in. clear on each side along the tie from such a laterally supported bar. The clear distance between the longitudinal bars must not be less than 1.5 times the vertical bar diameter, nor less than 1 1/2 in. For typical tie arrangements refer to Figure 4.7. Checking the critical column faces with the four bars and the wider spacing, as shown in Figure 4.7, yields

$$[22 - 2(1.5 + 0.375 + 2(1.128))]/3$$
$$= 4.58 \text{ in} < 6 \text{ in. The spacing is satisfactory.}$$

Remember that when the ties also have to act as stirrups to carry excess shear, then $s_{max} = d/2$ must be considered! Usually, inadequate shear reinforcement in the columns is the primary cause of most failure in earthquake regions (i.e., shear failure in the columns).

In tall rigid frame buildings the columns must resist the lateral forces in shear wracking, that is, in bending, which is especially critical in the bottom portion of the building. In addition, slenderness considerations become important, since they magnify the moment action. The predominant column action of laterally braced buildings is replaced by beam-column behavior in nonbraced buildings!

Beam columns can be roughly designed for a typical reinforcement ratio of about 2.5 to 3 percent by treating the moment action, M_u, as an equivalent eccentricity of the axial force, P_u: $M_u = P_u(e)$, or $e = M_u/P_u$. The required cross-sectional column area can be derived for Grade 60 steel as follows:

- The required cross-sectional area for a column that carries an axial load and only a small moment, and with a steel ratio of nearly 2.5 percent, can be derived from Equation 4.57 as approximately

$$A_g = 1.5P_u/f_c', \quad \text{for } e \simeq 0.1t \tag{a}$$

- For the special, so-called balanced, condition at which the column fails simultaneously in tension and compression due to axial action and extensive bending, the axial load, P_b, at which this type of failure occurs (see Problem 4.14) is approximately equal to

$$P_b = 0.3bdf_c'$$

By letting $P_u = P_b$ and $d \simeq 0.83t$, the balanced column area for this boundary stage can be roughly expressed as

$$A_g = P_u/0.25f_c' = 4P_u/f_c' \tag{b}$$

For a typical steel ratio of 3 percent, the tension failure occurs at about $e_b = 0.6t$ (see Problem 4.14).

- Now, the cross-sectional column area for a low but typical reinforcement ratio of about 2.5 to 3 percent can be approximated from Figure 4.7 by a straight-line interpolation between the upper and lower boundaries of the compression failure range of $1.5P_u/f_c'$ and $4P_u/f_c'$, respectively for Grade 60 steel (see Problem 4.15). This simplified version of the beam-column equation refers only to the special condition where the eccentricity does not exceed about six-tenths of the depth of the column, as based on 3 percent steel, keeping in mind that e_b varies with the steel ratio (e.g.: $\rho \simeq 2.5\%$, $e \leqslant t/2$).

$$A_g = P_u[1 + 5(e/t)]/f_c', \quad \text{for } e \leqslant 0.6t \tag{4.61}$$

It has been assumed that the bars are concentrated on the two end faces perpendicular to bending, which is reasonable when large bending moments are present. Should the bars be placed along all four faces, then the ones on the side faces parallel to the moment action (Fig. 4.7, top center) should be ignored for this preliminary design approach. Generally, when $e/t > 0.2$, the reinforcement may be placed most efficiently on the end faces of a rectangular column. It is apparent that a round column is inefficient for moment resistance.

For the condition where the moments are very large and the axial forces are small, so that tension failure controls the design of the column ($P_u < P_b$, $e > e_b$), the column may be treated as a beam (for preliminary design purposes) conservatively assuming $\phi = 0.7$.

$$M_u \simeq (\phi A_s f_y/2)0.9d \simeq 0.3A_s f_y d \tag{4.62}$$

In response to biaxial bending, the design of symmetrical square corner columns with bars along all four faces can be approximated (for preliminary design purposes) by simply adding the eccentricities $e_x + e_y$ or the moments M_{ux} and M_{uy}, rather than adding them vectorially.

As a rough first estimate of beam-column sizes, designers often use Equation 4.57, but ignore the capacity of the reinforcement. The effects of bending and slenderness will later be accommodated with the longitudinal bars in the final design stage.

$$A_g = P_u/0.5f'_c = 2P_u/f'_c \qquad (4.63a)$$

Notice that this expression can also be obtained from in Equation 4.61 for the special condition of $e = t/5$. Should this eccentricity be exceeded, but the steel ratio of 3 percent be kept, then the cross-sectional column area may be estimated as (2.5 to 3) P_u/f'_c, depending on how large the eccentric force action is. When the working stress method is used, Equation 4.63a can be replaced by the following expression for rough estimation purposes:

$$A_g = P/0.22f'_c \simeq 5P/f'_c \qquad (4.63b)$$

This equation represents a load capacity of 40 percent of that computed in accordance with Equation 4.63a, as is required by the ACI Code.

Example 4.11

An unbraced concrete column has an unsupported length of $l_u = 10$ ft; it carries $P_u = 640$ k and $M_u = 320$ ft-k. Determine the column size using a typical steel ratio of 3 percent so that secondary load effects can be covered with additional reinforcement. Use $f'_c = 4000$ psi and $f_y = 60$ ksi.

The equivalent eccentricity of the axial force is

$$e = M_u/P_u = 320(12)/640 = 6 \text{ in.}$$

As a first estimate from Equation 4.63a

$$A_g = P_u/0.5f'_c = 640/0.5(4) = 320 \text{ in.}^2$$

which yields about 18×18 in.; use Equation 4.61, and try $t = 18$ in.

$$A_g = P_u[1 + 5(e/t)]/f'_c = 640[1 + 5(6/18)]/4 = 427 \text{ in.}^2$$

This yields about 18×22 in. Try $t = 22$ in.

$$A_g = 640[1 + 5(6/22)]/4 = 378 \text{ in.}^2$$

Use an 18×22 in. column, $A_g = 396$ in.^{2}, $e/t = 6/22 = 0.27 < 0.6$

The approximate longitudinal steel, using conservatively $\rho_g = 3\%$, is

$$A_{st} = \rho_g A_g = 0.03(396) = 11.88 \text{ in.}^2$$

Try 10 # 10, $A_s = 12.70$ in.^{2} placed on the two end faces, and #3 ties @ 18 in. (see Example 4.10).

Check the slenderness:

$$l_u/t = 10(12)/20 = 6 \text{ in. (see Equation 4.54b)}$$

Hence, depending on the lateral stiffness of the building, the slenderness may not be critical for this case, since it is equal to the limiting value of six, which was, however, based on a nearly equal column-to-beam stiffness, or rigid beam action.

TENSION MEMBERS

In high-rise construction, tension members are found as primary structural elements in trusses, and for suspension buildings, where they are used as vertical hangers and catenaries. They also occur as diagonal lateral wind bracing, as vertical supports for wall girt systems, as cable stays for girders, guy wires for hoists and derricks, and as cable supports for elevators. They are usually steel, but can also be prestressed concrete, as is discussed further in Example 7.1. The typical steel tension members are cables, rods, bars, and rolled shapes, including built-up members.

Single Structural Shapes and Built-up Members

When some rigidity is required to resist bending and a reversal of loading, as for some truss members, rolled sections are selected, since cables and rods are flexible. It is clear that the structural shapes must possess a minimum slenderness to be considered rigid. The typical rolled member shapes are angles, channels, tubes, and wide-flange sections. Built-up shapes are assembled from single members and tied together with lacing, solid or perforated plates, or tie plates to form, for example, closed box shapes or open shapes like double angles and star forms. The connection details for tension members are most important as they often are the weakest part. Typical connections are of the welded or bolted angle or channel type.

Usually, the shape of the cross section has little effect on the tensile capacity of a member, that is, not considering the connection influence. Tension tends to straighten the member, in contrast to compression, which causes the member to laterally displace and buckle.

The preliminary design of tension members (with the exception of pin-connected members) is based, conveniently, on yielding of the entire gross area of the section, rather than fracture at its weakest effective net area at connection points, which obviously must also be taken into account in the final design, when the connection detail is known.

$$f_t = P/A_s \leq F_t = 0.6F_y, \quad \text{or } A_s = P/0.6F_y \qquad (4.64)$$

To provide some minimum stiffness for tension members other than rods, and to avoid undesirable lateral movement and vibrations, the AISC Specs recommend a maximum slenderness ratio of 240 for main members and 300 for lateral bracing and other secondary members. Hence, the minimum radius of gyration for main tensile members is

$$r_{min} = \sqrt{I/A} = L/240 \qquad (4.65)$$

When tension members are also subject to bending (in this case about one axis only) they can be proportioned according to the following familiar expression

$$A_s = (P + BM)/F_t = (P + BM)/0.6F_y \qquad (4.66)$$

Refer to the discussion of steel column design (Equation 4.52) for the derivation of the equation and the definition of the terms.

Round and Square Rods and Flat Bars

Rods and bars are used as wind bracing, hangers, sag rods for roofs, and as support for girts. To limit the slenderness of rods, often a minimum diameter of 5/8 in. or L/500 is taken. They may be pretensioned, to prevent the sagging of diagonal rods under their own weight due to a lack of stiffness. Rods are connected by employing welding or threading the ends and using bolts, turnbuckles, or clevises, while flat bars may be welded, bolted, or pin-connected (e.g., eyebars). The AISC specs allow for the reduced area through the threaded portion of a rod by conveniently incorporating the gross area A_s. The allowable tension stress for threaded rods is equal to $0.33F_u$, where the tensile strength, F_u, for A36 steel may be taken as 58 ksi.

$$f_t = P/A_s \leq 0.33F_u, \quad \text{or } A_s = 3P/F_u \qquad (4.67)$$

In this case, the tension force, P, should be at least 6 kips.

Cables (Wire Ropes and Bridge Strands)

Cables are used for hoists and derricks, possibly for suspension buildings, and occasionally as diagonal wind bracing, where they are pretensioned so that they can accept compression and will not vibrate. Special fittings are required for the connection of cables.

Since the capacity of cables is roughly five times higher than that of mild steels (with an ultimate tensile strength, F_u, in the range of 200 to 250 ksi, depending on the coating class) the resulting material will be a minimum. This yields small cross-sectional cable areas and, together with a lower effective modulus of elasticity of 20,000 ksi for ropes to 24,000 ksi for strands, causes cables to be flexible and form-active, and makes them vulnerable to excessive elongations, $\Delta L = PL/AE$. The cable capacity can be obtained from the manufacturer's catalogs. However, for rough preliminary design purposes, an allowable tension stress, F_t, of about $F_u/3$, together with a resisting metallic cable area, A_n, of roughly two-thirds of its nominal gross area, A_s, may be assumed. Hence, the required nominal cross-sectional cable area is in the range of

$$A_s \simeq 1.5A_n = 1.5(P/F_t) \simeq 4.5 \ P/F_u \qquad (4.68)$$

The corresponding cable diameter can be easily obtained from $A_s = \pi d^2/4$, as

$$d = \sqrt{4A_s/\pi} = 1.13\sqrt{A_s} \simeq 2.4\sqrt{P/F_u} \qquad (4.69)$$

PROBLEMS

4.1   Derive the equations that approximate the concrete beam size and the moment and shear reinforcement.

4.2   A concrete slab is supported by a rectangular beam that must resist a maximum shear force of $V_u = 30$ k. Use $f'_c = 4000$ psi and $f_y = 40,000$ psi. Assume the ratio of beam width to beam depth as about 1/2. Find the beam size, b/t, and the stirrup spacing, if required, for the following cases: a) no shear reinforcing, b) minimum shear reinforcing, c) absolute smallest section as based on d/2 maximum spacing.

4.3   A continuous concrete girder (b/d = 14/24) with a clear span l_n = 18 ft supports floor beams at the one-third points. These beams each cause concentrated loads of P_u = 80 k. For preliminary design purposes, the weight of the girder stem is already included in the single loads. Use f_c' = 4000 psi and f_y = 60,000 psi. Determine the spacing of the #3 stirrups.

4.4   A continuous beam (b/d = 12/24) with a clear span of 21 ft supports a uniform ultimate floor load of w_u = 8 k/ft. Determine the #3 stirrup spacing for f_c' = 3000 psi and f_y = 60,000 psi.

4.5   Derive the rules of thumb for the quick sizing of simply supported W-sections under a uniform load using A36 steel.

4.6   A welded steel plate girder (A36) with a span of 60 ft supports single loads of 150 k at the one-third points, in addition to a uniform load of 2 k/ft, which includes the girder weight. Do a preliminary design of the girder.

4.7   Approximate the allowable stress formula for A36 intermediate-long steel columns by assuming a straight-line relationship between allowable stress and slenderness. Use the range from Kl/r = 30 to 120.

4.8   Show that it would not have made much difference for the preliminary design of the column in Example 4.6 if an estimated effective length of the x-axis, for instance K_x = 1.8, rather than the known length of the y-axis, would have been used.

4.9   Show that the bending factor B_x for a W-section is roughly equal to 2.5/t, and explain the meaning of the bending factor.

4.10 Estimate the size of an interior 11-ft long W12 column (A36) that carries a concentric axial load of 600 k. The column is braced about its weak axis, but is free to sway about its strong axis.

4.11 Estimate the size of a 12-ft W14 column (A36) that supports 1000 k and a wind moment of 275 ft-k. Consider the column to be braced about both axes.

4.12 Estimate the size of a corner column in a skeleton building that is braced in both directions. The column has an unbraced length of 12 ft with respect to both axes, and is subject to the following loads: P = 500 k, M_x = 60 ft-k, M_y = 10 ft-k. Use A36 steel.

4.13 Determine the minimum dimensions for rectangular concrete columns in laterally braced and nonbraced buildings so that slenderness does not have to be considered in the structural design. Also derive the column equation Equations 4.57–4.59.

4.14 Determine the approximate magnitude of the balanced axial force P_b at which the column fails in tension and compression. Assume the placement of the reinforcement to be parallel to the bending axis and symmetrically arranged. Use f_y = 60 ksi.

4.15 Derive the beam-column equation (Equation 4.61) that represents the compression failure range by using a straight-line approach for reinforcement ratios between about ρ_g = 2.2 percent and 3.0 percent, as based on f_c' = 4000 psi and f_y = 60,000 psi.

4.16 An interior column of a braced flat plate building with a clear height of 9 ft, 6 in. carries an ultimate load of P_u = 400 k. The flat plate design requires a 12 × 12 in. column. Determine the approximate amount of steel needed, using f_c' = 4000 psi and f_y = 60,000 psi.

4.17 Select the size of a square, tied concrete column that must carry P_D = 200 k, P_L = 125 k, M_D = 50 ft-k, M_L = 25 ft-k. Use f'_c = 4000 psi and f_y = 60,000 psi. Also estimate the steel considering sideway not to be critical.

4.18 Size the column in Problem 4.17 so that it acts primarily in an axial action, and find the reinforcement.

4.19 What column sizes do you select for the case in Problem 4.17, if the eccentricity of the axial load increases to e = 8 in. and then to e = 12 in.?

4.20 Determine the preliminary size of an interior concrete column with about 2 percent steel for a laterally braced building. The column has an unsupported height of 8 ft, 6 in. and carries axial loads of P_D = 300 k and P_L = 200 k. Use f'_c = 4000 psi and f_y = 60,000 psi.

4.21 Estimate the size of an exterior column at the first floor level of a 50-story laterally braced building. The column supports a typical floor area of 10 × 20 ft, an average dead load of 165 psf, and a live load of 80 psf. Assume a live load reduction of 60 percent. The average floor height is 12 ft, 6 in. Use f'_c = 5000 psi and f_y = 60,000 psi. Keep in mind that the unbalanced gravity moments must be taken care of in the final design.

4.22 How much does the column in Problem 4.21 shorten due to elastic deformations under the full loading and due to creep of approximately 0.1% in./in.? Do you foresee any structural problems?

4.23 Find the size for an interior column of a laterally braced rigid frame concrete building at the top floor. Assume that the column has a clear height of l_u = 10 ft, and carries P_D = 100 k and P_L = 40 k. The live load moments due to gravity balance each other. Use f'_c = 4000 psi and f_y = 60,000 psi, and use the minimum amount of reinforcement.

4.24 Repeat Problem 4.23, but consider the column not to be laterally braced and part of a rigid frame.

4.25 The typical interior bays of a 12-story flat plate apartment building are 18 × 18 ft, with a typical story height of 9 ft. The lateral forces are resisted by shear walls. Assume a live load of 40 psf and a dead load of 100 psf, which includes the weight of the lightweight 7 1/2 in. concrete slab, the partitions, and columns. Estimate the size of a typical interior column at the first floor level. Use f'_c/f_y = 5/60 and ρ_g = 2% of reinforcement.

4.26 Determine the typical interior column size at the first floor level for an unbraced five-story flat plate building, using a story height of 12 ft. The three bay frames, with each bay 20 ft wide, are spaced 20 ft apart; thus each interior column supports a 20-ft square bay. The loading for each floor, including the roof, is: 8 in. concrete slab (100 psf), column weight assumed as 10 percent of slab weight, partition (20 psf), 50 psf live load, and 30 psf for the exterior wall. Assume a constant wind pressure of 20 psf. Use f'_c/f_y = 4/60.

4.27 A 12-story frame-shear wall building is subdivided into three bays of 22 ft in the cross direction and seven bays of 26 ft in the long direction. Two shear walls are provided in the transverse direction. The typical story height is 12 ft. Estimate the interior column size for the rigid frame at the second floor level. Use an average dead load of 160 psf, which includes the weight of the partitions (20 psf), ceiling and mechanical (10 psf), columns, beams, and slabs. The live load is 50 psf, which may be reduced by 60 percent. Keep in mind that the axial gravity action may be assumed to control the preliminary design of the column. The gravity moments can be ignored because of the interior position of the columns. The lateral force action is probably less critical, since the building is not very high and, in addition, the lateral shear may be assumed to be resisted primarily by the shear walls near the base of the building; it may, therefore, be concluded that the column moments are relatively small. Use $f_c'/f_y = 4/60$ and about 2 percent of reinforcement.

4.28   Do a preliminary design of an interior square column with a clear height of 12 ft at the second floor of an 11-story rigid frame office building in concrete. The loading is as follows: $P_D = 350$ k, $P_L = 130$ k, $M_D = 45$ ft-k, $M_L = 35$ ft-k. Use $f_c'/f_y = 4/60$.

4.29   A square, rigid frame concrete building with a central structural core is subdivided into nine structural bays, 36×36 ft each. The building is ten stories high, where the typical floor height is 13 ft and the bottom floor equal to 15 ft. Determine the size of a typical interior facade column at the first floor level, using Grade 60 reinforcing and 5 ksi concrete. Use an average dead load of 170 psf of floor area and an exterior wall load of 30 psf of wall area. The roof live load is 20 psf, while the typical office live load is 50 psf, and the partition live load is 20 psf. Reduce the live load by 60 percent for the column design. For this preliminary design approach assume that, for the given low building height, the wind loading case is not critical, especially since the concrete structure provides such a high gravity load. Further, ignore the gravity moment due to the unbalanced beam moment which the exterior column must carry; additional reinforcing will resist this moment.

4.30   Composite simply supported floor beams are spaced 8 ft apart and span 30 ft. Estimate their size, assuming a total floor load of 175 psf, A36 steel, and 4000 psi concrete.

4.31   A shallow 14×60 in. post-tensioned continuous beam supports a 30-ft, 7 1/2 in.-thick one-way concrete slab on each side. The beam span from center to center of the 16×16 in. columns is 30 ft. The superimposed dead loads consist of 10 psf for the ceiling and 20 psf for partitions, and a reduced live load of 50 psf. Estimate the size of the prestress tendons and check the concrete stresses. Use $f_c' = 4000$ psi and Grade 160 prestress bars. For this preliminary investigation, ignore the effective flange width of the slab and treat the beam as an isolated rectangular section.

4.32   A 20-in. wide $\times$ 30-in. deep, 24-ft span continuous concrete beam is to be post-tensioned with parabolic cables and a sag of 6 in. For this case, about 40 percent of the total 3 k/ft load is assumed to be balanced. Determine the required prestress force and the necessary tendons, and check the stresses. Use Grade 270 strands and 4000 psi concrete. Ignore the effect of the slab flanges or T-beam action.

4.33   Determine the upward loading for the cantilever and end span conditions of the beam in Figure 6.4b due to the prestress force found in Problem 4.31, which was derived from the typical interior span of a continuous beam. Assume that the interior beam span conditions do not change; use a cantilever length of 10 ft and an exterior span equal to the interior one. Draw your conclusions.

5 THE VERTICAL AND HORIZONTAL BUILDING PLANES

Buildings generally consist of parallel horizontal planes or floor structures and the supporting vertical planes, such as walls and/or frames. Gravity and lateral forces are dispersed through the floor structure to the vertical planes, and from there to the ground. The magnitude, direction, and type of action of the static force flow depend on the geometry and stiffness of the vertical planes and on their arrangement within the building volume. The distribution of the lateral forces along the floor diaphragms to the vertical resisting structures is here studied for typical conditions.

The floors are composed of the structural and possibly mechanical spaces. In this context, the structure portion consists of the floor slab, which may have to be supported by a framing system. Ceilings are directly attached to the structure or suspended from it. Various floor framing systems and the structural design of several types of reinforced concrete slabs, including typical stair slabs, are investigated.

THE DISTRIBUTION OF LATERAL FORCES TO THE VERTICAL RESISTING STRUCTURES

The basic behavior of the building structure has already been introduced (see fig. 2.14). The external lateral force action due to wind, and the inertial forces generated within the building by the vibration of the mass due to earthquakes and possibly also wind, must be distributed by the floor framing to the vertical resisting structures. Floors in high-rise buildings act generally as rigid horizontal diaphragms, which are assumed to be supported by elastic vertical elements that can be shear walls, braced frames, or tubes that behave like *flexural cantilevers* when they are slender, but they can also be supported by unbraced frames, which act like *shear cantilevers*. These lateral-force resisting structures can be visualized as vertical cantilevers fixed at the base with concentrated horizontal loads applied at the floor levels. The floors, in turn, act as horizontal deep beams, distributing the lateral forces among the vertical resisting elements to which they are properly connected in proportion to their relative shear and flexural stiffnesses.

The layout of the resisting vertical systems can take many different forms, varying from symmetrical to the asymmetrical arrangements, or ranging from a minimum of three planes to a maximum of a cellular wall subdivision. The resisting system may

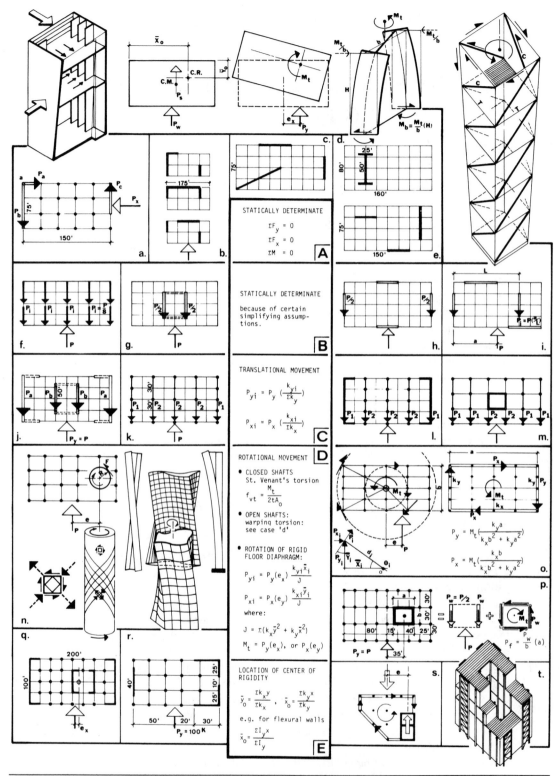

Figure 5.1. *The distribution of lateral forces to the vertical resisting structures.*

be located within the building as a single spatial core unit, or as separate individual planes, or it may be located along the periphery (see fig. 2.4 and fig. 5.1).

In a symmetrical building with a regular arrangement of vertical structures, where the line of action of the resultant of the applied lateral loads passes through the center of rigidity (center of resistance), the supporting systems deflect equally in a purely translational manner. In a symmetrical building, the geometric center coincides with the center of mass and the center of rigidity; it is located at the intersection of the axes of symmetry. However, even if a symmetrical ordinary structure should be loaded eccentrically, the torsional effects are relatively small and may be ignored for preliminary design purposes.

Asymmetry in buildings is caused by geometry, stiffness, or mass distribution—the applied resultant load does not act through the center of rigidity. In this case, the rigid floor diaphragms not only translate, but also rotate in the direction of the lateral load action. In eccentric buildings, it is essential that the vertical elements are arranged to provide adequate torsional stiffness as, for instance, along the periphery for curtain wall construction, to control rotational deflections. In asymmetrical building shapes of stepped or sloped configurations, torsional response is generated that varies with height and corresponds to the change of plan forms. It is beyond the scope of this investigation to deal with the effect of hybrid, compound building shapes, and the complexity of composite, irregular plan and elevation forms, as well as their response and their effect upon lateral load action. In this work, only basic concepts of action and reaction of simple, but typical, rectangular building forms are briefly investigated, to develop some feeling for the behavior of buildings.

The following simplifying assumptions are used for preliminary design purposes:

- The rigid diaphragm action of floors; they must be stiff enough to go through rigid body displacement.
- The bending stiffness of the planar, vertical structure about its weak axis (i.e., slab action) is neglected.
- The torsional resistance of the individual planar vertical elements is neglected.
- Intersecting planar units are treated as separate; there is no shear flow around the corners.
- The shear flow along shear walls is considered constant, as based on average shear stresses.
- Hinged connections between the diaphragms and the lateral force resisting vertical planes are assumed (i.e., the interaction of floor systems and shear walls is ignored).

Although floor structures for high-rise buildings can generally be treated as rigid diaphragms, there are situations where this may not be the case, such as:

- Closely spaced shear walls in relatively narrow buildings are stiffer in comparison to the floor diaphragms
- Floor diaphragms in long, narrow buildings with deep beam proportions of greater than approximately 3:1 that span large distances across the building
- Floor diaphragms that are weakened by cutouts and openings for atriums, escalators, stairs, etc.

For further discussion of floor diaphragms, refer to the next section in this chapter.

The following presentation of typical cases is approached from the point of view of an increase in the degree of complexity by, first, dealing with statically determinate conditions, proceeding with symmetrical buildings, then introducing torsion, and finally discussing irregular structures with mixed construction. This discussion follows the organization of Figure 5.1.

Statically Determinate Conditions

When a skeleton building is laterally braced by not more than three planar, lateral-force resisting systems, then the lateral force distribution to these elements can be treated as statically determinate, as based purely on equilibrium conditions, and independent of the stiffness of the resisting systems.

$$\Sigma F_x = 0, \qquad \Sigma F_y = 0, \qquad \Sigma M = 0$$

It should be noticed that there is no need to distinguish between direct and torsional force action, and no need to determine the location of the center of resistance. The three planes can be arranged in the building in any fashion, as long as they are neither parallel, since they could not resist perpendicular loads, nor is their line of action concurrent, since they could not provide rotational stability.

The typical placement of the vertical elements is along the perimeter, or as a core in the interior, or any other combination (Fig. 5.1a to e). Should the three structural planes form a spatial unit, such as a channel or wide flange section, or should the arrangement consist of an L- or T-section together with a separate plane, then the assembly can be treated (for preliminary design purposes) as consisting of separate planes, as shown in Figure 5.1b, and as is further discussed in the subsection here on "Torsion." Treating a spatial section as consisting of separate elements is usually conservative, because the additional strength due to continuity is ignored. However, it must be kept in mind that an earthquake may cause much larger forces in a connected system than in one where the planes are separated by joints that allow freedom of independent movement! Furthermore, it should be kept in mind that the flanges of T-, U-, and I-sections improve only the flexural rigidity, but hardly offer any shear rigidity, as is needed for lower buildings.

Example 5.1

For the 20-story building in Figure 5.1a, with an average floor height of 10 ft (or a building height of 200 ft), the force distribution to the walls is found as caused by an average wind pressure of 30 psf.

The total lateral wind pressure against the narrow face is

$$P_x = 0.030(200 \times 75) = 450 \text{ k}$$

Horizontal equilibrium in the x-direction gives the force in wall A at the base of the building.

$$\Sigma F_x = 0 = 450 - P_A, \qquad P_A = 450 \text{ k}$$

Rotational equilibrium about point *a* yields the wall forces that resist the torsion.

$$\Sigma M_a = 0 = 450(75/2) - P_c(150), \qquad \text{or} \qquad P_c = 112.5 \text{ k}$$

$$\Sigma F_y = 0 = 112.5 - P_B, \qquad \text{or} \qquad P_B = 112.5 \text{ k}$$

The total wind pressure against the broad face is

$$P_y = 0.030(200 \times 150) = 900 \text{ k}$$

Horizontal equilibrium in the y-direction shows that wall A does not resist any forces.

$$\Sigma F_y = 0 = P_A$$

Because of symmetry, walls B and C share the load equally.

$$P_B = P_C = 900/2 = 450 \text{ k}$$

Statically Determinate Conditions due to Symmetry

Even when there are more than three braced planes, the lateral force flow may still be treated as statically determinate for certain conditions of symmetry of plan layout, structure systems, material, and force action. Some examples are shown in Figure 5.1b.

For a regular layout of rigid frame bents or *widely spaced* cross walls with force action through the center of resistance, so that there is no torsion, and where the floor diaphragms are stiffer than the resisting elements, then all of the identical n walls carry an equal portion of the total load, P.

$$P_i = P/n$$

However, for many *closely spaced* cross walls, as shown in Figure 5.1f, with a much larger stiffness than that of the diaphragms, the floor diaphragm is treated as flexible with unyielding vertical supports, hence the lateral loads are distributed in proportion to the tributary facade area.

For the four walls along the periphery (Fig. 5.1h), or the interior rectangular core (Fig. 5.1g), it may be assumed that only the two parallel web walls resist the loads in each direction; because of symmetry, each of the walls resists one-half of the respective loads, as indicated. However, for Figure 5.1i, the load flow to the web walls must be found by rotational equilibrium. Naturally, the core (Fig. 5.1g) may also act as a three-dimensional tubular cantilever beam with respect to flexure, where the flanges provide most of the rotational resistance, while the webs resist the shear.

Indeterminate Force Distribution for Conditions of Symmetry

The lateral force distribution becomes indeterminate when more than three vertical planes are stabilizing the building and when special conditions of symmetry, as in the preceding subsection, are not available.

In this section, the special case of a symmetrical layout of different structure systems (Fig. 5.1c) that are not identical, as in Figure 5.1f, is investigated. In this instance, the loads cause only translational movement; in other words, the center of rigidity coincides with the resultant load action and the center of mass. The designer must distinguish, however, the difference between the construction that employs only one structure system of the same material such as the bearing wall building of Figure 5.1j, or the rigid frame structure of Figure 5.1k, and mixed construction such as, for example, the rigid frame braced by concrete walls shown in Figure 5.1l and m.

The horizontal loads cause the rigid floor diaphragms to laterally displace, thereby forcing all of the vertical structural planes to deflect by the same amount under direct load action. Therefore, the lateral forces are distributed in proportion to the relative stiffness of the supporting structural systems. It is apparent that the stiffer structure attracts a larger load than the more flexible one, since it takes more force to deflect the more rigid element by the same amount as the less rigid one.

The stiffness is a measurement of the resistance of the structure to lateral deformation. It is defined by a spring constant, k, as a function of the lateral deflection, Δ, caused by a unit force, P = 1; in this case, the stiffness is the reciprocal of flexibility.

$$k = P/\Delta = 1/\Delta \tag{5.1}$$

For the condition of n-connected parallel lateral-force resisting structure systems, where each one has its own absolute lateral stiffness, k_i, the lateral sway can be expressed as

$$\Delta = P/\Sigma k_i \tag{5.2}$$

It is shown in the derivation of Equation 7.22 that the lateral force, P, is distributed, at each floor level, to the cross walls and frames in direct proportion to the relative stiffness of the resisting systems. Therefore, a structural system, i, will resist the following portion of the total load $P_y = P$.

$$P_i = P(k_i/\Sigma k) \tag{5.3}$$

where: P = entire lateral load at given floor level

P_i = the lateral load to be resisted by the structure system i

k_i = absolute stiffness of system i

$k_i/\Sigma k$ = relative stiffness of system i

The above approach was assumed in the y-direction. A similar approach is used for the force action in the x-direction.

Since the distribution of the lateral force is not dependent on the absolute, but rather the relative stiffness of the connected elements, Equation 5.3 can be further simplified. However, first some of the approximate absolute stiffnesses for solid walls, rigid frames, and braced frames will be briefly defined.

In multistory wall structures, the story walls behave as though fixed at top and bottom. For this type of lateral displacement due to a unit force, P, at the floor level, the solid panel stiffness in response to flexure and shear action according to Table A.14, is

$$k = Et/[(h/L)^3 + 3(h/L)] \tag{5.4}$$

Refer to Equation 6.13 for a definition of terms.

Similarly, the approximate stiffness for ordinary story frames due to critical shear action, in response to the total story shear (according to Equation 7.15c), is

$$k = (12E/h)/(L/\Sigma I_G + h/\Sigma I_{ci}) \tag{5.5}$$

The stiffness of an x-braced frame panel, as based on web drift due to the tensile diagonal only (according to Equation 7.18), is

$$k = (EA_d \cos^3\theta)/L = (EA_d \sin\theta \cos^2\theta)/h \tag{5.6}$$

As has already been stated, it is not the absolute stiffness given above, but the relative stiffness, that is needed, because the various systems are only compared to each other. Simpler expressions are now developed for the relative stiffness of typical cases.

The deflection of a wall structure can be that of a flexural cantilever, similar to a shallow beam that is proportional to the bending moments, or it can be that of a shear cantilever, similar to a deep beam where the slope of the curvature is proportional to the shear force, or it can be a combination of the above.

For the distribution of the lateral forces to the long cross walls of a low-rise building (see Equation 6.14), the flexural deflections can be neglected, and analysis can be solely based on the shear deformations. Therefore, the relative rigidity of the wall is then directly proportional to its cross-sectional web area, A_i, assuming the same wall material (E = constant); or, for equally thick walls, the lateral forces can be distributed simply in proportion to the wall lengths, L.

$$P_i = P(k_i/\Sigma k) = P(A_i/\Sigma A) = P(L_i/\Sigma L) \qquad (5.3a)$$

For high-rise wall structures, where the solid walls are at least three times higher than wide (see Equation 6.14), the shear deflections may be neglected, so that the wall stiffness is proportional to the moment of inertia when E = constant; or, for equally thick walls, the wall stiffness is proportional to the cube of the wall length. Therefore, the lateral forces can be distributed as follows

$$P_i = P(k_i/\Sigma k) = P(I_i/\Sigma I) = P(L_i^3/\Sigma L^3) \qquad (5.3b)$$

Should different materials be used, such as concrete block together with clay brick walls, then the above equation becomes

$$P_i = P[(EI)_i/\Sigma(EI)] \qquad (5.3c)$$

The distribution of lateral forces to coupled shear walls is quite complex and depends on the evaluation of the stiffness of the wall, as is discussed further in Chapter 6.

Example 5.2

For the building in Example 5.1, but with a different shear wall layout, as defined in Figure 5.1j, the lateral force flow to the walls is determined as caused by the wind pressure against the long facade. The resistance of the flange walls perpendicular to the wind is conservatively ignored.

The total lateral wind pressure is

$$P_y = 0.030(200 \times 150) = 900 \text{ k}$$

The moment of inertia of wall A is expressed in terms of that of wall B.

$$I_A/I_B = L_A^3/L_B^3 = 75^3/50^3 = 3.375, \qquad \text{or} \qquad I_A = 3.375I_B$$

The total flexural stiffness is

$$\Sigma I = 2(I_A + I_B) = 2(3.375I_B + I_B) = 8.75I_B$$

The lateral force, P_y, is distributed to the cross-walls in direct proportion to their relative stiffness.

$$P_{yi} = P_y(k_{yi}/\Sigma k_y) \qquad (5.3b)$$

$$P_A = 900(3.375I_B/8.75I_B) = 347.14 \text{ k}, \qquad P_B = 900(I_B/8.75I_B) = 102.86 \text{ k}$$

Check: $\Sigma F_y = 0 = 900 - 2(347.14 + 102.86)$ OK

Torsion

In the following studies, torsion will be included in the analysis of the lateral force distribution as caused by asymmetry in geometry, stiffness, or mass distribution in the building plan, or as caused by eccentric force action.

It has been shown (in fig. 2.14) that torsion is generated by asymmetrical buildings, eccentric and asymmetrical layout of the bracing systems, when the center of gravity of the floors does not coincide with the center of rigidity, or by the direction of the load when the resultant force action at the center of wind pressure does not pass through the center of rigidity, that is, by eccentric force action. In addition, torsional effects are caused by nonuniform wind pressure, by accidental seismic eccentricities, and by torsional ground motion. Torsion is most effectively resisted at points furthest away from the center of twist, such as at the corners and perimeter of the building. Closed tubular shapes are clearly efficient because of their inherent torsional stiffness. It should also be kept in mind that asymmetry of irregular building plans can be avoided by separating individual wings with joints so that the building does not act as a unit, but rather as an assembly of separate symmetrical structures.

In the first two parts of this section, simple core structures are investigated, where the center of rigidity is known—in this instance, emphasis is placed on the behavior of the building core under torsion. In the two last parts, the force distribution to the various resisting structure systems of walls, frames, and cores, due to torsion, is studied, as based on rigid floor systems behaving as diaphragms.

As already mentioned in Chapter 2, cores can form closed or open shafts and single or multicell systems. Typical core cross sections are round, square, or rectangular, or they are open tubes composed of wide flange and channel shapes. The rigid floor diaphragms stiffen the cores and are assumed to prevent the distortion of the cross section at each floor level. In this simplified investigation, the effect of wall perforations and coupling between the various structure systems is ignored. For the approximate lateral force distribution to the core components, it will be distinguished between closed and open shafts.

Closed Shafts

Typical closed shafts are round, rectangular, or any other polygonal shape; they can form single or partitioned cells. The closed circular tube, because of its complete symmetry, is the ideal shape for resisting torsion. For a solid shaft, the torsional shear stress at any point is proportional to its distance from the axis of rotation; hence, the resulting stress state for a thin-walled hollow tube is one of pure uniform shear. When a hollow shaft is twisted, it does not laterally displace and change its cross section. The most convincing way for understanding the stress flow is achieved by twisting the shaft and noticing that the original straight vertical lines are forced into a helix (Fig. 5.1n); hence, the principal stresses due to torsion form helices in compression and tension. These normal stresses can be transformed into pure shear at 45° to the inclined planes, yielding uniform shear stresses in the transverse and vertical directions of a thin-walled shaft.

The moment due to the constant torsional shear flow, q, in Figure 5.1n balances the twisting.

$$M_t = qR(2\pi R), \quad \text{or} \quad q = (M_t/R)/2\pi R = M_t/2A_o$$

The corresponding uniform torsional shear stress, also often called *St. Venant's torsion*, after the French engineer Saint-Venant, who developed the theory of pure torsion in 1855, is

$$f_{vt} = q/A = q/L(t) = (M_t/R)/2\pi Rt \qquad (5.7a)$$
$$= M_t/2\pi R^2 t = (M_t/R)/A = M_t/2tA_o$$

This expression can also be derived from the more familiar general torsion equation, using the torsional moment of inertia.

$$f_{vt} = M_t c/J = M_t R/2\pi R^3 t = M_t/2tA_o \qquad (5.7b)$$

where: J = polar moment of inertia = sum of the rectangular moments of inertia

$$= I_x + I_y = 2I = 2\pi R^3 t$$

A = cross-sectional area of the thin-walled tubular wall = $2\pi Rt$

A_o = area enclosed by the centerline of the wall = πR^2

c = distance from center of the section to outer fiber $\simeq R$

The maximum shear stress due to the *direct force action* occurs at the neutral axis and can be roughly approximated for thin-walled round shafts as

$$f_{vmax} = 2P/A \qquad (5.8)$$

Also, in noncircular, closed, thin-walled tubes of any shape, the torque is resisted primarily by the shearing stresses along the walls of the tube, though a noncircular section (when twisted) will not only rotate, but also warp, due to the deviation of the cross section from the circle (Fig. 5.1n). However, in a thin-walled, closed tube, the bending of the thin walls in slab action is so insignificant that the deformation of the cross section can be neglected. Because of the low thickness-to-width and width-to-height ratios of the building core walls, the tube can be treated as a thin-walled beam. It can be assumed that the twisting is primarily resisted by the torsional shearing stresses flowing in a continuous fashion like a liquid along the closed ring. This shear flow can be taken as nearly uniform, as expressed by Equation 5.7a, letting A_o be the area enclosed by the centerlines of the tubular walls.

$$f_{vt} = M_t/2tA_o \qquad (5.9)$$

For a square tube (Fig. 5.2) the torsional shear stress, according to Equation 5.9, is

$$f_{vt} = M_t/2tA_o = M_t/2ta^2$$

In this case, the torsional resistance or effective polar moment of inertia, is $J = ta^3$.
The corresponding total shear force per wall is

$$P_t = f_{vt}(t)a = (M_t/2)/a$$

This result is also easily obtained from statics (see fig. 5.1p), where the torsion is balanced equally by the shear capacity of the four perimeter walls of length, a.

$$M_t = 2[P_t(a)], \quad \text{or} \quad P_t = M_t/2a$$

The translational forces are assumed to be resisted entirely by the web walls parallel to the force action.

$$P_w = P/2$$

The maximum shear force due to translation and rotation occurs along the inside wall, and (according to the expression from above) is equal to

$$P_{max} = (P + M_t/a)/2 \tag{5.10}$$

Alternatively, the corresponding maximum shear stress due to translation and rotation is

$$f_{vmax} = P/2ta + M_t/2tA_o \tag{5.11}$$

In a similar fashion, the approximate torsional forces for a rectangular shaft, a × b, can be determined. According to Equation 5.3, the uniform shear stress is

$$f_{vt} = M_t/2tA_o = M_t/2tab$$

Correspondingly, the shear force in the web wall is

$$P_w = f_{vt}(t)b = M_t/2a$$

This result is obtained from statics in the same fashion as before.

$$M_t = P_w(a) + P_F(b) \tag{a}$$

However, for constant torsional shear stresses along the thin-walled tube, the wall forces are proportional to the wall lengths, assuming a uniform wall thickness.

$$f_{vt} = P_w/bt = P_F/at, \quad \text{or} \quad P_w/b = P_F/a \tag{b}$$

Substituting Equation (b) into (a) yields, as before,

$$P_w = (M_t/a)/2, \quad P_F = (M_t/b)/2$$

Therefore, the maximum shear force due to direct and rotational action occurs along the inside web wall (Fig. 5.1p).

$$P_{max} = (P + M_t/a)/2 \tag{5.10}$$

The maximum shear stress is given again by Equation 5.11, letting a = b.

Torsion flows easily around a space truss (Fig. 5.1, top right). It can be treated in the familiar fashion, assuming a hinged truss and by balancing the forces at the joints. The space truss analogy is also used for the design of concrete beams that must resist torsion. In this case, the concrete beam consists of the tensile steel cage and the compressive diagonals provided by the concrete. The vertical corner bars and other vertical bars distributed around the section resist the vertical tension, and the horizontal bars (closed stirrups) cover the horizontal tension. In other words, the steel bars cover the diagonal tension components generated by the torsion.

In conclusion, it must be emphasized that cores are weakened by openings and thus can rarely be considered as truly closed systems. Therefore, St. Venant's torsion may be quite inadequate for analysis, and axial stresses caused by bending torsion may have to be taken into account, as is discussed in the next section. Furthermore, with the opening up of the core, the shear center moves outside (as, for example, for the symmetrical building in fig. 2.14k) and the lateral load will act in an eccentric manner, thus generate twisting of the core in addition to bending.

Open Shafts

For open building shafts such as cantilevering I, E, H and U-sections, the twisting moment, M_t, is no longer transmitted primarily in simple uniform torsional shear, as for closed tubes, but in *bending torsion* together with *nonuniform torsional shear*.

Open shapes cause significant warping under torsion thereby generating large torsional bending and shear stresses. Visualize a wide flange section that is not restrained at its boundaries to be twisted in such a manner that the section is free to warp (i.e., rotated flanges form straight lines and are not bent), then the member is subject to uniform warping and only torsional shear stresses (see fig. 5.1n). However, when this beam is fixed at one end, similar to a vertical building cantilever shaft restrained by the foundations, then the warping is prevented at the fixed end and the torsion is entirely resisted by the bending resistance of the flanges, which is called warping torque—in this case, the St. Venant's shear stresses are zero. The effect of the end restraint reduces as one moves towards the free end or top of the core. In other words, the rate of twist varies along the member and increases from zero at the base to a maximum at the top, indicating that warping is nonuniform, and that the torsional resistance of the section changes from pure warping torque to almost pure torsional shear at the free end.

Though pure torsional shear may be predominant near the top of a building, it is also relatively small and less critical for preliminary investigation purposes. The twisting is primarily resisted by the bending of the flanges in the lower building portion due to the restraint of the foundations and the coupling to the adjacent building components. The torque is assumed to be replaced by a force couple along the flanges, ignoring the bending of the web slab due to its lack of stiffness. It may be concluded that the twisting is resisted by the lateral bending of the flanges with the corresponding shear and normal stresses. It is beyond the scope of this investigation to further discuss the complexity of the interaction of St. Venant's torsion and warping torsion along the building between the extremes of fixed and free ends, particularly the effect of access openings.

When a single rectangular solid plate is twisted, it deforms nonuniformly, causing the warping of the cross section with a corresponding nonuniform shear stress distribution. As mentioned before, when the member is unrestrained from warping, then only St. Venant's torsion occurs. This behavior of the solid rectangular plate is quite in contrast to that of the circular shaft, which only rotates, but does not warp because of its perfect symmetry, allowing a uniform shear stress distribution. Since, for the solid rectangular plate, the plane sections are not maintained after twisting, the maximum shear stress does not occur at a point farthest removed from the axis of rotation, as at the corners of a stiff square shaft, but (surprisingly) at the midpoint of the long side of the rectangle, with no shear stress at the corners. St. Venant showed that the maximum shear stress for a thin plate of width b and thickness t is equal to

$$f_{vt} = M_t t/(bt^3/3) = M_t/(bt^2/3)$$

The simple torsional shearing stresses for thin-walled open sections that are made up of a combination of thin rectangular plates have been found to be nearly equal to the sum of the torsional resistances of the individual rectangular plates. The simple torsional shear distribution can, therefore, be roughly approximated by treating the section as separate individual plates. Hence, the maximum shear stress for open, thin-walled shafts with $b/t \geq 10$ can be obtained from the following expression by treating the section as an assembly of rectangular plates

$$f_{vt} = M_t t/J \tag{5.12}$$

where the effective torsional resistance or equivalent polar moment of inertia for the thin-walled open section, by evaluating each of the rectangles, is

$$J \simeq \Sigma bt^3/3$$

For example, the torsional resistance for a wide flange or channel section that consists of the sum of the torsional resistances of the two flanges and the web, is

$$J = \Sigma bt^3/3 = (2b_f t_f^2 + b_w t_w^3)/3$$

or, the equivalent polar moment of inertia for a slotted square tube, letting $b = a$, is

$$J = \Sigma bt^3/3 = 4(at^3)/3$$

Comparing the shear stresses of a closed square shaft with an open one clearly demonstrates the poor response of an open section to torsional shear flow and the small torsional moment of inertia for an open section.

The torsional shear stresses for the closed tube are

$$f_{vtc} = M_t/2tA_o = M_t/2ta^2$$

The maximum torsional shear stress for the slotted tube is

$$f_{vto} = M_t t/J = M_t t/(4(at^3)/3) = 0.75M_t/at^2$$

The ratio of the torsional shear stress of the open tube to the closed one is

$$f_{vto}/f_{vtc} = 1.5(a/t)$$

Using the shaft width-to-wall thickness of Example 5.3, yields

$$f_{vto} = 1.5[30(12)/18]f_{vtc} = 30f_{vtc}$$

Therefore, the torsional shear stress in the open slotted shaft is 30 times larger than that in the closed tube!

The torsional shear stress in open sections will not be dealt with further in this context. It is assumed, for fast approximation purposes, that the torque in open shafts is resisted by the shear and axial stresses due to bending torsion by neglecting St. Venant's twist. In contrast, however, the twisting moment in closed tubes is assumed to be resisted entirely by simple torsional shear by ignoring the warping of the walls and the effect of openings.

In open cores with at least two parallel walls, such as wide flange and channel shapes (e.g., H-, I-, U-shapes), these walls effectively resist the torsion by bending. Sections arranged without parallel walls, such as +-, L-, Y-, and T-sections, have a low torsional stability. It should be remembered that, by placing individual structural systems further away from the center of twist, the larger lever arms will provide a better resisting moment capacity! For conditions where the rotation around the core is too large, it may be more effective to use the eccentric core only for direct force action, while the perimeter frame, similar to a tube, provides the torsional stiffness (see fig. 5.1s). This approach may be especially important for curtain wall construction under seismic loads, where a minimum stiffness of the periphery may be necessary.

Because bending torsion controls the design of open shafts, the intersecting planar core walls can be treated as separate elements for determining the approximate distribution of the lateral forces, as well as for finding the warping shear and normal stresses. Naturally, with respect to the flexural beam stresses due to direct load action, the continuity of the shape may have to be considered, as is discussed in other parts of this book. The warping stresses are superimposed upon the direct stresses, as explained in Example 5.3.

Example 5.3

The average lateral wind pressure of 35 psf against the long facade of the 360-ft high building in Figure 5.1p is resisted, first, by using an eccentric 30-ft square core with 18 in. thick concrete walls, and then by using a 40 × 30 ft rectangular core. The approximate maximum stresses are found, first, based on a closed tube, and then on an open one, as caused by the perforations of the elevator and door openings.

The core must resist a total wind pressure of

$$P_y = 0.035(360 \times 160) = 2016 \text{ k}$$

This pressure causes a twisting of the core of

$$M_t = P_y(e_x) = 2016(35) = 70{,}560 \text{ ft-k}$$

The maximum wall force, according to Equation 5.10, along the inside web of the square core is

$$P_{max} = (P + M_t/a)/2 = (2016 + 70{,}560/30)/2 = 1008 + 1176 = 2184 \text{ k}$$

Should the core width be increased to 40 ft, a smaller critical force will occur in the web because of the larger lever arm, which provides more rotational resistance.

$$P_{max} = (2016 + 70{,}560/40)/2 = 1890 \text{ k}$$

The following critical stress conditions will occur:

- *Closed tube:* In this case, the twisting is assumed to be resisted entirely by simple torsional shear. In addition, the webs will resist nearly all of the direct shear. Therefore, the maximum total shear stress due to direct and torsional force action in the square core is approximately

$$f_v = P_{max}/bt = 2{,}184{,}000/18(30)12 = 337 \text{ psi}$$

The flexural stresses will only occur due to the direct force action of $2016/2 =$ 1008 k, which may be resisted by the full tubular cross section.

- *Slotted tube:* In the open tube, the torsion is assumed to be resisted entirely by the bending of the walls—the nonuniform torsional shear of the open section is ignored. Therefore, the maximum shear stresses are due to bending torsion and direct force action, and the flexural stresses will not only be due to the direct force action as for closed tubes, but also due to twisting.

   The slotted square tube is assumed to consist of two channel sections, where the torsion is resisted entirely by the two parallel web walls that have no penetrations; the effect of the two slotted flange walls is conservatively neglected. Therefore, the maximum wall force due to direct and torsional action is

$$P_{max} = 2016/2 + 70,560/30 = 1008 + 2352 = 3360 \text{ k}$$

This force causes a moment about the base of

$$M = 3360(360/2) = 604,800 \text{ ft-k}$$

The corresponding maximum shear and bending stresses in the critical wall are

$$f_v = V/A = 3,360,000/[18(30)12] = 519 \text{ psi}$$

$$f_b = M/S = \frac{604,800,000(12)}{18(30 \times 12)^2/6} = 18,667 \text{ psi}$$

The bending stresses in the wall are so high that they must be resisted by steel. Naturally, a more accurate solution should take the exact nature of the core into account, and will be somewhere between the two cases. This however, requires a complex three-dimensional computer analysis.

Symmetrically Arranged Multiple Lateral-Force Resisting Systems

The distribution of lateral forces to the various resisting structure systems of walls, frames, and cores, as based on a rigid diaphragm action of the floors, will now be briefly investigated. First, buildings with a symmetrical layout are studied. For example, for the rigid frame building in Figure 5.1o, under symmetrical loads, the lateral deflection is constant so that the loads are simply distributed in proportion to the stiffness of the widely spaced bents. In this case, identical bents are assumed, so that each of the five bents carries an equal portion of the total load: $P_i = P/5$.

Although the centers of mass and rigidity coincide in the given example, an accidental torsion may have to be taken into account. For instance, a minimum torsional eccentricity, e, equal to 5 percent of the maximum building dimension at the level to be be investigated, is required for seismic design. This torque, $M_t = P(e)$, causes a torsional shear force in all of the columns of a given story—in this case, the column sizes are assumed all equal. The shear in any column can be easily determined from the following familiar expression, assuming that the floor structure remains rigid when it rotates and the building cross section is maintained

$$f_{vt} = M_t c/J = M_t d_i/(I_x + I_y) = M_t d_i/(A\Sigma d^2) \tag{5.7b}$$

where: A = cross-sectional column area (assumed constant for all columns)

$\quad\quad$ J = polar moment of inertia

$$= I_x + I_y = A\Sigma d^2 = A\Sigma(\bar{y}^2 + \bar{x}^2)$$

$\bar{x}$ or $\bar{y}$ = the perpendicular distance from the center of rigidity to the axis of system i.

The torsional shear force that a column must resist is

$$P_i = f_{vi}A = M_t d_i/\Sigma(\bar{y}^2 + \bar{x}^2) \quad\quad\quad (5.13a)$$

The resultant torsional force, P_i, can be conveniently resolved into $\bar{x}$-, and $\bar{y}$-components (as shown in fig. 5.1o).

$$\tan\theta_i = \bar{y}_i/\bar{x}_i = P_{xi}/P_{yi}, \quad\quad \text{or} \quad\quad P_{xi}/\bar{y}_i = P_{yi}/\bar{x}_i \quad\quad\quad (a)$$

Rotational equilibrium yields

$$P_i d_i = P_{xi}(\bar{y}_i) + P_{yi}(\bar{x}_i) \quad\quad\quad (b)$$

Substituting Equation (a) into (b) gives

$$P_i/d_i = P_{yi}/\bar{x}_i = P_{xi}/\bar{y}_i \quad\quad\quad (c)$$

Substituting Equation (c) into 5.13a yields the torsional component forces

$$P_{yi} = M_t \bar{x}_i/\Sigma(\bar{y}^2 + \bar{x}^2), \quad\quad P_{xi} = M_t \bar{y}_i/\Sigma(\bar{y}^2 + \bar{x}^2) \quad\quad\quad (5.13b)$$

Should the stiffness of the various resisting columns be different, then Equation (d) changes to the general form of

$$P_{yi} = M_t k_{yi}\bar{x}_i/J, \quad\quad P_{xi} = M_t k_{xi}\bar{y}_i/J \quad\quad\quad (5.14)$$

where: J = rotational stiffness of entire floor level in a story

$$= \Sigma(k_x\bar{y}^2 + k_y\bar{x}^2)$$

$\bar{x}$ or $\bar{y}$ = perpendicular distance from the center of rigidity to the axis of the wall or frame.

$\bar{x}_i$ or $\bar{y}_i$ = perpendicular distance from the center of rigidity to the axis of the particular wall or frame system, i, to be investigated.

This equation is also used as a simplified method for structures with an asymmetrical arrangement of shear walls, as is discussed in the next section.

The design of the rigid frame structure in Figure 5.1o, with respect to torsion, is usually based on the assumption that the perimeter frames resist the entire twist and thus offer more stiffness to the curtain wall.

$$M_t = P_y(a) + P_x(b)$$

The perimeter frames, framed tubes, or other mixed perimeter structures may, however, have different stiffnesses, as identified in Figure 5.1o, for a rectangular plan form. Letting $\bar{x} = a/2$ and $\bar{y} = b/2$ yields (according to Equation 5.14) a torsional stiffness of

$$J = 2[k_x(b/2)^2 + k_y(a/2)^2] = (k_xb^2 + k_ya^2)/2$$

Substituting this equation, together with $\bar{x}_i = a/2$ and $\bar{y}_i = b/2$, into Equation 5.14, shows how the torque is resolved into forces around the periphery in accordance with the shearing resistance of each of the plane frames.

$$P_y = M_tk_ya/(k_xb^2 + k_ya^2), \qquad P_x = M_tk_xb/(k_xb^2 + k_ya^2) \qquad (5.15)$$

If it is assumed (for preliminary design purposes) that the webs are far stiffer, and therefore resist all the torsion, then the following simple expressions are obtained.

$$M_t = P_y(a), \qquad \text{or} \qquad P_y = M_t/a, \qquad P_x = 0$$

The torsional forces are added to the direct forces.

Asymmetrically Arranged Multiple Lateral-Force Resisting Systems

For the case where the lateral-force resisting systems are arranged in an asymmetrical fashion, the *center of rigidity* must first be found. Its location depends not only on the layout but also on the stiffness distribution of the resisting systems. When the resultant lateral load acts through the center of rigidity at any angle, there will be no torsion or rotational deformation!

The location of the center of rigidity can be derived from the fact that the rotation caused by the resultant force $\Sigma P_i = P$ about any arbitrary reference plane $\bar{x}_o$, such as wall axis A (Fig. 5.2) must be equal to the sum of the moments about that axis, due to the individual shear forces in the structural planes replacing the resultant force by assuming rigid diaphragm action of the floor systems.

$$P(\bar{x}_o) = \Sigma P_i(x) \qquad (a)$$

Substituting Equation 5.3, $P_i = P(k_i/\Sigma k)$, the location of the $\bar{x}_o$- and $\bar{y}_o$-axes can be found. The intersection of these shear axes is the shear center, which identifies the center of rigidity or the center of stiffness, as expressed by the terms in the following equations. In this context, these centers (including the center of twist) are treated as interchangeable, by just considering a single story and disregarding the change of stiffness and change of behavior of the various lateral-force resisting structural elements, with respect to each other, along the building height, and thus neglecting the corresponding variation in the locations of the center of rigidity and shear center at every floor level.

$$\bar{x}_o = \Sigma k_yx/\Sigma k_y, \qquad \bar{y}_o = \Sigma k_xy/\Sigma k_x \qquad (5.16a)$$

For *flexural cantilevers*, the moments of inertia are used to determine the shear axes and hence the torsion, but the cross-sectional areas of the individual structural elements are used to find the neutral axes and the corresponding flexural stresses under direct force action.

$$\bar{x}_o = \Sigma I_yx/\Sigma I_y, \qquad \bar{y}_o = \Sigma I_xy/\Sigma I_x \qquad (5.16b)$$

For *shear cantilevers* with equally thick walls and of the same material, the wall lengths are used to find the shear axes.

$$\bar{x}_o = \Sigma L_yx/\Sigma L_y, \qquad \bar{y}_o = \Sigma I_xy/\Sigma L_x \qquad (5.16c)$$

When the location of the center of rigidity is known, then the lateral forces can be transferred to this center, together with the rotation

$$M_t = P_y(e_x) + P_x(e_y)$$

Now, the direct and torsional forces can be distributed to the various structural elements according to Equations 5.3b and 5.14. For example, for the y-direction (e.g., P_y in Fig. 5.2), when the translational and torsional shear act in the same direction,

$$P_{yi} = P_y(k_{yi}/\Sigma k_y) + M_t(k_{yi}\bar{x}_i/J) = P_y[(k_{yi}/\Sigma k_y) + e_x(k_{yi}\bar{x}_i/J)] \qquad (5.17)$$

When (for approximation purposes only) the structural planes parallel to the force action are considered because of their larger stiffness and the contribution of resistance by the walls/frames normal to the load directions is ignored, then the expression for the rotational stiffness (see Equation 5.14) simplifies to

$$J = \Sigma(k_x\bar{y}^2 + k_y\bar{x}^2) \simeq \Sigma k_y\bar{x}^2 \qquad (5.18)$$

Should the asymmetrical building structure consist of mixed construction, for instance, of coupled shear walls or braced frames together with rigid frames and perforated cores, obviously a three-dimensional computer analysis is required. The problem becomes even more complex when the building shape becomes irregular.

Example 5.4

For the asymmetrical arrangement of the cross-walls of the high-rise building in Figure 5.2, the lateral force distribution is determined. All of the walls are assumed to be of the same thickness and material. The resistance of the flanges, that is, the walls perpendicular to the force action, is ignored for fast approximation purposes.

First, the center of rigidity must be found, realizing that it must be located on the x-axis, which is an axis of symmetry. To simplify the calculations, the moments of inertia of walls A and C are expressed in terms of that of wall B.

$$I_A/I_B = L_A^3/L_B^3 = 40^3/20^3 = 8, \quad \text{or} \quad I_A = 8I_B$$

$$I_C/I_B = L_C^3/L_B^3 = 30^3/20^3 = 3.375, \quad \text{or} \quad I_C = 3.375I_B$$

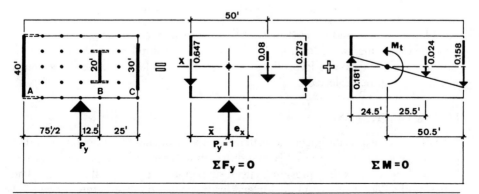

Figure 5.2. *Analysis of cross-wall structure in Example 5.4.*

The total flexural stiffness is

$$\Sigma I_y = I_A + I_B + I_C$$

$$= 8I_B + I_B + 3.375I_B = 12.375I_B$$

Therefore, the location of the shear axis, as defined in Fig. 5.2, is obtained by taking moments with the moments of inertia of the walls about wall A.

$$\bar{x}_o = \Sigma I_y x / \Sigma I_y \tag{5.16b}$$

$$= [I_B(50) + 3.375I_B(75)]/12.375I_B = 24.50 \text{ ft}$$

Therefore, the eccentricity of the resultant lateral force is

$$e_x = 37.5 - 24.50 = 13 \text{ ft}$$

The corresponding torsional moment is

$$M_t = P_y(e_x) = 13P_y$$

The rotational stiffness of all the walls, ignoring the flanges, is

$$J = \Sigma k_y \bar{x}^2 = \Sigma I_y \bar{x}^2 \tag{5.18}$$

$$= I_B[8(24.5)^2 + 1(25.5)^2 + 3.375(50.5)^2] = 14059.34I_B$$

Now the magnitude of the forces that each of the walls must resist can be determined by considering the direction of the rotational shear in relation to the transitional shear.

$$P_{yi} = P_y(k_{yi}/\Sigma k_y) + M_t(k_{yi}\bar{x}_i/J) \tag{5.17}$$

$$P_A = P_y(8I_B/12.375I_B) - 13P_y(8I_B(24.50)/14059.34I_b)$$

$$= 0.647P_y - 0.181P_y = 0.466P_y$$

$$P_B = P_y(I_B/12.375I_B) + 13P_y(I_B(25.50)/14059.34I_B)$$

$$= 0.081P_y + 0.024P_y = 0.105P_y$$

$$P_C = P_y(3.375I_B/12.375I_B) + 13P_y(3.375I_B(50.5)/14059.34I_B)$$

$$= 0.273P_y + 0.158P_y = 0.431P_y$$

Check: $\Sigma F_y = 0 = 0.466P_y + 0.105P_y + 0.431P_y - P_y$ OK

Notice that walls A and C resist most of the load, since wall B lacks stiffness and is located close to the axis of rotation. Hence, for approximation purposes, not only the structure components with low stiffness, but also the ones located close to the center of twisting, can be neglected.

FLOOR DIAPHRAGMS

It has been shown in the previous section that the floor framing functions as a diaphragm tying the building together and distributing the lateral forces to the vertical bracing elements, as well as providing lateral support to hinged vertical elements. The behavior of a floor diaphragm under lateral loads depends on its structure. In contrast to the thin skin construction for many roofs of low-rise buildings, the typical concrete slabs for high-rise buildings can be considered rigid. A floor diaphragm acts as a deep beam at each story with respect to lateral force action, and can be visualized as the thin web of a huge flat floor girder primarily resisting shear, while the boundary members (e.g., spandrel beams) perpendicular to the load action carry the rotation in an axial manner, similar to girder flanges. For instance, a typical floor diaphragm in Example 5.1 must transfer a wind pressure against the broad face of

$$w = 0.030(10) = 0.3 \text{ k/ft}$$

The load causes the following maximum moment:

$$M_{max} = wL^2/8 = 0.3(150)^2/8 = 843.75 \text{ ft-k}$$

This rotation is resisted along the diaphragm edges perpendicular to the load action in compression and tension. If spandrel beams are present and properly shear-connected to the slab, then they resist the maximum chord forces, P, at midspan in column action.

$$P = M/D = 843.75/75 = 11.25 \text{ k}$$

The diaphragm must carry the following maximum shear:

$$V_{max} = 0.3(150/2) = 22.5 \text{ k}$$

A collector or drag member may be necessary to transfer part of this shear to wall c, which has a length of 50 ft, and thus uses only a portion of the diaphragm depth of 75 ft. Assuming a uniform shear distribution of $22.5/75 = 0.3$k/ft, the collector reinforcement in the concrete slab should resist the following force of

$$P = 0.3(25) = 7.5 \text{ k}$$

For ordinary conditions, the slab easily absorbs the diaphragm stresses and is stabilized by the floor framing. The application of composite construction naturally improves the diaphragm action. However, there are situations where the floor framing must be strengthened and stiffened to perform properly as a diaphragm. These situations occur when the floor is weakened by numerous openings or abrupt changes of form where lateral forces have to be redistributed, or when stress concentrations occur due to sudden changes of mass and structure systems. Some of these conditions are briefly discussed in Figure 5.3.

When floors are constructed from precast concrete decks, the panels must be connected so that the entire floor can act as a unit, without allowing any slippage between the panels and to achieve diaphragm action. By using the truss analogy in Figure 5.3a and b, and the arch analogy in Figure 5.3c, one can develop some understanding and a model for the intensity of force flow for which the joints in precast panel construction must be designed. For example, in Figure 5.3d, the floor structure is tied together by continuous bond beams around the perimeter and across the building, possibly together with a cast-in-place topping to form cell-like compartments.

When facade columns are not laterally supported by floor beams, and when it is necessary to increase the torsional resistance of the spandrel beams, floor bracing may be employed along the periphery of the building. This condition occurs in Figure 5.3j, where the beams stiffen the outside columns against buckling, and in Figure 5.3k, where a system of horizontal bracing connects the wall trusses with the floor system. In this instance, the horizontal bracing provides lateral support for the wall and allows the transfer of wind loads along the floor diaphragm to the other peripheral walls. It may be necessary to tie separate vertical bracing systems

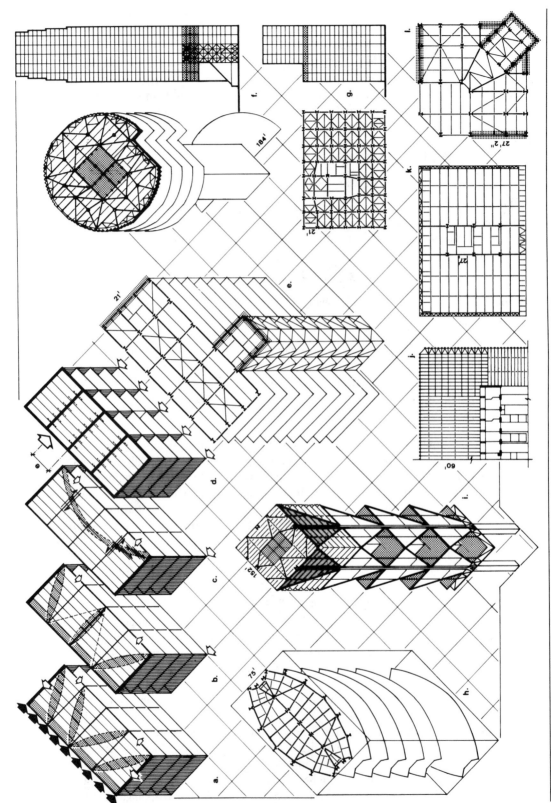

Figure 5.3. *Rigid diaphragm action of floor structures.*

together, particularly in composite buildings. In Figure 5.3 l, the five vertical trusses are connected by a horizontal bracing system at every third floor; a similar situation exists in Figure 5.3e.

When structure systems within a building change, horizontal transition systems may be required, as for Figure 5.3f, g, and i. In Figure 5.3i, the lateral forces are transferred at every eighth floor from the core to the exterior trusses. In Figure 5.3g, a ten-story rigid frame sits on top of a 15-story base, where only the peripheral frame resists the lateral force action. This change from the lateral-force resistance of two-way rigid frame bents to peripheral rigid frames necessitated extensive horizontal bracing of the two floors at the transition zone. In Figure 5.3f, tubular action occurs down to the notch, where a five-layer transfer structure ties the perimeter to a beefed-up core. The five transfer floor levels above the notch are trussed to transfer the lateral loads from the perimeter to the core, which obviously required extensive additional cross-bracing. Similarly, in the Sears Tower (see fig. 7.33) two-story trusses are placed at the locations where the tubes drop off. In this case, the trusses distribute the wind and gravity load to the greater number of tubes below.

FLOOR FRAMING SYSTEMS

In addition to providing the necessary diaphragms to stabilize the building volume and the transfer of lateral forces, the floor structures must carry the gravity loads to the vertical structural elements. The floor structure consists of the slab, possibly supported by framework. The floor framing is composed of the primary beams (girders), which span between the columns or walls, and the secondary beams (stringers, filler beams), which are supported by the primary beams; the secondary beams carry the slab. Various support conditions are shown in Figure 5.4b, ranging from one-way or two-way slabs that span directly to the primary beams, or slabs that are supported by secondary beams arranged in different patterns, to a two-way slab on shallow beams or a slab that is directly supported on columns. This hierarchy of flexural members, as reflected by the direction of gravity load flow from the slab to the beams, and then to the vertical structural elements, is most important for the understanding of the various framing layouts.

The arrangement pattern of the members is directly related to the shape of the building (plan form); the location of the vertical bearing elements, which are organized on a basic dimensional gird; and the efficiency of the load flow. The vertical bearing elements establish structural bays or basic building modules. A structure may consist of multiple bays of even or uneven spacing, it may just represent a single-bay long-span system, or it may form a radial structure, as for many tower buildings. Typical framing patterns are shown in Figure 5.4a for a square building plan, ranging from tower structures with central cores, and even two-directional column layouts or circumferential layouts, to one-directional systems of cantilevering beams spanning the full width of the building, or to central corridor systems. Naturally, there is an infinite variety of possible combinations yielding regular or irregular arrangements.

Corner conditions (Fig. 5.4c) may create interesting framing situations, since it is often desirable to equalize the load flow to the exterior columns and to control the differential axial shortening of the columns. For buildings with central cores, the floor framing spans directly from the perimeter to the core in a one-directional fashion, but this is not possible for the corner bay. Possible solutions for the distribution of the

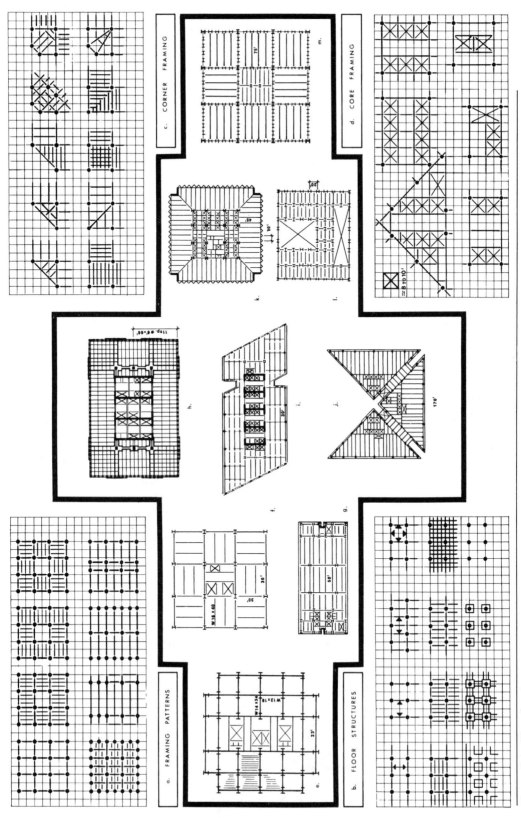

Figure 5.4. *Floor framing systems.*

force flow include: continuing the one-way span of the beams, but reversing the direction on alternate floors so as to achieve two-way distribution; using a two-way beam grid; placing a diagonal girder from the corner of the core to the building corner, thereby attracting a major load to the corner column; or using any of the variations indicated in Figure 5.4c.

The regularity of the floor framing patterns is disturbed by openings, as for the plumbing and exhaust shafts, as well as the elevator and stair shafts in the core area; here, in addition to the floor construction, the beams must support the enclosure walls. Typical framing layouts for elevator shafts are shown in Figure 5.4d.

The potential expression of floor framing systems in concrete can only be suggested in Figure 5.5. The plastic character of concrete, either cast-in-place or prefabricated, allows for a rich formal treatment and total integration of the floor structure components, as the various cases demonstrate. The plan transformation (center bottom of Fig. 5.5), from column-slab units of a four-leg table to an umbrella unit placed adjacent to each other, indicates that the spaces between the units can be bridged with intermediate floor slabs. It shows that, from a geometrical point of view, an infinite number of layouts are possible, particularly when polygonal forms are included. The visual transformation of a two-way slab on stiff beams to a waffle slab (Fig. 5.5, top left) demonstrates other possible solutions of support.

The organization of floor structures for high-rise buildings, according to the materials of concrete and steel, can be delineated as follows:

- Floor structures in lightweight or normal-weight *concrete:*
    One-way systems: cast-in-place solid slabs, joist floors, precast concrete slabs (solid and hollow panels, T-panels), composite slabs (steel decking, precast components), and post-tensioned slabs.
    Two-way systems: two-way slabs (on stiff beams, or on shallow beam bands), flat slabs/plates, waffle slabs, post-tensioned slabs, precast slabs and shells.
  The concrete slab systems are discussed further in the section in this chapter on "Reinforced Concrete Slab Systems."
- Floor structures in *steel* consist of the floor slab and the steel framework. They can act separately or in a composite manner.
    The *skin systems* may be: cast-in-place concrete slabs, cast-in-place concrete slabs on a metal deck (composite or noncomposite), precast concrete planks, ceramic slabs (usually in older buildings, and other countries).
    The type of *horizontal-span members* may be: rolled beams, open-web steel joists, plate girders, latticed girders, trusses, Vierendeel girders, castellated beams, stub girders, tapered or haunched girders, staggered trusses.

The steel floor framing must be fire protected by using either contact fireproofing or using the ceiling as a membrane fire protection. The ceiling, as a fire-resistant barrier, can protect the floor structure and the ducts, pipes, and other mechanical equipment from fire damage. Direct-applied fire protection, such as sprayed contact fireproofing, can be applied to the supporting steel beams (see fig. 1.22), and to the underside of the steel decking, when it is used structurally.

Some typical floor systems are shown in Figure 5.6.

(a) Steel beams support precast concrete planks and a suspended, fire-rated ceiling. The planks are tied together with a structural topping of lightweight concrete

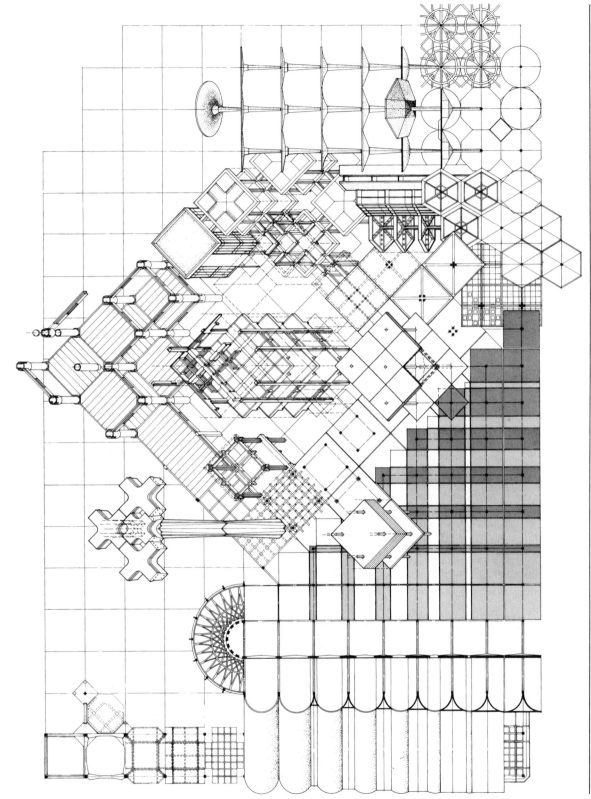

Figure 5.5. *Concrete floor framing systems.*

reinforced with welded wire fabric, and may be anchored with shear connectors to the beams.

(b) Open-web steel joists with a 2 1/2-in. concrete slab on steel forming and a suspended fire-rated ceiling.

(c) Built-up steel beams support a hollow-core precast concrete plank floor.

(d) Dyna-Frame precast concrete framing system (Strescon Industries, Inc., Baltimore, MD, 1984).

(e) Long-span, haunched steel girders, as part of rigid frame construction, accommodate ductwork at the midspan range.

(f) 4 1/2-in. concrete slab on permanent steel forming composite with steel beams, and a suspended fire-rated ceiling, rather than fire-protected beams with nonrated ceiling.

(g) Composite stub-girder floor system with cantilever cross-beams: This system was developed by J. P. Colaco in the early 1970s. In this case, stub beams are welded on top of the primary wide flange beams and act together with a composite floor slab. From a structural point of view, the stub girder behaves in a manner similar to a Vierendeel truss; its design is based on the assumption of shored composite construction. The apertures in the girder profile permit mechanical ducts to penetrate, and also allow the cantilever floor beams to pass through.

The depth of the floor structure depends on the bay sizes and the magnitude of the loads; office buildings must support roughly twice as much live load as residential buildings. Typical spans for high-rise buildings are in the range of 35 to 42 ft, with span-to-depth ratios for floor framing roughly from 20 to 24. Rules of thumb, as often used in practice by designers for first approximation purposes of member sizes, are given in other parts of the book. The floor framing depth is a critical design consideration; an increase of depth will result in a decrease of floor weight, especially when trusses become competitive, that is, when their higher unit costs can be offset by the decrease in weight. The weight of the floor framing per square foot of floor area is independent of the number of floors—it is nearly the same for low and high-rise buildings—but, as just mentioned, it is a function of the floor depth. Flat slab construction uses the least floor height, especially when the slab serves as the ceiling, keeping in mind that steel construction generally needs ceilings. Composite action of steel framing and concrete slabs also reduces the depth of the floor structure, as does high-strength steel, which yields shallower beam sections, as long as strength and not stiffness controls the design. The floor-to-floor height can also be minimized by cambering the beams upward so that the beam deflects downward into a level position under the weight of the concrete slab, thus also providing a slab of constant thickness.

The overall depth of the floor is crucial, since this also must accommodate the horizontal mechanical space; this floor depth may be in the range of 3 to 4 ft for standard tall office buildings. Since, in residential buildings, the services are routed vertically, the floor depth is naturally much less. The structural support of the horizontal distribution of the services in office buildings can be achieved in two ways. The ductwork and piping is either suspended from the floor structure and runs below it over the suspended ceiling, or it is directly supported by it. Direct support is provided by beams with penetrations, such as by trusses, framed girders, castellated beams, or stub-girders; in this case, the mechanical distribution system is integrated with the structural system. This approach is in contrast to the suspended support of the mechanical ductwork and piping, where the mechanical distribution system and the structural system are separated. Rather than placing the pipes and ducts together with the electrical and communication services *below* or *within* the floor framing structure, they also can be placed *above* the structural slab by providing a raised floor (see the section in Chapter 1 on "Mechanical and Electrical Systems"); the recommended height above the finished floor slabs for computer rooms is at least 12 inches.

Although deeper floors may be lighter than shallow ones, they also add to the overall height of the building, which is critical for tall structures. Not only will the costs for ducts, cladding, elevators, and other materials in the vertical planes be increased, but also the lateral force action, in addition to the vertical loading, will increase. One may conclude that bay sizes should be selected to yield a minimum story height and still provide the required space usage. To obtain an optimum floor depth, the following basic construction systems must be investigated for various floor framing layouts: (1) A shallow, heavy floor framing with ductwork below; (2) A deep, light floor framing with services passing through openings in the beams.

For very slender buildings, large bay sizes with long-span floor systems may be required to transfer all of the gravity loads to certain locations at the building perimeter, so that the entire gravity load of the building can act as a stabilizer with respect to lateral force action.

Although it seems to be efficient to reach a high live load to dead load ratio, as based on strength and deflection criteria, a floor structure may experience vibration problems. Flexible, long-span floor structures of minimum weight, such as steel joist-concrete slab floor systems or long-span precast concrete slabs, may be set in motion by the impact loads of pedestrian traffic, not to mention dancing or athletic activities. Although the floors are structurally safe, the oscillations may be disturbing. This problem can be solved by changing the natural frequency by increasing the floor stiffness (reduction of span or increasing the moment of inertia of the floor beams), and/or by increasing the mass for natural damping purposes, and by additional damping. Increasing the mass, that is, the thickness of the concrete slab, results in an increase of natural damping. By increasing the stiffness, the fundamental period is lowered so that it is below the one of the exciting force. However, to increase the stiffness may require deeper beams with larger moments of inertia, or a composite construction may have to be selected. The deeper floor structure, in turn, increases the height and costs of the building. Restrained support connections of the floor beams (on the other hand) increase the stiffness, but also help to transmit vibration to the adjacent structures, which may not be desirable. Hence, it may be necessary to add artificial damping such as partitions, rugs, ceilings, vibration absorbers attached to the joists, or floor-to-ceiling posts with neoprene pads, to quickly reduce the amplitude of vibration.

Some typical floor framing layouts for various building types are studied in the center portion of Figure 5.4. They range from the simple framing of a four-story office building with a braced core (Fig. 5.4f), where the filler beams support a 3 1/4-in. lightweight composite concrete topping over a 2-in. composite steel deck, to the 110-story Sears Tower (Fig. 5.4m). In this case, 40-in. deep long-span Warren trusses, in composite action with the floor slab, spaced on 15-ft centers, bridge the 75-ft square bay. The direction of the truss spans alternates in the corner tubes every six floors, to equalize the loading on the tubular frame. The floor slab consists of the composite assembly of 3-in. metal decking and 2 1/2-in. lightweight concrete on top.

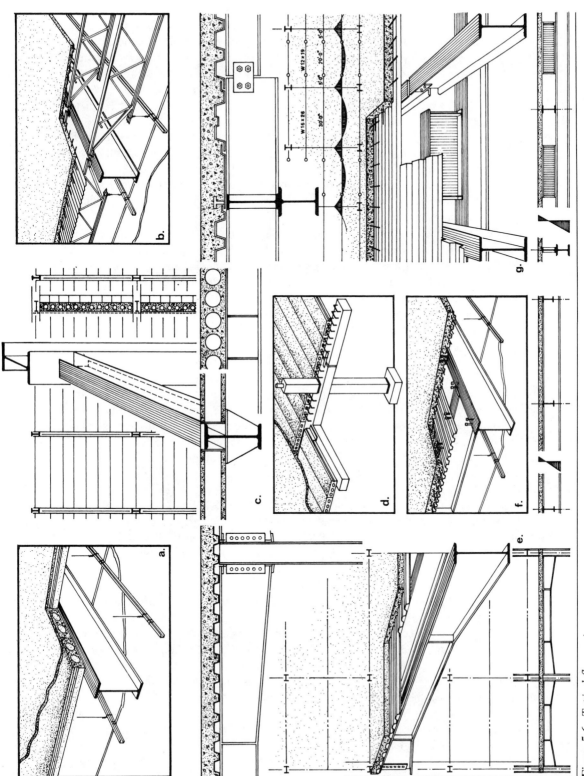

Figure 5.6. *Typical floor systems.*

In the 11-story rigid frame apartment building (Fig. 5.4e), the bays are bridged by open web steel joists and steel decking, with a 2 1/2-in. reinforced concrete floor slab. For the 17-story building with braced end cores (Fig. 5.4g) only four central columns were allowed, thus requiring two main long-span beams with cantilevers and long-span cross beams. The beams act composically with the floor slab, which consists of a 3-in. composite metal deck with a 2 1/2-in. normal-weight concrete topping.

For the 52-story One Shell Plaza building (Fig. 5.4h) in Houston, one-way 40-ft span concrete joist slabs are used between the core and exterior columns, and a two-way waffle system at the corners. Since the two-way slabs increase the loads on the beam strips spanning from the core corner across to the exterior columns, these columns get a larger share of the floor loads so that they had to be made larger by letting them project on the outside to form a visual expression.

The 1136-ft high AMOCO Building in Chicago (Fig. 5.4k) is a simple tubular steel structure which consists of 5-ft wide, V-shaped, folded plate exterior columns spaced 10-ft on center that are rigidly connected to 5.5 ft-deep channel-shaped spandrel beams. The floor system is made up of 38-in. deep simply supported trusses spanning 45 ft between the core and perimeter. The trusses are spaced 10 ft apart and are staggered from floor to floor so that they can frame into the alternate plates of the exterior V-columns. The floor deck consists of 1 1/2-in. composite steel deck topped with 5000 psi lightweight concrete of 4 in. thickness.

The cases in Figure 5.7 identify some typical floor framing layouts for tower structures of various forms, such as rectangular, truncated rectangular, triangular, trapezoidal, round, and other hybrid forms. Here, the primary beams span the shortest distance from the interior core to the exterior perimeter structure, and thus allow flexible open space around the central core area. The special problem in the corners has been discussed (see fig. 5.4c). The primary beams are either directly supported by the exterior columns, as for tubes with closely spaced columns (Fig. 5.7b, c, d, g, l, etc.) and other structure types (Fig. 5.7a, e), or they are carried by spandrel beams (Fig. 5.7f, h, i, j).

For the round Marina City Towers in Chicago (Fig. 5.7e), the concrete columns along the outer 109-ft diameter circle are connected to the columns on the inner 47-ft diameter circle directly by the 31-ft span beams that support the concrete slab. In contrast, for the octagonal tower (Fig. 5.7h), massive deep spandrel beams connect the eight columns around the perimeter and support the radial 37-ft span primary beams.

For the octagonal tube (Fig. 5.7c), the floor beams are spaced 10 ft on center and span an average of 41 ft from the exterior columns to the core. They act composically with the slab, which consists of a 2-in. composite metal deck with a 3 1/2-in. lightweight concrete topping. The beams are cambered for dead load deflection. For each of the trapezoidal towers (Fig. 5.7a), 42.5-ft stub girders span from each of the exterior columns, which are spaced 30 ft on center, to the interior core columns and support the secondary cantilever beams; the floor structure acts composically.

For the parallelogram-shaped tower (Fig. 5.7j), 33-ft concrete joist slabs span from the core to the exterior spandrels. For the concrete tower (Fig. 5.7i) with a pinwheel arrangement of the floor framing, the approximately 35-ft span girders are supported along the perimeter by the columns of the Vierendeel trusses.

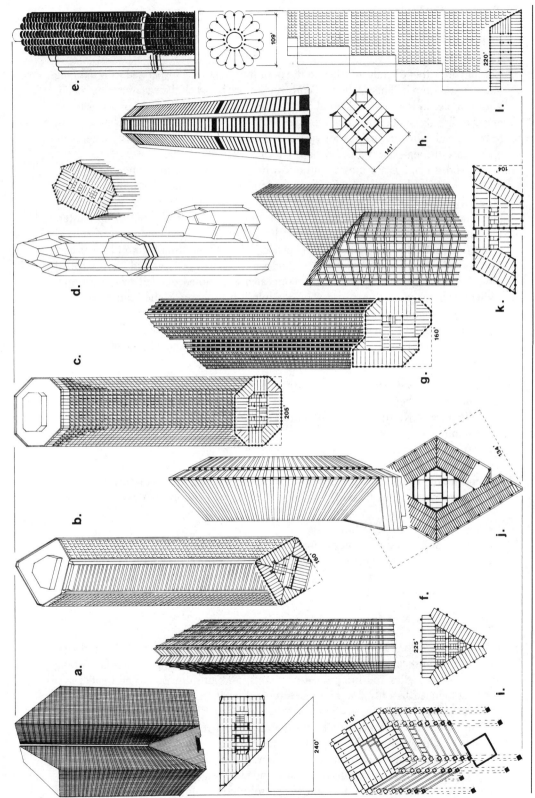

Figure 5.7. *Floor framing structures.*

CEILINGS

Ceilings are either directly attached to the floor framing or are suspended with hanging rods or wires from it. The methods of construction are based on the wet or dry approach. For example, in high-rise residental construction, plaster, spray finish, or acoustic tile may be directly applied to concrete slabs, or solid sheets, like the familiar gypsum board, metal lath, or furrings for plaster are fastened to open-web joists where the open space through the webs may be used for ducts and piping.

Suspended ceilings are most common in high-rise construction. They range from unique, sculptural types in public areas to plain functional ones that provide just the necessary ceiling space. They are either of the *solid* or of the *tile type*; occasionally, they may be fiberglass or reinforced gypsum moldings to imitate ornamental plaster. The more common ceiling types are the linear metal ceilings composed of spliced sections of metal pans or planks and the plaster ceilings used usually for public spaces. Also, office environments use the plaster or gypsum board ceilings and modular tiled acoustical ceilings, which have either concealed or exposed one- or two-directional gridworks that support the panels. The module is a function of the span capacity of the panels, ranging from tiles to large-scale panels spanning several feet.

From a material point of view, there are many different suspended ceiling systems. Their selection depends not only on cost, but also on performance. Some of the design criteria, depending on the building type, are

- acoustical (resistance to sound transmission and noise reduction through sound absorption), thermal and fire barrier
- a light reflective surface to maintain light levels (smooth, white)
- durability
- seismic resistance
- moisture resistance
- positioning of light fixtures and incorporation of air supply and return diffusers
- linkage to partitions by providing special guide rails (flexibility of office arrangement)
- accessibility to the plenum
- ceiling space for ducts and piping, and possibly as plenum for return air
- energy conservation: phase change materials above the ceiling serve as a heat sink, absorbing solar heat from daylight, and then radiating heat back when the temperature drops
- visual properties
- cost, etc.

In the *integrated ceiling*, either of the linear type with planks or of the grid type with lay-in acoustical panels, the various components of ceiling support system, ceiling material for acoustics and fire protection, lights, and air diffusers are all treated as a whole, produced by only one manufacturer, rather than as an assembly of separate parts controlled by the various trades.

Ceilings provide sound control by using soft, absorptive surfaces for noise reduction and hard, reflective surfaces for resistance to sound transmission. They must be fire rated if the structural floor framing above is not. Relief joints should be provided along the perimeter and wall junctions of the suspended ceiling to allow for building movement; they may also have to function as control joints for fire protection. To prevent independent swaying during an earthquake, the ceiling should be tied horizontally to the structure.

REINFORCED CONCRETE SLAB SYSTEMS

The primary action of the slab is generally in response to gravity loading and thus, for preliminary design purposes, can be treated locally without having to consider the entire building structure. The effect of differential vertical movement (i.e., support settlement) upon slab behavior in the upper floors of tall buildings, or the stressing due to lateral force action, is discussed elsewhere in this book.

The skin-like concrete slabs cover or directly form the floor framing. They are surface structures allowing the loads to flow in many directions, in contrast to beams, which are line elements permitting force flow along one direction only; beams usually have much more depth than width, as to efficiently resist flexural rotation. In contrast, slabs are relatively thin, two-dimensional elements supporting mostly small loads; their continuous surface provides a high redundancy with much more reserve strength, where the stresses are easily redistributed.

The behavior of slabs can be derived from the ultimate failure mechanisms, where the collapse loads of slabs are found (yield line theory of slabs), or from the traditional elastic plate theory, which will be used here. The response of slabs to loading is rather complex and influenced by the following criteria:

- *Slab form*: rectangular, round, triangular, trapezoidal, polygonal, free-form, etc.
- *Type of boundary condition and location of support*: degree of fixity versus simple support versus free edge, stiff beam/wall versus flexible beam band supports
- *Type and location of static loads*: point versus line versus surface loads
- *Material characteristics*: creep, shrinkage, composite action
- *Nature of slab*: isotropic versus anisotropic behavior, solid versus hollow versus corrugated versus ribbed, prestressed versus non-prestressed, one-directional versus two-directional, normal weight versus lightweight, etc.
- *Scale*: strength versus flexibility
- *Disturbances*: openings, mass concentration, etc.

The effect of support location upon force flow for rectangular slabs is exemplified in Fig. 5.8b. The location of the beam/wall supports can be along all four sides, three sides with one free edge, two parallel sides, two perpendicular sides (cantilever), column supports at corners, and so on. These supports may be continuous with the slab, as in monolithic construction, or discontinuous, so that the slab is free to rotate, as for example for a slab on a masonry wall.

Most slabs are designed for uniform dead and live loads; larger point and line loads may be assumed to be supported directly on beams. The loads are transferred by the slab in bending, shear, and torsion to the supports. For the case of two parallel walls the slab spans in one direction and is called a one-way slab. When the slab is supported along its periphery, it spans in two principal directions and is called a two-way slab. One-way construction includes solid, hollow, and ribbed slabs, while two-way construction encompasses two-way slabs on stiff beams and on shallow flexible beams, flat plates, flat slabs, and waffle slabs.

The effect of the slab proportion upon force flow and the transition from one-way

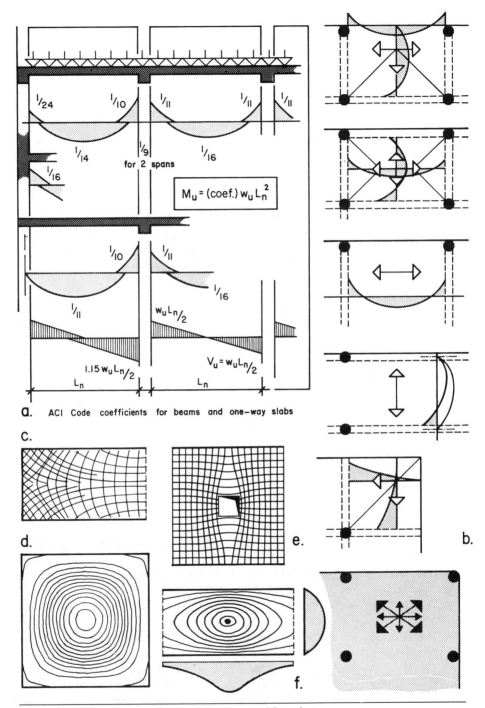

a. ACI Code coefficients for beams and one-way slabs

$$M_u = (\text{coef.}) \, w_u L_n^2$$

for 2 spans

$$w_u L_n / 2$$

$$V_u = w_u L_n / 2$$

$$1.15 \, w_u L_n / 2$$

c.

d.

e.

f.

b.

Figure 5.8. *Slab structures: the effect of support and boundaries.*

behavior to two-way behavior becomes quite apparent from Figure 5.8c, where a narrow slab is supported on three sides. The principal moment contours clearly express the one-way slab character for roughly one-half of the plate length of the slab, while the three-way span in the end zone is quite pronounced.

The behavior of a slab can best be visualized by its deformation, as the deflection contours for the two-way square slab indicate (Fig. 5.8d); notice the transition from the circle to the square with rounded corners in response to the rectangular slab form.

A rectangular slab may be considered to be subdivided by a rectangular grid system, or from a behavioral point of view, to be intersecting slab beams. To have a rough, first idea of the magnitude of stresses under uniform loading, the critical one-foot-wide center strips may be treated as isolated, independent shallow beams, taking into account the boundary conditions of the supports. Notice the moment diagrams for the various slabs shown in Figure 5.8b. This *strip method* of design obviously ignores the continuous character of the surface. Plates resist forces not only in bending but also in twisting, as well as in membrane action, if they deform into nondevelopable dished surfaces, as is the case for two-way slabs. In a cylindrically bent (i.e., developable) plate (such as a one-way slab), however, membrane stresses are only generated when the supports do not allow horizontal movement. Since the strip method only considers bending, it is apparent that it can only be used as a first preliminary estimate of force intensity. Generally, the loads tend to take the shortest and stiffest route to the supports.

The behavior of the slab becomes much more complicated when the usual uniform load is replaced by a concentrated force. The one-way slab under uniform load takes the shape of a suspended cylindrical surface, thus reflecting the one-way curvature and one-way slab action. However, when this same slab carries a point load (Fig. 5.8f), the force seems to fan out and causes the slab to respond with a dish-like deformation, indicating two-way action. It is obvious that the application of the one-directional strip approach here would be an extremely conservative solution, by ignoring the effect of the adjoining strips.

Openings in slabs, depending on their relative size, may disturb the continuity of the surface. Small openings, as for mechanical shafts, are usually not critical. The intensity of force flow increases adjacent to the holes (Fig. 5.8e). It is sufficient to add, along the sides of the opening, straight bars equal in number to the primary reinforcement cut off. Large openings, as for stairways and elevator shafts, must be framed by beams.

In this discussion, isotropic plate behavior has been assumed, which presumes the slab to be properly reinforced, so that it can carry the forces according to the geometry of the support layout. In this case, only the response of the slab to gravity loading is briefly investigated. For the effect of lateral load resistance, where a portion of the slab forms part of the continuous vertical frame, the reader is referred to the section on frame analysis.

The types of concrete that are used for slab construction are normal or lightweight, conventionally reinforced or prestressed, cast-in-place or prefabricated, or any combination of the above. From a construction point of view, concrete slabs can be organized as

- Cast-in-place concrete
- Precast concrete forms or metal deck, with or without the composite action of steel beams, composite with cast-in-place concrete

- Precast concrete decks with cast-in-place topping in composite action
- All precast concrete

Economical span ranges for cast-in-place reinforced concrete floor systems are:

20–25 ft:	Flat plate construction for typical high-rise buildings
25–30 ft:	Banded slab construction
	Flat slab construction for heavy loading conditions, such as found in industrial buildings and warehouses
35–40 ft:	One-way beam and slab construction
	Joist slab and skip-joist construction
	Waffle slab construction when exposure is desired or heavy loads must be supported
	Two-way beam and slab construction for heavy loading conditions
L > 40 ft:	Post-tensioned flat slab construction
	Post-tensioned waffle slabs
	Post-tensioned floor framing

Typical span-to-depth (L/t) ratios for continuous floor framing members are 36 for flat slabs, 28 for one-way slabs, and 21 for beams and joists. These ratios are increased for post-tensioned members to approximately 45 for slabs, 30 for joist slabs, and 24 for beams.

One-Way, Solid Slabs

Most floor slabs are of the one-way type. They span from cross-beam (filler beam) to cross-beam, which in turn are supported by the frame girders. In cast-in-place concrete construction, the slabs form a monolithic whole with the floor framing. They are narrow panels with the long sides at least twice the length of the short sides. They bend into a nearly cylindrical surface under uniform gravity loading, with the single curvature reflecting the one-way action of the slab. The dual function of the slab should be remembered: it not only spans from beam to beam, but also forms the flanges of the beams in the perpendicular direction (T-beam action).

In load-bearing masonry buildings, the slabs may span from wall to wall in a continuous fashion if cast in place, or they may be precast units with simple spans. Wall footings and retaining walls are other examples of primary one-way slab action, as are the curtain panels with respect to lateral force pressure.

The one-way slab is treated as a series of independent one-foot-wide shallow beams, the design of which has been discussed in the section in Chapter 4 on "Concrete Beams." The primary flexural reinforcement is positioned along the beam action, but the secondary steel must be placed perpendicular to this to control cracks due to shrinkage and temperature drop, as well as to redistribute possible concentrated loads. This so-called temperature reinforcement is also the minimum flexural steel required for the slab. It is equal to:

$$A_{smin} = 0.002bt \qquad \text{for} \qquad f_y = 40 \text{ and } 50 \text{ ksi} \qquad (5.19)$$

$$A_{smin} = 0.0018bt \qquad \text{for} \qquad f_y = 60 \text{ ksi}$$

This temperature reinforcement is located on top of the bottom bars but under the top bars, and shall not be placed further apart than 5t, or 18 in.

The primary flexural reinforcement in one-way slabs and walls other than concrete joists, shall not be spaced further apart than 3t, or 18 in.

In two-way slabs, the spacing of the bars shall not exceed 2t at critical sections, but otherwise 3t. The normal span range for one-way concrete slabs is *10 to 20 ft, with a typical slab thickness of 4 to 7 in.*

The slab thickness for residential and office construction is usually controlled by deflection limitations and by fire resistance requirements; flexural and shear stresses rarely control, except for heavy loading conditions. For a preliminary estimate of the slab thickness (in.), t, as a function of the center-to-center span (ft), L, use

simply supported: $t = L/2 \geq 4$ in. for fireproofing
cantilever: $t = L \geq 4$ in.
continuous both ends: $t = L/3 \geq 4$ in.

The continuous slab for exterior bays may have to be larger. The above values are based on Grade 40 steel and normal-weight concrete ($w_c = 145$ pcf). For Grade 60 steel, as for 90 pcf lightweight concrete, the slab thickness should be increased by 25 percent; for 110 and 120 pcf lightweight concrete, the slab thickness should be increased by 10 percent. However, it must be kept in mind that a thinner slab may be more economical, though deflection may have to be checked.

The monolithic character of concrete structures makes the design of slabs highly indeterminate and requires a frame analysis. To simplify the design of continuous beams and one-way slabs under gravity action, the ACI Code permits the use of approximate moment and shear coefficients (Fig. 5.8a), taking into account the critical live load arrangement for following conditions:

- There are two or more spans
- Spans are approximately equal, with the larger of the two adjacent spans not greater than the shorter by 20 percent
- Loads are uniformly distributed
- Unit live load does not exceed three times unit dead load

The coefficients are given in terms of clear span, l_n, at critical locations for various boundary conditions in Figure 5.8a. A typical factored moment is

$$M_u = (\text{coef.})w_u l_n^2 \qquad (5.20)$$

Example 5.5

A six-story concrete frame office building consists of 30 × 34-ft bays with a floor framing as shown in fig. 4.3. The concrete slab supports 5 psf for ceiling and floor finish and a 20 psf partition load, as well as a live load of 80 psf. In this instance, a typical inferior slab is investigated; for the design of the beams and girders, refer to Example 4.1. Use $f'_c = 4000$ psi and $f_y = 60,000$ psi.

The clear span of the typical interior bay, using a supporting beam width of 12 in., is

$$l_n = 15 - 12/12 = 14 \text{ ft}$$

The slab thickness is estimated as

$$t = (L/3)1.25 = (15/3)1.25 = 6.15 \text{ in.}$$

Use a 6 1/4-in. slab.

For a one-way slab there is no live load reduction. The uniform ultimate load that the slab must carry is

$$w_u = 1.4w_D + 1.7w_L$$

$$= (1.4(6.25(150/12) + 5 + 20) + 1.7(80))/1000$$

$$= 1.4(0.103) + 1.7(0.080) = 0.28 \text{ ksf} \quad \text{or} \quad \text{klf/ft of slab}$$

The critical moments at support and midspan are

$$- M_u = w_u l_n^2/11 = 0.28(14)^2/11 = 4.99 \text{ ft-k/ft}$$

$$+ M_u = w_u l_n^2/16 = 0.28(14)^2/16 = 3.43 \text{ ft-k/ft}$$

The corresponding slab reinforcement is

at support: $-A_s = M_u/(0.85f_yd) = 4.99(12)/[0.85(60)5.25] = 0.224 \text{ in.}^2/\text{ft}$

at midspan: $+ A_s = 3.43(12)/[0.85(60)5.25] = 0.154 \text{ in.}^2/\text{ft.}$

$$\geq A_{smin} = 0.135 \text{ in.}^2/\text{ft}$$

Temperature reinforcement:

$$A_{smin} = 0.0018bt = 0.0018(12)6.25 = 0.135 \text{ in.}^2/\text{ft}$$

Select the slab reinforcement as follows (or see Table A.2)

Top bars at support: $12/0.224 = s/0.2$, $s = 10.71$ in.
  use #4 @ 10 1/2 in. o.c., $A_s = 0.229$ in.^{2}
Bottom bars in the field: $12/0.154 = s/0.2$, $s = 15.58$ in.
  use #4 @ 15 1/2 in. o.c., $A_s = 0.155$ in.^{2}
Temperature steel: $12/0.135 = s/0.11$, $s = 9.78$ in.
  use #3 @ 9 1/2 in. o.c., $A_s = 0.139$ in.^{2}

The critical steel ratio is: $\rho = A_s/bd = 0.229/12(5.25) = 0.36\% < 1.6\%$
Obviously, the steel ratio for an ordinary slab is very low!
In one-way concrete slabs shear is rarely a problem. For example, the maximum shear in this exercise, conservatively ignoring the location at d distance from the beam face, is

$$V_u = 0.28(14/2) = 1.96 \text{ k/ft}$$

The shear strength of the concrete is

$$\phi V_c = 0.11b_wd = 0.11(12)5.25 = 6.93 > 1.96 \text{ k/ft} \tag{4.12}$$

The shear is easily resisted by the concrete.

Precast Concrete Slabs

There are basically two different precast concrete slab systems. One uses thin, precast slabs as formwork for cast-in-place concrete; in the final service stage, this system acts as a composite slab. In the other system, planks are laid on floor joists, or decks span directly from beam to beam or wall to wall. The mass-produced hollow-core decks and double-tee sections are quite popular in residential and office construction. Most are pretensioned and have a thin, cast-in-place concrete topping from 2 to 3 1/2 in. thick, which will act compositely with the precast slab, since its rough surface will provide the necessary shear resistance. The topping helps the floor structure develop the necessary diaphragm action with respect to the lateral force distribution. The typical 4-ft wide hollow-core slabs range in thickness from 4 to 12 in., and have a simple span range from 15 to about 38 ft. The maximum recommended span-to-depth ratio is 40 or

$$t = L/3.33 = 0.3L$$

The weight of hollow core slabs without topping varies with the manufacturer. For preliminary design purposes, a 45 psf slab weight may be assumed for 6 in. hollow core slabs, and 57 psf for 8 in. slabs.

Joist Floor

As the span increases, so does the thickness of the solid slab, resulting in much more dead load, so that the slab will have to be designed primarily for its own weight rather than the superimposed loads. On the other hand, the joist slab, ribbed slab, or pan joists, when poured in place, provide the necessary depth for strength and stiffness without the high dead load-to-live load ratio. The standard size pans for the voids between the ribs are 20 to 40 in. wide and 6 to 24 in. deep. The common spacing of the joists is 2 to 3 ft, with the width of the ribs varying from 5 to 6 in. and the slab thickness from 2 to 3 in. for normal loading conditions. The typical economical span range is from 35 to 40 ft; spans may be increased by about 30 to 40 percent through post-tensioning. For the preliminary estimate of the slab thickness (in.), t, as a function of the span (ft), L, use:

simply supported: $t = 3L/5 = 0.6L$
cantilever: $t = 1.2L$
continuous both ends: $t = L/2.2 \approx 0.5L$

The continuous slab for exterior bays may have to be larger. The above values are based on Grade 40 steel and normal weight concrete ($w_c = 145$ pcf). For Grade 60 steel, as for 90 pcf lightweight concrete, the slab thickness should be increased by 25 percent; for 110 and 120 pcf lightweight concrete, the slab thickness should be increased by 10 percent. However, it must be kept in mind that, for unreinforced webs, the shear may control the depth of the joist slab.

According to the ACI Code, the ribs shall not be less than 4 in. wide in joist construction, and shall have a depth of not more than 3-1/2 times the minimum width of the rib. The clear spacing between the ribs shall not exceed 30 in. Joist construction not meeting these limitations shall be designed as slabs and beams.

Since the ribs are so closely spaced, the thin slab only constitutes the flange portion of the joists (T-beams). The flexural stresses in the slab perpendicular to the ribs are easily resisted by the plain concrete for normal loading conditions, thus only temperature reinforcement (e.g., welded wire fabric) normal to the joists is required in the slab.

The shear stresses in the joist slab are usually not critical, so that stirrups are not used. If the shear should be larger than the capacity of the concrete, bent-up flexural bars or widened ribs can be provided at the supports. *Load distribution ribs* perpendicular to the joists are used for spans larger than 20 ft. Usually, one rib at center for spans up to 30 ft and two at third points for spans larger than that is adequate.

Where non-fire-resistant lightweight ceilings are used, the slab thickness may have to be increased to 4 or 4 1/2 in., so as to achieve the necessary fire rating. Since this slab can span much further than the typical 3 ft between the joists, wider joist spacing of 6 ft or 8 to 10 ft with special pan forms may be introduced to improve economy. Naturally, the behavior of this wide-pan, or skip-joist, slab approaches that of the slab on beams; hence the popular 30-in wide pans may often be replaced by 66-in. pans.

The slab thickness for concrete on open-web steel joists with a maximum spacing of about 2 ft is usually 2 to 2 1/2 in.

Example 5.6

A multiple bay, continuous concrete joist slab with clear spans of 20 ft and center-to-center spans of 21 ft, and with 5 in. wide, 10 + 3 in. deep ribs spaced at 35 in. on center, is to be investigated (Fig. 5.9). The preliminary design is based on the critical end span, where the slab is restrained by the spandrel beams. The slab must carry a live load of 80 psf and a superimposed dead load of 20 psf for the partitions and 10 psf for bridging, floor finish, and mechanical loads. Use $f_c' = 4000$ psi and $f_y = 60,000$ psi.

So that deflection does not have to be computed, the overall depth of the joist slab should be

$$t = 1.25(L/2.2) = 1.25(21/2.2) = 11.93 \text{ in.} \leqslant 13 \text{ in.}$$

The minimum reinforcement for the 3-in. slab is

$$A_{smin} = 0.0018bt = 0.0018(12)3 = 0.065 \text{ in.}^2/\text{ft}$$

$$s_{min} = 5t = 5(3) = 15 \text{ in.} \leqslant 18 \text{ in.}$$

Use #3 @ 15 in. o.c., $A_s = 0.11(12)/15 = 0.088$ in.^{2}/ ft, placed perpendicular to the joists and usually on centerline of the slab, to resist both positive and negative moments.

By using an average width of 5.83 in. for a standard taper of 1/12, as shown in Figure 5.9, the joist slab weighs

$$w = 3(150/12) + 150(5.83 \times 10/12^2)/(35/12) = 58.32 \text{ psf}$$

The ultimate load, with no live load reduction allowed, is

$$w_u = 1.4(58.32 + 30) + 1.7(80) = 260 \text{ psf}$$

Each joist carries a uniform load of

$$w_u = 0.260(35/12) = 0.76 \text{ k/ft}$$

The critical moments and the corresponding reinforcement, using $d \approx t - 1.25 = 13 - 1.25 = 11.75$ in., are at the

Interior support: $-M_u = w_u l_n^2/10 = 0.76(20)^2/10 = 30.4$ ft-k

$$-A_s = M_u/4d = 30.4/4(11.75) = 0.647 \text{ in.}^2$$

Use 2 #4 and 1 #5 truss bar, $A_s = 0.71$ in.^{2}, distributed across the effective flange width of $L/10 = 21(12)/10 = 25$ in. < 35 in. The steel ratio is $\rho = A_s/b_w d = 0.71/5(11.75) = 1.21\% < 1.6\%$

Similarly, the reinforcement for the exterior support can be found.

$$\text{Midspan:} \quad + M_u = w_u l_n^2/14 = 0.76(20)^2/14 = 21.71 \text{ ft-k}$$

$$+ A_s = 21.71/4(11.75) = 0.462 \text{ in.}^2$$

Use 1 #4 bar and 1 #5 truss bar, as shown in Figure 5.9, $A_s = 0.51$ in.^{2}

At the face of the first interior support, the maximum shear is $1.15 w_u l_n/2$, where the increase of shear is due to the effect of continuity.

$$V_u = 1.15(0.76)20/2 = 8.74 \text{ k}$$

The maximum shear at d-distance from the face of the beam is

$$V_{umax} = 8.74 - 0.76(11.75/12) = 8.00 \text{ k}$$

The shear strength for the ribs may be taken as 10 percent greater, as for the beams under the given ACI Code conditions. Therefore, according to Equation 4.12

$$1.1\phi V_c = 1.1(0.11 b_w d) = 1.1(0.11)5(11.75) = 7.11 \text{ k} < 8.00 \text{ k}$$

The capacity of the concrete is not enough. Normally, the ends of the ribs are widened to increase the shear strength of the concrete, or one bar is bent up to provide the additional shear capacity. In skip-joist construction, the required higher shear capacity may be covered with single-leg #3 stirrups located at the joist centerline.

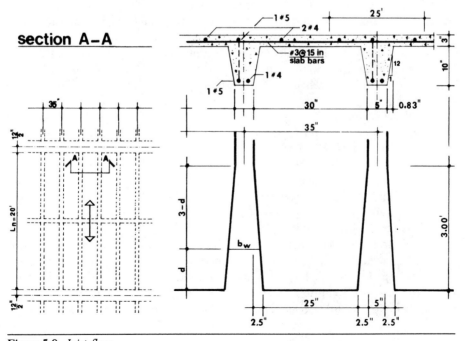

Figure 5.9. *Joist floor.*

Ignoring the bent-up bar, but taking into account the taper at the support as shown in Figure 5.9, the beam width is

$$b_w = 5 + \left(\frac{3-d}{3}\right)5 = 5 + \left(\frac{3-0.98}{3}\right)5 = 8.37 \text{ in.}$$

Hence, the shear strength of the concrete is

$$\phi V_c = 7.11(8.37/5) = 11.90 \text{ k} > 8.00 \text{ k} \qquad \text{OK}$$

Composite One-Way Slab Systems

The cast-in-place concrete can act together with steel decking or precast concrete components. Steel decking floors are most common in high-rise steel construction. A nearly unlimited variety of profiles and sizes of decks are commercially available. They range from corrugated forms to ribbed or folded forms to flat steel sheets with reinforcing ribs; they vary from closed to open profiles, and from narrow ribs to wide ribs. The most common steel decks for floors are 1 1/2 and 3 in. in depth, with rib spacings of 6 to 12 in., but also 2-in. deep decks are available—they range between 0.0596 in. (16 gauge) and 0.0295 in. (22 gauge) in thickness. Their capacity depends on the depth, profile, and metal thickness and type of support.

The steel decking may just be in-place decks to provide a form for the concrete (i.e., form deck) and a working platform. For this condition, the slab behaves somewhat like a ribbed floor with conventional reinforcement. The advantage of this approach lies in the rapid construction, since the deck does not have to be fireproofed. In steel deck composite slabs, the interface of floor deck and concrete must resist the horizontal shear; the materials are bonded together by a combination of chemical bond and physical connecting devices. Deep-embossed indentations in the flanges or webs of the deck, or transverse wires welded to the top of the deck, or other mechanical devices provide this horizontal shear resistance and prevent vertical separation. In composite floor decks, the cold-formed steel deck provides the dual role of serving as a platform during the construction period, and as a positive moment reinforcing for the one-way slab during the final service stage. Only welded wire fabric is needed across the interior supports as shrinkage reinforcement for simple spans, or special welded wire mesh as negative reinforcement if continuity across the interior beam supports is desired and sufficient slab thickness is available.

Since the cellular decks have considerable strength, lightweight concretes of less strength (e.g., 110 pcf) are usually used, thus providing a lightweight floor system. The minimum overall slab thickness should be 3 1/2 in. with a minimum concrete cover above the deck of 2 in.; this concrete cover varies typically from 2 1/2 to 4 in. Economical span ranges for 1 1/2-in. composite-cellular rib decks are up to about 10 ft and, for 3-in. decks, up to about 15 ft. Most composite concrete floor decks are treated as simply supported slabs. For this condition, the minimum slab thickness should be

$$t \geq L/2 \geq 3 \ 1/2 \text{ in.}$$

The flexural stresses for the steel deck during the noncomposite construction stage under the dead load, w_1, which includes the wet concrete and construction loads of 20 psf, are

$$f_b = M_1/S_d = (w_1L^2/8S_d)12 = 1.5w_1L^2/S_d \leq 0.6F_y$$

Should the stresses be larger than allowable, either a stronger deck must be selected or shoring must be provided during construction to reduce bending. However, the choice of metal deck not only depends on the loads and span conditions, but also whether floor electrification is desired.

Example 5.7

A 4-in. composite lightweight concrete slab with a 1 1/2 inch, 22 gauge steel decking spans 7 ft. Check whether temporary shoring must be provided if it supports, in noncomposite action, a load of 55 psf during construction. Use $S_d = 0.21$ in.³/ft and $F_y = 33$ ksi.

$$f_b = 1.5(0.055)7^2/0.21 = 19.25 \text{ ksi} \leq 0.6(33) = 19.8 \text{ ksi}$$

No shoring is required. Since an unshored floor deck carries the slab weight and some superimposed loads, the slab only needs to support a portion of the live loads.

In the final, composite stage, the simply supported slab can be considered as over-reinforced, with the concrete as the weaker element under bending. However, it has been found in numerous tests that the load-carrying capacity of the composite slab depends on shear rather than flexure. For the design of composite deck slabs, the reader should refer to the respective steel deck manufacturer's catalogs for the permissible loads. In concrete construction, there are two composite slab systems:

- Precast, pretensioned joists act compositely with the cast-in-place concrete slab. Shear reinforcement protrudes from the joists to provide composite action.
- Pretensioned, precast thin slabs (2 to 3 in. thick) serve as formwork for the cast-in-place concrete to form a composite slab system. The system is applicable to larger spans of beyond 25 ft. During construction, the thin precast plates must be shored to support the wet concrete. For longer spans, hollow-core precast sections are used, in order to reduce the dead weight.

Two-Way Slabs

The ACI Code treats two-way slabs in cast-in-place concrete construction as a continuous part of the three-dimensional structure. Visualize a frame building sliced vertically along floor panel centerlines, that is, midway between columns, first in the short direction and then perpendicular, in the long direction (Fig. 7.12, bottom left). Thus, the spatial building is subdivided into equivalent parallel planar rigid frame bents interlocking with the perpendicular frames. The beams in these frames vary in flexural stiffness; they may be shallow, wide slab-beams, as for flat slab/plate floors, or they may be deep T-beams, as for two-way slabs on beams.

The ACI Code uses this equivalent frame concept in both of its methods of design,

- Direct-Design Method:
  a semi-empirical method for gravity loading only,
- Equivalent-Frame Method.

For further discussion of the frame method, the reader may want to refer to the section on rigid frames in Chapter 7.

The equivalent frame concept is not further dealt with because of its complex analytical nature, which makes it difficult to develop a feeling for the physical behavior of the two-way action of continuous horizontal floor slabs. Further, it is assumed that the lateral forces are resisted solely by the rigid frames for buildings with two-way slabs on stiff beams, and that stiff shear walls or perimeter tubes resist the lateral forces of flat slab buildings, so that the design of two-way slabs is controlled by gravity loading. It should be realized that the assumptions above apply to tall structures, since lateral force action does not usually control the design of ordinary medium-rise buildings. Therefore, it is reasonable to consider only gravity loads for the preliminary design of the slabs.

In the past, the design of slab systems for gravity loads was done by either elastic analysis, usually by considering each floor slab with its columns above and below fixed at their remote ends, or it was done by empirical methods. Until 1971, the ACI Code clearly distinguished between the design of two-way slabs on stiff beams and the design of two-way slabs supported directly on columns. Since (in this context) only these extreme, but typical cases (and not the condition between, that of a slab supported on flexible shallow wide beam bands) are investigated, the empirical methods may be used, rather than the unified generalized approach of the current ACI Code, particularly to develop a sense for the response of horizontal surface plates to gravity loading.

Two-Way Slabs on Stiff Beams or Walls

As rectangular one-way slabs deform under uniform gravity loads, they take the shape of a cylindrical surface. All of the loads are transferred to the rigid supports in one direction, as reflected by the uniform single-curvature deflection. On the other hand, an isolated square slab, for instance, supported along its four sides (Fig. 5.10), takes on a dish-like deflected shape, which can be visualized as an inverted shallow dome along a nearly circular base (Fig. 5.10e), and a negative curvature in the corner zones, similar to the pendentives that allow the transition from a circular dome to a square base. This two-way action of the slab is rather complex, as is discussed later in this section. Rather than attempting to predict the true elastic behavior of the slab, an empirical method is used, where the coefficients do not reflect stress variations, but are constants, providing uniform envelopes of average moments; they are only concerned with the overall safety of the slab. For preliminary design purposes, Method 2 of the ACI 318-63 (Table A.7) will be used here; the ACI Code still permits the use of this method of analysis, as long as the slab is supported on sufficiently stiff beams on all four sides. This method is the simplest and shortest, but also the least economical one, because it does not take inelastic redistribution effects into account; however, it does give an appreciation for the construction of two-way slabs on beams. In the moment coefficient method, the two-way slab is subdivided in each direction into a middle strip of one-half panel width, symmetrical about the centerline, and a

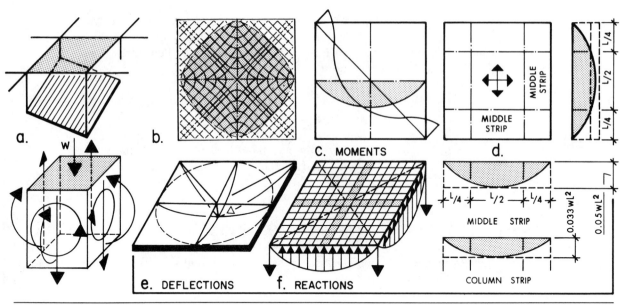

Figure 5.10. *Introduction to two-way slabs on rigid supports.*

column strip of one-half panel width, occupying the two quarter-panel areas outside of the middle strip (Fig. 5.10d). The average moments per foot of width in the column strip are taken as two-thirds of the corresponding moments in the middle strip, thus reflecting the decrease of the bending moments towards the edge zones.

To develop some understanding for slab behavior, first let a square slab be subdivided into parallel one-foot-wide strips in the longitudinal and transverse directions (Fig. 5.10e). When isolating the two imaginary perpendicular strips along the axes of symmetry, each of the shallow slab beams behaves as approximately a one-way, single curvature system, sharing the load equally. Thus, the maximum moment at the center in each direction is

$$M_{max} = (w/2)L^2/8 = 0.0625wL^2$$

In reality, however, the strips are not unconnected and isolated, they are continuous with their boundaries, which provide torsional resistance to the bending. In the empirical moment coefficient method (Table A.7), the approximate maximum moment for this case is

$$M_{max} = 0.05wL^2$$

Therefore, along the center strips, 80 percent of the rotational moment is resisted by bending, and 20 percent by the twisting—the effect of membrane action is ignored. It is interesting to note that the maximum moment for the simply supported square plate is approximately equal to the maximum moment for a simply supported circular plate, due to the fact that the stiff corners of the square slab hardly deflect. Based on a similar reasoning, the maximum bending moments and shear forces for simply supported polygonal slabs, including triangular slabs, can be roughly approximated by equivalent inscribed circular plates. It is also interesting to note that the maximum support moments for continuous or fixed circular and square slabs are nearly the same.

Should the simply supported slab be of rectangular shape, then the portion of the load carried in the short and long directions can be approximately found by equating the maximum deflections of the imaginary center strips, since, at their intersection (the slab center), the displacement must be the same (Fig. 5.10e).

$$5w_sL_s^4/384EI = 5w_LL_L^4/384EI, \quad \text{or}$$

$$w_L = w_s(L_s/L_L)^4 = w_sm^4$$

As the ratio $m = L_s/L_L$ decreases, more and more load is carried in the short direction and less in the long direction. Let (for instance) $m = 0.5$, then $w_L = w_s(0.5)^4 = w_s/16$, but

$$w = w_s + w_L = w_s(1 + 1/16), \quad \text{or} \quad w_s = 0.94w$$

Therefore, 94 percent of the load is carried in the short direction, and hardly any in the long direction. Obviously, the slab can be considered a one-way slab spanning in the short direction when the long side is more than twice the short side. Typical two-way slab proportions are in the range of $m = 1$ to 0.7. From this discussion, it is quite apparent that more loads are carried in the short direction, causing larger moments than in the long direction. Similarly, more loads are attracted to the stiffer, continuous boundaries of slabs than to the simply supported edges. Slab proportions and boundary conditions therefore determine the load distribution of the isotropic plate.

The strips parallel to the center ones not only bend, but also twist. The displacement of a slab element (Fig. 5.10a) indicates that, in addition to bending causing shear (V_x, V_y) and moments (M_x, M_y), the twisting introduces torsional shear and torsional moments (M_{xy}, M_{yx}).

As the designer progresses with imaginary parallel strips from the center to the supports, the bending moments decrease (Fig. 5.10e), while the twisting increases, or the effect of the diagonal increases, becoming quite pronounced in the stiff corner zones, where the slab tends to span diagonally. The corners prevent the slab from freely rotating, causing beam action along the slab diagonal similar to a fixed beam (Fig. 5.10b, c). To resist the uplift in the corner zones of noncontinuous edges, the slab must be anchored and the negative moments must be covered with diagonal top reinforcement.

This complex action of the slab in response to uniform loading is convincingly reflected by the maximum and minimum principal moment contours with no torsional stresses, indicating the bending stresses in the radial and circumferential directions (Fig. 5.10b). In this case, the continuous solid lines represent the tension at the bottom of the slab (maximum positive moments), and the dashed lines represent the tension at the top face (minimum negative moments), with the dash-dotted line indicating the change of curvature.

To follow these tension trajectories with the reinforcement is clearly impractical, though Pier Luigi Nervi, the famous Italian designer, did exactly that (see fig. 5.11g). By curving the ribs along the path of the principal stress flow for a simply supported plate, Nervi was able to uncover unexpected beautiful curvilinear patterns, which bring the material alive by letting the slab respond to external force action in a dynamic, organic manner.

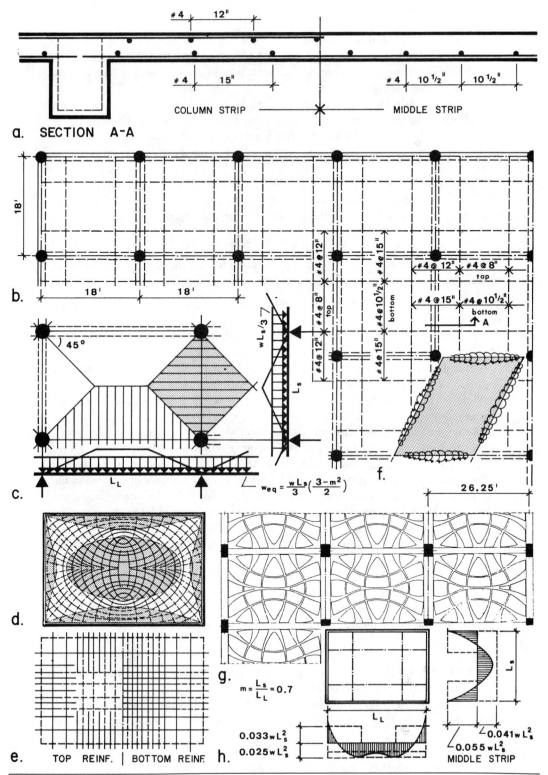

a. SECTION A-A

#4 12"

#4 15"   #4 10 1/2" 10 1/2"

COLUMN STRIP ——————✕—————— MIDDLE STRIP

b.

18'

18'   18'

#4@12" #4@12" #4@15"
#4@8"   #4@10 1/2"
top   bottom
#4@12" #4@15"

#4@12" #4@8" top
#4@15" #4@10 1/2" bottom
↑A

c.

45°

$w_{eq} = \frac{wL_s}{3}\left(\frac{3-m^2}{2}\right)$

$wL_s/3$

L_s

L_L

f.

d.

26.25'

e. TOP REINF. | BOTTOM REINF.

g. $m = \frac{L_s}{L_L} = 0.7$

h.

0.033wL_s^2
0.025wL_s^2

L_L

L_s

0.041wL_s^2
0.055wL_s^2
MIDDLE STRIP

Figure 5.11. *Design of two-way slabs on stiff beams.*

In contrast to the simply supported slab, there are hardly any corner moments in the continuous slab; here, the stiff beam intersection resists all of the rotation. The complex stress variations for an isotropic continuous plate are suggested by the principal moment contours of Figure 5.11d.

For two-way slabs on stiff beams, the critical moments may be assumed along the centerlines of the panels, at the middle region of the slab for the positive moments and along the beam edges for the negative bending. A similar approach can be used for visualizing the behavior of other panel types; always take the imaginary center strips as equivalent one-foot-wide shallow beams.

The moment coefficients for the various panel types and panel proportions, expressed as $m = L_s/L_L$ (the ratio of the short span to the long span), are given in Table A.7. The bending moments, M_u, for the middle strip are computed as

$$M_u = (\text{coef.})w_u L_s^2 \tag{5.21}$$

where: L_s = length of the short span, usually taken to be from center to center beam or wall (ft)

Notice that the coefficients for the long-span moments are multiplied by the square of the short span!

The average moments per foot of width in the column strip are taken as two-thirds of the corresponding moments in the middle strip.

Remember that the moment coefficients are based on slabs built monolithically with the supporting beams or walls, otherwise, for simple supports along the exterior building bays, positive field moments will have to be increased because of less restraining moments at the supports.

The flexural reinforcement is placed two ways along the imaginary one-foot-wide beam strips, corresponding to the assumed moment action as shown in Figure 5.11e, for a typical interior panel using bent-up bars. According to the ACI Code, the spacing of the main reinforcement in the critical field strips shall not exceed two times the slab thickness. At other locations, it is taken as three times the slab thickness, which corresponds to an increase of bar spacing by fifty percent in the column strips; but the spacing cannot exceed 18 in.

The reactions of the two-way slab cannot be constant, as is easily seen from its deflection (Fig. 5.10e, f). Hence, the corresponding shear distribution along the slab edges cannot be uniform either. Generally, the load to the beams is based on the subdivision of the floor panel into triangular tributary floor areas for the short beams, and trapezoidal areas for the long beams (Fig. 5.11c). For this condition, the maximum shear in the slab of a typical interior bay occurs at the ends of the center strips.

$$V_{umax} = w_u L_s/2 \tag{5.22}$$

In contrast to flat slabs, the shear in two-way slabs on beams is usually not critical; the beams eliminate the shear problem.

The loads on the supporting beams can be transformed into equivalent uniform loads (see Problem 5.33), so that the approximate bending moments can be found (see fig. 5.8a).

The thickness of the two-way slabs on stiff beams can be estimated for any type of steel, and independent of concrete strength, according to ACI 318-63, as

$$t = L/3.75 \geq 4 \text{ in.} \tag{5.23}$$

where: t = slab thickness (in.)

L = average slab span $(L_s + L_L)/2$,   (ft)

Since a steel ratio close to $\rho_{max}/2$ is used, the deflection should be within limits.

Example 5.8

The concrete floor framing of an office building is organized on a beam grid of 18 × 18 ft; two-way concrete slabs cover the bays. The live loads are 80 psf, and the superimposed dead load for floor finish is 10 psf and 20 psf for partitions. Use f'_c = 4000 psi and f_y = 40,000 psi. Do a preliminary design of a typical interior panel.

The slab thickness can be estimated as

$$t = L/3.75 = 18/3.75 = 4.8 \text{ in.,} \quad \text{use a 5-in. slab}$$

The uniform ultimate load is

$$w_u = 1.4(5(0.150/12) + 0.030) + 1.7(0.080) = 0.27 \text{ ksf}$$

There is no live load reduction, since the bay size is less than 400 ft².

The moment coefficients for the middle strip of an interior panel are found from Table A.7.

The negative support moments along the slab edges are

$$M_u = (\text{coef.})w_u L_s^2 = 0.033(0.27)18^2 = 2.89 \text{ ft-k}$$

The positive field moments at mid-span are

$$M_u = 0.025(0.27)18^2 = 2.19 \text{ ft-k}$$

The corresponding flexural reinforcement, using $d = 5 - 1.5 = 3.5$, is

$$A_s = M_u/(0.85f_y d) \tag{4.7b}$$

$$-A_s = 2.89(12)/[0.85(40)3.5] = 0.291 \text{ in.}^2/\text{ft}$$

$$+A_s = 2.19(12)/[0.85(40)3.5] = 0.221 \text{ in.}^2/\text{ft}$$

$$A_{smin} = 0.0020bt = 0.0020(12)5 = 0.120 \text{ in.}^2/\text{ft}$$

Select the reinforcement for the middle strip (Fig. 5.11a, b, e). The top reinforcement along the slab edges at the beams is

$$s = (0.2/0.291)12 = 8.25 \text{ in.}$$

Use #4 @ 8 in. o.c.

The bottom reinforcement in the field is

$$s = (0.2/0.221)12 = 10.86 \text{ in.}$$

Use #4 @ 10 1/2 in. o.c. both ways.

The average moment in the column strips is two-thirds of the corresponding moments in the middle strips, which is equivalent to increasing the bar spacing, s, by 50 percent; but the spacing should be not more than $3t = 3(5) = 15$ in.

Top steel at support: s = 1.5(8) = 12 in., use #4 @ 12 in. o.c.
Bottom steel spacing in the field: s = 1.5(10.5) = 15.75 > 15 in., use #4 @ 15 in. o.c.

Rather than using straight bars as top and bottom reinforcement, alternate field bars could be bent up (Fig. 5.11e).

The maximum shear occurs at the end of the center strips at the beam face (Fig. 5.10f).

$$V_u = w_u L_s/2 = 0.27(18 - 1)/2 = 2.3 \text{ k}$$

The shear capacity of the concrete is

$$\phi V_c = 0.11 b_w d = 0.11(12)3.5 = 4.62 \text{ k} > 2.3 \text{ k} \qquad (4.12)$$

Shear is rarely a problem for two-way slabs on beams.

For the preliminary flexural design of the continuous beams, the triangular slab loads can be transformed into equivalent uniform loads (Fig. 5.11c) so that the ACI moment coefficients for beams and one-way slabs can be used. When ignoring the live load reduction, the equivalent uniform beam load is

$$w_u L_s/3 = (0.27(18)/3)2 = 3.24 \text{ k/ft}$$

The stem weight and possibly a wall load must be added to this load.

Through the two-way action of the slab, reinforcement is saved, as compared to the one-way slab. However, the beam supports increase floor-to-floor heights and only allow passage of ductwork below the bottom of the beams. Because of these disadvantages, the flat slab concept has widely replaced two-way slabs on beams in typical high-rise building construction; it also allows simpler formwork and reinforcing steel layout, so that it can be constructed in less time with less labor costs.

Flat Slabs and Flat Plates

From a behavioral point of view, flat slabs are highly complex structures. The intricacy of the force flow along an isotropic plate in response to uniform gravity action is reflected by the principal moment contours in Figure 5.12d. In this case, the main moments around the column support are negative, and have circular and radial directions, while the positive field moments basically connect the columns linearly. The patterns remind us of organic structures such as the branching grids of leaves, the delicate network of insect wings, radial spider webs, and the contour lines of conical tents, realizing a similar relationship between cable response and loading as well as the corresponding moment diagram. Pier Luigi Nervi, for the Gatti Wool Factory [1953] in Rome, Italy (Fig. 5.13e) actually followed the principal bending moments with the layout of the floor ribs. Centuries earlier, however, the late medieval master builders had already intuitively developed patterns for ribbed vaulting predicting these tensile trajectories; the fan vaults of the Tudor period in England are a convincing example (see fig. 5.12e).

The common two-way slab systems on columns are

- *Flat plate:* The slab, with a typical economical span range of 20 to 25 ft, is directly supported on columns and possibly walls. The system is adaptable to an irregular

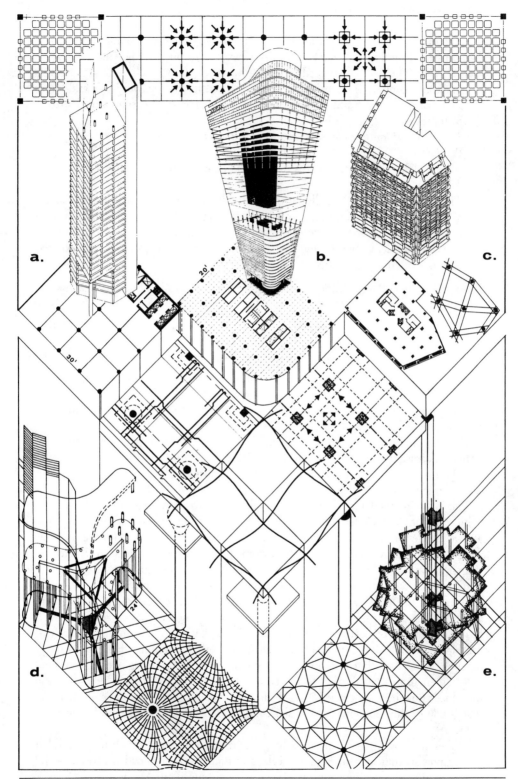

Figure 5.12. *Flat slab building structures.*

support layout. It is popular for apartment buildings, hotels, and dormitories. The thin slab beamless construction allows a low overall building height.

- *Flat slab:* The slab is thickened at the columns (i.e., drop panels), and the columns may have capitals. This beamless system is used for the heavier loads and larger spans of industrial buildings, warehouses, and garages. The economical span is in the range of 25 to 30 ft.
- *Two-way slab on flexible, shallow, wide beam bands*
- *Waffle slab:* There are two systems:
  A solid slab with waffle patterns only in the center portion
  Two-way ribbed slab
  The two-way joists are usually formed with standard square domed forms of about 30 in. square and 20 in. deep; they are omitted around the columns to form solid heads to resist the shear. The economical span range for waffle slabs is 35–40 ft, and are found in libraries and first floor levels for public spaces where exposed ceiling structures are desired.

For preliminary design purposes and light residential loading, the slab thickness can be estimated, as derived from deflection control (see Problem 5.17), as

$$\text{Flat slab:} \qquad t \geq L_L/3.33 \geq 4 \text{ in.} \qquad \qquad \textbf{(5.24a)}$$

$$\text{Flat plate:} \qquad t \geq L_L/3 \geq 5 \text{ in.} \qquad \qquad \textbf{(5.24b)}$$

Where L_L is the larger span from center column to center column in feet, and the slab thickness, t, is in inches. Generally, however, the slab thickness is controlled by the punching shear around the column. Typical flat slab/plate thicknesses in high-rise construction are 5 to 10 inches.

The values are for typical interior panels; for corner and side panels with spandrel beams, increase the slab thickness by about 10 percent. Further, the values are based on 60 ksi steel, they may be decreased by 10 percent for 40 ksi steel and by 5 percent for 50 ksi steel.

The following model for the preliminary design of flat slabs is taken from the earlier edition of the ACI Code (ACI 318-63). This empirical method compares in some ways with the Direct Design Method, which is currently in use. As with the previous design of two-way slabs on beams, it is assumed that the lateral forces are resisted by stiff shear walls, so that the gravity loads control the design of the flat slab/plate.

A two-way slab on rigid beams carries the load to its perimeter beams, which in turn transfer the loads to the columns. In the flat slab/plate construction, where there are no beams, the beam action must be taken over by the slab; visualize hidden slab beams that span from column to column, and the plate to behave similarly to a two-way slab on beams. It is apparent from Figure 5.13a that part of the slab spans directly parallel, while the other portion spans perpendicular, to the imaginary slab beams, which then must span parallel to the columns again. It may be concluded that the total load must be fully carried in each direction of the slab. This indicates that the magnitude of the moments for flat slabs must be in the same range as for one-way slabs.

The entire load of the floor panel wL_2 is carried in one-way action to the hidden slab beams, as indicated in Figure 5.13b, and then the entire load is carried in the transverse direction to the columns. The total static moment, M_o, is

$$M_o = (w_u L_2)l_n^2/8 \qquad \qquad \textbf{(5.25)}$$

where: l_n = length of clear span face to face columns or capitals of span L_1

L_2 = length of span transverse to span L_1

For a typical interior span, assuming fixed beam conditions, 2/3 of the total static moment, M_o, is negative and 1/3 is positive. According to the ACI Code, 65 percent of the total static moment is distributed to the support, and the rest (35 percent) is placed in the field. For the different distribution of an end span, the reader should refer to the code.

In response to the reserve strength and high redundancy of two-way slabs, the ACI Code allows a 10 percent redistribution of negative and positive moments, as long as the total static moment, M_o, is not altered.

In the next step, the slab is subdivided into column strips and middle strips in both the longitudinal and transverse directions, to take into account the variation of stress intensity (Fig. 5.13c). *The column strip has a width on each side of the column centerline equal to one quarter of the smaller of the slab dimensions L_1 and L_2* (see fig. 5.13b). The middle strip is defined by the portion between the column strips. Hence, for a uniform square slab layout, the middle and column strips are all of one-half panel width in both major directions.

According to the ACI Code, the stiffer column strips resist 75 percent of the support moment $0.65M_o$

$$M_1 = 0.75(0.65M_o) = 0.49M_o \qquad (5.26)$$

The more flexible middle strip resists the remaining 25 percent.

$$M_3 = 0.25(0.65M_o) = 0.16M_o \qquad (5.27)$$

Similarly, but with a smaller portion of the total, the column strip resists 60 percent of the positive field moment $0.35M_o$

$$M_2 = 0.60(0.35M_o) = 0.21M_o \qquad (5.28)$$

The remaining 40 percent is carried by the middle strip.

$$M_4 = 0.40(0.35M_o) = 0.14M_o \qquad (5.29)$$

The absolute largest moment or the flexural capacity of the slab, is at the column support. For a square panel, letting the clear span be conservatively equal to $l_n = 0.95L$ as for a flat plate, the total negative moment for the column strip is

$$M_1 = 0.49M_o = 0.49(w_uL)(0.95L)^2/8 = 0.055w_uL^3$$

or, the support moment per foot of width is

$$M_1 = 0.055w_uL^3/(L/2) = 0.111w_uL^2 \simeq w_uL^2/9 \qquad (5.30)$$

Notice that the magnitude of the moment is in the range of one-way slabs as mentioned before. The corresponding field moment is

$$M_4 = 0.14M_o = 0.111w_uL^2(0.14/0.49) = 0.032w_uL^2 \qquad (5.31)$$

Notice that the maximum field moment for the two-way slab on beams is about 20 percent less because of the stiffer boundary conditions.

For a square slab, the same reinforcement as for the longitudinal direction is placed in the transverse direction. The maximum spacing of all bars cannot be larger than two times the slab thickness.

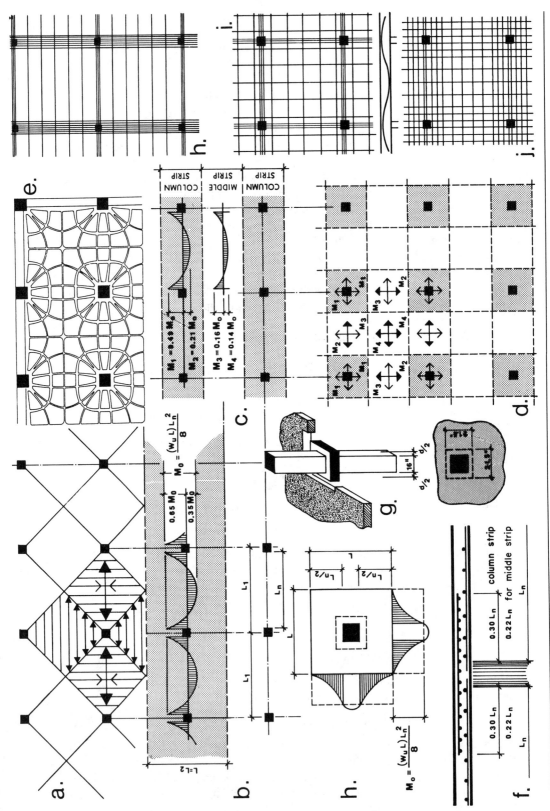

Figure 5.13. *Design of flat plates and post-tensioned slabs.*

One should keep in mind that, for a rectangular flat plate, the long span along the column strip generates the larger moments, in contrast to two-way slabs on beams, where the short span along the middle strip causes the larger moments.

In two-way slabs on beams, the shear is transferred from the beams to the columns; the shear from the slabs to the beams is usually negligible. However, in flat slab/plate construction there are no beams; in this case the columns tend to punch through the slab—this punching shear is very critical. The shear capacity of the slab can be increased by using column capitals, drop panels, or shearhead reinforcement in the slab. The shear strength of the concrete, in two-way action at d/2 distance around the column (Fig. 5.13g), is

$$\phi V_c = \phi(4\sqrt{f_c'})b_o d \tag{5.32}$$

where:

$$\phi = 0.85$$

$$b_o = \text{perimeter of critical section (in.)}$$

$$f_c' = \text{concrete strength (psi)}$$

$$d = \text{effective depth of slab (in.)}$$

The highest stress intensities occur at the columns in shear and bending (i.e., tension over columns).

As capitals and/or drop panels are added to the columns, not only does the shear capacity of the slab increase, but also the effective slab span decreases, resulting in smaller plate moments. On the other hand, the increase in stiffness around the column will attract a slightly higher portion of the total static moment, M_o, thus decreasing the field moment, M_4. In general, it has been found that the effect of the capitals only slightly decreases the negative moment.

Other examples of flat plate construction are found in foundation design. Foundation mats are treated as inverted two-way slabs; the column loads are in balance with the assumed uniform soil pressure. The single column foundation is a two-way slab structure, as are combined footings. The behavior of the single column footing may be visualized as an inverted, isolated flat plate uniformly loaded and supported by a central column. Also, in this instance, the entire load is carried in one-way action in both directions, requiring two identical layers of reinforcement for a square foundation placed perpendicular to each other at the bottom (Fig. 5.13h). The total maximum moment is

$$M_o = w_u L(l_n/2)l_n/4 = (w_u L)l_n^2/8 \tag{5.25}$$

where l_n = cantilever span of footing from column face = $(L - b)/2$

The critical punching shear occurs, as for the flat plate, at d/2 distance from the concrete column faces.

Example 5.9

Flat plate construction on 16 × 16 in. columns is used for a high-rise apartment building that is organized on 18 × 18 ft bays. The lateral forces are resisted by shear walls. The live loads are 40 psf and the superimposed dead load is 20 psf for partitions. Use $f_c' = 3000$ psi and $f_y = 60,000$ psi. Do a preliminary investigation of a typical interior panel.

The slab thickness is estimated as derived from deflection control, and it is assumed that the exterior panel determines the slab thickness of the entire floor.

$$t = 1.1(L/3) = 1.1(18/3) = 6.6 \text{ in.}, \quad \text{use a 7-in. slab.}$$

There is no live load reduction since the panel size is less than 400 ft^2. The ultimate slab load is

$$w_u = 1.4(7(0.150/12) + 0.020) + 1.7(0.040) = 0.22 \text{ ksf}$$

The effective depth, d, to the centroid of the upper layer of the steel, which is conservative, is estimated as

$$d = 7 - 1.5 = 5.5 \text{ in.}$$

The slab thickness must be checked for the punching shear at d/2 distance around the column face (Fig. 5.13g). The total shear is

$$V_u = 0.22(18^2 - [(16 + 5.5)/12]^2) = 70.57 \text{ k}$$

The shear capacity of the concrete against punching failure is

$$\phi V_c = \phi(4\sqrt{f_c'})b_o d$$

$$= 0.85(4\sqrt{3000})4(21.5)5.5/1000 = 88.09 \text{ k} > 70.57 \text{ k}$$

The 7-in. slab can resist the punching shear.
The total static moment for the interior panel is

$$M_o = (w_u L)l_n^2/8 = 0.22(18)(18 - 16/12)^2/8 = 137.5 \text{ ft-k/18 ft}$$

The column strip negative moment carries 49 percent of the total moment.

$$M_1 = 0.49(137.5) = 67.38 \text{ ft-k/9 ft}$$

The column strip positive moment resists 21 percent of M_o.

$$M_2 = 0.21M_o = 0.21(137.5) = 28.88 \text{ ft-k/9 ft}$$

The middle strip negative moment is

$$M_3 = 0.16M_o = 0.16(137.5) = 22 \text{ ft-k/9 ft}$$

The middle strip positive moment is

$$M_4 = 0.14M_o = 0.14(137.5) = 19.25 \text{ ft-k/9 ft}$$

The top reinforcement for the maximum moments over the columns can be determined approximately as

$$A_{s1} = M_u/(0.85f_y d) = 67.38(12)/[0.85(60)5.5] = 2.88 \text{ in.}^2/9 \text{ ft} = 0.32 \text{ in.}^2/\text{ft}$$

The total #4 reinforcement for the column strip negative moment is 2.88/0.20 = 14.4 bars, or s = (0.2/0.32)12 = 7.5 in.

Try #4 @ 7 1/2 in. o.c.

The bottom reinforcement for the column strip in the field is

$$A_{s2} = 28.88(12)/[0.85(60)5.5] = 1.24 \text{ in.}^2/9 \text{ ft} = 0.138 \text{ in.}^2/\text{ft}$$

$$A_{smin} = 0.0018bt = 0.0018(12)7 = 0.151 \text{ in.}^2/\text{ft} > A_{s2}$$

The required #4 bar spacing is:

$$(0.2/0.151)12 = 15.89 \text{ in.} > 2t = 14 \text{ in.}$$

Try #4 @ 14 in. o.c.

See Figure 5.13f for layout.

A similar approach can be used for finding the reinforcement in the middle strip.

Post-Tensioned Slabs

Precast, prestressed floor panels, such as pretensioned double-tees and hollow-core decks, have been used for many decades in high-rise construction. But the in-place unbonded post-tensioning of slabs on a large scale is a more recent development. The economy and advantage lies in the large span and shallow slab section. Since prestressed members are uncracked at service loads, they are much stiffer than nonprestressed reinforced concrete, hence they can be shallower, as reflected by the minimum depth requirements for deflection control. They span larger distances than conventional reinforced concrete slabs, but have the same thickness, thus allowing more flexibility in laying out interior spaces.

For example, prestressed flat plates for apartment buildings may efficiently span from 24 to 30 ft, or prestressed flat slabs for office buildings from 35 to 45 ft. This indicates that the span range for slabs can be increased by about 30 to 40 percent when post-tensioning, rather than ordinary concrete construction, is used. While the typical span-to-depth ratio for nonprestressed one-way and flat slabs or flat plates is in the range of 36, it is 45 for post-tensioned slabs.

For the continuous slab construction, the parabolic tendon profile is normally used. Since slabs are generally designed for uniform loads, the load balancing method offers itself conveniently as a basis for design. It has been shown in the section in Chapter 4 on "Prestressed Concrete Beams" that the prestress forces generate upward loads that balance a portion of the gravity load. For slab design, the balanced load is often taken as equal to the dead load, when the dead load is roughly equal to the live load, so that the dead load deflection is then equal to the camber due to prestressing, thus resulting in no deflection. The bending stresses, as caused by the live loads, are now superimposed upon the uniform axial prestress stresses.

One-way slabs are treated as continuous, one-foot-wide beams, hence the prestress tendons are placed perpendicular to the supports, while the nonprestressed temperature steel runs parallel (Fig. 5.13h, Example 4.3). The supporting beams may also be post-tensioned (see Problem 4.31), particularly when they form wide shallow bands; the depth of these beams can be estimated as $t = L/2.5$ to $L/3$.

Though the concept of two-dimensional load balancing (see fig. 4.6a) is rather similar to the one-directional beam approach, the two-directional tendon arrangement becomes an important criterion. The various tendon layouts for two-way slabs are indicated in Figure 5.13h–j. The pattern that responds best to two-way slabs on rigid beam supports seems to be the prestressed membrane network with an evenly distributed rectangular grid (Fig. 5.13i).

Post-tensioning is most common in flat plate construction either for cast-in-place slabs or lift-slab techniques. The tendon layout may directly respond to the continuous beam concept of flat slab design as shown in Figure 5.13j, where the denser column bands clearly indicate the higher stress concentrations along the column strips. The narrow banded tendon distribution in Figure 5.13h is quite popular for flat slabs; it greatly simplifies detailing and installation. This post-tensioned system creates a pre-stressed one-way slab supported on prestressed slab beams.

The typical span-to-depth (L/t) ratio for prestressed solid slabs is in the range of 45, while for prestressed joist slabs it is about 30. For the design of prestressed, continuous floor slabs, the following rules of thumb may be used, where the span, L, is in feet and the slab thickness, t, in inches.

One-way solid slab:	t = L/3.5 to L/4.2
One-way joist slab:	t = L/2 to L/2.5
Two-way slab on rigid supports:	t = L/3.75 to L/4.5
Flat slab:	t = L/3.75 to L/4
Flat plate:	t = L/3.5 to L/3.75
Waffle slab:	t = L/2.5 to L/3

For preliminary design purposes, the values above yielding the thicker slabs may also be applied to simple span floor structures.

Example 5.10

Do a preliminary design of a post-tensioned flat plate that forms a typical 30 × 30 ft interior bay; it is supported by 16 × 16 in. columns. The slab carries a partition dead load of 20 psf and a reduced live load of 40 psf. Use f'_c = 4000 psi concrete and Grade 270 strands. Base this first investigation on the assumption that 80 percent of the dead load is to be balanced, since the live load is much less than one-half of the total load.

Estimate the slab thickness as t = L/3.5 = 30/3.5 = 8.57 in.; try an 8 1/2-in. slab. Determine the loads.

Slab weight: (150/12)8.5 =	106 psf
Partitions:	20 psf
Dead load:	126 psf
Reduced live load:	40 psf
Total load:	166 psf

The loading for the continuous slab beam of 30 ft strip width is

$$w_D = 0.126(30) = 3.78 \text{ klf}$$

$$w_L = 0.040(30) = 1.20 \text{ klf}$$

$$w_s \qquad\qquad = 4.98 \text{ klf}$$

For a parabolic tendon layout, the approximate cable drape is

$$e = t - 2d' = 8.5 - 2(1.5) = 5.5 \text{ in.}$$

Assume an equivalent prestress load of

$$w_p = 0.8w_D = 0.8(3.78) = 3.02 \text{ klf} \tag{4.44a}$$

Therefore, the required prestress force to balance this load, according to Equation 4.39, is

$$P = w_p L^2/8e = 3.02(30)^2/[8(5.5)/12] = 741.27 \text{ k/30-ft strip}$$

This causes an average precompression stress, considering that 75 percent of this force will be acting on the 15-ft column strip, of

$$f_a = P/A = 0.75(741270)/[8.5(15)12] = 363.37 \text{ psi}$$

This value is within the typical range of 175 to 400 psi for slabs.
The bending due to the net loading is

$$w_{net} = w_s - w_p = 4.98 - 3.02 = 1.96 \text{ klf}$$

$$M_o = w_{net} l_n^2/8 = 1.96(30 - 16/12)^2/8 = 201.34 \text{ ft-k/30 ft}$$

The critical moment in the column strip at the column is

$$-M_s = 0.49 M_o = 0.49(201.34) = 98.65 \text{ ft-k/15 ft}$$

The corresponding flexural stresses are

$$f_b = \pm M/S = 98650(12)/[15(12)8.5^2/6] = 546.16 \text{ psi}$$

The combined compressive stresses are

$$f_c = f_b + f_a = 546.16 + 363.37 = 909.53 \text{ psi} \leqslant 0.45 f_c' = 0.45(4000)$$
$$= 1800 \text{ psi} \tag{4.42}$$

The compressive stresses are low as compared to the allowable ones.
The tensile stresses are also satisfactory, according to

$$f_t = f_b - f_a = 546.16 - 363.37 = 182.79 \text{ psi} \leqslant 6\sqrt{f_c'} = 6\sqrt{4000}$$
$$= 380 \text{ psi} \tag{4.43}$$

Since the tensile stresses are so low, the concrete is not utilized efficiently, and thus the prestress tendons will have to carry more loads.
The required strand area, according to Equation 4.41, is

$$A_s = P/0.56 f_{pu} = 741.27/0.56(270) = 4.90 \text{ in.}^2/30 \text{ft}$$

According to the ACI Code, 75 percent of the moment, or the corresponding reinforcement, is concentrated in the column strip

$$A_s = 0.75(4.90) = 3.68 \text{ in.}^2/15 \text{ ft} = 0.245 \text{ in.}^2/\text{ft}$$

The equivalent tendon spacing in the column strip, for a strand with 0.6 in. diameter (see Table A.5), is

$$s = (12/0.245)0.216 = 10.58 \text{ in.} < 4t = 4(8.5) = 34 \text{ in.}$$

Use Grade 270 unbonded stress-relieved 7-wire strands ASTM A416 with a diameter of 0.6 in. in each direction along the column strips at a spacing of 10 1/2 in. o.c.
In the field strip, 25 percent of the total moment is concentrated, which is equivalent to three times the spacing of the bars in the column strip

$$s = 3(10.5) = 31.5 \text{ in.} < 6t$$

The maximum tendon spacings of 4t and 6t, as used above, are based on the suggestion of the ACI 318-77 commentary. Because of the use of unbonded prestress tendons, the ACI Code requires that a minimum of bonded reinforcement must always be provided on the tensile face in the negative moment areas.

The steel weight per square foot of slab is

$$w_s = A_s\gamma_s = A_s(490/12^2) = A_s(3.4 \text{ lb/in.}^2/\text{ft}) = 3.4(4.9/30)2 = 1.11 \text{ lb/ft}^2$$

This value is high in comparison to the usual values for slabs of below 0.8 lb/ft^2; the result was predictable because of the low concrete stresses. The load-balancing forces should be revised so as to approach the net flexural tension stresses, which would result in smaller prestress forces, as is discussed further at the end of the section in Chapter 4 on "Prestressed Concrete Beams" (see also Problem 4.31). In the final design, the slab must also be checked for the initial design stage and for deflection. It must be checked for flexure, as a cracked section for the condition of ultimate strength, and it must be especially investigated for punching shear.

Stair Slabs

Stairs have often been a vital part of total architecture—they do not serve solely functional purposes. The powerful symbolic meaning of the monumental stairways of the Baroque period, or the daring construction of the spiral shell staircases with thin tiles in Catalonia, or some of the constructive experiments of the free-standing sculptural stairs of the modern movement, clearly prove this point. Some of this formal complexity of stairs is addressed in Figure 5.14.

From a geometrical point of view, there are an endless variety of stairs (Fig. 5.14h, i). They may be straight, curved, spiral (helical), and any combination of these; they may be continuous or broken up, one-directional or multidirectional, they may have platforms or transitional wedge-shaped wider steps.

The most simple stairs are straight, leading from one level to the next, with or without intermediate landings. The straight stairs can form L-shapes with one landing, Z-shapes with two turns, and U-forms, either as parallel or scissor stairs.

The structural action of the stair may be based on the step as a structural element supported by beams (stringers), or individually by hangers, or they may directly cantilever from a wall. Here, the treads act like horizontal, transverse beams with various possible support conditions (Fig. 5.14e, f). The treads may also form a continuous stepped, inclined concrete slab spanning vertically between the landings. Most concrete stairs, including landings, act as folded plates, where the steps are sitting on top of the slab and are considered nonstructural.

The stairs that are briefly investigated in this context are of a purely functional nature and public use, the type that provides the only means of access for low-rise buildings, say up to four floors, and stairs for high-rise buildings that function as escape routes, where access is provided by elevators. The minimum stair width for two people is usually 3 ft, 8 in., with a typical maximum vertical distance for a single flight between landings of 12 ft in high-rise construction. The rise of every step in a stairway should not be less than 4 inches nor greater than 7 inches, and the tread depth (run) should not be less than 11 inches. Exceptions are permitted for private stairways and stairways of other forms. In this context, only cast-in-place U-type or scissor concrete

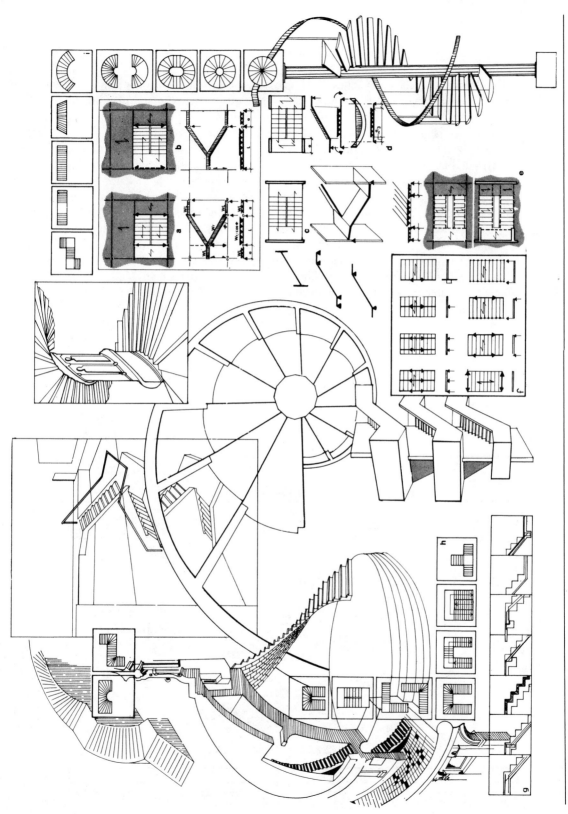

Figure 5.14. *Stairs.*

stairs are further investigated. Precast concrete and steel stairs are usually provided by manufacturers as a component package.

For the typical rectangular stairwells (Fig. 5.15), the inclined stair slabs may span directly from landing beam to landing beam (Fig. 5.15a), where the main landing is part of the floor framing, and the intermediate landing is hung from above or supported on struts from below. The landings above and/or below may be an integral part of the stair slab, such as for the cantilever stair system (Fig. 5.15c) and the hinged folded slab system (Fig. 5.15b). In Figure 5.15d, the inclined slab, together with the landings, form a simply supported folded slab, where the supporting beams are located along the edges of the stairwell. To reduce the span of the stair slab, the landing slab may also span crosswise to the flight (Fig. 5.15e) to form a wide slab beam. Should the U-shaped staircase be exposed, it may represent a free-standing cantilever stair with unsupported landings, where the flights are attached at floor levels only; obviously, this is a rather complex three-dimensional structural problem.

Concrete stairs for the typical conditions in Figure 5.15 are usually designed as simple-span slab beams under gravity loading, conservatively ignoring their additional strength because of the three-dimensional folds. While the live load is already given by codes on the horizontal projection, the dead load, w', along the flight must be transformed into an equivalent load on the horizontal projection, $w'/\cos\theta$, where θ is the angle of inclination of the slab. For these conditions, the folded slab can be treated as an equivalent wide beam on the horizontal projection, carrying the same load as the inclined beam, since the moment action controls the design, and shear and axial forces are usually not critical and can be ignored.

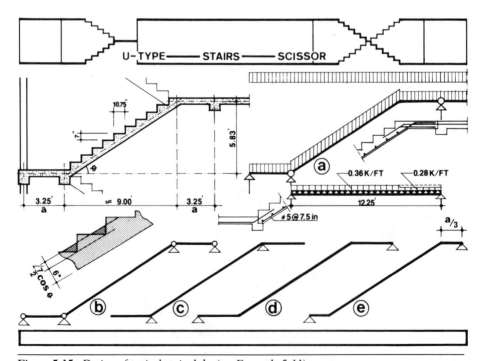

Figure 5.15. *Design of typical stair slabs (see Example 5.11).*

Example 5.11

Design the typical stair bent in Figure 5.15a, for a live load of 100 psf. Use $f'_c = 4000$ psi and $f_y = 40,000$ psi.

The slab thickness is estimated as

$$t = L/2 = 12.25/2 = 6.13 \text{ in.}$$

Since the deflection of this folded slab is obviously less critical than for a horizontal slab, investigate a 6-in. stair slab. The slope of the slab is: $\tan \theta = 5.83/9 = 0.648$ or $\cos \theta = 0.839$.

The average stair weight along the inclined portion is found by treating the treads as an equivalent inclined slab thickness of one-half the riser height multiplied by $\cos\theta$, which is added to the slab thickness

$$[(\cos \theta)7/2 + 6]150/12 = 36.71 + 75 = 111.71 \text{ psf}$$

The load on the inclined stair portion along the horizontal projection is

$$w_u = 1.4(111.71/\cos \theta) + 1.7(100) = 356.40 \text{ psf} = 0.36 \text{ ksf}$$

The load on the landing portion is

$$w_u = 1.4(75) + 1.7(100) = 275 \text{ psf} = 0.28 \text{ ksf}$$

Since the smaller landing load will not much decrease the maximum moment, as caused by a uniform load, the stair load is conservatively assumed to be uniform across the entire beam, for which the moment is quite familiar.

$$M_u = w_u L^2/8 = 0.36(12.25)^2/8 = 6.75 \text{ ft-k/ft}$$

The corresponding flexural reinforcement is

$$A_s = M_u/(0.85f_y d) = 6.75(12)/[0.85(40)5] = 0.477 \text{ in.}^2/\text{ft}$$

The required spacing for #5 bars is:

$$s = (12/0.477)0.31 = 7.80 \text{ in.}$$

Use #5 @ 7 1/2 in. o.c.; for the bending and lapping of the bars around the corners refer to Figure 5.15.

$$A_{smin} = 0.002bt = 0.002(12)6 = 0.144 \text{ in.}^2/\text{ft}$$

$$s = (12/0.144)0.11 = 9.17 \text{ in.}$$

Use #3 @ 9 in. o.c. temperature steel in the transverse direction on top of the main bottom reinforcement.

PROBLEMS

5.1 A 25-story building, 300 ft high, is laterally supported by a channel-shaped core structure or its equivalent (see fig. 5.1b). If an average wind pressure of 28 psf is assumed to act against the long facade, determine the approximate magnitude of the forces each wall will resist. What wall thickness is needed if the allowable shear stress for the 4000 psi core concrete is 210 psi?

5.2   For the various building layouts in Figure 5.1c to i, determine at the base of their first floor level the approximate force distribution to each of the shear walls for a 200-ft high building as caused by 30 psf of wind pressure. Use the plan dimensions of Figure 5.1e for Figure 5.1f to i. Assume the walls to have the same thickness and the same material; consider wind force action for both major directions.

5.3   Treat the open core structure (in fig. 5.1d) of Problem 5.2 as a closed rectangular shaft. Determine the approximate force flow.

5.4   Replace the square core in Example 5.3 with a round tube of the same circumference. Find the required wall thickness, as based on an allowable shear stress of the reinforced concrete of, for instance, $4.75\sqrt{f_c'} = 4.75\sqrt{4000}/1000 \simeq 300$ psi. Neglect the perforations in the wall.

5.5   Compare the simple torsional shear stresses of a closed circular shaft to an open one (i.e., slotted tube), using a ratio of the shaft radius to the wall thickness of $R/t = 15$. Do the same check for the rectangular tube in Example 5.3. Draw your conclusions.

5.6   Determine the approximate maximum stresses for the open core (fig. 5.1d) in Problem 5.2, for the wind action against the long facade, assuming 20-in. thick flange walls and a 15-in. thick web wall. Also, roughly check the maximum simple torsional shear stresses for the upper building portion; state your assumptions.

5.7   Determine the forces that each of the equally thick shear walls in the 200-ft high building of Figure 5.1q must resist. Use a uniform lateral wind pressure against the long facade of 25 psf. Ignore the resistance of the flange walls perpendicular to the force action. Assume flexural stiffness to also control the design of the end walls (for preliminary design purposes).

5.8   For the rigid frame building of Problem 3.10, determine the column that carries the highest shear as caused by translation and rotation. For this approximate approach assume the columns to be all of the same size. What is the magnitude of the force? Then assume the torsion to be only resisted by the perimeter frames parallel to the force action. Find the maximum column shear force for this condition.

5.9   Determine the approximate lateral force distribution, at a given floor level of 12-ft height, to the five planar rigid frames in Figure 5.1k using a lateral pressure of 25 psf. Assume 14 × 14 in. columns, 15-ft perimeter beams of 12 × 18 in. cross section, interior 30-ft girders of 12 × 24 in. cross section, and 4000 psi concrete.

5.10 Treat the structure in Problem 5.9 as a single-story building with a flexible roof diaphragm. Determine the force distribution to the frames.

5.11 Study the lateral force distribution for the cross-wall structure in Example 6.3 and the non-load-bearing shear walls in Example 6.4.

5.12 For the lateral-force resisting envelope structure with variable stiffness in Problem 3.15, Figure 3.13c, and Figure 5.1o, determine the additional seismic forces that must be resisted as caused by the minimum torsional eccentricity of 5 percent of the maximum building dimension of 150 ft at the story level to be investigated. Use a torsional moment of 7650 ft-k and a ratio of rigid frame to braced frame stiffness of $k_x/k_y = 0.35$.

5.13 Determine the lateral force distribution for a high-rise skeleton structure (see fig. 5.1r) that is stabilized along the perimeter by shear walls of equal thickness. Assume a total lateral pressure of $P_y = 100$ k.
- a) Treat the structure as statically determinate by disregarding the flanges.
- b) Consider the torsion to be resisted by all of the walls, but treat it as constant along each of them.
- c) Compare the solutions.

5.14 Consider the structure in Problem 5.13 to be a low-rise building, where the walls behave as shear panels. Determine the force distribution. Compare the results.

5.15 Treat the building in Example 5.4 as a low-rise shear panel structure. Determine the lateral force distribution.

5.16 Treat the lateral loading of Example 5.2 as a seismic load.
- a) Determine the additional forces due to torsion by letting the seismic shear act with an eccentricity of 5 percent of the long building dimension; ignore the torsional capacity of the flanges. Give the total result due to translation and rotation.
- b) Neglect the effect of the core in resisting the torsion. Draw your conclusions.
- c) Ignore the core completely for resisting the lateral loads.

5.17 Show how the rules of thumb for finding the thickness of nonprestressed slabs have been derived.

5.18 Do a preliminary design of the concrete slabs for the apartment areas of the cross-wall building in Example 6.4. Assume a concrete strength of 4000 psi and Grade 40 steel.
- a) Assume solid, single-span prefab concrete slabs of normal weight (150 pcf) to span between the walls. Add 10 psf for the floor finish and ceiling.
- b) Assume a cast-in-place continuous solid slab. Investigate a typical interior span, and add 10 psf, as for case a.

5.19 Design approximately the typical interior 12-ft span for a continuous one-way concrete slab supported by 10-in. wide floor beams. It carries a live load of 80 psf, besides its own weight. Use $f'_c = 4000$ psi and $f_y = 40$ ksi. Select the slab thickness so that no deflection check is required.

5.20 Determine the preliminary size of a continuous 20-ft concrete slab. Because of the very large span, use lightweight concrete (110 pcf) of $f'_c = 4000$ psi and $f_y = 40,000$ psi. For this concrete, the required depth (as based on no deflection check for normal-weight concrete) must be increased by 10 percent. What will be the maximum steel ratio if the slab must support a live load of 80 psf and a ceiling load of 5 psf? Assume the beam stems to be one foot wide.

5.21 Investigate the exterior span of a continuous one-way concrete warehouse slab that rests on 12-in. wide beams that are 12 ft apart; the slab is supported along the facade by spandrel beams. Assume a dead load of 25 psf, in addition to the slab weight, and a live load of 200 psf. Allow an extra 1/2 in. of slab thickness to serve as a wearing surface. Use $f'_c = 4000$ psi and $f_y = 40,000$ psi.

5.22 Investigate the stair in Figure 5.15 by applying the structure concepts of cases b-e in Figure 5.15. Notice that the stair was already designed in Example 5.11 for case a in the figure. The live load is 100 psf. Use $f_c' = 4000$ psi and $f_y = 40,000$ psi.

5.23 The floor structure of an apartment building is organized on a 20 × 20 ft structural grid using two-way slabs on stiff beams. Do an approximate design of a typical interior slab. Use a concrete strength of $f_c' = 4000$ psi and Grade 40 steel. Show how you place the reinforcement. Assume the following loads: floor finish 5 psf, 20 psf for partitions, and 40 psf live loads.

5.24 Design the corner slab of the building in Problem 5.23.

5.25 Redesign the two-way slab in Problem 5.23 as a flat plate, assuming 16 × 16 in. columns.

5.26 Design a typical interior bay of 18 × 22 ft for a two-way concrete slab on beams. Consider the following loads for this office building: 80 psf live load and 20 psf superimposed dead load for floor finish and ceiling. Use $f_c' = 4000$ psi and $f_y = 40,000$ psi.

5.27 Investigate a typical interior bay of 18 × 14 ft of a flat plate structure; assuming 16 × 16 in. columns. The loads for this apartment building are 40 psf live loads and a superimposed partition dead load of 20 psf. Use $f_c' = 3000$ psi and $f_y = 60,000$ psi. Determine the approximate reinforcement above the columns only.

5.28 Investigate a typical interior two-way beam floor framing system on a 48 × 48 ft grid for an office building. The post-tensioned 12-in. wide concrete beams are spaced 12 ft apart in each direction. The two-way slabs on top of the beam grid must support a reduced live load of 50 psf and a partition live load of 20 psf. Use $f_c' = 4000$ psi and Grade 160 prestress bars. Determine the prestress steel, assuming zero deflection under dead load.

5.29 A 30-ft one-way concrete slab for an office building is post-tensioned, and supported by post-tensioned 16 in. wide beams. The slab must support a superimposed dead load of 10 psf for ceiling and lights, as well as 20 psf for partitions; the live load is 60 psf. Use $f_c' = 4000$ psi and Grade 160 prestress bars. Do a preliminary design of the slab.

5.30 Investigate a typical interior post-tensioned two-way slab on stiff beams. The plate is 30 ft square and carries superimposed dead loads of 20 psf for partitions and about 7 psf for misc., and a reduced live load of 50 psf. Use $f_c' = 4000$ psi and Grade 270 unbonded strands. Base your design on the assumption of zero deflection under full dead load. Draw your conclusions about the solution.

5.31 Investigate a typical 24 × 25 ft interior bay floor framing for an office building. The one-way concrete slabs span 12 ft from beam to beam (Problem 5.19). The 25-ft beams, in turn, are supported by the 24-ft girders; hence, each girder supports, at its center span, one set of beams. Assume a live load of 80 psf. Use $f_c' = 4000$ psi and Grade 60 steel.
a) Design the beam, assuming a beam width of 10 in.
b) Design the 10/20 girder. Check whether the depth is adequate.

5.32 Derive a general expression for the equivalent uniform loads for a continuous beam, first with equal single loads at the one-third points, and then at one-quarter points, so that the ACI Code moment coefficients can be used.

5.33 The loads on supporting beams for a two-way rectangular panel may be assumed to be the load within the tributary areas of the panel bounded by the
intersection of 45-degree lines from the corners, with the median line of the
panel parallel to the long side. The bending moments may be determined
approximately by using an equivalent uniform load per linear foot of beam
for each panel supported as follows:

for the short span: $wL_s/3$

for the long span: $(wL_s/3)(3 - m^2)/2$

a) Derive the equivalent uniform load for the short beam, as caused by
uniform loads acting on the two-way slab; do not include the weight of
the beam.

b) Derive the equivalent uniform load for the long beam, as caused by
uniform loads acting on the two-way slab; again, do not include the weight
of the beam.

5.34 Determine whether the size of the interior beam $b/t = 10/15$ of Problem 5.23
is sufficient, and find the reinforcement for a typical interior span beam. Use
Grade 60 steel and a concrete strength of 4000 psi. Assume 10×10 in.
columns, and gravity loading controls the design.

5.35 Determine the approximate size of the beams that are spaced 8 ft on center
and span 36 ft from the building core to the facade frame. The simply supported
beams act compositely with the 4-in. slab of 3000 psi normal-weight concrete.
Assume A36 steel and the following loads: 100 psf live loads, 20 psf partition
loads, and 8 psf ceiling loads. Temporary shores are not used during construction stage. Also, determine the number of 3/4-in. diam studs. How much do
you save in comparison to the noncomposite approach?

6 WALL STRUCTURES

Wall structures are of many various types and may be organized as exterior and interior, as load-bearing or non-load-bearing, and as prefabricated or built in place walls. The interior walls may be non-load-bearing partitions that subdivide the functional space, (i.e., space dividers) or they also may be load-bearing walls that must carry the floor loads and act as shear walls to resist lateral forces, thus these load-bearing walls represent the building structure. The bearing walls may form the exterior perimeter and may be part of the so-called long-wall or cellular building structure systems; they may be solid, perforated, or coupled walls. The structural walls may just be isolated, stand-alone elements, or they may be part of a spatial all-bearing wall building. On the other hand, these exterior walls may only have to provide the protective environmental shield and serve as envelopes or non-load-bearing curtain walls.

Some wall characteristics are identified in Figure 6.1. Typical wall arrangements (Fig. 6.1a) and wall configurations (Fig. 6.1g), as related to high-rise buildings, are shown. The load action (Fig. 6.1b) determines how the wall behaves, and is treated in this context as a predominant action, in comparison to other secondary forces.

- Under concentric gravity loads, it acts as a *wall* in compression, where buckling and the corresponding boundary conditions (Fig. 6.1e) must be considered! The local moments induced due to an eccentric bearing of floors or due to continuity can be ignored for preliminary design purposes.
- Under lateral loads perpendicular to its surface, it bends about its weak axis and acts like a *slab*. This condition is typical for curtain walls that must resist wind and seismic forces, or for elevator shaft walls, which must withstand cyclic air pressures due to moving high-speed elevator cabs, and is typical for cantilever retaining and basement walls that must resist lateral earth and/or water pressure.
- Under lateral load action parallel to its surface, the wall bends about its strong axis and behaves like a *deep cantilever beam* (e.g., non-load-bearing shear wall).
- When the wall is supported by columns or single (rather than continuous) footings, it acts as a *wall beam*.
- For the general condition, where the wall resists the combined action of gravity and lateral loading, it is called a *plate* (e.g., load-bearing shear wall).

As the height of a plain wall increases, it may become necessary to stiffen it with piers or buttresses (i.e., stiffeners), or by using spatial geometries such as cellular,

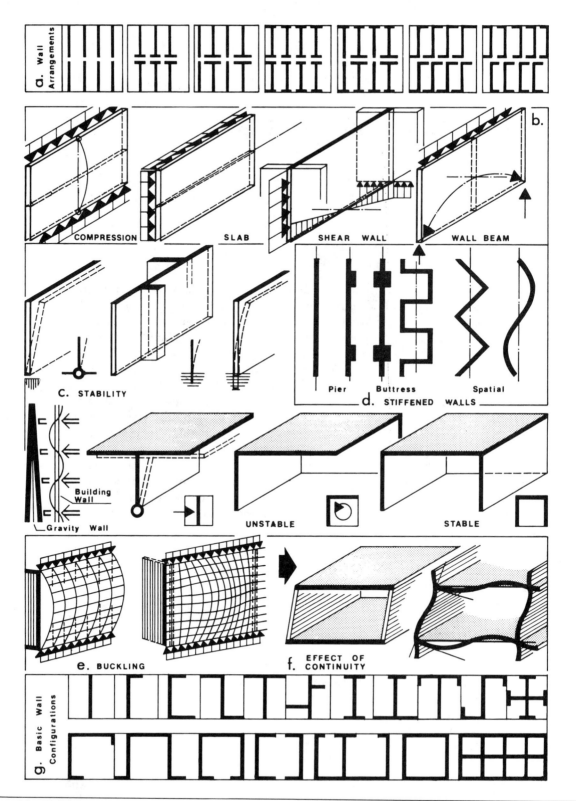

a. Wall Arrangements

COMPRESSION   SLAB   SHEAR WALL   WALL BEAM   **b.**

c. STABILITY

Pier   Buttress   Spatial

d. STIFFENED WALLS

Building Wall

Gravity Wall

UNSTABLE   STABLE

e. BUCKLING   **f. EFFECT OF CONTINUITY**

g. Basic Wall Configurations

Figure 6.1. *Wall behavior.*

corrugated, or other undulating forms (Fig. 6.1d). A free-standing flat wall (Fig. 6.1c) can be stabilized by fixing it to the ground so it behaves like a vertical cantilever slab, otherwise (and possibly more effectively) it can be stabilized and its capacity considerably increased by utilizing other intersecting structure elements (Fig. 6.1g) to form a wall assembly, as is the case for a cellular building, which does not consist of individual walls, but rather a spatial continuum. It has been shown (in fig. 5.1a) that, for hinged boundary conditions (Fig. 6.1f), a minimum of three properly arranged walls are necessary for a system to be stable (Fig. 6.1c). Naturally, when the wall-floor assembly is continuous (Fig. 6.1f), stability is provided.

The exterior walls are most challenging to the designer, since not only must they provide functional integrity, but at the same time they must represent an architectural expression. The term *facade wall* forms a contrast between this and the other exterior walls of a building, which may not give the primary identity to the building, but just provide protection; although an isolated glass-wall building revealing its interior may not be perceived as having facade walls.

The term *wall* reminds one of solidity, as expressed in nature by a rock formation; it exposes weight and strength, as developed over time from materials like stone, masonry, and concrete; it demonstrates a continuous surface character. In this surface, holes, which seem like the eyes of the building organism, are punched into the wall to place the windows. There can only be a maximum of fenestration such as does not destroy the nature of the wall as a surface structure. As the wall proportions become more slender, its behavior approaches that of a column. The spirit of the wall as an integral part of the building structure is contrasted by the envelope that forms an attached skin, and which typically consists of vision and spandrel panels.

The evolution of exterior wall construction is closely related to the development of glass. While the Syrians seem to have invented the art of glassmaking and introduced it to Egypt in about 1500 B.C., the Romans were probably the first to have used it as window glass. Employed in this manner since the first century, they were still not able to produce clear glass, which was finally accomplished in the Middle Ages. From Rome, glassmaking spread to northern Europe, where it reached perfection as a form of art and integral part of architecture in the stained-glass windows of the Gothic cathedrals. However, it was not until the late seventeenth century that plate glass was introduced in France and later produced in England as a cheap material to be used for the homes of the ordinary people and to replace wooden shutters, oiled paper, or muslin. Finally, Joseph Paxton's Crystal Palace of 1851 is a celebration of glass as a primary building material and form determinant, initiating an explosion of glass-roofed structures. At the same period, frame construction started to replace the conventional masonry bearing wall as the major support system. In 1891, the 16-story Monadnock Building in Chicago clearly demonstrated the limit of the traditional bearing wall structure, with more than 6 ft of wall thickness at its base (see fig. 1.1e).

The contribution of glass as an integral part of high-rise architecture, together with the development of the skeleton, occurred at the beginning of this century with the modern movement, possibly at the time when Gropius, Le Corbusier, and Mies van der Rohe worked for Peter Behrens in Berlin, although the separation between skin and building structure had already occurred before this time. Glass was no longer treated as a window in an opening protected by the wall, but as a solid and primary structure in its own right. Glass, together with the new architectural materials of steel and concrete, as well as daring structural engineering inventions, caused the solid wall to be dematerialized and to be replaced by the frame and glass curtain in the search for lightness and airiness (i.e., less is more), thereby exposing the inside of the buildings as for the Fagus Factory [1911] by Gropius and Meyer, in Alfeld, GDR. It led to the resolution of the solidity of the wall with a long horizontal ribbon window, and to the separation of the wall from the ground by lifting the building upon columns and treating it as a spatial object, thereby causing the facade to disappear, as so masterfully expressed in the Villa Savoye [1930], by Le Corbusier and Pierre Jeanneret, in Poissy, France.

The articulation of the glass wall in high-rise construction reached a level of maturity in the late 1950s with the pure architecture of glass and steel by Mies van der Rohe, as represented by the Seagram Building, [1958] in New York (see fig. 1.2). After this final perfection of the glass-metal curtain wall, the glass began to displace the supporting mullion frame in the search for making the curtain disappear. The lightness of the transparent clear glass does not seem to require any structural support, so that the mullion framing only represents a geometrical grid. This search for airiness of the skin may just be a result of the endeavor to minimize the weight and volume of the entire building. Some of the mirror buildings of the 1970s use monolithic all-glass facades with hidden mullion framing, thereby expressing the seemingly weightless nature of the skin, which is simply wrapped around the building.

Certain features of the bearing wall were rediscovered in the 1960s by the plastic and sculptural qualities of reinforced concrete, particularly in precast concrete construction. In this context, for example, small window units were placed to form a dense beam-column facade grid, having the appearance of a perforated wall.

Today, the post-modernists have brought back a language largely neglected by the modern movement. They are inspired by the past and tradition, by symbolism, imagery, metaphors, analogies, contextural reference, allusions, ornamentation, and other personal values. The rigid confinement of the pure boxy building has been freed and has resulted in compound hybrid free forms (see fig. 1.4). Designers have introduced the facade again; they experiment with its surface, and may richly ornament it with a variety of colorful materials, such as stone, tile, metal, and glass. They may reconnect with the past by giving it illusion, with their thin-walled construction, of the solid massive wall with punched openings, as for instance, in a Renaissance load-bearing wall that bears a decoration derived from the classical columnar order. The thin stone cladding of the 1980s has replaced the glass skins of the 1970s, although designers may still experiment with minimal geometry, including skins, to express formal purism.

The curtain wall, however, may not just represent facade architecture, with applied ornamentation or represent minimal skins. The appearance of the building may reveal an interplay of skin, structure, and ornament as an integral interaction, or it may express an exchange of the outside with the interior organism of the building. Naturally, there is no limit to expressing the exterior of a building, as can only be suggested here (see fig. 1.9).

CURTAIN WALLS

Exterior walls may function merely as envelopes or skins to protect the interior environment from the outside. They may be thin and nearly weightless, like a membrane,

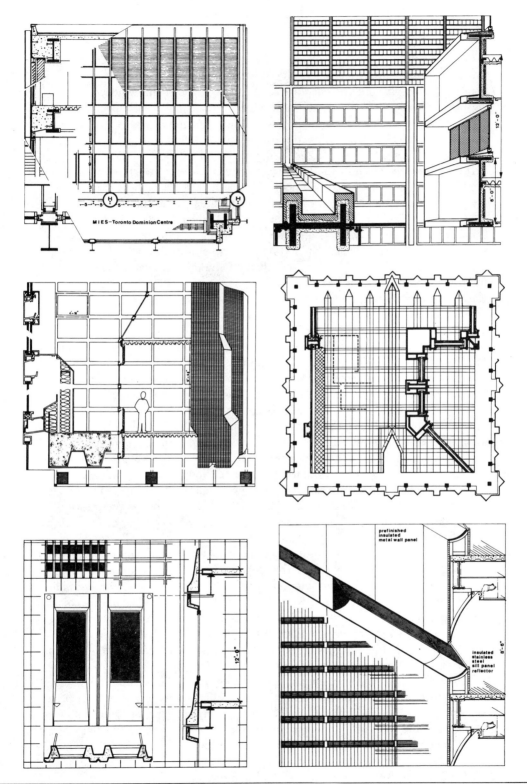

Figure 6.2. *Some typical curtain walls.*

and hang like a curtain from the building structure. Quite in contrast are the heavy load-bearing walls that must also support the floor loads and lateral loads, and must carry these forces directly down to their own foundations. Although a masonry or concrete wall with punched openings for the placement of the window units may appear as a bearing wall, it may still function only as a curtain that supports the wind loads like a vertical slab. The curtain must be thick enough to serve its functional and structural requirements, as discussed in the following sections.

William Jenney's ten-story Home Insurance Building [1885] in Chicago (see fig. 1.1b) is considered not only the first high-rise metal skeleton structure, but it is also the first to employ curtain walls, although during the first few decades of this century the stone wall continued to be used to hide the building frame, and often to articulate the sculptural quality of load bearing. It was not until the late 1940s that the structural cladding of massive brick and stone construction began to be replaced by the non-structural curtain wall, where skin and building structure are separated, reviving the early experiments of the Chicago School. Pietro Belluschi's Equitable Building [1948] in Portland, Oregon, must be considered one of the pioneer cases where the frame, clad in sheet aluminum, is exposed and clearly supports the infill aluminum spandrels and tinted glass. Slightly later, the building frame of the Lever House [1952] in New York, by SOM, is moved back from the curtain wall and concealed as much as possible with glass spandrels between window bands using only slender continuous mullions. Finally, Mies van der Rohe brought the curtain wall to maturity by reinforcing the structural articulation and modular organization of the facade with projecting mullions and spandrel cladding, which recreated the image of the supporting structure (Fig. 6.2, top left). Quite in contrast are the glass membranes of the late 1960s and the mirror walls of the 1970s. These curtains are purely skin: they may be nondirectional, antigravitational, and antistructural—the glass seems to float and dematerialize the wall. The slender, minimal projection mullion framing seems to be non-loadbearing and reflects solely a geometrical grid which has no relationship to the building organism (Fig. 6.2, second row). In the extreme, the curtain framing is hidden, so that a glass wall becomes a pure seamless taut membrane that appears as though prestressed from the inside, similar to a pneumatic skin.

Facade architecture has lately become much more decorative. The designer not only experiments with minimal austere geometrics of glass facades, or with functional concerns, such as the energy-conserving metal-clad panel wall with only 20 percent glass area (Fig. 6.2, bottom right), but also has rediscovered the qualities of solidity, weight, and strength of the bearing wall, as expressed in stone cladding or in decorative tile veneer. The trend toward thin stone cladding, especially granite-clad buildings, introduces new design approaches different from the minimal glass skins of the 1970s. Also, the sculptural possibilities of precast concrete are taken advantage of by some architects to create a wide range of complex geometric shapes.

Rather than separating structure and enclosure by clothing or skinning the building, the structure may be revealed by simply cladding the columns and spandrel beams and using infill panels within the frame openings. This can apply to frames with wide column spacing and deep long-span girders (Fig. 6.2, top right), or to typical rigid frame construction (Fig. 6.4, left), or it can apply to the closely spaced columns of the tubular wall (Fig. 6.4, right). Similarly, the precast concrete window wall panels (Fig. 6.2, bottom left) clearly express the structure of the frame facade.

Facade types may be organized according to the exterior *material* used, which

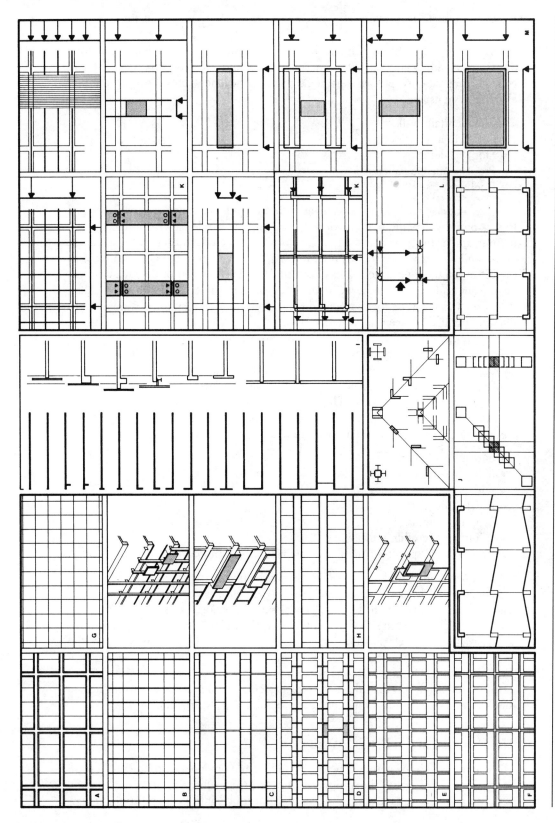

Figure 6.3. *Construction of curtain walls.*

includes metal (e.g., aluminum, carbon steel, stainless steel, bronze), glass, stone (limestone, marble, granite, slate, sandstone, travertine, etc.), ceramic tile, clay and concrete unit masonry, precast concrete, stucco, plastics and so on. The most common cladding systems are:

- Metal curtain walls
- Glass curtain walls
- Stone cladding
- Precast concrete panels
- Brick veneers

A first understanding of the organization of building enclosures can be developed from a purely geometric point of view by studying the facade patterns formed by the framing and/or the glass and solid panels, as well as the relationship of the envelope to the building structure. In Figure 6.3i and j, the location of the curtain wall with respect to the building structure has been studied. The position of the envelope can be in front, as is typical for facade architecture, it can be integrated with the building structure, and it can be located behind the exposed structure by using window walls that are installed between the floors.

The envelope is defined by the solid-to-transparent ratio. It may have punched openings, as for masonry walls, which may be arranged in various patterns, or the envelope may consist of continuous rows of horizontal or vertical fenestration. Naturally, in the all-glass wall, one may not be able to distinguish between the vision and insulated spandrel glazing. The exact opposite approach to the all-glass wall is column and spandrel cladding, thereby clearly identifying the building structure. From a *geometric* point of view, curtain walls may be organized as

- punched hole walls
- framed walls
- horizontal band facades
- panel walls
- continuous skins
- column and spandrel cladding
- hybrids (e.g., spandrel panels, window panels, column covers)

In framed walls, the fenestration is clearly defined by the exposed mullions. In horizontal band facades, the spandrel bands (narrow or wide, load-bearing or non-load-bearing) alternate with the glass bands, such as banded skins of stone or metal with mirror glass, thereby clearly identifying the location of the floor structure. In panel facades of discrete small or large story-high panel units, the subdivision of the stories is expressed, which is not, however, true for the continuous cladding of a gridded membrane or a seamless skin.

In the following sections, the nature of the curtain wall is briefly investigated in more detail as related to its function, construction, and its structure.

The Function of the Exterior Wall

The exterior wall is the barrier between the indoor and outdoor environments. It must be weatherproofed against the infiltration of water, snow, dust, air, and vapor, as well as sound and heat transmission; condensation must be controlled, and sunlight penetration must be regulated with the fenestration design. For reinforced concrete members, the width of hairline cracks must be limited.

Control of Air Leakage

Air infiltration due to wind induces additional thermal loads on the HVAC system. The lack of airtightness also allows water vapor and noise to penetrate into the building, as discussed further under the other functional criteria. The building envelope must form an airtight surface and barrier against air infiltration and exfiltration.

Control of Water Leakage

Wall cladding can be made watertight by fully sealing it using gunnable or solid preformed (e.g., gaskets, cellular tapes) elastomeric sealant materials. But in large-scale buildings, it is difficult to inspect the performance of the sealants and to repair them. The continuous movement of the facade elements, together with the effects of weathering, will wear the sealant down over time. High wind forces will drive the rain through the tiniest cracks in any direction, a state that is further compounded by the pressure differential developed between the inside and outside during this time. In addition, the vertical gravity flow of water along the outside surface may enter through these cracks, and surface tension and capillary action may also pull water through the openings. It is apparent that water penetration will reduce thermal resistance, cause freeze-thaw damage, and corrode the structural materials, in addition to other effects. Rather than sealing the outer wall, a double wall or *rainscreen wall* has been developed that is based on pressure equalization in the void between the outer and inner wall by allowing the water to penetrate through openings in the front wall, and is then drained with internal gutters. In this case, the seal is provided at the inside or interior face of the wall.

Condensation Control

Generally, condensation is possible in climates requiring winter heating. While protection against direct moisture penetration from the outside is provided by a water barrier, protection against condensation of water vapor produced on the inside may be a more serious problem. Indoor moisture is introduced by people, kitchens, humidifiers, and so on. However, the amount of water vapor that air can contain decreases with the air temperature, or, as the temperature decreases, the relative humidity rises until it reaches the saturation or dew-point temperature. Therefore, for any temperature and pressure, there is a maximum amount of water vapor that the air can hold; the temperature at which the vapor condenses is called the dew point. This condensation may occur in winter on cold interior wall surfaces and windows, for instance on the inside face of glass in bathrooms or on the interior surfaces of metal members that are highly conductive and form thermal bridges. It may also occur somewhere within the wall, since the warm air with high humidity flows from the inside to the cool outside with low humidity, causing the water vapor to be cooled to the point where it condenses to water at a certain location in the wall. It may be concluded that, when warm air cools to its dew point as it comes in contact with a cold surface or air, the vapor that it contains turns into water (e.g., 30 percent relative humidity at 68°F, but 100 percent relative humidity at 35°F).

The control of condensation is essential in order to prevent swelling of the materials and the corresponding stresses or cracks, to prevent deterioration due to freeze and thaw cycles, staining, and corrosion, the loss of thermal resistance due to disintegration of the insulation caused by trapped moisture, and other damage to the wall. Condensation can be controlled by keeping the indoor relative humidity within certain limits and by providing:

- an impervious vapor barrier such as plastic films, metal sheets, coated paper, etc., on the warm inside face, so that the cold surface cannot be reached;
- vented space on the cold side to relieve air and vapor pressure, possibly coupled with a vapor barrier on the warm side;
- a thick enough insulation that is air-permeable but water repellent;
- a thermal break in the mullions so that the surface temperature stays above the dew point temperature of the air with which they are in contact.

Fire Resistance

The exterior wall must have a certain hourly fire-resistance rating, as specified by the respective building code jurisdiction. Thin metal curtain walls may require independent back-up walls of noncombustible or fireproof material; gaps between the exterior wall and slab must be closed so that a flue cannot be created. The fire protection of exterior exposed steel members is discussed elsewhere in this book (see fig. 1.22).

Noise Control

Sound absorption and sound insulation are the basic means for controlling airborne and impact sound. Resistance to *airborne* sound transmission may be achieved through reflection by using hard, dense materials, such as provided by the mass of the nonglazed wall area, or through insulation by using soft, porous materials or multiple shell wall systems, where the shells are not rigidly connected and where each one has a different stiffness. Thin glass is a poor acoustic insulator, hence resistance to sound is achieved by reducing the window area, and by providing airtight joints. Noises from the outside may be isolated by absorbing the sound with materials having porous surfaces rather than reflecting the sound waves back into space. *Impact* noise is reduced by interrupting the sound path with insulation layers and/or air spaces.

Thermal Resistance

The building envelope constantly exchanges heat, which flows from the warm to the cool side; the flow varies in direction daily and from season to season. There are many variables that influence thermal transmission through the wall, including the type of wall construction, air infiltration through cracks, material properties, thermal mass with its delay effect, thermal bridging within the wall, wall orientation, air temperature, direct solar radiation, shading conditions, direction and speed of wind, type and area of glass, nature of occupancy, and so on. The surface properties of color and texture, as well as exposure conditions, determine if solar radiation is *absorbed* or *reflected*. For example, a metallic coating may act as a reflector, whereas a black surface captures solar heat.

Rather than using a dynamic model of heat flow analysis, which includes the complex interaction of the variables above and the effect of time, the heat transfer is usually simply estimated from static considerations of the steady-state heat flow concept, which is independent of time and obviously only rarely exists. In this case, heat is assumed to be transmitted through the wall solely in the three basic modes of *conduction*, *convection*, and *radiation*, assuming that there is an air space within the wall (which, by the way, is a good insulator), since otherwise the heat passes through a solid wall just by conduction because the materials are all in contact with each other. The composite total transmission of heat from one surface to the other is determined by the overall heat transmission coefficient or U-value. The resistance of a material to heat flow, R, is the reciprocal of the heat transmission; the thermal resistance $R = 1/U$ represents the degree of resistance to heat flow. The estimation of heat loss and heat gain (Btu/hr, or watts) depends on the surface area, the transmission coefficient, and the temperature differential between the outside and inside.

In lightweight construction, such as cladding, the *resistance insulation* is the primary heat controlling agent, while in heavy construction, the *capacity insulation* becomes an essential consideration. Ignoring the thermal storage capacity of massive wall systems, such as masonry and concrete, as an effective control of the indoor temperature in the single-factor static model of thermal performance, is quite unrealistic. In this instance, the mass absorbs and stores heat and, because of the thermal time lag, delays its migration. This thermal mass effect is well known from adobe architecture. The complex dynamic model of thermodynamic behavior not only includes the thermal mass, but also surface roughness and reflectance, besides the other variables mentioned before.

Heat flow through opaque curtain walls is controlled by sprayed-on or board-type insulation, which should be placed as near as possible to the outside face. It may be applied directly to the back of the panel by mechanical fixing or by adhesive, or it may be an independent back-up barrier for fire-rating purposes. The insulation levels vary with the wall orientation and temperature difference between outside and inside. The building envelope should efficiently absorb the winter sun while controlling heat loss, and resist heat gain in summer; thermal bridges must be avoided when detailing connections.

The crucial element in the wall is the window area, with its high heat transmission, as expressed by the *solid-to-transparent ratio*. Transparent materials reflect, absorb, and transmit solar radiation. Glass walls may cause a *greenhouse effect* by admitting the short-wave radiation of the heat from the sun, but then act more as opaque surfaces with respect to the reradiated longer wave lengths from the inside, so that the heat builds up, thus requiring larger cooling loads. Although fenestration effectively serves as a source of light and, in winter for brief periods, as a source of heat gain from the sun, the critical considerations are, indeed, its *heat gain in summer* and its *heat loss in winter*. It is apparent that a larger glass area allows for less electric lighting because of the increased natural daylighting, but the greater thermal skin loads must also be offset by heating and cooling systems. Hence, it should be quite clear that thermal and daylighting analyses are closely linked with each other and cannot be treated separately.

Special types of glass have been developed for curtain wall construction to optimize light intake while controlling heat flow. In addition, an effective control of thermal transmission through transparent glass panels in summer can be provided by exterior and interior shading devices such as overhangs, fins, and canopies; for smaller buildings, an independent screen wall can be attached in front. The north side of high-rise buildings may have the highest percentage of glazing to permit maximum day-lighting (though heat loss must be limited), while the other sides have the least to minimize the solar heat gain on the hot exposure sides, and thus lower the cooling load; it must be kept in mind that the high-rise buildings are heat dominated. For example, the north side of a tall building in a temperate to cold climate may have 75 percent glazing, while the other sides only have 50 percent. Should the windows be evenly distributed across the building faces, the south side efficiently lets the winter sun enter the building, thus possibly reducing the heating load along the perimeter zone, while overhangs or horizontal baffles during summer cast the necessary shadow to reduce the cooling load; along the east and west sides vertical baffles cast shadows during summer.

Control of Light

Ideally, glass should control the amount of light and solar heat admitted to the interior of buildings, as well as the indoor heat transmitted to the outside, as has just been discussed. Ultraviolet wavelengths of sunlight should be kept away from the interior, and the sunlight should not cause any glare. Since thin glass is not a good thermal insulator, double glazing was developed in the 1940s. Today, insulating glass is double and triple glazed with a dry air space between the glass panes to effectively control heat gain in summer and reduce heat loss in the winter, remembering that high-rise buildings are heat dominated and require high cooling loads. However, as the heat transmission is reduced, visible light may also be reflected, which (in turn) increases the building's lighting (electrical) demand and thus the cooling load. Therefore, the relationship between daylighting, heating, and lighting must be optimized. The de-velopment of a glass that minimizes heat loss in winter while allowing natural light and solar heat to enter is currently challenging the glass industry.

Glass may be organized according to its strength, its insulating quality, and its use as vision or nonvision glazing, besides formal and aesthetic considerations. From a visual point of view, a variety of glass colors and surface patterns on opaque ceramic-coated glass are available—glass may even look like stone. Any glass can be bent into cylindrical curves, double or triple reverse curves, possibly together with tangential flat parts, and it can be bent into irregular curves. For the discussion of glass types, from a material strength point of view, refer to the sub-section in this chapter on "The Exterior Wall as Structure."

Insulating glass consists of two or three glass panes hermetically bonded (sealed) at the factory with an air space between these panes. Any combination of glass type and thickness are possible. The monolithic, laminated, and insulating glass types are employed as vision and nonvision (e.g., spandrel) glass, and for other applications, such as sloped glazing. The vision glass types differ in the amount of light and heat transmitted. From an energy performance point of view, the various vision glass types are organized as clear, tinted, or coated.

Clear glass is a poor insulator, it lets in a lot of light, but also heat. It is usually used for residential applications, where interior shading devices such as draperies, roller shades, and venetian blinds control the heat flow. Although clear double-insulating glazing has about 80 percent transmission of visible light, it also has as much as 65 percent transmission of the invisible heat in the form of shortwave infrared radiation.

Tinted or heat-absorbing glazing was developed in the 1960s. It reduces the direct heat transmission significantly by absorbing heat (i.e., similar to a radiator). This is especially true for dark glass, but it thereby also heats the inside of the building, if not double glazed, which may be critical for warm climates. While the dark gray- and bronze-tinted glasses often have a higher transmission of heat than light, the opposite is true for blue-tinted glasses.

Reflective glazing (coated glass) became available in the 1970s. Its mirror-like coating bounces back some of the sun's rays and heat. It is more effective than tinted glass, since its surface reflects rather than absorbs the heat. Although it transmits less heat than gray- or bronze-tinted glass, it also reflects much of the visible light. Coated glass, especially the type resembling mirrors, may reflect heat onto neighboring buildings and cause a heat load problem there. Further, the glare reflecting off buildings constitutes a potential traffic hazard.

Selective, nonreflective coated glazing, developed in the 1980s, allows more visible light to be transmitted. The coatings can be applied to clear or tinted glass, and hence either concentrate on light transmission or insulation. Low-emissivity glass consists of two panes of glass, where the inner surface of one of the two panes is coated, thereby allowing the transmission of much more natural light and providing, at the same time, an effective thermal insulation.

Responsive glazing will be one of the "intelligent" products of the future. It will have a sophisticated switchable coating that physically changes as it responds to light, heat, and possibly to electricity. In this context, the coating may lighten, darken, change the color in response to light, as has been used in sunglasses, but (in addition) will also control the heat flow. The coating may be connected to low-voltage electrical distribution systems and be controlled with the building's thermostat, which will automatically be switched on and off as it adjusts to the heat requirements—the skin will have begun to become an integral part of the intelligent building.

Installation Methods

From an erection point of view, the organization of curtain walls depends on the degree of prefabrication, or the number of parts that must be assembled, ranging from frame to panel construction. In other words, the cladding may be installed piece by piece or as prefabricated units. In contrast to the curtain wall approach is the direct cladding of the structure, together with fill-in glazing units.

In conventional construction, concrete block or stud walls are used as back-up and are built in place; they fill the perimeter frame and also stiffen it. Against this back- up, brick, stone, or other veneers like tile, metal, or plastics are set. Veneers form a nonstructural facing: they can consist of various materials and are used as an exterior surface for protection, insulation, and appearance. They are adhered or anchored to the structural or nonstructural wall system. In *adhered veneers* (e.g., ceramic tiles, natural stone), plates are directly affixed to the backing with mortar or an adhesive such as silicone. In *anchored veneers*, metal grids are anchored to the structure and thin stone slabs or other panel types are inserted. Stones may also be attached to the back-up using clip angles (continuous or intermittent depending on loading conditions), kerfs, and pins. Thicker slabs, such as brick veneers or precast concrete panels, are supported on shelf angles at floor lines and anchored to the back-up (see fig. 3.16).

The prefabricated curtain wall systems for high-rise construction are custom designed in contrast to the industrial and commercial panels often used for low-rise buildings. The classification of curtain walls according to their method of installation is as follows:

Framing Systems

These systems use exposed or hidden mullion framing to support enclosure panels. The layout of the mullions (e.g., steel, aluminum, glass) may be a function of the structure and/or partitions, or it may reflect any desired grid pattern. Aluminum is quite popular because it is easily extruded and shaped in fabrication—a wide variety of mullion types is available. In the all-glass walls, the panels may be structurally adhered to hidden interior mullions with silicone. This approach is contrasted by the exposed vertical steel mullions on a 5-ft module, and the steel spandrel plates that enclose the aluminum glazing frame with the bronze-gray tinted glass (see fig. 6.2, top left). The steel cover plates for the spandrel beams and columns are anchored to the lightweight concrete, which was required as fire protection for the steel frame; the steel curtain wall is painted black. The regular rectangular aluminum framing in Figure 6.2, middle left, supports blue-green reflective glass; it seems not to have any relationship to the building structure from an appearance point of view, although each office floor, for instance, is composed of four bands (two clear and two insulated). Similarly, for the case of Figure 6.2, middle right, where the aluminum mullion framing supports the clear, reflective double-glazed vision glass (60 percent) and the reflective spandrel units with fiberglass backing. The following three framing systems reflect the degree of field assembly:

- *Stick or grid system* (Fig. 6.3b): This system is constructed in steel or aluminum. The vertical mullion members are attached to the frame, and together with the horizontal rail members form the supporting framework. They are assembled piece by piece in place to support the vision and spandrel units (e.g., glass, metal, or stone panels). This system is usually used for low-rise buildings.
- *Unit system*: In this case, the components of the stick system are prefabricated in the shop and installed as large-framed units—they may form single-story or multistory panels.
- *Unit-and-mullion system*: This system consists of a combination of the systems above, where the mullions are installed, but the panels are preassembled and placed as a framed unit between the mullions.

Column and Spandrel Cover Systems

As shown in Figure 6.3a and c, these express the layout of the building columns and spandrel beams by covering them with a cladding material that is directly attached to those members. This system therefore consists of column cover sections, spandrel units, and the infill glazing units. For the case in Figure 6.2, top right, the frame is steel-clad with the exception of the girder webs, and encloses horizontal bands of mullionless tinted glazing. Similarly, for the frames in Figure 6.4, where the rigid frame (left case) is cladded with aluminum that is coated white. In this instance, the

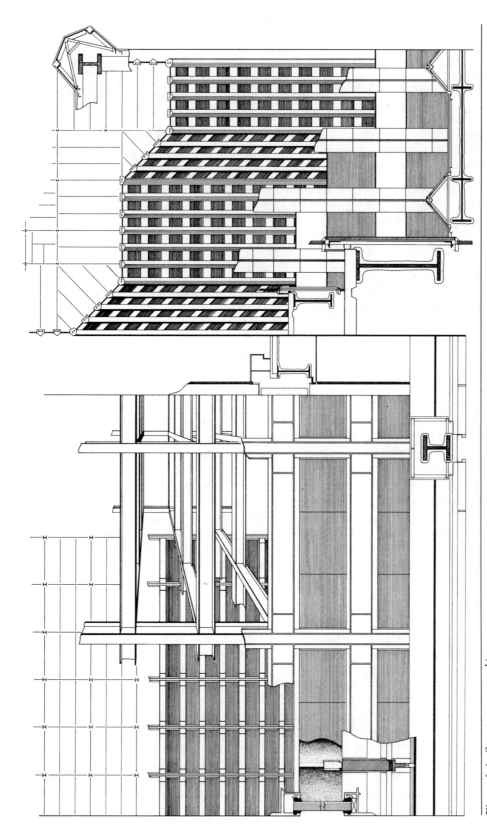

Figure 6.4. *Structure versus skin.*

clear glass spans vertically between the spandrel beams; there are no mullions, but the partitions must help to support the glass weight.

Precast Concrete Cladding Units

These units may also be used as formwork for cast-in-place concrete, or they may form load-bearing units. The typical precast concrete cladding units are usually of the larger, framed panel types.

- *Window wall panels* have a frame-like appearance similar to the case in Figure 6.2, bottom left. They may form large bay-size panels or small story-high panels.
- *Spandrel-unit-panel*
- *Mullion wall unit* (see Fig. 6.3d)

Panel Systems

These form integral units of a more composite surface action, in contrast to the framed unit systems. They are preassembled units formed from sheet metal, or formed as castings (e.g., aluminum, concrete, plastics) with a minimum of internal joints. They may be of full-story height and span from floor to floor as large (see fig. 6.3f) or small (see fig. 6.3e) panels, or they may form deep spandrel panels, possibly combined with column covers spanning from column to column, but here may require additional gravity supports at the spandrel beams, when the column spacing is too large. They may have openings for glazing units or may not. They can be of heavy-weight or lightweight construction, and they can be load-bearing or non-load-bearing. The organization of the panel systems depends on the material, weight, size, span, and structure. The preformed panels are usually of a multilayered construction, with an exterior facing backed by structure and insulation (thermal, acoustical, and fire-resistant), and possibly an interior finish. In this case, the structure may be integrated, as for plastic and metal sandwich panels, or it may form a separate component, such as steel diaphragm panels (e.g., stud walls with steel deck or lath), a steel truss, a prestressed stone panel or a reinforced concrete plate, to which for instance, stone pieces are fastened, using dry or wet methods. Most lightweight back-up systems encompass spandrel panels that form separate units but are integrated with window glazing components and column covers. Typical examples of panel construction are:

1. Precast concrete wall, possibly with window openings or spandrel panels: they can be flat, ribbed, or of any other sculptural configuration. The concrete can be exposed or may have a veneer facing. For example, a concrete-backed stone panel may have a 1 1/4-in. marble veneer directly affixed to the concrete panel. In contrast to heavy-weight concrete panels, precast glass fiber reinforced concrete cladding is thin and of lightweight construction.
2. Veneers attached to steel framework or trusses. For example, granite slabs (approximately 2 in. thick) fastened directly to the steel, or ceramic tile or stucco (e.g., cement plaster, cementitious coatings) first placed over metal lath or rigid board insulation and then connected to the steel back-up; clay brick that is veneered to steel stud construction.
3. Fiberglass-reinforced plastic panels and metal panels of aluminum or steel, may possibly use composite construction. Metal panels may also employ other assemblies, such as metal studs, insulation, and liner panels. Cast aluminum

and plastic panels are usually used as flat or moulded shapes. In composite construction, skins are laminated with an energy-efficient foam-filled insulating core or an aluminum/paper honeycomb. For example, in Figure 6.2, bottom right, the curtain wall consists of aluminum panels with an insulating foam core and clear double glazing that is slanted inward from the top. The curved insulated stainless steel sill reflector guides the light upward through the angled glass. The curtain panels are hung from a steel framework, which is bolted to the building frame. The building's dark gray north and east walls absorb heat, while the silver south and west walls reflect it.

4. The most common panel types used for the currently popular thin stone veneers typically incorporating 5/8-in. to 1 1/2-in. slabs (besides steel diaphragm panels, aluminum stick-framed panels, precast concrete, and glass-fiber-reinforced concrete panels), are the steel-truss-framed panels. Here, for example, four to five stone slabs are attached to the truss by means of clip angles and pins at regular intervals. Wind loading and stone strength determine the stone thickness and anchor spacing.

Composite Skins

These (see fig. 7.38b) are part of the building structure, such as in tubular construction, where the stressed skin provides additional stiffness against the building drift.

The Curtain Wall as Structure

In the previous section, the various curtain wall structure systems of frame construction (hinged, rigid or braced frames, stud wall, truss, etc.), surface construction (flat or folded plates, shells), or any combination thereof, where panels are supported by mullions or other panels, were discussed. Basically, non-load-bearing facade walls for tall buildings consist of individual panels, whatever their structure system—they are separated by horizontal and vertical soft joints, so that the panels can behave as independent elements. Lack of adequate separation between cladding and structure is the source of many problems.

There are various ways for supporting the cladding, depending on the system (as shown in fig. 6.3k, l, m). The cladding can span vertically, horizontally, and bidirectionally. Usually, solid panels span vertically between spandrels and, possibly, girts which are secondary horizontal members (for instance, channels with sag rods), acting as intermediate support; thus, the panels behave as one-way slabs. Mullions spanning vertically from floor to floor act primarily as beams by supporting the fill-in panels. The cladding may be suspended from the floor(s) above, or it may be bottom-bearing on the floor(s) below, in column action. Usually, panels are supported at two points: vertically and horizontally on the lower floor, with two tiebacks at the upper floor; therefore, they do not hang from the upper floor as a curtain.

The cladding may span horizontally, as in band facades, where the spandrel panels may be load-bearing or non-load-bearing. Load-bearing spandrel bands may form beams (trusses, plate girders, concrete beams) that span directly from column to column, thereby avoiding the problem of being carried primarily in torsion by the spandrel beams, although lateral stability may require intermediate diagonal and/or horizontal lateral bracing. The spandrel panels can support the vision and infill panels that span vertically (see fig. 6.3m). The cladding can span bidirectionally, as may be the case for a moulded panel, which can be supported along all four edges of columns and spandrel beams.

The structural designer of curtain walls must be primarily concerned with the: 1) strength and stiffness of the various wall components; 2) movement of these components to avoid the generation of additional stresses; and 3) weathertightness to prevent the deterioration of the structural materials.

The wall structure must resist the vertical loads due to its own weight, lateral loads, and hidden loads caused by movement. The wall attachments transfer these loads to the main building structure; their design is most critical, since they very often cause the failure of the curtain wall—in other words, the type and location of connection points for cladding are most important. The wall panel carries its own weight, which includes the window sash and glass, as well as objects attached to it, such as (possibly) the window cleaning equipment. Some typical weights for curtain walls are:

- 15 psf for lightweight wall panels (e.g., steel studs, steel framing, or trusses, for instance, with a stone finish)
- 35 psf for thin stone or precast concrete panels
- 50 psf for metal studs insulation and 4-in. brick facing
- 75 psf for 4-in. face brick with 4-in. concrete block back-up and insulation

The reduction of the cladding weight through lighter back-up and finish material is an important design consideration.

The gravity loads are less critical in comparison to the other design loads because they are transferred back to the building structure at frequent intervals. The lateral loads, as caused by the external wind pressure, internal pressure, and possibly seismic forces, are the primary loads which the non-load-bearing exterior wall must resist. In the section in Chapter 3 on "Wind Loads," it was shown that the overall wind pressure values for the design of the whole building are not the same as the local wind loading for the design of the facade wall. High local pressures can develop, especially at corners and edges; usually the largest pressures are negative (i.e., suction) and may be about 2.5 times or more as compared to code values. The lateral seismic forces in a major earthquake zone can be estimated to be 30 percent of the panel weight.

It was discussed, in the section in Chapter 3 on "Hidden Loads," that (for the design of curtain walls) the movement of the individual wall components and the differential movements between wall components, as well as between wall and building structure, must be taken into account. The various types of movement (which should not interfere with each other) are: story drift, including torsional movement, building structure shortening, spandrel deflection, foundation settlement, thermal movement, and any combination of these.

The facade panels must, in some manner, be allowed to float independently of each other, so that the hidden loads due to the prevention of differential movement will not be absorbed by the panels themselves. Since one of the primary reasons for the failure of cladding is as a result of overstressing, as caused by additional loads that were transferred from the building, one must consider the compatibility of the exterior wall and the building structure, which includes the movement of the curtain itself and that of the supporting structure—that is, the relationship of the building flexibility and the panel rigidity. A flexible structure, such as a skeleton building, will tend to transfer loads to the exterior wall, but the relief joints must take into account the translational and rotational movements due to the vertical and lateral loads, as well as the volumetric material changes (temperature, moisture, creep, and shrinkage). Some basic movements of the building structure, as those caused by gravity loads, lateral loads, and the buckling of mullions, as well as their effect upon some cladding systems, are shown in Figure 6.5.

Gravity loads cause the spandrel beam to deflect. It is conservative to use the maximum live load deflection for estimating the differential vertical movement between two floors. For example, for a 25-ft span the deflection is $25(12)/360 = 0.83$ in., which the horizontal joint must be able to take, or the spandrel beam deflection must be reduced by increasing the beam size or by cambering it. Furthermore, in order not to stress the cladding, the anchoring system must be flexible enough to allow for this vertical movement or, alternatively, vertical, slotted holes must be provided at one end. In addition, the eccentric position of a heavy panel weight may have to be resisted by the spandrel beam in torsion, so that the torsional rigidity of the beam must be large enough so as to prevent any stresses to the cladding caused by torsional deflection. It should be kept in mind that typical web connections of simply supported spandrel beams have little torsional stiffness.

The differential vertical movement due to material volume changes, particularly for heavy wall construction (e.g., masonry cladding), may become quite critical. It is shown (in fig. 3.16 and Problem 3.32) that, as the concrete columns shorten due to elastic deformations, creep, and shrinkage, or as a steel frame shortens due to significantly larger elastic deformations, while at the same time the veneer possibly expands due to temperature and moisture increase, the combined movements may result in the stressing or buckling of the cladding, if the horizontal joints at the shelf angles are not wide enough.

The horizontal loads cause a lateral building sway, that is, a horizontal displacement of one floor relative to the others, which is called the *story drift*; this interstory drift must be accommodated. For an allowable wind drift of h/500 and a typical story height of 12.5 ft, the horizontal movement of the wall perpendicular to the building structure is 0.3 inches. Should the anchorage system be not flexible enough to prevent stressing of the wall, then horizontal transverse sliding connections may have to be provided, or smaller panels hinged to each other have to be used. With respect to building deflection in the other direction parallel to the cladding, longitudinal slotted holes may be needed to prevent the wracking of the wall panel; this is particularly critical for rigid frame action in this direction. In addition to the lateral drift, torsional interstory movement may also have to be added.

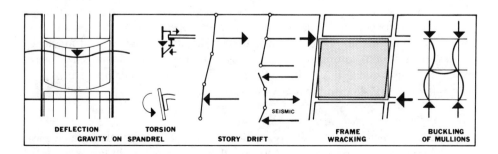

Figure 6.5. *Movement of structure.*

The floor diaphragms provide lateral support to the exterior walls. Should the floors be flexible, which may be the case (particularly for lower buildings), then they displace laterally, and in turn cause the curtain walls to deflect, if proper connections are not provided. To prevent the overstressing of the curtains, the stiffness of the horizontal diaphragms must be controlled. Some codes recommend maximum span-to-width ratios to control the diaphragm deflection of the floors.

The movement within the exterior wall itself may be as crucial as the effect of the movement of the building structure. The wall is exposed to variations in climate, which affects each of its components differently. It is apparent that exposed mullions expand or contract more than hidden ones, as was discussed with respect to exposed columns (see fig. 3.15). The glass, as a structure, must be treated like a curtain panel— it floats in the glass-holding frame opening, so that it is not thermally and mechanically stressed. The glazing system must adequately support and cushion the glass so that the loads due to various types of movement are not transmitted to the glass, thus requiring sufficient clearance between glass and frame. Naturally, there must be window or curtain wall expansion joints to control stress transfer to the glass-holding frame members. The floating glass panels may be laterally supported on all four sides and act like membranes with respect to wind pressure or suction, or they may only be supported on two opposite sides of the frame and behave like one-way slabs. The glass thickness depends on its strength, the magnitude of wind forces, its span, and the supporting conditions; for large panels, deflections and vibrations due to wind gusts must also be considered. Larger spans require thicker glass, especially for purposes of stiffness, since otherwise mullions are necessary to provide lateral support for the large lights of glass. All-glass walls are often found at the base of buildings because facades are not extended to the ground. With respect to their vertical weight, these structural glazing systems can be efficiently suspended from support structures at the top or they can be floor-supported, if they are not too high. The lateral loads are carried by the glass plates to the vertical tempered glass mullions (at each vertical joint), which are located behind the panels and act as beams. Horizontal glass mullions may be introduced to reduce the number of vertical mullions. Metal clips and clear structural silicone secure each glass mullion (fin) to its respective panel.

Because of the relatively large deflections of glass membranes under lateral loads, traditional methods of analysis, as based on small deflection theory, cannot be applied. Furthermore, it must be kept in mind that glass is a brittle material that exhibits a complex mechanism of breakage. Therefore, the structural design of window glass is usually done with empirically based design charts, which are provided by the manufacturers, as derived from tests. Typical glass thickness may vary from approximately 5/16 in. to 3/8 in. and 1/2 in, with the larger sizes along corners and edges of buildings.

In contrast to vertical glazing, for the design of sloped glazing, dead, snow, and projectile loads must also be considered. Usually, laminated glass is used for sloped planes, since it remains in place after breakage—it does not break into small pieces when it fractures due to missile impact (e.g., falling objects, windborne debris).

From a material strength point of view, glass is classified as

- *Regular* or *annealed glass.*
- *Heat-strengthened glass,* which is about twice as strong as normal annealed glass.
- *Fully tempered glass,* which is approximately four times as strong as annealed glass, but it is not much more rigid—it fractures into small fragments.

- *Laminated glass*, which may be comprised of any of the monolithic glass lights above, bonded together with a plastic film; it is usually used for sloped surfaces.
- *Wired glass*, which is not a reinforced glass—the wire mesh holds the glass pieces together at breakage and prevents it from falling—it does not increase the glass strength.

Failures of the currently popular thin stone veneers (e.g., fracture of granite, degradation of limestone through acid rain and bowing of marble into dish shape because of repeated thermal cycles) keep on reminding us that much must still be learned about this new construction technique. The stone strengths listed can only be considered approximate, since they are affected by weathering (e.g., freeze/thaw and thermal cycling, chemical attack), moisture content, surface finish, sawing and drilling, besides being dependent on the orientation of the grain structure relative to the direction of the load. Furthermore, critical are also attachment stresses and stresses due to handling are also critical. All of these unknown conditions clearly indicate that safety factors for thin stone design must be higher than for more traditional construction techniques.

At the interface of the cladding and building structure, attachments transfer the loads. *Load-bearing connections* support the gravity action, while *lateral connections*, such as horizontal ties perpendicular to the wall, resist wind suction and rotation due to gravity. The connection can be provided directly, as for precast concrete panels with brackets, or by using connector elements like stainless steel or aluminum anchors, or plates and angles, which should be corrosion proof. Bearing locations should be at points of minimal deflection, usually close to the columns. Connections should be stiff against lateral pressure, but must be flexible to allow for anticipated relative movements between the structure and cladding; in other words, they must *bend* or *slip* under lateral movement, since otherwise forces are transferred to the skin that it cannot resist. Longitudinal, transverse, and vertical slotted holes in the connections are not only necessary for freedom of movement, but also for accommodating the manufacturing, fabrication, and erection tolerances between the structure and cladding, and allow for adjustments in unpredictable field conditions. Naturally, if the skin is strong and stiff enough, as well as being properly connected to the primary structure, it becomes a composite part of the building structure, as is the case in the tubular stressed-skin concept. The gunnable or solid sealants form connections. They are most crucial, since they must assure weathertightness, offer adhesive qualities, allow for movement, retain their resiliency with age or exposure, be durable, and must be of high strength to transfer structural loads, for example, from the glass to the supporting perimeter frame. The most common sealants for curtain walls are silicone-based. In the structural silicone glazing, silicone sealants even replace the mullions. In this case, the colorless translucent sealants act as bonding agents to join panes of glass that span vertically, and to free them visually from the supporting mullions.

Deriving the movements of structure and exterior wall as independent, isolated cases are quite unrealistic; the combined action or superposition of all of these movements generates a new situation. Besides, other effects must be taken into account, such as the flexural and torsional stiffness of the building, which is determined by its shape and structure system, or the fact that the wall connections may not be as flexible or may not slide as assumed, or that the sealants between the panels may fail and lose their weathertightness over time and allow water to enter, thereby causing gradual deterioration of the anchorage system and pressure due to the freezing of water. This complex interaction of so many variables makes full-scale testing of curtain panels for high-rise buildings generally necessary to verify that anchors hold, and that windows will not pop out or break.

PARTITION WALLS

Partitions are interior walls that divide space to provide a visual, acoustic, and possibly fire barrier. They are non-load-bearing, they only carry their own weight and must resist a minimum lateral load of 5 psf, but they do not support floor loads. Sometimes, fixed partitions are designed as shear walls to provide additional lateral stiffness to the building. Partitions are classified as: 1) movable partitions; and 2) fixed partitions.

While permanent partitions enclose the space in the core area such as elevator, stairwell and air shafts, and machine rooms, open office spaces may be subdivided by movable partitions. These movable partitions are carried by the floors and may not be sitting directly on beams. They can take any position, and may be rearranged or relocated, and their weight is therefore treated as an equivalent uniform dead load of not less than 20 psf, except in the case of light partitioning, but that depends on the code. The weight of permanent partitions can be considered by using either the actual weight or equivalent uniform loads, but for shaft walls the actual weight is taken. The design of elevator shaft walls is much more critical, since they must resist positive and negative air pressures of up to about 15 psf, as caused by high-speed elevator cabs moving up and down the shaft, creating a pumping action. Similarly, shaft walls may have to resist lateral air pressure due to pressurization to keep the shaft free of smoke in case of fire. Hence, they may have to withstand extensive flexural stresses, and their lateral deflection should be less than 1/240 of the partition height.

Permanent partitions are either constructed of masonry (e.g., brick, lightweight concrete, or gypsum block), or they are framed partitions finished on each side with a single layer or multilayer skin system, such as plaster and/or gypsum board(s). Demountable partitions consist of prefabricated units; they are usually patented and hence much more expensive. They are organized according to their support structure as: frame plus infill panels; frame plus overlay panels; or panel systems.

The material for the frames is either metal or wood. In contrast to masonry partitions, framed partitions are of lightweight construction (roughly from 8 to 16 psf) and they are easier to assemble, which is an essential consideration for tall buildings. Typical framed partitions consist of stud drywall construction where fire- and sound-resistant sheet materials, possibly in several layers, are attached to the studs, and sound insulation blankets are installed in the cavities between the studs. For example, a movable partition may consist of gypsum panels, steel H-studs, the floor runner, and the flanged top rail.

Generally, partitions, in a manner similar to glass, must float in a frame opening so they will not be stressed by the vertical beam deflection and crack; also, loads must not be transferred through lateral frame wracking due to wind and earthquake action or by vertical frame wracking in the upper floors due to temperature differences (see fig. 3.19g). Special joints are required along the sides and top of the wall to allow for building movement. More sophisticated joints, like slip joints, are required for seismic design.

MASONRY WALLS

Masonry and unbaked earth construction have served men for several thousand years. While stone is the oldest raw building material, sun-dried or hard-burnt bricks are the oldest man-made ones. Precast concrete blocks, on the other hand, are a more recent development—by the 1870s they had become quite popular in Chicago as so-called artificial stones. Inexpensive *adobe* bricks, also often named sun-dried mud-bricks, are used in many regions. In this case, earth is mixed with water and vegetable fibers, such as chopped straw, shaped in molds, and then dried in the sun. In the *pisé de terre* method, hard earth walls are constructed by ramming earth within formwork. Rather than fitting bricks or blocks precisely with dry joints, as the Greeks did for their stone temples and the Incas for their stone walls, it is surely easier to use them together with mortar joints. The Egyptians used gypsum as mortar. Lime mortar was already used by the Minoans on Crete in 2000 B.C., and the Greeks discovered pozzolan cement c. 600 B.C., which was further perfected by the Romans.

In high-rise building construction, the traditional massive walls, where the weight had to suppress the tensile stresses due to lateral load action, have been replaced by the engineered thin-wall construction of the present. For the preliminary structural design of modern tall brick or concrete block buildings, it is necessary to first introduce some general principles of masonry technology before discussing wall types and wall behavior under load action. The most common masonry units in high-rise wall construction are either solid or hollow. They are: *brick*, made from clay, shale, or concrete; *concrete block*, with properties dependent on the aggregate of concrete mix; and *clay tile*, which is seldom used in the United States.

Bricks are available in many different sizes. Most of them are produced in modular sizes where the nominal dimensions are equal to the manufactured dimensions plus the thickness of the mortar joint (3/8 in. or 1/2 in.). For example, the nominal modular size of a standard brick is 4 in. wide, 8 in. long, and 2 2/3 in. high. It is laid in three courses to 8 in. height; this dimension is a multiple of the 4-inch basic dimensional reference grid of the building. Solid and hollow clay brick are graded as SW (severe weather), MW (moderate weather), and Grade NW (no exposure, as the least durable, for interior use only).

There are also many different sizes and shapes of concrete blocks. They are considered hollow blocks when they are less than 75 percent solid, otherwise they are treated as solid block units. Most concrete block units are of the stretcher type, and have either two or three cores. A typical nominal size for a 3/8-in. thick mortar joint is of 16 in. length, 8 in. height, and of 4 in., 6 in., 8 in., 10 in., or 12 in. in depth. The nominal size of the regular block will fit the basic 4 inch dimensional reference module. The section properties of the masonry units are given in Table A.8. Typical facing units are split blocks, split ribs, fluted blocks, and many other variations.

The compressive strength of masonry is defined as f'_m and is controlled by the compressive strength of the mortar (m_o), the strength of the masonry units, and the workmanship; it is given in Table A.10. The allowable stresses for clay brick and concrete masonry, as derived from the masonry strength, are given in Table A.9 for inspected construction, where the quality of work is in accordance with the contract drawings and specifications. Uninspected work typically requires a 33 to 50 percent reduction of allowable stresses in most codes. Clay brick and concrete masonry do behave differently. Clay brick has a higher modulus of elasticity and is stronger than the corresponding concrete units. While clay brick changes with temperature and expands because of moisture, concrete masonry shrinks and creeps after construction.

In addition to the mortar bond, the masonry units must be interlocked and tied together for the wall to act as a structural unit. This may be achieved by overlapping the masonry units, by metal ties, by joint reinforcement, by grout, or by horizontal bond beams, possibly combined with vertical core concrete columns. In traditional solid masonry construction, the bond is obtained by overlapping the units, where *stretchers* tie the wall together in the longitudinal direction and *headers* in the transverse direction. There are many different bond patterns, such as the familiar *English bond* and its *cross* variation with alternating courses of headers and stretchers (see fig. 6.7, top). In concrete block construction, the most common bond pattern is the *running bond*. Metal-tied masonry bonding is generally used for cavity walls, composite walls, double-wythe solid walls composed only of stretchers, and solid walls using the *stack bond* with no overlapping of units, where bond beams are often used to tie the wall together. In grouted wall construction, grout is placed in the vertical cores of the concrete blocks; it is basically a concrete made with small-size aggregates.

The strength of the masonry wall is dependent on the mortar strength, as well as on the strength of the masonry units. It is therefore apparent that strong bricks should not be used unless the mortar is sufficiently strong. Today, the advantage of the workability of the lime mortar of the past is combined with the strength of cement mortar. A cement-lime mortar is composed of cement, lime, sand, water, and possibly admixtures. Depending on the type and proportion of the cementing agents and sand, the mortars are classified in four grades according to their strength:

- *Type M mortar* (1: 1/4: 3, Portland Cement to lime to sand) has the highest compressive strength.
- *Type S mortar* (1: 1/2: 4 1/2) has a high tensile bond strength and a reasonably high compressive strength; it is recommended for exterior masonry below grade.
- *Type N mortar* (1: 1: 6) is a medium strength mortar.
- *Type O mortar* (1: 2: 9) is used for non-load-bearing walls.

Other mortar types have recently been developed to improve the tensile capacity. Usually, the bond of the adhesive material to the masonry is stronger than the unit itself. Examples are:

- *High-bond mortar*, such as Sarabond, developed by Dow Chemical Company. It is an organic additive that greatly improves the mortar's bonding capacity to the masonry and allows the designer to take advantage of the inherent tensile strength of the masonry units. The allowable flexural tensile stress can be taken as $0.3f'_t$, or 112 psi, without testing. It should be remembered that this allowable stress can be increased by one-third to 150 psi without testing when considering wind and earthquake; here f'_t is the ultimate flexural tensile stress.
- *Organic mortars* provide a high bond. The mortar joints are quite thin (about 1/8 in.) and require dimensional precision of the masonry units. The mortar is applied with a caulking gun.
- *Surface coat* of fiber-reinforced conventional mortar applied to the faces of the masonry wall. The masonry units are dry stacked without mortar between them, and then plastered along each face with the surface bonding mortar.

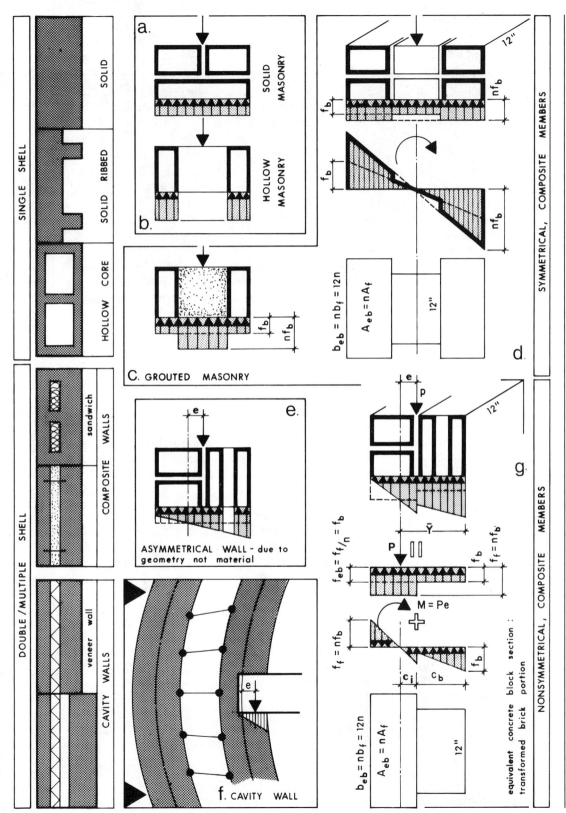

Figure 6.6. *The response of various wall types to gravity loading.*

Masonry Wall Types

Masonry walls can be load-bearing, non-load-bearing, or can be veneers—they form either single or multiple wythe wall systems (Fig. 6.6). Single shell walls can be either solid, hollow, ribbed, or curved. Typical multiple shell systems are cavity walls, veneer walls, or composite walls. Masonry walls may be reinforced, partially reinforced, or nonreinforced.

The stress distribution due to axial action and bending depends on the wall type. A wall that is symmetric from a geometric standpoint, and is constructed of only one material, can be treated as any other cross section of homogeneous material. Typical examples are solid brick and hollow-block concrete walls (Fig. 6.6a, b), or cavity and composite walls with identical facing and back-up wythes. In this case, the familiar axial stresses are $f = P/A \pm M/S$.

However, when *solid and hollow* masonry units of the same width and material are used (Fig. 6.6e), the section is asymmetric from a geometric point of view. Simple axial action at the centerline will be eccentric with respect to the centroidal or neutral axis, and thus will cause bending. When, in addition to asymmetry of geometry, different materials are used, such as the typical clay brick facing and concrete block backing (Fig. 6.6g), then the section also becomes asymmetrical from a behavioral (elastic) point of view, since the moduli of elasticity of the two materials are usually different. For this condition of $E_f > E_b$ (i.e., $E_{facing} > E_{backup}$), the centroidal axis moves even further to the left as compared to Figure 6.6e. Stresses in composite sections are determined by mathematically transforming or replacing one material (facing, clay brick) by the other material (backup, concrete block), so as to generate a homogeneous cross section. In the design process, first the minimum wall thickness, together with the lowest mortar strength, is assumed. Should this not be satisfactory, the mortar strength is first increased, and then the wall thickness. In the following paragraphs, the various wall types are briefly introduced (see also fig. 6.10).

Solid Masonry Wall

Masonry is considered solid when the units have less than 25 percent cells. Solid masonry may be single- or multi-wythe walls; multiple wythes are bonded with masonry units, metal ties, or metal joint reinforcement.

Hollow Masonry Wall

Typical single-wythe load-bearing concrete block walls are 6 in., 8 in., 10 in., or 12 in. thick. The stress diagrams for full mortar bedding of face shells and webs is similar to solid unit construction. The stress diagrams for *face-shell mortar bedding* are different, however, since the webs are not in contact and thus do not transfer stress. Therefore, the moment of inertia of an 8-in. concrete block wall using the minimum net section, as based on a 12-in. wall length, is

$$I_b = 12(7.625^3 - 5.125^3)/12 = 309 \text{ in.}^4/\text{ft}$$

This value is given in Table A.8.

Grouted Masonry Wall

The construction of grouted masonry walls is basically the same as for reinforced masonry walls (see fig. 6.7), with the exception of the steel bars.

Patented Masonry Wall Systems

There are many patented bearing wall systems on the market, for example, the Ivany concrete block, developed by George Ivany of Cleveland. In this case, single core, lightweight hollow concrete blocks have special slots in the webs to accommodate horizontal steel reinforcing bars, while the vertical bars are placed along grooves in the cores. Concrete is cast into the cores, embedding the two-way bar grid.

Cavity Wall

The typical cavity wall consists of two wythes of masonry separated by an air and/or insulation space of 2 to 3 inches; the cavity also acts as a moisture collector. Metal ties or steel rods are the load transfer mechanism—they join the facing and backing wall so that they can support each other and share the loads, but they do not act compositely. Usually, a 4-in. exterior brick facing acts purely as a veneer, or resists the wind pressure, while the interior back-up (e.g., clay brick, concrete block, or stud wall) may also have to carry the floor loads in addition to the horizontal loads (Fig. 6.6f). It is not necessarily always the case that only the inner wythe carries the floor loads, however, as both wythes may be used for load-bearing. Each of the two walls will resist a certain portion of the lateral load, w, with the stiffer wall attracting more of the load. For the two walls to act together, though, a sufficient number of ties must be properly placed and have the necessary capacity in tension and compression. It is assumed that both walls deflect equally in simple beam action under the lateral load, so that the following relationships can be derived:

$$\Delta_f = \Delta_b, \text{ or } 5wh^4/(384EI)_f = 5wh^4/(384EI)_b$$

This yields

$$(w/EI)_f = (w/EI)_b, \text{ or } w_b = w_f(EI)_b/(EI)_f = w_f(I_b/nI_f) \qquad \text{(a)}$$

where:
$$n = E_f/E_b = \text{modular ratio} \qquad (6.1)$$

However, the total load, w, is shared by the walls:

$$w = w_f + w_b \qquad \text{(b)}$$

Substituting Equation (a) into (b) yields the loads that each of the walls will carry; in other words, the portion resisted by the facing wythe, w_f, and the back-up, w_b.

$$w_f = \frac{w(EI)_f}{(EI)_f + (EI)_b} = \frac{wnI_f}{nI_f + I_b}$$

$$w_b = \frac{w(EI)_b}{(EI)_f + (EI)_b} = \frac{wI_b}{nI_f + I_b} \qquad (6.2)$$

For identical facing and back-up walls, each of the walls obviously carries one-half of the load. Notice that the expression above is similar in nature to Equation 5.3.

Example 6.1

Determine the flexural stresses in a 14-in. non-load-bearing cavity wall at the top floor of a five-story building. The wall consists of a continuous 4-in. clay brick facing with a compressive strength of $f'_m = 2800$ psi, and an 8-in. hollow concrete block back-up with $f'_m = 2000$ psi built between the floors. The wall height between the supports is 9 ft, and the wind pressure is equal to 23 psf. The relatively small axial stresses due to the wall weight are conservatively ignored.

From Table A.8:

$$I_f = 48 \text{ in.}^4/\text{ft}, \ S_f = 26 \text{ in.}^3/\text{ft}, \ I_b = 309 \text{ in.}^4/\text{ft}, \ S_b = 81 \text{ in.}^3/\text{ft}$$

From Table A.9:

$$E_m = 1000 \ f'_m$$

The modular ratio is: $n = E_f/E_b = 2800(1000)/2000(1000) = 1.4$
The total transformed moment of inertia is

$$nI_f + I_b = 1.4(48) + 309 = 67.2 + 309 = 376.2 \text{ in.}^4/\text{ft}$$

Each wall carries the following portion of the wind load.

$$w_f = 23(67.2/376.2) = \ \ \ 4.11 \text{ psf} \tag{6.2}$$

$$w_b = 23(309/376.2) \ = \ \underline{18.89 \text{ psf}}$$

$$w = \ \ \ \ \ \ \ \ \ \ \ \ \ \ \ 23.00 \text{ psf}$$

Both walls are assumed to behave as simply supported slabs, realizing that the brick facing is continuous and thus actually provides a higher stiffness.

$$M_f = 4.11(9)^2/8 = 41.61 \text{ ft-lb/ft}, \quad M_b = 18.89(9)^2/8 = 191.26 \text{ ft-lb/ft}$$

The flexural stresses in the walls, and the corresponding allowable tension stresses from Table A.9, are

$$f_f = M/S = 41.61(12)/26 = 19.21 \text{ psi} < 36(1.33) = 47.88 \text{ psi}$$

$$f_b = 191.26(12)/81 = 28.34 \text{ psi} < 23(1.33) = 30.59 \text{ psi}$$

The wall is satisfactory.

Veneer Walls

In this context, only some of the *anchored* veneers are of interest, but not the *adhered* veneers, such as tiles. A typical example is a 4-in. masonry veneer attached with metal ties to a relatively flexible back-up system, such as light gauge metal stud walls. Though veneers in general are nonstructural, and are assumed to transfer wind forces directly to the back-up system, in the case of the flexible backing, compatibility of stiffness between the two wall systems must be considered to prevent excessive deflection and cracking of the masonry wall (see Problem 6.10). Another example of a veneer wall is the exterior concrete masonry wall of double wythe cavity construction with insulation on the interior wythe, which consists of regular block back-up and split-ribbed and split-face solid veneer block.

Composite Walls

Composite members are nonhomogeneous—they are composed of various materials. Composite walls form multiple shell systems, usually of two materials, where the components act together as a unit. Typical examples are grouted masonry walls, reinforced masonry walls, and brick-block walls. Examples of nonmasonry composite walls are plastic or metal sandwich panels, composite skins, and so on. The composite masonry walls are further investigated here. They may be symmetric, such as the grouted masonry block wall (Fig. 6.6c), or the wall composed of clay brick facing with interior concrete bricks (see fig. 6.6d), or they may be asymmetric, such as the wall with brick facing and concrete block back-up (see fig. 6.6g). It should be remembered that asymmetry is not only caused by the different geometries of the outer and inner shells, but also by the different materials. Again, the two shells cannot act independently as in cavity walls, but must be bonded or tied together by mortar and ties, wall reinforcing, or headers. The general case of brick facing and hollow concrete block back-up (see fig. 6.6g) is conceptually investigated now.

In the elastic approach to the design of composite members, it is common practice to transform the composite system into an equivalent homogeneous material and then to proceed with the design in the conventional fashion. In the bending theory for solid beams, linear strain behavior is assumed; that is, the axial deformations of the longitudinal fibers are proportional to the distance from the neutral axis. The same assumption is applied to the composite action by not allowing slippage between the different materials, thereby clearly pointing to the importance of bonding. Hence, for a double-wythe wall, the strain at the interface of the facing and back-up shells must be the same.

$$\varepsilon_f = \varepsilon_b, \text{ or } (f/E)_f = (f/E)_b$$

From this expression, the relationship between the stresses in the two wythes can be found. For the case where $n > 1$, the face material is stiffer and attracts more stress.

$$f_f = f_b(E_f/E_b) = nf_b \qquad (6.3a)$$

Now the facing shell is transformed into an equivalent back-up material by dividing the stress in the facing, f_f, by the modular ratio, n, to yield a stress, f_{eb}, of the equivalent back-up material, thereby achieving a homogeneous wall.

$$f_{eb} = f_f/n \qquad (6.3b)$$

Since the facing portion of the wall has been transformed into an equivalent back-up material, both materials must carry the same force at a given point.

$$P_b = P_f, \text{ or } f_bA_b = f_fA_f = nf_bA_f = f_bA_{eb}, \text{ or}$$

$$A_{eb} = nA_f \qquad (6.4a)$$

Therefore, the facing shell is transformed into an equivalent back-up material, A_{eb}, by multiplying its cross-sectional area, A_f, by the factor, n. By assuming that the depths of the facing and back-up shells do not change, only dimensions of the cross section parallel to the neutral axis have to be altered. Thus the equivalent width, b_{eb}, of the transformed facing is

$$b_{eb} = nb_f \qquad (6.4b)$$

Once the cross-sectional areas of the various materials of a composite member have been transformed into an equivalent material, the stresses can be determined by the familiar design approach for homogeneous members, keeping in mind that the location of the centroidal axis for asymmetrical sections must be found first. Assuming the facing material to be transformed yields the following expressions for the combined stress in the extreme fibers of the back-up wall,

$$f_b = P/(A_b + nA_f) \pm M\bar{y}/I \qquad (6.5a)$$

and for the combined stress in the extreme fibers of the facing shell,

$$f_f = P/(A_b + nA_f) \pm nM(t - \bar{y})/I \qquad (6.5b)$$

Notice that the stresses break at the contact surfaces of brick and block (Fig. 6.6c, d, g), clearly supporting the fact that, under constant strain, the larger stresses develop for the stiffer material (see Problems 6.11 and 6.22).

It should also be noticed that the wall weight causes bending. While in the cavity wall each of the shells can shorten independently, this is prevented in the composite wall. The difference in shortening will cause bending, which is, however, small and usually neglected.

Reinforced Masonry Walls

Typical reinforced brick and concrete block walls are shown in Figure 6.7; they are used for large loading conditions. Because of the lack of ductility of masonry bearing-wall buildings, they must be reinforced in critical earthquake areas.

- *Reinforced Brick Walls:* This wall type is usually a double-wythe construction similar to a cavity wall with a 1 in. to 6 in. space, in which vertical and horizontal reinforcement is placed before grout is poured into the cavity to form a composite structure. The minimum thickness of reinforced brick walls, as based on construction requirements, is: 8 in. for brick walls with horizontal reinforcing placed in horizontal mortar joints (i.e., 2(3.625) + 3/4 in. for grouting); and 9.25 in. for brick walls with horizontal steel placed in the 2 in. grout space (i.e., 2(3.625) + 2).
- *Reinforced Concrete Block Walls:* The single-wythe hollow-block masonry units are aligned so that the cores are vertically continuous and the reinforcement can be placed in some of them, as required; the cores can then be grouted. The masonry, together with the grout and steel form a composite structure similar to reinforced concrete members.

Partially reinforced masonry walls are often used in areas where seismic action is not critical. They are designed as plain masonry walls, except that reinforcing is used to resist the flexural tensile stresses; the minimum steel area (ρ_{min}) is not applicable to partially reinforced masonry.

The design of the composite interaction of masonry and steel is analogous to the working-stress approach for reinforced concrete, where the steel is transformed to obtain a homogeneous material of an all-concrete member. In this context, the following approximations for the *preliminary design* of reinforced masonry walls are used. It is distinguished here between the design of walls that also act as vertical slabs resisting lateral loads and/or eccentric gravity loads, and between shear walls that act as deep cantilever beams.

When the wall acts as a *vertical slab* resisting the lateral forces perpendicular to its

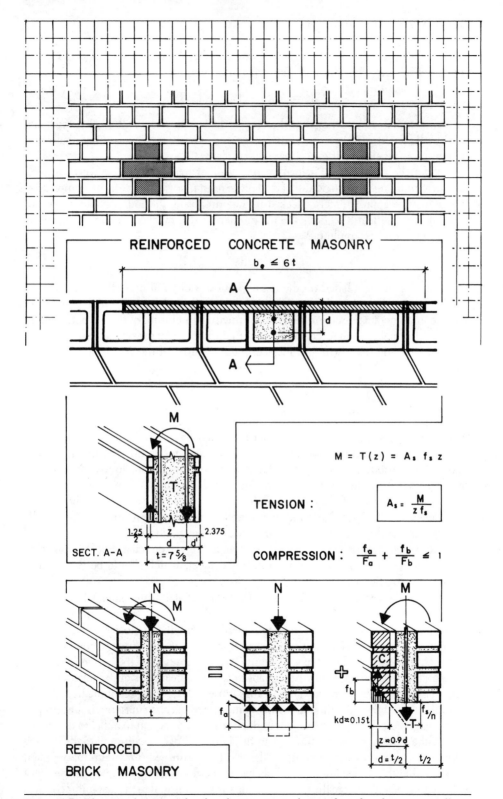

Figure 6.7. *The typical masonry bond and approximate design of reinforced masonry walls.*

surface, and when the tensile stresses, f_t, exceed the allowable ones, as based on an uncracked section, then the wall may have to be reinforced.

$$f_t = M/S - P/A > 1.33F_t$$

For the condition of a non-load-bearing curtain wall, the small axial stresses due to the wall weight may be conservatively ignored, and it may be assumed that all of the tension is resisted by the steel bars (Example 6.2).

Also, for the condition of a load-bearing wall, it may be conservatively assumed that the reinforcing bars provide all of the tensile resistance. The maximum compressive stresses may be approximately checked by superimposing the independent actions of the axial load on the full section, and the moment on the cracked section, as explained in Figure 6.7 and Example 6.2.

$$f_a/F_a + f_b/F_b \leq 1.33(1)$$

This unity equation is based on the temporary loads of wind or earthquake.

When the wall acts as a *shear wall* resisting the lateral loads parallel to it, a homogeneous, uncracked section may be assumed for preliminary design purposes. In this case, the total tension force, as found from the combined stress diagrams, is resisted by the reinforcing steel, which is located at the centroid of the tensile stress diagram (Example 6.5).

Exterior Non-load-bearing Walls

Walls that support no vertical loads other than their own weight are considered non-load-bearing, although exterior walls must resist lateral forces due to wind and earthquake. The various curtain systems have been introduced previously. It has been discussed that the wind loads for the design of the curtain walls are usually not the same as for the design of the entire building. The local wind pressure will be affected by corners and other peculiar building geometries; high local pressure, and especially suction, can develop at critical locations. Further, the additional effect of the internal building pressure must be considered. For approximate design purposes, the wind pressure values in Figure 3.8 are used in this context, which should be conservative for ordinary conditions.

Curtain walls can be visualized as vertical slabs that transfer lateral loads to the primary structure. Should these slabs behave like two-way systems, then part of the load is carried vertically and the remainder horizontally; the load flow depends on the nature of the material, the plate proportions, and the boundary conditions (see fig. 5.11). The magnitude of the loads transferred horizontally to the columns or cross-walls, and transferred vertically to the slabs or spandrel beams, is quite indeterminate, particularly when the effect of wall openings must be taken into account. In addition, the evaluation of the stress flow is further complicated because masonry is anisotropic; it has a larger tensile bending strength in the horizontal direction (parallel to the joints) than in the vertical direction (normal to the joints). Since openings partially destroy the two-way action of the masonry walls, it can be assumed (for preliminary design purposes) that the wall spans vertically between the floors, similar to parallel one-foot wide shallow beams, where the flexural tensile strength normal to the bed joints controls the design of the wall.

It has been discussed previously that the compatibility of the building structure flexibility and the curtain wall rigidity, that is, the differential movement between the two systems (which becomes more critical with an increase of building height), must be taken into account. It is apparent that a flexible structure such as a skeleton building will tend to transfer more loads to the exterior non-load-bearing walls than a stiff, bearing-wall structure.

A brick curtain may be assumed to support its own weight on the foundations up to roughly 100 ft height. For this condition, the veneer is continuous and hence acts as a continuous multiple-span vertical slab with respect to the lateral loads. It is laterally supported by the floors (Fig. 6.8), and is tied to the spandrel beams, columns, or bearing walls; this anchorage must be flexible enough to permit differential, longitudinal movement. Under certain conditions, self-supporting curtain walls may not be possible because of the location of openings; in this case, the curtain must be supported directly by the building frame.

For tall buildings, the curtain can no longer be self-supporting; it consists of discrete panels supported at each story, which are separated by horizontal and vertical joints,

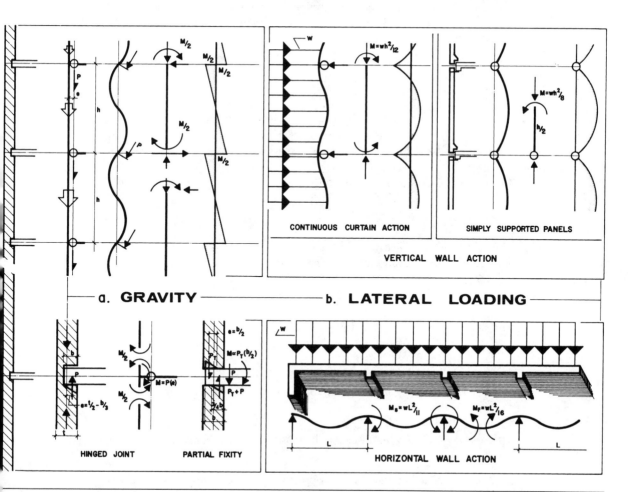

Figure 6.8. *The response of exterior brick walls to lateral and gravity loading.*

and thus behave as simply supported vertical slabs. The weight of the brick veneer is carried at the bottom of each floor level on shelf angles that are attached to the building structure (see fig. 3.16); beneath the angles are the horizontal joints. Hence, the masonry panels are only self-supporting between the stories. The connections at the top of the panel act as lateral ties only. The proper attachment of the curtain to the back-up structure is crucial; a sufficient number of ties and bearing connections are necessary. They must allow the differential movement of the curtain and the supporting structure, as has already been discussed. Rather than hand-laying the brick one at a time in place, self-supporting prefabricated brick panels may be used, which do not need any back-up structure. They could be high-bond brick masonry panels, or reinforced masonry panels where the hollow bricks contain the vertical bars.

Occasionally, the brick wall may span horizontally between the piers, columns, or bearing walls (Fig. 6.8, Problem 6.8). For example, in partially reinforced 4-in. brick walls, horizontal reinforcement of two #2 bars are laid along the horizontal joints, every second course depending on the magnitude of the moment, to resist the tensile stresses due to the lateral load action.

Usually, the structural design of the brick curtain is controlled by the tensile stresses at the top floor of the building, where the wind pressure is largest and the axial loads are small. The wall self-weight hardly offsets the flexural tension, however, and can be conservatively ignored.

Example 6.2

An exterior, non-load-bearing masonry wall is investigated at the critical top floor of a 15-story building for a wind pressure of 34 psf and a critical seismic location. The typical vertical span of the simply supported wall panels is 12 ft, but the top story panel has a 2 1/2-ft cantilever parapet.

The seismic loads that act against the 8-in. wall panels, which weigh $(120/12)8 = 80$ psf, are determined first. The lateral pressure against the parapet (Table 3.5) is

$$F_\rho = ZIC_\rho W_\rho = 1(1)0.8(80) = 64 \text{ psf} > 34 \text{ psf} \qquad (3.28)$$

The seismic load is clearly larger than the wind pressure, even if the wind loading should be increased because of the edge conditions. This load causes a maximum cantilever moment of

$$M_{max} = 64(2.5)2.5/2 = 200 \text{ ft-lb/ft}$$

The lateral load for a typical story-high panel is

$$F_\rho = ZIC_\rho W_\rho = 1(1)0.3(80) = 24 \text{ psf} < 34 \text{ psf.}$$

Therefore, the wind pressure controls. The corresponding maximum moment for the simple span panel by ignoring the wind action on the cantilever portion, is

$$M_{max} = 34(12)^2/8 = 612 \text{ ft-lb/ft}$$

This moment is larger than the cantilever moment, and hence controls the design of the wall panel.

a) Determine whether an 8-in. solid brick wall with S mortar and inspection is satisfactory. For the wall properties, refer to Table A.8. Assume the weight of the panel to be supported at the bottom.

Because of the beneficial effect of the parapet, the critical flexural tensile stress occurs at midheight of a simply supported typical wall panel (Fig. 6.8), at the second floor from the top.

$$f_t = f_b - f_a = M/S - P/A$$

$$= 612(12)/116 - 80(12/2)/92 = 63.31 - 5.22 = 58.09 \text{ psi}$$

The allowable tension stress is

$$F_t = 1.33F_t = 1.33(36) = 47.88 \text{ psi}$$

The wall is not satisfactory. Notice that the axial compression stress due to the weight of the panel at the point of maximum moment is small, and can be neglected for preliminary design purposes. To increase the tensile capacity of the masonry wall, either a larger concrete block back-up must be used or the wall should be reinforced.

b) Determine the approximate amount of steel required for an 8-in. reinforced brick wall. Use 8000 psi brick units with S mortar and inspection, and Grade 40 steel with $F_s = 20$ ksi.

The compressive strength of the masonry is $f'_m = 2400$ psi, according to Table A.10. The steel is assumed to resist all of the tension caused by the maximum moment, similar to the derivation of Equation 4.7c for reinforced concrete beams; the effect of the compressive stress is ignored.

$$M = T(z) = A_s f_s(z) \tag{a}$$

The internal lever arm, z (as identified in fig. 6.7), is approximated as

$$z = 0.9d, \quad \text{where } d = t/2 \tag{b}$$

Hence, from Equations (a) and (b), the required reinforcement can be found

$$A_s = M/(zf_s) = M/(0.45tf_s) \tag{6.6}$$

The approximate steel area needed to resist the wind moment in this example is

$$A_s = 612(12)/[0.45(8)1.33(20000)] = 0.077 \text{ in.}^2/\text{ft}$$

where the minimum area of reinforcement should not be less than 0.07 percent of the gross cross-sectional area of the wall.

$$A_{s \text{ min}} = \rho_{min}bt = 0.0007(12)8 = 0.0672 \text{ in.}^2/\text{ft} < A_s$$

Select #4 bars spaced at $(12/0.077)0.2 \simeq 31$ in.

The approximate compressive stresses due to bending are derived as follows and identified in Figure 6.7.

$$M = C(z) \tag{c}$$

where the resultant compression force is equal to

$$C = b(kd)f_b(1/2) \tag{d}$$

In this case, the stress block depth can be derived from the location of C (at 0.1d from the face), as

$$kd = 3(0.1d) = 0.15t, \quad \text{where } d = t/2 \tag{e}$$

Substituting Equations (b), (d) and (e) into (c), yields

$$M = bf_b(0.03375t^2)$$

Therefore, the maximum compressive stress due to bending can be approximated as

$$f_b \simeq M/[bt^2/30] \leq F_b \qquad (6.7)$$

Should the section be treated as *homogeneous*, then the bending stress is 20 percent less, as compared to the cracked section.

$$f_b = M/S = M/[b(t/2)^2/6] = M/[bt^2/24]$$

Hence, the approximate maximum compression stress in the brick wall is

$$f_b \simeq 612(12)/[12(8)^2/30] = 286.88 \text{ psi}$$

The allowable compression stress, from Table A.9, is

$$F_b = 1.33(0.33f'_m) = 1.33(0.33)2400 = 1053.36 \text{ psi} > f_b$$

The flexural compressive stresses are easily resisted by the brick wall, although the small axial compression stresses must be added.

c) Now, for the sake of exercise, the vertical reinforcement for an 8-in. reinforced concrete block back-up is determined, using 4000 psi concrete block units with S mortar and inspection, and Grade 40 steel with $F_t = 20$ ksi.

The reinforced concrete block wall under bending in its cracked state forms a T-section consisting of the grouted core together with the shell wall (see fig. 6.7). Because of the large compression area available, which (in turn) moves the neutral axis close to the shell walls, it is conservative to assume that the resultant compression force acts at the center of the thin shell wall. The d'-distance is taken as 2.375 in. for 8- and 10-in. blocks, and 2.625 in. for 12- and 16-in. blocks.

Similar to the reinforced brick masonry case and Equation 4.7c, the tension generated by the moment is fully resisted by the reinforcement.

$$A_s = M/(zf_s)$$

Therefore, for the given conditions, the following reinforcement is required.

$$A_s = 612(12)/[4.625(20000)1.33] = 0.0597 \text{ in.}^2/\text{ft}$$

where the internal lever arm is equal to

$$z = 7.625 - 2.375 - 1.25/2 = 4.625 \text{ in.}$$

Checking the minimum steel ratio by considering two bars

$$\rho = A_s/A_c = 0.0597(2)/48 = 0.0025 > \rho_{min} = 0.0007$$

Try #4 bars spaced at $(12/0.0597)0.2 \simeq 40$ in., where the cores are spaced at 8 in. on centers. Use one #4 @ 40 in. near each of the outside shells to take into account seismic action in both directions, as well as wind pressure and suction. Note that it may be more economical to place the reinforcing at the center of the grouted core for the condition of force action in both directions; in this instance, the design is analogous to the reinforced brick wall design.

Since the extensive compression flange area is available, the T-section can be considered underreinforced, and the compression stresses may be considered not critical for preliminary investigation purposes. The *compression stresses* can be roughly checked by treating the wall as not reinforced, realizing that the stresses will be slightly higher because of the cracked section, and ignoring the weight of the back-up wall and veneer for this fast check.

$$f_b = M/S = 612(12)/81 = 90.67 \text{ psi}$$

$$F_b = 1.33(0.33f'_m) = 1.33(0.33)2000 = 877.80 \text{ psi} > f_b$$

Bearing Walls

The bearing wall is part of the main building structure; it supports floor loads and lateral forces. Typical wall arrangements have been introduced at the beginning of this chapter (see fig. 6.1), and the many possible layouts of bearing wall buildings are discussed further in Chapter 7 (see fig. 7.2). As a rough, first classification, the following basic groups can be identified:

- Long-wall systems
- Cross-wall systems
- Cellular wall systems

In the long-wall system (see fig. 6.11), the main bearing walls run parallel to the length of the building, and form the exterior facade walls and interior corridor walls. The floor slabs span perpendicular to them, hence allowing cantilever action at the exterior. Besides carrying the floor loads, the facade walls also must act as curtain walls (i.e., vertical slab action) to distribute the wind perpendicular to them to the floor diaphragms, which then transfer the lateral forces to the short, transverse shear walls, which may be nonbearing. These shear walls run in the cross direction and are usually the end facade walls and elevator shafts. With respect to the less critical lateral loads parallel to the long dimension of the building, the long walls act as load-bearing shear walls.

The cross-wall system consists of parallel walls running perpendicular to the length of the building (see fig. 6.15), and hence do not interfere with the treatment of the facade. These transverse load-bearing walls also act as shear walls with respect to the lateral load action parallel to them, that is, perpendicular to the long direction. The floor slabs span between the cross walls. Stability in the long direction is provided by the elevator shafts or the corridor walls.

The cellular wall system consists of bearing walls in the transverse and longitudinal directions, thus representing a combination of the long-wall and cross-wall systems. Depending on the layout of the walls, the floor slabs may be two-way slabs. The infinite combinations between the basic groups above can only be suggested by the hybrid cases or layouts in Figure 7.2.

In this section, only some characteristics of bearing walls as related to concentric and eccentric gravity loading and curtain action are investigated, while shear walls are discussed in the next section. The floor load distribution to the bearing walls is relatively simple by using the approximations discussed in the section in Chapter 3 "Dead and Live Loads" (see fig. 3.2).

To develop some understanding for the axial force flow in a wall and the effect of openings, some visual studies have been done in Figure 6.9. The nature of this force flow can be visualized as the flow of water, which is distributed when an object is submerged in the uniform current thereby displacing the flow lines. The resulting pattern of the flow net depends on the shape of the object or the type of opening in the wall and support conditions. Naturally, the streamlined tear-drop shape provides the least resistance to the original parallel flow lines and would provide a perfect opening. The degree of disturbance, that is, the crowding of the stream lines, indicates the increased speed or the corresponding intensity of load action.

In Figure 6.9, various wall conditions are described by treating the wall as an ideal, homogeneous material, which is obviously not true for masonry walls and not exactly the case for reinforced concrete walls. In this case, the wall surface is disturbed by various types and shapes of openings at various locations. The loads may act at the top or bottom as uniform or concentrated forces. The overall response of the wall plate to all of these effects of geometry, load, and support conditions may be (in the extreme) either an axial wall system or a wall beam structure.

When an opening is placed in the wall, the simple parallel vertical compression stress trajectories (solid lines) are pushed aside, thus causing lateral thrust and tensile trajectories (dashed lines) perpendicular to them. In other words, the uniform axial stresses are replaced by a complex state of varying stresses. When adjacent openings are placed in such a way that only a narrow space between them is available, then the compressive force lines must be funneled through it. Since the path of the straight compression lines is diverted, curved lines are generated that must be kept in equilibrium by tensile lines at right angles to them, hence the corresponding stream line pattern or flow net consists of a set of tension and compression arches that balance each other. The degree of disturbance is reflected by the steepness and concentration of the arched tensile and compressive stress trajectories.

For example, when a simply supported wall beam is loaded at the top, most loads are carried directly by compression arches to the supports—the thrust is balanced by relatively flat tensile stress trajectories throughout the beam. In contrast, when this wall beam is loaded at the bottom, then the loads are hanging from steep tension lines that are anchored in the compression arches. It is apparent from Figure 6.9 that sharp corners and geometrical irregularities cause the stress trajectories to be crowded, and thus give rise to local stress concentrations. When a concentrated load bears on a small area of the wall (Fig. 6.9, bottom), it spreads out in the wall, thereby developing tensile stresses. It may be concluded from this discussion that it is helpful in the general conceptual design stage to visualize the varying stresses in a member in terms of the principal stress trajectories, so as to develop a feeling for flow patterns, and hence stress concentrations, rather than using the mathematical model that analyses the stresses by subdividing the structural element into a rectangular grid.

The transfer of the floor loads to the wall depends on how the floor structure is supported on the wall. Details for various wall types with different floor systems are identified in Figure 6.10. They range from cast-in-place concrete slabs, precast concrete planks, and joist floors in steel and prefabricated concrete, to patented systems. The precast concrete connection (Fig. 6.10d, j) is improved by letting looped bars protrude from the concrete member(s) and by adding transverse reinforcement; the connection is completed when the joint concrete has been cast. It should also be kept in mind that the floors act as horizontal diaphragms that carry the lateral forces to the

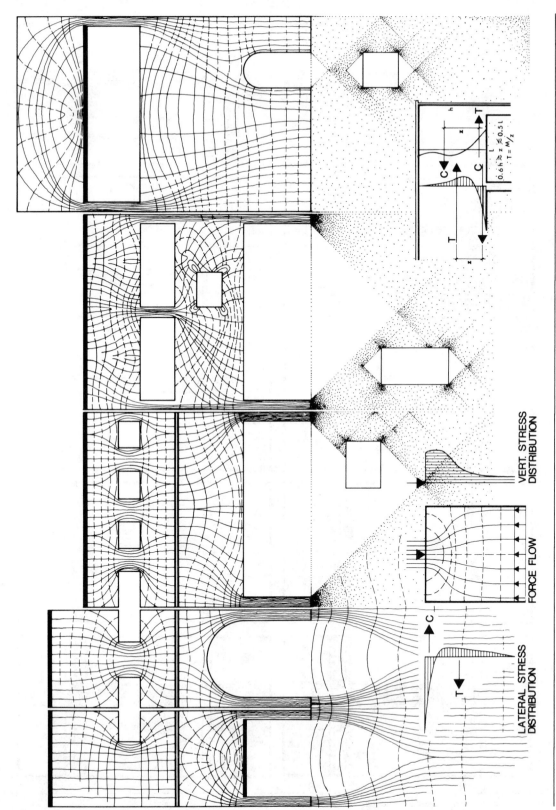

Figure 6.9. *Gravity force flow along walls.*

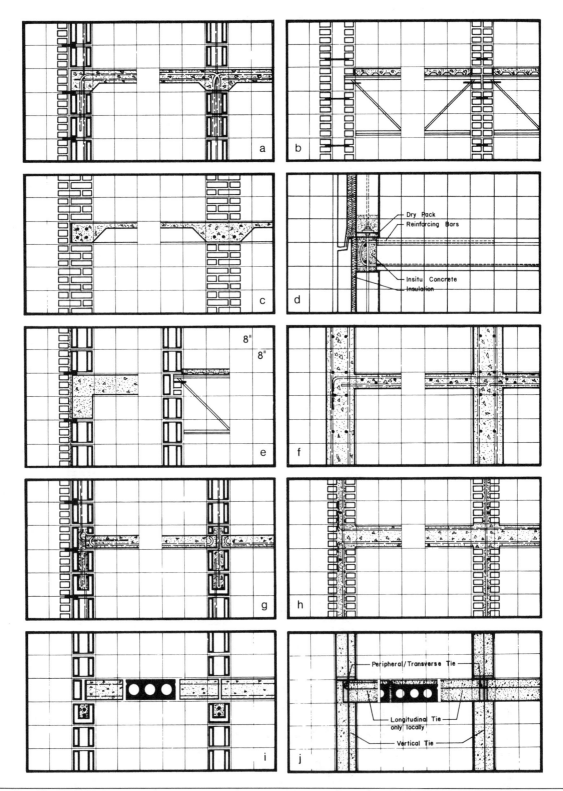

Figure 6.10. *Typical bearing wall and floor intersections.*

shear walls, thus requiring that the connections between the floors and walls be able to transfer the horizontal shear and tie forces, in addition to the forces due to gravity action on the floors.

For a symmetrical layout condition, the interior transverse load-bearing shear walls may be assumed to be fully stressed by the total floor loads on each side, thus causing only axial action in the wall, and no rotation, as in the asymmetrical live loading case (see fig. 3.3). However, when a wall supports a floor load on one side only, as is the case for an exterior long wall, then the floor reaction will cause the wall to rotate, since the center of action rarely coincides with the centroidal axis of the wall (see fig. 6.8a).

In addition to the *eccentric force* action, the *clamping action* of the wall, that is, the *degree of restraint*, must be evaluated, which is a difficult task. Typical floor structures like steel joists with concrete deck and precast single and double T-sections, where only the stems sit on the wall, can be considered as hinged floor-to-wall connections. Generally, a triangular stress distribution is assumed under the bearing. In this case, the location of the resultant eccentric floor reaction, P, due to dead and live loads, as derived from the triangular stress distribution under the bearing area of width, b (see fig. 6.8a), causes the following moment, M,

$$M = P(e) = P(t/2 - b/3) \tag{6.8}$$

For precast concrete plank construction, the sequence of erection must be taken into account. As the plank is placed upon the masonry wall or concrete wall panel, it rotates freely, and its weight, P_D, causes a moment of

$$M_D = P_D(e) = P_D(t/2 - b/3) \tag{a}$$

Should the topping be placed after the wall above has been constructed, then, with respect to the superimposed loads of topping weight and live loads, clamped conditions occur. In this instance, it is assumed that the masonry is placed directly on top of the slab, and thus does not allow free rotation. This fixed-end moment is a function of the slab span, L, and may be roughly evaluated as

$$M_L = w_L L^2/18 \tag{b}$$

Hence, the resultant moment due to eccentric dead load of the plank and the fixed-end moment due to the superimposed loads is

$$M = P_D(t/2 - b/3) + w_L L^2/18 \tag{6.9}$$

For a cast-in-place concrete slab system, rotational restraint is provided by the masonry wall, since the wall above the slab is usually constructed before the slab shoring is removed. In the upper portion of a multistory building, it may be argued that, due to the absence of large axial forces providing clamping action and due to joint slippage, hinged conditions may still be assumed. For the behavior of a continuous, cast-in-place concrete bearing wall structure, refer to the discussion of rigid frame buildings in the section in Chapter 7 on "Skeleton Buildings."

The moments in the wall due to eccentricity of the floor reactions are reduced to a minimum by extending the slab to the face of the wall (see fig. 6.8a), but the clamping action is also very much increased. The clamping action provided by the wall can be roughly approximated as

$$M = P_T(b/2) \tag{6.10}$$

From this expression, it is quite obvious that, in the lower portion of a high-rise building, the clamping action of the wall due to the large total axial forces, P_T, from above is quite high. The wall can be visualized as prestressed by the weight, thereby suppressing the tensile stresses due to the floor connections.

The bending moment, M, at the floor-wall connection is distributed to the walls above and below in proportion to their relative stiffness, I/h (see also Equation 7.6). Since, for typical conditions, the wall stiffness does not change (i.e., wall thickness and wall height), it can be assumed that one-half of the bending moment, M, is carried upward and one-half downward by treating the wall as continuous. Naturally, this is not true for the top story, where all of the roof moment must be carried downward. Further, the moments induced at the floor levels only occur locally and will not accumulate, since the floor diaphragms will provide the necessary horizontal reactions for the unbalanced moment, and transfer these forces to the shear walls.

The design of high-rise bearing walls is usually based on compressive stress, although it may happen in the upper stories of an exterior wall under wind pressure that the roof/floor dead loads may not be sufficient to suppress the tensile stress due to vertical slab action (i.e., curtain action).

For preliminary design purposes, it may be helpful to use the rather arbitrary minimum wall thickness values provided by some codes as a function of the unsupported span, h (i.e., critical clear height or length), of the masonry wall. All masonry walls should have a minimum thickness of h/36, but nonreinforced load-bearing walls must have a minimum thickness of h/20. For fast approximation purposes, it may be easier to remember the following expression for the minimum masonry wall thickness of nonreinforced bearing walls for t (in.) and h (ft), as

$$t_{min} = 0.6h \tag{6.11}$$

Basic concepts for the preliminary design of typical bearing walls are discussed in the following example.

Example 6.3

A ten-story apartment building of the spine wall type (a typical long-wall structure) in Figure 6.11 will be investigated. In this case, the load-bearing walls are parallel to the longitudinal building axes, and the floor structure spans between them; non-load-bearing shear walls are placed perpendicular to resist the wind in the primary direction (see Example 6.5).

The capacity of the 8-in. facade brick wall is checked first. Notice that the wall has only 50 percent bearing area, since the perforated wall is assumed to consist of a series of individual 5-ft-wide wall strips. The floor system consists of steel joists with a 4-in. bearing on the walls. The typical unsupported wall height is 9 ft. The building must resist a wind speed of 80 mph. Use 8000 psi brick units with type N mortar and inspection (from Table A.10: f'_m = 2000 psi). Assume the following loads:

   45 psf roof dead load, 30 psf roof live load;
   55 psf floor dead load, which includes partition weight;
   40 psf floor live load for apartment area only; and
   80 psf wall weight, which includes the window weight.

The critical compressive stresses are investigated at the building base just below the second floor structure level. It is assumed that the first floor framing sits on the concrete

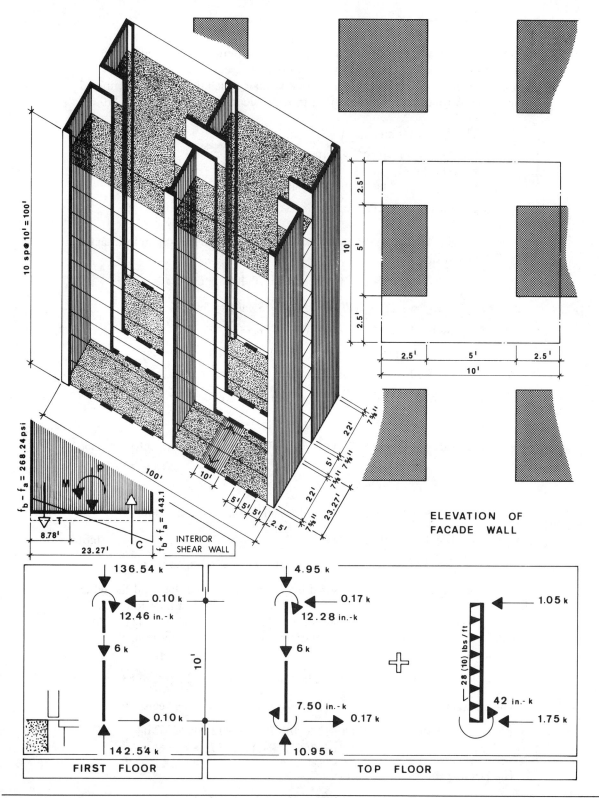

Figure 6.11. *Typical long-wall structure.*

block basement walls, which form an independent substructure. The critical tensile stresses in the brick wall are checked at the top floor level.

First, the gravity loads just above the second floor structure level are determined (Fig. 6.11). A typical 5-ft wall supports a tributary floor area of $10 \times 11 = 110 \text{ ft}^2$.

The dead loads are:

Roof loads:	0.045×110	$= \quad 4.95 \text{ k}$
Floor loads:	$8(0.055 \times 110) = 8(6.05)$	$= \quad 48.40 \text{ k}$
Wall loads:	$9[(10^2 - 5^2)0.080] = 9(6.00)$	$= \quad 54.00 \text{ k}$
Total loads:	ΣP_D	$= 107.35 \text{ k}$

The live load reduction for the apartment area, according to the section in Chapter 3 on "Dead and Live Loads" is

$$R_1 = 0.08\%A = 0.08(110)8 = 70.40\% > 60\%$$

$$R_2 = 23(1 + (D/L)) = 23(1 + 55/40) = 54.63\%$$

Therefore, the live loads can be reduced by 55 percent. The live loads are

$$P_L = 0.030(110) + 8(0.45 \times 0.040)110 = 3.30 + 8(1.98)$$
$$= 3.30 + 15.84 = 19.14 \text{ k}$$

The total load just above the second floor structure level is

$$P_T = 107.35 + 19.14 = 126.49 \text{ k}$$

For a live load reduction of $0.08 (110) = 8.8\%$, the wall must carry a typical floor load of

$$P_F = [0.055 + 0.91(0.040)]110 = 10.05 \text{ k}$$

This floor load causes a moment upon the wall due to eccentric action of

$$M = P(e) = P(t/2 - b/3) \tag{6.8}$$
$$= 10.05[(7.625/2) - 4/3] = 10.05(2.48) = 24.92 \text{ in.-k}$$

One-half of this moment will be resisted by the wall above the floor and the other half by the wall below. The critical stresses at the top of the first floor level can be determined now (Fig. 6.11). The axial forces at that location are due to the total load just above the second floor structure and due to the floor load causing the local moment. The axial stresses are

$$f_a = P/A = (126.49 + 10.05)1000/[5(92)] = 136540/[5(92)] = 296.83 \text{ psi}$$

The allowable compressive stresses (Table A.9) are approximately equal to

$$F_a \simeq f'_m(0.233 - h/300t) = 2000\{0.233 - 9(12)/[300(7.625)]\}$$
$$= 466 - 94.43 = 371.57 \text{ psi}$$

The bending stresses are equal to

$$f_b = \pm M/S = (24920/2)/[5(116)] = \pm 21.48 \text{ psi}$$

The allowable flexural compressive stresses are approximately equal to

$$F_b \simeq 0.33f'_m = 0.33(2000) = 660 \text{ psi}$$

Checking the beam-column interaction yields

$$f_a/F_a + f_bF_b \leq 1$$

$$(296.83/371.57) + (21.48/660) = 0.799 + 0.033 = 0.832 < 1 \qquad \text{OK}$$

This clearly shows that axial action controls the design so that local bending can be ignored for preliminary design purposes. In addition, the effect of slenderness can be neglected for the design of typical bearing wall buildings during the first approximation stage.

The following simplified approach can be used as a fast check of the critical compressive stresses in the bearing walls at the building base or other levels where the wall thickness or material strength change.

$$f_a = P/A \leq F_a \simeq 0.2f'_m \tag{6.12}$$

Hence, the critical axial compressive stresses can easily be checked at the first floor level at the bottom of the wall rather than at the top, as before.

The respective axial dead and live loads are:

$$P_D = 4.95 + 9(6.05) + 10(6.00) = 119.40 \text{ k}$$

$$P_L = 3.30 + 9(1.98) = 21.12 \text{ k}$$

$$P_T = P_D + P_L = 119.40 + 21.12 = 140.52 \text{ k}$$

The corresponding axial compressive stresses are

$$f_a = P/A = 140520/[5(92)] = 305.48 \text{ psi} < 400 \text{ psi}$$

where the approximate allowable compressive stresses are

$$F_a \simeq 0.2f'_m = 0.2(2000) = 400 \text{ psi}$$

The shear stresses in the wall due to the end moments are small and can be neglected. The loading case, where the wind acts perpendicular to the wall together with the gravity loads, rarely controls at the bottom portion of the typical high-rise building (see Problem 6.15).

Now, the wall is checked at the top floor level for the critical tensile stresses, since, for this condition, the axial forces due to dead load action are small, but the wind pressure is high. At the top floor level, the wind pressure may be assumed to be 28 psf (see fig. 3.8), ignoring the larger pressures along the roof corners.

The maximum wind moment at the bottom of the continuous wall, by conservatively taking the wall span as equal to the floor height, is

$$M_w = wh^2/8 = (28 \times 10)10^2/8 = 3500 \text{ ft-lbs} = 42 \text{ in.-k}$$

The maximum wind shear at the bottom of the wall is

$$V_w = 0.028(10 \times 10)/2 + 3.50/10 = 1.75 \text{ k}$$

Since the weight counteracts the wind stresses, only the dead load is considered.

The moment in the wall, due to eccentric floor action at the bottom, is

$$M_{Db} = 6.05(2.48)/2 = 7.50 \text{ in.-k}$$

However, the gravity moment at the top must be fully balanced by the wall.

$$M_{Dt} = 4.95(2.48) = 12.28 \text{ in.-k}$$

The corresponding shear is

$$V_D = [(12.28 + 7.50)/12]/10 = 0.17 \text{ k}$$

Now, the critical tensile stresses at the bottom of the top floor level can be checked. The axial forces at this location are due to the wall and roof weight. Notice that the floor rotates the wall in a direction opposite to the wind moment (Fig. 6.11).

$$f_t = M/S - P_D/A$$

$$= [(42.00 - 7.50)/5(116) - (6.00 + 4.95)/5(92)]1000 = 59.48 - 23.80$$

$$= 35.68 \text{ psi}$$

The allowable tensile stresses normal to the bed joints are

$$F_t = 1.33(28) = 37.24 \text{ psi}$$

Hence, the type N mortar is satisfactory. Should the wall, for fast approximation purposes, be treated as non-load-bearing, and the axial compressive stresses be neglected, then it would not be satisfactory.

The shear stresses are equal to

$$f_v = V/A = (1.75 - 0.17)1000/[5(92)] = 3.43 \text{ psi}$$

The allowable shear stresses, neglecting the beneficial effect of the weight, are

$$F_v = 0.5\sqrt{f'_m} = 0.5\sqrt{2000} = 22.36 \text{ psi}$$

Increasing the allowable stresses for the wind condition yields

$$F_v = 1.33(22.36) = 29.74 \text{ psi} > 3.43 \text{ psi} \text{ OK}$$

For ordinary loading conditions, the shear stresses do not have to be checked during the preliminary design stage! Refer to Problem 6.16 for the investigation of the full gravity loading case at the top floor level.

The exterior long wall must also be checked as a shear wall with the gravity loading and the wind action parallel to the wall. For this condition, the wall is subject to biaxial bending. However, since the wind pressure is relatively small and the moment of inertia of the long walls is quite large, this case may be ignored for preliminary design purposes.

Shear Walls

In a building, the vertical structural planes are tied together by the horizontal floor planes. The lateral forces from wind or earthquake are distributed along the floor framing, which act as deep horizontal beams, to the vertical shear walls, which are assumed to be located parallel to the direction of the lateral loads. The shear walls, in turn, carry the loads to the foundations and ground. The name *shear wall* may be confusing, since it may be a flexural beam, and not a shear beam where shear action controls the behavior of the vertical cantilever wall. In cross-wall buildings (see fig. 6.15), the shear walls are also bearing walls; their arrangement can be parallel, in-line, or staggered, and sometimes randomly distributed. The typical spacing of the parallel cross-walls in residential construction is between 12 and 24 ft. In long-wall buildings (Fig. 6.11), the shear walls may not be load-bearing, similar to the shear walls that stiffen skeleton structures. The typical shear walls in bearing wall buildings are constructed from cast-in-place concrete, precast concrete, and masonry.

In large-panel concrete construction, the shear walls may form simple planar surfaces, while in masonry buildings they usually intersect with other walls, which provide single or double returns at the ends and a certain effective width in resisting lateral forces, assuming proper bonding at the intersection points. Hence, masonry shear walls may be visualized as vertical cantilever plate girders over the full building height, with the walls representing the web, the floors the web stiffeners, and the end returns the flanges. It is quite clear that the capacity and stiffness of the tall flanged wall is considerably increased because of the increase of the moment of inertia of the wall, in comparison to the plain wall without returns. The lateral stiffness can be further improved by coupling collinear shear walls, which also may help to reduce possible tension stresses, as is discussed in more detail later in this subsection.

The force flow in response to vertical and lateral loading along a simple solid cantilever shear wall without any openings is familiar from basic mechanics. It is discussed later (see fig. 6.14, top); the lateral shear and gravity flow increase linearly towards the base, while the moment increases much more rapidly, possibly causing tension on the windward side. Typical failure mechanisms are shown in Figure 6.12. The geometric proportions of the shear wall will determine whether its strength is controlled by shear or flexure, as has been examined in Chapter 4, in the discussion of shallow and deep beams.

In order to design a shear wall, its loading must be known. With respect to lateral loading, the building is assumed to deflect as a unit, because it is tied together by the rigid floor diaphragms. Therefore, each of the shear walls must take the same deflected shape under symmetrical conditions, and thus will attract a certain portion of the total lateral load in proportion to its stiffness. It is assumed that the connections of floors and walls are adequate, so horizontal and vertical forces can be transferred. When the building consists of a combination of different shear wall types, the stiffest system will attract most of the loads, and the most flexible one the least. Once the loading for each of the walls has been determined, their behavior can be investigated separately.

The stiffness or rigidity of the shear wall is measured by its deflection (see the subsection in Chapter 5 on "Indeterminate Force Distribution for Conditions of Symmetry"). The type of deflection, in turn, depends on the wall proportions of height-to-width (H/L), as indicated in the top middle of Figure 6.13, by treating the wall as solid and the foundations as rigid. The lateral deflection of the solid shear wall due to flexure and shear, as based on uniform loading according to Table A.14, assuming the shearing modulus, G, for concrete and masonry as 40 percent of its modulus of elasticity, E, is

$$\Delta = \Delta_f + \Delta_s$$

$$= wH^4/(8EI) + 6wH^2/(10GA), \ \text{let } G \simeq 0.4E, \text{ and } V = wH$$

$$\Delta = 1.5V[(H/L)^3 + H/L]/Et \tag{6.13}$$

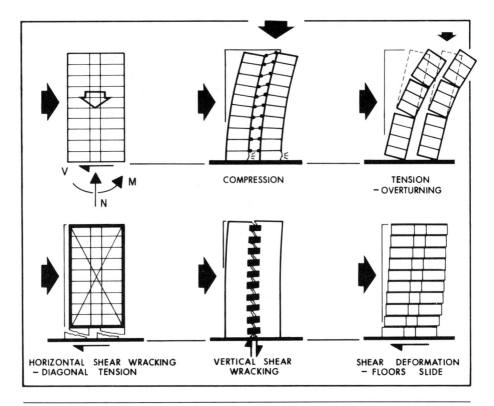

Figure 6.12. *Typical failure mechanisms in cantilever shear walls with openings.*

The lateral sway due to a triangular load as for earthquake action is roughly 50 percent larger. The respective ratio of the lateral shear deflection to the flexural deflection is

$$\Delta_s/\Delta_f = (L/H)^2 \tag{6.14}$$

From this expression one may draw the following conclusions:

- Long, solid walls that are at least three times longer than high ($L/H \geqslant 3$) behave like stiff shear panels where shear deflections control. For this condition, the flexural deflection may constitute only about 10 percent of the total deflection and can be neglected for preliminary design purposes. The walls may be treated on a story by story basis.
- In contrast, for tall, slender, solid walls that are at least three times higher than wide ($H/L \geqslant 3$), the shear deformation may represent about 10 percent of the total deflection and be neglected for preliminary design purposes. Naturally, wall openings may reduce the resisting shear area so that shear deformations can no longer be ignored. Tall, slender, solid walls are obviously more flexible than shear panels; they are moment systems and act as deep cantilever beams where flexural deflection clearly controls.
- The intermediate solid walls represent the transition stage from the long to the high slender wall. For this condition, shear together with flexure cause the lateral deflection.

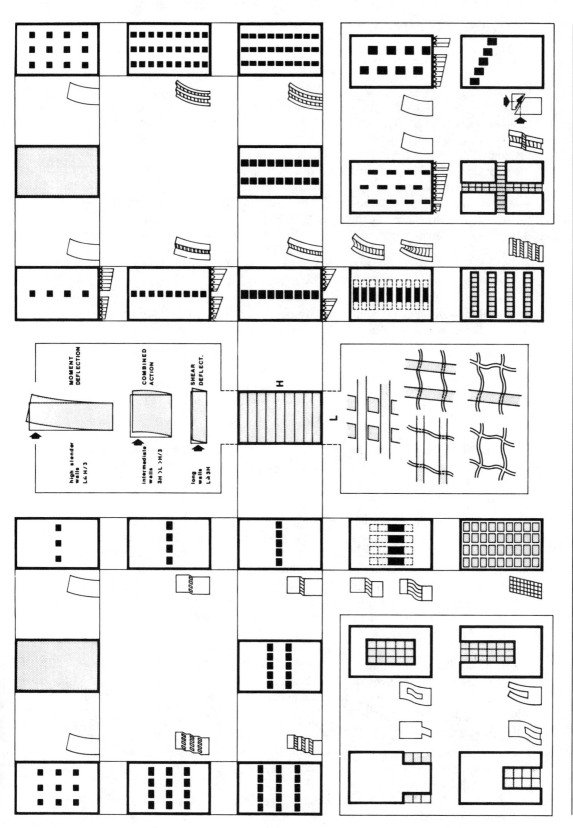

Figure 6.13. The effect of lateral load action upon walls with openings.

Typical solid shear walls may occur as load-bearing interior cross-walls for single-loaded corridors, and together with longitudinal corridor walls for double-loaded corridors.

Naturally, openings will influence the behavior and stiffness of the wall; their size, location, shape, and arrangement are critical, since they disrupt the continuity of the load paths. Also, abrupt changes in the outline and cross section of the wall may have the quite negative effect of stress concentration (see fig. 6.9). The reader may want to study Figure 6.13, where the effect of various opening patterns upon stress distribution and lateral deformation are investigated. Large openings in walls may result in predominant frame action, while small openings may not affect the overall behavior of the wall at all. One large opening may have less influence than a narrow window strip, which may subdivide the wall. Similarly, a staggered opening arrangement may not interrupt the force flow as much as a regular window pattern. It is evident that the wall behavior is not only related to the percentage of wall perforations, but also to the arrangement of the openings.

There are an infinite variety of possible patterns for wall openings, particularly for exterior shear walls; they range from uniform, to grouped, to random arrangements. Some of the more common uniform window wall patterns of regular spacing are:

- *Vertical window slots with small spandrels and large piers*, approaching the coupled shear wall situation (which is discussed later).
- *Horizontal window bands with heavy spandrels and flexible piers*: Because of the large stiffness of the deep spandrels, they hardly deform, whereas the columns bend over the height of the opening in a way similar to the behavior of an interstitial structure.
- *Pierced walls*: As the size of the openings on a regular grid increases, the behavior of the pierced wall changes from the stiff solid wall, with large piers and deep spandrels, to the flexible frame.
- Combinations of the above.

The typical story wracking for various structural systems is shown at the center bottom of Figure 6.13; depending on the member proportions, that is, their relative stiffness, either shear or flexural distortions are indicated. The lateral deflection of individual story walls or frames of high-rise buildings may be treated as fixed on top and bottom because of the rigid diaphragm action of the floors and the weight from above. Naturally, the lateral deflection of the entire building is that of a cantilever.

The condition where individual shear walls are connected or coupled is quite common. Frequently, vertical rows of windows, doors, or corridors divide the wall into two or more walls. The coupling of the collinear shear walls is provided by beams, plates, lintels, floor slabs, or any combination of the above. The determination of the degree of coupling is quite complex. But a sense for their behavior can be developed by dealing with the following extreme conditions, using collinear shear walls of equal stiffness and proportions (Fig. 6.14).

1) *Flexible coupling*: In this case, the openings are so large that the connecting beams are thin and very flexible, or the lintels are only simply supported. For this condition, the coupled shear walls may be treated as consisting of separate walls, each acting as independent vertical cantilevers. The connecting beams do not resist any vertical shear ($V_G = 0$), they simply act as hinged struts that transfer a portion of the lateral loads in axial action, similar to the ties in cavity walls. The

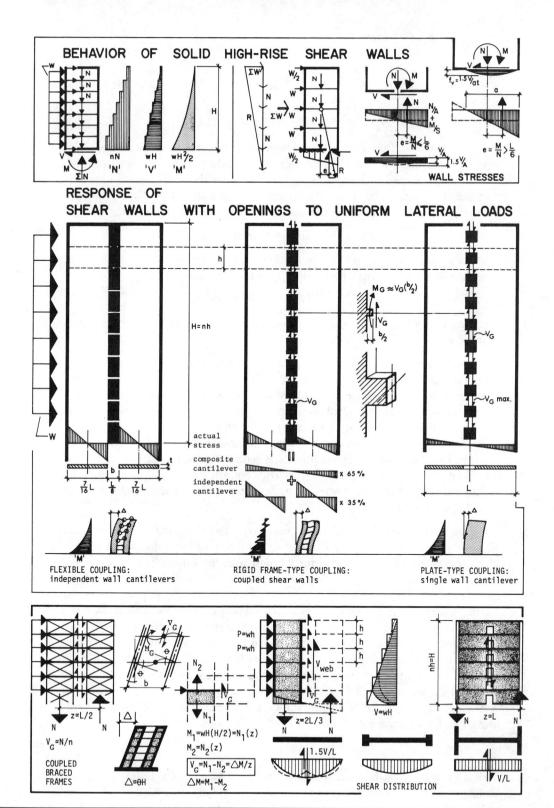

Figure 6.14. *The behavior of ordinary shear walls.*

moment of inertia provided by two collinear, identical walls separated by a corridor width of b = L/8 is: $I = 2[t(7L/16)^3/12] \simeq tL^3/72$. The corresponding maximum flexural stress is

$$f_b = (M/2)(7/32)L/[tL^3/144] \simeq 16M/tL^2$$

2) *Plate-type coupling:* For this condition, only small openings pierce the wall, so that deep beam sections can be provided. In other words, the deep spandrel beams are so rigid that the openings do not influence the single cantilever action of the shear wall. The moment of inertia of the wall is $I = tL^3/12$ and the corresponding lateral building deflection, ignoring the vertical loading, is

$$\Delta = wH^4/8EI$$

It is apparent that this wall system is about six times stiffer, compared to the very flexible coupling system. The maximum flexural stresses are

$$f_b = M(L/2)/[tL^3/12] = 6M/tL^2$$

Hence, the stresses are roughly only one-third as large as for the individual action of the walls with flexible coupling.

3) *Rigid frame-type coupling:* In general, the connecting beams are neither infinitely flexible nor infinitely rigid; the collinear walls do interact because the beams do transfer some shear and moment. The lateral sway of the coupled wall system is very much dependent on the stiffness of the linking beams, which resist the vertical shear wracking; for slender walls, they may determine the lateral building sway. The assessment of the flexural stresses in the walls is quite complicated and beyond the scope of this investigation. For the specific case in Figure 6.14, it can be visualized that the behavior consists of the combined action of approximately 65 percent composite cantilever action plus 35 percent independent cantilever action.

For the preliminary design of a rigid-frame type coupling, one may conservatively disregard the coupling effect with respect to the design of the walls, and just use the simple cantilever mode; however, for the design of the linking beams, the vertical shear wracking must be taken into account to eliminate possible crack formation in the beams.

The magnitude of the vertical shear at midspan of a connecting beam can be estimated by assuming the beam to be rigidly connected to a composite wall system. It is known from the strength of materials that the maximum shear stress for a rectangular beam section at the centroidal axis is $f_v = 1.5(V/A)$. In other words, the maximum horizontal shear force per unit length, and the vertical shear per unit height, of a solid cantilever wall is $P_v = f_v(t) = 1.5 V/L$, where the cross-sectional area is $A = tL$. It may be concluded that the vertical shear force to be resisted by a rigid coupling girder for a constant story shear is

$$V_G = 1.5(V/L)h \tag{a}$$

Here, V is the total horizontal shear at the floor below the coupling beam level. Usually, the shear walls are flanged. However, for large flanges the web shear is constant, $f_v = V/A$. Therefore, the coupling girder shear approaches

$$V_G = (V/L)h \tag{b}$$

Equations (a) and (b) can also be easily derived by balancing the flexural stresses above and below the story with the vertical shear, V_G, as indicated at the bottom of Figure 6.14. The magnitude of the girder shears given in Equations (a) and (b) are reasonable for walls with relatively small openings; however, for large openings, the beam shear approaches zero ($V_G = 0$). In other words, as the openings increase, more independent cantilever action of the walls takes place, thus causing less girder shear; that is, the bending stresses in the walls at the center increase, while the shear stresses at that location decrease, as shown in Figure 6.14. One may conclude that it is generally conservative (considering the effect of the flanges and the effect of the openings) to use, as a first estimate of girder shear,

$$V_G \simeq (V/L)h \qquad (6.15)$$

Hence, for approximation purposes, a constant vertical shear flow or unit shear is assumed, so that the total shear increases in a stepped triangular fashion towards the base.

Nearly the same result is obtained for the case of two identical individual trussed steel frames coupled together, as shown at the bottom left of Figure 6.14. The rotation, M, due to the lateral force action at the level to be investigated is resisted by axial forces, N, located at the center of resistance of the walls: $M = N(z)$; the internal lever arm between these forces is roughly $z = 0.56L$, or approximately $z = L/2$. The total vertical web shear must resist the axial force N, or $N = V_{web}$. For the ductile material, steel, it may be assumed that the web shear is equally resisted by all n coupling girders, $V_G = V_{web}/n$

$$V_G = (M/z)/n = [wH(H/2)/L/2]/n = (V/L)h$$

where:
$$H = nh$$

The maximum girder moment at the walls due to the vertical shear wracking can be approximated as

$$M_G \simeq V_G(b/2) \qquad (6.16)$$

The lateral deflection of a coupled shear wall system that consists of slender walls and long-span coupling girders, may be roughly approximated by conservatively assuming that the lateral deflection is primarily due to the vertical shear wracking of the horizontal coupling beams (see Equation 7.12). It is shown (in fig. 6.14 and fig. 7.17) that the maximum lateral sway can be estimated as

$$\Delta \simeq \theta H = \frac{HV_G b^2}{12EI_G} = \frac{HM_G b}{6EI_G} \qquad (6.17)$$

It should also be remembered that, for tall walls ($H/L \geq 3$), the effect of the flanges very much increases the lateral stiffness of the building, which is not, however, the case for long walls in low-rise buildings, where the shear deformation (i.e., the web area) controls the lateral rigidity.

In precast large-panel wall construction, the shear resistance of the vertical joints can be compared with the rigidity of the beams in coupled wall structures. In this case, the wall-to-wall connections may be open, and hinges may only be provided at floor levels, so that vertical shear forces cannot be transmitted and sliding is not prevented, thus resulting in independent action of the panels and a reduction of overall stiffness. In the rigid connection system, full interaction of the walls is provided, while in the semirigid joint, some vertical slippage does occur, similar to the rigid-frame type coupling.

In the following, some typical building cases are investigated, to develop a basic understanding for shear wall design. The primary conclusions are:

- Usually, the design of load-bearing flanged shear walls in cross-wall buildings is controlled by the compressive stresses, since the tension due to lateral force action is suppressed by the weight. Shear stresses rarely control.
- The design of non-load-bearing shear walls, on the other hand, may be controlled by stability against overturning and tensile stresses. This is especially true when there are only a few walls, which are spaced far apart, so that they must resist large lateral forces. It may be necessary to tie the shear walls to the intersecting bearing walls to control the overturning moment. Also, the shear stresses may be critical.

Example 6.4

A typical interior, 8-in., transverse brick wall of a ten-story apartment building shown in Figure 6.15 (a typical cross-wall structure), with a regular story height of 10 ft, will

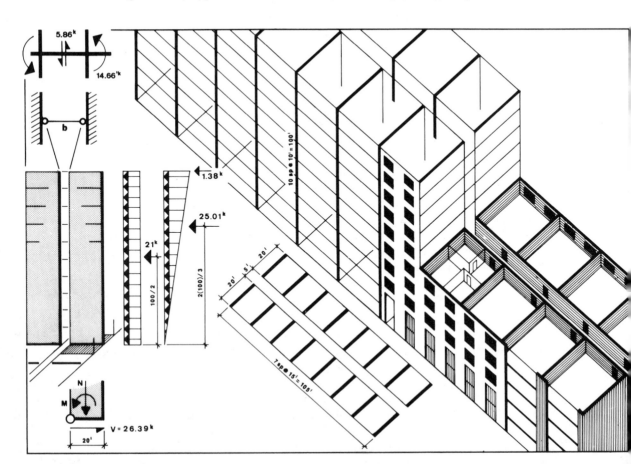

Figure 6.15. *Typical shear wall structure.*

be investigated. The cross-walls are load-bearing walls, which support the floor loads, but also act as shear walls to resist the wind and seismic forces against the long building face. The parallel pairs of transverse walls are 15 ft apart; the flexural coupling across the corridor is treated as representing hinged struts. Only the webs of the load-bearing shear walls are assumed to resist the forces (see fig. 2.16), and the flanges are conservatively neglected for this preliminary check; all of the shear walls are assumed to provide the same stiffness. The lateral forces against the short building face are resisted by a system of longitudinal shear walls along the facades and the interior corridor.

The following loading conditions are used by considering the roof dead load to be equal to the floor dead load: 65 psf floor dead load, 40 psf floor live load, 30 psf roof live load, and 80 psf wall weight. The additional corridor live loads are assumed to be carried by the longitudinal corridor walls, which have been conservatively ignored. The basis of design for the lateral loads is an 80 mph wind speed and seismic zone 2. 8000 psi brick units with type N mortar and inspection are to be used. The unsupported wall height is taken as 9 ft, 6 in.

A typical 20-ft interior bearing wall carries, at the first floor level, the following dead load due to the floor and wall weight:

$$P_D = [0.065(15 \times 22.5) + 0.080(10 \times 20)]10 = 379.38 \text{ k}$$

The live load reduction is 0.08 percent per square foot of area supported, except roof loads, but should not exceed

$$R = 23(1 + (D/L)) \leq 60\% \tag{3.1}$$

The typical reduction per floor is

$$R = (15 \times 22.5)0.08 = 27\%$$

The live load reduction for this investigation at the first floor level is

$$R_1 = (27\%)9 = 243\%$$

$$R_2 = 23(1 + 65/40) = 60.38\% > 60\%$$

Therefore, the live loads are to be reduced by 60 percent.
The axial live loads at the first floor level are

$$P_L = [0.040(0.4)9 + 0.030](15 \times 22.5) = 58.73 \text{ k}$$

The total axial loads at the first floor level are

$$P_T = P_D + P_L = 379.38 + 58.73 = 438.11 \text{ k}$$

With respect to the lateral force action, the many closely spaced cross-walls give a very rigid support to the floor diaphragms, which can thus be treated as flexible. Therefore, the lateral wind pressure is distributed to the interior walls as derived from the tributary areas, regardless of stiffness conditions. However, the rigid floor diaphragm concept may still be used conservatively for the load distribution to the end walls.

First, the wind pressure is investigated. It is safe to use a uniform wind pressure of 28 psf (see fig. 3.8) for this preliminary investigation, rather than taking the true wind distribution.

An interior shear wall must resist a total wind shear of

$$V_w = [0.028(15 \times 100)]/2 = 21 \text{ k}$$

This load causes the following moment about the building base

$$M_w = 21(100/2) = 1050 \text{ ft-k}$$

The seismic action for zone 2 is checked now. First, the total building weight is found; for the longitudinal walls along the facades and corridor only 50 percent of the wall area is assumed to account for openings.

The long walls weigh:	$[0.080(105 \times 100)0.5]4$	$= 1680 \text{ k}$
The cross walls weigh:	$[0.080(20 \times 100)]16$	$= 2560 \text{ k}$
The floors weigh:	$[0.065(45 \times 105)]10$	$= 3071 \text{ k}$
The total building weight is:		$= 7311 \text{ k}$

In the absence of a more accurate information on the period of the building, it may be found from

$$T = 0.05h_n/\sqrt{D} = 0.05(100)/\sqrt{45} = 0.745 \text{ s} \tag{3.21}$$

The seismic coefficient is

$$C = 1/[15\sqrt{T}] = 1/[15\sqrt{0.745}] = 0.077 \leqslant 0.12 \tag{3.19}$$

The numerical coefficient for the site-structure resonance is not established.

$$CS = 0.077(1.5) = 0.1158 \leqslant 0.14$$

The framing coefficient for a box structure is $K = 1.33$. The importance factor is $I = 1$, and the zone factor for zone 2 is $Z = 3/8$. The total lateral seismic force acting on the building is

$$V = ZIKCSW = (3/8)1(1.33)(0.1158)7311 = 422 \text{ k} \tag{3.18}$$

Therefore, each of the 16 cross walls carries a shear of

$$V_s = 422/16 = 26.39 \text{ k} > 21 \text{ k}$$

Since the period $T \geqslant 0.7s$, the whiplash force at the top of the building is

$$F_t = 0.07TV = 0.07(0.745)26.39 = 1.38 \text{ k} \tag{3.24}$$

The rotation due to the seismic force action, assuming a uniform weight distribution, that is, a triangular load distribution, is

$$M_s = 1.38(100) + (26.39 - 1.38)100(2/3) = 1805 \text{ ft-k} > 1050 \text{ ft-k}$$

The seismic shear and rotation are both larger than the respective wind values, hence control the design. Notice that the additional torsional shear forces due to the minimum seismic eccentricity of $(5\%)105 = 5.25$ ft has been ignored!

First, the stability against overturning is checked. The resisting dead load moment is

$$M_{res} = P_D(20/2) = 379.38(20/2) = 3794 \text{ ft-k}$$

The safety factor against overturning is

$$\text{S.F} = M_{res}/M_{act} = 3794/1805 = 2.10 > 1.5 \qquad \text{OK} \tag{2.9}$$

The axial compressive stresses at the first floor level, due to the dead and live loads, are

$$f_a = P_T/A = 438110/[7.625(20)12] = 239 \text{ psi}$$

$$F_a \simeq 0.2f'_m = 0.2(2000) = 400 \text{ psi} > f_a \qquad \text{OK}$$

Notice that the effect of the wall slenderness upon the allowable compressive stress has been ignored (see Example 6.3). The flexural stresses due to lateral force action are

$$f_b = \pm M/S = 1805000(12)/[7.625(20 \times 12)^2/6] = 296 \text{ psi}$$

$$F_b \simeq 0.33f'_m = 0.33(2000) = 660 \text{ psi}$$

Now, the combined axial and flexural compressive stresses due to the dead, live, and seismic loads are checked.

$$f_a/F_a + f_b/F_b \leq 1(1.33)$$

$$239/400 + 296/660 = 0.60 + 0.45 = 1.05 < 1.33 \qquad \text{OK}$$

The critical tensile stresses due to lateral force action and dead load reaction are

$$f_t = f_b - f_a \leq 1.33F_t$$

$$= M/S - P_D/A = 296 - 379380/[7.625(20)12]$$

$$= 296 - 207 = 89 \text{ psi} > 1.33(28) = 37 \text{ psi} \qquad \text{NG (No Good)}$$

It is quite clear that the weight of the flanges and their bending resistance must be taken into account. For the condition where there are no flanges available, reinforcing bars should be provided (Example 6.5), or the two collinear shear walls should be coupled with stiffer beams, or the wall thickness should be increased, possibly together with a higher quality mortar.

The shear stresses may be assumed to be uniformly distributed along the length of the wall because of the flanged shear wall shape.

$$f_v = V/A = 26390/[7.625(20)12] = 14.42 \text{ psi}$$

$$F_v = 1.33(0.5\sqrt{f'_m}) = 1.33(0.5\sqrt{2000}) = 29.74 \text{ psi} > f_v \qquad \text{OK}$$

In this example, the connecting slabs have been treated as flexible with respect to the approximate design of the walls. For the sake of the exercise, it is assumed now that the slabs in the corridor area are sitting on continuous concrete beams that provide rigid frame-coupling to the walls. Although the approach used for the checking of the walls is still alright (though conservative), the coupling beams will resist part of the lateral load in bending and may crack if they cannot carry the moments.

The critical seismic wall shear at the base of the building is $V_s = 26.39$ k. The corresponding beam shear is estimated as

$$V_G \simeq = (V/L)h = 2(26.39/45)10 = 11.73 \text{ k} \qquad (6.15)$$

Hence, the beam at the first floor level must be designed for the following moment.

$$M_G \simeq V_G(b/2) = 11.73(5/2) = 29.32 \text{ ft-k} \qquad (6.16)$$

Example 6.5

In Example 6.3, the exterior walls of the long-wall structure shown in Figure 6.11 have been investigated. Here, the interior non-load-bearing 8-in. brick shear walls across the short direction of the building (a typical non-load-bearing shear wall) will be checked with respect to lateral wind action; for the loading conditions refer to Example 6.3. The walls are assumed not to be coupled across the central corridor space.

It has been shown, in the subsection in Chapter 5 on "Indeterminate Force Distribution for Conditions of Symmetry," that the lateral forces are distributed to the vertical shear walls in proportion to their relative stiffnesses, as based on rigid diaphragm action of the floors. The stiffness, k, is the inverse of the lateral deformation, Δ, due to a unit force, P.

$$k = 1/\Delta \tag{5.1}$$

Since the approximate wall proportions are $H/L \simeq 100/23 = 4.35 > 3$, moment deflections clearly control. Therefore, the moment of inertias of the walls' cross section determine the distribution of the lateral wind pressure, P. Because of the symmetric wall arrangement and building elevation, only translational forces occur; however, torsion must be considered for seismic action. Each of the shear walls will resist

$$P_i = P(k_i/\Sigma k) = P(I_i/\Sigma I) \tag{5.3b}$$

The following moment of inertias have been determined in Problem 2.21 (as related to fig. 2.16f):

- each of the end walls: $I_e = 1118 \text{ ft}^4$
- each of the interior walls: $I_i = 1616 \text{ ft}^4$

Hence, the total lateral stiffness provided by all of the shear walls is

$$\Sigma I = 4(1118) + 2(1616) = 7704 \text{ ft}^4$$

Each of the four end walls resists the following proportion of the total lateral load, as based on their relative stiffness

$$I_e/\Sigma I = (1118/7704)100 = 14.51\%$$

Each of the two interior walls resists

$$I_i/\Sigma I = (1616/7704)100 = 20.98\%$$
$$\text{Check: } 4(14.51) + 2(20.98) = 100\% \qquad \text{OK}$$

Since it is time-consuming to find the moment of inertias of the shear walls for preliminary estimation purposes, the designer may want to treat the relatively long spans of the floor diaphragms between the shear walls as flexible. In this case, the wind pressure can be distributed according to the tributary facade area, that is, each of the end walls resists 12.5 percent and each of the interior walls, 25 percent of the total lateral force. Notice that this approach would be unconservative with respect to the design of the end walls, each of which should carry 14.51 percent!

For the condition where the shear walls are closer together, one may assume (for preliminary design purposes) the same stiffness for all of the shear walls, together with rigid diaphragm action. For this example, assuming that the conditions above would be applicable, then each wall would have to resist $100/6 = 16.67\%$ of the total load.

Rather than altering the wind pressure with height, an equal pressure of 28 psf is conservatively assumed. The total wind pressure on the building (see fig. 3.15), is

$$P_H = 0.028(100 \times 100) = 280 \text{ k}$$

Therefore, each of the interior shear walls must resist

$$P_w = 280(20.98\%) = 58.74 \text{ k}$$

This lateral force causes a rotation about the building base of

$$M_w = 58.74(100/2) = 2937 \text{ ft-k}$$

The resisting weight is conservatively based on a symmetrical shear wall with 7.5-ft flanges at each end, as derived from an average flange width at the facade (see fig. 6.11, right); the floor loads supported by the flanges are ignored.

web weight:	$0.080(22 \times 100)$	$= 176 \text{ k}$
exterior flange weight:	$0.080(7.5 \times 100)$	$= 60 \text{ k}$
interior flange weight:		$= 60 \text{ k}$
total wall weight:		$= 296 \text{ k}$

The wall provides the following resisting moment:

$$M_{res} = 176(11 + 0.635) + 60(0.635 + 22 + 0.635/2) + 60(0.635/2)$$

$$= 296(0.635 + 22/2) = 3444 \text{ ft-k}$$

The safety factor against overturning is

$$\text{S.F.} = M_{res}/M_w = 3444/2937 = 1.17 < 1.5 \qquad \text{NG}$$

The structure is not safe if it is not anchored to its base or to the ground. It is apparent that the floor loads on the flanges must be considered; see Problem 6.26 for a more precise investigation.

The shear is resisted primarily by the web of the flanged wall in a relatively uniform manner, and yields the following shear stress

$$f_v = V/A = 58740/[7.625(23.27)12] = 27.59 \text{ psi}$$

$$F_v = 1.33(0.5\sqrt{f'_m}) = 1.33(0.5)\sqrt{2000} = 29.74 \text{ psi} > f_v \qquad \text{OK}$$

Since it is the intention here to demonstrate how to find the steel reinforcing to resist the high tensile stresses, only the wall web is considered in the following calculations.

The axial compressive stresses, as based on web weight, are

$$f_a = P/A = 176000/[7.625(22)12] = 87.43 \text{ psi}$$

$$F_a \simeq 0.2f'_m = 0.2(2000) = 400 \text{ psi, ignoring the effects of wall slenderness.}$$

The flexural stresses due to the lateral force action are

$$\pm f_b = M/S = 2937000(12)/[7.625(23.27 \times 12)^2/6] = 355.67 \text{ psi}$$

$$F_b \simeq 0.33f'_m = 0.33(2000) = 660 \text{ psi}$$

The combined axial and flexural compressive stresses do not present a problem.

$$f_a/F_a + f_b/F_b \leq 1(1.33)$$

$$87.43/400 + 355.67/660 = 0.76 < 1.33 \qquad \text{OK}$$

The tensile stresses are far beyond the masonry capacity.

$$f_t = f_b - f_a \leq 1.33F_t$$

$$= 355.67 - 87.43 = 268.24 \text{ psi} > 1.33(28) \qquad \text{NG}$$

It is obvious that steel reinforcing is required for the assumed conditions.

For fast approximation purposes, a homogeneous, uncracked wall section is assumed where the tension resistance is provided by the steel bars. The reinforcing is considered to act at the centroid of the tensile stress triangle, and to be strained to its allowable stress state. These assumptions are conservative since, in reality, the steel bars are located closer to the wall end, thus providing a larger lever arm, in turn requiring less steel for the resulting lower forces; therefore, the actual steel stress is below the allowable as based on elastic material behavior and linear variation of the strain from the neutral axis (N.A.). Refer to Figure 6.11 for the stress diagram.

From similar triangles, the length, x, of the tensile stress block can be found as

$$23.27/[443.10 + 268.24] = x/268.24, \text{ or} \qquad x = 8.78 \text{ ft}$$

The tensile force, T, is equal to the contents of the tensile stress block.

$$T = 268.24(8.78/2)12(7.625)/1000 = 107.75 \text{ k}$$

The required steel area to cover this force is approximately equal to

$$A_s \simeq T/1.33F_s = 107.75/1.33(20) = 4.05 \text{ in.}^2$$

Try 4 #7 and 4 #6, $A_s = 4.16 \text{ in.}^2$

For a more precise approach towards finding the steel reinforcing, refer to Problem 6.29.

Bridging the Wall Opening

As the regular load flow in a wall due to the wall weight and other superimposed uniform loads is interrupted by an opening for a window or door, the compression lines are pushed aside (Fig. 6.16b) and lateral thrust is generated (as has been discussed in fig. 6.9). The crowding of the displaced flow lines adjacent to the opening reflects the stress concentration at these locations, which is especially apparent for sharp corners, in contrast to smooth transitional curves. The uniform flow lines, as generating an imaginary funicular arch within the wall around the opening, which causes a lateral thrust that must be resisted by the adjacent masonry, or by the floor slab acting as a tie, or by other tension-resisting devises, can be visualized. When the opening is too close to the wall edge, there may not be enough shear resistance in the wall to counteract the lateral thrust, so that cracks may form.

It may be concluded that, for a rectangular opening, this arch action transmits most of the wall load adjacent to the opening, so that only a relatively small portion of the load below the imaginary arch must be carried by a beam, called a lintel. It is common

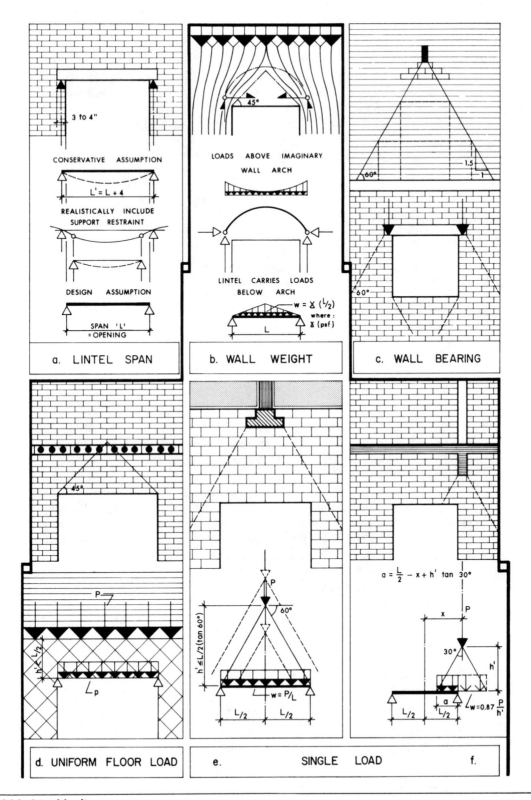

Figure 6.16. *Lintel loading.*

practice to treat this rather indeterminate beam load due to the wall weight as triangular, as based on a crack formation at 45°, with the height of the triangle equal to one-half of the lintel span; this type of crack formation has been observed after the lintel has settled. Naturally, this assumption is not true for a wall laid in stack bond, or where there is not enough wall height to provide arch action—here all of the loads above the lintel must be carried directly in beam action.

The bridging of a wall opening can be achieved by the following means:

- *Lintels*
- *Beams* (e.g., bond beams, large-scale lintels)
- *Arches*: major or steep arches (e.g., round, semicircular, pointed), corbelled or false arches, minor arches (jack or flat arches, segmental arches)
- *Wall beams*

Here, lintels and beams are further investigated. The various lintel types are:

- *Steel lintels*: for example, angles for cavity or veneer walls, I-, T-, or U-beams for solid walls, flat bars or plates for small openings, or built-up members, such as double angles.
- Precast or cast-in-place *reinforced concrete beams*.
- Prefabricated or cast-in-place *concrete masonry lintels* (e.g., channel shaped or solid cross section).
- *Reinforced masonry brick lintels* or brick beams (hidden beams).

The size of the opening, the loading conditions, and possibly aesthetic conditions determine the lintel type to be used. Lintels may be hidden within the wall or they may be exposed as structural elements. Lintels may be supported on the wall, or they may be hung from the spandrel beams to carry masonry cladding.

Should the lintel be simply supported on the wall and allowed to rotate freely, then its effective span would be approximately equal to the distance from center to center of bearing. Since the wall provides some clamping action, however (Fig. 6.16a), the simple-span portion of the beam will fall within the clear span of the opening. It may be concluded that it would generally be conservative to approximate the lintel span as the clear span or opening width. For long spans, where concrete or masonry lintels carry heavy loads, negative moment top steel at the supports is recommended, to avoid crack formation; for this condition and uniform loading, use: $-M = wL^2/10$.

The typical design loads for which a lintel may have to be designed are (Fig. 6.16c–f):

- Wall weight, as represented by a triangular load.
- Uniform load, as due to a floor slab.
- Uniform load over a portion of the beam, as may be caused by a concentrated load above (e.g., beam reaction).
- Concentrated load(s) due to direct application of beam and/or column loads.

The maximum moment due to the triangular wall weight, $W = wL/2$, on top of the lintel (Fig. 6.16b), is

$$M_{max} = wL^2/12 = WL/6 \qquad (6.18)$$

where:

$$W = \gamma[L(L/2)/2] = \gamma L^2/4$$

Uniform loading is not only caused by the lintel itself, but also possibly by the floor above. Should the uniform floor loads act below the apex of the imaginary 45° triangle (Fig. 6.16d), that is, at a height above the lintel of less than L/2, then it may be conservatively assumed that the uniform load is carried directly by the lintel, yielding the familiar moment of

$$M_{max} = pL^2/8 \qquad (6.19)$$

However, one must keep in mind that, when the floor load acts at or above the apex, then the imaginary arch carries the load, but will cause thrust, which must be resisted. Further, temporary shoring must be provided during construction, until the masonry has cured and will provide the arch action!

The force distribution of a concentrated load due to a beam or column within the wall depends on the nature of the wall (e.g., mortar, bonding, type of masonry units, material). Usually, an angle of distribution of 30° with the vertical or 60° with the horizontal (see fig. 6.16e, f) is recommended. Therefore, the single load disperses in a triangular fashion as a uniform load at the base of the triangular flow. Depending on the location of the concentrated load in relation to the lintel, only a portion of the uniform load may act on the lintel (Fig. 6.16f).

The load, w, due to the single load, P, distributes directly above the lintel over a width of $\tan 30° = (b/2)/h'$, or $b = 2h' \tan 30°$, here h' is the vertical distance of the load above the lintel (Fig. 6.16f). Hence, the load intensity at that level is

$$w = P/b = P/[2h' \tan 30°] = 0.87\ P/h' \qquad (6.20)$$

The portion *a* of the uniform load that must be carried by the lintel, by defining the location of P from the centerline of the opening as x, is

$$a = (L/2) - (x - b/2) = (L/2) - x + h' \tan 30° \qquad (6.21)$$

The crowding of the compression flow lines in a wall (i.e., stress concentrations) occurs at points where single loads must be distributed within the wall, as is the case for floor beam and lintel supports (Fig. 6.16c). It may be necessary at these points of support to use bearing plates, possibly together with stronger masonry units below (to disperse the reaction force), so that the masonry is not overstressed. It is customary to consider the load as equally distributed under the bearing plate, or as an average stress, rather than the actual stress, that may have a true stress distribution closer to a trapezoidal or triangular configuration. As a minimum bearing length for lintels, usually 3 to 4 inches is used.

The allowable bearing stress for masonry is $0.25f_m'$ for bearing on full area, but where the seat occupies one third or less of the bearing area, the allowable bearing stress is $0.3f_m'$ (Table A.9).

Example 6.6

Do a preliminary design of an 8-in. reinforced brick lintel (a reinforced brick beam) that must span a window opening of 5 ft and must support a uniform floor load of 1 k/ft applied at 2 ft above the window opening, in addition to the wall weight. Use a compressive strength of brick masonry, $f_m' = 2500$ psi for inspected workmanship, and an allowable steel stress of $F_s = 20,000$ psi.

Since the uniform floor load is located below the apex of the imaginary 45° triangle, and $h' = 2ft < L/2 = 5/2$ ft, the load must be carried by the lintel. For a masonry weight of 125 pcf or a wall weight of $(125/12)8 = 83.33$ psf, the total triangular lintel load is

$$W = 83.33[5(5/2)/2] = 520.83 \text{ lb}$$

The maximum moment due to the uniform and triangular loads is

$$M_{max} = wL^2/8 + WL/6$$

$$= 1000(5)^2/8 + 520.83(5)/6 = 3559.03 \text{ ft-lb.} = 42{,}708 \text{ in.-lb}$$

The maximum shear at the reaction, by conservatively ignoring the controlling shear at d-distance from the face of the support, is

$$V_{max} = 520.83/2 + 1000(5/2) = 2760 \text{ lb}$$

The depth of the brick beam is selected as based on shear requirements, so no web reinforcement is needed.

$$f_v = V/(bjd) \simeq V/[b(0.9)d] \leq F_v = 0.7\sqrt{f'_m}, \qquad \text{where: } z = jd \simeq 0.9d$$

$$2760/[8(0.9)d] = 0.7\sqrt{2500}, \qquad \text{or } d = 10.95 \text{ in.}$$

Select a beam depth according to the nearest brick module of about $5(2\ 2/3) - 3/8 \simeq 13$ in. (Fig. 6.17), so that $d = 13 - 1.5 = 11.50$ in. For concrete masonry lintels, the effective depth may be taken as $d = t - 3.5$ in.

The required steel to cover the moment (see fig. 6.7), is

$$A_s = M/zf_s \simeq M/[f_s(0.9d)]$$

$$= 42.708/[20(0.9)11.50] = 0.21 \text{ in.}^2$$

Place one #5 horizontal bar, $A_s = 0.31$ in.^{2}, at the bottom of the central grout core (Fig. 6.17).

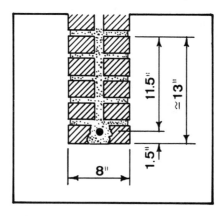

Figure 6.17. *Brick lintel.*

The flexural compressive stresses in the brick are approximately checked by treating the beam as a solid (rather than a cracked) section, which is not conservative (see Equation 4.3).

$$f_b \simeq M/S \leq 0.33f_m'$$

$$= 42708/[8(11.5)^2/6] = 242.20 \text{ psi} \leq 0.33(2500) = 825 \text{ psi}$$

Since the allowable stress is so much higher than the estimated actual one, the section can be considered satisfactory.

REINFORCED CONCRETE WALLS

There is a substantial difference between the monolithic, cast-in-place concrete structure and the precast concrete structure. The monolithic structure allows a continuous interaction between the vertical and horizontal planes, thereby providing more stiffness and redundancy in comparison to the structure built from prefabricated elements, such as large panels or boxes, which depends on its connections for integrity and thus may not provide much reserve strength. Often, the degree of continuity is difficult to evaluate, similar to the partial restraint between floors and masonry walls (see fig. 6.10).

While masonry, with its low tensile capacity, must express the character of the wall as an element of compression, reinforced concrete allows, in addition, the wall to act as a beam, and thus provides much more potential with respect to space manipulation. Some of the terraced buildings in Figures 1.19 and 1.20 express the flexibility of layout.

In Figure 6.18, various three-dimensional precast rectangular units or box kits further exemplify the many ways that space can be defined. It must be realized, however, that the investigated case in Figure 6.18 is a special one, and is only part of the infinite number of space-filling or non-space-filling cellular (polyhedral) systems.

From a formal point of view, the kit characteristics can be described as follows:

Tubular units	long-narrow shapes	upright (Fig. 6.18, 2a) horizontal (Fig. 6.18, 2b)
Closed-wall units	short-wide shapes	upright (Fig. 6.18, 3a) horizontal (Fig. 6.18, 3b)
Wall-floor shapes	long-narrow (Fig. 6.18, 4a,c) short-wide (Fig. 6.18, 5a,c) tilted: upright (Fig. 6.18, 4b) sideways (Fig. 6.18, 4d, 5b)	
H-shaped units (Fig. 6.18, 6a,b)		
Sliced units	for all shape forms (e.g., Fig. 6.18, 2c, 3c)	

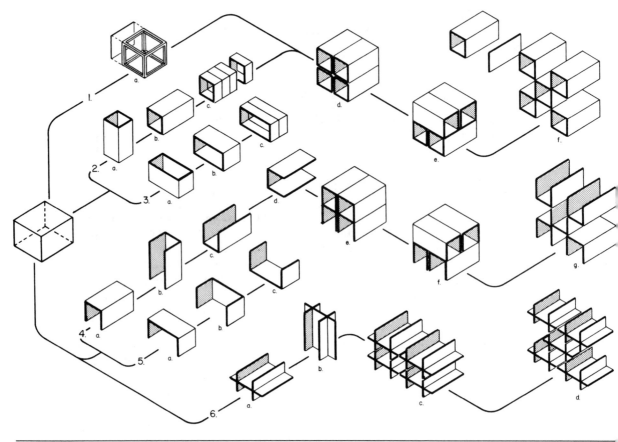

Figure 6.18. *Box types. Reproduced with permission from* High-Rise Building Structures, *2nd ed., Wolfgang Schueller, copyright © 1986 by Robert E. Krieger Publ. Co.*

There are three basic forms, from a spatial point of view:

- Long, horizontal, closed shapes forming sequential spaces, thereby generating a tunnel effect.
- Vertical, closed shapes forming two-story space interaction.
- Wide, narrow, sectional, or sliced units, allowing more freedom in subdividing space.

The panel systems may use dry, semidry, or wet joints. Wet joints are preferred for large scale structures; they utilize interlocking hooked bars or a longitudinal and transverse tie system.

The typical application of large precast concrete panels is as load-bearing shear walls, although they can be used for any other layout, as has been discussed in the subsection in this chapter on "Bearing Walls." Here, the common cross-wall structure in Figure 6.19 is briefly investigated. Usually, large panels are of single-story height and have at least the dimensions of a room; they range in length from 10 ft to 45 ft, are 8 ft to 10 ft high, and 6 in. to 12 in. thick. Common wall sections were shown earlier (see fig. 6.6, left); external walls may be of the sandwich type, since they may require insulation. Typical floor planks are solid, hollow core, or ribbed units.

When (in 1968) a gas explosion on the eighteenth floor of the 22-story Ronan Point apartment building in London, England, blew out an exterior load-bearing panel, it triggered a progressive collapse of the walls above and below, resulting in the destruction of an entire corner of the building (Fig. 6.19, top right). An investigation into the structural integrity of large-panel buildings was thereafter initiated in many countries; it was concluded that a large-panel structure must not only be designed for the typical primary loading conditions of the vertical and horizontal loads, but also for progressive collapse.

To reduce the risk of progressive collapse due to local failure, the various building elements must be tied together by tensile ties in the vertical, peripheral, longitudinal, and transverse directions, as shown in Figure 6.19; ties (that is, reinforcement) must be placed in the horizontal connections in the in situ joint material. Should a local failure occur, the structure must be capable of providing an alternate path of load flow, so that a progressive collapse cannot occur. This can be achieved through the following means (Fig. 6.19, left):

- beam/cantilever action of walls
- catenary behavior of slabs
- vertical suspension of walls
- diaphragm action of floors

The reinforcing steel ties the various building elements together. It provides a certain degree of continuity between the structural components across the joints and increases the ductility within the joints, which is needed to sustain deformations; it also increases the shear resistance of the joints and helps to redistribute the loads. It is apparent that the ties are essential in severe earthquake areas, in conditions of other abnormal loads such as blast, and for control of the differential settlements to which the relatively stiff panel structures are vulnerable. The main purpose of the various ties is as follows:

- The continuous prestressed or unstressed *vertical ties* in the walls assure clamping and dowel action in the horizontal joints for shear friction to develop cantilever and beam action, should panels below fail. They provide vertical suspension of ineffective panels and they assure resistance against the "kicking out" of the walls laterally.
- The continuous *peripheral ties* form ring beams along the periphery. They hold the floor panels together, with the help of the cross ties (see fig. 5.3d), and thus allow the necessary diaphragm action of the floors to transfer lateral forces and to act as the flexural reinforcement for the deep floor beams. They also allow membrane action in the corner slabs, should the supporting wall panels become ineffective.
- The local, unstressed *longitudinal ties* in the floor joints over the wall supports provide continuity between adjacent slab spans, so that they can develop the necessary catenary action, should the wall panels below fail.
- The *transverse ties* (prestressed or unstressed) within the floor level over the supporting walls are anchored to the peripheral ties. They provide resistance to slippage in the horizontal and vertical wall joints, and assure horizontal beam action in case of panel failure below (Fig. 6.19, top left), which includes that the ties may also be used as flexural reinforcement for individual panel behavior.

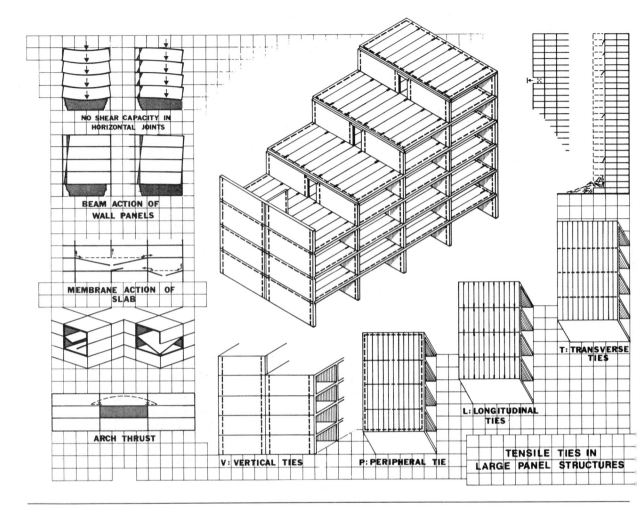

Figure 6.19. *Large panel construction.*

The structural behavior of large-panel structures, such as the typical building in Figure 6.19, center, is highly indeterminate. The load capacity of the whole structure depends on the degree of interaction between the individual wall panels, as well as the interaction between the walls and floors. It depends on the strength and stiffness of the joints (where the stress concentrations occur) rather than the capacity of the panels. The precast slabs resist the vertical loads, either in simple span action or as continuous slabs with cast-in-place topping, which includes negative reinforcing over the wall supports. Continuous slab action can also be provided by reinforcing steel loops protruding from the slabs and walls to interlace within the joints. With respect to lateral loads, the floors can be assumed to act as rigid diaphragms when the floor panels are properly tied together.

Under lateral force action, the large panel walls may be treated (for preliminary design purposes) as monolithic independent vertical cantilevers. However, when they are composed of multiple panels, for example, for the typical case of a double panel wall assembly, it has been shown in the subsection in this chapter on "Shear Walls," that the shear stiffness of the vertical wall-to-wall shear key joints (usually of the castellated type) behave in a manner similar to the flexible coupling of two walls (see fig. 6.14), and hence determines the stress distribution in the shear walls. For preliminary design purposes, the interaction between the panels may be disregarded, and the vertical semirigid connections treated as open.

The behavior of the shear wall depends on its height-to-width ratio. Flexural considerations will control the tall, slender wall, while shear governs the long, massive wall often found in low-to-moderately high buildings. For slender shear walls (H/L > 2.5), the vertical shear reinforcement, A_v, is less effective in resisting shear than the horizontal, A_h (e.g., stirrups in beams), and therefore a minimum of vertical shear reinforcement should be used. For long shear walls, however (L/H > 2), the vertical shear reinforcement is more effective and is taken as being equal to the amount of the horizontal reinforcement. For this condition, the uniformly distributed vertical shear reinforcement will usually also be sufficient for providing the necessary moment strength. However, for tall, slender shear walls, the flexural reinforcement, A_s, must be concentrated in the extreme ends (Fig. 6.20)!

According to Equation 4.16, minimum shear reinforcement is required when the ultimate design shear force is less than the shear strength of the concrete. But there is no minimum shear reinforcement required (according to Equation 4.15) when $V_u \leq \phi V_c/2$, however, the minimum wall reinforcement in both the horizontal and vertical directions must still be provided to limit crack widths (see Example 6.7).

For the approximate structural design of reinforced concrete walls in general, the following conclusions can be drawn:

- Non-load-bearing walls such as curtain walls, partition walls, cantilever retaining walls, basement walls for low-rise buildings, and the exterior walls of multistory underground structures, are primarily flexural systems, where the horizontal forces act perpendicular to the plane of the wall and the relatively small axial forces may be ignored. Therefore, a reinforced concrete curtain panel, for example, can be treated simply as a vertical slab resisting wind or seismic loads, where the effect of the axial load due to the panel weight on the moment capacity of the wall can be neglected.
- Many load-bearing walls, on the other hand, may be treated as predominantly axial systems, where the resultant compressive forces fall within the kern of the wall section (e $\leq$ t/6), and the effect of bending may be neglected. This situation is typical for the relatively short-span walls of cellular bearing wall buildings, where lateral force action is not critical, so that shear wall action can be ignored, as has been shown in the section on masonry walls. For this condition, the ACI Code provides the empirical design method, as is further discussed in Example 6.7.
- For the preliminary design of non-load-bearing reinforced concrete shear walls refer to Problem 6.40.
- For the approximate structural design of wall beams refer to the section in Chapter 4 on "Concrete Wall Beams."
- For the general condition where a reinforced concrete section is subject to both bending and axial loads, rational methods of design must be applied, similar to the requirements for concrete column design.

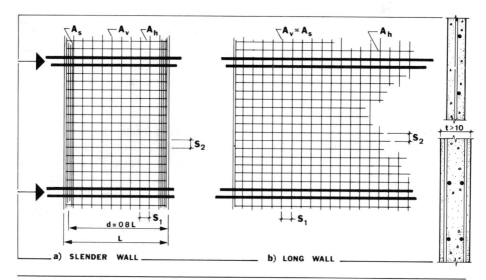

Figure 6.20. *Reinforced concrete walls.*

Example 6.7

Preliminary design of a reinforced concrete bearing wall: In a five-story apartment building with typical floor heights of 8 ft, 8 in., 8-in. thick, cast-in-place, reinforced concrete cross walls are spaced 30 ft apart and support 8-in. precast, hollow-core floor slabs. The loading conditions are taken as follows:

Floor/roof dead load, which includes 5 psf for
   mechanical loads and 5 psf for built-up roof: 60 psf
Partition dead load: 10 psf
Floor/roof live load, where the reduction of the
   floor live load is conservatively neglected: 40 psf

Normal-weight concrete of $f'_c = 4000$ psi and $f_y = 60$ ksi is used. Because the building is relatively low, it is assumed that the combined loading case of lateral force action and gravity is not critical, so that the shear wall action of the cross walls can be neglected for this preliminary investigation.

The overall thickness of the bearing walls according to ACI 318-83, Section 14.5.3.1 cannot be less than 1/25 the unsupported height or width, whichever is shorter, nor less than 4 in., when the empirical design method is used.

$$t \geq l_c/25 \tag{6.22}$$

$$= 8(12)/25 = 3.84 \text{ in. } \leq 8 \text{ in.} \qquad \text{OK}$$

The gravity load action at the first floor level is:

Dead load: $30[0.060(5) + 0.010(4)] + 8(0.150/12)8.67(5) =$
   $10.2 + 4.34 = 14.54$ k/ft

Live load: $30(0.040)5 = 6.0$ k/ft

The factored load is

$$P_u = 1.4P_D + 1.7P_L = 1.4(14.54) + 1.7(6) = 30.56 \text{ k/ft}$$

The ACI empirical design method for walls can be used here, since the resultant of the loads is located within the middle third of the overall thickness of the typical interior wall. Due to the fact that the lateral loads are resisted by the corridor walls, the effective eccentricity of the resultant vertical load is due only to the eccentric support action of the precast slabs, which can be considered to be negligible ($e \leqslant t/6$, see also Example 6.3). Consequently, the wall can be treated as carrying a reasonably concentric vertical load.

The design strength of the wall or its axial load capacity, according to the code, is

$$\phi P_{nw} = 0.55 \phi f'_c A_g [1 - (kl_c/32t)^2] \tag{6.23}$$

where: ϕ = the strength reduction factor = 0.70
$\quad$ k = the effective length factor
$\qquad$ = 2.0 for sidesway conditions
$\qquad$ = 1.0 for laterally braced conditions, but unrestrained against rotation at both ends
$\qquad$ = 0.8 for laterally braced conditions, but restrained against rotation at one or both ends
$\quad$ t = thickness of wall (in.)
$\quad$ l_c = vertical distance between supports (in.)
$\quad$ A_g = gross area of wall section (in.^{2})
$\quad$ f'_c = concrete strength (psi)

Since the building is laterally braced by the shear walls along the interior corridor, and conservatively assuming that the typical wall is unrestrained against rotation at both ends, its slenderness ratio can be taken as

$$kl_c/t = 1.0(8 \times 12)/8 = 12$$

Hence, the capacity of the wall, according to Equation 6.23, is

$$\phi P_{nw} = 0.55(0.7)4.0(8 \times 12)[1 - (12/32)^2]$$

$$= 147.84(0.86) = 127.05 \text{ k/ft} \geqslant 30.56 \text{ k/ft}$$

The wall capacity is by far larger than the factored load, and thus has a sufficient margin for a possibly larger load eccentricity. It also shows that it was reasonable to ignore the live load reduction for the floors. Further, notice that the effect of slenderness results in a 14 percent reduction of the wall capacity.

The minimum vertical and horizontal reinforcement, according to the ACI Code, Section 14.3, assuming deformed bars not larger than #5 and Grade 60 reinforcement for one foot width of wall, is:

Vertical: $A_v = 0.0012bt = 0.0012(12)8 = 0.115 \text{ in.}^2/\text{ft}$
Horizontal: $A_h = 0.0020bt = 0.0020(12)8 = 0.192 \text{ in.}^2/\text{ft}$

The maximum spacing of the reinforcement is

$$s \leqslant 3t \leqslant 18 \text{ in.} \tag{6.24}$$

$$s = 3(8) = 24 \text{ in} \geqslant 18 \text{ in.}$$

Select from Table A.2 and refer for arrangement to Figure 6.20b,

#4 @ 18 in. vertical bars, $A_v = 0.13$ in.^{2}/ft

#4 @ 12 in. horizontal bars, $A_h = 0.20$ in.^{2}/ft

Since the wall is not more than 10 in. thick, the reinforcement can be placed in a single layer at the center of the wall, rather than as a double layer, such as #3 @ 13 in. horizontal bars and #3 @ 18 in. vertical bars along each face. The same uniform bar arrangement would be used for precast wall panels, but the vertical continuous tensile ties must be added. However, should the wall panels not require uniformly distributed vertical reinforcement for strength, then the panels may be designed as peripherally reinforced with the continuous vertical tensile ties, and the horizontal reinforcement at the top and bottom of the panels.

For conditions where the wall is subject to concentrated loads, the length of wall considered as effective in carrying the load must not exceed the center-to-center distance between the loads nor the width of bearing plus four times the wall thickness (see Problem 6.39).

LOAD-BEARING FRAME WALLS

Multistory load-bearing walls in timber or steel of up to four stories are permitted by most major codes. They use platform frame construction, but diagonal bracing is required to resist wracking due to lateral force action. Critical in the structural design of the wood stud wall is the compression perpendicular to the grain and the shrinkage characteristics of the wood. Here, due to the cross-grain wood for the stud supports at the floor levels, considerable cumulative vertical shrinkage occurs due to the loss of moisture. This shrinkage, together with creep, will not be even and creates differential vertical movement that must be controlled; this is necessary so that nonstructural elements like plaster, doors, and windows are not stressed and do not buckle or crack, or that openings will not be distorted. It is apparent that the thickness of the cross-grain wood should be minimized and should be equal along all the walls to keep the shrinkage uniform. In addition, expansion joints should be employed at critical areas at each floor level. As has already been discussed, the differential movement between the relatively rigid masonry veneer and the wood frame, particularly at openings, must be considered. Further, at the lower floor levels, the compression perpendicular to the grain may control the stud sizes, due to end bearing requirements on the wall plates! The problems above could be reduced by using the more expensive balloon framing, where the studs are continuous over the full building height.

The construction of load-bearing steel stud walls is similar to wood stud walls. Channel-shaped studs are roll-formed from corrosion-resistant steel and are spaced 16 to 24 inches apart; they are anchored to top and bottom channel-shaped steel tracks.

STEEL-PLATE SHEAR WALLS

The concept of steel-plate shear walls as an economical method of construction was developed in the late 1970s, although it has been used for years in the shipbuilding industry. Here the steel plates are stabilized by vertical and horizontal stiffeners, such as T-sections and channels. Tubular beams are welded to the plates at the floor levels to provide stiff torsional restraints, thereby decreasing the effective buckling length of the plate. These thin walls not only furnish the necessary stiffness, but also ductility with a minimum of material, an important feature for seismic loading. In addition, the steel-plate walls can be prefabricated efficiently, using shop welding and field bolting. The design of stiffened steel-plate walls (possibly with window openings) is based primarily on buckling as caused by axial, moment, and shear actions. Here, overall buckling of the story panels between floor and column lines, as well as local buckling of the plates between the stiffeners, must be considered. It should, however, be realized that plates anchored along their edges are capable of carrying loads substantially higher than buckling loads.

PROBLEMS

6.1 Determine the thickness of a typical 11 × 5 ft curtain limestone panel for a wind pressure of 20 psf and a vertical span of 9 ft. Assume an allowable flexural tensile capacity of 125 psi; the weight of the material is 144 pcf.

6.2 Assume that the limestone fascia of the curtain in Problem 6.1 extends 3 ft past the roof line, forming a parapet. Determine whether the panel thickness is still alright.

6.3 A non-load-bearing 4-in. curtain brick wall of a multistory building is supported laterally at 8 ft. intervals by the floor structure. Determine whether the wall is adequate for a wind pressure of 24 psf. Assume Sarabond mortar with an allowable tensile strength in bending of 112 psi is used. The brick weighs 120 pcf.

6.4 Determine whether a 12-in. non-load-bearing continuous brick wall at the upper portion of a building is adequate with respect to seismic zone 4 and a wind action of 20 psf. The facade wall is supported laterally by the floor structure at 20 ft intervals. Use M or S mortar with inspection.

6.5 Check whether a 4-in. prefabricated 10-ft high non-load-bearing brick panel is satisfactory to resist a wind pressure of 26 psf. Use $f_m' = 8000$ psi units and Sarabond mortar with an allowable tensile strength in bending of 112 psi and a shear strength of 100 psi.

6.6 Find the required wall thickness for the conditions in Problem 6.5 by using clay brick with N mortar.

6.7 Check whether a 10-in. cavity wall consisting of two wythes of 4 in. brick units, is satisfactory for the conditions in Problem 6.5. Determine the mortar type.

6.8 A typical exterior non-load-bearing masonry wall at the top floor of a building spans 16 ft to the center of the pilasters and it cantilevers vertically 10 ft, 8 in. Although the plate proportion is 1.5, the fixed horizontal beam conditions versus the vertical cantilever conditions will cause most of the loads to flow horizontally, hence assume conservatively only one-way horizontal slab action.

  a. Check whether the 12-in. hollow concrete masonry wall is adequate for a wind pressure of 20 psf. Use N mortar with inspection.

       b. Determine the critical forces that the pilaster must support, assuming fixed bottom and simply supported top conditions.

6.9   Do a fast check of a brick/concrete block cavity wall at the top floor of a 10-story building. The arrangement of the wall openings forces the wall to span vertically between the floors, which are spaced 8 ft apart. The design wind speed for the building location is 70 mph. The strength of the 6-in. hollow concrete block back-up is $f'_m = 2000$ psi, and $f'_m = 4000$ psi for the 4-in. clay brick units. Type M mortar is used with inspection.

6.10 For the conditions in Problem 6.9, determine the necessary stiffness of a metal stud wall back-up so that the 4-in. brick veneer is not overstressed.

6.11 Do a fast check of a continuous composite 4-in. brick, 6-in. hollow concrete block wall, with a 3/8 in. collar joint between the two wythes, at the top floor of a 12-story building, for a wind speed of 80 mph. Assume one-way slab action and an unsupported span of 8 ft height and a story height of 9 1/2 ft. Use 4000 psi concrete block units and 10,000 psi brick units with M mortar and inspection. Assume the allowable tensile stress for the concrete wall as 23 psi, and ignore the wall weight.

6.12 A 5-in. precast concrete sandwich curtain panel has a 2 in. insulating core with the shells on each side of the same thickness. Investigate the wall first as a composite section with the necessary and adequate shear connectors, and then as a noncomposite system similar to a cavity wall. Determine the difference in stiffness; assume an uncracked section so that the effect of the shell reinforcement can be safely neglected for this preliminary investigation.

6.13 Determine the spacing of the grouted cores with the necessary reinforcing for an 8-in. hollow, 16-ft high simple span concrete block wall that is located in the most critical seismic zone and must resist a wind pressure of 20 psf. However, first check whether the wall can resist the lateral forces without any reinforcement. Use a masonry unit strength of 2000 psi with a density of 120 pcf and S mortar, and Grade 40 bars with an allowable tensile stress of $F_s = 20$ ksi.

6.14 A 25-ft high reinforced brick wall must resist a wind pressure of 30 psf. Determine the minimum wall thickness and the required reinforcement using Grade 40 steel with $F_s = 20$ ksi. Assume simple beam action of the wall, and 8000 psi brick units with S mortar.

6.15 Investigate the loading case of full gravity plus wind perpendicular to the wall for the bottom floor of Example 6.2.

6.16 Investigate the loading case of full gravity loading for the top story of Example 6.2.

6.17 Add a 3-ft parapet wall to the wall that has been investigated in problem 6.16. Ignore the higher wind pressure for this approximation.

6.18 Investigate the exterior wall of an eight-story long-wall structure; the spacing between the exterior facade and interior corridor walls is 16 ft. Assume a typical floor height of 9 ft center to center floor structure with an unsupported wall height of 8 ft. For the precast concrete floor deck that consists of wide units with no topping, a 4-in. bearing is assumed. Estimate the required masonry capacity for the 8-in. hollow concrete block wall. The wall openings are rather small, so that their effect upon stress distribution may be ignored.

Assume the following loads for this apartment building:

roof: 45 psf dead load and 30 psf live load
floor: 60 psf dead load, which includes partitions
40 psf live loads for the apartment areas
wall: 65 psf, which includes the nonstructural veneer

Use a wind speed of 70 mph as the basis for determining the wind pressure.

6.19 At the base of a multistory building, an 8-in. brick wall with an unsupported height of 8 ft, but an actual story height of 9 ft, carries an axial load of 12k/ft from the floors above. Added to this load is the floor load of 1.4 k/ft due to a precast concrete joist slab with a minimum bearing of 4 in. Find the necessary unit strength of the brick if type N mortar is used without requiring any special inspection.

6.20 Find the wall moments for the conditions in Problem 6.19, if concrete planks are used as floor structure. Assume they span 28 ft, and the live loads are 40 percent of the given load. Study the effect of the clamping action.

6.21 A 9-in. reinforced brick wall must support an axial load of 5 k/ft that acts 3 in. eccentric with respect to the centroidal axis. If 4000 psi brick units are used, together with type N mortar, and no special inspection is required, find out whether the wall is satisfactory. Determine the approximate reinforcing (F_s = 20 ksi), if needed. Assume an unsupported wall height of 10 ft.

6.22 An exterior 8-in. composite wall that consists of a 4-in. brick veneer, a 4-in. hollow concrete block back-up and a 3/8 in. collar joint between, supports 8 floors of an apartment building. The wall height between the floors is 8 ft. The precast deck floors, consisting of wide units without topping, sit on the full depth of the concrete block.
  a. Select the mortar type based on the stress state at the roof level, where the floor load is equal to 1 k/ft. Assume the modular ratio of elasticity n = 1.2.
  b. Select the compressive strength of the concrete block units based on the load conditions at the first floor of the building, where the axial load along the centroidal axis is 12 k/ft and the floor load is 1 k/ft. Check the assumed n-value of 1.2.
  c. Show the stress diagrams for the first floor and roof level conditions.

6.23 A nine-story apartment building with a crawl space below the first floor is constructed based on the cross-wall principle, with a layout similar to Example 6.4. The transverse walls are spaced 28 ft on centers and are 23 ft wide; the corridor is 5 ft wide. The typical story height is 8 ft, 8 in. with an unsupported wall height of 8 ft. Determine the required strength of the typical interior 12-in. hollow concrete block shear wall at the critical location below the first floor framing level in the crawl space. Assume the following loads: floor and roof dead loads = 70 psf, roof live loads = 30 psf, floor live loads = 40 psf, interior 12-in. hollow block wall with plaster = 75 psf. Design for an 80 mph wind speed.

6.24 Use an exterior, 6-in. hollow concrete block, self-supporting curtain wall for the ten-story cross-wall structure in Example 6.4. The wall weight is 60 psf, which includes a ceramic veneer. Assume only 50 percent of wall area available for curtain action.

6.25 Determine the number of stories, if the self-supporting masonry curtain in Problem 6.24 should be designed for compression at the building base rather than tension at the top of the building. Use f'_m = 2000 psi, and ignore the window openings.

6.26 Investigate the rotational stability of the long-wall structure of Example 6.5, that is, check the rotation of the cross-walls by taking into account their flanges.

6.27 An apartment building is 100 ft long, 100 ft high, and 60 ft deep. A uniform wind pressure of 30 psf against the short facade will be resisted by a central non-bearing wall spine, which can be considered as three 25-ft independent walls. Check the capacity of the 12-in. brick wall, and use Grade 40 steel bars if required. The masonry strength is f'_m = 1400 psi with type S mortar and inspection. The weight of the wall is 120 psf, and it has an unsupported height of h = 9 ft.

6.28 A five-story cross-wall apartment building is constructed from large, 8-in. thick concrete panels. They are 8 ft high, spaced at 30 ft, and support prestressed 8-in. hollow core slabs. Assume the following loads: live load for roof and floors = 40 psf, dead load of slab = 55 psf, partition load = 10 psf, mechanical load = 5 psf, and built-up roof = 6 psf; wind loads can be ignored for the preliminary investigation of this low-rise building. Use normal-weight concrete of f'_c = 4000 psi and f_y = 40 ksi as reinforcing steel. Consider the unsupported panel height also equal to 8 ft. Check the wall capacity and determine the reinforcing steel.

6.29 Be more realistic with respect to determining the steel reinforcing in Example 6.5, by taking into account the actual location of the steel resultant.

6.30 Determine the height of the interior shear wall in Example 6.5, at which the masonry can take the wind tension by itself without the help of the steel bars.

6.31 Do an approximate check of the moments induced in the 6-in. thick corridor concrete slab (f'_c = 4000 psi) of Example 6.4 due to the story drift at the critical top floor level. Assume safely an allowable story drift index of 1/400 for this unreinforced masonry building.

6.32 In case d of Figure 6.16, the lintel not only must carry the 12-in. brick wall weighing 120 psf, but also the uniform floor load of 1 k/ft at 3 ft above the opening. Determine the maximum beam moment, assuming a door opening of 12 ft.

6.33 Assume that the lintel in Problem 6.32 must support, in addition, a concentrated column load of 5 k at the floor level located at 1.73 ft from the centerline of the opening. Find the maximum moment due just to the additional load. Select a W or M section using A36 steel by considering the entire load.

6.34 An angle lintel spans a 6-ft window opening and must support a 4-in. brick curtain weighing 120 pcf. Determine the lintel size using A36 steel.

6.35 Use a segmental brick arch for the wall opening in Problem 6.34, with a maximum rise-to-span ratio of 0.15; for this case select a 5-in. rise. Determine the arch depth, conservatively assuming the tension stresses as zero. Use a masonry strength of 2000 psi with type N mortar and no inspection.

6.36. Determine, first, the size of a T-section, and then the sizes of two angles (A36 steel) to act as lintel for a 10-ft opening in an 8-in. thick brick wall (120 pcf).

6.37 Use an 8-in. reinforced grouted brick lintel for the 10-ft opening in Problem 6.36. Determine the approximate amount of steel reinforcement for the 8 × 10.5 in. beam using Grade 40 steel. Check also whether the shear and compressive capacity of the brick is satisfactory, using a masonry strength of 8000 psi and type N mortar with no inspection.

6.38 Determine the approximate amount of reinforcing bars (F_s = 20 ksi) for an 8 × 16 in. concrete masonry lintel that must carry a uniform floor load of 0.5 k/ft, including its own weight, and a 2 k single load at midspan, both located at h' = 2 ft above the lintel; the clear span of the opening is 6 ft. Assume the weight of the masonry wall as 80 psf, and f'_m = 2000 psi.

6.39 A 14-ft high reinforced concrete bearing wall that is laterally braced, and has its bottom end fixed against rotation, must support a floor system of precast single tees spaced 8 ft on centers with a stem width of 8 in. Assume a floor reaction of the T-beams of P_D = 30 k and P_L = 15 k; ignore the wall weight for this preliminary investigation. Use f'_c = 4000 psi and f_y = 60 ksi.

6.40 Investigate the apartment building in Example 6.7 as a skeleton structure. Assume the volume of 120 ft × 65 ft × 50 ft height to be stabilized across the short direction by two non-load-bearing concrete shear walls of 10 ft length arranged symmetrically. Do a preliminary design of the shear walls for a uniform wind pressure of 30 psf, using the minimum shear reinforcement; also determine the approximate flexural reinforcement. Ignore the weight of the wall for this preliminary design approach. Use f'_c = 3 ksi and f_y = 60 ksi.

7 BUILDING STRUCTURE SYSTEMS

The common support structures for high-rise buildings will now be discussed; they have already been introduced conceptually (in figs. 2.3 to 2.7).

A building must resist the primary loads of gravity and lateral force action. It is shown in Figure 7.1 that, with respect to gravity loads, the weight of the structure increases almost linearly with the number of stories. In this context, it is the weight of the vertical structural elements (columns, walls) that increases roughly linearly with height, since their weight is proportional to the axial stresses that determine the vertical member sizes, while the weight for each floor remains constant. It is also interesting to note that the floor weight, hence floor cost, constitutes more than one-half of the cost of the entire structure, for buildings not higher than the 30- to 40-story range. However, with an increase in building height and slenderness, the importance of lateral force action rises in a much faster nonlinear fashion, and becomes dominant. Therefore, the material needed for the resistance of lateral forces increases as the square of the height, that is, at a drastically accelerating rate, as is expressed by the upper curve in the weight-to-height diagram in Figure 7.1.

For typical *medium-rise* structures in the 20- to 30-story range, the vertical load resistance nearly offsets the effect of the lateral forces; according to Figure 7.1 only about 10 percent of the total structure material is needed for lateral force resistance. The late eminent structural engineer, Fazlur Khan of SOM, has shown that it may be economical to select (for a tall building) a structure system in which the bending stresses due to lateral force action do not exceed one-third of the axial gravity stresses, so that the effect of the lateral forces, shown by the upper curve in the diagram of Figure 7.1 can be ignored (see also Equation 2.8). At a certain height, however, the lateral sway of a building becomes critical, so that considerations of stiffness, rather than the strength of the structural material, control the design. The degree of stiffness depends on the building shape and the spatial organization of the structure.

It has been a challenge to building designers to keep the difference between the two upper curves in the weight-to-height diagram to a minimum, that is, to minimize the effect of the horizontal forces by developing optimum lateral-force resisting structure systems. The efficiency of a particular system is directly related to the quantity of material used, at least for steel structures. It is measured as the weight per square foot (psf)—that is, the weight of the total building structure divided by the total square footage of the gross floor area. Therefore, optimization of a structure for given spatial

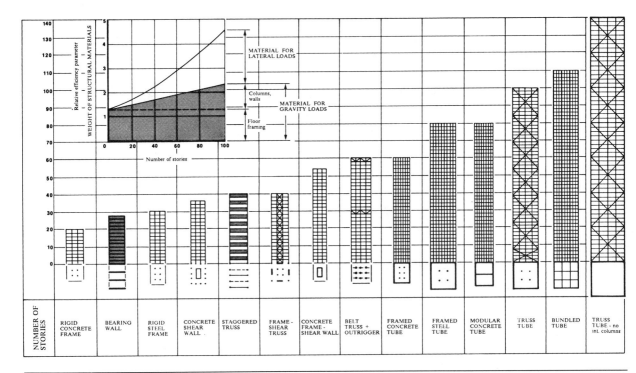

Figure 7.1. *Structure types.*

requirements should yield the maximum strength and stiffness with the least weight. This results in innovative structure systems applicable to certain height ranges. Naturally, it must be kept in mind that not only material costs, but also fabrication costs and erection time must be considered.

Fazlur Khan argued, in the mid-1960s, (and was supported by the development of computer simulations), that the rigid frame that had dominated high-rise building construction was not the only system associated with tall buildings. Due to a better understanding of the mechanics of materials, and how materials and members interact, the structure could now be treated as a whole, or the building form as a three-dimensional unit. Khan later proposed a range of structure systems for office buildings of ordinary proportions and shapes that are appropriate for certain heights, as is partially reproduced in Figure 7.1. Table 7.1 indicates that steel structures range from 29 psf for a 100-story building to 6.3 psf for a 10-story structure.

It must be realized, however, that the various systems applicable to certain heights only represent a rough rule of thumb, since many factors could not be considered in this oversimplified approach. Factors such as the building shape, the building slenderness (height-to-width ratio), the functional requirements (e.g., percentage of office space for multiuse buildings, or type of function), particular exterior loading conditions (e.g., environmental context), and many other variables have important bearing on the selection of the appropriate structure. However, it may be concluded that there is an optimum solution for any building type for a given situation, although it may not necessarily be based on Figure 7.1.

Table 7.1. Typical Weight of Highrise Steel Structures

Building Cases	Year	Stories	Height/Width	psf	Structure System
Empire State Building, New York	1931	102	9.3	42.2	Braced rigid frame
John Hancock Center, Chicago	1968	100	7.9	29.7	Trussed tube
World Trade Center, New York	1972	110	6.9	37.0	Framed tube
Sears Tower, Chicago	1974	109	6.4	33.0	Bundled tubes
Chase Manhattan, New York	1963	60	7.3	55.2	Braced rigid frame
U.S. Steel Building, Pittsburgh	1971	64	6.3	30.0	Shear walls + out-
I.D.S. Center, Minneapolis	1971	57	6.1	17.9	riggers + belt trusses
Boston Co. Building, Boston	1970	41	4.1	21.0	K-braced tube
Alcoa Building, San Francisco	1969	26	4.0	26.0	Latticed tube
Low Income Housing, Brockton, Mass.	1971	10	5.1	6.3	

Further, Frank Lloyd Wright's experiments, as early as the 1920s, with slender tree-like concrete cantilever structures that were quite opposite in nature to the traditional skeleton construction, should not be forgotten. This approach has recently lead to buildings more than 50-stories high, where large concrete cores alone provide lateral force resistance. Many of the concrete structures of the 1960s exposed the cores, in order to articulate the strength of the three-dimensional support structure (often of the bridge-type) and to express the servicing as clearly separated from the served spaces. In this case, the design philosophy is quite different from most of the steel skeleton structures of the same period. The efficiency of a concrete structure is evaluated (to a great extent) in terms of the process of construction, in addition to the quantities of materials used (roughly between 0.5 to 1.0 ft^3/ft^2 concrete and reinforcing steel of 2 to 4 psf), in contrast to steel, which considers only the quantity of material used.

By comparing the building cases (Table 7.1) of the transition period of the late 1960s with the ones built before, the progress made in the development of new structure systems becomes apparent. The 102-story Empire State Building, with its rigid frame-shear wall system, uses 42.2 psf of structural steel, while, in contrast the John Hancock Center in Chicago, with its trussed tube at 29.7 psf, requires about 30 percent less structural material. The 60-story Chase Manhattan Bank Building in New York, with a braced long-span rigid frame structure, is the least efficient (with 55.2 psf of structural steel), as compared to the slightly lower 54-story IDS Building in Minneapolis, with only 17.9 psf, using shear walls with outriggers and belt trusses. It is apparent from these comparisons that each structure system is only applicable within certain height limits; in other words, as the scale changes, different structures are required, as was already suggested by Myron Goldsmith of SOM in the early 1950s, as he derived from animal structures and trees (see the subsection in Chapter 2 on "Basic Concepts of Structural Language").

As a rough rule of thumb for the preliminary estimation of the weight of ordinary steel structures, designers often divide the number of stories, N, by three and add the number 6. For example, for a 40-story building, the weight of the steel structure should be roughly in the range of

$$w = 6 + N/3 \qquad (7.1)$$

$$= 6 + 40/3 = 19.33 \text{ psf}$$

The total steel weight for typical mixed steel-concrete building construction systems is in the range of 40 to 60 percent of that for all-steel buildings.

Many of the tall buildings of today no longer represent the pure shapes of the 1960s and early 1970s. Compound, hybrid building shapes have become fashionable. With the aid of computers, in response to complex geometric shapes, a wealth of new structure layouts have been made possible, which basically consist of combinations of the fundamental structure systems identified by Fazlur Khan in Figure 7.1. In addition, wind tunnel studies have become quite accurate in evaluating the response of a building to wind flow. Despite this increased sophistication of structural analysis and design, however, the fundamental fact should not be overlooked that the material and layout of the structure should not provide the stiffness solely by themselves, the *form* of the structure must also be searched for, with the help of computers, so as to efficiently reduce the use of materials.

The new structure systems reflect optimum solutions for given complex building shapes, which include composite structures and the mixed construction of concrete and steel. Large buildings are broken down into smaller zones; megaframes will give support to the supertall buildings of the future. It is hoped that the architect will use the potential richness of structure to express its power and its purpose as support, rather than just letting the engineer plug the structure into a form that was derived, independent of its nature. In the following sections, the most common structure systems are briefly investigated.

BEARING-WALL STRUCTURES

Traditional bearing-wall construction is directly associated with brick or stone masonry, which can be traced back thousands of years. The nature of bearing is probably most convincingly demonstrated by the sculptured stone walls of the Incas in the Peruvian Andes where each block was shaped and fitted with incredible precision. The daring use of stone is proudly expressed by the tall, medieval cathedrals of Europe. Brick is the oldest manufactured building material: the Romans raised the art of brickmaking, as had been brought to them from Egypt and the Ancient Near East (i.e., Syria, Babylon, and Persia), to a high level of sophistication—brick had become a common building material in Rome. But with the fall of the Roman Empire, the art of brickmaking died out in Europe, until it was revived during the Romanesque and Gothic periods, when it was increasingly substituted for stone. Inspiring examples are the Gothic brick buildings of the Hanseatic towns along the Baltic Sea, such as the St. Mary's Church in Lübeck, FRG. Also mentioned should be the majestic Gothic brick Cathedral in Albi in southern France, where the pier buttresses are projected into the interior, thereby allowing the exterior to give a heavy-walled, fortress-like appearance.

The bearing wall was the primary support structure for high-rise buildings before the steel skeleton was introduced in the 1880s. In the past, bearing-wall structures have been of thick, massive masonry wall construction. Their heavy weight and inflexibility in plan layout made them rather inefficient for multistory application, and their use declined rapidly in the late nineteenth century, when they were replaced by the steel skeleton structure. The limits of this type of construction became apparent with the 16-story Monadnock Building [1891] in Chicago, which required the walls to be more than 6 ft thick at the base, decreasing gradually to 30 inches at the upper story. Here, the design of the wall was based on a rule-of-thumb method that required a minimum wall thickness of 12 in., with an increase of 4 in. for every story below the top. Therefore, the effect of increasing building height requiring a corresponding substantial increase of the percentage of wall area relative to floor area became quite critical.

During the first half of this century, masonry was used in high-rise construction primarily as facing, in-fill, and fireproofing. It was not until after World War II, when European designers introduced thin-walled masonry construction, that the use of bearing walls for high-rise buildings was revived. They supported 16-story buildings on only 6-in. thick nonreinforced masonry walls. For example, an 18-story apartment building [1957] in Schwamendingen, Switzerland (Fig. 7.2c) has interior brick walls of 5 to 10 in. thickness, and exterior load-bearing brick walls of only 15 in., which were, however, based on thermal insulation rather than structural requirements. Thin-walled construction is based on a rational design approach, as derived from working-stress methods, and also treats the building as a whole by letting the walls and floors work together, that is, by using the floors as diaphragms to carry lateral loads to the shear walls. This approach is quite in contrast to the traditional, outmoded empirical methods, which employ a minimum wall thickness and maximum height, and which treat the building as consisting of individual walls similar to an assembly of cantilever retaining walls.

In the 1960s, the engineered design approach was introduced in the United States. An early example of brick bearing walls is the 23-story Penn Circle Apartment Building [1967] in Pittsburgh, by the architect Tasso Katselas and the structural engineer Richard Gensert. In contrast, reinforced masonry construction was already being used on a larger scale in the 1930s, primarily in California, to provide the necessary resistance to earthquakes.

Besides masonry, large-panel concrete high-rise construction for residential use greatly developed in Europe after World War II because of a severe housing shortage. These industrialized concrete systems not only allowed for a rapid erection of multiple housing units, but also claimed a cost advantage over conventional construction. Many of these European systems were introduced into the United States in the 1960s, especially in the later years, as encouraged by HUD's "Operation Breakthrough." The large-panel concrete systems came from France (e.g., Cebus, Coignet, Tracoba, Sectra), England (e.g., Bison, Camus, Wates, Tersons), Denmark (e.g., Larsen & Nielson, Jesperson), Canada (Descon/Concordia), and from other countries. Two of the American large-panel concrete systems were, first, the Sepp Firnkas System in the early 1960s and then the Dillon System, at the beginning of 1970. Other American companies of that period concentrated on concrete box-modules (e.g., Shelley System, Zachry System), rather than on the open and more flexible panelized systems, possibly incorporating utility modules. For example, the Zachry System was used for the 21-story Hilton Palacio Del Rio Hotel [1968] in San Antonio, where the precast concrete boxes were assembled in a checkerboard pattern. Safdie's Habitat for Expo 67 in

Montreal (see fig. 1.19) is a visual manifestation of architecture as a process of building, as expressed by the acrobatic assembly of concrete box units that form a terraced, pyramid-like urban building. In the end, the European large panel systems of the late 1960s proved to be more expensive than the traditional types of construction in the United States because of the large floor panels and the number of walls per square foot of floor area. In addition, they did not allow much flexibility in layout, since the apartments had to be confined between the relatively narrow spaces of the walls.

Bearing-wall construction is mostly used for building types that require frequent subdivision of space, such as for residential application. The bearing walls may either be closely spaced, say 12 to 18 ft and directly define the rooms, or they may be spaced say 30 ft apart and use long-span floor systems that support the partition walls subdividing the space. Bearing-wall buildings of 15 stories or more in brick, concrete block, precast large-panel concrete, or cast-in-place reinforced concrete are commonplace today; they have been built in heights up to the 26-story range.

The bearing-wall principle is adaptable to a variety of building forms and layouts, as is suggested by the cases in Figure 7.2. Plan forms range from slab-type buildings and towers of various shapes to any combination. The wall arrangements can take many different forms. Some basic wall layout systems are identified by the abstract diagrams in Figure 7.2 as

- Cross-wall system
- Long-wall system
- Combination: e.g., double-cross-wall system (see fig. 3.13a)
- Tubular system
- Cellular system
- Radial system

Naturally, there are an endless variety of hybrid systems possible just by combining the cases above. The walls may be continuous in nature and in line with each other, or they may be staggered; they may intersect or they may function as separate elements to form individual wall columns. The cellular organization of the residential buildings according to apartment units and function, rather than according to the bearing-wall pattern, was identified earlier (see fig. 1.18).

The dense, cellular arrangement of the walls for many of the cases in Figure 7.2 indicates an inherent compactness of bearing-wall structures. In the most efficient layouts, the walls are oriented to provide lateral stability in the primary directions and to brace the volume internally, which is necessary for slender buildings. The strength and stiffness of a building, however, not only depends on the wall layout in plan, but also on the degree of continuity between the horizontal floor planes and the vertical wall planes. It also depends on the degree of continuity *within* the walls, which (in turn) is a function of the wall material, the type of joints, and the wall penetrations, which not only include the number but also the size, shape and arrangement of openings (see fig. 6.13). It was discussed in the section in Chapter 6 on "Masonry Walls" that walls may be coupled or be separate, and they may act together with intersecting walls to form flanged walls, if they are properly connected.

As the name suggests, high-rise bearing-wall structures do act primarily in compression. This is especially true for the dense cellular wall layouts of typical apartment buildings and hotels. Here the self-weight is so large that it easily offsets the flexural tension due to lateral force action and due to other secondary effects, so that only

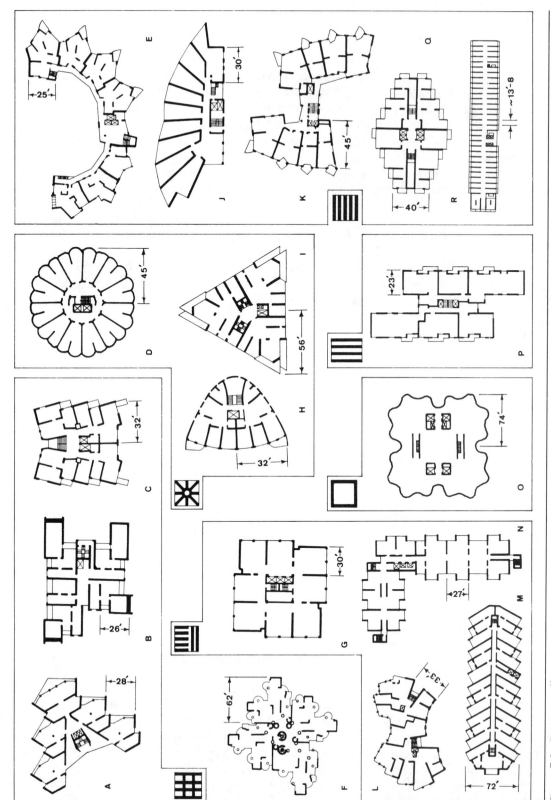

Figure 7.2. *Bearing-wall structures.*

gravity loads have to be considered, at least for preliminary design purposes. It is assumed here that the stresses induced by the lateral loads do not exceed one-third of the compressive stress caused by gravity loading (Equation 2.8). Although compression governs the design of interior bearing walls, tension may control the design of exterior bearing walls in the upper stories, where the slab action of the wall as a curtain resisting wind is predominant because the axial loads are small.

The reader may refer to Chapter 6, where the behavior of walls under load action and wall types has been discussed in more detail. Perforated walls in steel and concrete for tubular construction are discussed later in this chapter. Also mentioned in this context should be Bertrand Goldberg's innovative experiments with the nature of the concrete wall. For example, in the 13-story St. Joseph Hospital in Tacoma, WA (Fig. 3.5a and Fig. 7.2o), the unusual elliptical window openings keep the stress flow in the walls uniform and continuous and prevent the possible cracking at corners as may be the case for square windows. In this structure, the tower's 10-in. undulating perimeter concrete wall does not sit on the ground, but acts more like a deep beam or shell that is supported by sixteen, 3-ft diameter columns with conical capitals that connect into arches along the corrugated shell wall.

CORE STRUCTURES

The linear bearing-wall structure works quite well for residential buildings, where functions are fixed and the energy supply can be easily distributed vertically. In contrast, office and commercial buildings require maximum flexibility in layout, calling for large open spaces subdivided by movable partitions. In this case, the vertical circulation and the distribution of other services must be gathered and contained in shafts, and then channeled horizontally at every floor level. These vertical cores may also act as lateral stabilizers for the building.

The many ways in which a building may be stabilized by shear walls, trusses, or frames has been suggested (see fig. 2.4 and fig. 5.1). These lateral-force resisting vertical planes may be arranged as separate elements (as is discussed further at the beginning of the section in this chapter on "Braced Frame Structures"), or they may form spatial core units that usually consist of either braced steel frames or cast-in-place concrete. The core(s) should be located so as to generate a minimum eccentricity for the lateral forces. A typical interior concrete core is weakened by penetrations for door openings and holes for the service systems. When the total area of penetrations is below about 30 percent of the core wall area, and the openings are small and arranged in a staggered fashion, then the effect of the openings upon the closed tubular behavior of the whole core may be ignored for preliminary design purposes (see the subsection in Chapter 5 on "Torsion"). Should the openings occur in a vertical row, however, then the core may be treated as an open tube by ignoring the coupling effect of the link beams. Often, the core is broken up by the vertical rows of openings into individual uncoupled planar wall elements that may occur as a series of wide-flange or channel sections. In tall buildings, elevator banks are terminated at certain levels, thereby changing the vertical profile of the core, which (in turn) may result in an eccentric core with a corresponding generation of torsional forces.

The behavior of the core depends on its slenderness, similar to the wall action

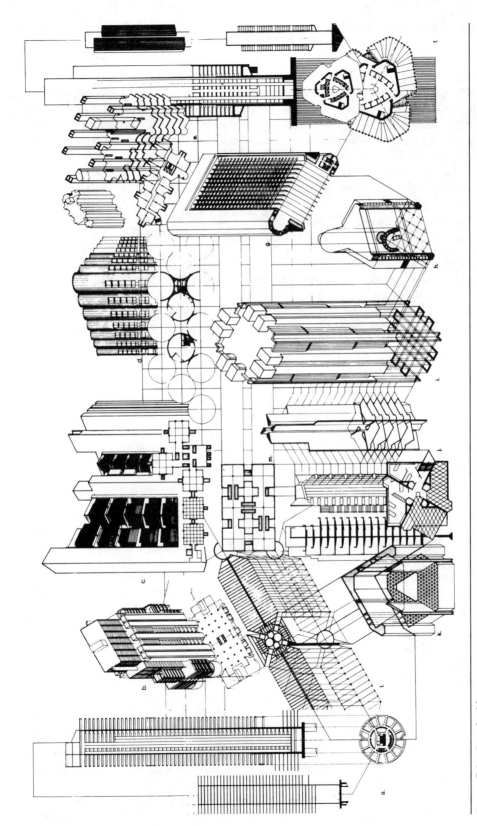

Figure 7.3. Core buildings.

shown earlier (see fig. 6.13). The short, massive core acts as a shear tube, where the web or webs provide most of the resistance; here bending can be neglected. When the shaft has a height-to-width ratio of approximately 3:1 or more, only bending needs to be considered for preliminary design purposes; in this case, the flanges provide most of the resistance. For slender towers with an aspect ratio of 7 or higher, however, flexibility becomes the dominant consideration. The infinite variety of possibilities as related to the shape, number, arrangement, and location of cores can only be suggested by the cases discussed (see fig. 2.4 and fig. 7.3).

From a structural point of view, the core or cores may constitute the sole vertical supporting elements in a building, or they may act together with other structure systems that either carry only gravity loads or share the lateral force resistance with the cores. Therefore, the designer may want to distinguish between the following cases:

- Single core structure: the entire building is supported solely by the core
- Multiple core structure, possibly of the bridge-type: the entire building is supported solely by the cores
- Core in combination with columns and/or shear walls:
    Cores together with wall beams form megaframes to support secondary buildings, possibly of the skeleton type or suspension type
    Core(s) act as the stabilizing element(s) to hinged frames such as precast concrete skeletons
    Cores act together with shear walls, etc.

In this section, emphasis is placed on the more or less pure core structure, where the cores provide the primary connection to the ground, with the exceptions of the bridge structures and the suspension systems, which are discussed later. The combination systems of core and frame are investigated in other sections of this chapter in more detail.

To develop some initial understanding for this interaction of the core(s) with other structure systems, various types of single-core structures have been investigated in Figure 7.4. In this instance, the structural response of the different core structure systems to vertical and lateral loading is described; most cases are self-explanatory, with the exception of the more complex composite systems, which are discussed in the later portion of this chapter.

Some basic core structures are identified in Figure 7.4 and in the following paragraphs:

1) Single-core Structure
In buildings where the center core is the only support structure (Fig. 7.4, center top, a, c), the core is a compression member, so that reinforced concrete usually provides the most efficient solution; this, however, requires a core of sufficient size.

- *Core with cantilevered floor framing* (Fig. 7.4, center top). Since all the vertical and horizontal loads are carried solely by the core, it should be weakened by only a minimum of penetrations; the openings may be arranged in a staggered fashion so as to have the least effect. Although the bending stiffness is not as high as it would be if columns were also used along the perimeter, the shear rigidity is much higher because of the concentration of all of the gravity loads. It must be kept in mind that the entire building weight prestresses the core and tends to keep it under

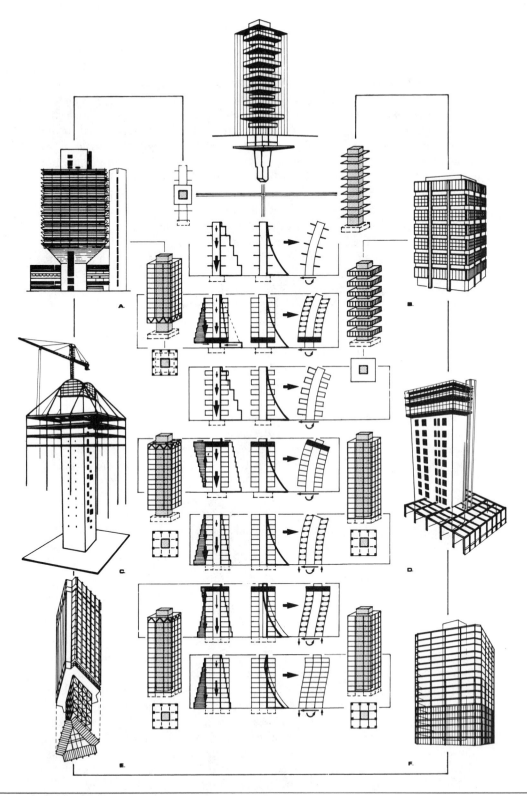

Figure 7.4. *Central core structures.*

compression, thus yielding a higher stiffness, since the concrete section is not cracked due to the tension stresses. Obviously, the attraction of all of the loads within the building to a relatively small core area requires exceptional soil conditions with a high bearing capacity.

The response of a core to lateral loading is dependent on its shape, the degree of homogeneity and rigidity, and the direction of the load, as discussed in the subsection in Chapter 5 on "Torsion." The structural strength of a single, large concrete-core cantilever structure is clearly demonstrated by the 52-story Singapore Treasury Building (see fig. 7.3a), which has 40-ft cantilever steel floor framing and was designed by the noted structural engineer, William J. LeMessurier.

Frank Lloyd Wright is credited with the first tree-like construction of a high-rise building that integrates the nature of concrete into the design. The 15-story laboratory tower of the Johnson Wax Company [1950] at Racine, Wisconsin (Fig. 7.4, top) seems to grow out of the earth with the central shaft extending far into the ground, similar to the deep roots of a tree. With the Price Tower [1956] in Bartlesville, Oklahoma, (see fig. 7.3j), Wright finally realized a 1929 proposal for the St. Mark's Tower in New York City. The 16-story cantilevered construction uses cross walls as the interior core; in this case, the peripheral core wall is replaced by an open cruciform concrete core that encloses four shafts. The plan organization is based on a rotated square configuration. The cellular, tapered floors are cantilevered from the four secondary hollow cores.

Kenzo Tange's Shizuoka Building (see fig. 7.5g) in Tokyo, Japan, is more representative of a giant sculpture, or an abstraction of his urban megastructure concept, by emphasizing the power of the cylindrical tower with a minimum of cantilevered capsule-like units. When the effect of cantilevering the floor structures becomes critical because of the magnitude of the negative moments and the lack of rigidity, it may be advantageous to tie two floors together along the perimeter by using a system such as a Vierendeel truss belt (Fig. 7.4b) in order to share the loads and stiffen the cantilever.

- *Core with massive base cantilever.* A structure possibly of the mushroom type or the inverted truncated pyramid upon which the external columns rest. In this instance, the floor loads are not all transferred at every floor level directly to the core, but instead are first partly collected by story-high cantilevers at one or several levels, and then brought to the core. Egon Eiermann's Olivetti towers (Fig. 7.4a) of seven- and nine-story height, in Frankfurt, FRG, are a fine example of this principle. The irregular central concrete core of the 52-story National Westminster Bank in London supports three cantilevered office wings, each of which are carried by massive, cellular, haunched cantilevered slabs located at different levels above the ground (see fig. 7.3f). The typical floor framing for the wings consists of steel beams with a span range of about 23 to 30 ft, anchored to the core at one end and connected to steel columns at the perimeter; the exterior columns, which resist only gravity, are supported by the huge base cantilever brackets. The 40-story Rainier Bank Building in Seattle uses the pedestal design principle (Fig. 7.5a). Here, however, a 30-story steel-framed 140-ft square tower sits on top of an outward sloping concrete pedestal-shaped base, where all of the lateral loads upon the tower are resisted by the perimeter rigid steel frame.
- *Core with large top and possibly intermediate cantilevers* (Fig. 7.4c and Fig. 7.5h, i). In this instance, the perimeter columns act as tensile members or hangers

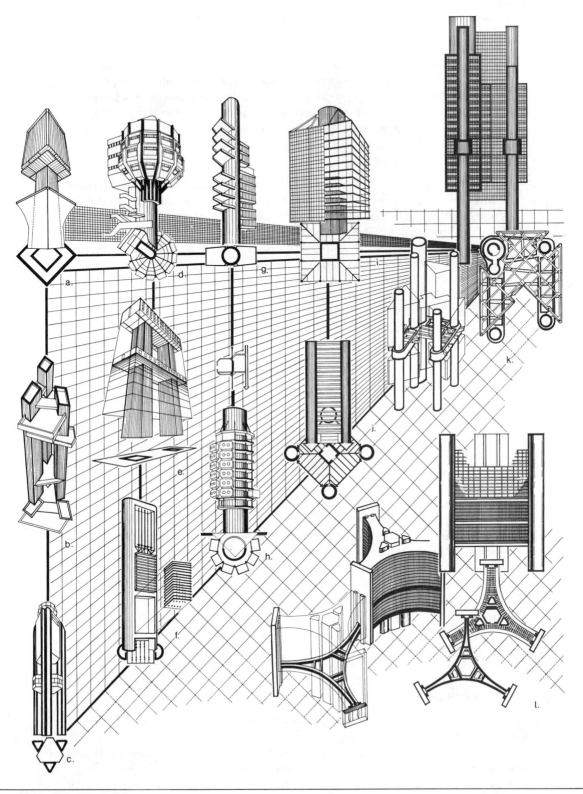

Figure 7.5. *The building core.*

suspended from the story-high outriggers to support the floors. In contrast to the other systems, the gravity flow takes a long route, since it must first be brought up in tension, possibly to the top of the building, and then flows along the core down to the foundations. This excess load can be beneficial, because it prestresses the concrete core directly at the top and keeps it under compression, thereby increasing its stiffness, since the reinforced concrete core will not behave as a cracked cantilever member when it responds to lateral loads.

2) Core with Hinged Frame

In this case (see fig. 7.4d), the core alone resists the lateral forces, while the frame carries only the gravity loads directly to the foundations. By providing lateral stability entirely within the central building core, the perimeter structure can be as light as possible. The shear core walls can be dropped off or stepped back at the termination of the elevator banks, which correspond with the lateral strength and stiffness requirements that reduce with building height. For buildings up to about 20 stories, the central concrete core that encloses the elevator and stair shafts is usually sufficient to resist all of the lateral loads. Naturally, for the tall 59-story Tour Maine Montparnasse (see fig. 7.27b) in Paris, France, a huge spinal reinforced-concrete core was required to provide the lateral force resistance, while the perimeter steel framing and the hinge-connected steel floor beams carry only the gravity loads. When a centrally braced core is too slender, a high steel-weight will result, because drift and acceleration must be controlled, which may make this structural concept uneconomical or impossible. In addition, large uplift forces on the foundation will require extensive tiedown systems.

3) Core with Rigid Frame

In this context (see fig. 7.4f), core and perimeter frame share the lateral force resistance, which can be visualized as a core within a core (see also fig. 7.29).

4) Cores with Shear Walls

In this category, reference should be made to one of the first important high-rise reinforced concrete structures, the 34-story Pirelli Building [1959] in Milan, Italy (Fig. 2.1 top, right). The narrow, slender tower, designed by the famous structural engineer Pier Luigi Nervi, is laterally stabilized by a pair of triangular cores at each end and by a massive pair of interior coupled-wall columns. These vertical structures support the ribbed floor framing of approximately 80-ft span at the center and 42.5-ft spans at the ends. The layout was designed so as to concentrate most of the gravity loads on the interior coupled shear walls for purposes of stability. The thickness of these wall columns and the connecting floors reduces with building height as shear and compressive strength requirements decrease.

5) Core with Outriggers and Belt Trusses

In this case, (see fig. 7.4e), a cap truss is used as a horizontal extension of the core to tie the exterior columns to the core, and thus forces the entire structure to react as a spatial unit.

The concern for making the functioning of space possible through structure by not separating the structure from the building or the *total architecture*, was a challenge to the early structuralists of the 1960s. Some of the architects of this early period who articulated these issues by using the core principle in the context of the bridge-concept, or megaframe, besides the Metabolists or the English Archigram with their utopian schemes, were Louis Kahn, John Andrews, Kevin Roche, I. M. Pei, and Paul Rudolph, to name just a few.

Many multicore buildings, with their exposed service shafts, have been influenced by the thinking of the Metabolists of the 1960s, who clearly separated the vertical circulation along cores and the served spaces. According to Kenzo Tange: "Buildings grow like organisms in a metabolic way." Their urban clusters consisted of vertical service towers linked by multilevel bridges which, in turn, contained the cellular subdivisions. Kisho Kurokawa expresses this relationship in the individual building with his capsule architecture, where prefabricated steel boxes are clipped onto the exposed concrete service shafts (see fig. 7.41).

In the Richards Medical Laboratory [1961] in Philadelphia, PA (see fig. 7.3c), Louis Kahn attaches tower shafts to each of the building clusters, which serve them and express them as dominant form generators. The 23-story Knights of Columbus building [1969] in New Haven, CT (Fig. 7.5j), is supported by four cylindrical concrete towers at the corners and a central square elevator core. Steel girders of 72-ft length and 36-in. depth span as bridges between the corner shafts and carry the floor beams. Since the spandrel girders and part of the floor beams are exposed and outside of the glass line of the building, they were placed on frictionless pads to allow movement due to the wide temperature variations. The towers were post-tensioned to keep them in compression when they act as vertical cantilevers under lateral force action.

The truncated triangular plan floors of the 34-story King George Tower (see fig. 7.3h) in Sydney, Australia, designed by John Andrews, are serviced by three concrete cores at the apexes—one large elevator core and two smaller stair towers. The gravity loads are carried by conventional flat slabs on columns. The three towers of Figure 7.5b are jointed at the cantilevering observation platforms. A cluster of grain silos (see fig. 7.3d) has been transformed into an eight-story convention hotel. The Faculty of Sciences Complex of the University of Paris (see fig. 7.3l) is an example of an urban megastructure. The site is subdivided into 25 rectangular bays, each one about 180 ft × 220 ft in size. The 59-ft wide buildings are placed along the grid lines, to form net-like chains that enclose interior courtyards. At the intersection points, cylindrical service shafts are located; these 31-ft concrete cores provide lateral stability to this urban complex.

The 52-story OCBC Center [1976] in Singapore (Fig. 7.5f), by I. M. Pei, consists of the primary concrete megaframe, which supports the secondary, plug-in, 14-story stacks of conventional concrete frame construction. The two semicircular cores, together with the approximately 115-ft span prestressed concrete girders and composite steel trusses, interact to resist the vertical loads like a ladder and the lateral loads like a superframe. The Hypobank [1981] in Munich, FRG (Fig. 7.5k), represents a bridge-type concrete structure of 21 stories. Four cylindrical towers, together with a story-high platform at the 11th floor service level (consisting of three rigidly connected prestressed boxlike girders), form an irregular spatial rigid megaframe. This primary structure supports the secondary 15-story structure above and the hanging 6-story portion below. Clearly expressed are the rigid connections of the platform with the collars around the tubular towers.

Another bridgelike structure is the Y-shaped plan 24-story building of the UN City complex [1978] in Vienna, Austria (Fig. 7.5l). In this structure, the 235-ft long prestressed story-high curved concrete box girders form central spines and bridgelike

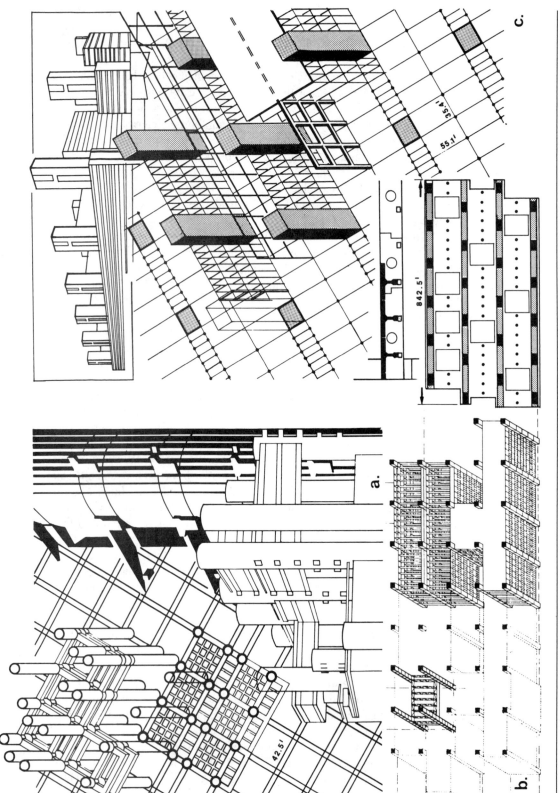

Figure 7.6. *Megastructures.*

platforms at three levels. They span between the six outer hollow legs at the end of the three wings and the central triangular cores, to produce a spatial megaframe with respect to gravity action; the curved box spines support up to 13 floors. Torsional forces on the curved girders are eliminated by the balanced symmetry as they lean on one another and the central core. Lateral loads are resisted by the central core and the three outside stairwell towers adjacent to the pairs of legs that carry the spinal girders. The bridgelike platforms laterally brace and tie the outer and central towers together.

The Metabolists used the bridge concept to articulate the urban building or megastructure as an organism, where energy and services were placed along the vertical cores to the multilevel horizontal bridges, thereby expressing functional diagrams and the zoning in the building's massing. A realization of this concept is Kenzo Tange's Yamanashi Communications Center [1967] in Kofu, Japan (Fig. 7.6a). In this case, four functional spatial blocks are supported by 16 hollow concrete shafts; the closely spaced cores along the center provide a higher resistance to earthquakes. The clear span between the towers is 42.5 ft, using a 10.625 ft module throughout the plan.

Other examples of the bridgelike megastructures are to be found in the hospital planning of the 1970s. A typical structural module of the Health Sciences Centre in Hamilton, Canada, consists of long-span, two-way truss grids of about 8.5 ft depth, which span 73.5 ft to the steel-framed service shafts; they resemble a flat plate in behavior. Similarly, for the Health Sciences Complex of the University of Minnesota in Minneapolis (Fig. 7.6b), 10-ft square service shafts occur at the corners of the square modules. In this instance, however, the primary twin steel girders span 49.33 ft between the concrete towers in only one direction, to support the secondary 36-in. deep steel trusses perpendicular to them.

Probably one of the mightiest megastructures ever built is the University Clinic of Aachen, FRG (Fig. 7.6c). In this case, also, the service towers dominate and order the huge complex with overall plan dimensions of 842.5 ft × 441 ft. Twenty-four 177-ft high concrete towers provide lateral stability to the building. They are arranged in parallel rows of six with the alternating rows staggered and forming open courtyards between them. The prefabricated concrete skeletons between the cores (along the core strips) form spines. The parallel spines give support to the prestressed long-span concrete joist floors, with the help of columns midway between.

BRIDGE STRUCTURES

This topic is introduced only as a special theme to develop some awareness for the use of long-span structures in high-rise buildings. Closely related to the bridge concept is the core structure of the previous section, where many of the buildings formed megaframes to support, in bridgelike fashion, secondary building packages. Similarly, several of the suspension structures in the next section are based on the bridge principle, as are supertall buildings that use megaframes or superdiagonals to gather the building weight to certain points for the purpose of stability.

The idea of the bridge structure had been vitalized by the designers of the 1960s, who were concerned with large scale urban architecture and wanted to separate the ground and services from social activities. These megastructures, or urban structures, were proposed by the Metabolists in Japan, Archigram in England, and designers such as Yona Friedman in France and Eckhard Schulze-Fielitz in the Federal Republic of

Germany, with their horizontal space frame structures. The infinite number of possibilities with which the long span from vertical support to vertical support can be achieved, and as it has been expressed in architecture, is studied as a visual experience in Figure 7.7; many other examples of bridge buildings are given elsewhere in this book. Space can be bridged by using any of the following basic structural concepts:

- Rigid frame beams (Vierendeel trusses)
- Trusses
- Arches
- Suspended arches, possibly post-tensioned
- Wall beams, possibly post-tensioned

These structural elements, in turn, can be arranged in many ways along the exterior of the building, or they can be hidden in the interior. The whole facade can be built as one wall beam, or the entire building can form a bridge. Story-high deep beams can be arranged to generate free spaces one or more stories in height. They can support additional floors on top or at the bottom, through suspension. These deep-beam structures may form interstitial or staggered wall systems, as discussed later. Early examples in high-rise construction include the 21-story Earth Sciences Building [1964] at MIT in Cambridge, MA in concrete, where 42-in. deep girders spaced 9 ft on centers span directly 48 ft from facade column to facade column, and the 60 story First National Bank of Chicago [1969] in steel (Fig. 7.7b). This 844-ft high building forms a megaframe in section by transferring the vertical loads of the two rows of interior columns to the sloping, curved exterior columns by using story-high steel trusses at the mechanical floor levels. The stepped V-plan of the Cambridge Marriott Hotel [1987] uses staggered trusses with triangulated cantilever extensions to span 42 ft across the building to the perimeter columns (Fig. 7.7i).

In Boston's 33-story Federal Reserve Bank [1976], massive twin 36-ft deep transfer trusses near the tower base span between the two braced end cores to carry the weight of 30 floors above (see fig. 7.25g). In contrast, for the 50-story 609-ft high Seattle-First National Bank [1969], exterior shell walls stiffened by two 2-story girders at different levels span between the four corner columns (Fig. 7.7h). The gravity loads are carried by the steel Vierendeel walls, which consist of 3.83-ft deep spandrels and columns spaced 4.67 ft on center, down to the sixth floor, where they are transferred to the huge corner box columns; the box columns continue to the thirty-fifth floor, where they change to W14 sections. The lateral shear due to wind and earthquake is also transmitted to the central core at the sixth floor level by horizontal trusses within the floor structure. For the bank in Lugano, Switzerland (Fig. 7.7f), the perforated concrete wall along the facade forms a deep cantilever Vierendeel beam spanning between the heavy wall columns.

The superstructure of the 21-story Hong Kong Club and Office Building (Fig. 7.7a) in Hong Kong consists of the exposed service core on the East facade and four huge S-shaped corner columns bridged on the Western facade with 112-ft curved prestressed concrete girders that are shaped according to the intensity of the force flow. They are of T-shaped cross section at midspan, so that the flange at the top can efficiently resist the compression due to bending, and gradually change the profile to a vertical rectangle at the ends, as controlled by shear. They support the prestressed concrete T-beam floor of about 59-ft span bridging the column-free interior space.

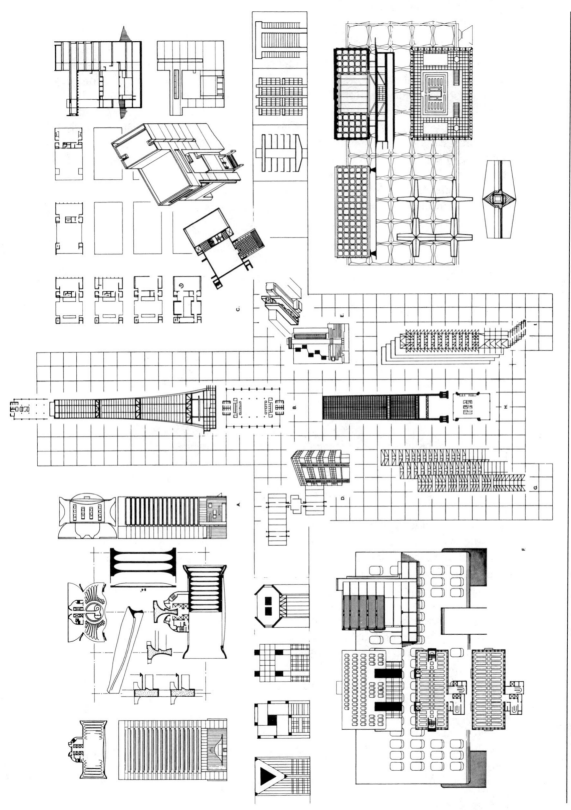

Figure 7.7. *A visual experience of bridge buildings.*

The enclosing walls of Yale University's Rare Book Library (Fig. 7.7j) consist of five-tier 49.5-ft-high all-welded exterior Vierendeel trusses with 1.25-in. thick white marble infill panels; the trusses are supported by heavy concrete columns at the corners. The longer frame walls span 130 ft and the ones along the short sides, 86.67 ft. The tapered box-section web members are fabricated of individual steel-framed cross-shaped units on a 8 × 8 ft grid, and are faced with granite. The box girder bottom chords of each truss carry the mezzanine floor framing one story above grade, while the top chords support the roof framing. The six-story stack of library shelves in the central part of the building is supported separately. Quite in contrast to the more minimal structure approach is the Johnson Art Center at Cornell University (Fig. 7.7c), which resembles a Constructivist sculpture, where the various masses, including the bridge, articulate the functioning of the building organism.

SUSPENSION BUILDINGS

The application of the suspension principle to high-rise buildings rather than to roof structures is essentially a phenomenon of the late 1950s and 1960s, although experiments with the concept go back to the 1920s. The structuralists of this period discovered a wealth of new support structure systems in the search for minimizing the material and to express antigravity, that is, the lightness of space and openess of the facade, by allowing no visual obstruction with heavy structural members. The fact that hanging the floors on cables required only about one-sixth of the material compared to columns in compression, as in skeleton construction, provided a new challenge to the designers. In addition, this type of structure allowed a column-free space at the base.

The tree-like buildings with a large central tower, from which giant arms are cantilevered at the top to support the tensile columns at their ends, are quite common today. The characteristics of this building type have been briefly described in the section in this chapter on "Core Structures" (Fig. 7.4c), where the stresses are primarily in compression and tension, by calling attention to the detour of the force flow up to the rigid support rather than down directly to the ground.

Tensile members can be of absolute minimum size from a strength point of view. There is no need to decrease the allowable stresses due to instability considerations as in members under compression. The cross-section is fully used in resisting the loads, directly in contrast to bending members. Further, high tensile strength materials can be used with a capacity of roughly six times higher than ordinary structural steels. Although high strength cable steels are extremely efficient from a minimum material point of view, the resulting lightness also results in a lack of rigidity and consequent large deformations for long tensile members, as is discussed later in this section. Furthermore, when hangers are exposed to the change of weather, additional movement due to temperature variations is generated. The large differential movements between the suspended envelope structure and the core are essential design considerations.

Typical tension members in steel were discussed in the section in Chapter 4. They range from rolled shapes and built-up members to cables, rods, and flat bars. The large elongations of the steel hangers are contrasted by the behavior of the prestressed concrete tension columns, which combine the high-tensile strength of the prestress tendons with the rigidity of the concrete; they also do not have to be fire-protected, as with steel members.

A possible organization of suspension buildings is demonstrated in Figure 7.8. The major subdivision is based on the primary support structure, but realizing that there are an infinite number of geometric layouts. The floors or spatial units (e.g., capsules, entire building blocks) are suspended from these support structures by using either vertical or diagonal tensile members.

- *The rigid core principle*: single cores with outriggers (e.g., frames, trusses, prestressed concrete wall beams, horizontal or diagonal compression struts) at the top and possibly at intermediate levels, depending on the height of the building; multiple cores connected by deep beams, trusses, arches, space frames or catenaries, to form bridge-type structures; megaframes and megatrusses (e.g., tree-type) of various forms.

The following support structures have only been investigated on an experimental level:

- *The guyed mast principle*, used as the primary support structure for the suspended floors: In this instance, the masts are stabilized and compressed by prestressing the guyed cables and anchoring them directly to the ground or other boundary systems. The entire primary structure is pretensioned so that the guy cables can absorb the lateral forces and provide the necessary stability.
- *The tensegrity or spacenet principle*, used as the primary support structure: The tensegrity structure is a closed three-dimensional system consisting of continuous tensile elements and noncontinuous compression members that keep each other in equilibrium when the entire system is prestressed. In contrast, prestressed spacenets consist only of tensile elements that must be stabilized by fixed exterior boundary conditions.

In the following paragraphs, some typical suspension buildings are briefly discussed. The 35-story Standard Bank Centre [1970] in Johannesburg, South Africa (Fig. 7.8, top right), resembles a tree-like structure with massive branches cantilevering at three levels from the central trunk of four coupled concrete tubes that form the core, to support three stacks of ten floors each. These branches consist of story-high four-way brackets, double cruciform in plan, which support eight prestressed concrete hangers to which the perimeter floor beams are connected, which (in turn) carry the precast lightweight concrete double T-beam floor panels. The three-level suspension system was necessary to control the size of the cantilever brackets.

The four hanging cylinders of the 20-story BMW Administration Building [1972] in Munich, FRG (Fig. 7.8, center top), clearly express the suspension principle. The four central hangers are suspended from the post-tensioned bracket cross, which cantilevers from the central concrete core at the top of the building, to support the clover-shaped floors at the center of the circles. While the huge prestressed concrete hanging columns, with an average diameter of 2.6 ft, give the main support, secondary columns along the perimeter of the drums are carried by radial story-high cantilevers at the mechanical floor level. In this instance, the seven floors above are carried in compression, and the eleven floors below in tension.

The primary structure of the 11-story Federal Reserve Bank Building [1973] in

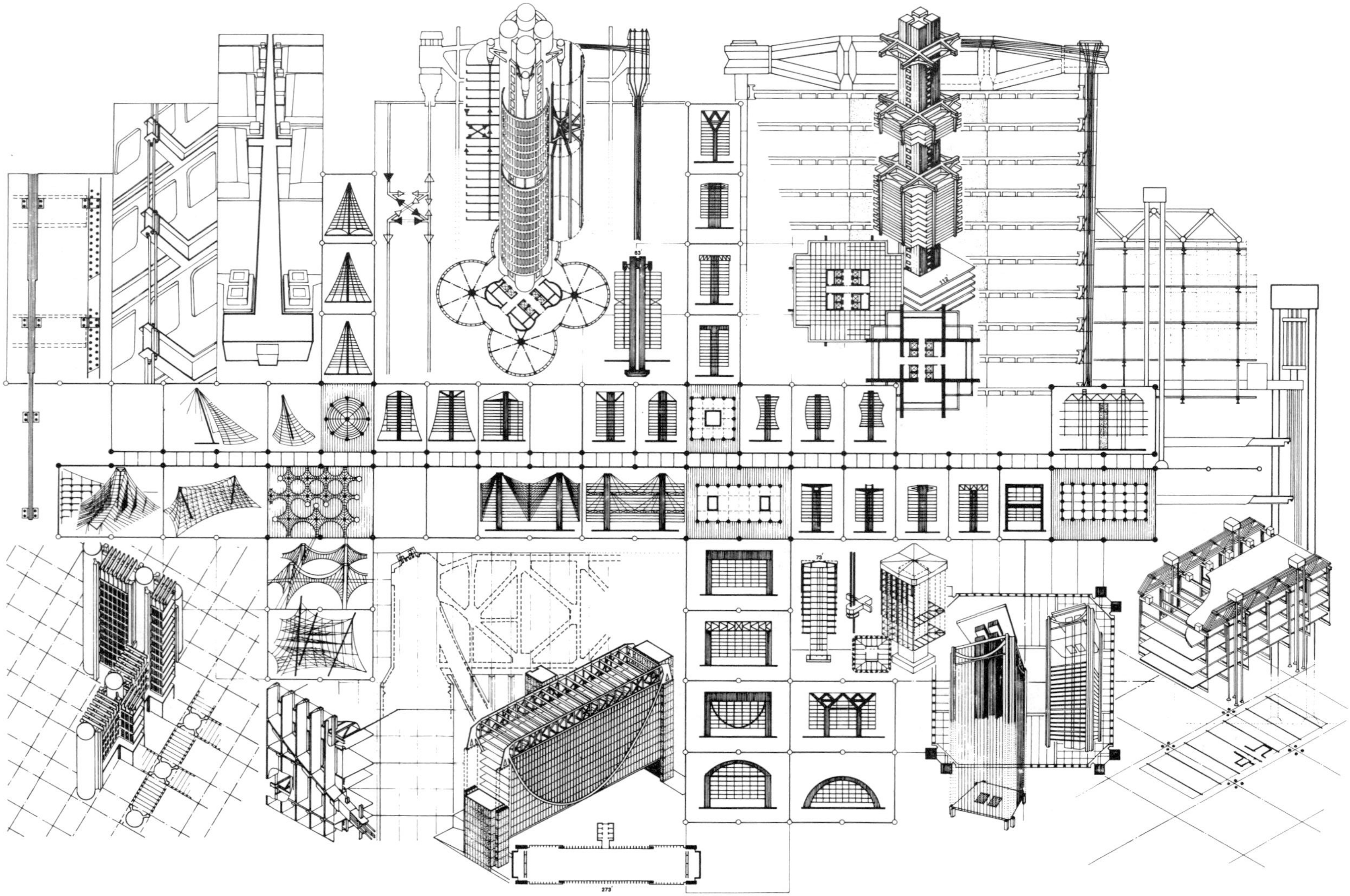

Figure 7.8. *Suspension structures.*

Minneapolis (Fig. 7.8, center bottom), is of the bridge-type, where two suspended catenaries, one along each of the facades, are supported by four corner piers and 30-ft deep 273-ft-span trusses at the top, which prevent overturning of the piers. The future 6-story addition on top will be carried by arches, causing an outward thrust in the truss system which will be counteracted by the compression due to the suspension system. The catenaries each consist of a 3-ft-deep welded wide flange member, aided by 4-in. diameter cables, which vary from eight cables at the support portion to two at the midspan range. The W8 columns on top of the catenaries are supported in compression, while the 1 × 8 in. plate hangers are attached to the bottom. These tension and compression columns support the 60-ft floor trusses, which span across the full width of the building.

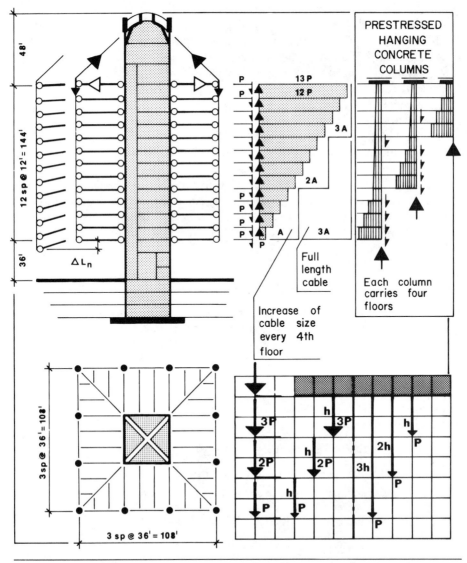

Figure 7.9. *Typical suspension structure: investigation of tension members.*

Four floors of the Narcon Building [1984] in Hannover, FRG (Fig. 7.8, right), are suspended from a set of five exposed adjacent cantilever trusses above the roof level in front of each of the long facades. The trusses are carried by a cluster of four exterior tubular steel columns close to each end of the building. Each of the approximately 54-ft long floor beams that bridge across the full width of the structure are supported by individual circular suspension steel tubing. The hollow steel pipes for the columns and hangers are filled with water for fire-protection purposes. The concrete service core provides lateral stability to the building.

Louis Kahn's proposal for a 32-story Kansas City office building (Fig. 7.8, bottom right) uses double-corner columns with a prestressed inverted arch-type concrete head structure, from which the floors are suspended. In the 43-story Hongkong Bank Building (see fig. 7.37), double-story cantilever trusses at five different levels are supported by towers. They each carry three vertical hangers, one inside at midspan and one at each cantilever end, to support the floors below.

The floors of the 12-story Westcoast Transmission Building [1969] in Vancouver, Canada (see fig. 7.5i and fig. 7.9), are suspended by six sets of continuous cables 36 ft o.c., which are draped over the curved concrete core walls and the two diagonally intersecting concrete arches. Each set contains a pair of 2 7/8 in. diameter galvanized bridge strands (A586, F_y = 240 ksi) for the face cables, and a pair of 2 3/4 in. diameter strands for the corner cables. They are joined at the roof level by a pair of 2 1/2 in. diameter cables to carry the roof weight and the additional components of stress due to the cable inclination; the inclined cables place the roof beams in compression. The cables are fireproofed with vermiculite plaster and enclosed in channels. They are protected by the interior environment so that they do not expand with temperature change. This type of structure is considered very effective in withstanding earthquakes because it is allowed to move more freely similar to trees which are rarely damaged during an earthquake. The building is further investigated in Example 7.1 and in Problem 7.1.

Example 7.1

A typical tension hanger of the 12-story core-supported building in Figure 7.9 is briefly investigated. The floor dead load is 50 psf, the live load is 80 psf, and the facade wall load is taken as 15 psf, which includes the hanger weight. For a more detailed discussion of the building refer to Problem 7.1.

A face hanger supports a perimeter floor area of approximately

$$(18 \times 18) + (2/3)(18 \times 36)/2 = 540 \text{ ft}^2$$

The live load reduction is

$$R = 23(1 + D/L) = 23(1 + 50/80) = 37.38\%$$

The total cable load per floor due to floor and wall weight, is

$$P_T = 540(0.050 + 0.63(0.080)) + (36 \times 12)0.015 = 60.70 \text{ k}$$

The maximum cable force at the top of the building is

$$P_{max} = 12(60.70) = 728.40 \text{ k}$$

In the following paragraphs, some typical cable systems are investigated.

1) Two high-strength steel cables with an allowable tension stress of $F_t = 80$ ksi are considered. The following approximate nominal cross-sectional area, according to Equation 4.68, is required:

$$A_s \simeq 1.5(P/F_t) = 1.5(728.40)/80 = 13.66 \text{ in.}^2$$

The corresponding cable diameter is obtained from

$$A_s = \pi d^2/4 = 13.66/2, \text{ or} \quad d = 2.95 \text{ in.}$$

Try two 3 in. ϕ cables.

Now, the designer may want to determine at what story only one cable is required. The flexibility of the cable hangers must also be investigated, as is further discussed in the solution to Problem 7.1.

2) A Grade 50 steel W-section and a built-up member are considered as a hanger. The following cross-sectional area is needed:

$$A_s = P/F_t = 728.40/(0.6 \times 50) = 24.28 \text{ in.}^2$$

Try W14 $\times$ 90, $A_s = 26.5$ in.^{2}, or try four L 6 $\times$ 6 $\times$ 9/16, $A_s = 4(6.43) = 25.72$ in.^{2}, so the number of angles can be reduced at certain floor levels.

3) Three prestressed concrete columns are used to carry each of four floors with $f'_c = 5000$ psi and a stress relieved strand of $f_{pu} = 270$ ksi. Any load increase due to the heavier concrete construction is ignored for this approximate investigation.

The hangers are treated as concrete columns under enough precompression so that the tension due to the external loads can be carried without cracking of the concrete. The concrete columns are treated as short members because the tension tends to straighten the member and counteract buckling. Any bending of the hangers is ignored.

Each concrete hanger carries the following maximum load

$$P = 4(60.70) = 242.80 \text{ k}$$

The required strand area for prestressing, according to Equation 4.41, is

$$A_s = P/0.56f_{pu} = 242.80/(0.56 \times 270) = 1.61 \text{ in.}^2$$

Try fourteen 0.438 in. diameter Grade 270 strands, from Table A.5,

$$A_s = 14(0.115) = 1.61 \text{ in.}^2$$

Estimate the preliminary column size according to Equation 4.42, as

$$A_g = P/0.45f'_c = 242.80/(0.45 \times 5) = 107.91 \text{ in.}^2$$

Try a 12 $\times$ 12 in. column, $A_g = 144$ in.^{2}

It is assumed that additional nonprestress conventional reinforcing will control concrete cracks due to overloads. Further, it should be noticed that there will be much less deformation, as compared to the hanging cable!

Rather than using a continuous full-length hanger of the same size, the sizes could be chosen in accordance with strength and stiffness requirements. For example, tension column sizes could be increased at every second floor, or they could be grouped as in Figure 7.9, where there are three hangers at the top of the building. In this case, one carries the upper four floors, one carries floors five through eight, and one carries floors one through four.

For the preliminary design of the concrete core, refer to Problem 7.1. The compressive stresses in the core from the gravity loads of the building relieve the tensile stresses due to the overturning wind forces. Therefore, the section is not cracked, so that the entire cross section and the full moment of inertia is effective in resisting the lateral forces (see also Example 7.8).

Of critical concern is the large elongation of the hangers, recognizing that the constant dead load extension can be offset during construction by inclining the floor beams upward, so that the floor framing is level when it is in place; this, however, requires the connection points to be calculated for each floor. Similarly, the time when a concrete hanger is prestressed should coincide with the application of the dead load, which also prevents overstressing the concrete; it may, therefore, be advantageous to apply the prestress in stages.

As mentioned before, the high-strength steel cables require much less cross-sectional area to resist a given force, compared to structural steel, and are particularly prone to large elongations as long tension members. For example, the instantaneous unit deformation for structural steel under an allowable stress of 22 ksi, according to Equation 7.1, is

$$\varepsilon = f/E = 22/29,000 = 0.00076$$

The corresponding strain for a typical cable steel under a stress of 80 ksi is roughly

$$\varepsilon = 80/23,000 = 0.0035$$

The unit shortening of a typical prestressed concrete hanger under a stress of, for example, 800 psi, is

$$\varepsilon = 0.8/4000 = 0.0002$$

Hence, the unit elongation for the cable steel is about 5 times higher than for structural steel and 17 times more than the unit shortening for the prestressed concrete member. The beneficial teamwork of the tensile strength of steel and the rigidity of concrete for prestressed concrete tensile columns is apparent.

Decisive in the design is the additional deformation due to the superimposed dead and live loads—especially the latter, which cause a back and forth movement. In this instance, the prestressed concrete members offer a distinct advantage, since the steel tendons are stressed to offset the dead and live load elongation. The deformation in prestressed tensile concrete columns is small; the shortening due to creep is gradually recovered under the sustained external loads. The deformation can be further reduced by using a concrete with a higher modulus of elasticity, besides a larger member size.

The total elongation of an axial member of length, l, with variable cross-sectional area, A(x), modulus of elasticity, E, under a constant axial force, P, is known from the basic mechanics of materials as

$$\Delta l = (P/E) \int_o^l dx/A(x), \text{ but when A = constant, then} \qquad (7.1)$$

$$\Delta l = Pl/AE = fl/E = \varepsilon l$$

The total elongation of a hanger supporting the gravity loads of n floors is simply the sum of the elongations of each of the story-high hangers above the level to be investigated (Fig. 7.9),

$$\Delta L_n = \Delta l_1 + \Delta l_2 + \ldots + \Delta l_n = \varepsilon_1 l_1 + \varepsilon_2 l_2 + \ldots + \varepsilon_n l_n \quad (7.2a)$$

When assuming the same story heights $h = l_1 = l_2 \ldots = l_n$, and the floor weights to remain constant $P = P_1 = P_2 = \ldots = P_n$, then Equation 7.2a, for the same hanger material, simplifies to

$$\Delta L_n = (Ph/A_1 + 2Ph/A_2 + \ldots + nPh/A_n)/E$$

$$= Ph(1/A_1 + 2/A_2 + \ldots + n/A_n)/E \quad (7.2b)$$

Assuming that the cross-sectional area of the hangers remains constant throughout the building height $A = A_1 = A_2 = \ldots = A_n$, then Equation 7.2b takes the following simple form:

$$\Delta L_n = Ph(1 + 2 + \ldots + n)/AE = Ph[n(1 + n)/2]/AE \quad (7.3)$$

For instance, at the lowest floor (Fig. 7.9), the maximum elongation of the W14 × 90 section in Example 7.1 is

$$\Delta L_{12} = \Delta L_{max} = 60.70(12)12[12(1 + 12)/2]/26.5(29,000) = 0.89 \text{ in.}$$

By far more critical is the elongation of the high-strength cables because of their much smaller required cross-sectional area and the smaller modulus of elasticity (see Problem 7.1).

SKELETON BUILDINGS

When William Jenney, in the 10-story Home Insurance Building [1885] in Chicago, used iron framing for the first time as the sole support structure carrying the masonry facade walls, all-skeleton construction was born. It was preceded by cage construction, where the interior frame bore only the weight of the floors and the exterior masonry walls were self-supporting and provided lateral stability to the building. For a short period, the Chicago frame came to symbolize the new direction in building design, towards integrating engineering, construction and architecture; its spatial network, requiring only a minimum amount of material, lent itself towards expressing the purity of form. Steel frame construction had been the primary structure system for high-rise buildings in this country until the development of software simulations for computers, in the 1960s, made many other structure systems possible.

For post-World War II architecture, the skeleton again became a central theme of the modern movement in the search for merging technology and architecture, thereby reviving the tradition of the Chicago School. SOM's Lever House in New York [1952] and Mies van der Rohe's two 860–880 Lake Shore Drive Apartment Buildings in Chicago [1951] (see fig. 7.11b) became famous landmarks. They have been most influential to the subsequent generation of designers; they symbolized, with their simplicity of expression, the new spirit of structure and glass. The frame seems to have reached its limits with the unprecedented 87-ft girder spans of the 662-ft high Chicago

Civic Center [1965] (see fig. 7.11c); in this structure, in the lower half of the building, the rigid frame had to be stiffened by shear trusses in the core area.

Although the pure, boxy shapes of the 1960s are closely associated with skeleton construction, as derived from Miesian minimalism, other high-rise building forms (based on quite different design philosophies) have been built, such as the unusually hammer-shaped Velasca Tower [1957] in Milan, Italy (Fig. 7.10, top left). In this instance, the perimeter columns of the 26-story reinforced concrete structure fold outward at the eighteenth floor level to support the larger mass above, reminding us of a medieval fortified tower.

Today, there seems to be no limit to the variety of buildings' shapes. The skeleton, as an organizing element for this new generation of hybrid forms, is investigated and freely interpreted in Figure 7.10. Odd-shaped towers, possibly with tapered frames, reflect the change of irregular plan forms with height. Skeleton buildings may be stepped at various floor levels, where large set-back terraces may be fully landscaped. In the Lloyd's of London Building, the braced perimeter concrete frame (see also fig. 5.1t) is surrounded by six satellite service towers, while the internal perimeter columns carry the elaborate central atrium structure. Kisho Kurokawa articulated the regularity of the three-dimensional grid and its adaptability to growth and change by constructing the Takara Beautillion for Osaka's Expo '70 from single six-pointed spatial cross units. The typical unit is made from twelve steel pipes bent into 90° angles and welded to steel plates; a unit measures about 22 ft in each of the main directions. The open, airy skeleton is contrasted with the perforated tubular walls of the Olympia Center in Chicago, shown on each side at the bottom of Figure 7.10.

The skeleton, as seen in the plan, has been further investigated in Figure 7.11. The organization of typical column layouts, as related to building form, are shown alongside the study of some plans from a geometric and functional point of view. The morphology of the plans is distinctly different from that of the core buildings with their concentrated masses. The column layouts range from the parallel two-way frames based on regular 21 ft square bays (Fig. 7.11b) to the huge repetitive bays of 87 × 48 ft for Chicago's Civic Center, which required large courtrooms (Fig. 7.11c); the column spacing ranges from the short-span apartment building framing of about 15 ft, to the longer spans for office buildings. In Figure 7.11a and d, rectangular bays of various sizes have been used, in contrast to Figure 7.11h, with a perimeter frame and columns on a diagonal grid in the core area. Figure 7.11i and j have a parti that departs widely from the more standard rectangular grids just discussed; in this case, the irregular column layouts lend themselves to flat-slab construction.

One of the three staggered, stepped towers for the U.S. Embassy Housing (Fig. 7.11g) in Tokyo, Japan, is organized on a 30 ft square grid in a zigzag fashion, separated by the 7.2 ft wide corridors; in this structure, each square unit is surrounded by eight columns. The terraced housing development for Metastadt (Fig. 7.11f) in Wulfen, FRG, is based on a staggered spatial network of 4.20 × 4.20 × 3.60 m (approx. 14 × 14 × 12 ft), which is defined by the steel skeleton. The column arrangement for Figure 7.11e is based on a square grid that is set at 45 degrees to the main axis of the V-shaped plan. The eight-story, transverse rigid concrete frames for the UNESCO Secretariat (Fig. 7.11k) in Paris, France, are spaced at about 20 ft and consist of cantilever girders supported by only two columns.

A typical skeleton structure consists of parallel plane frames in the longitudinal and

Figure 7.10. *Skeleton structures.*

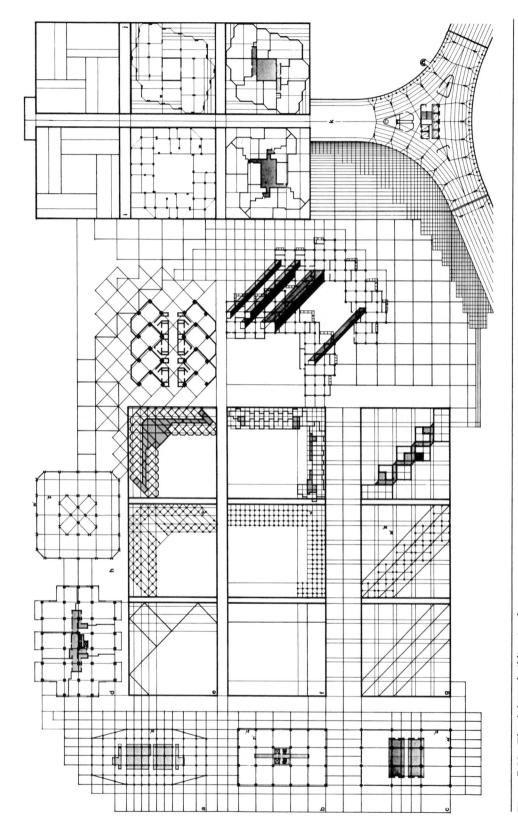

Figure 7.11. *The skeleton building in plan.*

transverse directions (Fig. 7.12, bottom left). The frames may be organized as discussed in the following paragraphs.

1) *Continuous rigid frames* in steel or cast-in-place reinforced concrete generally consist of a rectangular grid of horizontal beams and vertical columns connected together in the same plane by means of rigid joints. They resist lateral forces primarily by the bending of members, hence their efficiency is dependent on bay spans and member depths. Rigid frames are quite flexible with respect to lateral loading, so that sidesway must be considered in the structural design.

Flat slab buildings can be treated as consisting of rigid frames with imaginary wide shallow slab beams, which may not, however, provide much lateral stiffness as they increase in span.

The rigid frame principle seems to be economical for steel framing up to about 30 stories, and for concrete framing up to 15 stories for office buildings and 20 stories for residential construction.

The manipulation of form for tapered frames with sloping exterior columns results not only in a reduction of wind pressure but also in an increase in strength and lateral stiffness.

2) *Hinged frames* in steel and precast concrete consist of basic component kits (see fig. 7.14, top) such as frame units or beam-column components that are pin-jointed, remembering that pin-joints transfer only shear and axial forces. Depending on the arrangement and number of hinges, the frame may resist lateral forces, although it is quite flexible because of the lack of continuity of its members. It allows for free movement under temperature changes, in contrast to the restraint provided by the rigid frame. Often, the hinged frame only carries gravity loads and is stabilized by rigid frames or other lateral force resisting structural systems. Besides steel and concrete, heavy-timber framing can occasionally be found. Stanford's University Terman Engineering Center [1977] (Harry Weese, Architect) uses a five-story glulam superstructure. Here the glulam beams and columns support only gravity loads. The 3-in. tongue-and-groove laminated lumber deck, together with 1/2-in. plywood sheets, act as the floor diaphragm to carry the lateral forces to the shear walls.

3) *Braced frames* may be a more economical solution, or they may be necessary because of lateral stiffness requirements. Any of the frames just discussed may be laterally braced by shear walls (e.g., vertical trusses) or other methods, as discussed in the following sections of this chapter.

4) Some *combinations* of the rigid frame and hinged frame are shown in Fig. 7.10 middle, bottom.

Typical arrangements of lateral-force resisting frames are identified in Figure 7.12, top left. The lateral force resistance may be provided by the perimeter frame alone, which may be closed or open, while the interior skeleton only carries the gravity loads. For reasons of stiffness or strength, it may be necessary to internally brace the perimeter frame by using cross-frames. Naturally, the entire skeleton may represent a two-way moment-resisting rigid frame system. The frames in elevation (Fig. 7.12, right) reflect the plan layout. They may form equal bay systems, central corridor systems, or long-span frames, possibly of the cantilever type. They may be tapered, internally or externally stepped, or they may act as beams. The number of columns are limited for internally located frames for functional reasons, while the design of exterior frames

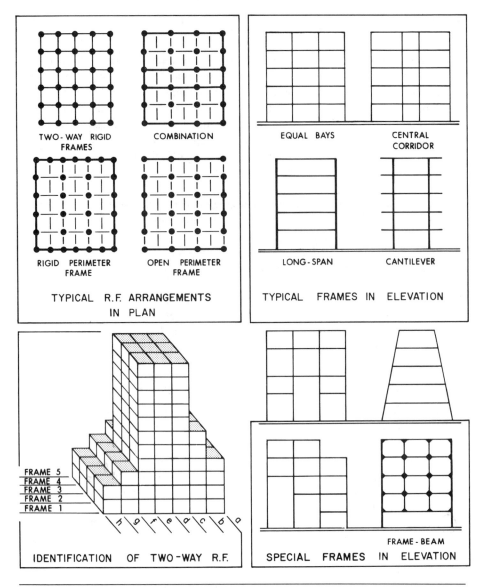

Figure 7.12. *Introduction to skeleton structures.*

has often been a design opportunity, from a visual and/or behavioral point of view. Various framing patterns are shown at the top (left and right) of Figure 7.10. They range from long-span, deep girder systems to perforated walls, that is, from open frame systems to wall-like frames.

Typical connection systems for steel frames are identified in Figure 7.13. They include:

- beam-to-beam connections
- beam splices
- beam-to-column connections

- column splices
- beam-to-concrete wall connections in mixed construction

Some other connections, such as of the hanger type or of the sloped type for trusses, are shown elsewhere in this book. Connections can be one-sided (eccentric) or two-sided (symmetric). The connector types are usually bolts and/or welds. The mechanical fasteners are common bolts (A307) and high-strength bolts (A325, A490); rivets (A502) have become close to obsolete. Bolted connections are either of the bearing-type or the friction-type. A typical construction approach is often, shop welded and field bolted. The fasteners are stressed in tension, shear, or a combination of the two. For example, splices and gussets usually subject fasteners to shear. The basic distinction between the connection systems is according to the degree of continuity, that is, their degree of rotational restraint. They are classified and described in Figure 7.13 as listed in the following paragraphs.

- *Flexible connections*—also called simple, hinged, or shear connections. These connection types are so flexible in acceptance of end rotation that they are only capable of transferring shear. Typical connections are: web connections (e.g., double angles, single or double plates); seated connections; and end-plate connections. To achieve a moment connection, the flanges must be connected.
- *Semirigid beam connections* usually consist of web angles, and top and seat angles. They do allow some end rotation because of the flexibility of the angles, and thus develop partial-end moments. In this case, the web connection is assumed to resist the shear, while the top and bottom angles transfer the moment. Semirigid joints allow the redistribution of the moments, so that it is reasonable to design a joint for a support moment equal to the field moment of

$$M_s = (wL^2/8)/2 = wL^2/16$$

so that the section modulus for the beam is one-half of that for a simply supported beam, and 75 percent of that for a fixed beam.
- *Rigid connections* are provided by the nondeformability of the beam-column joints; for example, a beam connected to a column at 90 degrees remains at 90 degrees during deformation under loading. The necessary rotational resistance is obtained by stiff flange connections; therefore, a rigid frame analysis can be carried out. Some of the typical moment connections in Figure 7.13, which may be fully welded, fully bolted, or a combination, possibly shop welded and field bolted, are: end-plate connections; T-stub connection; two-sided beam-stub bracket; flange plate connection; and directly welded connection.

The joining of members in cast-in-place concrete structures is, by their very nature, continuous; monolithic beam-to-column joints are briefly discussed later (see fig. 7.20). The joints for precast concrete frames are usually hinged; moment connections in the field are expensive. Continuous connections can be achieved by post-tensioning the precast elements together, or by using precast shell units with cast-in-place cores, where the shells are the formwork for the wet concrete. Typical soft or pin beam-to-column connections for multistory precast concrete construction are:

- Haunches projecting from the columns, possibly using elastomeric bearing pads
- Embedded steel shapes (e.g., vertical plates, W-sections), projecting from the column
- Seated connections using horizontal plates or angles

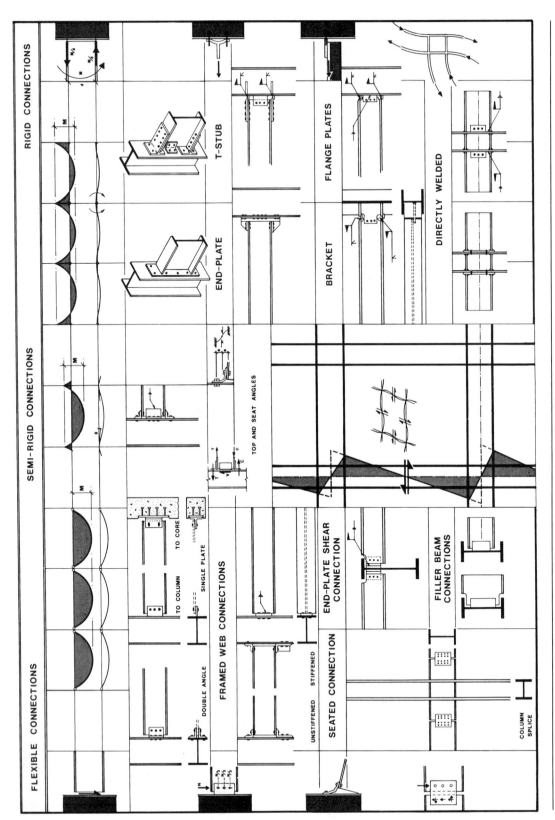

Figure 7.13. *Steel connections for frames.*

Joints may be organized according to the forces that they transfer as compression, tension, and shear joints. They are either of the dry or wet type. The dry or cold joints, reinforced with mild steel, usually form hinged joints for member splices. Wet splices, possibly with interlocking hooked bars, will provide moment resistance.

Designers have often articulated the member forms, not only to respond to the intensity of force flow (see fig. 7.15) but also to consider the ease of assembly. Precast concrete frame construction developed a wealth of assembly patterns in the 1960s. At the top of Figure 7.14, various basic frame components are shown, ranging from planar to spatial kits, from single-story to multistory units, from separate pieces for beams and columns to beam-column frame components. Not only in precast concrete construction, but also in conventional steel construction, the hinging of perimeter frame units is quite common. Since the primary action of tall skeleton structures is in response to lateral force action, prefabricated column trees can be used as assembly units for the perimeter structure. In this context, the continuity at the beam-column joint is easily achieved by welding in the shop, while simple hinged field splice connections at the inflection points (i.e., theoretical minimum moment points) at midspan of the members allow fast erection without disrupting the total frame action.

Some typical frame components are:

- Separate beam and column kits, such as continuous columns with beams, or tree columns with branch-like beam projections and fill-in beams.
- Single-bay or multibay frame units, such as portal frames, U-units, planar or spatial ring units, or units with cantilevers.
- Open frame units, such as T-, L-, H-, cross-, and star-shaped kits.
- Panel-frame units (see fig. 5.5), such as skeleton plus two-way slabs, (e.g., mushroom-type columns).
- Special facade framing systems (see the subsection in Chapter 6 on "Installation Methods").

These units are either pin-connected at the midspan of beams and/or columns, where the moments due to lateral force action are close to zero, or they are placed according to the gravity load flow, with the beam hinges close to the column supports (e.g., tree columns).

Naturally, the hinges may be located at the floor level as for the two-hinged portal frames stacked on top of each other, or they may be placed at other locations so that they are accessible and the units are easy to transport and to assemble. Simple connections are not only more economical, in terms of fabrication and erection, but they can also readily accommodate relative movements.

The various types of beam-column joints are identified in Figure 7.14, top, where the hinged joint transfers shear and axial forces while the rigid joint also resists bending. The members may all be connected rigidly or be pin-jointed; the beam, column, or frame unit adjacent or below may be continuous with the other members or hinged to them, and so on. In the field, pin-joint connections are preferred; moment connections—especially for frame assemblies composed of precast concrete elements—are complex and expensive, and are therefore avoided.

The provision for lateral stability is of critical concern for hinged frame construction (Fig. 7.14, left). Some of the common methods for stiffening a building are listed in the following paragraphs.

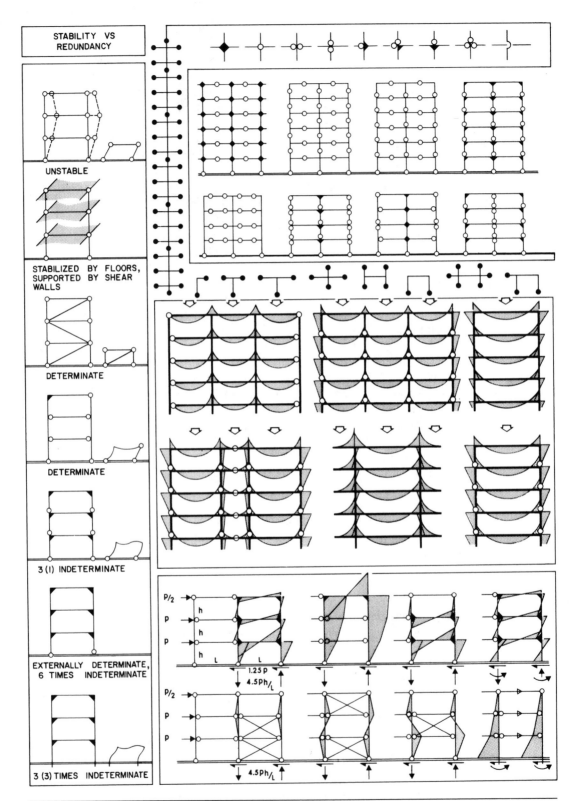

Figure 7.14. *Various frame types and their response to loading.*

1) One way in which a building may be stiffened is by stabilizing the hinged frame in its own plane, so that it can resist lateral forces by using:

- Continuous columns fixed to the foundations (similar to flagpoles), which may only be reasonable up to about four stories.
- Frame action, by properly arranging the frame units, such as the stacking of portal frames, a cross-unit assembly responding to lateral load action, tree-columns together with short-span beams responding to gravity loading, or H-shaped units with short beam cantilevers, which tend to reflect the combined loading case.
- Post-tensioning to tie the frame together in a continuous fashion.
- Lateral bracing, for instance, for one vertical bay, to laterally support the adjacent hinged frames.

In high-risk seismic zones precast concrete framing is not widely used as a primary lateral-force resisting system. Since the reliability of the connections becomes of critical importance, the number of connections tends to be reduced to a minimum where precast concrete is employed.

2) Hinged frames with many joints that consist of separate beam and column components, or small open frame units together with individual members, may not be able to resist lateral loads, but instead carry only gravity loads. These types of skeleton buildings are stabilized by the rigid floor diaphragms, which are laterally supported by shear walls, such as a slipformed concrete core or a concrete masonry core.

The stability of hinged frames during the erection stage is an important consideration; under these circumstances, temporary bracing is required. The response of various hinged and rigid frames to gravity and lateral loading, as expressed by the moment diagrams, is illustrated in Figure 7.14.

The various skeleton systems in Figure 7.15 can only suggest the infinite variation in which the linear beam and column elements can be formed and related to one another. They range from aluminum cladding, which reflects the continuous rigid frame steel structure (Fig. 7.15, top left), to the individual two-bay ladder frames (Fig. 7.15, top middle) and the assembly of single concrete tree columns (Fig. 7.15, bottom right), where the branchlike beam projections at each story support the single-T floor panels. Other examples show continuous cruciform columns, with capitals at the floor levels to support the primary beams upon which the floor panels rest; T-sections used as columns and floors; ladder columns with haunches on each side carrying perforated girders; or double columns that allow the radiating cross beams to project out between them, thereby articulating the joints. The sculptural possibilities of precast concrete are challenged by the interwoven facade members of the U.S. Embassy in Dublin, Ireland. In this case, only the straight portion of the vertical twisted unit acts as a column, while the curved remainder is treated as wall cladding.

Approximate Design of Frames

While planar hinged frames may be statically determinate (see fig. 7.14, left) and can be analyzed with only the three equations of equilibrium, multistory rigid frames are highly indeterminate, requiring other methods of analysis to determine the magnitude of force flow along the members. In order to find the degree of indeterminacy of a rigid frame, the frame is transformed into a determinate one by, for instance, cutting

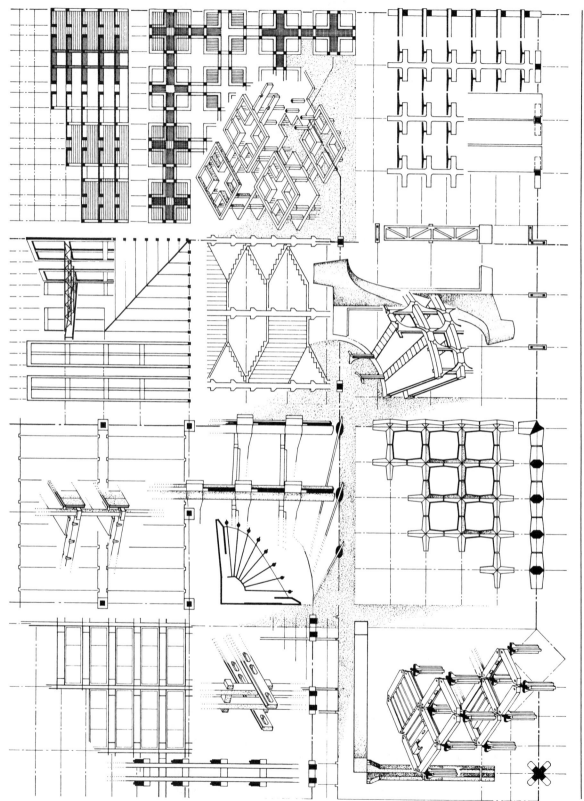

Figure 7.15. *The skeleton as assembly.*

vertically through the girders in each bay so as to generate independent statically determinate column trees. By cutting a girder, three unknown conditions have been removed ($\Sigma M = 0$, $\Sigma F_x = 0$, $\Sigma F_y = 0$); the summation of all unknown conditions is equal to the degree of indeterminacy of the building structure. For example, a ten-story, three-bay, rigid frame (see fig. 7.18) has $10(3) = 30$ girders, where each girder has three unknown forces, so that the building frame is $3(30) = 90$ times indeterminate. This clearly indicates that the force flow in a rigid frame is quite complex; the intensity of the stresses will depend on the relative stiffness of the members, which will not, however, be known at the beginning of the analysis.

Because of the high degree of indeterminacy of rigid frame buildings, approximate methods of analysis have been developed. Since the rigid frame will respond to gravity loading in a distinctly different manner than to lateral force action, as is expressed in the deformed configurations of the frames, each of the loading cases will be investigated separately. These deflected frame shapes will be the basis for the following approximate analysis, by making assumptions as to the locations of the inflection points where the moments are zero, thereby transforming the indeterminate structure into a determinate one.

For preliminary design purposes, and to aid in developing some feeling for the distribution of loads, only regular (but typical) frame layouts are considered.

Gravity Load Action

It has been shown that, for regular frames under gravity loads, the effect of sidesway can be ignored. A frame can thus be treated as a subassembly of single story frames, each of which consists of the continuous beam and the columns, which are assumed to be fixed at the floors above and below. The discussion of gravity load action is exemplified in Figure 7.16.

With respect to the design of the beams, the location of the inflection points depends on the degree of boundary restraints, the live load arrangement, and the frame layout. In this case, the beam spans are assumed to be nearly equal, with the larger of two adjacent spans not greater than the shorter by more than about 20 percent, and the live loads are not excessively high, such as for storage and equipment. Under uniform gravity loading, the location of the inflection points in a beam depends on the boundary conditions, that is, primarily on the stiffness of the members directly connected to it.

- For fixed boundaries, the zero-moment points occur at roughly 0.2L from the beam ends. For instance, this condition may occur for an interior beam where the support moments balance each other and do not rotate the column, or where the connected members are, together, at least five times stiffer (e.g., wall column) than the beam.
- Hinged conditions occur, for example, when a flexible facade column with a total stiffness ($\Sigma I/h$) from above and below of less than one-half of the beam stiffness (I/L) cannot provide any restraint to a stiff beam; in this instance, the point of contraflexure is moved to the support.
- Under conditions of partial fixity, the inflection point location will vary between the fully fixed and hinged case; for instance, should the member be fixed only to 50 percent, then the inflection points appear at 0.1L from the beam ends. One may conclude that, for conditions of partial fixity, including joint slippage, imaginary hinges for a uniformly loaded beam are located between 0.2L and 0.0L.

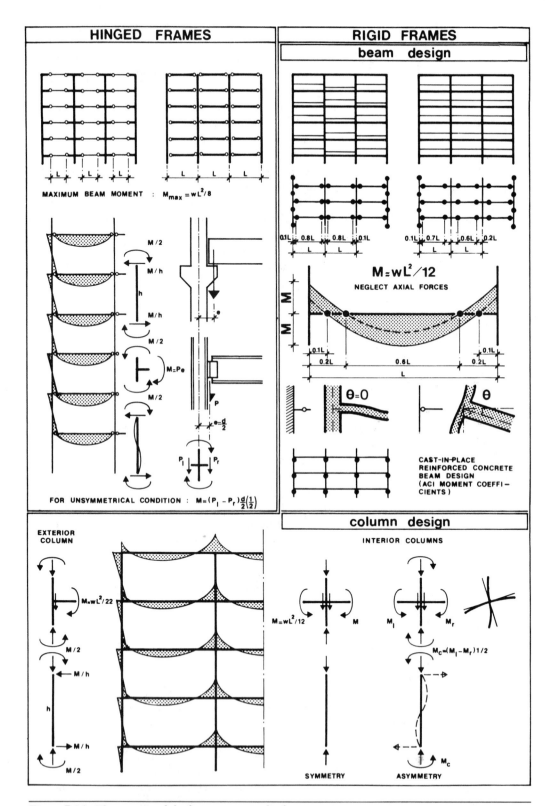

Figure 7.16. *The response of the frame to gravity loading.*

Considering the effect of the critical live load arrangement in Figure 7.16, rather than the effect of the boundary restraints, an approximate inflection point location of 0.1L may be assumed. The effect of live load arrangement for concrete buildings may be less critical because the dead load is usually much higher than the live load.

As based on the foregoing discussion, it may be concluded that beam inflection points for gravity loading of frames may be assumed to be between 0.1L and 0.2L from the beam ends. The maximum field moment occurs for inflection points at 0.1L, which can easily be determined by visualizing a simply-supported beam of span 0.8L between the imaginary hinges.

$$M_F = w(0.8L)^2/8 = 0.08wL^2$$

For fixed conditions, the maximum support moments occur for inflection points at 0.2L, that is

$$M_s = wL^2/12 = 0.08wL^2$$

Nearly the same result is obtained from the ACI coefficient method (see fig. 5.8), using a maximum moment of $M_s = wl_n^2/10 \simeq w(0.9L)^2/10 \simeq 0.08wl^2$

Semirigid connections allow some joint rotation, and thus cause moment redistribution. For flexible connections, the support moments are zero, and the beam is simply supported. Under symmetrical loading conditions, the slopes of the elastic curve at the reactions (using the conjugate-beam method) are:

$$\theta = \pm \frac{2}{3} \left(L \frac{wL^2}{8} \right) \frac{1}{2} \frac{1}{EI} = \pm \frac{wL^3}{24EI} \quad \text{(radians)}, \qquad M_s = 0 \qquad \text{(a)}$$

In contrast, for fixed boundary conditions, the end rotations are zero, but the support moments are of maximum value.

$$\theta = 0, \qquad M_s = wL^2/12 \qquad \text{(b)}$$

It may be concluded that, as the end slopes increase (i.e., semirigid connections), the support moments decrease. The following equation can be derived from relationships (a) and (b) (see also Table A.14) as

$$M_s = wL^2/12 - 2EI\theta/L \qquad (7.4a)$$

In general, it is reasonable, for preliminary design purposes of continuous beams, to use a maximum moment in the field and at the support of

$$M = \pm wL^2/12 \qquad (7.4b)$$

Any larger moment may be assumed to be redistributed by the frame because of its ductile character. It should be kept in mind that the absolute maximum beam moment is $M_{max} = wL^2/8$, when the entire moment is concentrated in the field because of the lack of continuity at the supports. For plastic design, refer to Equation 3.52.

For the design of continuous concrete beams under gravity loads, the ACI coefficient method (see fig. 5.8) may be used. Further, it may be assumed, for the preliminary design of beams under gravity loading, that the axial forces are zero.

For the preliminary design of the columns, uniform gravity loading on the entire building may be assumed. Therefore, interior columns will carry only axial loads for relatively symmetrical layout conditions, since the balanced loading causes the beam moments at the columns to cancel each other. However, the situation is different for the exterior columns, which are rotated by the beams. For the exterior bay beam, with its partial outside boundary restraint, imaginary hinge locations at 0.1L are conservatively assumed at both ends. Therefore, the corresponding support moment at the exterior column is equal to

$$M_s = (wL^2/8) - (w(0.8L)^2/8) \simeq wL^2/22 \qquad (7.5)$$

This beam moment must be balanced by the columns above and below (see Fig. 7.16) in proportion to their stiffnesses, which (in high-rise construction) may be assumed in proportion to the column lengths above and below. For the typical condition of equal story height, one-half of the beam moment will be resisted by the column above and the other half by the column below.

$$M_{cl} = M_s \left[\frac{(I/h)_1}{(I/h)_1 + (I/h)_2} \right] \simeq M_s \left[\frac{h_2}{h_2 + h_1} \right] = M_s/2 \qquad (7.6)$$

Therefore, for typical conditions of symmetry, inflection points are assumed at mid-heights of exterior columns and no flexure of interior columns.

Lateral Load Action

Lateral loads due to wind and earthquake action are reduced to concentrated forces applied to the frame at each floor level. In other words, the lateral loads are assumed to be carried by the curtain to the spandrel beams, which (in turn) transfer them to the columns. A thorough examination of lateral load action is presented in Figure 7.17.

The lateral deformation of a rigid frame depends on the relative stiffness of its members. For the extreme condition of much stiffer columns than beams, as may be the case for a tall flat plate building, the frame may be visualized as consisting of individual vertical cantilevers similar to flagpoles. In this instance, the entire moment must be resisted by each of the column cantilevers, obviously producing large column moments at the building base. As the columns rotate, they induce bending in the shallow slab-beams; for this case, inflection points are formed only at the midspan of the beams.

For the other extreme condition, where the floors are rigid and the columns are flexible, as for an interstitial structure, the horizontal shear wracks the columns, thereby forming inflection points at the midheight of the columns. The rigid frame lies some-where between these extreme cases; here inflection points form near the center of each member. For the preliminary design of typical, regular rigid frames, equal beam and column stiffnesses may be assumed with points of contraflexure at midlength of all members.

The portal and cantilever methods may be used for the quick approximate analysis of rigid frames. In the *cantilever method*, the frame is treated as a flexural cantilever controlled by bending action, whereas in the *portal method* it is treated as a shear cantilever dominated by shear wracking (Fig. 7.17, top left). Both methods assume points of contraflexure at midlength of all members. In addition, the cantilever method assumes the axial unit stress in each of the columns to be proportional to the distance from the neural axis of the bent (or to the centroidal axis, if column areas in a story are all the same); the portal method assumes a certain shear distribution, as will now be discussed. The foregoing assumptions in the two methods are sufficient to reduce

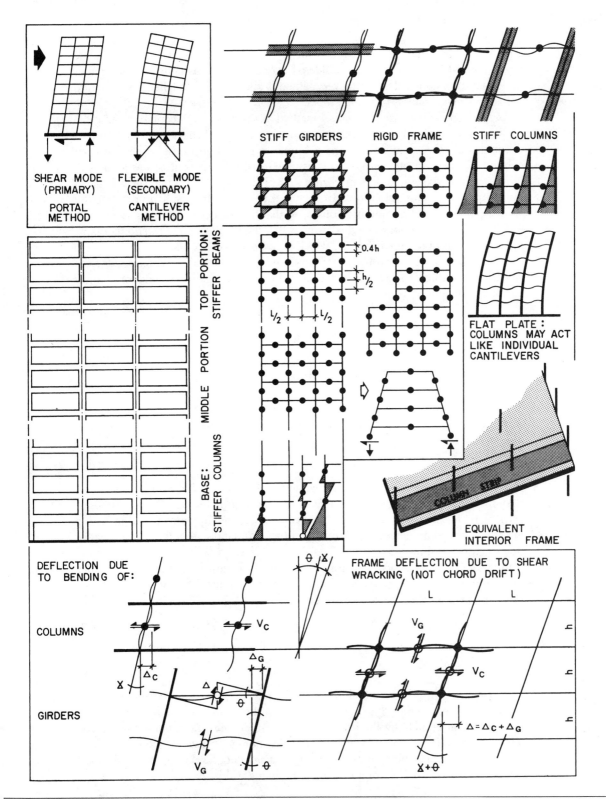

Figure 7.17. *The response of the frame to lateral loading.*

a highly indeterminate structure to a statically determinate one. In this context, the portal method is used, which is considered reasonable for conventional rigid frames up to about 25 stories. The cantilever method is often applied to taller frames and to the approximate analysis of framed tubes.

As mentioned previously, the portal method assumes points of contraflexure at the midlength of all members and assumes that the total lateral shear upon the frame is distributed at every floor level to the columns in proportion to the width that each column supports. In other words, it is assumed that a frame bent acts as a series of independent portals, as shown in Figure 7.18, where the total frame shear is taken by each portal in proportion to its span. For equal bays, this results in dividing the shear equally among the number of bays, so that the interior columns carry twice as much load as the exterior ones; this assumes that the stiffness of the exterior columns is one half that of the interior ones, which is not necessarily true.

For a building frame with n equal bays, and a total lateral pressure, W, for the given level (see fig. 7.18, top), the typical column shear against an interior column at story midheight is

$$V_{ci} = W\left[\frac{(L_1 + L_2)/2}{B}\right] = W/n \tag{7.7a}$$

The corresponding shear for an exterior column is

$$V_{ce} = V_{ci}/2 \tag{7.7b}$$

This horizontal shear at midheight (h/2) wracks the column and causes moments at the top and bottom of each column. For example, the maximum moments for an interior column are

$$M_c = \pm V_{ci}(h/2) \tag{7.8}$$

The vertical shear is resisted by the girders in bending. It is found by isolating a story-high frame, as demonstrated in Example 7.2. For preliminary design purposes, particularly for tall buildings, the girder shear can be quickly approximated by letting the girder and column shears balance each other in rotation around a typical interior story-high cruciform module, as defined by the inflection points at midlength of the columns and girders; in addition, it is conservatively assumed that there is no change in column shear for the given story, and hence no axial force flow in the girders, by using the column shear below the girder level.

$$V_{ci}(h) \simeq V_G(L), \quad \text{or} \quad V_G \simeq V_{ci}(h/L) \tag{7.9a}$$

A similar moment equation can be set up for the exterior beam-column joint:

$$V_G \simeq V_{ce}(2h/L) \tag{7.9b}$$

According to the portal method, the interior axial column forces of the independent portals cancel each other (Fig. 7.18, top), so that the rotation of the frame is resisted in axial action entirely by the exterior columns; therefore, the interior columns do not carry any vertical forces. In other words, the moment caused by the lateral forces acting about the inflection points at midheight of any floor level is resisted by the axial forces, N_c, in the exterior columns across the frame width, B.

$$N_c(B) = M, \quad \text{or} \quad N_c = M/B \tag{7.10}$$

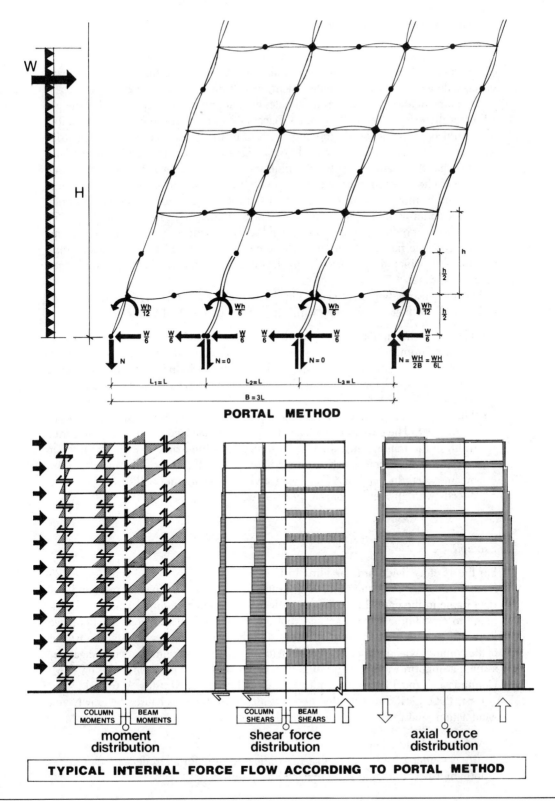

PORTAL METHOD

TYPICAL INTERNAL FORCE FLOW ACCORDING TO PORTAL METHOD

Figure 7.18. *Typical force flow according to portal method.*

The assumptions regarding the locations of the inflection points and the shear distribution to the columns in the portal method reduces a highly indeterminate structure to a statically determinate system, as is further explored in Example 7.2.

The portal method generally overestimates the interior column moments, but under-estimates the exterior ones. It becomes inaccurate in the assumption of the location of the inflection points at the top and the base of the frame (see fig. 7.17), that is, using a regular frame with no setbacks or significant changes in member stiffness. It assumes that the beam and column stiffnesses are roughly equal to each other, but this does not hold true for the upper stories (with smaller columns), where the inflection points tend to move downward (e.g., 0.4h from the base), and for the lower stories where the columns are larger and the points of contraflexure move upward (e.g., 0.6h from the base for ordinary conditions), possibly even into the next story; this is especially true for concrete frames. Naturally, for the condition of small foundations on com-pressible soil, which do not provide much restraint to rotation, the imaginary hinges move downward.

The force distribution due to lateral load action along a regular rigid frame is shown in Figure 7.18, bottom. The following characteristics are identified:

- The frame rotation causes an axial force flow in the exterior columns, which increases from the top to the bottom of the building; the interior columns are assumed not to carry any axial forces. The axial forces in the beams are considered not to be critical for typical conditions, and are neglected for the preliminary design.
- The column shear increases from the top of the building to the base, as does the beam shear. The beam shear at each floor level remains constant across the entire width of the frame, since there are assumed to be no axial forces in the interior columns. The lateral wracking of the frame due to the shear forces is resisted by the members in bending. The beam and column moments therefore increase directly with an increase of the shear forces.

Example 7.2

For the 15-story rigid frame building in Figure 7.19, with the frame bents spaced 25 ft apart in the long direction and 20 ft in the short direction, a typical interior 3-bay rigid frame in the cross-direction is investigated. The floor framing spans from cross-frame to cross-frame. The floor and roof dead loads are 80 psf each, the floor live load is 80 psf, the roof live load is 30 psf, and the curtain load is 15 psf. The weight of the columns is assumed to be equivalent to 14 psf. For this preliminary investigation, assume a uniform wind pressure of 26 psf, as based on a 70 mph wind (see fig. 3.8). Columns 1–5 and 13–17, and girder 17–18, will be designed using A36 steel.

First, the critical forces on the members are determined, as caused by gravity and wind action, and then the members are designed.

A) *Gravity Loads.*

A.1) Girder 17–18. Notice that the vertical loading for all girders at any level and any bay is the same. The live load reduction is

$$R_1 = 0.08\%A = 0.08(20 \times 25) = 40\%$$

$$R_2 = 23(1 + D/L) = 23(1 + 80/80) = 46\%$$

$$R_3 = 60\%$$

There will, therefore, be a live load reduction of 40 percent. The girder must carry the following uniform dead and live load.

$$w = 25[0.080 + 0.6(0.080)] = 3.2 \text{ k/ft}$$

The corresponding maximum girder moment is

$$M_{max} = wL^2/12 = 3.2(20)^2/12 = 106.67 \text{ ft-k}$$

A.2) *Columns.* Each of the exterior columns above and below must resist one-half of the girder moment $M = wL^2/22$, as has been discussed earlier (see fig. 7.16).

$$M_c = (wL^2/22)/2 = 3.2(20)^2/44 = 29.09 \text{ ft-k}$$

Notice that, due to the symmetrical layout conditions, the interior columns do not carry any gravity moments, assuming unbalanced live loading is not critical.

The live load reduction for the columns is

$$R = 46\% < (40\%) \times (\text{n-floors}).$$

The curtain load per floor is

$$P_c = 0.015(12 \times 25) = 4.5 \text{ k/floor}$$

The exterior column carries the following roof load

$$P_R = 25 \times 10(0.030 + 0.080 + 0.014) = 31 \text{ k}$$

The interior columns carry twice as much roof load.

The exterior column carries the following floor load

$$P_F = 25 \times 10[0.080 + 0.014 + 0.54(0.080)] = 34.3 \text{ k/floor}$$

The interior columns carry twice as much floor load. From the foregoing, we find the columns to be investigated must resist the following axial loads:

$$N_{1-5} = 31 + 14(34.3) + 15(4.5) = 578.7 \text{ k}$$

$$N_{2-6} = 2[31 + 14(34.3)] = 1022.4 \text{ k}$$

$$N_{13-17} = 31 + 11(34.3) + 12(4.5) = 462.3 \text{ k}$$

B) *Wind Loads.* Because the cross-frames are all rigid, and because of their large number, it is conservatively assumed that each interior frame supports the uniform wind pressure from the tributary facade area extending to the centerlines of the adjacent bays, which is equal to

$$w = 0.026(25) = 0.65 \text{ klf}$$

B.1) *Columns.* To find the horizontal shear forces that the columns must resist, a section is cut through the inflection points at midheight of the floor level where the column to be investigated is located. At those points of contraflexure, the columns resist only shear and axial forces. First, a horizontal section is taken at

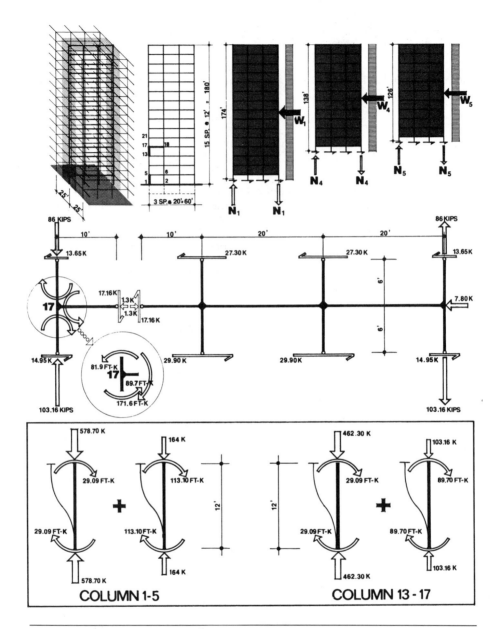

Figure 7.19. *Investigation of rigid frame structure in Example 7.2.*

the first floor level, as shown in Figure 7.19. From this free body, the following results are obtained.

The total wind shear is

$$W_1 = 0.65(174) = 113.1 \text{ k}$$

Because of the three equal bays, each of the interior columns resists one-third of the total wind shear.

$$V_{ci} = W_1/n = 113.1/3 = 37.7 \text{ k}$$

The exterior columns resist one-half of this value

$$V_{ce} = V_{ci}/2 = V_{1-5} = 37.7/2 = 18.85 \text{ k}$$

The corresponding column moments, due to shear wracking, are

$$M_{1-5} = 18.85(12/2) = 113.1 \text{ ft-k}, \qquad M_i = M_{2-6} = 2(113.1) = 226.2 \text{ ft-k}$$

The rotation due to the wind pressure is assumed to be resisted entirely by the external columns. Taking moments about the exterior windward column yields

$$N_{1-5}(60) = 113.1(174/2), \qquad \text{or} \qquad N_{1-15} = 164 \text{ k in compression}$$

A similar approach is now used for finding the column forces at the fourth and fifth floor levels, with the corresponding free bodies shown in Figure 7.19.
Fourth floor level:

$$W_4 = 0.65(138) = 89.7 \text{ k}, \qquad M_{13-17} = 14.95(12/2) = 89.7 \text{ ft-k}$$

$$V_{13-17} = (89.7/3)/2 = 14.95 \text{ k}, \qquad N_{13-17} = 89.7(138/2)/60 = 103.16 \text{ k}$$

Fifth floor level:

$$W_5 = 0.65(126) = 81.9 \text{ k}, \qquad M_{17-21} = 13.65(12/2) = 81.9 \text{ ft-k}$$

$$V_{17-21} = (81.9/3)/2 = 13.65 \text{ k}, \qquad N_{17-21} = 81.9(126/2)/60 = 86 \text{ k}$$

B.2) *Girder 17–18.* To find the girder forces due to wind action, the multistory rigid frame can be visualized as consisting of single-story frames with their columns cantilevering to midheight above and below the floor levels, as based on the inflection point location assumed for the portal method. The free body of the story-high frame for the fifth floor level of this example is shown in Figure 7.19. Indicated are the respective axial and shear forces in the columns at the fourth and fifth floor levels, as have already been determined.

When isolating the T-shaped free body at the midspan of girder 17–18, that is, at the point of contraflexure, the axial and shear forces in the girder can easily be obtained by statics.

Vertical equilibrium of forces, the difference in axial forces, is provided by the girder shear.

$$V_{17-18} = 103.16 - 86 = 17.16 \text{ k}$$

This vertical shear wracks the girder and causes the following moments at the girder ends.

$$M_{17-18} = 17.16(20/2) = 171.6 \text{ ft-k}$$

It is now checked to verify whether the moments around joint 17 are in rotational equilibrium.

$$\Sigma M = 89.7 + 81.9 - 171.6 = 0 \qquad \text{OK}$$

Horizontal equilibrium of forces, or the difference in column shears, yields the axial force in the girder.

$$N_{17-18} = 14.95 - 13.65 = 1.3 \text{ k}$$

This force is insignificant for the preliminary design of the girder, and will be ignored.

For preliminary design purposes, the girder shear could have been conservatively approximated, as based on the column shear for the story below, as

$$V_G = V_{ci}(h/L) = 29.90(12/20) = 17.94 \text{ k} \qquad (7.9)$$

C) *Member Design.*

C.1) *Girder 17–18.* The combined loading case of gravity and wind, for which codes allow a load reduction of 25 percent or an increase in allowable stresses of 33 percent, is clearly larger than the gravity case.

$$M = 0.75(106.67 + 171.60) = 208.70 \text{ ft-k} > 106.67 \text{ ft-k}$$

The axial forces in the beam are neglected for this preliminary design, since they are insignificant. The required section modulus, according to Equation 4.23, is

$$S = M/F_b = 208.70(12)/24 = 104.35 \text{ in.}^3$$

Select, as a trial section: W24 × 55

$$A = 16.2 \text{ in.}^2, \ I_x = 1350 \text{ in.}^4, \ S_x = 114 \text{ in.}^3$$

C.2) *Column 1–5.* Since the lateral forces in the critical short direction of the building are resisted by the columns in bending, the columns are oriented so that they bend about their strong axis (see fig. 7.19). The moment is transformed into an equivalent axial force by assuming a bending factor of $B_x = 0.18$ for W14 sections (see Equation 4.52). At the building base, the combined loading case clearly controls the column design.

$$P + P' = P + B_x M_x = 0.75[578.7 + 164 + 0.18(29.09 + 113.1)12]$$

$$= 787.37 \text{ k} > 578.7 + 0.18(29.09)12 = 641.53 \text{ k}$$

For this laterally nonbraced building, a stiffness factor of $K = 1.8$ is assumed as a first trial, as based on the discussion of steel columns in the subsection in Chapter 4 on "Steel Columns." According to the column tables in the AISC Manual for $KL = 1.8(12) \simeq 22$, the following section is selected as a first trial:

$$W14 \times 145, \qquad P_{all} = 718 \text{ k}$$

$$B_x = 0.184, \ A = 42.7 \text{ in.}^2, \ I_x = 1710 \text{ in.}^4$$

It should be remembered that not the section required, but the next smaller one, is selected as a first trial.

This section could also have been obtained without the use of the column tables in the AISC Manual, as follows,

$$A_s = \frac{P + B_x M_x}{F_a} \simeq \frac{P + 2.5(M_x/t)}{23 - 0.1(KL/r)} \qquad (4.53)$$

where the allowable axial compression stresses are approximated as

$$F_a \simeq 23 - 0.1(KL/r) = 23 - 0.1(1.8(12)12/4) = 16.52 \text{ ksi}$$

Therefore, the required cross-sectional area, using $B_x = 2.5/t = 2.5/14 = 0.18$, is

$$A_s = 787.37/16.52 = 47.66 \text{ in.}^2$$

Selecting the next smaller section yields the same member as before.

C.3) *Column 13–17.* Using a similar design approach as for the other column, one finds that this column, at the fourth-floor level, is also controlled by the combined loading case (see fig. 7.19).

$$P + P' = 0.75[462.3 + 103.16 + 0.18(29.09 + 89.7)12]$$

$$= 616.54 \text{ k} > 462.3 + 0.18(29.09)12 = 525.13 \text{ k}$$

From the column tables in the AISC Manual for KL = 1.8(12) ≃ 22 the following first trial section is obtained

$$W14 \times 120, P_{all} = 578 \text{ k}$$

$$B_x = 0.186, A = 35.3 \text{ in.}^2, I_x = 1380 \text{ in.}^4$$

C.4) *Interior columns* (see Problem 7.5). For ordinary multistory rigid frame buildings up to about 15 stories, with regular layout and normal loading conditions, the preliminary design of an interior column may be based on axial gravity action only, since the additional moment contribution due to the combined loading case is usually below 10 percent.

The preliminary design of moment-resisting concrete frames can be approached in a similar manner as for steel frames; for the design with respect to gravity loading, refer to Figure 5.8a. Special attention, however, must be given to the design of concrete frames for seismic resistance, since concrete (as a brittle material) lacks the ductility of steel, a property that is necessary for the relatively flexible frames in order to deal with the unknown, randomly reversing inertia forces. Codes allow yielding to take place in some members for buildings located in regions of moderate to high seismic risk, naturally without causing a collapse of the structure. It has been discussed, in the section in Chapter 3 on "Seismic Loads," that ductile moment-resisting frame structures have the lowest horizontal force factor of K = 0.67, in contrast to the much stiffer bearing-wall systems (lacking the ductility), which have the highest factor of K = 1.33, since they are designed for a nearly elastic response. In the past, concrete buildings on the West Coast never went above 20 stories, but *ductile concrete design*, which considers inelastic deformations, has now made buildings of 30 stories and more possible.

Concrete frames must provide both the ductility, which tolerates large inelastic deformation reversals, and the energy dissipation capability of steel frames, without significant loss of strength; brittle failure (compression crushing, shear failure) of the concrete must be controlled. This ductile flexural hinging is achieved by keeping the tensile reinforcement ratios well below the balanced ratios, to allow for yielding under bending and for the redistribution of overloads, keeping in mind that high-strength steels with limited ductility should be avoided. However, closely spaced transverse reinforcement is required and is distributed over regions where inelastic action is expected to confine the concrete and maintain lateral support of the bars; in other words, it acts to maintain the member's axial load and shear capacities. Therefore, large quantities of confining closed-hoop ties (spirals) and stirrups are needed at the critical stress locations of the beam-to-column joints (Fig. 7.20) to increase the member ductility and rotational capacity; the steel ties raise the compressive strain capacity of the concrete and prevent buckling of the longitudinal bars in compression.

Critical, with respect to strength and especially to ductility, is the transmission of forces through the connection points of beams and columns. The continuous force

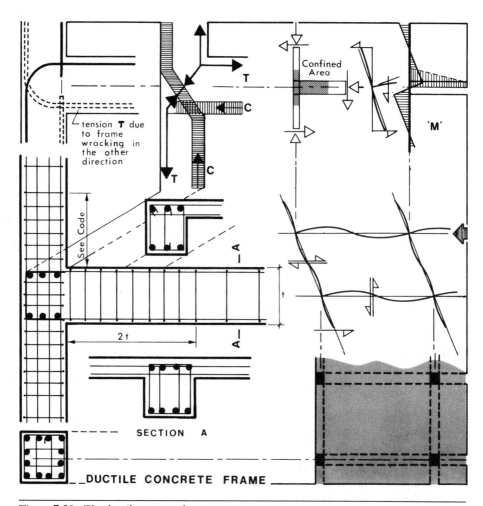

Figure 7.20. *The ductile concrete frame.*

flow within the joint is illustrated in Figure 7.20 in schematic fashion. Diagonal forces occur in the beam-to-column joint core due to the change from tension to compression; in other words, shear is generated within the joint core by the flexural reinforcement, which must be laterally restrained with closed horizontal ties that also provide the necessary confinement to the compressed concrete. It is apparent that the longitudinal bars should run through the joint without interruption, otherwise attention must be given to special anchorage. Lap splices should only be within the center half of the member lengths. For the specific design requirements, the reader should refer to the ACI Code Appendix A.

The control of the inelastic deformations is achieved by deliberately making certain members weaker than others. Recommended by codes is the *strong column-weak beam* potential collapse mechanism (see fig. 3.21), because it has less effect on the structural stability and beams can be repaired more easily than columns; furthermore, axial compressive loads tend to reduce available ductility. Plastic hinges form at the ends of beams during an earthquake, absorbing energy and dissipating it through large deformations; they redistribute forces to the stronger columns, which are kept elastic throughout their seismic response. Hence mechanisms associated with a *soft story* concept (see fig. 3.21) should be avoided; only under controlled conditions should hinges be allowed to form in specific columns. Designers seem to agree that ductile concrete frame construction can be economical for buildings up to the 40-story range. However, it should not be overlooked that ductile concrete frames require more reinforcement than conventional frames, thereby creating a congestion at the joints, especially for higher buildings, and require more laborious detailing so that the bars do not conflict.

Codes discourage the use of precast concrete ductile frame construction because of the difficulty in evaluating the behavior of the connectors. Although it seems to make sense to place the connection at the location where a ductile hinge forms, designers avoid this situation, since it is extremely complex to evaluate the capacity of a ductile energy-dissipating connection. It may be concluded that plastic hinge location and joint location should not coincide! By providing a non-yielding continuous connection, a plastic hinge is forced away from the joint. Continuous connections in precast concrete can be achieved by precast shells with cast-in-place concrete cores; in this case, the precast shells are the formwork for the wet concrete. Continuity can also be achieved by assembling precast elements by grouted post-tensioning, which may respond as a ductile energy-absorbing connection; this will not be true, however, for ungrouted post-tensioning. Cold joints reinforced with mild steel form hinged connections that are placed in regions of low moment demand close to where inflection points occur.

Approximate Lateral Deflection of Rigid Frames

It has been discussed in Chapter 2 that the lateral drift of a building must be limited to ensure not only the comfort of the occupants, but also to maintain the structural integrity of the connections and to minimize nonstructural damage of curtain walls, partitions, ceilings, and elevator shafts.

In the previous section, it was shown that the lateral drift of the frame is due to the shear wracking of the columns and girders and also due to the cantilever behavior of the frame as a whole (i.e., due to the axial deformations of the columns). Whether *web drift* (shear wracking) or *chord drift* (cantilever bending) controls the lateral deflection depends on the height-to-width ratio of the building and the stiffness of the frame, which is proportional to the member sizes, and inversely proportional to the member spacing; it will be shown that the lateral sway of ordinary planar frames is primarily due to web drift rather than due to chord drift. The effect of joint stresses and joint slippage is ignored in the following investigation. Also, the effect of sidesway due to asymmetrical placement of the gravity loading is considered not to be critical because of sufficient building stiffness and balance of loading from other floors. The terms for the derivation of the lateral building deflection were defined earlier (see fig. 7.17, bottom).

The web drift is a shear-type deflection caused by the shear forces that bend the columns and girders. The *column sway* is derived by considering the column boundaries (i.e., floor girders) as rigid. The lateral deflections, Δ_c, due to the shear, V_c, at midheight can then be obtained by visualizing two vertical cantilevers each of $h/2$ length. The lateral sway due to column bending is thus

$$\Delta_c/2 = \frac{V_c(h/2)^3}{3EI_c}, \quad \text{or} \quad \Delta_c = \frac{V_c h^3}{12EI_c} \tag{7.11}$$

In other words, the relationship between the column rotation, γ, and the deflection is

$$\tan \gamma = \Delta_c/h \simeq \gamma = \frac{V_c h^2}{12EI_c}$$

The lateral deflection of the frame due to the girder rotation, θ, is derived in a similar fashion, by assuming the girder boundaries (i.e. columns) as rigid.

$$\tan \theta = \frac{\Delta/2}{L/2} \simeq \theta = \frac{V_G(L/2)^3}{3EI_G L/2} = \frac{V_G L^2}{12EI_c}$$

Therefore, the lateral sway due to girder bending is

$$\Delta_G \simeq \theta h = \frac{V_G L^2 h}{12EI_G} \tag{7.12}$$

It may be concluded that the total story drift due to shear wracking is

$$\Delta_T = \Delta_G + \Delta_c = (\theta + \gamma)h$$

$$\Delta_T = \frac{V_G L^2 h}{12EI_G} + \frac{V_c h^3}{12EI_c} \tag{7.13}$$

The total lateral frame sway is obtained by assuming a linear drift of the entire building with n stories.

$$\Delta = (\theta + \gamma)hn = (\theta + \gamma)H$$

$$\Delta = \left[\frac{V_G L^2}{12EI_G} + \frac{V_c h^2}{12EI_c} \right] H \tag{7.14}$$

This equation can be further simplified for rigid frames with nearly equal bays by expressing the girder shear in terms of the column shear (assuming the difference in column shears above and below the floor level to be insignificant). The interior column shear, V_{ci} according to Equation 7.9a, is

$$V_G = V_{ci}(h/L) \tag{7.9a}$$

Substituting this expression into Equation 7.14 yields

$$\Delta = \frac{V_{ci}hH}{12E} \left[\frac{L}{I_G} + \frac{h}{I_{ci}} \right] \tag{7.15a}$$

Using the exterior column shear, V_{ce}, of Equation 7.9b yields

$$V_G = V_{ce}(2h/L)$$

Substituting this expression into Equation 7.14 yields

$$\Delta = \frac{V_{ce}hH}{12E}\left[\frac{2L}{I_G} + \frac{h}{I_{ce}}\right] \tag{7.15b}$$

Several column and girder combinations satisfy the relationships in Equations 7.15a and b.

Often, a similar expression as above for approximating the frame drift is used, but in terms of the total horizontal story shear, V, as is derived in Problem 7.14.

$$\Delta = \frac{VhH}{12E}\left[\frac{L}{\Sigma I_G} + \frac{h}{\Sigma I_c}\right] \tag{7.15c}$$

For preliminary design purposes, rigid frames with H/B ratios of less than 3 may be assumed to be dominated by web drift (i.e., shear wracking) so that chord drift can be ignored. For more slender frames, the deflection due to the axial deformations of the exterior columns can be approximated as shown in the following paragraphs.

Since the axial gravity stresses in the exterior columns increase in an approximately linear manner towards the base, it is assumed that the wind forces also increase linearly from zero at the top to a maximum at the bottom. The resultant wind pressure, $W = V = wH$, is therefore assumed to act at the building top. The familiar lateral deflection of the vertical cantilever frame is

$$\Delta = \frac{VH^3}{3EI} \tag{a}$$

According to the portal method, the resisting moment of inertia is provided by the exterior columns with their cross-sectional areas, A_c, which are separated by the distance, B (i.e., the frame width). The equivalent moment of inertia is, therefore,

$$I = 2[A_c(B/2)^2] = A_c B^2/2 \tag{b}$$

Since the exterior columns resist all of the rotation, the following relationship holds true:

$$N(B) = V(H), \quad \text{or} \quad V = N(B/H) \tag{c}$$

Substituting Equations (b) and (c) into (a) yields the chord drift:

$$\Delta = \frac{2VH^3}{3EA_cB^2} = \frac{2NH^2}{3EA_cB} = \frac{2f_cH^2}{3EB} \tag{7.16a}$$

where: $f_c = N/A_c =$ maximum axial stresses in the exterior columns at the base due to the lateral forces

B = width of bent

H = height of frame

E = modulus of elasticity

For the earthquake action, the chord drift is increased by 50 percent to

$$\Delta = \frac{VH^3}{EA_cB^2} = \frac{f_cH^2}{EB} \tag{7.16b}$$

Notice that the ratio of the lateral deflection to the total axial deformation of the chord is equal to the aspect ratio of the frame: $\Delta/(f_c H/E) = H/B$

Example 7.3

The approximate lateral sway of the rigid frame in Example 7.2 is determined.

Since $H/B = 180/60 = 3$, the total deflection due to shear and chord drift for wind action is determined according to Equations 7.15b and 7.16a.

$$\Delta = \frac{V_{ce}hH}{12E}\left[\frac{2L}{I_G} + \frac{h}{I_{ce}}\right] + \frac{2f_c H^2}{3EB} \tag{7.17}$$

The various terms are obtained from Example 7.2; $f_c = N_c/A_c = 164/42.7 = 3.84$ ksi.

$$\Delta = \frac{14.95(12 \times 12)}{12(29,000)}(180)[2(20 \times 12)/1350 + 12(12)/1380]$$

$$+ 180\left[\frac{2(3.84)180(12)}{3(29,000)60(12)}\right]$$

$$= 0.396 + 0.116 + 0.048 = 0.512 + 0.048$$

$$= 0.56 \text{ ft} = 6.72 \text{ in.} \geqslant H/500 = 180(12)/500 = 4.32 \text{ in.}$$

The frame is too flexible and must be stiffened! In this case 91 percent of the lateral building sway is caused by the web drift (i.e., 70 percent by girder bending and 21 percent by column bending) and only 9 percent by the chord drift or axial deformations of the columns. The shear deflection clearly controls, so that (for preliminary investigations) the bending deflections can be ignored, as was expected because of the low H/B ratio. Plane rigid frame structures with reasonably wide bays are often treated as if 90 percent of the total deflection is due to shear wracking, realizing that the stiffness of the frame girders controls, to a large extent, the lateral sway of the building. In contrast to rigid frames, cantilever beams such as slender braced frames are dominated by axial column deformations—in this case, more than 90 percent of the total deflection may be due to chord drift.

Now the critical bending stiffness of the girder at the fifth floor level is found, as required to satisfy the allowable story drift, by considering only shear deflections and neglecting axial deformations (Equation 7.15b).

$$\Delta = \frac{14.95(12 \times 12)}{12(29,000)}(12)[2(20 \times 12)/I_G + 12(12)/1380] = 12/500$$

$$I_G = 2192.26 \text{ in.}^4$$

Try W24 × 84, $I_G = 2370 \text{ in.}^4$

FLAT SLAB BUILDINGS

Flat slab buildings consist of horizontal planar concrete slabs directly supported on columns, thus eliminating the need for floor framing. This results in a minimum story height—an obvious economic benefit that is especially advantageous for apartment buildings. Drop panels and/or column capitals are frequently used because of high shear concentrations around the columns. Slabs without drop panels are commonly called flat plates. The system is adaptable to an irregular support layout.

Some of the disadvantages of the flat slab structure are:

- It is a relatively short-span system, limited to building types with relatively light loading and frequent partition layout, as for residential buildings.
- As a heavy-weight system, it requires appropriate foundation conditions.
- As a flexible system, it may be necessary to increase its lateral stiffness by adding shear walls.

Two-way slabs are discussed in more detail in the section in Chapter 5 on "Reinforced Concrete Slab Systems." It is shown that a flat slab structure may be idealized by dividing it into a series of individual orthogonally crossing two-dimensional frames, consisting of the columns and the shallow slab-beams. For lateral force resistance, the effective slab-width concept is often used in rigid frame analysis. As the slab spans increase, less lateral stiffness can be provided by the slabs, thereby causing the columns to approach the behavior of individual cantilevers (see fig. 7.17). Frame action can only give adequate lateral stiffness for buildings up to about ten stories.

When the flat slab structure is stiffened by an interior core or a perimeter tube, it may be assumed that the much stiffer rigid shear walls resist the lateral loads entirely. This is particularly true for the lower portion of the building, hence the flexible flat slab structure will carry only gravity loads. However, it must not be forgotten that the lateral forces in the upper portion of the structure are primarily resisted by frame action, which may be assumed not to be critical for preliminary design purposes.

The 20-story Hartford Building [1961] in Chicago (SOM) is a fine example of the expression of the flat slab principle in the tradition of the Chicago School. The exterior concrete frame on an 11.5 × 22 ft grid articulates the thin horizontal edges of the flat slabs and the rounded haunch connections at the columns, quite in contrast to the more familiar wide bands of the dropped spandrel beams. Several other flat slab buildings have been shown (see fig. 5.12).

INTERSTITIAL STRUCTURES

The interstitial design concept provides intermediate or interstitial spaces between column-free working floors, so that they can be independent of each other. In the interstitial space, story-high long-span trusses (Fig. 7.21) of the bridge-type accommodate ducts, plumbing, distribution systems, service piping of many types, and wiring; they contain transformer rooms and the complex support equipment especially required by hospitals (and sometimes by laboratories), so that the utilities for the working spaces can be distributed through the ceilings and floors. The separation between the servicing and served spaces is essential; the maintenance of the mechanical spaces should not disturb or interfere with the free movement of personnel on the service floors or with the operational functions. Hospitals are highly susceptible to rapid obsolescence due to changing medical technology. Flexibility of plan layout and structural versatility, including adaptability to alterations and spatial changes, are met by the interstitial

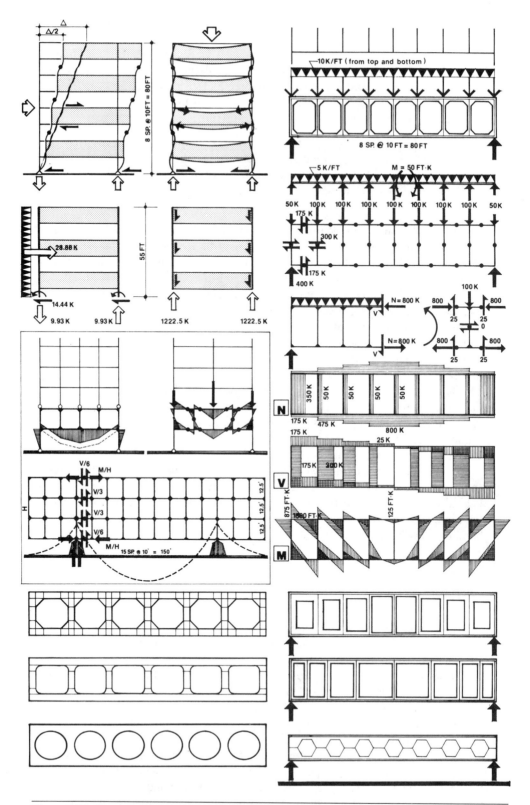

Figure 7.21. *Vierendeel beams.*

design concept. Although the interstitial space adds more volume to a building, it requires hardly any plumbing, light, or heat for its own use.

The concept of sandwiched spatial organization allowing free-space platforms was introduced by the structuralists of the 1960s, for instance, as found in the urban megastructures. An early example is Louis Kahn's Salk Institute Laboratories of 1965 and then the many hospital structures of the 1970s (see fig. 7.7).

In this section, the *Vierendeel truss*, in the context of the interstitial design principle, is investigated because the reader may not be familiar with the structure concept. The Vierendeel truss avoids the diagonal members of conventional trusses, thus providing continuity of space and allowing freedom of movement. It is basically a rigid frame that is used as a beam, although it is much more compact than a rigid frame building structure. Since the horizontal frame-beam resists uniform gravity loads, similar to the way that the vertical cantilever frame resists lateral loads, the portal method of analysis can be used for preliminary design purposes, assuming uniform member distribution. The force flow can thus be approximated by estimating inflection points at the midlength of all members for each bay of the Vierendeel truss. It should be realized, however, that the frame-beam is considerably less efficient compared to an ordinary truss, since the shear must be resisted in bending by the members rather than directly in axial action by the truss diagonal members.

The reader is possibly familiar with the use of Vierendeel trusses for pedestrian bridges or the application of the principle to stub-girders and castellated floor beams. Several examples of Vierendeel trusses in buildings have been shown (see figs. 7.4b, 7.7, and 7.21).

Example 7.4

An eight-story interstitial frame building (Fig. 7.21) with Vierendeel trusses spaced 25 ft apart and spanning 80 ft from outside column to outside column is investigated. A typical truss carries a uniform gravity load of 5 k/ft at the top and bottom chords each; the curtain load is taken as 15 psf. Grade 50 steel is used for the preliminary design of some of the critical truss members and for the design of the building column at the third floor level.

1) Preliminary investigation of the Vierendeel truss.

The top chord acts as a continuous beam that transfers, in bending, the uniform load of 5 k/ft to the truss joints; the secondary effects of truss deflection are ignored (i.e., support settlement) because the truss is considered to be of sufficient stiffness. The critical negative chord moments at the joints are conservatively taken as

$$M = wl^2/10 = 5(10)^2/10 = 50 \text{ ft-k}$$

The top and bottom chord beam reactions are transferred to the top joints. Therefore, the single loads at the top chord joints are

$$P = [5(10)]2 = 100 \text{ k}$$

The Vierendeel truss as a whole resists the uniform load of 10 k/ft, which causes the following maximum moment at center span

$$M_{max} = wL^2/8 = 10(80)^2/8 = 8000 \text{ ft-k}$$

This moment, in turn, is resisted by the top and bottom chords in axial action.

$$N(h) = M, \quad \text{or} \quad N = M/h = 8000/10 = 800 \text{ k}$$

This approach is conservative, since the axial chord forces will be slightly smaller. The truss chords should have been investigated at the center of the innermost bays adjacent to midspan, where the location of the inflection points are assumed, according to the portal method (see Problem 7.15).

Isolating the I-shaped freebody of the truss at midspan (Fig. 7.21) clearly shows that the 100 k joint load is resisted in shear evenly by the chords: $V = 100/4 = 25$ k. These shears bend the chords and cause the following moments

$$M = 25(5) = 125 \text{ ft-k}$$

Therefore, the chords at center span must be designed for an axial force of $N = 800$ k and a moment of $M = 50 + 125 = 175$ ft-k.

Since the chords are fully laterally supported by the floor framing, the allowable axial stresses may be assumed as controlled by yielding with $F_a = 0.6F_y = 0.6(50) = 30$ ksi, rather than buckling about the weak axis; the displacement of the chord about the strong axis is considered not to be critical for this preliminary investigation. Hence, the required cross-sectional area for a W14-chord section is approximately

$$A_s = [P + 2.5(M/t)]/F_a \quad\quad\quad\quad\quad (4.53)$$

$$= [800 + 2.5(175/14)12]/30 = [800 + 375]/30 = 39.17 \text{ in.}^2$$

Try: W14 × 132, $A_s = 38.8 \text{ in.}^2$

The next larger section may have to be selected because the predominant action of the member is axial since the moment uses only about 30 percent of the member's capacity.

The chords adjacent to the supports will be much larger—they will behave primarily as *beams* rather than *beam-columns*, as at midspan, where the truss moment is large but the shear wracking is small. They must resist large bending moments because of the high shear adjacent to the support, but only a little axial force due to the small truss moment. The chord size can be quickly estimated by simply treating it as a beam and adding about 10 percent to its required section modulus to take axial action into account; refer to Problem 7.15 for a more accurate solution.

The shear in the top and bottom chords each resist $(400 - 50)/2 = 175$ k; hence the chord moment, due to shear wracking adjacent to the supports, is

$$M = 175(5) = 875 \text{ ft-k}$$

Adding the local beam moment due to the uniform load yields

$$M_T = 875 + 50 = 925 \text{ ft-k}$$

The required section modulus is

$$S \simeq 1.1(M/F_b) = 1.1[925(12)/33] = 370 \text{ in.}^3$$

Try W33 × 130, $S_x = 406 \text{ in.}^3$

The truss columns act primarily as *beams*, especially adjacent to the support, where the shear is maximum. The horizontal shear wracks the columns and causes large moments. For the approximate design of the truss columns as based on the portal method, refer to Problem 7.15.

It is beneficial to build up the sections so that the inertia is concentrated at the joints where the moments are maximum (Fig. 7.21). Further, by decreasing the bay sizes towards the supports, the chord spans become smaller and more columns resist the horizontal shear, thereby decreasing the effect of shear wracking; for this condition, the chords and web columns tend to be of a constant size, as shown in Figure 7.21.

> 2) The design of the column at the third floor level is clearly controlled by the large gravity loads, as investigated in Problem 7.16. The axial load due to the facade and truss loading is

$$N = 0.015(25 \times 60) + 3(10 \times 40) = 1222.5 \text{ k}$$

The columns bend when the truss deflects under the gravity loading (Fig. 7.21, top, left). There is no accumulation of these secondary gravity moments along the columns, since the floor diaphragms are assumed to take the moments out by resolving them at each floor level into an axial load couple that is carried to the shear walls (e.g., elevator core or facade cross walls). If, during the construction stage, the truss chords are connected to the columns with slotted holes, then the dead load rotation of the truss will not introduce secondary moments in the columns. Once the truss is in place, the chords can be rigidly connected to the columns so that only the superimposed loads will cause rotation. Truss cambering can also be introduced to relieve the dead load deflection by making the top chords longer than the bottom chords. In this context, the secondary moments due to the truss rotation are ignored. It is further assumed that the laterally swaying truss does not allow rotation of the columns so that they may be treated as fixed at top and bottom with a theoretical K-value equal to one for first approximation purposes. Using $K = 1.2$ as recommended by the AISC specifications, one obtains, for $KL = 1.2(10) = 12$ from the column tables of the AISC Manual, the following section:

$$\text{W14} \times 159, \qquad P = 1232 \text{ k}, \qquad I_x = 1900 \text{ in.}^4$$

The lateral deflection of this frame due to wind action is not critical for this low building; it is approximately checked in Problem 7.16 by using the column sway due to shear wracking, as expressed by Equation 7.11, together with rigid plate action of the trusses (see fig. 7.21).

STAGGERED WALL-BEAM STRUCTURES

In this innovative structure system, story-high wall-beams span across the full width of the building on alternate floors of a given bay and are supported by columns along the exterior walls, as shown in Fig. 7.22; there are no interior columns. The wall-beams are usually steel trusses, but can also be pierced reinforced concrete wall beams. The steel trusses are concealed within the room walls. Conventional gypsum wallboard is mounted on steel studs and runners to provide a fire-resistant envelope.

In the interstitial system, wall beams are used at every other floor to allow for uninterrupted free flow in the floor space between, while in the staggered wall-beam system, the wall-beams are used at every floor level, but arranged in a staggered fashion

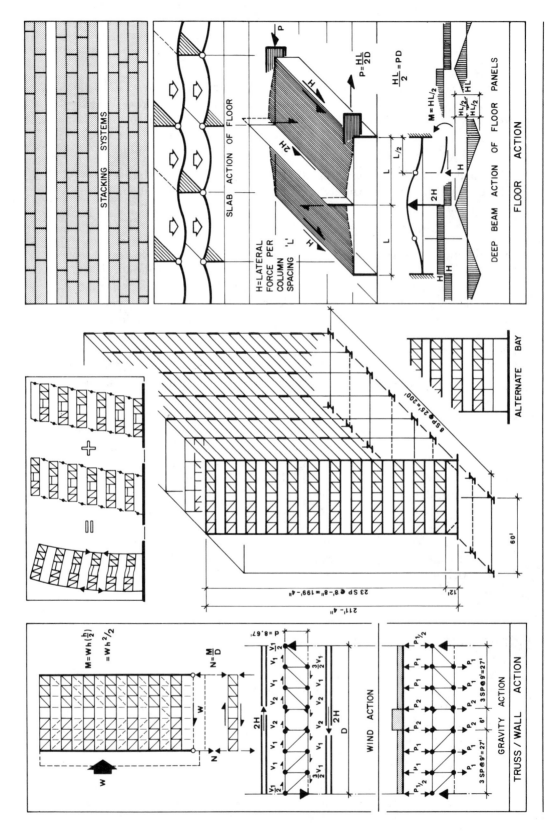

Figure 7.22. *Staggered truss buildings.*

between adjacent floors. The arrangement of the story-high members depends on the layout of the functional units. Apartment units to be contained between the wall-beams and to be vertically stacked to resemble masonry bond patterns can be visualized as indicated in Figure 7.22, right. As the unit sizes change, the spacing of the wall-beams may be adjusted or additional openings may be provided. The most common system of organization is the running bond or checkerboard pattern.

Since alternate walls at each floor level are eliminated by staggering the wall-beams from floor to floor, the room width is doubled without having to increase the floor span. In this case, the slabs or floor framing sit in bearing on top of one wall-beam and hang from the bottom of the next.

The staggered steel truss system was developed in the mid-1960s by a team of architects and engineers from the Departments of Architecture and Civil Engineering at M.I.T.; the principle was first used in 1967. The system is best suited for narrow structures of the slab-type in the 60-ft width range, especially for hotels and apartment buildings with central corridors. The staggered wall-beam principle can be used in rectangular, circular, and curvilinear configurations, as well as in many other plan forms. A more flexible plan layout, which has the appearance of a stepped V-shape, has been shown (see fig. 7.7i); in this instance, the trusses have triangular cantilever extensions. The staggered truss principle has typically been used for structures in the mid-rise range of 15 to 20 stories, but can be appropriate for tall, narrow buildings as well. Currently, the world's highest staggered-truss structure is the 44-story Resorts International Hotel in Atlantic City.

In the following discussion, steel truss construction will be used. The Pratt truss principle is applied, since the diagonals efficiently respond in tension under gravity action. They are usually placed at 45 to 60 degree angles, and thus determine the panel widths of a truss. Diagonals are eliminated at certain locations, to permit openings for the corridors or other doorways, although rectangular penetrations should be kept to a minimum and placed close to the centerspan of the truss where the shear is lowest, so as not to weaken the truss action. The spans for typical Pratt trusses are about 60 ft, with an approximate steel framing weight of 6 to 8 psf. The lighter-weight staggered truss framework results in savings of foundation costs; in addition, foundations are only needed along the exterior column lines.

The structural design of the members in the staggered truss system for gravity loading is rather conventional, as is demonstrated in Example 7.5, which is not, however, true for the lateral loading case. Although the trusses do not lie in one plane, with respect to lateral loading, they can be treated as if they form a continuous web because of the rigid diaphragm action of the floors. The behavior is similar to a vertical I-beam, where the trusses form the stiff web to resist the shear, while the perimeter columns represent the flanges to carry the rotation in axial action. Since all of the gravity loads are concentrated along the edges, the entire building is activated efficiently in counteracting overturning.

In the staggered truss system, the floors span between the trusses from the top chords to the bottom chords. They not only carry the gravity loads to the wall-beams, but must also act as horizontal shear diaphragms with respect to lateral loading, since the trusses are staggered and do not lie in one vertical plane. The lateral forces flow into the top chord of each truss and leave through the bottom chord, as shown in Figure 7.22. However, since the vertical continuity is interrupted, they must travel along the floor structure to enter the top chords of the adjacent trusses. While this process continues, additional lateral forces keep on accumulating at each floor level, requiring a truss to carry the cumulative lateral load from the building portion above over a two-bay width when a checkerboard pattern of vertical stacking is being used. Therefore, with respect to horizontal load action, the floor structures must act as diaphragms capable of transferring the wind or seismic shear to the truss chords. This shear is considered to be distributed as concentrated horizontal loads acting at the truss panel points and resisted entirely by the diagonals, thereby introducing only axial forces into the columns. The floor diaphragms, acting as deep beams, must resist the in-plane bending, which could, however, be provided by flange action in the exterior walls.

The interaction of wall-beams and floors results in a stiff three-dimensional structure when reacting to lateral load action. This spatial behavior of the structure can be visualized by the superposition of two adjacent interstitial systems (Fig. 7.22), with their separate modes of deformation connected by the rather rigid floor plates, to yield the final deflection of the building—that of a single, stiff vertical cantilever similar to a tube, where the shear is efficiently resisted by the stiff, continuous (but staggered) trussed web, and the rotation by the exterior columns. Therefore, the columns in this system are primarily axial members rather than beam-columns, as in rigid frames where the columns are bent by shear. The lateral building sway is controlled primarily by the drift of the perimeter columns. Because of the predominant axial action of the columns, they can be oriented with their strong axis in the longitudinal direction. With respect to seismic action, the staggered truss structure is treated as a shear wall system (K = 1.33), but the chord members in the open corridor panel will be able to develop the beneficial ductility and energy absorption capacity.

When the number of trussed panels is decreased, the effect of the central Vierendeel panel becomes more pronounced. While the typical five- and seven-panel trusses behave similarly to a braced frame or bearing-wall structure, a three-panel truss behaves more like a frame truss (i.e., braced rigid frame, see fig. 7.7g).

Example 7.5

In a 24-story apartment building, 8 ft, 8 in. deep Pratt trusses span 60 ft and are spaced at 50 ft on alternate floors (a staggered-truss structure) as described in Figure 7.22, center; along the central corridor, the trusses have a Vierendeel panel. Assume 105 psf floor dead load and 40 psf live load for the apartment area; the concrete block facade walls weigh 50 psf. The loading conditions are described in more detail in Problem 7.18. Because the design of steel-frame staggered-truss buildings is usually governed by strength rather than stiffness, it is generally economical to use high-strength steel. In this case, A572 Grade 50 steel is used for the columns and truss chords, and A36 steel for the remainder.

For the preliminary investigation of the floor structure refer to Problems 7.18 and 7.19.

> 1) *Chord design.* The truss design is based on continuous chords and pinned web members, causing no eccentricity at the joints. Since the staggered truss system resists the lateral forces primarily in direct stress, the preliminary design of the trusses in this example can be based on gravity for the normal loading and height conditions, as well as for the upper half of very tall buildings. It must be kept in mind, however, that the vertical wracking of the Vierendeel panels due to wind

may control the design of the central chord portion at the lower floors of a tall building. For the investigation of the combined loading case refer to Problem 7.19.

Since the truss supports a large floor area, a maximum live load reduction of 60 percent is allowed.

$$R_1 = 0.08\%A = 0.08(60 \times 25) = 120\%$$

$$R_2 = 23(1 + D/L) = 23(1 + 105/40) = 83.38\% \qquad (3.1)$$

$$R_3 = 60\%$$

Hence, the uniform load along each of the chords at top and bottom, is

$$w = 25(0.105 + 0.4(0.040)) = 3.03 \text{ k/ft}$$

For fast estimation purposes of member sizes, the truss loading, including the corridor area, is considered uniform. Notice, however, in the solution of Problem 7.18, that the corridor loading is 3.41 k/ft.

The corresponding truss reactions, which are considered to act at the top chord, are

$$R_A = R_B = 2(3.03)60/2 = 181.8 \text{ k}$$

The moments at the nodal points of the continuous chord, assuming a truss of sufficient stiffness, are conservatively taken as

$$M = wl^2/10 = 3.03(9)^2/10 = 25.54 \text{ ft-k}$$

In the truss design, the chords basically resist the rotation and the web members the shear. Thus, the critical location for the web member design is close to the column supports where the shear is largest, and at center-span for the design of the chords, since here the maximum moment occurs. The effect of the central Vierendeel panel, where the corridor passes through the truss, is ignored; in this instance, diagonals are assumed across the opening. Actually, the shear must be resisted by the chords in bending because the diagonal is not available to carry it directly. The shear forces due to asymmetrical gravity and lateral loading, which would be carried by the diagonals, are then applied to the chord members at the inflection points, to be resisted in bending. Since the shear is small at midspan, however, the effect of chord bending may be ignored for preliminary design purposes.

The maximum truss moment, by ignoring the small restraint of the exterior columns, is

$$M_{max} = wL^2/8 = 2(3.03)60^2/8 = 2,727 \text{ ft-k}$$

This moment is resisted axially at midspan by the chords:

$$N = M/h = 2,727/8.67 = 314.53 \text{ k}$$

As a conservative approximation, the longer 9-ft chord span adjacent to the corridor is assumed to be critical and its local bending moment is used, together with the maximum axial forces at truss midspan, for this preliminary investigation. The chords are laterally braced by the floor framing, so that the allowable axial stresses may be based on yielding for this approximation. Assuming a bending factor of $B_x = 0.26$ for a W10 section yields a required cross-sectional area of

$$A_s = (P + B_x M_x)/F_a \tag{4.53}$$

$$= (314.53 + 0.26(25.54)12)/0.6(50)$$

$$= (314.53 + 79.69)/30 = 13.14 \text{ in.}^2$$

As a first trial, the following chord member is selected as based on predominant axial action, which is not necessarily conservative. It should be kept in mind that the next lower section may still be alright.

$$W10 \times 45, \quad B_x = 0.271, \quad A_s = 13.30 \text{ in.}^2$$

This section is also conservatively used for the continuous tensile bottom chord. For the preliminary design of the web members, which are assumed pinned, refer to Problem 7.18.

2) *Column design.* Since the lateral forces will cause only axial forces, and no bending of the columns, and since the gravity loads are so large because they are concentrated entirely along the edges, the combined loading case of gravity and lateral load action can be ignored for preliminary design purposes, assuming ordinary building loading and height conditions. The gravity loads are assumed to cause only direct axial stresses because the trusses are connected directly to the column webs. The secondary flexural stresses due to truss deflection are considered negligible, since the bottom chords are connected to the columns only after erection (see fig. 7.21, top). In other words, slotted holes are used in the bottom chord connections to the columns to relieve the rotation due to dead load deflection. The reaction forces of the trusses are assumed to act at the top chord. Therefore, the column loads at the tenth floor level due to the truss loads and the curtain wall, considering the roof load equal to the floor load, are

$$N = 181.8(8) + 25(8.67 \times 15)0.050 = 1616.96 \text{ k}$$

For this laterally braced structure, the effective column length is equal to the unbraced story height; it is assumed that the building is also braced in the longitudinal direction. According to the column tables in the AISC Manual for $KL = 1(8.67) \simeq 9$, the following section is selected as a first trial.

$$W14 \times 211, \quad P_{all} = 1709 \text{ k}$$

For the investigation of the combined loading case refer to Problem 7.19.

Because of the inherent stiffness of the staggered truss building, the drift is quite small, at least for buildings below 20 stories under normal loading conditions. The lateral deflection can be estimated by the column drift formula of Equation 7.16. This approach is not necessarily conservative, since the shear wracking of the Vierendeel frame in the corridor has been ignored.

BRACED FRAME STRUCTURES

The concept of resisting lateral forces through bracing is the most common construction method—it is applied to all types of buildings, ranging from low-rise structures to

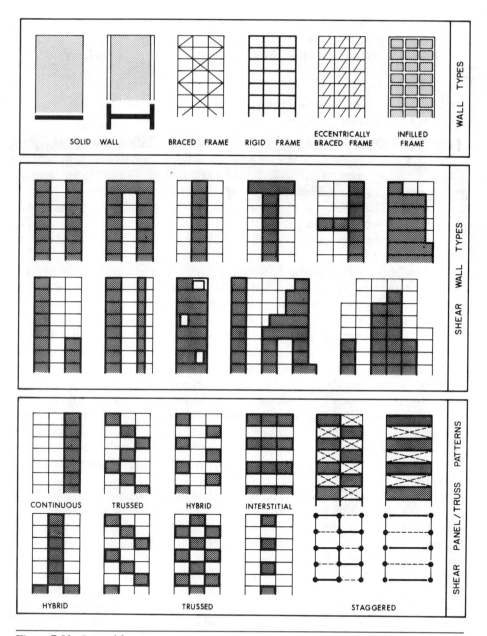

Figure 7.23. *Lateral bracing systems.*

skyscrapers. The emphasis in this section will be on truss frames. However, some basic concepts related to the bracing of frames to increase their lateral stiffness will first be briefly reviewed in Figure 7.23. The various bracing systems are:

- Rigid connections of beam-column joints (i.e., rigid frames, framed tubes)
- Truss frames
- Infilled frame, composite construction (e.g., masonry, concrete, light-gauge partition panels)

- An infill panel in a frame acts as a compressive diagonal strut
- Slotted shear walls behave as ductile panels (see fig. 7.33)
- Solid shear walls (e.g., cast-in-place or precast concrete walls, masonry walls)
- Perforated walls
- Steel plate shear walls
- Composite rigid frame and cladding
- Mixed construction

Usually, the term "braced frames" is not used in the general sense of stiffening, but in place of truss frames, while the term "shear walls" often refers not only to solid walls but, in more general terms, to the stiff lateral-force resisting structure system, as contrasted with the flexible rigid frame. In this context, shear walls are therefore assumed to include braced frames. Shear walls may resist lateral forces in both shear and rotation, as well as supporting gravity loads, they may resist only the lateral shear and gravity, or they may solely resist the lateral forces primarily in shear, similar to shear panel buildings.

The possible layouts of the lateral-force resisting systems within the building have been investigated earlier (see figs. 2.4 and 5.1, and elsewhere). They may be part of the exterior perimeter structure and may form continuous or isolated elements, or possibly the resistance may be concentrated in the interior slip-formed concrete core to which a pin-connected or rigid steel frame is attached. The various shear wall forms in Figure 7.23 range from individual walls, tied walls, coupled walls, trussed walls, walls with outriggers, walls that form a superframe, stepped walls, perforated walls, to any combination of the above. The wall does not necessarily form a continuous vertical element—discrete panels can be arranged in a truss-like configuration or they can be arranged as interstitial or staggered shear panel or wall-beam systems. A braced system may be more efficient than a rigid frame to support a building's many setbacks and cantilevers. The interaction of braced frames and rigid frames with respect to lateral force resistance will be discussed in the section in this chapter on "Frame–Shear Wall Interaction."

A general introduction to trusses in high-rise buildings is presented in Figure 7.24. Trusses not only constitute support structures hidden within the building, but also may be revealed. One of the earliest examples of skeleton buildings is the Chocolate Factory [1872] at Noisiel-sur-Marne near Paris by Jules Saulnier, where the walls consist of exposed trussed iron framework. This method of construction was surely inspired by trussed bridge construction, as well as by the timber framing that first occurred in Europe during the Middle Ages, where each region developed its own distinct pattern of braced wall heavy timber framing, with the space between the timber members infilled with masonry or other material mixtures.

An early example of high-rise braced frame construction is Gustave Eiffel's interior-braced iron skeleton for the 151-ft high Statue of Liberty [1886] in New York. He also designed the braced skeleton wrought iron structure of the Eiffel Tower [1889] in Paris (see fig. 8.2), the tallest building of its time at almost 1000 ft. This first modern tower became a symbol for a new era with its daring lightness of construction.

The elaborate tops of the skyscrapers of the early part of this century required complex bracing systems. For example, a high spire structure with a needlelike termination was designed to surmount the dome of the Chrysler Building [1930]. Currently, the post-modern building tops with their spires and pinnacles revive ornamentation and

Figure 7.24. *Trussed frames.*

the architectural styles of the past. Intricate braced frames are required for the various roof shapes such as pyramids, domes, spirals, and gabled, stepped, folded, or arched forms. These structural complexities are not only found in the roof spires but also lobby entrances and atria of high-rise buildings.

The basic bracing types for frames are identified at the bottom of Figure 7.24. They are: single diagonal bracing, X-bracing, K-bracing, lattice bracing, eccentric bracing (single diagonal or rhombic pattern), knee-bracing, and combinations. When the diagonal members have to be kinked for the placement of openings, then they must be stabilized by additional members. The most common vertical trusses use X-bracing, K-bracing, or vertical K-bracing (i.e., the Warren truss).

The plane frame may be braced across the full width of the building or possibly only certain bays or portions of the building may be trussed in a regular or irregular fashion. The bracing may span across several stories and bays to form superdiagonals, or it may be confined to each story and bay to be arranged in a repetitive pattern; vertical K-bracing may also be used only along the columns. The bracing systems, in other words, may be narrow or wide, and may be arranged on top of each other or in a staggered fashion. The width of the bracing is an essential consideration with respect to stability. The selection of the bracing type to be used in a particular circumstance is a function not only of strength, stiffness, and economy, but also of the size of wall openings.

When the diagonal members have to carry just tension and act as ties, then X-bracing is required, with relatively simple connections; when the bracing has to resist both tension and compression, however, then only a single stiff strut may be selected. In an X-bracing system, just one diagonal member is activated, thus always making one redundant, therefore more material is generally required. Further, the X-bracing of a bay does not allow door openings, in contrast to other bracing systems; it also does not give support to the floor beam, thus resulting in larger beam sizes.

The typical member shapes for light diagonals are single angles, double angles, or channel shapes connected to single gusset plates, while the shapes are W-sections with double gusset plates for heavy bracing members. In most trusses, the chords (columns) are continuous, while the diagonal members are separate and hinge connected. Typical connections for concentrically and eccentrically braced frames are shown in Figure 7.24.

In braced structures, the truss action eliminates the bending of the columns and girders as occurs in frames. In this instance, the truss members respond primarily in axial action, assuming that the major loads are transferred to the truss joints. The lateral drift is very much reduced because the rigidity of the structure depends on the cross-sectional area of its members, and not on their moments of inertia.

In earthquake regions, it may not be economical to use braced frames alone. To provide a minimum overall structural ductility during a severe earthquake, a secondary back-up ductile moment-resisting frame may be used to resist at least 25 percent of the base shear: in this case, the K-factor may be reduced to 0.8 from 1.33. Another way to achieve the necessary ductility is for the beam-to-column connections of the braced bays to be made moment-resistant, so that, when a brace is overstressed and buckles under severe seismic loading, rigid frame action takes over.

The conventional concentrically braced frames are very stiff because the thrustline passes through the centerlines of the beam and column intersection; it is well known that the triangle is inherently stiff in contrast to the rectangle. Under severe cyclic loading, however, the braces may buckle because of the lack of ductility and cause an unbalance of loading, which may result in failure if the joints are not made moment-resistant.

In the late 1970s, Professor E. P. Popov of the University of California at Berkeley showed that when a brace is at least connected at one end eccentrically to the beam, rather than concentrically to the beam-column intersection, then the diagonal axial force is transmitted in shear and bending along the beam segment, forming a ductile link that behaves in a manner similar to a stiff piece of rubber under severe cyclic loading (see fig. 3.9). Depending on the length of the link, it behaves either as a plastic moment hinge for a long link or as plastic shear hinge for a short link. The clear eccentricity for shear links is on the order of two times the beam depth. The strength of the link is chosen so that its large plastic deformations prevent the accumulation of axial loads and keeps the compression brace from buckling (see fig. 3.21).

Lateral bracing for high-rise buildings need not be restricted to internal cores, shear walls, and outrigger beams—it may be expressed on the exterior facade, serving aesthetic as well as structural functions. The bracing may be more easily accommodated on the exterior since, in the interior, it may restrict the free flow of the activity space. The major exterior X-bracing of superdiagonals may be supplemented by interior minor secondary bracing. The arrangement of the braced bays along the perimeter also efficiently solves torsional problems, particularly associated with asymmetrical buildings. For mid-rise buildings, perimeter bracing may also be used solely for the purpose of resisting torsion, while the direct action is carried in the interior of the building.

The cases in Figure 7.25 reveal various exterior truss applications. The facade trussing may be:

- part of the facade wall and fully visible (Fig. 7.25e)
- subtly contained within the wall surface (Fig. 7.25b)
- hidden behind the skin (see fig. 5.3i)
- in front of the skin (Fig.7.25m)

The facade trussing may define the shape of the building. The trusses may be planar, folded (Fig. 7.25j), or curved; they may form individual vertical cantilevers (Fig. 7.25f), spatial core units (Fig. 7.25h, g), bridges (Fig. 7.25k), partial tubes (Fig. 7.25d), a continuous spatial tubular unit (Fig. 7.25e, p), or they may form space trusses of a polyhedral nature. The trussing can be dense, as for lattice frames (Fig. 7.25l, n), or minimal, as for superdiagonals (Fig. 7.25a, b, c, etc.). In the 42-story First Wisconsin Center in Milwaukee (Fig. 7.25i), horizontal belt trusses at the bottom, middle, and top of the building are connected to interior outrigger shear trusses, thus stiffening the rigid frame perimeter structure. Eccentrically braced trusses for the proposed Century Tower in Tokyo (Fig. 7.25o) are expressed in two elevations. The A-shaped 25-story Nonoalco Tower [1964] in Mexico City (see fig. 2.21b) is laterally stabilized by sloping concrete side walls and braced by diagonal concrete struts in triangular faces 153 ft wide and 417 ft high.

In the 41-story Boston Company Building (Fig. 7.25b), about one-half of the gravity load and all of the wind forces are resisted by the diagonal facade members, and then transferred to the massive tapered corner columns. The perimeter framing represents vertical K-frame trusses, with respect to lateral loads, while it is comprised of three separate stacked K-brace units sharing the same four corner columns with respect to gravity loads. In this case, the vertical facade columns transfer the gravity loads to the

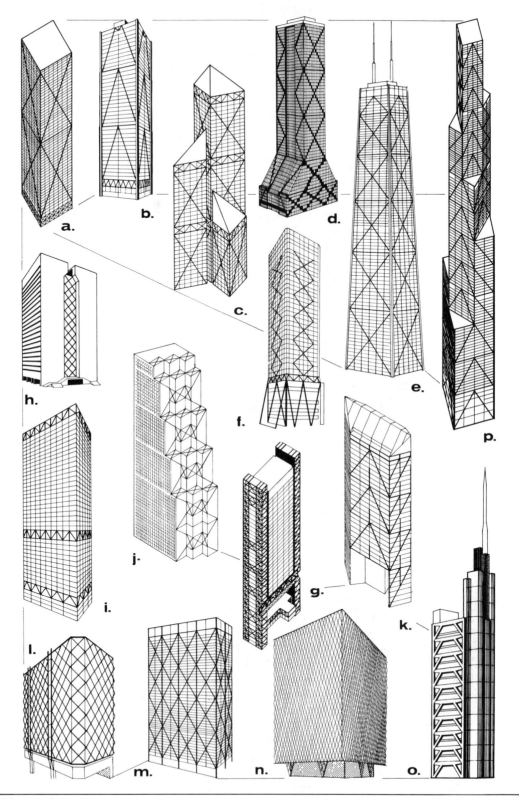

Figure 7.25. *Facade trussing.*

diagonal members in compression and tension. This 140-ft square, 601-ft high tower is not a pure tube structure, although it behaves as one. At the base, a two-story unit is hanging from 24-ft deep trusses that span the 140 ft distance between the corner columns.

Boston's slender steel-framed Federal Reserve Bank building (Fig. 7.25g), with an extremely large height-to-width ratio of 6.82, stands on two end cores. Two huge 36-ft deep transfer trusses at each facade near the tower base carry the entire gravity load of the 30 floors above like bridges across the 143-ft span to the flanking core walls. The two braced end cores also resist the entire wind load, although most is resisted by the two outside walls that form X-braced supertrusses.

The 390-ft high, 25-story Hennepin County Government Center in Minneapolis (Fig. 7.25h) consists of two slab towers linked by a central atrium where bridges tie the two towers together at various floors. The towers are coupled to the atrium space frame, which forms a large X-braced open core or cage around the atrium periphery and resists all the lateral forces.

Most of the other cases in Figure 7.25 are braced tubes, which are discussed in the section in this chapter on "Tubular Structures" and "Megastructures."

In Example 7.6 and the corresponding Problems, typical truss frames using X-bracing, K-bracing, and knee-bracing are briefly investigated (Fig. 7.26). These systems are economical for mid-rise buildings; for these conditions, only the vertical trusses resist the lateral forces. All girder and diagonal members are pin connected so that the vertical single-bay truss can be treated as a determinate structure, even though the column chords are continuous; the secondary stresses can be neglected. Only the knee-braced frame represents a more complex structure system, since the diagonals bend the beams and columns, thus generating a system between rigid frame and braced frame. With the exception of X-braced frames, all truss frames carry gravity loads. Because of the shorter length of the diagonals in the K-bracing system which produce less axial deformations, there will be less building drift as compared to the diagonally braced bent case.

Example 7.6

An eight-story skeleton building is laterally stabilized by K-bracing along the facades in the short direction and by rigid frame systems, also along the facades, in the long direction; for the layout of the framing refer to Figure 7.26a. For this preliminary investigation of the braced frames, where the vertical trusses primarily resist the lateral loads, assume the following loading conditions:

- Floor dead load, including partitions: 75 psf
- Equivalent floor load for girders, columns, and spray-on fire proofing: 10 psf
- Live load: 40 psf
- Cladding: 15 psf
- Wind load: 28 psf
- Treat the roof load as a floor load

Since the lateral forces are rather large because only two vertical cantilever trusses resist the entire wind pressure, while the reduced live loads are comparatively small, it is reasonable to assume (for a fast approximation of the truss member sizes at the building base) that the combined loading case with the 33 percent increase of allowable

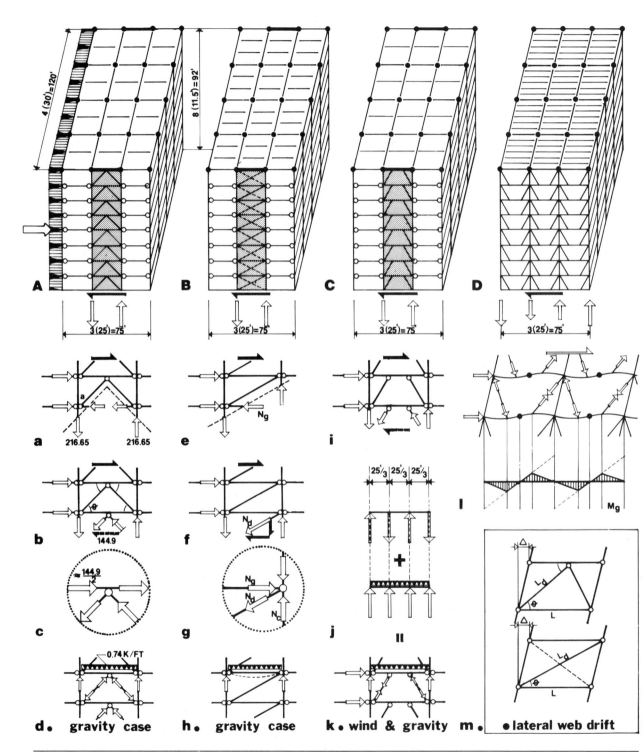

Figure 7.26. *Investigation of typical truss frames.*

stresses does not control. Therefore, only the dead load case, together with wind, is investigated here.

First, the member forces due to wind action are determined. The wind pressure against the entire building of 7.5 floors, to be resisted at the first story as based on single load action at the floor levels, is

$$0.028[11.5(7.5)]120 = 289.8 \text{ k}$$

Each of the vertical cantilever facade trusses will resist one-half of this load

$$289.8/2 = 144.9 \text{ k}$$

The wind shear is resisted by the web members. First the length and inclination of the diagonals are found.

$$l_d = \sqrt{(25/2)^2 + 11.5^2} = 16.99 \text{ ft}$$

$$\cos\theta = 12.5/16.99 = 0.736, \quad \sin\theta = 0.677$$

The horizontal components of the diagonals in Figure 7.26b resist the wind shear in tension and compression.

$$\Sigma F_x = 0 = 144.9 - 2(N_d/16.99)12.5 = 144.9 - 2N_d \cos\theta, \quad N_d = 98.47 \text{ k (T,C)}$$

For fast approximation purposes, it is assumed that one-half of the shear is resisted by the girder in tension on the leeward side and one-half on the windward side in compression (Fig. 7.26c).

$$N_G \simeq 144.9/2 = 72.45 \text{ k (C,T)}$$

The lateral forces that must be resisted in rotation by the truss chords or interior columns at the first floor level (Fig. 7.26a) are

$$0.028[11.5(6.5)]120/2 = 125.58 \text{ k}$$

The lever arm of the resultant lateral pressure to the point of rotation at a is

$$[6.5(11.5)/2] + 11.5/2 = 43.13 \text{ ft}$$

Rotational equilibrium yields the column forces

$$\Sigma M = 0 = 125.58(43.13) - N_c(25), \quad N_c = 216.65 \text{ k (T,C)}$$

The spandrel girder carries the following uniform gravity load due to floor and curtain action.

$$w_D = 0.075(15/2) + 0.015(11.5) = 0.74 \text{ k/ft}$$

This load causes a maximum moment at the interior support of this two-span continuous girder of

$$M_D = 0.74(25/2)^2/8 = 14.45 \text{ ft-k}$$

The first floor column must support the following dead load:

$$N_D = 0.085(25 \times 15)8 + 0.015(25 \times 92) = 289.5 \text{ k}$$

The total axial column load, due to wind and dead loads, is

$$N_T = 289.50 + 216.65 = 506.15 \text{ k}$$

The slenderness ratio of this laterally braced column can be estimated, for a W14 section with $r \simeq 4$ in., as

$$Kl/r \simeq 1(11.5)12/4 = 35$$

The corresponding allowable axial stresses are roughly

$$F_a \simeq 23 - 0.1Kl/r = 23 - 0.1(35) = 19.5 \text{ ksi} \qquad (4.47)$$

Therefore, the required cross-sectional area for the column is

$$A_s = N/F_a = 506.15/19.5 = 25.96 \text{ in.}^2$$

Try W14 × 90, $A_s = 26.5$ in.^{2}, $r_y = 3.70$ in.

The same section can be obtained from the column tables in the AISC Manual for $Kl = 1(11.5)$.

Notice that the gravity load of 289.50 k is not that much more than the tension force of 216.65 k due to wind, clearly indicating that the resisting moment due to dead weight is not 50 percent larger than the acting moment. Therefore, the braced frame must be anchored to the ground against uplift forces by using caissons or piles, or by bracing the first three story bays to increase the dead load and the resisting moment.

The diagonals also carry a little bit of the gravity loads, since they provide support to the spandrel girder at midspan (Fig. 7.26d). The beam reaction at this point is

$$P_D = 0.74(12.5)5/4 = 11.56 \text{ k}$$

Summing up the vertical forces at the joint yields the compressive forces in the diagonals due to the dead load.

$$\Sigma F_y = 0 = 11.56 - 2(N_d/16.99)11.5 = 11.56 - 2N_d \sin\theta, \qquad N_d = 8.54 \text{ k (C)}$$

Thus, the critical compression diagonal must carry the following total load.

$$N = 8.54 + 98.47 = 107.01 \text{ k}$$

The minimum radius of gyration, as based on a maximum slenderness ratio of 200 for compression members, taking into account their actual unbraced lengths between the W12 spandrel beams, is

$$Kl/r \simeq 1(16.99 - 1)12/r = 200, \qquad r_{min} = 0.96 \text{ in.}$$

From the column tables of the AISC Manual for $Kl = 1(15.99) \simeq 16$, the following section may be obtained.

Try 2L 6 × 4 × 5/8, $P_{all} = 129$ k, $A_s = 11.7$ in.^{2}, $r_y = r_{min} = 1.67$ in.

The compression force of 8.54 k in the diagonals causes the following relatively small tension force in the continuous girder:

$$N = (8.54/16.99)12.5 = 6.28 \text{ k (T)}$$

The girder size is approximated as based on an axial compression force of $N = 72.45 - 6.28 = 66.17$ k and a moment of $M = 14.45$ ft-k. The required cross-sectional area for this laterally supported light W12 beam column (see Equation 4.53), with $F_a = 0.6F_y \simeq 22$ ksi for a mid-rise structure, is

$$A_s = [P + 2.6(M_x/t)]/F_a = [66.17 + 2.6(14.45(12)/12)]/22 = 4.72 \text{ in.}^2 \quad (4.53)$$

Notice that the predominant action is axial. Try the following section, although the next lower one may be satisfactory.

$$W12 \times 19, \qquad A_s = 5.57 \text{ in.}^2$$

The lateral deflection of the K-truss can be approximated as follows. The total building deflection is equal to the sum of the web drift and chord drift. The web drift can be estimated by adding the horizontal girder shortening and the elongation of the bracing (Fig. 7.26m).

The girder shortening simply consists of its horizontal displacement, that is, the elongation of the tie and the assumed equal shortening of the strut.

$$\Delta_1 = f_G(L/2)/E = f_G L/2E \tag{a}$$

The lateral displacement of the panel due to elongation and an equal shortening of the diagonals can be approximated by letting the tensile brace freely change in length to Δ_d, and then rotating it back to the horizontal position, but along an assumed straight line, ignoring the effect of the angle change.

$$\Delta_2 = \Delta_d/\cos\theta = (f_d L_d/E)/\cos\theta$$

$$\Delta_2 = \frac{f_d L}{2E \cos^2\theta}, \qquad \text{where } L_d = L/2\cos\theta \tag{b}$$

Therefore, the story web drift is equal to

$$\Delta_{web} = \Delta_1 + \Delta_2 = \frac{L}{2E}(f_G + f_d/\cos^2\theta) \tag{c}$$

The linear shear drift for the entire building with n stories is

$$\Delta_{web} = \frac{nL}{2E}(f_G + f_d/\cos^2\theta) \tag{7.18}$$

For single diagonal bracing, or X-bracing with an inactive compression diagonal, the linear shear drift is twice as much!

The approximate lateral wind drift due to chord sway has already been derived in Equation 7.16a; hence, the lateral sway of the vertical K-truss can be estimated as

$$\Delta = \Delta_{ch} + \Delta_{web} = \frac{2f_c H^2}{3EB} + \frac{nL}{2E}(f_G + f_d/\cos^2\theta) \tag{7.19}$$

where: $f_G = (N/A)_G$ = average stress in a horizontal member due to horizontal forces

$f_d = (N/A)_d$ = average stress in a diagonal member due to horizontal forces

$f_c = (N/A)_c$ = axial column stress due to lateral forces

n = number of stories

E = modulus of elasticity

The building deflection of slender trusses is controlled by chord drift (flexural deformations), while the deflection of low-rise truss panels is primarily due to web drift, that is, the elongation of the diagonals (shear deformations).

Until now, only single-bay trusses have been investigated. To determine the approximate force flow in multibay bracing systems (i.e., indeterminate trusses) with equal bay spacing, the total horizontal shear may simply be divided by the number of bays with respect to web drift. The investigation may then proceed as for the single-bay case. For multibay bracing with respect to chord drift, the truss may be considered to be a vertical beam with an equivalent moment of inertia of the column areas, A_c, times their distance, r, to the center of gravity squared, using a truss efficiency factor of 0.8 (see derivation of Equation 7.16a).

$$I \simeq 0.8 \Sigma A_c r^2 \tag{7.20}$$

Example 7.7

Determine the approximate lateral building sway for the trusses in Example 7.6.
The wind stresses in the members are

$$f_G = 72.45/5.57 = 13.01 \text{ ksi}$$

$$f_d = 98.47/11.7 = 8.42 \text{ ksi}$$

$$f_c = 216.65/26.5 = 8.18 \text{ ksi}$$

Substituting these results into Equation 7.19 yields

$$\Delta = \frac{2f_c H^2}{3EB} + \frac{nL}{2E}(f_G + f_d/\cos^2\theta)$$

$$= \frac{2(8.18)(92 \times 12)^2}{3(29,000)25(12)} + \frac{8(25 \times 12)}{2(29,000)}(13.01 + 8.42/0.736^2)$$

$$= 0.764 + 0.538 + 0.643 = 1.945 \text{ in.}$$

$$\Delta_{all} = 0.002H = 0.002(92)12 = 2.208 \text{ in.} > 1.945 \text{ in.} \quad \text{OK}$$

FRAME–SHEAR WALL INTERACTION

At a certain height, depending (among other criteria) on the building proportions and the density of frame layout, the rigid frame structure becomes too "mushy" and may be uneconomical, so that it must be stiffened by steel bracing, concrete shear walls, or some of the other methods discussed at the beginning of the section in this chapter on "Braced Frame Structures" (with respect to fig. 7.23). The bracing usually occurs in the core area if the building is relatively symmetrical. It must be kept in mind that the service shear core at the center of the building is flexurally weak beyond a certain height because of its location at the axis of rotation, although it does provide shear resistance, which (in turn) reduces the shear deformations of the rigid frame. It is shown later in this section that, for tall buildings, the majority of the lateral loads are carried by the shear walls in the lower portion of the building, and mostly by frame action in the top portion.

Although shear walls stiffen the structure and are necessary for the control of drift due to wind and moderate earthquakes, they also lessen its ductility—a property which

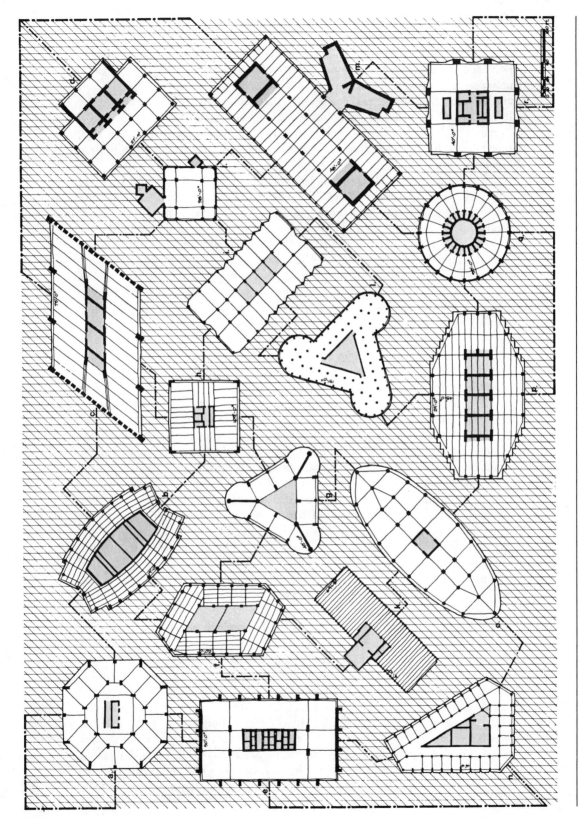

Figure 7.27. Shearwall/core-frame structure systems.

may be essential in order to deal with unpredictable severe cyclic seismic forces. To improve the ductility of concrete walls, shear walls may be coupled by diagonal strut reinforcement to cause a force flow, as in a truss, thereby improving the available rotational ductility. The beneficial aspect of eccentric bracing of trusses and slotted walls has already been discussed. To reduce the rigidity of a concrete core, the core walls may be split horizontally at each floor so as to be connected only by steel rods, so they take solely shear forces and no vertical loads, thus allowing the shear walls to absorb large seismic forces.

Currently, the world's tallest all-concrete building is the 75-story 946-ft high 311 South Wacker Drive tower [1990] in Chicago (see fig. 7.28b), where lateral resistance is provided by interior core shear walls and frame interaction. Typical frame–shear wall buildings are shown in the plan in Figure 7.27. In some of the buildings, the core is by far the stiffest element and resists all of the lateral loads (see Fig. 7.27b, k, l, m, r, or for the three separate concrete shear walls of Fig. 7.27g). In other buildings, the resistance to the lateral force action is shared. For example, in the 40-story concrete structure (Fig. 7.27e), 50 percent of the wind in the cross direction is resisted by the core and 25 percent by each of the end frames. A similar type of force distribution occurs in Figure 7.27c. The 13-story hybrid building structure of Figure 7.27h consists of the slip-formed concrete core, the steel floor beams, and the post-tensioned precast concrete perimeter frame, made up of massive L-shaped corner columns and 105-ft span deep girders. The rigid frame provides 60 percent, and the core 40 percent, of the wind load resistance. The 38-story Seagram Building in New York (Fig. 7.27d) is a rigid frame steel structure where some bents in the core area have wind bracing. The bracing is encased in 12-in. thick concrete shear walls up to the seventeenth floor, and then up to the twenty-ninth floor there is only the diagonal bracing, followed by rigid frame action above. In the cylindrical 51-story Australia Square Tower in Sydney, Australia (Fig. 7.27q), lateral stability is provided by tube-in-tube action. The concrete cores of the 60-story Marina City Towers in Chicago (see fig. 5.7e) take about 70 percent of the horizontal loads, while the rest is carried by frame action around it.

In Figure 7.28, various other types of frame–shear wall interaction structures are studied. The 27-story Gulf Life Tower in Jacksonville, FL (Fig. 7.28i), consists of a 53-ft square slip-formed concrete core and perimeter cantilever frames, with only two tapered columns on each face. By placing the columns at approximately one-third points, the spandrel girders must cantilever a daring distance of 42 ft; the girders are tapered according to the intensity of load flow. The major feature of the building is the precast perimeter frame, which was constructed with precast column shell sections filled with lightweight concrete, and 16 precast girder segments post-tensioned together to form the 133-ft spandrel girders. The joints were grouted and a compressible mastic gasket was used around the periphery of the joint. The floor framing consists of precast 40-ft span floor tees acting compositely with the concrete topping. To equalize the load distribution to the columns, the span of the T-sections is shifted to the opposite direction on alternate floors. All of the lateral loads are resisted by the central core.

The floors of the 40-story City Center office towers in Fort Worth (Fig. 7.28e) are pinwheel-shaped and supported by tripod and cluster-pair columns. Lateral stability is provided by the braced core and the welded perimeter frame. In the upper stories, the wind load is shared nearly equally by the two systems, while at the base the core takes nearly all of the lateral forces, since the tripod columns have little bending stiffness. The 73-story Raffles Hotel [1986] in Singapore (Fig. 7.28h) is of teardrop

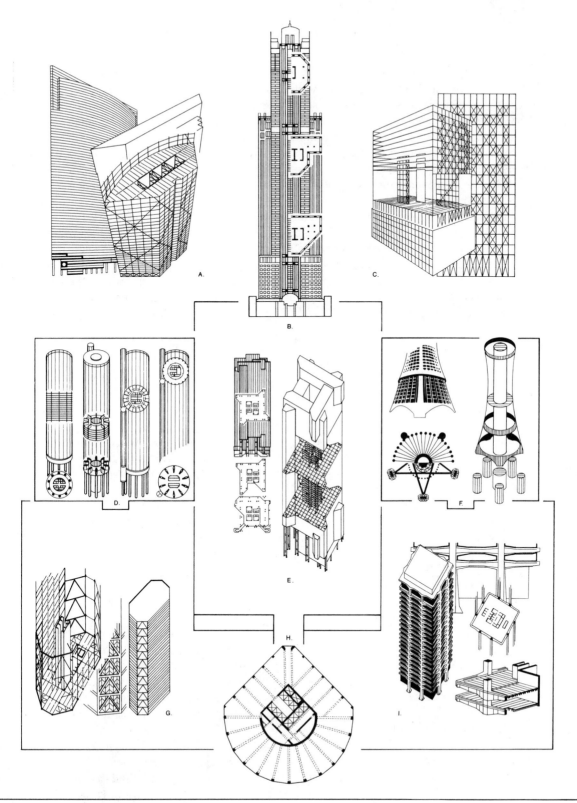

Figure 7.28. *Braced frames.*

shape in plan. This 741-ft tall concrete tower required the combining of typical floor framing systems for square and circular plan forms. The floor beams are directly supported by the perimeter columns, so that spandrel beams were not needed.

A 30-story tower in Rosslyn, VA is shaped like an airfoil, which causes large torsional wind forces. Exterior diagonal X-bracing on each of the two flat sides of the building resists these torsional loads and the wind on the narrow faces, and also provides the necessary stiffness along the perimeter walls; direct wind action on the long faces is resisted by interior core bracing. Composite frame construction was used for the eight lower levels, with conventional steel framing on top to achieve savings in erection time. Similarly, for the 36-story fan-shaped 333 West Wacker Building in Chicago (Fig. 7.28a), the exterior X-bracing along the side walls of the curved face resists the torsional wind loads, while the X-braced steel core carries the rest of the lateral forces. At the arcade level, the exterior X-bracing cannot continue to the foundations, so the floor diaphragms have to transfer the loads to the interior core, which is in concrete up to the third floor level in order to provide the necessary stiffness resisting the torsional forces. The 24-story apartment building in the Rue Croulebarbe in Paris (Fig. 7.28c) is a steel frame structure where tubular columns are filled with concrete. It is braced in the longitudinal direction along the central frames, in the transverse direction in each of the end walls of the lower portion by concrete shear walls, and by staggered tubular lattice bracing in the upper portion of the building.

The cylindrical 70-story Peachtree Plaza Hotel Tower in Atlanta (Fig. 7.28d) consists of three sections: the seven-story atrium base, the hotel tower, and the three-level restaurant on top. The hotel tower is supported over the atrium on ten massive perimeter columns and the core. Above the atrium, at the transfer level, massive box beams make the transition from the tower's radial wall construction to the base columns possible. The tower shape offers little resistance to the wind, and its cellular geometry provides excellent stiffness, although the attached exterior elevator shaft increases wind drag and also generates torsion. The coupled shear walls radiate from the 53-ft wide core and are tied together by the flat plate floor slab and the perimeter beams to form an interaction of core and walls like the spokes of a wheel in resisting the wind. At the base, the massive perimeter columns, similar to pin-connected members, do not carry any lateral loads—they are entirely resisted by the core from the top of the building down to the foundation mat. The narrow 35-story Mercantile Tower in St. Louis (Fig. 7.28g), looks like an elongated octagon in plan. Exposed vertical K-trusses, arranged in three-story tiers, are located on all four diagonal faces at each end of the building. To complete a channel-type bracing, spandrel beams are moment-connected to the trusses at the narrow end faces, and one-bay welded rigid frames are placed adjacent on the broad faces. All of the lateral forces are resisted by these huge channels at each end. This structural system forms a kind of partial tube with respect to lateral loads on the narrow face, but a frame–shear wall interactive system for force action on the critical broad sides. Because the trusses are on an angle at the corners, they are able to resist the lateral loads from both major directions. The interior framing carries only gravity loads.

The 38-story round, hyperbolic LUTH Tower in Kuala Lumpar, Malaysia (Fig. 7.28f), consists of a central circular concrete core and radial precast pretensioned floor beams, directly supported by perimeter columns with precast floor planks and cast-in-place topping. Below the story-high prestressed concrete transfer ring beams (which act like a belt) the columns continue in pairs until they reach a massive prestressed arched support system at the base, which serves to transfer the loads from the many small perimeter columns to the five massive base columns. The outward-sloping columns at the upper and lower portion of the building cause tension in the floor framing, while the inward-sloping portion causes compression. These lateral forces due to gravity action are particularly critical at the foundation level, where a severe outward thrust on the pile caps is caused by the sloping columns, which is balanced by prestressed tie beams linking the pile caps in a star-like pattern. The exterior frame wall and the core are connected by shear walls so that both structure systems act together to resist the lateral loads, similar to a tube-in-tube structure.

The 73-story First Interstate World Center [1989] in Los Angeles (Fig. 7.29) is currently the world's tallest office building in a most critical seismic zone. This 1018-ft tall cylindrical steel tower with multiple stepbacks consists of a flexible perimeter ductile moment-resisting frame and a stiff central braced core (formed by four massive 4 ft × 4 ft steel box columns using 7-in. thick plates welded together), which provides the necessary stiffness for most of the building height and carries nearly two-thirds of the building's vertical loads. The steel floor beams, supporting the concrete slab span, up to 55 ft from the inner approximately 74-ft square core to the perimeter frame.

The interactive behavior of the rigid frame–shear wall structure depends on the relative stiffness of the right frame to that of the shear walls (i.e., the relative stiffness of the various components of the lateral-force resisting structure). Usually, the shear walls provide most of the stiffness, as is the case for a tower structure with a central core and widely spaced perimeter columns, where the shear core may provide 90 percent of the lateral stiffness. For buildings of moderate height, and for the bottom portion of tall buildings, shear walls generally provide most of the lateral resistance. For preliminary design purposes, the effect of the rigid frame in resisting lateral forces can be ignored, as is demonstrated in Example 7.8; there must be a minimum of shear walls, however, to consider a rigid frame braced. According to the ACI code, a reinforced concrete frame is considered braced if the shear walls in a story provide a total stiffness of at least six times the sum of all of the column stiffnesses within the story (see Problem 7.23). Therefore, six-sevenths, or nearly all, of the lateral loads will be resisted by the shear walls,

$$\Sigma(EI/L)_{wall} \geq 6\Sigma(EI/L)_{col} \qquad (7.21a)$$

For constant material and story height, the equation can be simplified to

$$\Sigma I_w \geq 6\Sigma I_c \qquad (7.21b)$$

This consideration is most important for the design of columns (see the section in Chapter 4 on "Columns"). For laterally braced buildings, the complexity of column design is very much reduced, since the effect of slenderness may be ignored and since the relatively small building sway hardly effects the magnitude of the column moment.

For preliminary design purposes, the uncracked flexural stiffness for columns and walls may be used. Therefore, the required minimum length of the walls, L, for a selected wall thickness, t (assuming that all of the walls are of the same length), according to Equation 7.21b is,

$$\Sigma I_w = \Sigma t L^3/12 \geq 6\Sigma I_c, \quad \text{or} \quad L \geq \sqrt[3]{72\Sigma I_c/\Sigma t} \quad \text{(in.)} \qquad (7.21c)$$

Example 7.8

A 100 × 100 ft, 30-story, 360-ft-high reinforced concrete tower is supported by a central 40 ft square core (measured from centerline to centerline) and perimeter columns, which are spaced 25 ft apart. The assumed closed tubular shape of the core is obviously a simplification, since the wall perforations and inside cross-walls are neglected. For this preliminary investigation of the office building at its base, a floor dead load of 100 psf and a reduced live load of 50 psf, as well as a uniform wind pressure of 33 psf, are chosen; the roof loads are treated as floor loads for this high building. At the base of the building, 6000 psi concrete is used for the core; the 18-in. thick core walls are considered constant over the entire height (i.e., constant stiffness properties through the building height).

The core carries the following loads:

$$\text{wall loads: } 4(40 \times 360)1.5(0.150) = 12,960 \text{ k}$$

$$\text{floor dead load: } 0.100(70 \times 70)30 = \underline{14,700 \text{ k}}$$

$$P_D = 27,660 \text{ k}$$

$$\text{floor live load: } 0.050(70 \times 70)30 = \underline{7,350 \text{ k}}$$

$$P = 35,010 \text{ k}$$

The wind moment at the base is

$$M_w = 0.033(100 \times 360)360/2 = 213,840 \text{ ft-k}$$

The resisting moment provided by the core dead load is

$$M_r = 27,660(40/2) = 553,200 \text{ ft-k}$$

Therefore, the safety factor against overturning is

$$\text{S.F.} = M_r/M_w = 553,200/213,840 = 2.59 \geqslant 1.5 \qquad \text{OK}$$

The slender core behaves like a cantilever tube, where the lateral deformation is controlled by the moments; in other words, the bending stiffness is low while the shear stiffness is high. It is assumed that the core cantilever will always be in compression so that the section is not cracked under tension and can be fully utilized. Hence, for preliminary design purposes, the reinforced concrete can be treated elastically, and the contributing capacity of the reinforcing steel is conservatively ignored.

The sectional properties of the core tube are found by assuming the perforations not to be more than approximately 30 percent, so that the weakening of the shaft can be ignored for this first check.

The cross-sectional area is

$$A = 4[40(12)]18 = 34,560 \text{ in.}^2$$

The moment of inertia is

$$I = (41.5^4 - 38.5^4)/12 = 64,090 \text{ ft}^4$$

The section modulus is

$$S = I/(d/2) = I/41.5/2 = 64,090/20.75 = 3089 \text{ ft}^3 = 5,337,230 \text{ in.}^3$$

The approximation according to Equation 7.26 is

$$S \simeq 4td^2/3 = 4(1.5)40^2/3 = 3200 \text{ ft}^3$$

In the working-stress approach (ACI Alternate Design Method), the compressive stress in concrete for flexure without axial load is limited to $0.45f'_c$, while for walls with compression and not excessive flexure, it is taken as $0.22f'_c$, neglecting the effect of slenderness. Therefore, the allowable compressive stress for 6000 psi concrete is about $0.22(6.00) = 1.32$ ksi. Now the stress conditions at the building base can be checked.

The stresses due to gravity loading are:

$$f_c = P/A = 35,010/34,560 = 1.01 \text{ ksi} \leq 1.32 \text{ ksi}$$

The stresses due to gravity plus wind loading are:

$$f_c = 0.75(P/A + M/S)$$

$$= 0.75(1.01 + 213,840(12)/5,337,230)$$

$$= 0.75(1.01 + 0.48) = 1.12 \text{ ksi} \leq 1.32 \text{ ksi}$$

The stresses due to dead load plus wind loading are:

$$f_c = (27,660/34,560) + 0.48$$

$$= 0.80 + 0.48 = 1.28 \text{ ksi} \leq 1.32 \text{ ksi}$$

$$f_t = P_D/A - M/S$$

$$= 0.80 - 0.48 = 0.32 \text{ ksi; there is no tension.}$$

The core stays in compression so that the elastic approach can be used for fast investigation purposes. It has also been shown that the compressive stresses are kept within their allowable limits. The shear stresses are not considered critical.

The flexural sway at the building top can be roughly checked by ignoring the beneficial effect of the perimeter framing, and also the weakening effect of the wall openings, by simply treating the core as a vertical cantilever resisting the uniform horizontal wind load.

The modulus of elasticity of normal-weight concrete is

$$E_c = 57,000\sqrt{f'_c} = 57,000\sqrt{6000}/1000 = 4415.20 \text{ ksi} = 635,789 \text{ ksf}$$

The maximum building deflection is

$$\Delta = wH^4/8EI = MH^2/4EI = \frac{213,840(360)^2}{4(635,789)64,090} = 0.17 \text{ ft} = 2.04 \text{ in.}$$

$$\Delta_{all} = 0.002H = 0.002(360)12 = 8.64 \text{ in.} \geq 2.04 \text{ in.} \qquad \text{OK}$$

Alternatively, the maximum overall drift index, R, for the building is

$$\Delta/H = 0.17/360 = 1/2118 \leq 1/500$$

The overall drift index is well within the limits, but it must be kept in mind that the inter-story drift index may still be larger.

In a typical rigid frame–shear wall structure—that is, a laterally braced rigid frame—a much higher portion of the lateral loads is usually resisted by the stiffer shear walls as compared to the flexible frames. For rough preliminary design purposes, it may be assumed that each of the structure systems acts independently, though the deflection for both will be the same at the top of the building. It must be realized, however, that the flexural deformation of the shear wall is quite different from the linear shear deflection of the rigid frame for high buildings, and that the interaction of the two systems results in a completely new structure type. Therefore, it can only be very approximate to consider just the deflection at one location, the building top. Further, due to the interaction of the systems, forces are redistributed, not just distributed in proportion to their relative stiffness, as is discussed later.

$$\Delta_w = \Delta_f \tag{a}$$

The deflections of the two structure systems are expressed in terms of their stiffness, according to Equation 5.2.

$$P_w/\Sigma k_w = P_f/\Sigma k_f \tag{b}$$

Each of the structure systems carries a certain portion of the total load, P.

$$P = P_w + P_f \tag{c}$$

Substituting the results of Equation (b) into Equation (c) shows that the lateral forces are distributed through the rigid floor diaphragms in proportion to the relative stiffness of the supporting structural systems; this has already been shown in Equation 5.3.

$$P_w = P/(1 + \Sigma k_f/\Sigma k_w) = P[\Sigma k_w/(\Sigma k_w + \Sigma K_f)]$$
$$P_f = P/(1 + \Sigma k_w/\Sigma k_f) = P[\Sigma k_f/(\Sigma k_w + \Sigma k_f)] \tag{7.22}$$

Example 7.9

Rather than increasing the girder stiffness to control the lateral building sway of the rigid frame structure in Example 7.3, the structure is stiffened by two 16-in. thick shear walls across the short direction in the center bay at the perimeter. The building is considered to be a five-bay structure in the longitudinal direction.

The total wind shear against a typical interior frame, according to Example 7.2, is $W_1 = 113.1$ k. This lateral pressure causes a deflection of $\Delta = 0.56$ ft, as derived in Example 7.3.

Therefore, the corresponding frame stiffness is

$$k_f = P_f/\Delta_f = 113.1/0.56 = 201.96 \text{ k/ft}$$

It is now assumed that the two concrete walls resist the entire wind pressure by themselves. Each of the walls must carry the following load:

$$P = 0.026(180 \times 125)/2 = 292.5 \text{ k}$$

The walls each provide a moment of inertia of

$$I = (16/12)(20)^3/12 = 888.89 \text{ ft}^4$$

The modulus of elasticity for normal-weight 4000 psi concrete is

$$E_c = 57,000\sqrt{f'_c} = 57,000\sqrt{4000}/1000 = 3605 \text{ ksi} = 519,120 \text{ ksf}$$

Therefore, the maximum deflection of the cantilever wall is

$$\Delta_w = \frac{WH^3}{8EI} = \frac{292.5(180)^3}{8(519,120)888.89} = 0.462 \text{ ft}$$

The corresponding wall stiffness is

$$k_w = P_w/\Delta_w = 292.5/0.462 = 633.12 \text{ k/ft}$$

The ratio of the wall and frame stiffnesses, consisting of two walls and the four interior frames (end frames are not rigid), is

$$\Sigma k_w/\Sigma k_f = 2(633.12)/4(201.96) = 1.57$$

Hence, the frames resist, roughly, following portions of the wind load.

$$P_f = P/(1 + \Sigma k_w/\Sigma k_f) = P/(1 + 1.57) = 0.39P$$

The four rigid frames resist approximately 39 percent, while the shear walls resist 61 percent of the wind load. However, as has been emphasized, this percentage varies along the height of the building and cannot be constant!

The interaction of the rigid frame and shear wall is actually far more complex than this rough approximation and cannot be solved by just having compatibility between the two structure systems at the top of the building. It must also be kept in mind that the relative stiffness of each wall and frame may not be constant throughout the building height and may change abruptly (e.g., building geometry varies). It must be emphasized that the deformations of the individual structure systems simply do not add up in a linear fashion; the resulting stiffness of the combined structure is considerably higher than simply the sum of its components. In Figure 7.29, it is shown that the addition of the frame shear deformation and the bending deflection of the wall results, for tall buildings, in a flat S-curve. Because of the different deflection characteristics of each of the structure systems, the wall is pulled back by the frame in the upper portion, since the slope of the deflected wall is large there, while it is pushed forward in the lower portion, where the slope of the deflected frame is largest. It is as though a shear-type frame building sits on top of a flexural cantilever-type structure.

It is shown in Figure 7.29 that the rigid frames resist a portion of the total cantilever moment, primarily in axial action of their outside columns, while the remainder is resisted by the shear walls in bending. The design of the shear walls is clearly controlled by the large moments in the bottom half of the building; the reversal of the moment curvature in the upper half indicates the effect of interaction with the rigid frames. Critical, with respect to the design of the rigid frames, is the magnitude of the shear they must resist in bending. For the special condition in Figure 7.29, the frames must support a maximum shear of about 40 percent of the total cantilever base shear. Notice the shear distribution is relatively uniform over the upper building half, therefore resulting in frames of uniform stiffness, quite in contrast to free-standing frames. Most important, however, is the fact that the frames must carry shear forces at the top of

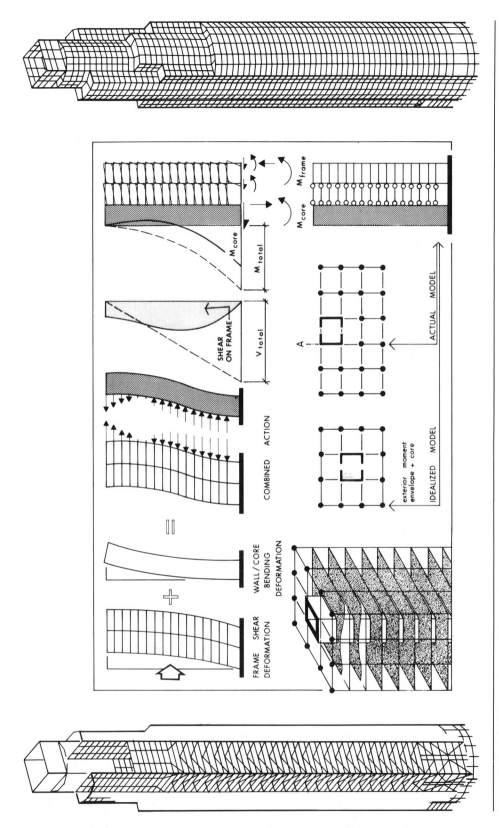

Figure 7.29. *Rigid frame shear wall interaction.*

the building that were generated internally through wall-frame interaction, even though the exterior shear is zero.

It may be concluded that, for *tall* slender rigid frame–shear wall buildings:

- The top portion consists of predominant shear beam action where the lateral forces are primarily resisted by the rigid frames. Due to the redistribution of the forces, however, the frame must resist a higher portion of the loads than just the exterior ones.
- The lower portion consists primarily of flexural beam action, where nearly all of the lateral forces are carried by the shear walls.

In contrast, for buildings of *moderate* height, it is reasonable to ignore the frame–shear wall interaction when the walls are at least six times as stiff as the total columns within a story (see Equation 7.21); in this case, the entire lateral load may be assigned to the shear walls, while the frame only carries the gravity loads. This is especially true for ordinary flat slab–shear wall buildings, where frame action offers very little stiffness.

SHEAR WALLS WITH OUTRIGGERS

At a certain height, the braced frame will become uneconomical, particularly when the shear core is too slender to resist excessive drift. In this instance, the efficiency of the building structure can be greatly improved by using story-high or deeper outrigger arms that cantilever from the core at one or several levels and tie the perimeter structure to the core by either connecting directly to individual columns or to a belt truss. This activates the participation of the perimeter columns as struts and ties, thus redistributing the stresses and eccentric loading, as is explained in Figure 7.30. In other words, the core is utilized in resisting the horizontal shears, and the outrigger arms are used in transferring the vertical shear from the core to the exterior columns; this makes the structure act as a spatial building unit similar to a cantilever tube-in-tube, but with no shear stiffness in the outer tube. The core with outriggers and verticals is somewhat similar in behavior to a sailboat mast with the spreader and shroud system. In addition, differential column shortening due to gravity, and a large portion of the thermal movement in fully or partially exposed exterior columns, is neutralized. The thermal motion along the perimeter must be controlled. For example, uniform seasonal expansion is prevented by stiff outrigger brackets, and thus causes additional compression in the columns (see fig. 3.19), while unequal daily thermal movement causes the buildings to bend.

Figure 7.30a and e demonstrate the lateral deflection of a free cantilever. When the outrigger arms couple the braced core to the exterior columns, and when the core (together with the arm extensions) attempts to rotate under the lateral loads, the exterior columns acting as tie-downs do not allow free cantilever rotation, but force upon the core an inflection point with a reversal of curvature (Fig. 7.30c, d, f). This rotational restraint, or partial fixity, reduces the moments of the core at the location of the outriggers by letting the axial forces in the columns (which form a couple) resist this portion of the rotation; this is obviously a more efficient use of the material. Therefore, the outrigger arms apply counteracting moments (Fig. 7.30h, i) or rotations to the core, which reduce the overall sway of the building. It is interesting to note that the

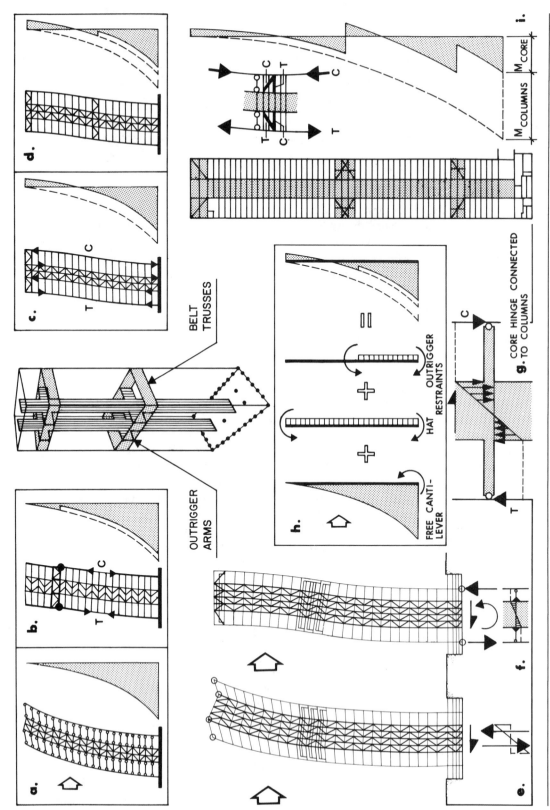

Figure 7.30. Hinged frame + core/outrigger building structures.

outrigger trusses do influence the drift beneath the trusses, but more so above (Fig. 7.30b). It may be concluded that the outrigger arms force the perimeter columns to participate in carrying the rotation or bending moments.

A building with a head structure (Fig. 7.30c, f) exhibits a reverse curvature. In this instance, the axial forces along the core due to the lateral loads change from tension to compression, forming a point of contraflexure at roughly 5/8 height, while the exterior columns carry the axial forces over the full height in tension on the windward side and in compression on the leeward side. As the outriggers become more flexible and rotate, the inflection point moves upward. Compared to an unrestrained cantilever, the overturning moment in the core of this structure is reduced by about 30 to 40 percent and the drift is also significantly reduced. The behavior of a building with a head structure can be simplified by treating it as a cantilever of uniform stiffness, EI, responding to a uniform wind pressure, w. The resulting maximum deflection is that of a free cantilever under lateral load action, Δ_w, reduced by that due to the moment restraint, Δ_M, as provided by the head outrigger arms and the exterior column ties.

$$\Delta = \Delta_w - \Delta_M = \frac{wH^4}{8EI} - \frac{MH^2}{2EI} = \frac{H^2}{2EI}\left(\frac{wH^2}{4} - M\right) \qquad (7.23)$$

For further discussion of the moment restraint, M, refer to Problem 7.30.

The outrigger arms are cantilevered from the core and hinged to the perimeter columns to better develop the moment-resisting capacity of the core (Fig. 7.30g) and to not induce any moments into the columns, but only axial loads, thereby increasing the axial capacity of the columns. If the cantilever trusses were to be rigidly connected to the columns, the entire system would act as a unit, thus utilizing only a small percentage of the moment-resisting capacity of the core, whose walls are relatively close to the neutral axis of the building. The floor beams are also pin connected to the exterior columns so that they do not cause any gravity moments in the columns.

The outrigger system may consist of reinforced concrete wall beams, steel trusses, I-shaped steel-plate wall girders, or diagonal struts (brackets), which may form super-diagonals (see fig. 7.39a). These diagonals will have to be designed for compression and tension because of lateral force reversal. In areas of strong seismic activity, an outrigger Vierendeel system cantilevering from an eccentrically braced central frame will provide the necessary ductility. The outriggers are placed at various crucial locations, which usually coincide with the mechanical floor levels. The lateral drift of the building depends on the stiffness, the number and location of the outriggers, the belt trusses, and the cores.

Some applications of the outrigger principle are briefly discussed now. The 47-story Place Victoria in Montreal (see, for concept, fig. 2.3) was, in 1964, the first reinforced concrete building to utilize the principle. Diagonal story-high X-braced girders tie the massive corner columns at four levels to the X-shaped core shear walls. The 64-story U.S. Steel Building in Pittsburgh (see figs. 5.7f, 7.4e, and 7.30f) consists of a braced triangular core, an exposed primary perimeter steel frame of hollow box columns, stub girders that tie the columns to the box spandrels at every third floor level, and a stiff space frame hat connecting the core to the 21 exterior columns, thus prohibiting top rotation. The primary framing carries the two-story secondary framing, which resists only gravity loads; the building can be visualized as a series of vertically-stacked three-story units. The response of the equilateral corner-notched triangular building substantially reduces the wind loading (see fig. 3.6). The exterior exposed tubular columns are filled with a fireproofing liquid, as is discussed in the section in Chapter 1 on "Fire Safety."

The 52-story IDS Building in Minneapolis (Fig. 7.30i) is a hybrid structure with a reinforced concrete core, a perimeter steel frame, and outrigger trusses at three levels tying the core to the exterior columns. Spanning the 45-ft free space from the core to the perimeter columns are 3-ft deep floor trusses on 29-ft centers. Filler beams of 16-in. depth span between the trusses to support composite flooring. In the 42-story First Wisconsin Center in Milwaukee (see fig. 7.25i), horizontal belt trusses at the midsection and top of the building, together with outrigger trusses cantilevering from the central core, stiffen the rigid frame perimeter structure with a 20 ft column spacing

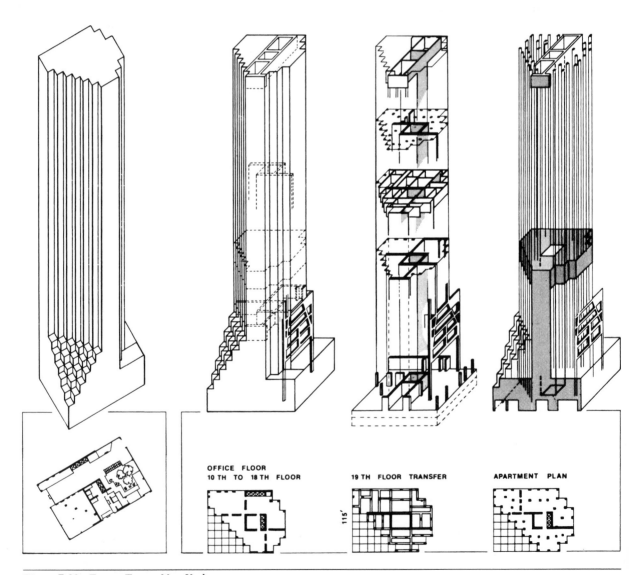

OFFICE FLOOR
10 TH TO 18 TH FLOOR

19 TH FLOOR TRANSFER

APARTMENT PLAN

Figure 7.31. *Trump Tower, New York.*

to attain a framed tube behavior under wind action. The belt trusses at the bottom of the building transfer the column loads to the wider column spacing at the base.

In the 61-story Trump Tower (Fig. 7.31) in New York, the pinwheel-shaped core structure is anchored at the top to two huge hat girders in the perpendicular direction and is tied into perimeter shear walls below the twentieth floor. In this case, the hat girders and the perimeter shear walls help to control the lateral displacement of the

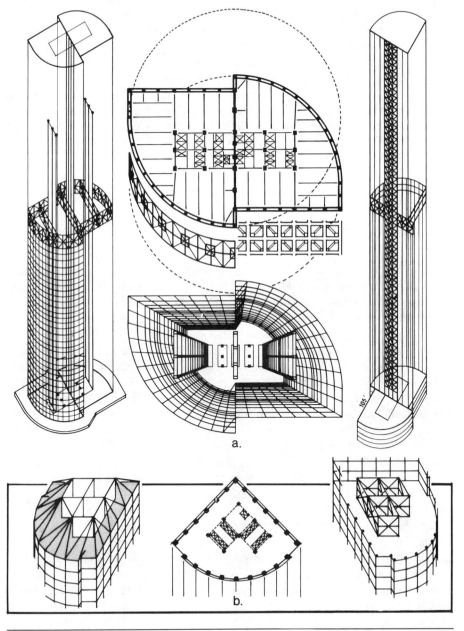

Figure 7.32. *Allied Bank Tower (Houston) and 17 State St. building (New York, bottom).*

core. For the upper 38-story apartment portion of the concrete structure, the vertical loads are carried by the core and columns. Since virtually no interior columns were allowed for the 13 floors of office space, a complex three-story deep transfer girder system at the mechanical floors at the nineteenth level was required to catch 52 columns and reduce them to only 8. However, the structure system also transfers a portion of the wind loads to the perimeter shear walls. Another series of transfer girders was needed to open up the atrium at the back of the building. In this instance, the loads of four columns at the eleventh floor level had to be transferred to two columns at the seventh floor level with the help of a post-tensioned A-frame.

The 71-story Allied Bank Tower in Houston (Fig. 7.32a) is a bundled tube structure. It consists of two quarter-cylindrical steel tubes placed back to back but offset. They are stiffened by two-story outrigger trusses at floors 34 and 35, which are linked to three vertical core trusses. Belt trusses around the perimeter distribute the loads more uniformly and maintain the building's stiffness. A secondary system of columns is placed where the tubes meet. The typical spacing of the perimeter columns is 15 ft. The curved form of the building reduced the wind loading by more than 25 percent. The 42-story 17 State Street building in New York (Fig. 7.32b) has a shape of a quarter circle in plan with a 123 ft radius. The hybrid structure for this steel-framed building consists of bundled, braced core-tubes linked to moment-resisting perimeter frames that resist torsion, and are coupled by an outrigger hat space truss.

TUBULAR STRUCTURES

As a building increases in height in excess of roughly 60 stories, the slender interior core and the planar frames are no longer sufficient to effectively resist the lateral forces. At this point, the perimeter structure of a building must be activated to undertake this task by acting as a huge cantilever tube. For this condition, the outer shell may act as a three-dimensional hollow structure, that is, as a closed box beam, where the exterior walls are monolithically connected around the corners and internally braced by the rigid horizontal floor diaphragms. The concept evolved out of the three-dimensional action of structure as found in nature and in the monocoque design of automobiles and aircraft. The dense column spacing and the deep spandrel beams also tend to equalize the gravity loads on all of the exterior columns, similar to a bearing wall, thereby minimizing column sizes. In addition, the closed perimeter tube provides excellent torsional resistance. The tubular concept has revived the bearing wall for tall building construction, but in steel, concrete, and composite construction, rather than in masonry. In this structure, the window lights can be placed directly between the columns of the punched wall, and hence the need for a separate curtain wall is eliminated.

The effectiveness of the behavior of the simple, vertical, hollow tube under lateral force action depends on the nature of its cross section (i.e., building plan) and the number of perforations in the webs. A *perforated shell tube* is considered to have less than 30 percent perforations, and may be assumed to behave as a perfect cantilever beam developing a uniform flange response. In contrast, a *framed tube* with opening sizes on the order of 50 percent of the wall surfaces cannot develop the tubular action or proper flange behavior because of the flexible nature of its soft webs. By substantially increasing the depth of the frame beams as in the *deep spandrel tube*, however, the lateral stiffness of the tube is very much improved. Rather than increasing the depth of all of the spandrel beams to improve the web stiffness of the soft framed tube, belt trusses at the mechanical levels can be provided, as in the *framed tube with belt trusses*. Naturally, true tubular action is achieved when sufficient diagonal braces are super-imposed upon the frame. In this instance, the tube is tied together to form a *braced tube*, where the diagonals resist the lateral forces efficiently in an axial manner. For this condition, the vertical columns can be spaced further apart, and thus allow a more open facade.

The wall perforations do not have to remain constant, though this is often done for reasons of economy. The window sizes may respond to the change of functional requirements or to the intensity of force flow. Because a constant wall thickness was used for the 590-ft high concrete tube of the 45-story Fiat Tower [1974] in Paris, France, the window size could increase in width with the height of the building so that the largest windows are at the top, where the least strength is needed. Similarly, for the 728-ft high, 64-story Olympia Center [1981] in Chicago (see fig. 7.10 bottom left and right), the concrete-framed tubular walls demonstrate different degrees of openness in response to change in occupancies. Various types of wall perforations and wall framing for tubes are shown in Figure 7.33H to O, which may serve also as a basic organization for tubular structures.

- Perforated shell tube (Fig. 7.33J)
  - concrete wall tube
  - stressed skin steel tube
  - composite steel-concrete tube
- Framed tube or Vierendeel tube (Fig. 7.33H)
- Deep spandrel tube (Fig. 7.33I)
- Framed tube with belt trusses (Fig. 7.33L)
- Trussed or braced tube (Fig. 7.33M)
- Latticed truss tube (Fig. 7.33N)
- Reticulated cylindrical tube (Fig. 7.33O)
- Combinations (e.g., Fig. 7.33K)

The simple perimeter tubular concept is ideal for compact, doubly symmetrical plan forms of round or nearly square cross section. Naturally, this cannot be true for narrow, elongated shapes, which must be stiffened in order to keep their tubular shape under lateral force action, so as to effectively resist twisting. In addition to attracting more lateral forces, elongated shapes have a very low shear resistance along their narrow faces, and full flange action along their long faces cannot be activated. For the condition where the perimeter of a building becomes too large for a single tube, it must be internally braced. Today, most of the plan configurations of tall buildings are not of regular form, and therefore provide less efficient tube action. When a plan configuration necessitates a perimeter wall with discontinuities, such as by offsetting two quarter circles (see fig. 7.32a), then there will be local stress concentrations at the abrupt changes of geometry, which must be superimposed upon the overall tubular action. Many hybrid building shapes have complex geometries with setbacks requiring transitions from one form to another, which necessitates various combinations of structure systems to fit the space. The resulting structure solutions are quite complex in contrast to the simple, pure prismatic cantilever forms of the 1960s. In response to these additional considerations of building form and size, tubular systems can be

Figure 7.33. *Tubular structures.*

further organized according to their cantilever cross section (i.e., plan forms) according to Fig. 7.33A to G, as

- Pure tubular concepts:
    Single perimeter tubes (Fig. 7.33G; e.g., punched, framed, or trussed walls)
    Tube-in-tube (Fig. 7.33E)
    Bundled tubes or modular tubes (Fig. 7.33D)
- Modified tubes:
    Interior braced tubes (Fig. 7.33F)
    Partial tubes (Fig. 7.33A)
    Hybrid tubes (Fig. 7.33B, C)

The invention of the tubular principle for tall buildings is reflective of the spirit of the 1960s, with the search for a clearer understanding of the behavior of structures and the optimization of structure systems. However, the development of tubular structures cannot be talked about without giving credit to the late distinguished structural engineer Fazlur Khan of SOM. His research, writings, and design of buildings had a profound effect on the field of high-rise construction, especially on tubular construction. He was deeply concerned with the efficiency of structure, to express its beauty as architecture, and to articulate the purity, honesty, and simplicity of structural form. Many of the concepts of minimal but essential structures with an emphasis on the tubular principle were investigated by Khan at I.T.T. in Chicago, as a teacher of architecture with his thesis students and the help of colleagues.

Among the first examples of tubular construction are the 13-story latticed steel tube of the IBM Building [1963] in Pittsburgh and the framed concrete tube of the 43-story DeWitt-Chestnut Apartment Building [1965] in Chicago, although the exterior framed concrete tube principle, stiffened by interior cores, had already been applied in 1964 to two 38-story office buildings: the Brunswick Building in Chicago and the CBS Building in New York. The rapid development of the tubular principle finally resulted in the new generation of superskyscrapers of that period. In 1968, the trussed steel tube of the John Hancock Building in Chicago reached 1127 ft. It was followed in 1971 by the tallest light-weight concrete building, the 714-ft high One Shell Plaza in Houston, using the tube-in-tube principle. In 1972, the framed steel tubes of the World Trade Center reached 1368 ft, and the composite perimeter tube of the 679-ft high, 51-story One Shell Square Tower in New Orleans was completed. In 1973, the 1136-ft high framed steel tube of the AMOCO building in Chicago (see figs. 2.21f, 5.4K) was finished. Finally, the bundled steel tubes of the 1454-ft high Sears Tower in Chicago were completed in 1974. This era of supertall skyscrapers of pure tubular forms was, in a way, concluded in 1976 with the then-tallest concrete structure, the 859-ft high Water Tower Place in Chicago (see fig. 2.17e), using a perimeter tube stiffened by an interior shear wall.

Perforated Shell Tubes—Framed Tubes

Tubes are three-dimensional hollow structures internally braced by rigid floor diaphragms. They cantilever out of the ground, and overturning is resisted by the entire spatial structure as a unit and not as separate elements. The moment is carried by the tube in tension and compression, while the shear is resisted primarily along the webs, that is, by the diagonals in trussed tubes and by the bending of columns and girders in framed tubes. The effectiveness of the cantilever action of the simple perimeter tube depends on the size and the shape of the wall penetrations. A windowless tube is obviously more rigid than one with many openings. It is apparent that small, round window openings interfere less with the force flow as compared to rectangular ones, which cause stress concentrations at the corners and along the edges (see figs. 6.9, 6.13). A tube with small openings, say less than 30 percent perforations, especially when the window openings are round, behaves like a perfect cantilever. However, as the rectangular perforations become larger, the perforated wall tube changes to a framed tube, which is soft and flexible and behaves like an open or partial tube consisting of channel-shaped rigid frames along the webs parallel to the lateral force action. By increasing the depth and stiffness of the spandrel beams in the framed perimeter tube, however, its stiffness and true cantilever action will be very much improved. This type of tube is called the deep spandrel tube.

The opening sizes for typical framed steel tubes may be on the order of 50 percent of the wall surfaces, while for punched concrete tubes it may be 30 to 40 percent. The wide columns are closely spaced and tied together with rigid connections to deep spandrel beams. The typical exterior column spacing for framed tubes ranges from approximately 4 ft to 10 ft and larger depending on the member proportions, and the spandrel beam depth ranges from 2 to 5 ft. Naturally, the proportions of the beam-column grid depend on the slenderness and height of the building, as well as the functional requirements and the building module. The extremely close column spacing in steel construction requires many joints and thus tends to be uneconomical, since the unit cost of steel is proportional to the number of joints. Since the primary task of the perimeter frame is to resist the lateral forces, it may be more efficient to use deep beam sections as columns, rather than the conventional W14 columns. In steel construction, all-welded column-trees are often shop prefabricated and then field assembled by bolting the spandrels at midspan and welding the columns' splices. The interior gravity steel framing in steel or composite tube construction is all pin connected, and the rigid floor diaphragm is hinged to the columns and beams of the perimeter tube. Usually, the floor structure consists of a composite floor deck, possibly with composite steel beams or trusses.

Some important buildings that show the development and describe the nature of the single tube are discussed first, before investigating the behavior of the tube from an analytical point of view.

The 43-story DeWitt-Chestnut Apartment Building [1965] in Chicago is the first single framed tube structure. The exterior column spacing for this slender 125 × 81 ft, 395-ft high concrete tube is 5.5 ft with 2-ft deep spandrel beams. The floor structure consists of a flat plate supported by interior columns spaced at approximately 20 ft.

The most famous example for the application of the framed tube principle is the 110-story World Trade Center [1972/73] in New York (see figs. 1.14a, 2.17f, 5.3j, and Problem 7.26). The 1368-ft high tower forms a 208-ft square perimeter steel tube; 14-in. square box columns are spaced at only 3.33 ft on center and are tied together by 4.33-ft deep plate spandrels, thus allowing narrow 19-in. wide by 78-in. high windows 14 in. above the finished floor. Floor trusses approximately 35 in. deep are spaced 6.66 ft apart and act compositely with the 4-in. thick 3000 psi lightweight concrete slab; they span 35 ft and 60 ft between core and exterior walls.

   The tubular concept is not necessarily associated only with tall, slender buildings; the system has also proven economical for mid-rise structures. For example, a 23-story 110-ft square steel skeleton may be organized on 22-ft square bays. By adding columns along the exterior at the midbays, a uniform column spacing of 11 ft is achieved. This close column spacing, together with stiff spandrel beams, forms a perimeter tube of a nearly square grid.

   The traditional pure rectangular perimeter tubes of the 1960s and early 1970s have been slowly replaced by free-form shapes. These new forms range from an 11-sided free-form for a 38-story folded perimeter tube [1978] (see fig. 5.7g), to the irregular-shaped octagon of a 50-story tower [1980] (see fig. 5.7c); both are steel tubes using a column spacing of 10 ft on center. The 145 × 205 ft octagonal plan, 692-ft high perimeter tube lacks symmetry, and thus generates additional lateral force action; in this instance, however, the torsion is efficiently resisted by the closed tube. For the 323-ft high framed concrete tube of rectangular plan form with a large chamfered corner [1981] (Fig. 7.33h), which resembles a multisided folded tube, the columns are spaced 15 ft on center and are tied together by 4-ft deep spandrel beams. For the 726-ft high, roughly 160-ft square tower with a diamond-shaped plan formed by intersecting squares [1982] (Fig. 7.33c), the composite columns of the tower portion are spaced 10 ft on center and are tied together by steel spandrel beams. The steel floor beams span 42 ft from the composite tube to the steel core.

   The 54-story, 727-ft high One Mellon Bank Center [1983] in Pittsburgh, of elongated octagonal plan (see figs. 5.7d, 7.38b), is an exposed steel-clad stressed-skin steel tube. The columns of the framed tube with belt trusses and thick infill plates at the corners are spaced 10 ft on center and are tied together by deep spandrel beams. The 1/4-in. to 5/16-in. thick A36 face steel plates act compositely with the frame to reduce wind sway; they were erected as 10-ft wide by 36-ft high 3-story stiffened panels. The facade panels had to be connected to the frame so that only shear, and not axial, forces are transferred. In addition, tiebacks were required and thermal expansion had to be relieved by attachments and joints between the panels. Since the stressed skin controls only wind drift, it does not need to be fireproofed. Core bracing had to be added at the upper building portion where the setbacks discontinue the perimeter columns.

   In the following discussion, the reaction to lateral loading of ordinary symmetrical tubes without any discontinuities is studied. First, *perfect tube* behavior, such as for a braced, deep spandrel, or perforated wall tube with less than approximately 30 percent perforations, is investigated. The effect of the ratio of wall thickness to the cross-sectional dimensions of the hollow perimeter tube (notice also the difference from the ratio for an interior core tube) is ignored because of the stiffening effect of the rigid floor diaphragms. This thin-walled beam behavior under lateral loading may, however, cause bending stresses that are no longer proportional to the distance from the neutral axis. For the ideal condition of a single cantilever box (Fig. 7.34, center) the familiar bending theory, or cantilever method, can be applied for preliminary design purposes. The closely spaced perimeter columns of cross-sectional area, A_c, which are spaced at L-distance, are replaced by an equivalent wall thickness, t:

$$A_c = tL, \quad \text{or} \quad t = A_c/L$$

Now, the moment of inertia of the rectangular tube can be found as

$$I = [bd^3 - (b - 2t)(d - 2t)^3]/12 \tag{7.24a}$$

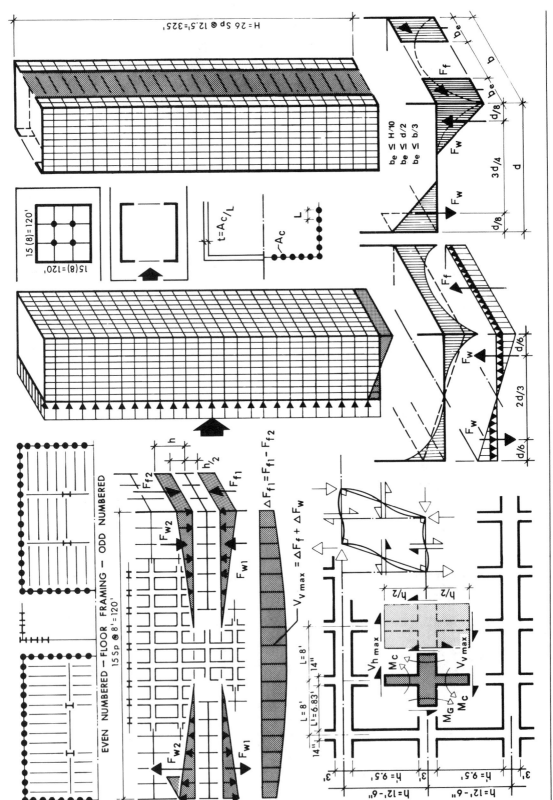

Figure 7.34. *The behavior of the cantilever tube.*

It may be more convenient, however, to treat the tube as an assembly of webs and flanges by using the dimensions to the centerlines of the walls and neglecting the moment of inertia of the flanges about their own axes, thereby obtaining a simpler expression for the moment of inertia of the tube.

$$I = I_w + I_f \simeq 2[td^3/12 + tb(d/2)^2] = td^2(d + 3b)/6 \qquad (7.24b)$$

The section modulus is

$$S = I/0.5d \simeq td(d + 3b)/3 \qquad (7.25)$$

For a square tube, b = d, the section modulus becomes

$$S = 4td^2/3 \qquad (7.26)$$

The extreme fiber stresses, f_z, in the square tube are equal to

$$f_z = M/S = 0.75M/td^2$$

or, the column forces in the flanges are

$$F_f = f_z bt = f_z dt = 0.75M/d, \qquad or \qquad F_f(d) = 0.75M \qquad (7.27)$$

It is thus shown that, in the perfect tube, 75 percent of the overturning moment is resisted by the flanges and the remaining 25 percent by the webs.

Therefore, each column along the windward and leeward faces carries the following load, N_f

$$N_f = f_z A_c = (0.75M/td^2)A_c = 0.75ML/d^2$$

Letting: n = number of bays = b/L = d/L yields

$$N_f = 0.75M/dn = F_f/n \qquad (7.28a)$$

However, when the moment is assumed to be resisted entirely by the flanges, then each column carries the same force:

$$N_f = (M/d)/(n + 1) = F_f/(n + 1) \qquad (7.28b)$$

The shear along the web faces must be resisted by the columns and girders in bending. For first estimation purposes, the horizontal shear stresses are taken as equally distributed, as derived from the closed tubular shape. In this case, the web walls are treated as rigid frames where, according to the portal method, the shear each interior column resists in bending at midstory height is

$$V_h = f_v A_c = (V/2)/n \qquad (7.29)$$

where: V = total lateral shear at given level

The girder shear can be approximately determined by subdividing the wall frame into cruciform modules (Fig. 7.34), as defined by the inflection points at the midheight of columns and the midspan of spandrels, or by their equivalent rectangular plate modules. The equilibrium of the shear forces along the module results in

$$V_v(L) = V_h(h), \qquad or \qquad V_v = V_h(h/L) \qquad (7.9a)$$

The corresponding column and girder moments at the critical girder and column faces are

$$M_c = V_h(h'/2), \qquad M_G = V_v(L'/2) \qquad (7.30)$$

It may be unrealistic for the preliminary design of *perforated wall tubes* to ignore the nature of the wall and use perfect tubular behavior associated with solid walls, so that the following assumptions may be made to reduce the cantilever effectiveness, as is shown in Example 7.10,1.

- The shear is considered uniform and carried by the web, that is, the walls parallel to the lateral force action (Equation 7.29).
- The bending is resisted by the flanges evenly, that is, the walls perpendicular to the lateral forces (Equation 7.28b).

The behavior of *framed tubes*, on the other hand, is quite different; it lies somewhere between a pure cantilever structure and a pure frame. Thin-walled beam behavior, together with the effect of large wall penetrations (i.e., window opening sizes on the order of 50 percent of the wall surface), makes the assumption of linear stress distribution quite unrealistic. The efficiency of the three-dimensional tube is limited by the flexibility of the web-frame, particularly the spandrel beams allowing a *shear lag* in axial load distribution. Due to the shear lag between the columns, much larger forces in the corner columns are generated, so that the columns away from the corners are much less effective (see fig. 7.35); therefore, the entire three-dimensional form of the tube is no longer effective in resisting the lateral forces. For preliminary estimates, the closed tube may be replaced by an open or partial tube consisting of two equivalent channels along the webs resisting the total overturning moment, M. In this case, the effective flange width, b_e, is approximated as not more than one-half of the web depth (d/2), one-third of the width of the tubular flanges (b/3), or 10 percent of the height of the building (H/10), keeping in mind that these values only represent a rule of thumb. Using the information for the channel dimensions of Figure 7.34, right, the moment capacity of the channels can be derived as

$$M = F_f(d) + F_w(3d/4)$$

$$= f_z(b_e t)d + f_z(3dt/8)(3d/4)/2 \tag{a}$$

The axial capacity of the flanges, F_f, can be derived from Equation (a) by letting

$$F_f = f_z b_e t, \qquad m = b_e + 9d/64$$

$$F_f = Mb_e/dm \tag{7.31}$$

The axial web force F_w is derived from Equation (a) by letting

$$F_w = f_z(3dt/8)/2, \qquad m = b_e + 9d/64$$

$$F_w = 3M/16m \tag{7.32}$$

The fact that the shear stresses are not constant along the web, but are largest at the neutral axis of the web (Fig. 7.34, left), is demonstrated in Example 7.10. It should be remembered that the larger shears in some columns and girders will also cause larger moments. To overcome the shear lag problem in the framed hollow tube, either deep spandrel sections or trussing can be used, as has already been mentioned.

The renowned structural engineer Fazlur Khan of SOM developed the " NO-PREMIUM FOR HEIGHT" concept for the design of the tubular bearing walls of the John Hancock Center and the Sears Tower in Chicago. In this case, the bending stresses, f_b, due to lateral force action, do not exceed one-third of the axial gravity stresses, f_a, or $f_b/f_a \leq 0.33$, so that wind does not cause an increase in the size of the supporting members (see Equation 2.8).

The dish effect of tall tubular structures has already been discussed in Chapter 3 under "Hidden Loads." There will be a differential settlement between the central core, which only carries gravity loads, and the exterior tubular wind-resisting system. The core will shorten in relation to the exterior wind columns, which will only be fully activated on rare occasions, thus causing wracking of the floors. To overcome this problem, the nonwind columns should be made longer and be of low-strength material, while the wind columns along the perimeter should be of high-strength material to provide less stiffness.

The lateral sway of an ideal tall and slender tubular structure of constant wall thickness is controlled by flexure. A uniform wind pressure will deflect the vertical cantilever roughly as

$$\Delta = wH^4/8EI \tag{7.33}$$

In this case, the moment of inertia is provided by the closed hollow tube, as defined by Equation 7.24b. In reality, the degree of flexibility of the tubular walls determines the lateral sway. For a single framed tube, the deflection is primarily due to the flexibility of the web-frame parallel to the lateral forces. Usually around 70 to 75 percent of the lateral drift is due to frame action of the webs in shear, while the rest is due to ideal tube deflection in compression and tension in the perimeter columns. For rough preliminary estimation purposes, the framed tube could be conservatively treated as a plane frame, where the two faces parallel to the wind act as independent rigid frames, according to the portal method (see Equation 7.17), realizing that the axial deformations (i.e., resistance) of the interior columns are ignored. For example, the lateral deflection of one of the slender World Trade Center towers in New York could be roughly estimated by considering only the chord drift of the webs (Equation 7.15a), as demonstrated in Problem 7.26.

Example 7.10

The 26-story, 120 × 120 ft framed tube structure in Figure 7.34 is 325 ft high, with an average story height of 12 ft, 6 in.; the columns are spaced 8 ft apart. The following load conditions are used:

floor/roof dead load, including ceiling and miscellaneous:	60 psf
partition dead load:	20 psf
reduced office live load:	50 psf
roof live load:	15 psf
wind load:	40 psf

A typical perimeter column is investigated at the building base, using A36 steel for drift control purposes.

In tubular structures, the exterior perimeter framing acts much like a bearing wall. Column groups help each other in distributing the gravity flow, so that average loads can be assumed; there is a reserve strength because stresses tend to redistribute from points of overload to areas of strength. However, it must be kept in mind that, when side walls are quite soft with respect to the vertical stiffness due to wide column spacing, then a large portion of the gravity loads may be transferred to the corner columns.

Therefore, circulating the axial gravity loads by the tributary area method, as is done in this example, and which is applicable to frame structures where the columns are spaced relatively far apart, should be conservative if the vertical stiffness is satisfactory.

The gravity flow to the columns is not uniform because of the floor framing layout, although redistribution of forces occurs. To minimize the difference in load distribution, the even-numbered floor framing systems are oriented orthogonally to the odd-numbered ones. It can be assumed that, with the exception of the eight columns supporting the girders and the four corner columns, the force flow to the rest of the columns is rather uniform. It is apparent that the girder columns carry more gravity loads and the corner columns less. However, it must be kept in mind that, even though the corner columns do not support much of the gravity load in comparison to the other columns, they may carry most of the wind load.

A typical facade column that is located on one of the centroidal axes of the building tube carries the following axial load at the ground level. Notice that, for the selected odd-numbered floor systems, the roof framing is not included.

$$
\begin{aligned}
\text{curtain load: } 0.015(8 \times 325) &= 39.0 \text{ k} \\
\text{floor dead load: } 0.080(8 \times 20)13 &= \underline{166.4 \text{ k}} \\
\text{Total axial dead load: } &= 205.4 \text{ k} \\
\text{floor live load: } 0.050(8 \times 20)13 &= \underline{104.0 \text{ k}} \\
\text{Total axial load: } &= 309.4 \text{ k}
\end{aligned}
$$

1) First, the framed tube is roughly checked by assuming (as a first approximation) that the shear is carried by the webs and the entire moment by the flanges.

The building must resist the following total wind shear

$$V_w = 0.040(120 \times 325) = 1560 \text{ k}$$

The overturning moment at the building base is

$$M_w = 1560(325/2) = 253,500 \text{ ft-k}$$

The tubular flanges, that is, the walls perpendicular to the wind direction, must resist the entire moment in tension and compression.

$$253,500 = N_T(120), \quad \text{or} \quad N_T = 2112.5 \text{ k}$$

Hence, each of the 16 columns on the windward and leeward faces carries, in tension or compression (depending on the wind direction),

$$N_c = 2112.5/16 = 132.03 \text{ k} \tag{7.28b}$$

The columns along the web faces parallel to the wind are assumed to support only shear, and not axial forces, According to the portal method, for 15 equal bay spaces and a uniform wind shear, disregarding the actual location of the shear at story midheight, each interior column must resist

$$V_h = 1560/2/15 = 52 \text{ k} \tag{7.29}$$

For fast approximation purposes, it is assumed that the column shear acts at midheight of the first story, the location of the inflection point, so that the critical moment at the girder face due to shear wracking is

$$M_c = \pm V_h(h'/2) = 52(9.5/2) = 247 \text{ ft-k} \tag{7.30}$$

The corresponding girder shear and moment are

$$V_v = V_h(h/L) = 52(12.5/8) = 81.25 \text{ k} \tag{7.9a}$$

$$M_G = V_v(L'/2) = 81.25(6.83/2) = 277.47 \text{ ft-k} \tag{7.30}$$

2) Now, the more accurate approach of a partial tube with equivalent channel sections is used to determine the approximate wind forces in the columns.

The building must resist the following total wind shear, considering the resultant wind forces to act at the floor levels.

$$V_w = 0.040(120 \times 318.75) = 1530 \text{ k}$$

The overturning moment at midheight of the first floor level is

$$M_w = 1530(318.75/2) = 243,844 \text{ ft-k}$$

The effective flange width is

$$b_e = b/3 = 120/3 = 40 \text{ ft} \leqslant d/2 = 60 \text{ ft}$$

$$b_e = H/10 = 325/10 = 32.5 \text{ ft, controls}$$

The factor, m, in Equation 7.31 is

$$m = b_e + 9d/64 = 32.5 + 9(120)/64 = 49.38 \text{ ft}$$

The total flange force is

$$F_f = Mb_e/dm = 32.5M/120(49.38) = 0.005485M \tag{7.31}$$

At midheight of the first floor, the total flange force is

$$F_{f1} = 0.005485(243,844)/2 = 668.74 \text{ k}$$

One floor above, the total flange force is

$$V_{w2} = 0.040(120 \times 306.25) = 1470 \text{ k}$$

$$F_{f2} = 0.005485(1470 \times 306.25/2)/2 = 617.32 \text{ k}$$

The total flange force is distributed over $b_e = 32.5$ ft, so that each column, for the column spacing of 8 ft, carries

$$N_c = 668.74(8/32.5) = 164.61 \text{ k}$$

When comparing this result with the one obtained by ignoring the shear lag in part 1, it is found that $x = 164.61/132.03 = 1.25$, or that the forces in the corner columns are actually 25 percent larger. Often, for fast approximation purposes, the axial wind forces in the corner columns are increased 50 percent, as compared to the ideal tube behavior, to take shear lag into account.

The difference between the axial forces at the base and one floor above is equal to the web shear, as reflected by the curvilinear portion of the shear diagram in Figure 7.34.

The minimum web shear at the building corner is

$$\Delta F_f = 668.74 - 617.32 = 51.42 \text{ k}$$

The total web force is

$$F_w = 3M/16m = 3M/16(49.38) = 0.0038M \qquad (7.32)$$

At the base, the total web force is

$$F_{w1} = 0.0038(243,844)/2 = 463.30 \text{ k}$$

One story above, the total web force is

$$F_{w2} = 0.0038(1470 \times 306.25/2)/2 = 427.68 \text{ k}$$

The difference of the axial web forces is equal to the additional web shear.

$$\Delta F_w = 463.30 - 427.68 = 35.62 \text{ k}$$

The maximum vertical shear at the neutral axis of the web consists of the sum of the flange and web shear.

$$V_{vmax} = \Delta F_f + \Delta F_w = 51.42 + 35.62 = 87.04 \text{ k}$$

The corresponding girder moment at the column face is

$$M_G = 87.04 \ (6.83/2) = 297.24 \text{ ft-k}$$

Again, the shear lag causes slightly larger moments, of about 7 percent, than was assumed in the more simplified approach.

The corresponding column shear and moment are

$$V_{hmax} = V_{vmax}(L/h) = 87.04(8/12.5) = 55.71 \text{ k}$$

$$M_c = 55.71(9.5/2) = 264.60 \text{ ft-k}$$

It is reasonable (for this preliminary design of the columns) to consider the tube stiff enough that the columns can be considered laterally braced. The unbraced length of $h_x = 9.5$ ft is also used for the weak axis of the section:

$$(Kh)_y = 1(9.5)$$

The typical web columns (not girder or corner columns) must resist the axial gravity load and the moment due to shear wracking, but the same columns must also act as flange columns when the wind acts orthogonally. For this first check, it is assumed that the critical typical column is located at the neutral axis of the tube. At this point, it must resist the largest shear but no axial wind forces, while the columns adjacent to the corner columns do carry axial wind forces, but also less shear. Notice that the columns are oriented with their strong axes as to efficiently resist shear wracking.

The various loading conditions are investigated for a W14 column with $B_x \simeq 0.18$.

Web action:

D + L + W: $\qquad P_{eq} = P + B_x M_x$

$$= 0.75[309.4 + 0.18(264.6)12] = 660.71 \text{k} \geqslant 309.4 \text{ k}$$

D + W: $\qquad P_{eq} = 205.4 + 0.18(264.6)12 = 776.94 \text{ k}, \qquad \text{clearly controls.}$

Flange action:

$$D + W: \quad P_{eq} = 205.4 + 164.61 = 370.01 \text{ k}$$

$$D + L + W: \quad \geqslant 0.75(309.4 + 164.61) = 355.5 \text{ k}$$

Notice that the design of the column is clearly based on web action, that is, shear wracking, rather than flange action!

From the column tables of the AISC Manual for KL = 1(9.5), try

$$W14 \times 132, \quad P_{all} = 772.5 \text{ k}, \quad A = 38.8 \text{ in.}^2$$

Alternatively, use the manual design approach with an approximate allowable compressive stress of

$$F_a \simeq 23 - 0.1KL/r = 23 - 0.1(1)9.5(12)/4 = 20.15 \text{ ksi} \qquad (4.47)$$

Therefore, the required cross-sectional column area is

$$A = N/F_a = 776.94/20.15 = 38.56 \text{ in.}^2 \leqslant 38.8 \text{ in.}^2$$

Braced Tubes

The inherent weakness of the framed perimeter tube lies in the flexibility of its spandrel beams. Its rigidity is substantially improved by adding sufficient diagonal members to minimize the shear-type deformations. In this case, the shear is primarily absorbed by the bracing directly in axial action, rather than in bending by the spandrel beams so, that nearly true cantilever action is produced. The diagonals effectively move the forces around the corners and tie the tubular shape together. Since the bracing absorbs the lateral forces efficiently, the exterior columns may be more widely spaced and thus allow a more open base.

Besides the popular *diagonally braced framed tube* using widely spaced diagonals, also found are the *trussed tube* (i.e., truss walls) and *latticed truss tube* (i.e., closely spaced diagonals with no vertical columns) systems, as described in the following examples.

The behavior of diagonally braced framed tubes is quite complex, due to the interaction of the spatial truss and framed tube systems. To develop some understanding of this, each of the systems may initially be visualized as acting independently. Under lateral loads, the truss webs (consisting of diagonals and horizontal ties) efficiently resist the shear in tension and compression, while the entire spatial truss-tube counteracts rotation (i.e., overturning); the corner columns function as the chords of the cantilever space-truss and carry the largest loads. The trusses transfer the loads, especially torsion, efficiently around corners, particularly when the sides are not perpendicular to each other. Under gravity loads, the diagonals do redistribute the loads and differential movements to equalize the wall stresses. Because they also act as inclined columns under gravity loading, the diagonals do not usually develop tension stresses. For very tall buildings, the diagonals should be at an angle of approximately 45 degrees to result in widely spaced crosses.

The 26-story Alcoa Building [1967] in San Francisco (see fig. 7.25m) is a powerful early statement of the trussed steel tube principle. The twelve exterior columns of the 100 × 200 ft, 398-ft high tube are spaced at 51-ft intervals and are tied together by X-braces to give the appearance of true vertical trusses. The wall members project nearly 3.5 ft beyond the curtain wall, but still required fireproofing. Slender tension members hang midway between the columns from the intersection points of the diagonal members to support the load of five floors, structurally making the building a stack of four 6-story buildings. Although the stiff perimeter tube resists all of the lateral loads, two additional interior rigid frames in each direction had to be provided to resist 25 percent of the seismic forces in order to achieve the necessary ductility.

The most common braced tubes are of the diagonally braced frame type, as boldly demonstrated for the first time by the 100-story John Hancock Center [1968] in Chicago (see fig. 7.25e). The diagonal members of the 1127-ft high truncated pyramid not only act as bracing, but also tie together all of the perimeter columns to make the entire steel building act as a hollow tube with an average cross section of 132 × 211 ft. The diagonal members will be in compression under almost all loading conditions. The reader may refer to Problem 7.27 for a further investigation of the building.

The first diagonally braced frame concrete tube is New York's 50-story 780 Third Avenue Office Building [1985] (see fig. 7.33, center bottom). For this extremely slender 570-ft high, 70 × 125 ft tube with a height-to-width ratio of 8:1, braced action is achieved by simply filling the window openings with concrete in a diagonal pattern and by running the reinforcement along the diagonals. Only single diagonals could be accommodated on the narrow faces, in contrast to the double diagonals on the broad sides. The exterior columns are spaced at 9.33 ft on center around the perimeter. In a way, the extremely slender building proportions called for a diagonally braced concrete tube to obtain both the necessary lateral stiffness and to efficiently use the weight of the concrete as a stabilizer and prestressing agent to suppress tension stresses and thus reduce crack formation. The interior core walls, which allowed for column-free interior space, have only a minimal influence on resisting the lateral forces because of their small size and central location. In addition, the eccentric location of the core causes an uneven distribution of the gravity loads to the exterior columns, with corresponding long-term effects of creep and shrinkage; the diagonals are therefore also required to effectively redistribute and equalize the column stresses.

In the pure latticed truss tube, the perimeter walls are made up of closely spaced diagonals acting as inclined columns to carry the gravity loads and brace the building against lateral forces; the diagonals may be tied together by the spandrel beams. The diagonals are efficient in responding to lateral forces, but they are less efficient than vertical columns in transmitting gravity loads. Furthermore, the large number of joints required between diagonals, and the problems related to window details, make the pure lattice truss impractical, especially in steel construction.

The perimeter tube concept was applied first in 1963 to the 13-story IBM Building in Pittsburgh (see figs. 2.17b, 5.3 k, 7.25n), by using four latticed truss walls connected at the corners to form a tube. The diagonal lattice of the 110 × 137 ft, 170-ft high steel tube is tied together by a system of horizontal plates; these plates also serve as a connection for the floor framing, which transfers about one-half of the total floor load and also transfers the lateral forces in diaphragm action. For other braced tube structures, the reader may refer to Figure 7.25.

Tube-in-Tube

To improve the shear stiffness of the framed perimeter tube, an inner braced steel or concrete tube may be added. Although the shear core at the center of the building is flexurally weak because of its location close to the axis of rotation, it does provide shear resistance, which reduces the shear deformations of the outer framed tube. It may be concluded that this interaction causes the outer tube to primarily resist the rotation, while the interior tube resists the shear; this may result in axial stresses in the inner braced tube as large as those in the outer framed tube. This interaction between the two structure systems generates a completely new structure system, the behavior of which can be visualized by superimposing the shear mode deformation of the framed tube upon the bending mode deformation of the shear walls. The result is flat S-curve (Fig. 7.35, left), where the framed tube reduces the lateral deflection of the shear core at the top and the core reduces the deflection of the frame near the base.

For preliminary design purposes, the outer framed tube may be replaced by its webs, that is, by equivalent plane frames parallel to the direction of the lateral force action. Assuming the floors to be pin connected to the tubes, so that they act only as axial members in transferring the lateral loads, the two systems are pushed apart in the upper portion of the building but pulled together near the base, as explained previously. It may be concluded that an approximate analysis can be based on a frame–shear wall interactive system (as shown in fig. 7.29). Naturally, when the floor framing is moment-connected to the inner and outer tubes, then the coupling effect makes the analysis much more complex.

In 1964, the first two tube-in-tube buildings, both in reinforced concrete and with nearly the same gross square footage, were completed. The exterior 155 × 125 ft tube of the 491-ft high CBS Headquarters in New York (see fig. 7.33, bottom right) encloses an interior tube of 85 × 55 ft. The exterior 5-ft wide columns are spaced 10 ft apart. The massive triangular (actually five-sided) columns allow an opening of only 5 ft, and thus give the appearance of a bearing column wall. In contrast, the perimeter frame wall tube of the 170 × 114 ft, 478-ft high Brunswick Building in Chicago, which encloses a 93 × 37 ft inner core, has 1.75-ft wide columns spaced at 9.33 ft all around the building (see fig. 2.6, center).

The tube within tube concept was first introduced on a large scale for Houston's 714-ft high One Shell Plaza [1971] (see fig. 5.4h). In this structure, the 192 × 132 ft exterior tube encloses a 98 × 56 ft interior tube. All exterior columns are 18 in.

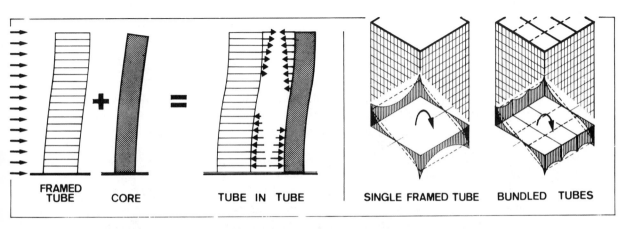

Figure 7.35. *Stiffening the perimeter tube with core or cross-bracing.*

wide, but vary in depth, and are spaced 6 ft on center. The floor structure is similar to that of the Brunswick Building in Chicago.

The 70-story MLC Centre [1976] in Sydney, Australia (see fig. 5.7h), is octagonal in plan, with only eight massive tapered perimeter columns tied together with 6-ft deep spandrel beams, which alternately span 35 and 60 ft around the perimeter. The 786-ft high reinforced concrete structure acts like a tube-in-tube, where the nearly 141-ft square, deep spandrel exterior framed tube encloses a 67-ft square interior tube.

The 60-story Sunshine 60 Office Building [1978] in Tokyo, Japan (see fig. 7.33, middle right), consists of a 143 × 234 ft, 742-ft high tube with a 41-ft wide interior spinal core. This structure uses a mixed system of a steel skeleton, where the exterior steel columns are spaced at about 10.5 ft on center, and slotted precast reinforced concrete fill-in wall panels in the interior core for the tower portion of the building, which resist 30 percent of the lateral forces and increase the building stiffness. The slits prevent the panels from absorbing too much energy, and thereby allow them to act like ductile shear walls. Their behavior is thus compatible with the flexible behavior of the steel skeleton under repeated cyclic seismic loading.

The Japanese took the tube-in-tube concept a step further by introducing the triple-tube for a 52-story office building [1974] in Tokyo (see fig. 7.33g). The tower is of triangular shape with large chamfered corners, where the three primary faces are 161 ft long and the three chamfered sides are 63 ft. In this structure, the exterior framed tube resists the entire wind load, while the three tubes together resist earthquake forces. All of the building columns are spaced 9.9 ft apart.

The free-form development of the new generation of skyscrapers is expressed by the plan form in Figure 7.33e, representing a parallelogram with truncated and re-entrant corners. This tube-in-tube structure consists of a reinforced concrete perimeter framed tube and an interior concrete core. Steel is used for the floor framing and interior columns. The perimeter tube consists of 3-ft wide columns spaced at 9.66 ft on center and 3.25-ft deep spandrel beams that increase in thickness from the top to the bottom of the building. The framed tube is completed around the corners by solid walls to allow the continuous transmission of forces.

Bundled Tubes

The bundled tube principle can be visualized not only as an assembly of multiple tubes of various heights placed directly adjacent to each other, but also as a large perimeter tube that is stiffened with interior framed webs to form a modular tube structure that behaves like a huge multicell perforated box girder cantilevering out of the ground. Since each tube appears to act independently under lateral force action, the exterior and interior webs parallel to the lateral force action generate points of peak stress at the exterior frame junctions (Fig. 7.35, right). The shear lag of the single perimeter tube is thus greatly reduced and replaced by a more uniform distribution of the column forces, thereby achieving a more effective flange action. Therefore, because of the three-dimensional action of the multicell cantilever tube, for preliminary design purposes a bundled tube of regular shape may be treated as a true cantilever by disregarding shear lag. Critical, with respect to the design of bundled tubes of different heights, is the differential column shortening due to gravity. For example, the incompatibility of a 100-story tube that shortens twice as much as an adjoining 50-story tube must be accommodated.

The bundled tube principle is not only effective from a strength point of view for very tall buildings, but also as a support structure for irregular plan and building form configurations. In this instance, irregular plan forms can be subdivided into individual cells of different shapes that form vertical tubes that can each be terminated at any desired level.

The first and most renowned bundled tube structure is the 110-story Sears Tower [1974] in Chicago (see fig. 7.33, center, right). This 1454-ft high tube is composed of nine 75-ft square framed steel tubes of varying heights bundled together to form an overall tube 225-ft square in the bottom portion; the adjacent tubes thus share common frame walls. The columns are spaced 15 ft apart, with larger columns at the corners in response to wind action. The columns are connected in the interior by composite 40-in. deep 75-ft span floor trusses (see fig. 5.4m). The perimeter of the structure was too large for a single tube—it had to be internally braced by cross-frames so that true tubular cantilever action with greatly reduced shear lag could be achieved. The efficiency of the structure is reflected by the low steel quantity of 33 psf. Because of the independent strength of the individual tubes, they are dropped off at different heights. At two setbacks immediately below the sixty-sixth and ninetieth floors, and at the twenty-ninth to the thirty-first floor mechanical levels, belt trusses are used as transfer systems. These trussed levels increase the lateral stiffness of the structure by tying the tubes together vertically and horizontally; they distribute the gravity and wind loads from the upper floors to the greater modules below, and reduce the dish effect, especially at the setbacks, due to differential column shortening caused by gravity. A preliminary investigation of the magnitude of the force flow is presented in Problem 7.28.

The 57-story One Magnificent Mile [1983] in Chicago (see fig. 7.33, middle, center left) is a bundled tube structure made up of three near-hexagonal reinforced concrete framed tubes, each one of a different height. The exterior column spacing responds to the different functions at various levels: for the office floors it is 10 ft; for the apartment levels, it varies from 2.5 to 9 ft; and for the commercial and parking levels, it is 20 ft. The interior tube column spacing is 10 ft. Flat plate construction is used for the apartment and office floors.

Modular tubes do not have to be of the same shape. For example, the 730-ft high composite bundled tube for the 53-story Southeast Financial Center [1984] in Miami (see fig. 7.33d) is composed of a rectangular tube and a stepped triangular tube with a common frame–shear wall between them. The interior gravity framing is all steel.

The Torre Banaven Office Tower, in Caracas, Venezuela (see figs. 5.4j, 7.25c), required a special type of bundled steel tube structure because of severe seismic activity. The tower consists of three independent triangular-plan trussed frame tubes, with the central tube being 680 ft high. At the two-story mechanical levels, the three tubes are tied together by horizontal steel trusses to activate the building as a whole with respect to lateral force action.

In the following sections, the pure tubular concepts just discussed are modified.

Interior Braced Tube

Elongated or complex geometric plan forms may make it impossible to obtain the proper web action of the perimeter tube, so that shear resistance cannot be developed if some sort of interior bracing is not provided. Similarly, compound interacting shapes may require an additional interior stiffening structure to allow the perimeter tube to act as an overall system. The following cases will identify some typical conditions.

An early example of the modified tube principle is the 859-ft high Water Tower Place [1976] in Chicago (see fig. 2.17e). The slender, 220 × 95 ft tower has exterior columns spaced at 15.5 ft and relatively shallow spandrel beams; this perimeter tube is thus very soft and had to be braced transversely by an internal shear wall that bisects the tube and takes 65 percent of the wind load. The transition of the 64-story tower to the double bay sizes of the 12-story base, where the wind resisting system changes to double cores, was made at the thirteenth floor with a grid of 10-ft deep concrete girders.

The more complex geometric shapes may not lend themselves to simple tubular action, as is demonstrated by the elongated 253 × 113 ft irregular hexagonal plan form of a 503-ft high tower (see fig. 7.33f). In this case, the concept of the single, framed perimeter tube with steel columns on 10 ft centers had to be modified because the pointed ends of the hexagonal plan exhibited excessive shear lag, which made it impossible to get an effective tubular response. However, by adding rigid frames in the transverse direction, the exterior walls were tied together and adequate flange action was achieved.

The 50-story Four Allen Center [1984] in Houston (see fig. 7.33, bottom left) is of an elongated rectangular plan form with semicircular ends. This slender 109 × 259 ft, 695-ft high tower, which has a narrow noncompact shape and a shallow core width of only 26 ft, required a hybrid steel structure consisting of a perimeter framed tube with columns spaced at 15 ft on center. It is braced internally along three lines by webs made up of the core shear trusses and a frame system on each side, as well as two-story subgrade trusses, to form a four-celled bundled tube. The cross-frames are composed of 40-ft span *tree-beams*, that is, floor girders to which short column stubs are welded above and below at center span; they improve stiffness by providing the necessary vertical shear resistance between the core and perimeter columns. The connections between the column stubs are nonbearing and do not transmit any vertical loads; the columns provide only flexural wind resistance by forcing intermediate inflection points into the girders, thereby greatly improving the stiffness and strength of the structure and permitting the building to act as a tube. Since the core shear truss/frame interaction is not needed at the top and bottom portions of the building, the frame columns are omitted at those levels. The typical 40-ft span W21 floor beams frame into each perimeter column; they act compositely with the 15-ft span composite floor structure consisting of a 3 1/4-in. lightweight concrete slab on a 3-in. deep metal deck.

The 970-ft tall bundled steel tube Allied Bank Tower [1983] in Houston (see fig. 7.32a) is composed of two quarter-circle tubes placed antisymmetrically back to back but stiffened in the transverse direction by braced shear walls with outrigger trusses and belt trusses.

Partial Tubes

Usually, the framed tube concept is associated with tall buildings. For mid-rise structures, however, it may not be necessary to develop the entire perimeter for lateral force resistance; instead, merely engaging incomplete or partial tubes of the C-type along the webs of a building may be sufficient. This approach corresponds directly with the approximate design of framed tubes, where the perimeter tube may be reduced to two equivalent channels resisting the overturning moment (see fig. 7.34, right). An early example of this concept is the 183 × 93 ft, 405-ft high Town Center [1975] in Southfield, Michigan. In this structure, partial steel tubes, that is, channel-shaped incomplete trussed frame tubes, were provided across the short end sides extending 45 ft along each of the long sides.

The 58-story Onterie Center [1986] in Chicago (see fig. 7.25d) is stabilized by two exterior diagonalized framed tubular channels at each end. The concrete framed tubes, with a column spacing varying from 4.5 to 6.5 ft on center, are braced by filling the window spaces with concrete in a diagonal pattern. These infill panels act not only as diagonal braces as for trussed steel tubes but also as shear panels. The interaction between the two open tubes is provided by the floor diaphragms and by spandrel beams across the open central bays. The resultant outward horizontal thrust at the bracing corners is restrained over three floor levels by spandrel beams continuous tension around the building.

The trend towards a new generation of skyscrapers and away from pure structure systems is expressed by Houston's 75-story Texas Commerce Tower. Completed in 1981, this 1002-ft high composite tube is 160-ft square in plan with a large 45 degree chamfer at one corner, thus forming a pentagonal-shaped configuration (see figs. 5.7b, 7.33a). The tower is a ruptured tube with four sides forming an open framed perimeter tube where the composite columns are spaced 10 ft on center. This incomplete tube is connected across the chamfered fifth side by 85-ft long-span steel girders. Along this open side, the lateral loads are transferred by link beams to the U-shaped composite shear core. The floors are steel framed.

Hybrid Tubes

Many complex building shapes are derived from intricate geometries and not from the nature of the support structure. Thus, the result is often not a straightforward pure structure, but a combination of structure systems in response to an economical solution in order to fit the geometric requirements. These mixed structure systems can be quite complex, as has already been indicated by some of the modified tubular systems. There are many possible combinations of tubes with other systems, as is discussed with the following building cases.

An early example is the 32-story building (see fig. 7.33b) of the mid-1970s. This irregular 181 × 100 ft mid-rise tower consists of two intersecting octagons. It is laterally stabilized by an interior tube with columns spaced at 10 ft on center and exterior channel-like wall frames at opposite ends with column spacings of 9.3 ft and 14 ft.

The 52-story Georgia-Pacific Center [1982] in Atlanta (see figs. 1.14c, 5.7l, 7.25j) has a plan form of a parallelogram, but with four large setbacks at various floor levels, and forms a five-sided sawtooth face on the inclined east facade. This 671-ft high and 123-ft wide steel structure consists of a framed tube with a column spacing of 10 ft on center on three sides, and is closed at the fourth faceted face by a doubly-folded truss wall. The tube had to be stiffened in the lower building portion by a set of three vertical interior core trusses with outriggers tied to belt girders.

The stepped configuration of the 57-story Republic Bank Center [1983] in Houston (see fig. 7.33, middle left) gives the appearance of three adjoining towers. Because of their varying height, each of the towers required a different lateral-force resisting system.

In order to minimize the overall twisting of the building due to unequal wind loading, the shear resistance had to be distributed to the three tower structures and optimized to locate the center of rigidity close to the center of the resultant lateral load when the wind blows against the 250-ft long face. The tallest tower is 780 ft, and consists of a perimeter tube closed on the inside with a Vierendeel hat truss following the gabled roofline that ties the braced frame of the interior core to the exterior tube. The intermediate tower is laterally stabilized by a channel-shaped partial tube with a stepped Vierendeel outrigger hat truss connected to the core, similar to the perimeter tube for the high-rise tower. The low-rise tower only required a planar welded frame along the end face. The welded perimeter steel structure of tubes and rigid frames consists of the columns spaced at 10 ft apart and the 43-in. deep channel-shaped flat-plate spandrels; the tree-columns were bolted at midspan. The floor construction consists of the primary composite beams spanning 42 ft from the core to the exterior columns and the 4 1/2-in. lightweight concrete slab on a 2-in. deep metal deck. The overall steel weight for the building structure is 25.2 psf.

The office tower of the 36-story AT&T building [1983] in New York (see fig. 7.33, top center) is supported on exterior stilt-like columns and a central core to allow for an open public plaza space of 134 ft height at the base. The tower portion consists of two vertical partitioned steel tubes at each end of the building, connected by 50-ft span hinged beams at the center; they are tied together at top and bottom by cross-braced horizontal tubes. Each of the vertical tubes is partitioned and internally braced by two I-shaped steel plate outrigger wall girders with door openings; located at every eighth floor, the girders extend from the vertical core trusses to the exterior columns. The transition of the lateral forces from the tower to the base structure of the exterior stilt columns and steel-plate shear tubes in the central core has been described earlier (see fig. 2.13).

The perimeter tube for the 200 × 135 ft, 797-ft high, 60-story Momentum Place [1987] in Dallas (see fig. 7.33, top right) consists of punched concrete walls at the four building corners, with infills between composite columns and steel spandrels. At the fiftieth level, the tubular walls around the corners could not continue because of the transition from the rectangular to the cruciform-shaped plan; at that level the building becomes all steel. There, two-story braced frames had to be added in the core, together with a strengthening of the floor diaphragms, to allow the transfer of wind shear from the core to the perimeter tube. In order to provide a column-free public space at the base (sixth floor level), the column spacing in the concrete walls changed from 5 ft to 25 ft along the long faces, and, on the short faces, the composite columns from above were transferred to concrete columns at 38-ft spacing below by deep concrete-encased steel plate girders.

The 60-story, 726-ft tall steel framed Allied Bank Tower [1986] in Dallas (see fig. 7.33, top left) is of an elaborate formal geometry. Its shape is composed of a main 40-story rectangular block with tetrahedron-shaped wedges carved out on each side; its plan thereby changes with height from a square at the twelfth floor level to a rhomboid at the forty-fifth floor; the top is capped by a 16-story triangular prism. The rhombus-shaped plan at the upper level is repeated by a 4-story high rhomboidal block at the base; in this structure, the overhanging corners of the square block above are supported on 30-ft wide pylons. It is apparent that this combination of different shapes causes unsymmetrical bending and torsion when acted upon by wind; also critical is the transition of the forces from above to the open base structure. In addition, the opposing slopes of the tetrahedrons cause twisting about the vertical axis under gravity loads. The perimeter trussed frame for the 40-story tower portion not only resists lateral loads, but the megatrusses on two sides also bridge the 156-ft span at the base, while the 8-story trusses on the two opposite sides span the 96 foot open space. The megatruss, in turn, consists of 8-story deep subtrusses with one-story Vierendeel girders acting as their chords to resist the unbalanced horizontal components of the diagonals at the floor midheight intersections. The structure for the prism at the top consists of a welded triangular hat-truss. The transition from the top to the trussed frame is provided by extending the moment-resisting welded frame to the thirty-sixth floor level. The typical floor construction consists of composite 50 ksi steel beams supporting a 2 1/2-in. normal-weight concrete slab on top of a 3-in. deep metal deck; the concrete is mesh-fiber reinforced to speed up construction. With only a 22 psf steel weight, the building structure proved quite economical.

MEGASTRUCTURES

Megaforms were the dominant visionary concept of architecture and urbanism in the 1960s. Cities were idealized, with the help of the exciting potential of technology, by proposing structural frameworks broken down into smaller modular components to allow for unlimited future growth and change. From a purely structural point of view, these megaforms ranged from large-scale functional organizations through massive multilayer long-span bridges connecting vertical cores (thus expressing the servicing and functioning of the urban organism), to the breaking down of the faceless mass into a more human scale through the spontaneous stacking of modular clusters or the plugging of units into space frames.

In this context, the term megastructure refers not to a visionary solution expressing the comprehensive planning of a community or even an entire city, but solely to the support structure of a building. However, this megastructure is still based on the basic concept of a primary structure that supports and services secondary structures or smaller individual building blocks. Some of the early examples that more or less express the megastructure principle are: the bridge structures of the Yamanashi Communications Center [1967] in Kofu, Japan (see fig. 7.6a) and the Knights of Columbus Building [1969] in New Haven, CT (see fig. 7.5j); the superdiagonals forming the tube of the John Hancock Center [1968] in Chicago (see fig. 7.25e); the U.S. Steel Building [1970] in Pittsburgh (see fig. 7.4e) where the primary framing supports vertically stacked three-story units; the Federal Reserve Bank [1973] in Minneapolis (see fig. 7.8), using the suspension bridge concept; and the megaframe of the OCBC Centre [1976] in Singapore (see fig. 7.5f) with its 14-story plug-in stacks.

The megastructures of today evolve out of structural efficiency in response to spatial requirements; in addition, sophisticated computers and software have now made it possible to understand the behavior of structures much better. In the current hybrid building forms, with their many setbacks, as well as in slender building shapes, the material must be arranged to efficiently resist overturning. Furthermore, the urban high-rise buildings are multiuse structures and are integrated (at the base) with the city fabric, therefore requiring a variety of spatial layouts, which include multilevel open atria stacked on top of each other. All of these reasons, and many others, have motivated the further development of the megastructure concept.

Surely one of the most important examples of this new breed of megastructures is the 59-story Citicorp Center [1977] in New York (Fig. 7.36, bottom, fifth from left), which is discussed in more detail later in this section. In this case, the renowned structural engineer William J. LeMessurier introduced a completely new way of thinking about structure with the eight-story series of chevron bracing, a structure that seems so simple and clear, but which is so much more complex and sophisticated than designers have been accustomed to.

In contrast to this development of exciting structural principles by engineers, and where the dynamics of the structure is usually hidden beneath the skin, because it is a servant to the stronger concerns of the facade (that is, to symbolism and iconography), another (more recent) trend is indicated by the 43-story Hong Kong Bank [1985] in Hong Kong (see fig. 7.37). In this instance, a more encompassing design dimension is introduced, which reminds the architect that there is a way of masterly integrating architecture, structure, and technology, although it does not express the predominance of structural efficiency or the logic and beauty of the pure minimal structure. Foster Architects, in a team approach with Ove Arup, Structural Engineers, and together with many other specialists, expressed (in an original and brilliant fashion) the current sophisticated level of technology, and here celebrated science and function as a work of art. The supporting bridgelike structure, which allows opening up of the central space, is articulated, and only escalator circulation is used, simply by placing the structural towers, the elevators, and the prefabricated mechanical modules containing the toilet facilities and individual air conditioning plants along the sides of the building; this is quite opposite in approach to the traditional central core idea of conventional high-rise buildings. The functioning of the building and the movement of people, the assembling of component parts, service modules and other prefabricated packages, the building intelligence, the sophisticated cladding systems, and the perfection of the detail are all articulated. The structure itself is discussed in more detail later in this section.

In addition to the development of sophisticated structures in response to spatial requirements and economical solutions, as well as to structural exhibitionism, the obsession in the United States with superskyscrapers during the 1980s has generated many new innovative structure concepts. America, being the birthplace of the skyscraper, has always been fascinated with, and felt challenged to conquer, height. In 1956, Frank Lloyd Wright proposed the mile-high cantilever Illinois Sky-City (Fig. 7.36, left). This 528-story steel tower, a tripod in plan, is vertically divided into five sections of 100 floors, and would house 130,000 inhabitants. It should also not be forgotten that there has been no dramatic increase in height since the 1250-ft high Empire State Building in New York was finished in 1931. The new breed of ultra-high-rise buildings (above 150 stories) that are being proposed form vertical cities, where the impact of scale can be visualized by looking to the World Trade Center in New York, which already must service 50,000 workers and 80,000 daily visitors. Naturally, the problems related to servicing a vertical city and its impact upon the surrounding city are immense, and far more complex than the design and construction of the support structure, which is of primary concern here.

With an increase of height, the building will become more slender, and hence more vulnerable to dynamic behavior. Drift and acceleration under wind become most critical, in addition to differential column shortening and temperature change, among other considerations. Because of the large slenderness, the stability of the

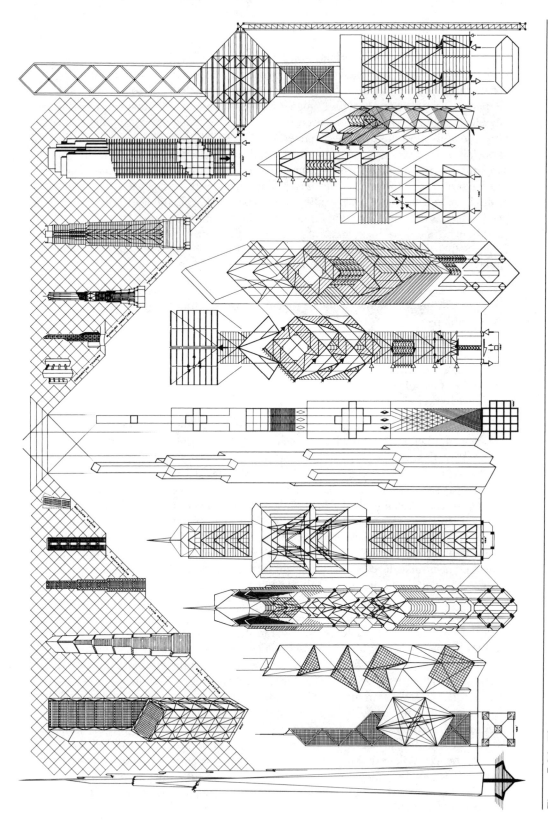

Figure 7.36. The megastructure.

building, as expressed by the relation of the height to the width at the base parallel to the wind, is of fundamental concern. For example, for a 200-story building of approximately 2400-ft height, the common maximum aspect ratio of from 6 to 8 must be increased to 10, to yield a 240-ft square tower at the base to accommodate a typical city block dimension and to allow occupants to be close enough to natural light and view. Not only must this extremely slender structure be stiff, but also, as much weight as possible must be collected to the perimeter and concentrated in some supercolumns at optimum locations, possibly as four massive corner columns, to effectively counteract the overturning due to wind (see fig. 2.1c). The supertowers are tapered from the top to the base, and their geometric outlines are shaped to provide the least resistance to wind, while at the same time helping to damp the vibrations.

Although superskyscrapers may not be built in the near future, the new structure concepts that they have generated may be applied to more conventional buildings. Among those new structure systems are:

- Double-layer space-frame braced tube
- Internally and/or externally braced superdiagonal structure
- Megaframe
- Telescoping megaframe
- Space-frame braced tube
- Guyed tower
- Hollow megatube
- Linked tower clusters

One of the pioneers in the development of supertall buildings is Myron Goldsmith of SOM. In his concern for the laws of scale in relation to structures, as early as 1953 (for his Master's thesis) he proposed an 86-story reinforced concrete superframe. Later, together with Fazlur Khan, as professors of architecture at I.I.T. in Chicago in the early 1960s, he carried out (with his students) numerous studies dealing with the design and construction of tall buildings that efficiently resist lateral forces. In 1969, Alfred Swenson, another professor at I.I.T., developed the concept of the superframe further with his 150-story fully-braced megatube (Fig. 7.36, top, second from the left). He used a double-layer perimeter spaceframe wall structure for this 1656-ft high tower. The building is subdivided vertically by eight trussed superfloors set 207 ft apart, which support 36 floors each; the upper half of each set of floors is suspended, while the lower half is column-supported on the superfloor below. All of the loads are brought to the perimeter megatube.

In the early 1970s, Fazlur Khan proposed to replace the multicolumn concept by the superframe, supporting four massive corner columns. By using supertransfer trusses at every 20 floors or so, on both the exterior and interior of the building, this would allow all of the gravity loads to flow to the four supercolumns. The megaframe idea used for the 70-story Dearborn Center [1988] in Chicago is the first example of this type of construction (Fig. 7.36, top, fourth from the left). The principle can be traced back to Khan's studies of superframes for multiuse urban skyscrapers, with the John Hancock Center in Chicago representing the forerunner of this idea. In the Dearborn Center, the legs of the exterior portal megaframes are efficiently concentrated at each corner of the building; they represent framed composite tubes that are linked at six levels by horizontal steel trusses. These superframes support the secondary frames within them, and allow the stacking of a series of 11 atriums varying in height from five to seven stories; this permits the admission of daylight and creates buildings within a building.

In the following paragraphs, some of the structure systems proposed in the mid-1980s for supertall buildings are briefly discussed.

Challenged by Frank Lloyd Wright's mile-high tower, the well-known structural engineer Joseph P. Colaco of Houston proposed an all-concrete tapered perimeter-trussed megatube with a 500-ft square base that is divided into twenty-one 100-ft square modules (Fig. 7.36, center). Interior columns are spaced 20 ft on center along the modular lines; openings through the building would minimize the accelerations due to wind. The building weight is about 25 pcf.

A further evolution of the megaframe principle for ultra-high-rise structures in the 150- to 200-story range is the telescoping superframe scheme that Fazlur Khan was interested in, and which was developed by Hal Iyengar of SOM for a 168-story building proposal. In this step-tapered tower, three stacks of superframes are telescoped inside of each other (Fig. 7.36, top, third from the left).

LeMessurier also studied the feasibility of a supertall building, which he called the Erewhon Center (Fig. 7.36, top right). He argues that there is no limit to height if the building's footprint grows in proportion to its height. However, since the occupants need to be reasonably close to natural light and views, he limited the building to a 220-ft square base and an aspect ratio of 12. For this condition, he proposed a 207-story tower (about one-half mile high), using a perimeter-trussed megaframe supporting a stack of 18-story modules. This 2760-ft high megaframe consists of four high-strength concrete supercolumns at the corners, measuring 40-ft square at the ground, and which are connected by external steel superdiagonals forming the web. Internal superdiagonals carry all of the gravity loads to the four massive corner columns.

The architect Harry Weese, together with the structural engineer Charles H. Thornton of Lev Zetlin Associates, proposed a 210-story guyed tower consisting of seven 30-story sections (Fig. 7.36, top, second from the left). The 2500-ft high twisted building shaft has a 300-ft square base, and takes a 45-degree turn as it tapers to the 200-ft square top. The four corners at the top are connected to the ones at the base with eight inclined supercolumns, thereby imitating guyed cable action. When, under wind pressure, the building moves laterally and is twisted, the guy columns pull the tower back to its vertical and correct position.

Kay Vierk Janis proposed a 142-story superskyscraper (in her M.Arch. thesis at I.T.T.), in 1986 (see fig. 7.25p). This 1745-ft high steel tube, with an aspect ratio of 8.3, has a 210-ft square base that rotates to smaller squares through the use of setbacks as the building rises. The exterior K-bracing carries through from the top to the bottom of the building without interruption. In the upper half of the structure, interior hanging trusses transfer all of the gravity loads, including the elevator core loads, to the perimeter every ten floors. The exterior diagonals collect all of the loads in the corner columns to efficiently stabilize the tower against overturning.

Among the many other proposals for the construction of megastructures is the linking of very slender buildings with sky bridges, not just for the purpose of sharing the lateral force resistance and dampening wind-induced vibrations much more efficiently, but possibly also to transfer some of the city functions to upper building levels. In the following paragraphs, some of the more outstanding newer megastructures are discussed in more detail.

The 59-story Citicorp Center [1977] in New York (see figs. 5.3i and 7.36) is indeed a unique structure. The 152-ft square, 914-ft high tower consists of giant trusses along the building perimeter that define independent 8-story stacks. At every eighth floor, the superdiagonals branch out from the supercolumns centered on each tower face. They frame into large spandrel beams at the corners to form three-dimensional 8-floor structure units (Fig. 7.36, center, right). One half of the building's weight is carried by the central core, and the other half by the trussed perimeter structure. Along the exterior, the gravity loads are first gathered by the diagonal columns in compression and stabilized by the horizontal spandrels, which act as tension belts. Then they are fed to the four 60-in. wide massive box columns, which vary in depth with the height of the building, and brought to the giant 9-story legs at the base. The lateral forces are resisted entirely by the trussed perimeter frame. In order to transfer the wind to this primary structure, however, the lateral forces upon the 8-story tiers must first be carried by the floors to the 68-ft square rigid core (see fig. 5.3i). This interior core acts like a vertical beam supported laterally at each eight levels by trussed floors, which (in turn) transfer the lateral loads back to the primary trussed exterior perimeter structure. Therefore, the wind shear forces are resolved axially by the truss, while the bending is resisted by the mast columns. Because the trusses do not continue to the base, the wind shear is transferred from the perimeter base trusses along trussed floor diaphragms to the central braced core, so that the 17.5-ft square giant legs only have to resist gravity and overturning due to wind (see fig. 2.13); the massive core must act as a shear tube to resist the wind shear of the entire building. The two-story base trusses not only support the slab, but also were used as a cantilever working platform structure, requiring no shoring during construction. The efficiency of the structural system, with 22 psf steel weight, compares rather favorably with 27 psf for most buildings of that height. The 160-ft crown houses the tuned mass damper (see fig. 3.5c). It was originally designed to accommodate a solar collector along the sloped face.

William J. LeMessurier designed another unique but complex steel structure, the Medical Mutual building of Cleveland [1983] (Fig. 7.36, bottom right). In this building, the open space at the base required a special transition system for the force flow. The critical wind loads perpendicular to the broad building face are transferred by the floor diaphragms to the central trussed core, which acts like a continuous vertical beam that is laterally supported at every fourth floor, where the loads are transferred from the interior core to the exterior trussed structure. The main supports, however, occur only at every eighth level, where the primary floor diaphragms link the core directly to the perimeter megatrusses.

The 70-story InterFirst Plaza Tower [1985] in Dallas (Fig. 7.36, top, second from the right) resembling a stretched hexagon in plan, is an extremely slender building with a critical height-to-width ratio of 7.24 to 1. Since the owner required an open perimeter, LeMessurier came up with an ingenious solution for this 921-ft high building, which is of mixed construction with exterior steel columns embedded in concrete and interior steel framing. For the purpose of stability, he supported the entire building on sixteen exterior columns, which are set 20 ft back from the curtain wall and are spaced 30 ft on center; they vary in size from 6 × 6 ft to 8 × 8 ft. Between these outer columns span two-way Vierendeel steel frames, which act as deep beams with respect to carrying the gravity loads, and as webs to resist the wind shear; the beams are 42 inches deep to obtain the necessary stiffness in the frame. The light steel core is hung from the frame, it does not rest on foundations.

Helmut Jahn's proposal for the Bank of the Southwest [1982] in Houston (Fig. 7.36, bottom, third from the left) is an obelisk-like 82-story square tower with chamfered corners. The slender, 1220-ft high structure tapers from a 165-ft square base to a 135-ft square plan at the top. Again LeMessurier came up with an ingenious support structure, which is not, however, integrated in articulating the building form. The entire building is supported by eight supercolumns, two on each side, which reduce in size from 10 × 15 ft at the bottom to 5 × 5 ft at the top of the building. Interior steel superdiagonals straddling the core cross the plan to connect the massive perimeter concrete columns on the opposite sides. Similar to a Greek cross configuration, they gather and then transfer the gravity loads at the base of each module as well as act as the web with respect to wind shear. The chevron configuration of the primary interior bracing is organized in nine-story modules.

The well-known structural engineer Leslie E. Robertson of New York developed another unique structure for the 72-story Bank of China Building [1988] in Hong Kong (Fig. 7.36, bottom, second from the left) consisting of four adjacent triangular prisms of different heights rising out of the square base. Currently the tallest building in Asia and the fifth tallest in the world, the 1209-ft high tower is a *space-frame braced tube* organized in 13-story truss modules, where the 170-ft square plan at the bottom of the building is divided by diagonals into four triangular quadrants. The space truss resists the lateral loads and transfers almost the entire building weight to the four supercolumns at the corners; the column at the center of the four quadrants is discontinued at the twenty-fifth floor, where it transmits the loads to the top of the tetrahedron, which carries them to the supercolumns. Midway through the 13-story truss modules, transverse trusses wrap around the building to transfer the gravity loads from the internal columns to the supercolumns at the corners; the horizontal trusses are not expressed in the facade. The loading conditions in Hong Kong, in contrast to the United States, are much more severe: the live loads and wind loads are twice those in New York, and the earthquake load is four times higher than in San Francisco. The superdiagonals are not directly attached to each other at the corners to form complex spatial connections, but are, instead, anchored in the massive concrete columns, thereby forcing the concrete to behave as a shear transfer mechanism. The mixed construction of the primary structure consists of the separate steel columns at the corners (to which the diagonals are connected), which are encased and bonded together by the massive concrete column. The giant diagonal truss members are steel box columns filled with concrete. The open space at the base of the building did not allow the diagonals to continue to the bottom; at the fourth level, a specially reinforced floor diaphragm was required to transfer the lateral shear to steel-plated core walls, which were designed as three-cell shear tubes.

The 574-ft high 43-story Hongkong Bank [1985] in Hong Kong (Fig. 7.37) is supported by a cluster of eight towers, where pairs of towers form four parallel megaframes. Structurally independent building stacks of varying height, separated by double-story spaces, are suspended from the megastructure, reflecting the 5-level vertical zoning of the building. The primary structure is made up of the four parallel fire-protected aluminum-clad steel megaframes, each consisting of two towers connected by two double-story pin-connected cantilever suspension trusses, which span 110 ft between the masts and cantilever 35 ft beyond. Three vertical steel tube hangers are suspended from the bridgelike trusses, one at midspan and one at each of the cantilever ends, to support the primary floor girders of as many as nine decks. For architectural

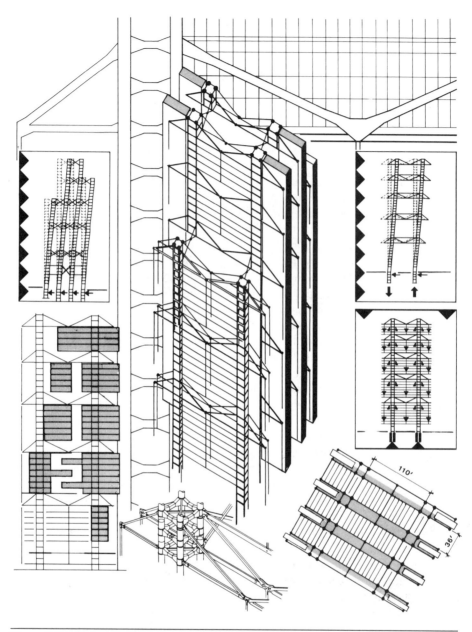

Figure 7.37. *Hongkong Bank, Hong Kong.*

reasons, the top chords of the suspension trusses between the masts are omitted on the facades. The efficiency of the gravity flow is convincingly manifested by the minimum of tensile material needed to bridge the (for buildings) unusually large span from tower to tower. The megaframe clearly expresses the character of suspension bridges vertically stacked on top of each other. The building, with rectangular plan dimensions of about 180 × 236 ft is divided horizontally by the four parallel mega-frames into vertical sections of 30, 37, and 43 stories. The megaframes also form the

16.7 ft wide circulation and edge zones, and support the 36-ft span composite floor beams, spaced at about 8 ft in the three general zones. The towers are linked at the truss levels by cross-bracing located on the inner line of the tubular tower columns. The lateral loads are resisted by the masts (together with the trusses and cross-bracing) in frame action, as shown in Figure 7.37 for the two principal wind directions, indicating that shear wracking of the columns in the multistory portal frames is predominant. The towers are composed of four corner supercolumns rigidly connected at every floor by haunched beams to form a Vierendeel box truss with plan dimensions of 15.8 × 16.7 ft, center-to-center. The columns are circular steel tubes that taper from 4.6-ft diameter, with a wall thickness of about 3.9 in. at the base, to 2.6-ft diameter and 1.6-in. wall thickness at the top.

COMPOSITE AND MIXED STEEL–CONCRETE BUILDINGS

The integral interaction of reinforced concrete and steel is found not only in the popular composite metal deck and floor framing systems, but can be seen on a much larger scale as well. It is not the composite action of the structure members—the slabs, beams, and columns—that is of interest here, but rather the combination or interaction of these members, which are blended into a single structure system. Typical *composite building types* that have developed over the last decade or so are

- composite framed tubes
- composite steel frames
- composite panel-braced steel frames
- composite interior core-braced systems
- composite megaframes
- hybrid composite structures

Concrete-encased steel frames have been used since the beginning of this century, although full composite action was only allowed in the late 1950s for floor framing systems using mechanical shear connectors. Composite construction takes advantage of the positive characteristics of both steel and concrete. The strength, light weight, speed of construction, and the flexibility of interior layout of steel are combined with the low material costs (which includes a steel reduction of around 50 percent), stiffness, larger mass for damping, fire resistance and moldability of concrete. Composite construction methods have helped to liberate the building form, according to the current trends, from its dependence on the pure, rectilinear organization of structure.

Recently, *mixed steel–concrete buildings* have also become popular. In a way, mixed construction has always existed in steel buildings by using concrete slabs and foundations, although the concrete played a secondary role. Combining major structure components of concrete, steel, or composite buildings is, however, a relatively new development. For example, it may now be economical to place a steel building on top of a concrete building or vice versa; alternatively, a central concrete core may be slip-formed to a predetermined height and then the steel frame built around it. Naturally, the steel frame can be erected first to a certain number of levels and then the interior core frame can be encased in concrete. Usually, the concrete is used for the vertical building elements and slabs, while the floor framing is steel. It is apparent that, in mixed steel–concrete construction, many practical combinations are possible.

Before studying composite and mixed building systems, the common composite member types are briefly reviewed by referring to the respective sections in this book.

Composite Members

To achieve full composite action between concrete and steel members, the bonding of the two materials must be such that no slippage occurs between them, so that they can behave as a new interlocking material similar to reinforced concrete. Composite behavior is automatically achieved for bending members (beams, beam-columns) when the steel section is fully and sufficiently embedded in concrete. However, when the steel is only partially encased, or when the members are only touching each other, then mechanical shear connectors are required. The purpose of this new composite material is to improve the strength and stiffness, among other criteria. The typical composite types are:

- Composite columns (see the subsection in Chapter 4 on "Concrete Columns")
  concrete-filled tubular columns, where pipes provide erection steel forming, as well as vertical reinforcing and horizontal ties
  concrete-encased steel columns: reinforced concrete columns, or only simple fireproofing—the steel core columns are small when used only for erection purposes.
- Composite beams (see the section in Chapter 4 on "Composite Beams")
  concrete-encased steel beams
  steel beams connected to the floor slab or metal deck with shear connectors
- Composite T-beams, castellated beams, trusses, girders, and open-web steel joists (see the section in Chapter 5 on "Floor Framing Systems")
- Stub-girders (see fig. 5.6g)
- Composite slabs (see the subsection in Chapter 5 on "Composite One-Way Slab Systems")
  composite metal deck
  precast concrete joists or thin slabs
  precast concrete floor panels
- Composite formwork cladding
- Composite shear walls (e.g., cores): see discussion below

Composite Building Structure Systems

In this exploration of the new composite building structure systems, the following organization may be used:

Composite Framed Tubes

- concrete-encased steel frame
- composite columns of rolled steel shapes embedded in reinforced concrete columns with steel spandrels
- composite columns with concrete spandrels
- concrete-filled tubular steel columns with steel spandrels

In composite tube construction of the concrete-encased steel frame type, the exterior composite framed tube is combined with simple gravity steel framing in the interior.

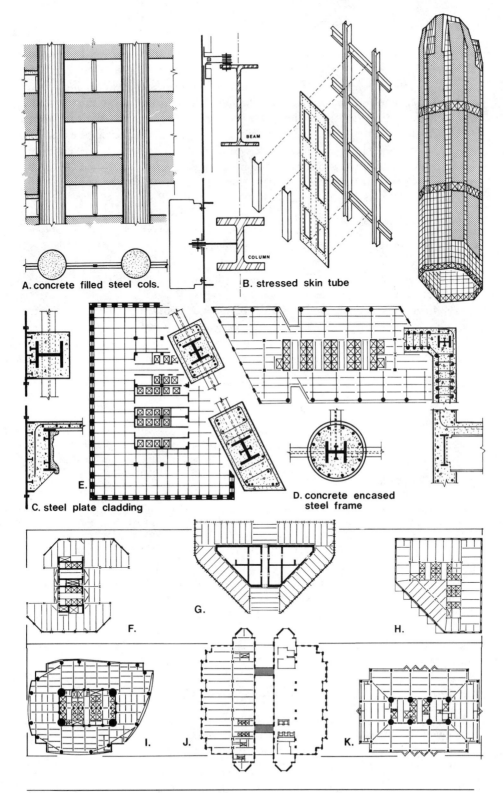

A. concrete filled steel cols.

B. stressed skin tube

C. steel plate cladding

D. concrete encased steel frame

E.

BEAM

COLUMN

F.

G.

H.

I.

J.

K.

Figure 7.38. *Composite building structures.*

The construction of the steel framing begins first and advances to a predetermined number of stories, often about ten stories ahead of the concrete construction. Usually, the steel framing is designed for gravity loading only, thus resulting in small column sizes; it must, however, be stabilized with temporary cable bracing and possibly with welded frame action during erection. When the floor structure has been finished, prefabricated reinforcing cages, and forms around the cages, are placed for the exterior columns and spandrels, so that the concrete can be cast. Permanent precast concrete forms for the facade may also serve as the final building skin. It may be concluded that, in this type of construction, the simple steel frame is laterally stabilized and stiffened by the exterior cast-in-place concrete punched-wall tube.

Among the first composite tubes is the 51-story One Shell Square Tower [1972] in New Orleans (Fig. 7.38e). This 130 × 180 ft, 697-ft high framed perimeter tube of the concrete-encased steel frame type uses 4-ft wide columns spaced 10 ft on center. Another typical design example is the composite tube of the 58-story, 775-ft high, First National Plaza [1981] in Chicago (Fig. 7.38h), with an exterior column spacing of 15 ft. Houston's 75-story Texas Commerce Tower [1981] (see figs. 5.7b and 7.33a) is a ruptured composite tube building with perimeter composite columns on four sides and a composite shear core behind the clear span of the fifth side. The potential adaptability of composite construction to shaping a building may be indicated by some of the cases in Figure 7.33, such as by the composite perimeter tube of Figure 7.33c, the composite tube-in-tube of Figure 7.33e, and the composite bundled tube of Figure 7.33d.

Composite Steel Frames

- concrete-encased steel frames
- composite columns of rolled steel shapes embedded in reinforced concrete columns with steel beams
- concrete-filled tubular steel columns with steel girders
- unencased composite steel frames: steel columns with composite beams and reinforcement in the negative moment areas.

An example of this type is a proposal, by Alexander Tarics of Reid & Tarics Associates in San Francisco, for a 50-story building with a composite exterior steel frame consisting of 5-ft diameter concrete-filled tubular steel columns spaced at 15 ft on center with the flanges of the steel spandrel girders protruding through the columns without interruption, so that only the girder webs have to be connected to the steel tubes (Fig. 7.38a). Deep, 6-ft girders are needed where lateral load resistance of the frame is required, while only shallow girders are used where the exterior frame carries only gravity loads.

Composite Panel-braced Steel Frames

- precast reinforced concrete shear wall panels
- slotted precast reinforced concrete shear wall panels
- steel-clad composite systems
- stressed-steel skin attached to steel tube

The bracing of steel skeletons with precast reinforced concrete infill panels has already been mentioned in the subsection in this chapter on "Tube-in-Tube" (see also fig. 7.33, center right). These panels can be placed in the interior core area or along the exterior as facade panels, or they can be arranged in a trusslike pattern to stiffen the frame. Mies van der Rohe was one of the first architects to introduce steel cladding as a facade treatment. In his 860-880 Lake Shore Drive Apartment Buildings in Chicago, he used 5/16-in. thick painted steel plates to cover the concrete that fire-proofed the steel frame. When the cladding is bonded with shear studs to the reinforced concrete (Fig. 7.38c), however, it serves not only as a protective skin and decoration, but it also provides structural resistance. Since the steel cladding is not fireproofed, building codes only allow it to carry lateral forces and not gravity loads. For the discussion of the stressed-skin principle, refer to the subsection in this chapter on "Perforated Shell Tubes—Framed Tubes" (Fig. 7.38b).

Composite Interior Core-braced Systems

- composite shear core walls with encased steel framing
- concrete-filled steel pipes in the core area (e.g., jumbo column scheme)
- steel frame with diagonal braces, web plates, or vertical truss webs embedded in concrete
- composite hybrid core (e.g., where steel trusses act as webs and are linked to concrete shear wall flanges)

An example of this type of construction is the 44-story Tower 49 [1985] in New York (Fig. 7.38f). This faceted crystal-like structure consists of a composite core assisted by channel-like exterior rigid steel frames on two opposite faces, while the core alone resists the lateral forces parallel to its long walls. Above the thirty-fifth floor, the composite core is discontinued and replaced by rigid frames. In this building, the steel frame was erected first, and subsequently the concrete shear walls were infilled.

Seattle's 58-story, 720-ft high Two-Union Square Building (Fig. 7.38i) is supported by four massive 10-ft diameter core columns and fourteen widely spaced perimeter columns ranging in diameter from 3 to 4 ft at grade level. In this case, the core resists the lateral forces and carries about 40 percent of the gravity loads. The jumbo core columns are connected to a braced moment-resisting frame; diagonal braces connect the core with the perimeter from the thirty-fifth to thirty-eighth floors. The composite concrete columns possess no conventional reinforcing, but rather a permanent 5/8-in. thick steel shell is lined on the inside with shear studs. The steel pipe columns are filled with 19,000 psi concrete, currently the strongest concrete mix ever produced in construction. The steel erection was two stories ahead of the concrete pumping. The floor structure consists of composite girders and beams, together with a composite deck. The 44-story Pacific First Center in Seattle (Fig. 7.38k) uses a similar structure system. In this instance, eight 7.5-ft core pipe columns and perimeter columns of 2.5-ft maximum diameter were used, both filled with 19,000 psi concrete.

Composite Megaframe

The huge legs of a megaframe form framed tubes, which may be in concrete or composite construction. These superlegs may be linked by steel trusses, and the interior frame may be all steel (see fig. 7.36). An example of a composite megatruss structure is the 1209-ft high Bank of China Building in Hong Kong, currently the world's tallest composite structure. It uses composite supercolumns at the four corners and giant steel box diagonals filled with concrete (see fig. 7.36, and the discussion in the section in this chapter on "Megastructures").

Hybrid Composite Structures

See the discussion in the following subsection, on "Mixed Steel–Concrete Building Construction Systems".

Mixed Steel–Concrete Building Construction Systems

Mixed steel–concrete building construction has become popular with the development of complex geometric forms and the corresponding necessity for hybrid structures. Sophisticated computer software allows the engineer to analyze structures never before possible, and to come up with the most economical solution for the construction of a building; this includes a wide selection and many possible combinations of major building elements borrowed from steel, concrete, and composite buildings. Vertically mixed systems may result from mixed-use buildings, where structure systems may alter with occupancies. For example, flat plate concrete construction for the upper residential portion of a building may change to long-span steel framing for the lower office levels. Typical examples of mixed steel-concrete construction, realizing that some may overlap with the composite structure systems, are:

- slip-formed concrete core surrounded by hinged or rigid steel frame
- all poured-in-place concrete up to a certain level, then only steel above
- all-steel structure up to certain level because of column sizes, then changing to composite construction
- composite framed tube (or concrete tube), plus interior steel framing
- composite shear walls, plus composite frames
- cast-in-place concrete building with steel floor framing

Typical examples range from the 60-story Momentum Place in Dallas (see fig. 7.33, top right) to the 52-story, 730-ft high Eau Claire Estate Tower [1983] in Calgary, Canada (Fig. 7.38g). In the latter case, the interior concrete core, together with the exterior perimeter rigid steel frame (with a typical column spacing of 10 ft along the folded faces), resist the lateral forces; the two structure systems are connected by composite trusses. This example probably best indicates the potential complex, hybrid nature of building construction. The 131 × 240 ft, 49-story First City Tower [1981] in Houston (Fig. 7.38d) is a parallelogram in plan with four 11-story staggered notches cut into the long faces, which rupture the continuity of the perimeter structure, therefore requiring a complex hybrid structure. In this 660-ft high mixed steel-concrete building, the lateral forces are resisted primarily by composite shear walls in the core, and secondarily by composite frames along the short sides at the building ends. The composite concrete core changes to rigid frame construction in the upper stories. The composite frame consists of composite columns of rolled steel shapes embedded in reinforced concrete columns and moment-connected steel beams. The composite columns along the broad face take only gravity loads. A composite stub girder system was used for the floor structure. The composite columns and shear walls were constructed 10 to 12 floors behind the steel frame.

The 42-story Cityplace Tower [1988] in Dallas (Fig. 7.38j) consists of two concrete tubes with highly articulated exterior walls and two projecting buttresses at each end.

The tubes stand side-by-side, 30 ft apart, and are joined at every fifth floor by a full atrium floor, and at other levels by two bridges, to form seven stacked five-story high atriums. The twin poured-in-place concrete tubes contain interior steel framing and are joined with steel. At the lower levels, most of the lateral forces are resisted primarily by the buttresses in shear wall action, while at the upper levels they are resisted by the tubes in frame action.

HYBRID STRUCTURES

The trend away from pure structure systems towards hybrid solutions, as expressed in geometry, material, structure layout, and building use, has already been emphasized in the previous sections. Interactive computer-aided design ideally makes a team approach to design and construction possible, allowing the designer to stay abreast of new construction technology at an early design stage. Due to the trend toward further automation and rationalization of the building process, the construction manager must already be involved in the project planning at that early stage. The design of complex structures not only includes efficient structure systems and construction techniques (including management), but also building intelligence for energy applications and the financial strategies of the developer, among other criteria. The recent development of complex building shapes with many setbacks, elaborate tops, and rich facade modulations (possibly with deep notches), in addition to multiuse spaces and interior atriums at various levels, has resulted in sophisticated structures to fit these complex spatial geometries; computerization has allowed engineers to cope with these novel situations. Naturally, it could be argued that a better understanding of structure at the early design stage, rather than using geometry for its own sake, would result in less conflicts and awkward solutions for many building cases.

In the search for more efficient structural solutions, especially for very tall buildings, a new generation of systems has developed with the aid of sophisticated computers and software, which have, in turn, an exciting potential for architectural expression. The new structures do not necessarily follow the traditional classification used in the previous sections of this text. Currently, the selection of a structure system, as based on the basic variables of material and the type and location of structure, is no longer a simple choice between a limited number of possibilities. Mathematical modeling with computers has made mixed construction (which may vary with building height) possible, thus allowing nearly endless possibilities that could not have been imagined only a few years ago. The computer software simulates the effectiveness of a support system, so that the structure layout can be optimized and nonessential members can be eliminated to obtain the stiffest structure with a minimum amount of material. Naturally, other design considerations besides structure will have to be included, but the design concepts can be tested quickly and efficiently by the computer program.

The discussion of some hybrid structures in Figures 7.39 and 7.40 attempts to identify the potential of new construction techniques. The sole intention here is to emphasize the hybrid character of a building as expressed by:

- the massing of a building
- the mixed construction
- the complex structure
- the effect of setbacks

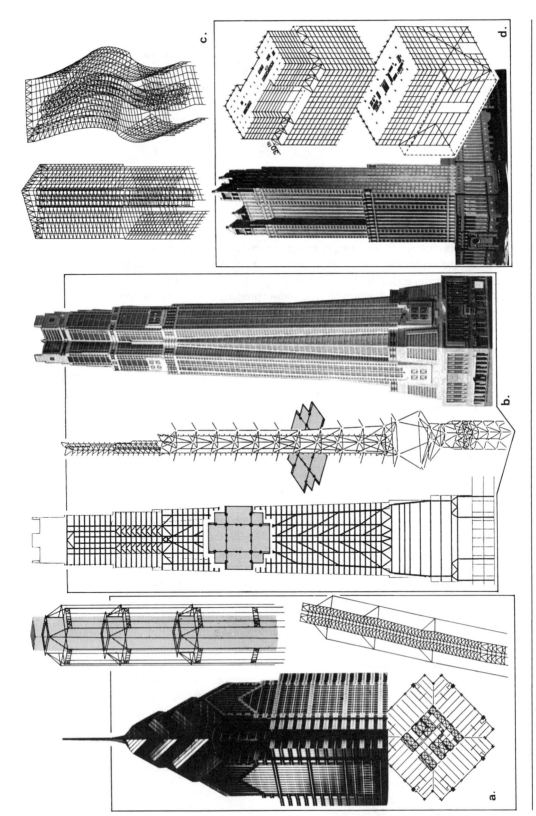

Figure 7.39. *Hybrid structures.*

An unusual condition exists for the Wilshire Finance Building [1986] in Los Angeles, where a 21-story triangular office tower sits atop an 11-story rectangular 176 × 150 ft parking structure (Fig. 7.39c). In this case, due to the radical change of mass and stiffness of the irregular building, the behavior during an earthquake, as well as under wind torsion, is most critical. The designers selected rigid perimeter frames for both shapes, supplemented by common rigid core frames. The core acts in a similar manner to a spinal chord to tie the wedge to the box below, as well as to tune and control the dynamic response, as is convincingly expressed by the computer drawing in Figure 7.39c for one of the deformed modes.

The 66-story 900 North Michigan Building [1989] in Chicago (Fig. 7.39d) is of multiuse and mixed construction, where a 36-story concrete building rests on top of a steel building at the thirtieth floor. The 30-story steel structure consists of a perimeter-framed tube with a column spacing of 15 ft starting at the ninth floor, while the structure on top for the residential portion is composed of two C-shaped partial framed tubes linked by frames, as shown in Figure 7.39d. The concrete construction and flat plate floor structure were found to be most economical for the 15 × 20 ft bay sizes. At the transition level, the closer interior column spacing of the concrete section is accommodated by a system of transfer girders.

LeMessurier proposed, for the 71-story 383 Madison Avenue Building in New York (Fig. 7.39b), a similar structure as for the Bank of the Southwest in Houston (see fig. 7.36). In this case also, the primary superstructure has a Greek cross configuration in plan; it is confined to the interior and crosses the building to define the service core, rather than being located along the periphery. The supertrusses resist wind, and the superdiagonals transfer gravity loads to the supercolumns at every eighth floor. In this structure, reinforced concrete will be used for the critical first ten stories above grade with clear spans, while steel is employed above and also below grade. The tapered 1040-ft high tower is of a tripartite vertical organization. At the transition from the middle to the top, at about two-thirds up the building, a shoulder truss acts as a platform to support the approximately 20 stories above and to tie the top together with the structure below.

The 61-story One Liberty Place [1987] in Philadelphia (Fig. 7.39a) is an example where computer optimization of the structure layout has obtained a most economical steel system, with 23 psf. The 165-ft square, 945-ft high tower is laterally stabilized by three sets of eight 4-story high superdiagonal outriggers, which cantilever from each face of the 70-ft square braced core to connect to single exterior columns. To control the uplift forces due to wind, most of the building weight had to be concentrated on the corner core columns and the outside outrigger columns. To achieve this, exterior trusses at three levels, spanning between the outrigger columns, had to collect and transfer the dead load to them.

The setbacks of building profiles, and possibly the cantilevering of a large portion of a building over an adjacent landmark building, have generated some unusual structure problems. New loading situations are created at building setbacks where vertical columns must be terminated. The columns are either picked up by a standard deep beam system, or the columns are sloped, and thereby generate lateral thrust forces from one plane to the next. Typical transfer systems (with some shown in Fig. 7.40) are:

- Individual floor beams cantilevering from interior columns
- Story-high Vierendeel trusses cantilevering from primary interior columns

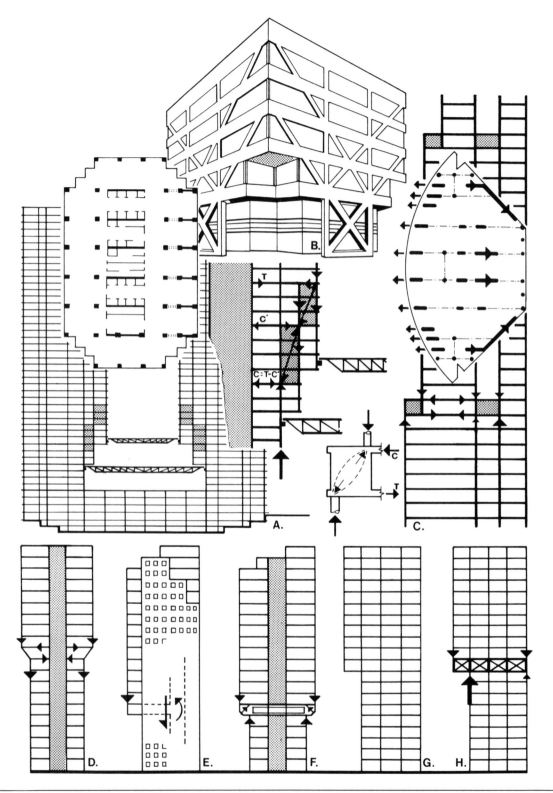

Figure 7.40. *Setbacks and overhangs.*

- An entire multistory rigid frame or huge Vierendeel truss cantilevering (Fig. 7.40g)
- A building cantilevering from a core or cores
- Cantilever steel trusses (Fig. 7.40h), girders, or reinforced deep concrete beams are used to support the overhanging building portion above
- By prestressing the tendons in a deep concrete beam, the columns are lifted upward (Fig. 7.40f)
- The exterior column is continued as an inclined column, supported laterally by the top beam as a tension tie and the bottom beam as a compression strut (Fig. 7.40d)

The exterior load-bearing braced steel frame walls of Figure 7.40b allow cantilever action by stepping backward at the corner to allow an inverted entryway.

The 37-story Centrust Tower [1986] in Miami (Fig. 7.40c) rests on top of an 11-story parking garage. It is a hybrid reinforced-concrete structure consisting of two C-shaped partial tubes, with 15-ft column spacing, which are linked by frames. Critical to the structural design were the three 15-ft setbacks along the circular wall. Conventional transfer girders were used at the top stepback, while for the others, eccentric one-sided panel brackets transfer the column loads from above, diagonally through the solid bracket, directly to the outside columns below. This shifting of the vertical loads causes rotation of the bracket, which is restrained by the upper floor in compression and by the lower floor in tension, and which therefore had to be prestressed. The thrust forces in the floor diaphragms are, in turn, laterally stabilized by interior and exterior vertical shear panels at the transition floor levels.

The twin 44-story office towers for the Chicago Mercantile Exchange [1984] (Fig. 7.40a) are linked near the base by 150-ft long, 9.5-ft deep steel roof trusses, and below by 175-ft long 14-ft deep floor trusses, to cover the column-free trading halls. The upper 30 stories of the two concrete towers are each cantilevered 32 ft over the main trading floor. A unique load transfer system was developed by Shankar Nair, the chief structural engineer of Alfred Benesch and Co. The exterior column loads, including the hanger columns that support the huge roof trusses for the trading hall, are transferred diagonally through seven-story high diaphragm walls, or shear panels, to large interior columns. The rotation developed by the cantilever action is resisted axially by the floors, which are anchored to the interior elevator core shear walls. In other words, eight floors act as lateral stabilizers for the transfer brackets; the upper floors act as tension ties and the lower floors as compression struts to transmit the thrust forces to the shear walls. Since significant bending is introduced in the shear walls, the towers were built out of plumb with the floors cambered to compensate for the horizontal deflection.

UNCONVENTIONAL BUILDING STRUCTURES

As illustrated in Figure 7.41, there are many structure types other than those covered in the previous sections. Some are experimental in nature, demonstrating fascinating new structural concepts, possibly motivated by more encompassing design philosophies. Quite often, these concepts relate to structures in nature or have evolved out of biology in the search for innovative solutions. On the one hand, they may directly reflect the cell structures of plant organisms or the uniform repetitive order of crystals,

Figure 7.41. *Unconventional building structures.*

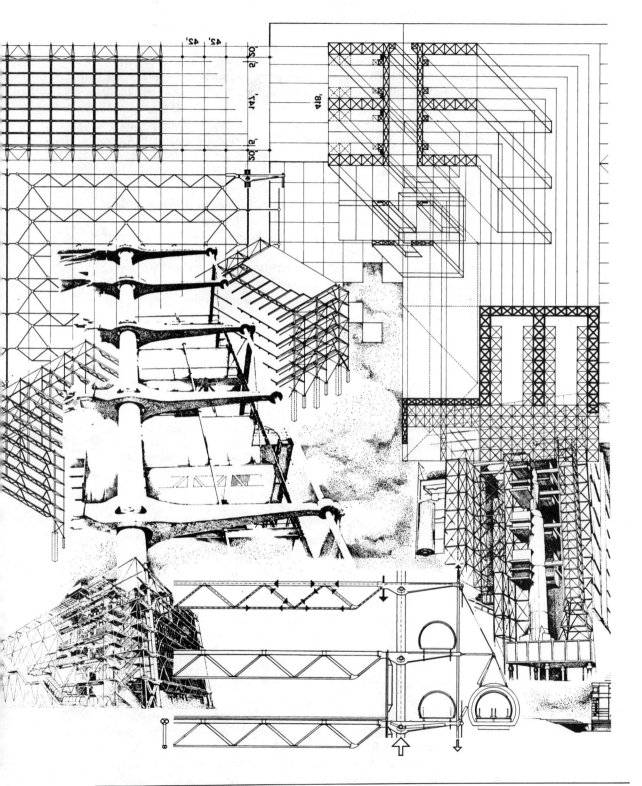

Figure 7.41. *(continued)*

to inspire the structure of space frames and the packing of polyhedra. On the other hand, they may represent visionary schemes, such as biogenetic models with self-generating environments, where designs initiate processes, such as chemical reactions in seawater causing coral-like growth, to enable the structure to build itself. Quite in contrast are Paolo Soleri's Arcologies, his utopian megacities of the 1960s, resembling organisms of gigantic scale and possessing amazing geometric complexity. Soleri sees Arcology as architectural ecology. He started work in 1970 on the desert city Arcosanti in Arizona, which will eventually house 2500 people. Soleri relies primarily on traditional ways and low-tech methods of construction for the curvilinear biological forms of concrete arches, vaults, and apses.

While Soleri believed that function follows form, Archigram represented, in contrast, responsive dynamic environments in their mechanized cities of the early 1960s, which celebrated the poetry of advanced industrialization. These hardware environments describe machine or robotlike organisms, which integrate various fundamental design concepts. Elements such as responsive kinetic structures, kits of parts, plug-in and clip-on capsule housing, variable and adaptable geometry, as expressed with space frames and polyhedra, as well as movement (including mechanical services) were all explored. This spirit is reflected by Peter Cook's Montreal Tower project of 1964 for Expo 67. To the central concrete shaft of this treelike 600-ft high tower, with gigantic exposed roots, he attached various entertainment modules and temporary exhibition elements.

The Pompidou Center [1977] in Paris (see fig. 7.4, right) is a realization of the spirit of the 1960s, as reflected by many of the Archigram schemes. The architects Piano and Rogers perceived the building as a huge adaptable enclosure that seems to be held up by adjustable scaffolding. All of the structure components are exposed, clearly expressing their structural responsibilities; the joints are especially articulated, and the innovative use of cast steel has opened new opportunities for the design of buildings. The structure is divided by parallel cross-frames into thirteen 42-ft wide bays. The typical six-story frame consists of 8-ft deep 147-ft span Warren truss beams, which are suspended from small cantilever beams or *gerberettes*. The gerberettes are small cast-steel beams pin connected to the cast-steel tubular columns. Their short arms cantilever about 5 ft from the columns to the interior in order to carry the trusses, while their long outriggers extend 20 ft to the outer face, where they are tied down by vertical tension rods. The gerberettes minimize eccentric loading under gravity, by placing the columns in compression and the vertical ties in tension. The system was also used in order to reduce the live load deflection of the long-span trusses, thus allowing their depth to be kept to a minimum. The gerberettes were named after the German bridge engineer Heinrich Gerber who, in response to settlement problems, pioneered the hinged cantilever beam system in 1866. The long-span Warren trusses of the Pompidou Center consist of continuous double tube compression chords at the top and continuous double solid round bars for the tension chords at the bottom. The diagonal members alternate as single compression tubes and solid round tension bars; they are connected by cast-steel joints. Standard rolled steel beams, spaced at about 10 ft on center, span between the trusses and support the concrete floor panels in composite action. The glass facade runs behind the columns and gerberettes. The building is laterally stabilized in the cross-direction by braced frame action along the short end facades, by linking the horizontal trusses with three pairs of tubular diagonal struts at the ends and midspan. Lateral stability in the longitudinal direction is provided along the two long outer faces by the post-tensioned diagonal steel rod bracing placed between the vertical ties, which are pretensioned by gravity. The horizontal forces are transmitted from the floor diaphragms to the outer horizontal tubes that connect the ends of the gerberettes via horizontal bracing located at alternate floor levels.

The following cases derive their structure systems from the tensegrity principle, space frames, box construction, or pneumatics.

Manfredi Nicoletti proposed, in the early 1970s, a 200-ft high helicoidal skyscraper, where each blade of the helix is shaped like a sail. The warped shape of the three sails is highly complex and seems to offer a minimum resistance to wind. The helicoidal form not only generates horizontal drag forces in the two major directions but also vertical lift forces. Nicoletti used the tensegrity principle for the tower structure by separating the primary compression and tension stresses. The three central cylindrical steel shafts are connected by bridges to form a rigid megaframe. This vertical compression superstrut system supports a series of diagonal stay-cables along the external surfaces of the sails, and wind brace cables along the exterior edges of the wings (i.e., exterior corners of each floor). The double-layer space frame floors are suspended from the stay-cables and held in compression against the cores by the stay and wind brace cables.

Carlo Moretti's proposal for a 50-story residential tower [1976] consists of a central cylindrical concrete utility shaft, to which (at every floor level) three independent apartment tubes are anchored, to form a large triangle in plan. These floor-tubes are of elliptic cross section and composed of identical upper and lower shells. They act as huge struts stabilized by steel cables along the edges, which wind in a screw-like fashion up to the top of the building, where they are anchored to core outrigger arms.

One group of designers has been experimenting with the spatial truss or space frame concept. This type of structure is more rigid than conventional cubical structures, since (because of the triangulation of space) the members respond directly to the loading, rather than in bending, and thus also require less material. In this case, the structure is formed by spatial frameworks or prestressed tension-compression networks, which enclose polyhedral chains; the formation of the geometric space can be visualized as the stacking of polyhedra. On the large scale of a superbuilding, the space frame consists of basic multistory megamodules of some polyhedral configuration; the long-span horizontal members and diagonal columns of the megamodules each also form space frames, which contain the actual habitable spaces.

In 1957, Louis Kahn, together with Anne Griswold Tyng, proposed the mega-spaceframe concept for a 616-ft high building. The primary floors are part of the space frame and are supported by 66-ft high tetrahedrons. At the nodal points where the sloped columns intersect, 11-ft deep capitals house the mechanical services. The height of the intermediate floors can be varied. Professor Francois Gabriel of Syracuse University proposed, in 1981, a basic building block for a housing cluster, in the form of a hexagonal prism; it is made of hexagonal floor caps at top and bottom, connected by six diagonal columns. These individual units are stacked to form a vertical space truss.

The Vehicle Assembly Building [1966] at Cape Kennedy, Florida, is one of the largest buildings in the world in terms of enclosed volume; it is as tall as a 50-story skyscraper. It is essentially a huge box big enough to shelter the assembly of the rockets, which are more than 30 stories high; it is so large that clouds sometimes form indoors and it rains. The building consists of two sets of three trussed towers, anchored to the ground with piles to withstand uplift forces, and arranged in plan as two E's placed back to back and tied together by the roof structure.

Rather than using a spatial framework, the same geometric subdivision can be achieved by stacking polyhedra on top of each other. The more common use of simple rectangular box units is just a special case of the infinite number of space-filling cellular systems. One of the most famous examples of this type of construction is Moshe Safdie's Habitat [1967] in Montreal, Canada (see fig. 1.19, top right). Prefabricated box units can also be plugged into a frame structure (see fig. 2.7a) or they can be clipped onto vertical utility shafts. Kisho Kurokawa's 15-story Nakagin Capsule Tower [1972] Tokyo (Fig. 7.41, center), consists of two steel-framed concrete cores 21 ft apart, with each one nearly 17 ft square. From the twin shafts, 175 ft and 150 ft high, cantilever 140 factory-produced capsules, which act independently of each other. The vertical and horizontal clearances between them are 12 in. and 8 in., respectively. Each capsule is 8 ft high, 8 ft wide, and 13 ft long and weighs nearly 4 tons without furnishings. The primary structure of the capsule consists of an all-welded steel-truss box. Each capsule is a self-contained one-man unit. It is attached to the core at four points (that is, four seats), at the corners: two steel-plate box seats at the lower corners, and two I-beams cantilevering 6 in. from the wall at the upper corners.

In the past, several people have investigated high-rise pneumatics. Jean-Paul Jungmann, in 1967, proposed a pressurized cellular construction for housing clusters, where the wall membranes consist of air members (that is, high-pressure tubes), and the floors of inflated mats. The bulbous multilevel aggregate of pneumatic cells is held up and stabilized by the internal low-air pressure.

In the early 1970s, Professor Jens G. Pohl of California Polytechnic State University at San Luis Obispo actually tested multistory air-supported buildings from a technical and functional point of view. It is interesting to investigate an eight-story inflated tube, so as to develop a rough idea of the forces involved. For a typical floor load of 140 psf, an internal pressure of 1 psi = 144 psf above atmospheric pressure would be required to support each floor in a building, hence 8 psi for all the floors; an internal pressure of 14 psi is considered a maximum for the human comfort range. This pressure of 8 psi should be increased by about 10 percent because of the reduction in strength of the membrane attributable to the slenderness ratio. For an eight-story tubular building of 50-ft diameter, the following hoop forces are generated by the internal pressure

$$T = pR = 1.1(8)(25 \times 12) = 2640 \text{ lb/in. of height}$$

For a membrane thickness of 0.13 in., the tensile stresses in the skin are

$$f_t = T/A = 2.64/(1 \times 0.13) = 20.31 \text{ ksi}$$

Using a safety factor of 2.2 yields a required material capacity of

$$f_{ut} = 2.2(20.31) = 44.68 \text{ ksi}$$

It is apparent that this high tensile strength requires the use of an external cable net, which supports the membrane.

It may be concluded from this discussion of a few cases of unconventional building structures that there is no limit to the creation of new structure systems, and that the imagination and ingenuity of designers knows no boundaries.

PROBLEMS

7.1 In a 12-story, core-supported building (see fig. 7.9), six sets of continuous cables are draped over the core to form twelve hangers carrying the perimeter weight of all twelve floors. At the roof level, a further set of cables is used to carry the additional forces caused by the roof weight and the change of direction of the cables. The core crown consists of two diagonal cross arches to support the corner cables. The main roof beams are in compression because they hold the main cables apart from the core, which is, in turn, stiffened by a compression ring. Assume the following loading:

typical floor weight:
$1 \frac{1}{2}$-in. composite deck with $2 \frac{1}{2}$-in. concrete + beams

spaced at 12 ft o.c.	35 psf
ceiling	7 psf
mechanical	5 psf
miscellaneous	3 psf
	50 psf

curtain wall, hangers and spandrels: (per wall area)	15 psf
wind load	24 psf
live load (office space):	80 psf

a. Determine whether the core has to carry tension stresses at plaza level. Assume the roof weight to be double the floor weight. Further, the loads that the core contains (e.g., elevators, utilities, heating, air conditioning equipment, washrooms, lobbies, and service rooms) are assumed equal to floor loads in this approximate approach. Neglect the increase of bending stresses due to lateral displacement of the core and any openings in the core walls that reduce the core strength.

b. Determine approximately whether the 10-in. concrete core walls are adequate in resisting the loading. Assume 5500 psi concrete with an allowable compressive stress of $0.22 f'_c = 1210$ psi, neglecting buckling criteria.

c. Determine the maximum elongation of a typical facade cable (not a corner cable), and also find the extension at the fifth floor level. Check whether the 2–2 5/8 in. ϕ galvanized bridge strands are all right. Each cable has the following properties: breaking strength 417 tons, 4.13 sq. in., modulus of elasticity $E = 23,000$ ksi after prestretching.

d. Determine the increase in the maximum elongation of the cable in (c), if the cable size is to be increased at every fourth floor level, corresponding to the load increase.

7.2 Check the tensile condition in the core for the building in Problem 7.1, if the facade hangers are replaced by columns that carry the gravity loads directly to the ground, but do not resist any lateral forces.

7.3 The two steel catenaries of the Federal Reserve Building in Minneapolis (see fig. 7.8) are suspended from 200-ft high piers, resembling a suspended bridge. They have a span of 273 ft, a rise of 150 ft, and are assumed to each support a load of 30 k/ft. Determine the approximate cross-sectional area of the ca-tenaries, consisting of the primary 3-ft deep welded wide flange members and the cables, assuming an average allowable tensile stress of 60 ksi, which takes into account control of flexibility.

7.4   Check the assumed K-factor of 1.8 at the fourth floor level in Example 7.2 by using the alignment chart in the AISC Manual.

7.5   Determine the size of a typical interior column at the first floor level for the building in Example 7.2. Use A36 steel.

7.6   Determine at which level the combined loading case starts to control the design of the floor girders in Example 7.2. Further check to discover what case does control the design of an interior W14 column at the top floor.

7.7   A 25-story rigid-frame concrete building forms a horizontal grid of 20 × 25 ft bays; five 20-ft spaces are in the short direction and seven 25-ft spaces in the long direction. The typical floor height is 12 ft, except the first floor and the top three floors, which are 15 ft high. A typical interior column at the first floor is to be investigated. An average dead load of 165 psf per floor may be assumed, which includes the heavier weight of the two upper mechanical floors; the partition load is 20 psf. The typical office live load is 80 psf, which may be reduced by 60 percent; the roof live load is 20 psf. The wind pressure for this preliminary investigation may be taken as 30 psf. Ignore the secondary effects due to building sway. Determine the preliminary column size by using 5000 psi concrete and Grade 60 steel.

7.8   Investigate the structural members of a typical interior frame at the first floor level in Problem 3.15 (see fig. 3.13c). Remember, this simple framing system carries only gravity loading; the floor live load is 50 psf. Use A36 steel.

7.9   A three-story rigid frame building consists of three 18-ft bays in the short direction and, in the long direction, of four 21.5-ft bays; assume each floor to be 14 ft high. Consider the filler beams to span parallel to the long direction and to be spaced 6 ft apart.

Live loads:
1st floor (retail occupancy):	100 psf
2nd & 3rd flr. (office use):	80 psf
roof:	40 psf

Wind loads
assume average pressure:	20 psf

Dead loads:
for typical floor and roof:
4 in. concrete slab:	48 psf
3 in. cinderfill:	15 psf
1 in. cement finish:	12 psf
suspended ceiling (metal lath and gypsum board):	10 psf
	85 psf
filler beams with fireproofing:	25 psf
	110 psf

Design a typical filler beam. Investigate one of the interior frames that is the exterior girder at the second floor level and the interior column at the first floor level. Consider the wind and gravity loading cases. Assume the girder weight as 200 lbs/ft, which includes the concrete fireproofing, and assume the column weight with fire protection to be equivalent to 10 psf of floor area. Use A36 steel, and assume $K_y = 1.3$ for the column design.

7.10 Since the building in Problem 7.9 is relatively low, fixed columns with the beams hinged to them may be employed as a structural system. Using the same loading and material as in Problem 7.9, investigate a typical filler beam and girder for the office space, and an interior column at the first floor level.

7.11 The 26-story office building for which the loads were determined in Problem 3.17 (see fig. 3.13d) is stabilized by four rigid frames across the short direction. Frame B will be investigated in the following exercises; assume that all four frames have the same stiffness.

  a. Select open web steel joists from the J-series, as based on 2-ft spacing.

  b. Set-up the moment diagrams for girders and columns due to gravity loading. Find the gravity load for the exterior column at the first floor level.

  c. Determine the shears, axial forces, and moments in the exterior column and outer girder at the first floor level, as caused by a wind pressure of 30 psf.

  d. Design the exterior column and outer bay girder at the first floor level using 50 ksi steel. Think about the impact of lateral force action, as related to building height.

7.12 Determine the moments in the columns of the building in Problem 7.11 on the first floor level due to wind action should the building be hinged to the base.

7.13 Design the interior column and the central bay girder at the thirteenth floor level (as shown in fig. 3.13d). Consider both gravity and wind action. Use A36 steel for columns, otherwise use data from Problem 7.11.

7.14 Derive the approximate lateral deflection for a rigid frame building by considering only the web drift (Equation 7.15c).

7.15 Do a more precise investigation of the force flow for the Vierendeel truss in Example 7.4. Determine the approximate chord and web column sizes adjacent to the exterior column supports. Use Grade 50 steel.

7.16 Show that the wind does not control the design of the column in Example 7.4. Further, determine how much the building laterally sways under the given wind loading of 21 psf.

7.17 A 40-ft span, 9-ft deep four-bay Vierendeel truss supports a building column at center span that transfers a load of 200 k; assume uniform bays of 10 ft each. Determine the approximate force flow using inflection points at mid-length of members.

7.18 In a 24-story apartment building, 8 ft, 8 in.-deep Pratt trusses span 60 ft; they are spaced 50 ft apart and are arranged in a staggered pattern at alternate floors. Along the central corridor, the trusses have a 6-ft wide Vierendeel panel, while the trussed panels are 9 ft wide (as shown in fig. 7.22). Assume the following loading:

floor weight:

8" precast hollow core concrete planks with 2" topping:	72 psf
steel framing:	8 psf
ceiling, flooring, mechanical, etc.:	5 psf
partitions:	<u>20 psf</u>
	105 psf

Consider the roof weight equal to the floor weight.

facade weight:

Assume average weight for windows and lightweight concrete block:	50 psf

live loads:

apartments:	40 psf
corridors:	100 psf
roof:	20 psf

Do a preliminary design of the major structural members (floor slab, trusses, columns) as based on gravity action only, which is reasonable for a staggered-truss structure. Use A572 Grade 50 steel for the columns and truss chords, but A36 steel for the remainder.

7.19 Investigate the staggered truss building of Problem 7.18 at its critical second floor level by considering a uniform wind pressure of 24 psf. Check the top truss chord and column, and think about the floor slab behavior.

7.20 Replace the K-bracing of the building in Example 7.6 by X-bracing (see fig. 7.26B). Find the truss member sizes at the first story level and determine the approximate lateral building sway. Compare the results of the two bracing systems. Use A36 steel.

7.21 Replace the K-bracing in Example 7.6 by full story knee braces (see fig. 7.26C). Estimate the member sizes and compare the results with the ones in Example 7.6. Use A36 steel.

7.22 A five-story flat slab concrete building is laterally braced by shear walls. It consists of five 24-ft spans in the long direction and three 20-ft spans across the short direction. The first story height is 15 ft, and the other floors are 12 ft high. Assume the following loading conditions; 122 psf dead load and 20 psf live load for the roof, and 136 psf dead load (which includes 20 psf partition load and 10 psf ceiling load) and a live load of 50 psf for the floors; consider the column weight as equivalent to 8 psf of floor area. Determine the reinforcing for a 14 × 14 in. typical interior column at the first floor level. Use Grade 60 steel and 4000 psi concrete.

7.23 Check whether the columns in Problem 7.22 can be considered braced across the short building direction, where two 8-in. thick, 8-ft long shear walls are located along each of the exterior frames. Assume that the eight inside columns are 14 × 14 in. and the twelve exterior columns are 14 × 12 in. Determine also the minimum length of the walls.

7.24 Assume the flat slab building in Problem 7.22 to be unbraced, so that the frame must resist the 25 psf lateral wind pressure. Determine the column size, and discuss how much reinforcing you may approximately provide.

7.25 Assume that the five-story flat slab building in Problem 7.22 is laterally braced by two interior channel-shaped concrete cores located at each end of the building. For a uniform wind pressure of 25 psf against the long facade, determine whether the 20-ft long 8-in. thick concrete walls (i.e., webs of channel cores) with a minimum amount of reinforcing are satisfactory, using 4000 psi concrete and 60 ksi steel. Consider the lateral forces to be resisted entirely by the walls, and the slab-column framing to carry only gravity loads; this assumption should be reasonable for this relatively low-rise building.

7.26 The 110-story twin World Trade Center towers form 208-ft square hollow tubes with 80 × 138 ft central cores, and cantilever 1350 ft. Determine the approximate axial forces in the perimeter columns at the base, using a building weight of 12 pcf for the typical perimeter columns and a wind pressure of 45 psf; assume the entire wind moment to be resisted by the tubular flanges only. At the street level, 80 columns, spaced at about 10 ft, support the tower; 300 columns are used for typical floors. Also estimate the lateral deflection, assuming a cross-sectional column area of $A_c = 408$ in.^{2}.

7.27 The double-tapered truncated pyramid of the 100-story John Hancock Center (1091 ft) in Chicago has an approximate base dimension of 164 × 262 ft tapering upward to 100 × 160 ft at the top floor level. The exterior column-diagonal tube resists all of the lateral loads, while the core only carries gravity loads; it is assumed that the core carries roughly 50 percent of the entire building gravity load. The diagonals tie the 24 perimeter columns together, that is, 7 columns in each face. Investigate a typical facade column on the long leeward side at the building base. Assume a dead load of 165 psf, a reduced live load of 50 psf for the perimeter loading, and a uniform wind load of 45 psf, taking into account the efficiency of the tapered building form. Use $F_y = 50$ ksi and ignore the effect of the building shape.

7.28 The 1454-ft Sears Tower in Chicago consists of nine 75-ft square bundled tubes (see figs. 5.4m and 7.33) up to the fiftieth floor, that is, the building is 225 × 225 ft square at the base. Stepbacks of the tubes occur at floors 50, 66, and 90. The columns are built-up wide-flange sections, which are spaced around each modular tube at 15 ft on center. With the exception of the larger corner module columns, which attract more load because of wind shear lag, the typical columns measure 39 in. from flange to flange and are 24 to 30 in. wide. Do a quick estimate of the typical column size (A36) at the base to resist a gravity load of 120 psf and an average wind pressure of 50 psf, assuming an equivalent height of 100 floors for wind pressure analysis.

7.29 A 20-story, 136 × 136 ft square, framed-tube structure with a 40 ft central core is 250 ft high. The columns are spaced 8 ft apart, and the average story height is 12 ft, 6 in. The floor framing is changed in direction at alternate floors (see fig. 7.34). Use the following loading conditions: 47 psf for light-weight concrete on steel deck for floors and roof, 20 psf partition dead load, 5 psf for ceilings, 5 psf miscellaneous, 30 psf for curtain wall, 50 psf live load for office space and 20 psf for the roof, and finally 40 psf wind load. Determine the size of a typical perimeter column, but not a corner or girder column, at the building base, using A36 steel for drift control purposes. Assume a column width of 14 in. and a girder depth of 3 ft.

7.30 Investigate the moment restraint, M, of the head outrigger system and the column ties in Equation 7.23.

8 OTHER VERTICAL, ARCHITECTURE-RELATED STRUCTURES

There are many other structures of the scale of high-rise buildings. They are not only land-bound, occurring above or below the ground surface, but also may be found in deep water and in outer space. Among the many examples are a 550-ton mobile crane used for the construction of Munich's Olympic Stadium, which stood 530 ft tall and had a maximum horizontal reach of almost 300 ft. A transmission line tower in Germany is 750 ft high, and a cooling tower is slightly over one-half the height of the Eiffel Tower. The giant Bullwinkle Platform (in the Gulf of Mexico) is, at 1615 ft, 365 taller than the Empire State Building, while a concrete chimney in Sudbury, Canada, is as tall as the Manhattan tower. Currently, the world's tallest structure is a guyed transmission tower in Warsaw, Poland, at 2108 ft. The world's deepest mine is 17,600 ft (Western Deep, South Africa), like fourteen Empire State Buildings stacked up on top of each other. Eventually, giant space satellites will dwarf the Earth's skyscrapers.

All of the current tallest buildings have been discussed elsewhere in this book; they are identified in Figure 8.1. The race for height started with the Eiffel Tower [1889] in Paris, and ended with the Empire State Building [1931] in New York, representing the second generation of skyscrapers. The third generation reached its high point with the world's currently tallest building, the Sears Tower [1974] in Chicago. All of the ten tallest buildings are constructed in steel or mixed construction. Several proposals for super-high-rises, in the 150-story range, are on the drawing boards. It seems that this fourth generation of tall buildings will not necessarily appear only in the United States.

The following brief excursion into tower, underground, hydrospace, and aerospace construction presents a new frontier, and should be valuable for developing a better understanding of the behavior of tall buildings.

TOWERS

Up to the last century, towers had been of heavy stone or masonry construction; they were square, round, or of some polygonal plan form. They were either free-standing or part of a building—be it a church, monastery, city hall, or palace—or they were strengthening walls that encircled cities and castles, possibly located at corners and

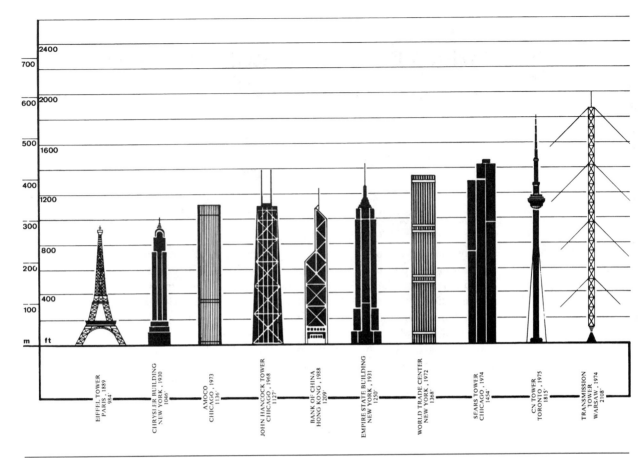

Figure 8.1. *The tallest building structures.*

gates. Towers served for observation or communication, but most often they were built for purposes of defence. Later, towers developed symbolic meanings besides their social, religious or functional purpose. Towers like Big Ben in London, the Eiffel Tower in Paris, the Washington Monument, the Leaning Tower of Pisa, the Gateway Arch in St. Louis, and possibly the Watts Towers in Los Angeles, just to name a few, have become landmarks of their cities.

The first historically known high-rise structures were probably the man-made mountainlike temples of the Sumerians in Mesopotamia (Ur, c. 3000 B.C.), which are similar in appearance to the stepped pyramids of Egypt and the Indian temple pyramids in Mexico, or the monumental terraced Buddhist temple-mountains in India. These so-called ziggurats may be considered as the forerunners of tall tower structures. Legendary is the "Tower of Babel" [c. 600 B.C.], which was rebuilt by the Neo-Babylonian king Nebuchadnezzar. The tapered brick structure supposedly had a height of about 300 ft, with a square base of nearly 300 ft with ramps winding around the tower towards the top platform in a rectangular spiral-like fashion. Famous also is the lighthouse at Pharos, near Alexandria, Egypt, one of the Seven Wonders of the Ancient World—this huge, 450-ft high masonry tower was built circa 280 B.C., during the Ptolemaic era.

Not only are the campaniles, the free-standing bell towers, a symbol of many Italian cities, but also the medieval tower houses with the appearance of castle towers, which were built for defensive purposes but also symbolized the power of the ruling families. Famous are the towers of San Giminiano in Tuscany, and the towers of Bologna, where (for instance) the Asinelli Tower [c. 1119] is 319 ft high and 29 ft square in plan. Ireland's ancient, sturdy, round stone towers, with hardly any openings, were built in conjunction with monasteries. They rise up to about 130 ft to their conical belfries, and are only up to 19 ft in diameter. They gave refuge to the monks when they were attacked by the Vikings. Among other tower structures of the past are the Egyptian obelisks; the lofty muslim minarets, with their exterior cantilevering platforms at the top from which the call to prayer was and continues to be made; and the Buddhist pagodas in China, Japan, Burma, etc.

In contrast to the many massive towers of the early Romanesque churches and abbeys, often part of fortifications, are the lighter and richly ornamented high towers of the Gothic era. The upper portions of the towers were left open to rain and wind, and the masonry spires often were constructed in a delicate skeletal fashion. The 466-ft high tower of Strasbourg Cathedral was the tallest structure of the Middle Ages and remained so for 450 years, until it was surpassed in 1880 by the tower of Cologne Cathedral at 515 ft (and in 1890 by the 528-ft high tower of Ulm Cathedral). In 1884, the 555-ft Washington Monument became the tallest structure (and still is the tallest load-bearing masonry structure in the world; see also Problem 2.6), and finally in 1889, the Eiffel Tower, the first truely modern high-rise structure, reached the unbelievable height of 984 ft. The transition from masonry bearing walls to the new generation of lightweight tower structures of this century was made possible by the development of the materials steel and reinforced concrete in the context of the changing political and social structure of society. The Eiffel Tower, built for the 1889 Paris Exhibition, may be considered to symbolize the beginning of this new era.

It was not until the 1950s, however, with the progress in the communication industry, that record heights were achieved. The 1572-ft high guyed Oklahoma City Television Tower, consisting of a latticed triangular steel mast, was erected in 1954. The steel towers of that period were contrasted by the concrete towers. In 1953, with the 710-ft high Stuttgart TV Tower in the Federal Republic of Germany, the eminent structural engineer Fritz Leonhardt pioneered the construction of slender self-supporting concrete cantilever towers. It served as a model for the TV tower wave in Europe, and set the basis for the record heights of Toronto's 1815-ft CN Tower in 1976, Moscow's 1762-ft Ostankino Tower, East Berlin's 1187-ft tower, and Frankfurt's 1086-ft communication tower in 1978. A concrete chimney built in 1970 at Sudbury, Ontario, is 1250 ft high (as tall as the Empire State Building); it tapers from its 116-ft base to a 52-ft diameter at the top. The world's highest structures are the guyed masts in Warsaw [1974] (Fig. 8.2), Poland, at 2108 ft, and in Fargo, North Dakota, at 2063 ft.

In contrast to the towers of the past, which were generally of solid shaft construction using stone or masonry, the towers of today may take any shape and can reach immense heights with a minimum of material because of new construction techniques and due to the strength and continuity of steel and reinforced concrete. The heavy weight *bearing wall gravity towers* have been widely replaced by light skeleton and shell construction for the building of the tall slender *cantilever towers*. The towers of today perform many more functions than the ones of the past, as can only be suggested by the cases in Figure 8.2. They may form support structures for buildings, bridges, offshore platforms and tanks, or they may simply carry floodlights, advertisement signs, flags, electrical transmission lines, or radio and television antennas. They may also be free-standing, for example: diving towers, ski jump towers, hoisting shafts for underground mining, drill towers, grain silos, cooling towers, cranes, chimneys, air traffic control towers, observation towers, elevator towers, air shafts for exhaust fumes, exhibition towers, a leaning tower supporting a tent roof over a stadium, and landmark towers such as the Statue of Liberty in New York.

The shape of the bridge towers in Figure 8.2 depends on how they are used as supports for the roadway. The tower of the cable-stayed Friedrich-Ebert Bridge in Bonn, FRG, is a single cantilever in response to the fan-shaped single plane cable arrangement. The tower of the suspended Humber Bridge in England, however, forms a transverse single-bay rigid frame. On the other hand, a twin trapezoidal box girder deck bridge is resting on two tower-like legs surging up from the canyon floor, which are connected by tie beams to provide portal frame action.

Tower structures may be organized as follows:

- Free-standing towers
- Guyed towers

Free-standing towers are self-supporting. Depending on their slenderness, they act as

- Gravity towers: stocky towers
- Cantilever towers: slender towers (e.g., chimneys, poles, communication towers)

They may be concrete towers, trussed steel towers, rigid frames, or metal poles. The common structure types are

- Closed tubes: single-cell and multicell towers
- Open tubes: single-cell or multicell towers
- Legged towers: three-legged or four-legged, possibly forming multiple shaft towers
- Bundled walls/frames with independent action of vertical structural planes
- Tree-form towers
- Experimental tower structures

In guyed towers, such as tall antenna structures, the poles are extremely slender, since the wind is not resisted by bending of the material as for cantilever towers, but by the couple formed between guys and mast or between the guys. Basic behavioral concepts such as stability, stressing under gravity and lateral force action, slenderness, flexibility, and dynamic properties have already been briefly introduced in the subsections in Chapter 2 on "Introduction to Buildings as Support Structures" and "Basic Concepts of Structural Language" (see also fig. 2.1).

The structural design of slender free-standing towers is based primarily on wind considerations, and consequently on the overall shape of the tower. For very tall towers, the circular or polygonal tapered tubular shafts seem to be well streamlined, since they provide the least resistance to varying wind directions and provide a high torsional stiffness. It should be noted that the wind distribution obviously is not constant, and the suctions perpendicular to the direction of the wind are larger than the direct pressures (see fig. 3.6). The solidity ratio of open trussed steel towers indicates

Figure 8.2. *Towers.*

the relation of surface area of exposed steel to envelope area, and thus the magnitude of wind action.

The slender tower acts as a vertical cantilever in bending to guide the wind forces to the ground (see fig. 2.1a), where the profile may be a curvilinear taper, possibly taking the efficient form of an exponential function (see Equation 2.7). In contrast to the free-standing tower, where the section modulus of the shaft provides the bending resistance, for the guyed mast the rotational resistance is provided by the cables. While the static wind pressure is the basis for stability considerations, the design of a slender flexible tower must also incorporate dynamic loading criteria (see fig. 2.1e), since turbulent wind gusts, and possibly seismic ground accelerations, are applied abruptly and change rapidly. In addition, the wind (with its longer periods) may come close to the natural period of a flexible structure, thereby causing resonant loading, where the oscillations and the corresponding lateral forces build up, depending on how much damping is present in the system.

The tower not only bends due to lateral force action but also due to eccentric gravity loading, such as live loads placed asymmetrically upon the head and platforms (see fig. 2.1b). Similarly, temperature differences between the sunlit and the shaded areas cause rotation of the shaft away from the sun. When a flexible tower sways excessively, additional moments are generated by gravity, especially the tower head (i.e., P-Δ effect). This is similar to the wheat head catching the wind and swaying the slender stalk, where the ear of grain causes additional bending due to its eccentric position (Fig. 8.3). In this case, the cross section of the grass blade responds to the intensity of lateral force flow: it is a tube at the base, V-shaped above, and flat and adjustable at the tip. If the grass blade would be flat over the full height, it could obviously not even support its own weight.

It may be concluded that stiffness must be considered in addition to stability, strength, and dynamic properties. Especially when the comfort of people at elevated restaurants and observation platforms must be considered, or when sensitive equipment is supported, then deflection and acceleration control is necessary. Towers must provide a certain amount of rigidity; they cannot deflect as structures in nature, which only want to provide a minimum resistance to lateral force action.

Various tower structure types are identified in Figure 8.3. They range from tapered free-standing shafts to laterally supported guyed masts, or from stocky solid gravity towers that seem only to support themselves, to light antigravity columns, and to slender cantilevers, possibly with large heads. Towers may be relatively stiff or quite flexible, such as guyed stacks, cabled trussed towers, tensegrity masts, or high-pressure pneumatic tubes.

The Viennese artist Walter Kaitna was fascinated by the visual equilibrium of force and deflection. He constructed a sculpture in 1962 by anchoring several vertical steel bars of different length in a base plate according to some pattern. Then he bent and twisted the bars so he could force them into position to unite them at the top, where they keep each other in balance. Now the bars are in a stressed state, transmitting the power of the locked-in forces. One senses the thrust for release, but which can only happen if they are disconnected, so they can snap back and straighten into their natural position.

The slenderness of a building partially dictates whether gravity and static loading, or lateral force action and possibly dynamic criteria control the design. It is apparent that the dome towers of Freiburg (FRG) and Pisa (Italy) are examples of gravity

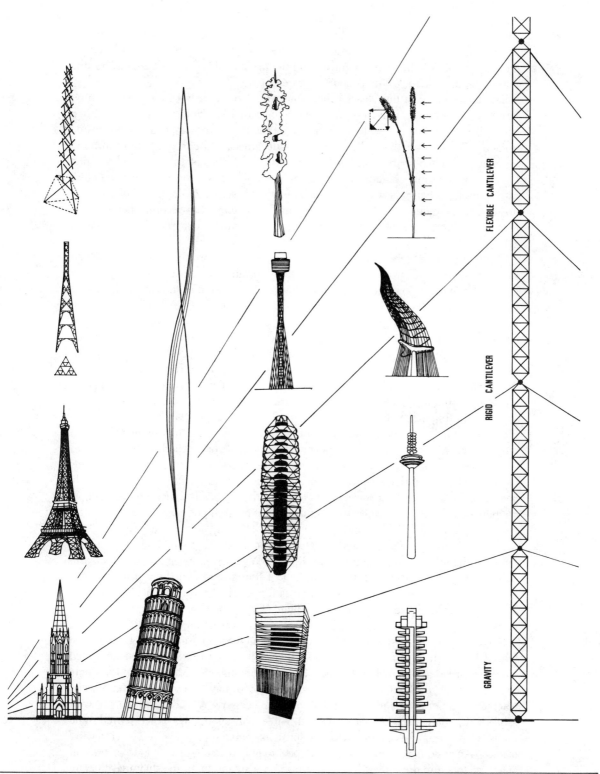

Figure 8.3. *Tower structures.*

structures, where the weight induces so much compression that it prestresses the building and thus suppresses tension due to lateral force rotation. In this instance, the resultant due to the vertical gravity and the horizontal wind loads falls within the middle third, or kern, of the supporting base (see fig. 2.1c).

The large spread of the 984-ft Eiffel Tower in Paris is clearly expressing stability. The weight of the wrought iron tower seems to easily balance the overturning due to wind pressure, which is kept to a minimum by the narrowing shape and open trussed surfaces. The structure basically consists of four latticed box corner columns forming exponential arched shapes, which lean against each other at the top, and are tied together by four square platforms at different levels. The tower form is nearly funicular with respect to a constant wind pressure. The four arches under the first platform have no structural function and are purely ornamental. Above the second platform, the four box columns are cross-braced to form a rigid truncated pyramid (see Problem 2.7).

Quite in contrast is the slender, 1086-ft Frankfurt telecommunication tower, which cantilevers out of the ground with the bending resistance hidden in the earth, similar to that of tree roots. The tubular concrete tower tapers from 65.6 ft at the base to 18.4 ft at the 968-ft mark, and its wall thickness decreases from 58.6 to 39.3 in. It supports an enormous 188-ft diameter head consisting of a six-story steel frame conical pod and seven cantilevered concrete platforms above. The head contains a revolving restaurant, an observation deck, antenna platforms, and a large operation floor for postal installations. Seattle's 600-ft "Space Needle" [1968] (see fig. 2.1d) is basically a steel tripod where three paired legs taper inward to a slim waist and then flare out to support a five-story building, including a revolving restaurant.

The 1815-ft high CN Tower [1976] (see figs 1.2 and 8.2, right) in Toronto, Canada, is currently the world's tallest self-supporting structure. It has a three-legged Y-shaped cross-section which tapers upward and supports a seven-story skypod acting as both an observation restaurant and service facilities for broadcasters; the concrete tower rests on an 18-ft thick Y-shaped post-tensioned cellular raft foundation. Fully vertically post-tensioning the slender tower adds greatly to the tower's life span and resulted in an increase of stiffness, because the member is not cracked and its entire cross section is available for resisting rotation.

Frank Lloyd Wright's 15-story laboratory tower structure [1950] (Fig. 8.3, bottom right) for the Johnson Wax Company in Racine, Wisconsin, appears as an abstraction of a tree with the central trunk growing out of the ground and the mushroom-type floors branching from it. Quite in contrast is the efficient form of the 140-ft square, 22-story tower in Figure 8.3 with a 70-ft square central core; the tower's post-tensioned framed cantilever floors are based on economy of construction.

Guyed masts (see figs. 2.1a and 8.3) are extremely slender and must therefore be braced by pretensioned cables; they are either hinged or fixed to the base and need ample ground space to anchor the guy tables. Depending on the tower height, a single cable set attached either at the top or below the top may be used for lower towers, whereas for taller towers multiple cable sets are necessary. A *three-cable system*, with cables 120° apart, or a *four-cable set* with cables 90° apart, may be employed. The masts act as continuous beams on elastic springlike cable supports with respect to wind loading and as compression struts with respect to gravity loads, as well as to pretension and wind forces in the cables. If the prestress forces are too high, the shaft may buckle, but will sway excessively if they are too low; the compromise results in a rather flexible guyed tower. It should be kept in mind that, due to prestressing of the cables, the tower is compressed and its stiffness is increased similar to the vertical prestressing of concrete towers. The cables of short towers remain basically straight when the masts deform, but for tall towers the cable sag is large and must be considered. Very slender structures are subject not only to static, but also to dynamic, wind action, causing additional flexure and torsion. The dynamic wind action causes the bending moments in a mast to change with different frequencies.

The 700-ft high Sydney tower [1971] in Sydney, Australia (Fig. 8.3), is laterally stabilized by a prestressed cable network that forms a hyperboloid of revolution, which is constructed as a ruled surface by straight diagonal cables. The 22-ft diameter steel shaft supports the head which contains two restaurants, observation decks, and a water tank which, together with hydraulic shock absorbers, is used as part of a damping system for wind gusts.

Robert Le Ricolais proposed a tower in the form of a funicular polygon of revolution in 1962 (Fig. 8.3). In this case, the cables form a spatial closed tensile network along the perimeter to give support to the floors, and are connected to the central compression spine; the funicular cables are rotated around the circular floor diaphragms. Stiffness and strength are provided by geometry with a minimum of weight. The architect Wolfgang Döring proposed, in the 1960s, the cable-trussing of open, flexible skeletons (see fig. 2.1d) by treating the columns and some of the floors as struts and using continuous cables to tie the building together, thereby providing the necessary lateral stiffness.

Designers have been intrigued by more dynamic support structures, in contrast to the rigidity provided by static one-material systems as in conventional construction. The human body is an example of a sophisticated tower-like structure, where the hinged unstable bone skeleton in compression is stabilized by the ligaments along the joints and the skeletal muscle and tendons in tension, allowing body movement. Frei Otto, the famous designer of tensile structures, in his search for optimizing movement and weight, studied the human spine, which consists of multiple articulated vertebrae (i.e., vertebral column) and is held together and upright by muscles and ligaments. In his *vertebrae mast* of 1963, he expresses free spatial movement in response to random pressure, together with the antigravity character of the composite action of tension and compression in this kinetic sculpture. Similar action can be found in cranes and derricks, or in the guying principle (as in the rigging of sailing ships).

Kenneth Snelson explored this spatial interaction of tension and compression in his *tensegrity masts*, such as his famous tapered tensile cable and tube sculpture, the Needle Tower [1968]. The term *tensegrity* was established by Buckminster Fuller as derived from tensional integrity. In contrast to conventional building structures, where the solid members form a continuous rigid skeleton, in tensegrity systems volume is defined by a tensile network of continuous cables or skins prestressed into shape by the individual, discontinuous compression struts, which never touch each other.

Typical foundations for towers have been shown (see fig. 2.1g). For smaller towers and chimneys they form flat slabs, while for legged towers individual footings are used, which may have to be anchored to resist uplift. For tall towers, shell foundations, such as single and combined truncated cones and other shells of revolution (with or without ribs), possibly together with ring footings, are employed. Depending on the bearing capacity of the soil, additional pile or caisson foundations may be required. It must be kept in mind that guy foundations must resist uplift and horizontal shear.

Quite in contrast to the typical flat root construction, Frank Lloyd Wright used, for the Johnson Wax Laboratory Tower (Fig. 8.3), a deep root approach, which is typical for pole construction.

UNDERGROUND STRUCTURES

People have used the earth as shelter for many thousands of years, beginning with the earliest cave dwellers. They have been living below ground, such as in the subsurface towns in the loess belt of northern China. They have carved buildings from stone and hollowed out spaces (e.g., Buddhist rock-cut architecture in India, rock-cut monasteries in Ethiopia). They have carved out of soft rock entire villages or towns along mountain slopes, such as in Anatolia (Turkey), Sicily, Tunisia, and many other places. Rock architecture has been practiced in many countries in the past, particularly in India, Egypt, and Assyria, by the Greeks in Lycia, by the Persians, as well as by the Romans, Arabs, and others at Petra in the Arabian desert (located in Jordan). Famous are the honeycombed sandstone cliffs of the southwestern United States, built by the Pueblo indians.

The self-preservation instinct has lead countless animals to live underground, often in apartmentlike multilevel dwellings with at least two exits. Underground structures have always been used during wars for protection and defense purposes; this includes bombproof shelters for civilians and factories. In the 1930s, France built the huge fixed frontier fortifications along the German border, known as the Maginot Line. While the fixed fortifications of World War I had been semi-underground, the main nerve centers of the Maginot Line were deep in the earth and could only be reached by a system of tunnels and elevators. Although the extensive underground urbanlike multilevel complexes did represent a significant physical design accomplishment, and a first realization of the utopian dreams about subterranean living by Jules Verne in the second half of the nineteenth century, and later by H. G. Wells, they eventually proved worthless as a military endeavor. At the end of World War II, Goebbels' propaganda machine claimed that an underground army stationed in a subterranean fortress, with the help of powerful new weapons, manufactured in underground factories, would liberate Germany from the occupying forces.

In many countries, buildings are designed for a dual purpose of serving their peacetime function as well as operating as civil defense shelters during a war. Subsurface shelters against nuclear attack have been built by countries like the Soviet Union, China, Switzerland, Sweden, and Norway. Other examples of military subsurface structures include land-based missiles that are placed in buried heavily reinforced concrete silos to protect them from nearby nuclear explosions.

While the reason for underground living in the past was primarily because of the climate, defense, and cultural context, the extensive subsurface developments of today, and particularly during the last two decades or so, have been due to a variety of other intentions. Underground urbanization has become necessary because of surface congestion. It includes mass transportation systems, utility lines, storage of life-supporting material (e.g., food, water, oil, gas, and other goods), parking garages, and disposal facilities; commercial, industrial, institutional, and some other special occupancies are invading the urban underground space as well. The Japanese have not only been experimenting with artificial islands in the sea, but are also designing subterranean cities to relieve the country's space crunch. In the early part of the next century, they expect to build an underground network of huge vertical cylinders (containing elevators, power generators, and other energy supply systems) connected by tunnels to spheres (accommodating commercial, industrial, institutional, and other facilities) but keeping residential construction above ground. In these underground environments, plants can be grown, sunlight can be reflected in through vents, and temperature and humidity can be controlled.

Energy conservation has popularized the building of earth-sheltered homes of the berm or dug-out types near the surface. The usage of deep underground space has emerged because of the necessity for the storage of toxic and radioactive materials, for producing and storing thermal energy, for hydropower and nuclear power plants, and (as previously mentioned) for military or civil defense purposes.

The development of underground space is based on the expertise acquired in *mining* and *tunneling* during many centuries. Over 4000 years ago, engineers built a tunnel beneath the river Euphrates. More recently, tunnels have been built for highways, railways, shipping, sewerage, subways, fuel storage, power plants, public utilities, and so on. Construction breakthroughs have been achieved by the 33.5 mile Seikan Tunnel, currently the world's longest underwater corridor, which connects Japan's main island of Honshu with Hokkaido. The experience gained in mining is often closely related to that of coal mines, which may be 3000 to 4000 ft deep, using vertical air and elevator shafts as access systems. The complexity and sophistication of operation required for the giant South African gold mines more than 12,000 ft below surface, employing 15,000 people or so can only be imagined and require enormous refrigeration plants for cooling.

Underground construction may be organized according to how closely the structure is located to the surface:

- Near-surface construction is more of a horizontal spread type, and consists of a man-made roof or cover for the underground structure. The following structure types are included:
  Earth-sheltered buildings
  Basement-type spaces
  Cut-and-cover techniques (e.g., for transportation systems)
- Underground, multilevel surface construction expands vertically, but is accessible by inclined roadways and escalators, as well as carved out courtlike cavities. In this case, the conventional construction for basements, or cut-and-cover trenches, may be used. For greater depths, however, earth or rock may provide a natural roof structure, so that one of the following procedures is applicable:
  Tunneling (e.g., for mass transportation, storage, or commercial facilities)
  Possibly mining (e.g., room and pillar mining)
- Deep underground space is only accessible by vertical shafts with elevators or hoists, so that mining techniques must be applied.

Tunnels are either driven or bored deep underground, so they do not interfere with the surface activities, the building basements and the energy network, or they are placed directly below the streets, close to the surface, using the cut-and-cover method by digging an open trench and then filling the upper portion of the excavation to form a roof.

Various applications of subsurface construction are briefly investigated in the following discussion of Figure 8.4. At the center of the drawing, an impression of the complex network of underground structures beneath cities is given. It reflects a massive root system that keeps the urban organism alive by providing physical support with the pile and shallow slab foundations, and by supplying the life-supporting energy through a feeder system to the basements of the buildings. An intricate distribution system with branches of many different sizes and types such as pipes, ducts, tubes, and cables follow the streets and supply water, gas, steam, electricity, and communications to the buildings, and also take care of drainage and sewage. The magnitude of this underground utility system is directly proportional to the size of the city. For large cities, it becomes quite congested and may have to be placed together in deep multi-purpose tunnels; but here, in addition, elaborate tunnel systems for roads, railways, and subways are required. People are distributed through a labyrinth of multilayer tubes, along halls, passages, stairs, escalators, and elevators. The multilayer network of subway tubes is serviced by underground stations and vertical ventilation shafts.

The large multiuse basement spaces beneath many urban skyscrapers are well known, but the extensive underground pedestrian networks, for example, as found in Toronto and Montreal, are unique. The Toronto underground mall system, possibly the largest of its kind in North America, consists of a labyrinth of tunnels, escalators, subway stations, basement connections, retail shops, restaurants, cinemas, gardens with fountains, and much more. The large underground urban complex at Les Halles in the heart of Paris not only represents a huge transfer station and junction for the subterranean transportation networks of subways and roads (including stations for buses and subways as well as parking lots), but also provides shopping facilities, swimming pools, theaters, etc. Sweden has a long tradition for underground placement of factories, power plants, warehouses, and sewage treatment plants, along with shopping malls.

Unique in the United States are the underground spaces created throughout Kansas City, Missouri, as a result of ongoing limestone mining operations. Here the *room and pillar* method of mining is used, where the various chambers are excavated by leaving pillars between to support the roof (Fig. 8.4, bottom right). The huge caverns remaining allow commercial use, such as for manufacturing and storage. At depths of 50 to 200 ft below the ground, the temperature remains nearly constant at 57°F, while humidity is a moderate 50 percent.

The University of Minnesota adopted energy conservation requirements in the mid 1970s because of the harsh environment and the few indigenous energy resources available. The temperature range is from about 100°F in summer to −30°F in winter. The ground temperature at a depth of 8 ft, however, never falls below freezing, and at a depth of 26 ft remains nearly constant at 50°F all year round, a mere 18°F colder than that required for comfort. It is apparent that underground structures can take advantage of nearly constant temperatures, so that energy consumption for heating and cooling is drastically reduced in comparison with exposed surface buildings. The university's Underground Space Center has become one of the leading groups in the United States to develop guidelines for earth-sheltered buildings. Williamson Hall [1977] at the Minneapolis campus, a courtyard-type building with three levels below grade, is an early example of the application of the principles, not to homes, but to broader use.

Other examples in Figure 8.4 refer to deep underground structures. For instance, underground storage of combustible, explosive, or poisonous material can be done in abandoned mines, salt caverns, and other mined spaces, such as horizontal tunnel-

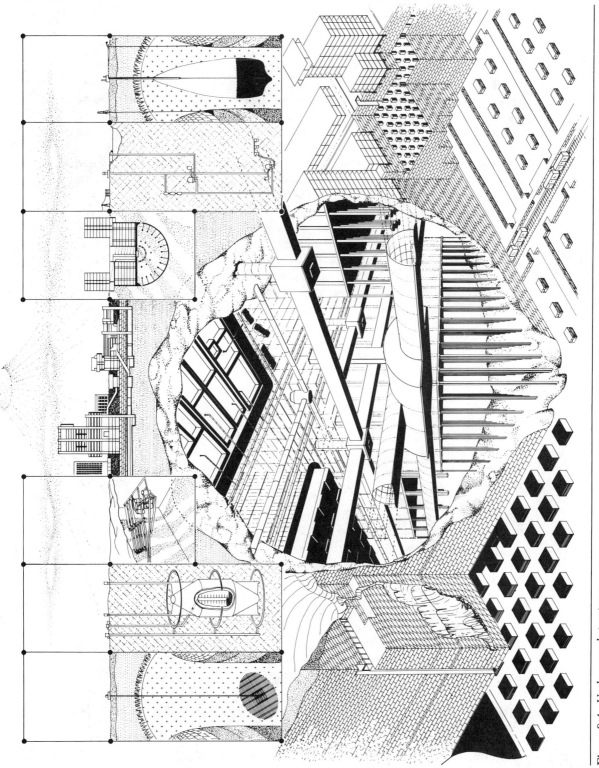

Figure 8.4. *Underground structures.*

shaped caverns, or in a multitank system with vertical rock caverns. Huge underground salt dome caverns are used as a storage reserve for crude oil (Fig. 8.4, top, right); here, borehole mining is commonly applied with water as the solvent. Mined underground chambers, as based on a room and pillar design, with a network of several thousand feet of tunnels, for instance, at 500 ft depth, are used to store liquefied petroleum gas (Fig. 8.4, bottom, left).

It makes sense to isolate radioactive waste deep underground in mined rooms or salt caverns, so that it does not become a threat to human health. A repository in a salt dome is shown in Figure 8.4 (top, left) with a central pillar that provides protection for several vertical shafts for access, ventilation, rock hoisting, and waste handling. An underground repository for reactor waste from nuclear power plants (Fig. 8.4, top, third from the left) may consist of various rock chambers which contain only waste with low activity content and are linked by a tunnel system, and of several concrete silos for storage of the majority of the more hazardous waste.

Underground hydroelectric and pumped storage powerplants discharge into manifold storage chambers. Energy is generated at peak times when the water flows through turbines to the lower reservoir in the cavern complex, and then spare electrical energy is used during off-peak periods to pump water from the lower reservoir to an upper reservoir at ground level. Shown in Figure 8.4 (top, second from the right) is a section through a proposed power installation for underground pumped hydro storage using two powerplants, one housing one-half the pumping and generating capacity at a depth of about 1800 ft and the other half at 3600 ft.

OFFSHORE ARCHITECTURE

People have always been living on water either in their houseboats, as in the Orient or in houses built on piles. The floating villages of the Urus on Lake Titicaca at the Peru-Bolivia border are supported by 6 to 7 ft thick mats made of totora reeds. In the past, water has provided a natural protective shield against invaders, as exemplified by the pre-Columbian Mexico City and Venice, both cities resting on piles above the water. Military pontoon bridges were used as early as 483 B.C. by the Persian king Xerxes during a war with Greece. However, it was not until the energy crisis of the early 1970s, which forced the oil industry to drill offshore oil wells into deep and hostile waters, that some of the largest man-made structures were built.

A huge underwater steel oil-storage tank [1972] at Dubai stands in 200 ft of Persian Gulf water; it weighs 15,000 tons and is as tall as a 20-story building, being 270 ft across its greatest diameter. The Bullwinkle Oil Platform in the Gulf of Mexico rises 1615 ft to the top of its drilling derricks. It is 161 ft higher than Chicago's Sears Tower, currently the world's tallest building. The Ninian Central Platform in the North Sea is a 560-ft high massive concrete gravity structure, weighing about 550,000 tons; quite a bit more than the dead weight of 325,000 tons for each of the World Trade Center towers in New York. The Statfjord B Platform off the coast of Norway, with a total displacement of 899,000 tons, is the heaviest object ever moved by man.

Today the oceans are no longer used solely for transport, fishing, or waste disposal, as in the past. To cope with land shortage, population growth, and increasing food demands, as well as dwindling energy resources on land, the oceans are being explored. They constitute a major biological and mineral resource; they can be harvested and the ocean beds can be mined. The fast rate of urban growth and the overcrowding of cities along coasts, including the lack of basic services they can provide, may force cities to expand onto the water to build artificial islands. The sea can be used as a site for self-sufficient communities and as a source of energy.

Marine structures may be above water, on water, under water, or on the seabed. They may be located in shallow or deep waters, and they may be fixed or movable. They may be rigid structures anchored directly to the oceanbed, or they may be floating. Marine structures are of many varieties, including landfill and dikes in shallow waters, pontoons for marinas, lighthouses, lightships, floating barge-like structures serving as bridges/buildings/airports/dry-docks, research platforms, floating or bottom-standing storage tanks, submarine stations, loading facilities, drilling vessels, oil rigs, etc. These ocean structures are made from steel, concrete, and prestressed concrete.

To gain some basic understanding for the behavior of offshore structures, it is essential to know how they respond to an extremely hostile environment and how they are constructed. The magnitude and type of loads that marine structures must resist depend on the locality, the depth of water, and the size (as well as type) of structure. It is clear that fixed structures have to withstand larger forces than floating ones. Permanent loads consist of gravity loading (dead load) and external water pressure. Other gravity loads include the live loads (e.g., ballast, deck loads). Most critical are the wind, and especially the wave impact loads that develop during a storm, with the exception of the U.S. West Coast, where earthquake action controls. Other loads to be considered are the current forces (e.g., tidal current, wind-driven current, river currents), impact loads (e.g., collision: ship, sea ice, iceberg; crane loads, docking impact, mooring forces), ice forces, earthquakes (tsunamis: long-period waves), fatigue loads, ocean bottom slides, thermal loads, transportation/assembly loads, etc.

Hydrostatic pressure acts uniformly in all directions and is proportional to depth. For seawater at ordinary conditions, with a density of $\gamma = 64$ lb/ft^3, it is equal to

$$p_i = \gamma h_i = 64h_i \qquad (8.1)$$

The resultant of all forces on a submerged body is an upward force or buoyancy due to the increase of hydrostatic pressure with depth. This buoyant force is equal to the difference between the vertical resultant forces on top and bottom surfaces, A_1 and A_2, of the body, which is equal, in turn, to the weight of the fluid displaced (Fig. 8.5).

$$\Sigma F_y = 0 = \gamma h_1 A_1 + W - \gamma h_2 A_2 \qquad (8.2)$$

Letting $A_1 = A_2 = A$, for purposes of simplicity, yields the following buoyant force

$$P_B = \gamma(h_2 - h_1)A = \gamma V, \qquad \text{where: } (h_2 - h_1)A = \text{volume} = V$$

Therefore, a body will sink until its weight, W, is balanced by the buoyant force, P_B: $W = P_B$. Should the buoyant force be greater than the weight, it will float on the surface. However, if the weight is greater than the buoyant force, the object will sink to the bottom. From this follows *Archimedes' principle*, that a body wholly or partly immersed in a fluid is buoyed up with a force equal to the weight of the fluid displaced by the body.

Most critical is the hydrodynamic wave pressure, which is greatest at the surface but which diminishes with depth. When the period of intense wave loading is close to the natural period of the platform structure, dynamic analysis is required. Resonant

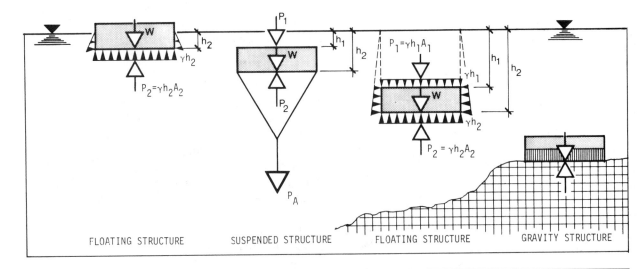

Figure 8.5. *Basic concepts related to marine structures.*

loading is generated by the many wave periods that happen to match the fundamental period of the structure. Although these forces may be small, they may add up over time and cause fatigue. Typical wave periods of 15 sec for ordinary braced steel platforms and 20 sec for gravity towers, as derived from a regular wave of about 100 ft height and 1300 ft length, are often used as critical for preliminary design purposes. In contrast, compliant structures have long natural periods of up to around 100 sec, hence well above the main wave periods. For example, a guyed tower with a natural period of 30 sec will resonate with fewer waves than a comparable fixed platform.

The ocean fiercely attacks anything standing in its way; the forces are by far more violent than the ones we are used to on land, with the exception of earthquakes. Stability of floating and fixed structures becomes of primary concern. The superstructures are positioned above the crest of the highest wave expected to come once in 100 years; this, for an example in the North Sea, is at least 80 ft.

In the following discussion, various proposed or built offshore structures are briefly investigated; some are shown in Figure 8.6. The futurists of the 1960s, in the search for more responsive urban and social forms, envisioned the ocean as having an exciting potential for new marine civilizations and habitats. Among the many designers of that era were Paul Maymont, who proposed earthquake-resistant floating towns for Japan. Buckminster Fuller and Shoji Sadao designed the floating "Triton City" [1968]. Stanley Tigerman's project "Instant City" is a floating town assembled from inverted tetrahedra. Rudolf Doernach and William Katavolos imagined self-built, floating, organic-like structures that grow through chemical reaction or like marine organisms. Some of the schemes of the Archigram group (1964) were derived from pure fantasy, such as Warren Chalk's "Underwater City" and Ron Heron's "Walking City on the Ocean." Edouard Albert and Jacques Cousteau proposed an artificial floating island [1966] in the Bay of Monaco, which has a plan form of a pentagon with a diameter of 722 ft. The tubular substructure is 164 ft high with roughly one-half of it under water (Fig. 8.6, top left).

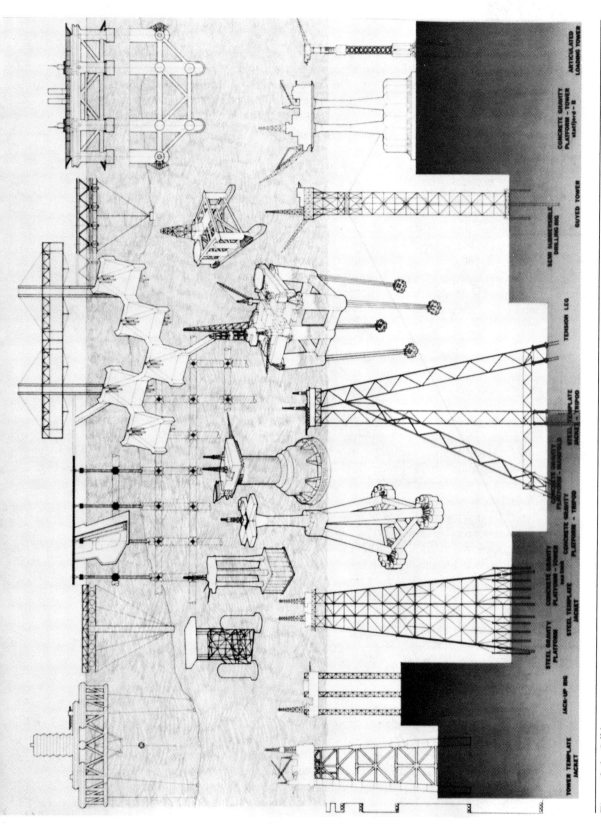

Figure 8.6. *Offshore structures.*

As early as in the late 1950s, Kiyonori Kikutake proposed several versions of a marine civilization concept. Finally, he realized his dream in 1975 with Aquapolis, a floating exhibition pavilion and the world's first self-contained aquatic city of this type. This huge floating and semisubmerged steel structure, which had to accommodate 2400 visitors at a time and which has an upper deck area alone of 2.5 acres, was designed for Okinawa's EXPO 1975 (Fig. 8.6, top right). Three square decks are supported by sixteen cylindrical columns, twelve of them 25 ft in diameter and four 10 ft in diameter. The columns are braced horizontally and diagonally by 10- and 6-ft diameter steel tubes to form an octagonal frame in plan view. They are resting on the substructure consisting of four huge hulls, 33 ft wide, 20 ft high, two of which are 341 ft long and two 184 ft long. Aquapolis is fixed and anchored in place by two chains from each of eight columns. The sixteen chains are fastened to 25-ft diameter reinforced concrete anchors sunk as much as 73 ft through bedrock. When a storm approaches, the structure is released from these anchors, towed further out and more water is let into the ballast tanks, so that the structure may draw as much as 66 ft and displace 28,000 tons, and the stability closely matches the sea conditions. It takes four hours to change from a semisubmerged draft of 66 ft to a normal display draft of about 20 ft, displacing approximately 19,000 tons. Kikutake has also proposed a much larger 150 acre, prestressed concrete version for 20,000 persons, in Tokyo Bay. The structure would consist of open-bottom prestressed concrete hulls $165 \times 660 \times 20$ ft high, where the buoyancy would be provided by air trapped under the hulls.

The structural engineer Lev Zetlin proposed, in the early 1970s, floating airports for deep waters using buoyant tubes to support the platform structure (Fig. 8.6, top, second from the left); in this case, the floating platforms are stabilized and supported by caissons and cables. For a concrete runway pier at LaGuardia Airport in New York, which is supported on steel piles, he proposed adding large diameter plastic air tubes to generate uplift, so that heavier airplanes can be supported (Fig. 8.6, top, third from the left). In another project (Fig. 8.6, top, second from the right), Lev Zetlin uses natural buoyant forces by prestressing the submerged platform with cables to the ground as well as floating it. Thomas R. Kuesel proposed, in 1985, a cable-stayed bridge for deep water crossings, where each of the towers would be founded not on a pier, but on a huge circular pontoon, which would be held in place by vertical and diagonal cables anchored to the bottom.

In 1982, a Singapore architect was commissioned to design a 50-story hotel that would float on a 6.5 acre concrete barge anchored to the seabed. A seven-story hotel was floated in 1987 from a Singapore shipyard to be anchored on a lagoon on Australia's Great Barrier Reef; the 9800-ton structure covers 322,000 ft² of the lagoon. The five-story superstructure is supported by a 292-ft long, 90-ft wide, and 19-ft deep steel barge, which contains two floor levels. The steel barge has a double skin, where the space between the side walls and bottom form water-tight tanks, which may be used to hold water as ballast. Similar to the anchoring of oil supertankers offshore, the floating hotel uses a single-point mooring system; the mooring can be adjusted to conform with changing wind patterns.

In Lake Ontario, off Toronto, five pavilions [1971] built on stilts are linked by bridges to form artificial islands (Fig. 8.6, center top). These units, 90-ft square in plan, are each supported by a central core consisting of four tubular corner columns that cantilever out of the water to carry the cable-stayed trusses at the top, from which are suspended vertical hangers that support the floor framing, together with diagonal cables.

Marinas are designed as fixed or floating structures. In locations where the spring tide is not high, and the bottoms are not too deep and rocky, fixed marine structures on piles may be built, while for a situation of high tidal range, a float system of barges, plastic tubes, metal drums or galvanized steel pontoons should be provided for all catwalks and main piers. Pontoons have a very wide range of applications, such as for swimming or work platforms, tennis courts, ferries, rafts, houseboats, fuel docks, and so on.

Large-scale offshore structures are closely related to the oil industry. As early as the turn of this century, the search for oil occurred off the coast of California. However, it was not until around 1930, in California and the Gulf of Mexico, that larger scale offshore developments by the oil industry took place. The early offshore platforms were mostly constructed with timber and supported on timber piles, which were braced only above the water; they were placed in water only about 15 ft deep. After World War II, the first self-contained steel platforms were built, and barge-mounted mobile drilling units were also introduced. Concrete platforms cast-in-place on driven concrete piles were already used on Lake Maracaibo, Venezuela, by the 1920s. In 1941, the British floated four complete concrete naval forts out to be sunk in the Thames estuary, this in a way portending the construction of the concrete gravity structures for the North Sea oil operations in the 1970s.

The early platforms were erected in fairly shallow waters close to the shore, and thus their construction was only faced with relatively modest technical problems. When, however, (in the 1950s) the oil operations moved into deep open waters and extremely hostile environments, the platforms had to grow in size and became complex structure systems, the construction of which became highly sophisticated. Finally, in the late 1970s, record-setting fixed platforms were constructed in water depths near 1000 ft! In the 1980s, new structural concepts have made the construction of platforms in extreme water depths of 5000 ft possible.

Offshore structures are fabricated on shore. The major parts or the entire structure are then transported on barges or floated and towed to the site. They are usually end-launched, that is, rotated into vertical position; for example, they may be lowered into place by slowly flooding the floatation tanks. Once the tower is anchored to the foundations, the superstructure or platform can be placed. Should the tower not have sufficient buoyancy in itself, buoyancy tanks must, in addition, be used to keep the structure afloat. The tanks are of cylindrical shape and must be capable of resisting the tremendous water pressure.

In contrast to a fixed platform anchored directly to the ground, a floating rig maintains its position in shallow to mid-depth water by means of a massive multiple anchoring system; mooring lines are anchored to the ground to prevent the rig from moving. In deeper waters, a floating rig may use a dynamic positioning system, such as motor-driven propellers (thrusters), to adjust the rig's position for the action of wind and waves. Instead of fixed moorings, single-point moorings are advantageous for the loading of tankers and barges in rougher seas.

Various large-scale offshore structures related to the oil industry are shown in Figure 8.6. Their basic organization is as follows:

- Exploration rigs
- Permanent production platforms
     fixed platforms
     compliant platforms

Mobile drilling rigs are usually used for exploration purposes. The most common types are:

- The *jack-up drill rig* is used in water depths up to about 400 ft, and rests on the ocean floor. The rig is towed like a barge, with the legs extended above the deck, or possibly with two-piece telescoping legs for greater depths. On-site, the legs are jacked downward through the water and into the sea floor.
- The *semisubmersible rig* is used in water depths from about 300 to 3000 ft. The working deck rests atop hollow columns, which are supported on giant pontoons. They float high in the water when the rig is moved, but the pontoons are flooded with sea water to lower the structure on-site. The submerged ballasted pontoons damp the sea's swells and give the rig its stability, so that the rig moves only slightly with wind and currents.
- The *drillship* is an ocean-going vessel or barge adapted for drilling work in water depths to about 8000 ft.

Permanent production platforms may be organized as fixed and compliant platforms. The major categories of *fixed platforms* are the lighter lattice-type steel platforms, the heavier concrete gravity towers, and the tripod platform structures.

The *steel template platforms* are the most common; they consist of the deck, jacket, and foundation. The jacket is a tower-like three-dimensional X- or K-braced frame, usually consisting of from four to twelve jacket legs that support the platform. Each leg foundation consists of several pile groups. The *steel tower platform* system has only a few large-diameter, nonbattered legs (e.g., four-leg platform), with fewer diagonal braces. Rather than employing a barge for launching, it can be floated using the buoyancy of its large legs. The steel-jacket platforms reached record-setting dimensions in the late 1970s. Exxon's Hondo Platform [1976] stands in 850 ft of water and rises 945 ft above the floor of the Santa Barbara Channel off California. This eight-leg platform has a jacket base of 235 × 170 ft. The jacket was fabricated in two sections, floated to the site, joined in a horizontal position, rotated by flooding the hollow jacket legs and buoyancy tanks, and then anchored to the sea floor. Shell Oil's Cognac Platform [1978] stands in 1025 ft of water in the Gulf of Mexico and rises 1407 ft to the top of its drilling derricks. The jacket is of enormous scale, it tapers from 84 × 164 ft at 14 ft above the mean water line to 380 × 400 ft at the mudline. It is supported by eight 7-ft diameter main legs and two additional legs of the same size at the lower tower portion. The jacket was transported in three sections and assembled vertically under water. First, the barge-carried 14,000-ton base section was secured to the ocean floor with twenty-four 7-ft diameter piles driven about 450 ft into the seabed, then the mid- and top sections were vertically joined to support the deck and drill rigs.

The *gravity platforms* provide enough weight to hold them down while the tubular steel template platforms rely on piles to stay in place. They are either massive structures or towers anchored to cellular foundations or caissons that are filled with sand or water ballast or are used as oil storage tanks to provide the necessary gravitational pull. Gravity structures are usually of concrete, although steel gravity platforms incorporating oil storage at the base can be found (e.g., Tecnomare platform).

The development of large concrete offshore structures began in 1973 with the Ekofisk oil tank, a huge and massive artificial island in 230 ft of water in the middle of the North Sea, which now serves primarily as a platform. It consists of a central cellular, nine-lobed oil-storage caisson about 170-ft square in plan and is 295 ft high, surrounded by a perforated breakwater wall of roughly 302 ft in diameter. Even larger is the Ninian Central Platform [1978] located in 456 ft of water in the North Sea. It has a single column of 148-ft diameter and 560-ft height, supporting a steel deck. It is surrounded by concentric curved walls joined by radial walls to form a complex honeycomb of cells or manifold structure. The enormous 455-ft diameter base consists of three concentric rows of cells that rise to different heights. The cells are not used for oil storage, as originally planned, but are filled with sand and water ballast. Rather than using a single massive shaft protected by a perforated wall around it, as in the so-called *manifold platform* concept, current designs favor the *tower concept* or the so-called *Condeep design*, which includes the Andoc and Sea Tank configurations.

The Condeep design consists of a wide cellular hexagon-shaped foundation of vertical cylindrical storage shells capped with spherical domes at the top and bottom. The cells provide buoyancy during towing, and then serve as oil storage tanks. Two to four shafts or towers extend from the caisson cells to form conical shells at the lower part and cylindrical shells at the upper portion. For example, the huge Statfjord B oil production platform [1981] in 490 ft of water in the North sea, has a 164-ft high and 555-ft diameter base, consisting of twenty-four storage cells. Four towers support the two-level, 96-ft high, steel deck structure with plan dimensions of 181 × 351 ft. The Andoc and Sea Tank configurations are based on the same design principles as the Condeep concept; they use a nearly square base divided into square cells.

Among the various concepts using concrete structures for oil production at greater water depths, of about 1000 ft, has been a proposal based on inclined hollow legs that form a tripod providing the necessary lateral stability. Similarly, the Mandrill 400 tripod steel structure concept was developed for depths of up to about 1600 ft.

In contrast to the fixed platforms, the *compliant structures* move with severe wave action rather than rigidly resisting it, thus reducing the magnitude of the lateral forces. The major representatives of the compliant platform group are described in the following paragraphs.

The *guyed tower* is a relatively slender column behaving in a manner similar to the guyed mast. It carries only gravity loads, since the lateral forces and the corresponding bending moments are resisted by guylines. Exxon's Lena [1985] in the Gulf of Mexico is 1305 ft high, standing in 1000 ft of water. The 120-ft square, single-piece jacket steel tower, with its three-level, 156-ft square platform, rests on a pile foundation and is laterally stabilized by an array of twenty 5-in. cables that extend radially to 200-ton articulated steel anchor weights connected to 72-in. diameter piles. During severe storms, the weights on the stormward side are designed to lift off the bottom, permitting the tower guying system to absorb large wave loads. The space frame tower has sixteen legs at the top to accommodate the deck and buoyancy tanks, and twelve legs in the lower portion. The buoyancy tanks were made a permanent feature of the tower to increase stability and reduce vertical foundation loads, and hence were not only used for tower installation.

The *articulated tower* may consist of a single hollow buoyant concrete tower shaft or a steel lattice structure with buoyant tanks; the tower is connected to the base by an articulated hinge, and the base (in turn) is anchored to the sea bed by piles or ballast. The tower is kept upright by buoyancy, and the articulation allows no bending moment to be transferred to the base. The tower behaves similar to an inverted pendulum with its period well above that of the waves.

The *tension leg platform* is a vertically moored floating platform similar to a semisubmersible mobile drilling barge with a taut mooring system. The superstructure consists of the deck and the buoyant hull held down by highly tensioned vertical tubular steel ties that are anchored to tension piles or a gravity foundation. The buoyancy of the platform creates an upward force that keeps the legs under tension and the platform in place. The system is especially applicable for deepwater structures.

While the design of the various large offshore structure systems just discussed is quite complex, their installation may be far more challenging. To float a concrete gravity structure weighing several hundred thousand tons, or to transport a single-piece 1078-ft long 120-ft square steel tower on a 580-ft long barge and to launch it sideways in 1000 ft of water (i.e., the Lena Platform) represents an extraordinary human achievement!

STARSCRAPERS

In the near future, giant structures will be fabricated and assembled in space that will dwarf the size of tall buildings on Earth. For example, a solar power satellite the size of Manhattan has been proposed, which would collect sunlight in space and beam it to Earth as microwave energy, where it would be converted to electricity. There has always been a fascination with space and its limitless potential—as reflected, for instance, by Jules Verne's famous space-travel novel of 1865 about man's first trip to the moon. The early space station designs of Hermann Oberth in Germany [1923] have been most influential. A significant landmark for the postwar development of space flight was the German V2 rocket, largely designed by Wernher von Braun, which was used towards the end of World War II. The romantic excitement of space living began with the launch of the satellite Sputnik 1 in 1957, and grew rapidly after the Soviets had launched the first man into orbit in 1961, and climaxed with the American Apollo spacecraft landing on the Moon in 1969.

The futurist designers of the 1960s dreamt about space colonies by envisioning them as a way out of the urban crisis on Earth; in a true pioneer spirit, the past could be left behind and one could start all over again. Paoli Soleri, fascinated with the city-as-space-satellite, designed the Asteromo orbital arcology in 1967. Lockheed proposed "Space City 1990," a giant revolving wheel containing the necessary functional modules. Similarly, in 1975, a group of engineers, scientists, and designers at the Ames Institute proposed a one-mile wheel-like orbiting structure as a space habitat for 10,000 people. In this case, the 427-ft diameter ring-shaped outer tube, where people would live and work, is connected by six 48-ft diameter tubular spokes that provide access routes to a central hub serving as a docking station. The structure orbits the Earth at a position equidistant from both Earth and Moon. The large annular torus rotates at one revolution per minute about the central hub to simulate the Earth's normal gravity.

The typical hardware envisioned to be employed reflected the spirit of the experimental architecture of the 1960s. Space architecture was first perceived as huge machine-line projectile-shaped vehicles, and then as cellular agglomerates hooked together, or as open framework facilities containing various types of modules. Concepts of minimum weight, space frames, polyhedral packing, assemblies of components and functional elements, and inflating units in space as well as the adaptable and expansive nature of megastructures to which capsules could be clipped on or plugged in were all articulated. Simulated gravity would be achieved through the centrifugal force of rotating elements.

Now, more than twenty years later, living in space is no longer a dream. The Soviet Mir space station is designed for permanent occupation, with its large winglike solar energy panels; it has already been in orbit since 1986. With their rocket, Energia, the Soviets can lift 100-ton payloads, and thus make deployment of larger manned space stations possible. Expendable heavy-lift vehicles, rather than recoverable heavy winged ones, have an even higher payload capacity.

There is no question that, in the near future, large satellites will orbit in space, be it huge multibeam communication antennas, microwave radiometers, which could be as large as nearly 1000 ft in diameter, or multifunctional manned platforms. These satellites will have forms of dishes, booms, spheres, cylinders, spokes, wheels, planes, and combinations thereof; they will be trussed and open-surface structures where containers or other elements will be attached. Because these orbital space structures are so large, they must be built in space with the help of construction platforms using components delivered by the space shuttle. In 1984, NASA developed a possible configuration for a first generation space station, which is called the "power tower" (Fig. 8.7), receiving its name from the 450-ft central spine space truss structure with cross arms. Modular units, which would be arranged in various ways, are attached to the tower. At the base of this linear truss structure are hung pressurized modules for living and working (i.e., habitat, laboratory, logistics modules) as well as other sub-systems, such as for utility and communications; platforms for servicing, repairs, and construction are attached to the spine, and solar panels are fastened at the top of the tower. Since man can live and work without artificial gravity for only limited periods in space (about 90 days), the space station will be manned with rotating crews.

Eventually, space-born megaspace stations, consisting of a vast multilayered truss-like latticework, will represent self-sufficient urban colonies containing all of the required services, including laboratories, factories, warehouses, service stations, parking areas, docking facilities for free-flying space vehicles, as well as shuttle and cargo transfer terminals associated with missions to other planets.

As mentioned earlier, huge satellites cannot be built on Earth and launched into orbit in their finished configuration, since they are, by far, too large (heavy) and too flexible. They must be assembled on construction platforms, which will include long booms for handling materials and equipment; they will serve as space factories on low Earth orbits (LEO), up to 500 miles distant, accessible by shuttle. A number of shuttle missions will be required to carry up the hardware, of several hundred tons or so. From there, the finished satellites can be ferried out by orbital transfer vehicles or they can be boosted, for example, 22,300 miles to a geosynchronous orbit (GEO) where the stations remain over the same spot on Earth.

Before any of these dreams are to become reality, however, we must first learn how to build large platforms of low mass in a weightless environment. Furthermore, it must be realized that the weight of the payload that can be transported by a shuttle vehicle to the required altitude, and the size of the cargo bay, are limited. Costly shuttle transport trips must be minimized, which indicates that minimum weight structures and efficient packaging become the most important design criteria. The construction of the basic structural framework can be based on a *deployable* or *erectable* approach, that is, either a low-density cargo may be used, which may consist of a

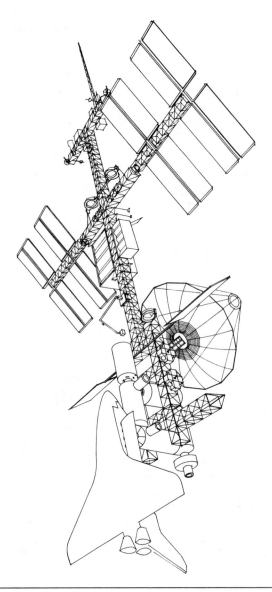

Figure 8.7. *Space station.*

single- or double-fold deployable system preassembled on Earth, which can then be unfolded like an umbrella, or structure components can be densely packed in the shuttle bay and then hand-assembled in space piece by piece, possibly aided by space cranes and by machines fabricating members such as triangular and circular beams (e.g., automatic beam builders). While the primary orbital support structure is assembled in space, the pressurized modules are fabricated on Earth, although some may be membrane structures taken as compacted packages into space, where they are inflated. Rather than letting astronauts assemble deployable and erectable space structures, this will, in the future, be done robotically. Furthermore, new materials will eventually be created, and new products manufactured, in space.

An entirely new approach is needed for the design of these orbiting platforms, since the space environment is radically different from Earth conditions. Gravity loads are nearly nonexistent, hence structures don't even have to support their own weight; wind and seismic forces are not present. Foundations are not needed, and plastic materials like concrete won't pour! However, the absence of gravity and weathering is countered by cyclic radiational heating and cooling for a temperature range, for instance, as wide as $-250°F$ to $250°F$. The effect of these extreme temperature variations upon materials (e.g., mechanical properties, degradation) and upon the structure (e.g., thermal loads due to induced structural deformations) become most critical. In addition to the thermal effects, enormous dynamic loads must be considered as generated by lift-off, transportation, and shuttle docking, as well as from control thrusters, which keep the structure in orbit and adjust position. Stiffness is the primary design determinant for the extremely light orbital structures. They must not be deformed during construction and must allow only a minimal deflection when they are in place. Space stations must also be shielded from orbiting debris (e.g., space litter from broken-up satellites). The modules for space living are pressurized to maintain an interior atmosphere and thus respond, from a structural point of view, like balloons, thereby pointing to cylinders, spheres, and certain polyhedra as efficient shapes.

The materials for the framework of the orbiting platforms will be either aluminum tubes or composite tubes (e.g., graphite epoxy), where thermal movement must be controlled. The materials must be light and strong; they must allow optimal packaging and ease of construction, including adaptable unit attachments with, for instance, one hand and no tools, and they must be reliable in their performance. Certain composite materials (e.g., polysulfone, a thermoplastic resin reinforced by a hybrid fiber of glass and graphite) have a high thermal insensitivity. They are designed to resist thermal expansion and contraction, keeping in mind that thermal changes could be as much as $700°F$. In thermoplastic composite materials, a temperature increase causes the glass to expand but the graphite to shrink; hence, the glass/graphite fabric can have a coefficient of thermal expansion equal to zero.

This brief introduction to space structures was primarily concerned with the constructive aspects of large platforms, and not with the complex issue of livability and behavior of people in zero-gravity space. Basic concepts of enclosure, as defined by top, bottom, and sides (as interpreted by earthbound thinking) now take on a completely different meaning, since all planes can be used as work surfaces. Humans must anchor themselves in order to work in a stationary position. Furniture becomes an integral part of the enclosing structure and is very differently organized; any small elements must be fastened so that they do not float. In a low-gravity environment, humans begin to lose muscle tone and calcium starts leaching from bones; they must do constant physical exercises. Naturally, it would be beneficial to generate an Earth-like gravity for the living quarters. At any rate, the planning of space environments will be an exciting part of the near future.

APPENDIX A: TABLES FOR STRUCTURAL DESIGN

TABLES FOR REINFORCED CONCRETE DESIGN

Table A.1. Total Areas for Various Numbers of Reinforcing Bars (in.^{2})

Bar Size	Nominal Diameter (in.)	Weight (lb/ft)	Number of Bars									
			1	2	3	4	5	6	7	8	9	10
# 3	0.375	0.376	0.11	0.22	0.33	0.44	0.55	0.66	0.77	0.88	0.99	1.10
# 4	0.500	0.668	0.20	0.40	0.60	0.80	1.00	1.20	1.40	1.60	1.80	2.00
# 5	0.625	1.043	0.31	0.62	0.93	1.24	1.55	1.86	2.17	2.48	2.79	3.10
# 6	0.750	1.502	0.44	0.88	1.32	1.76	2.20	2.64	3.08	3.52	3.96	4.40
# 7	0.875	2.044	0.60	1.20	1.80	2.40	3.00	3.60	4.20	4.80	5.40	6.00
# 8	1.000	2.670	0.79	1.58	2.37	3.16	3.95	4.74	5.53	6.32	7.11	7.90
# 9	1.128	3.400	1.00	2.00	3.00	4.00	5.00	6.00	7.00	8.00	9.00	10.00
#10	1.270	4.303	1.27	2.54	3.81	5.08	6.35	7.62	8.89	10.16	11.43	12.70
#11	1.410	5.313	1.56	3.12	4.68	6.24	7.80	9.36	10.92	12.48	14.04	15.60
#14	1.693	7.650	2.25	4.50	6.75	9.00	11.25	13.50	15.75	18.00	20.25	22.50
#18	2.257	13.600	4.00	8.00	12.00	16.00	20.00	24.00	28.00	32.00	36.00	40.00

Table A.2. Average Area per Foot of Width Provided by Various Bar Spacings (in.^{2}/ft)

Bar Size	Nominal Diameter (in.)	Spacing of Bars in Inches													
		2	2½	3	3½	4	4½	5	5½	6	7	8	9	10	12
# 3	0.375	0.66	0.53	0.44	0.38	0.33	0.29	0.26	0.24	0.22	0.19	0.17	0.15	0.13	0.11
# 4	0.500	1.18	0.94	0.78	0.67	0.59	0.52	0.47	0.43	0.39	0.34	0.29	0.26	0.24	0.20
# 5	0.625	1.84	1.47	1.23	1.05	0.92	0.82	0.74	0.67	0.61	0.53	0.46	0.41	0.37	0.31
# 6	0.750	2.65	2.12	1.77	1.51	1.32	1.18	1.06	0.96	0.88	0.76	0.66	0.59	0.53	0.44
# 7	0.875	3.61	2.88	2.40	2.06	1.80	1.60	1.44	1.31	1.20	1.03	0.90	0.80	0.72	0.60
# 8	1.000	——	3.77	3.14	2.69	2.36	2.09	1.88	1.71	1.57	1.35	1.18	1.05	0.94	0.78
# 9	1.128	——	4.80	4.00	3.43	3.00	2.67	2.40	2.18	2.00	1.71	1.50	1.33	1.20	1.00
#10	1.270	——	——	5.06	4.34	3.80	3.37	3.04	2.76	2.53	2.17	1.89	1.69	1.52	1.27
#11	1.410	——	——	6.25	5.36	4.69	4.17	3.75	3.41	3.12	2.68	2.34	2.08	1.87	1.56

Table A.3. Minimum Beam Width (in.) for Interior Exposure According to the ACI Code with #3 Double-Leg Stirrups

Bar Size	Number of Bars in Single Layer of Reinforcement						
	2	3	4	5	6	7	8
# 4	6.8	8.3	9.8	11.3	12.8	14.3	15.8
# 5	6.9	8.5	10.2	11.8	13.4	15.0	16.7
# 6	7.0	8.8	10.5	12.3	14.0	15.8	17.5
# 7	7.2	9.0	10.9	12.8	14.7	16.5	18.4
# 8	7.3	9.3	11.3	13.3	15.3	17.3	19.3
# 9	7.6	9.8	12.2	14.3	16.6	18.8	21.1
#10	7.8	10.4	12.9	15.5	18.0	20.5	23.1
#11	8.1	10.9	13.8	16.6	19.4	22.2	25.0
#14	8.9	12.3	15.7	19.1	22.5	25.9	29.3
#18	10.6	15.1	19.6	24.1	28.6	33.1	37.6

Table A.4. Maximum Reinforcement Ratio for Singly Reinforced Rectangular Beams: $\rho_{max} = 0.75\rho_b$

f_y (psi)	f'_c (psi)				
	3000 ($\beta_1 = 0.85$)	3500 ($\beta_1 = 0.85$)	4000 ($\beta_1 = 0.85$)	5000 ($\beta_1 = 0.80$)	6000 ($\beta_1 = 0.75$)
40,000	0.0278	0.0325	0.0371	0.0437	0.0491
50,000	0.0206	0.0241	0.0275	0.0324	0.0364
60,000	0.0160	0.0187	0.0214	0.0252	0.0283

Table A.5. Typical Characteristics of Stress-Relieved 7-Wire Strands (ASTM A416)

ASTM Type or Grade	Nominal Diameter		Nominal Area		Minimum Tensile Strength, f_{pu}	
	in.	mm	in.^{2}	mm^2	ksi	MPa
Grade 250	0.25	6.35	0.036	23.22		
	0.313	7.94	0.058	37.42		
	0.375	9.53	0.080	51.61		
	0.438	11.11	0.108	69.68	250	1725
	0.500	12.54	0.144	92.90		
	0.600	15.24	0.216	139.35		
Grade 270	0.375	9.53	0.085	54.84		
	0.438	11.11	0.115	74.19		
	0.500	12.54	0.153	98.71	270	1860
	0.563	14.29	0.192	123.87		
	0.600	15.24	0.216	139.35		

Table A.6. Reinforcing Steel Bars for Concrete and Masonry

Type of Steel	Bar Size Numbers	Grade/Yield Strength, ksi	Allowable Tensile Stress, ksi	Min. Tensile Strength, ksi
Billet steel	3–11	40	20	70
(ASTM A615)	3–11	60	24	90
	14 and 18			
Rail steel	3–11	50	20	80
(ASTM A616)	3–11	60	24	90
Axle steel	3–11	40	20	70
(ASTM A617)	3–11	60	24	90
Low-alloy steel	3–11	60	24	80
(ASTM A706)	14 and 18			

Table A.7. Moment Coefficients for Two-Way Slabs on Beams (ACI, Method 2)[a]

Moments	Short Span (Values of m)						Long Span, All Values of m
	1.0	0.9	0.8	0.7	0.6	0.5 and less	
Case 1—Interior panels							
Negative moment at—							
Continuous edge	0.033	0.040	0.048	0.055	0.063	0.083	0.033
Discontinuous edge	—	—	—	—	—	—	—
Positive moment at midspan	0.025	0.030	0.036	0.041	0.047	0.062	0.025
Case 2—One edge discontinuous							
Negative moment at—							
Continuous edge	0.041	0.048	0.055	0.062	0.069	0.085	0.041
Discontinuous edge	0.021	0.024	0.027	0.031	0.035	0.042	0.021
Positive moment at midspan	0.031	0.036	0.041	0.047	0.052	0.064	0.031
Case 3—Two edges discontinuous							
Negative moment at—							
Continuous edge	0.049	0.057	0.064	0.071	0.078	0.090	0.049
Discontinuous edge	0.025	0.028	0.032	0.036	0.039	0.045	0.025
Positive moment at midspan	0.037	0.043	0.048	0.054	0.059	0.068	0.037
Case 4—Three edges discontinuous							
Negative moment at—							
Continuous edge	0.058	0.066	0.074	0.082	0.090	0.098	0.058
Discontinuous edge	0.029	0.033	0.037	0.041	0.045	0.049	0.029
Positive moment at midspan	0.044	0.050	0.056	0.062	0.068	0.074	0.044
Case 5—Four edges discontinuous							
Negative moment at—							
Continuous edge	—	—	—	—	—	—	—
Discontinuous edge	0.033	0.038	0.043	0.047	0.053	0.055	0.033
Positive moment at midspan	0.050	0.057	0.064	0.072	0.080	0.083	0.050

[a]Reproduced from *Building Code Requirements for Reinforced Concrete (ACI 318-63)* courtesy ACI.

TABLES FOR MASONRY DESIGN

Table A.8. Section Properties of Masonry

Nominal** Wythe Thickness (in.)	Concrete Masonry Hollow Units*								Solid Units		
	Net Area (in.²/ft)	Wall Weight (psf)					Moment of Inertia (in.⁴/ft)	Section Modulus (in.³/ft)	Area (in.²/ft)	Moment of Inertia (in.⁴/ft)	Section Modulus (in.³/ft)
		Density (pcf)									
		60	80	100	120	140					
4	28	14	18	22	27	31	38	21	44	48	26
6	37	20	26	33	40	46	130	46	68	178	63
8	48	24	32	40	47	55	309	81	92	443	116
10	60	28	37	47	56	65	567	118	116	892	185
12	68	34	45	55	67	78	929	160	140	1571	270

*Section modulus and moment of inertia of hollow units are based on face-shell mortar bedding and the minimum face-shell thickness for load-bearing units (ASTM C90).
**The real wythe thickness of the units is based on: 3.625, 5.625, 7.625, 9.625, 11.625 in.

Table A.9. Selected Allowable Stresses (psi) for Masonry Construction with Inspected Workmanship (for preliminary design purposes only)* (Ref.: 1980 NCMA, 1969 BIA, 1976 UBC, ANSI A41-2-1960)

Type of Stress	Concrete	Brick
Compressive Axial (walls)	$F_a = 0.225 f'_m (1 - (h/40t)^3)$ ≤ 1000 psi	$F_a \simeq f'_m (0.233 - h/300t)$
Flexural	$F_b = 0.33 f'_m \leq 1200$ psi	$F_b \simeq 0.33 f'_m$
Shear (no shear reinforcement) Flexural members	$F_v = 1.1 \sqrt{f'_m} \leq 50$ psi	reinforced: $F_v = 0.7 \sqrt{f'_m} \leq 50$ psi nonreinforced: $F_v = 0.5 \sqrt{f'_m} \leq 80$ psi
Shear walls	multistory (h/2L < 1): $F_v = 2.0 \sqrt{f'_m}$ $\leq 40 (1.85 - h/2L)$	(56 psi for N mortar) reinforced: $F_v \leq 100$ psi
Tension (no tension reinforcement) Normal to bed joints	$F_t = 0.5 \sqrt{m_o} \leq 25$ psi $F_t = 23$ psi (M or S mortar) $= 16$ psi (N mortar) —— for hollow units ——	$F_t = 36$ psi (M or S mortar) $= 28$ psi (N mortar)
Parallel to bed joints	double the allowable stresses normal to bed joints	$F_t = 72$ psi (M or S mortar) $= 56$ psi (N mortar)

Table A.9. (continued)

Type of Stress	Concrete	Brick
Bearing On full area	$F_a \leq 0.25\ f'_m \leq 900$ psi	$F_a = 0.25\ f'_m$
On one-third area or less	$F_a \leq 0.375\ f'_m \leq 1200$ psi	$F_a = 0.375\ f'_m$
Modulus of elasticity Modulus of rigidity	$E_m = 1000\ f'_m \leq 2{,}500{,}000$ psi $E_v = 400\ f'_m \leq 1{,}000{,}000$ psi	$E_m = 1000\ f'_m \leq 3{,}000{,}000$ psi $E_v = 400\ f'_m \leq 1{,}200{,}000$ psi

*Where there is no inspection, reduce allowable masonry stresses in compression by one-third, and shear and tension by one-half.

where: f'_m = specified compressive strength of masonry, psi (Table A.10);

m_o = specified 28-day minimum required compressive strength of mortar;

L   = length of wall;

h   = clear height or length of wall;

t   = thickness of wall;

Table A.10. f'_m-Values for Masonry (Ref.: 1980 NCMA and 1969 BIA)

Compressive Strength of Masonry Units (psi)	Compressive Strength of Masonry f'_m (psi)				
	Concrete Masonry		Brick Masonry		
	Mortar Type				
	M and S	N	M	S	N
14,000 or more			4600	3900	3200
12,000			4000	3400	2800
10,000			3400	2900	2400
8000			2800	2400	2000
6000	2400	1350	2200	1900	1600
4000	2000	1250	1600	1400	1200
2500	1550	1100			
2000	1350	1000	1000	900	800
1500	1150	875			
1000	900	700			

TABLES FOR STEEL DESIGN

Table A.11. A Partial Selection of Weight Economy Sections[a]

S_x (in.^{3})	Shape	F_y' (ksi)	$F_y = 36$ ksi		S_x (in.^{3})	Shape	F_y' (ksi)	$F_y = 36$ ksi	
			L_c(ft)	L_u(ft)				L_c(ft)	L_u(ft)
1110	W36 × 300	—	17.6	35.3	107	W12 × 79	62.6	12.8	33.3
1030	W36 × 280	—	17.5	33.1	103	W14 × 68	—	10.6	23.9
953	W36 × 260	—	17.5	30.5	98.3	W18 × 55	—	7.9	12.1
895	W36 × 245	—	17.4	28.6	97.4	W12 × 72	52.3	12.7	30.5
837	W36 × 230	—	17.4	26.8	94.5	W21 × 50	—	6.9	7.8
829	W33 × 241	—	16.7	30.1	92.2	W16 × 57	—	7.5	14.3
757	W33 × 221	—	16.7	27.6	92.2	W14 × 61	—	10.6	21.5
719	W36 × 210	—	12.9	20.9	88.9	W18 × 50	—	7.9	11.0
684	W33 × 201	—	16.6	24.9	87.9	W12 × 65	43.0	12.7	27.7
664	W36 × 194	—	12.8	19.4	81.6	W21 × 44	—	6.6	7.0
663	W30 × 211	—	15.9	29.7	81.0	W16 × 50	—	7.5	12.7
623	W36 × 182	—	12.7	18.2	70.6	W12 × 53	55.9	10.6	22.0
598	W30 × 191	—	15.9	26.9	70.3	W14 × 48	—	8.5	16.0
580	W36 × 170	—	12.7	17.0	68.4	W18 × 40	—	6.3	8.2
542	W36 × 160	—	12.7	15.7	66.7	W10 × 60	—	10.6	31.1
539	W30 × 173	—	15.8	24.2	64.7	W16 × 40	—	7.4	10.2
504	W36 × 150	—	12.6	14.6	62.7	W14 × 43	—	8.4	14.4
487	W33 × 152	—	12.2	16.9	60.0	W10 × 54	63.5	10.6	28.2
455	W27 × 161	—	14.8	25.4	58.1	W12 × 45	—	8.5	17.7
448	W33 × 141	—	12.2	15.4	57.6	W18 × 35	—	6.3	6.7
439	W36 × 135	—	12.3	13.0	54.6	W14 × 38	—	7.1	11.5
414	W24 × 162	—	13.7	29.3	51.9	W12 × 40	—	8.4	16.0
411	W27 × 146	—	14.7	23.0	49.1	W10 × 45	—	8.5	22.8
406	W33 × 130	—	12.1	13.8	48.6	W14 × 34	—	7.1	10.2
380	W30 × 132	—	11.1	16.1	47.2	W16 × 31	—	5.8	7.1
371	W24 × 146	—	13.6	26.3	45.6	W12 × 35	—	6.9	12.6
359	W33 × 118	—	12.0	12.6	42.1	W10 × 39	—	8.4	19.8
355	W30 × 124	—	11.1	15.0	42.0	W14 × 30	55.3	7.1	8.7
329	W30 × 116	—	11.1	13.8	38.6	W12 × 30	—	6.9	10.8
329	W24 × 131	—	13.6	23.4	38.4	W16 × 26	—	5.6	6.0
329	W21 × 147	—	13.2	30.3	35.3	W14 × 26	—	5.3	7.0
299	W30 × 108	—	11.1	12.3	35.0	W10 × 33	50.5	8.4	16.5
299	W27 × 114	—	10.6	15.9	33.4	W12 × 26	57.9	6.9	9.4
291	W24 × 117	—	13.5	20.8	32.4	W10 × 30	—	6.1	13.1
269	W30 × 99	—	10.9	11.4	31.2	W 8 × 35	64.4	8.5	22.6
267	W27 × 102	—	10.6	14.2	29.0	W14 × 22	—	5.3	5.6
258	W24 × 104	58.5	13.5	18.4	27.9	W10 × 26	—	6.1	11.4
243	W27 × 94	—	10.5	12.8	27.5	W 8 × 31	50.0	8.4	20.1
231	W18 × 119	—	11.9	29.1	25.4	W12 × 22	—	4.3	6.4
227	W21 × 101	—	13.0	21.3	24.3	W 8 × 28	—	6.9	17.5

Table A.11. (continued)

S_x (in.^{3})	Shape	F_y' (ksi)	$F_y = 36$ ksi L_c(ft)	$F_y = 36$ ksi L_u(ft)	S_x (in.^{3})	Shape	F_y' (ksi)	$F_y = 36$ ksi L_c(ft)	$F_y = 36$ ksi L_u(ft)
222	W24 × 94	—	9.6	15.1	23.2	W10 × 22	—	6.1	9.4
213	W27 × 84	—	10.5	11.0	21.3	W12 × 19	—	4.2	5.3
204	W18 × 106	—	11.8	26.0	21.1	M14 × 18	—	3.6	4.0
196	W24 × 84	—	9.5	13.3	20.9	W 8 × 24	64.1	6.9	15.2
192	W21 × 93	—	8.9	16.8	18.8	W10 × 19	—	4.2	7.2
188	W18 × 97	—	11.8	24.1	17.1	W12 × 16	—	4.1	4.3
176	W24 × 76	—	9.5	11.8	16.7	W 6 × 25	—	6.4	20.0
171	W21 × 83	—	8.8	15.1	15.2	W 8 × 18	—	5.5	9.9
166	W18 × 86	—	11.7	21.5	14.9	W12 × 14	54.3	3.5	4.2
157	W14 × 99	48.5	15.4	37.0	13.8	W10 × 15	—	4.2	5.0
154	W24 × 68	—	9.5	10.2	13.4	W 6 × 20	62.1	6.4	16.4
151	W21 × 73	—	8.8	13.4	12.0	M12 × 11.8	—	2.7	3.0
146	W18 × 76	64.2	11.6	19.1	10.9	W10 × 12	47.5	3.9	4.3
143	W14 × 90	40.4	15.3	34.0	10.2	W 5 × 19	—	5.3	19.5
140	W21 × 68	—	8.7	12.4	9.72	W 6 × 15	31.8	6.3	12.0
134	W16 × 77	—	10.9	21.9	8.51	W 5 × 16	—	5.3	16.7
131	W24 × 62	—	7.4	8.1	7.81	W 8 × 10	45.8	4.2	4.7
127	W21 × 62	—	8.7	11.2	7.76	M10 × 9	—	2.6	2.7
123	W14 × 82	—	10.7	28.1	7.31	W 6 × 12	—	4.2	8.6
117	W18 × 65	—	8.0	14.4	5.56	W 6 × 9	50.3	4.2	6.7
117	W16 × 67	—	10.8	19.3	5.46	W 4 × 13	—	4.3	15.6
114	W24 × 55	—	7.0	7.5	4.62	M 8 × 6.5	—	2.4	2.5
108	W18 × 60	—	8.0	13.3	2.40	W 6 × 4.4	—	1.9	2.4

[a]Reproduced with permission from *Horizontal-Span Building Structures,* Wolfgang Schueller, copyright © 1983 by John Wiley & Sons, Inc.

Table A.12. Allowable Stresses for Columns of A36 Steel[a]

Main and Secondary Members Kl/r not over 120						Main Members Kl/r 121 to 200				Secondary Members* l/r 121 to 200			
$\frac{Kl}{r}$	F_a (ksi)	$\frac{Kl}{r}$	F_a (ksi)	$\frac{Kl}{r}$	F_a (ksi)	$\frac{Kl}{r}$	F_a (ksi)	$\frac{Kl}{r}$	F_a (ksi)	$\frac{l}{r}$	F_{as} (ksi)	$\frac{l}{r}$	F_{as} (ksi)
1	21.56	41	19.11	81	15.24	121	10.14	161	5.76	121	10.19	161	7.25
2	21.52	42	19.03	82	15.13	122	9.99	162	5.69	122	10.09	162	7.20
3	21.48	43	18.95	83	15.02	123	9.85	163	5.62	123	10.00	163	7.16
4	21.44	44	18.86	84	14.90	124	9.70	164	5.55	124	9.90	164	7.12
5	21.39	45	18.78	85	14.79	125	9.55	165	5.49	125	9.80	165	7.08
6	21.35	46	18.70	86	14.67	126	9.41	166	5.42	126	9.70	166	7.04
7	21.30	47	18.61	87	14.56	127	9.26	167	5.35	127	9.59	167	7.00
8	21.25	48	18.53	88	14.44	128	9.11	168	5.29	128	9.49	168	6.96
9	21.21	49	18.44	89	14.32	129	8.97	169	5.23	129	9.40	169	6.93
10	21.16	50	18.35	90	14.20	130	8.84	170	5.17	130	9.30	170	6.89

Table A.12. (continued)

Main and Secondary Members Kl/r not over 120						Main Members Kl/r 121 to 200				Secondary Members* l/r 121 to 200			
$\frac{Kl}{r}$	F_a (ksi)	$\frac{Kl}{r}$	F_a (ksi)	$\frac{Kl}{r}$	F_a (ksi)	$\frac{Kl}{r}$	F_a (ksi)	$\frac{Kl}{r}$	F_a (ksi)	$\frac{l}{r}$	F_{as} (ksi)	$\frac{l}{r}$	F_{as} (ksi)
11	21.10	51	18.26	91	14.09	131	8.70	171	5.11	131	9.21	171	6.85
12	21.05	52	18.17	92	13.97	132	8.57	172	5.05	132	9.12	172	6.82
13	21.00	53	18.08	93	13.84	133	8.44	173	4.99	133	9.03	173	6.79
14	20.95	54	17.99	94	13.72	134	8.32	174	4.93	134	8.94	174	6.76
15	20.89	55	17.90	95	13.60	135	8.19	175	4.88	135	8.86	175	6.73
16	20.83	56	17.81	96	13.48	136	8.07	176	4.82	136	8.78	176	6.70
17	20.78	57	17.71	97	13.35	137	7.96	177	4.77	137	8.70	177	6.67
18	20.72	58	17.62	98	13.23	138	7.84	178	4.71	138	8.62	178	6.64
19	20.66	59	17.53	99	13.10	139	7.73	179	4.66	139	8.54	179	6.61
20	20.60	60	17.43	100	12.98	140	7.62	180	4.61	140	8.47	180	6.58
21	20.54	61	17.33	101	12.85	141	7.51	181	4.56	141	8.39	181	6.56
22	20.48	62	17.24	102	12.72	142	7.41	182	4.51	142	8.32	182	6.53
23	20.41	63	17.14	103	12.59	143	7.30	183	4.46	143	8.25	183	6.51
24	20.35	64	17.04	104	12.47	144	7.20	184	4.41	144	8.18	184	6.49
25	20.28	65	16.94	105	12.33	145	7.10	185	4.36	145	8.12	185	6.46
26	20.22	66	16.84	106	12.20	146	7.01	186	4.32	146	8.05	186	6.44
27	20.15	67	16.74	107	12.07	147	6.91	187	4.27	147	7.99	187	6.42
28	20.08	68	16.64	108	11.94	148	6.82	188	4.23	148	7.93	188	6.40
29	20.01	69	16.53	109	11.81	149	6.73	189	4.18	149	7.87	189	6.38
30	19.94	70	16.43	110	11.67	150	6.64	190	4.14	150	7.81	190	6.36
31	19.87	71	16.33	111	11.54	151	6.55	191	4.09	151	7.75	191	6.35
32	19.80	72	16.22	112	11.40	152	6.46	192	4.05	152	7.69	192	6.33
33	19.73	73	16.12	113	11.26	153	6.38	193	4.01	153	7.64	193	6.31
34	19.65	74	16.01	114	11.13	154	6.30	194	3.97	154	7.59	194	6.30
35	19.58	75	15.90	115	10.99	155	6.22	195	3.93	155	7.53	195	6.28
36	19.50	76	15.79	116	10.85	156	6.14	196	3.89	156	7.48	196	6.27
37	19.42	77	15.69	117	10.71	157	6.06	197	3.85	157	7.43	197	6.26
38	19.35	78	15.58	118	10.57	158	5.98	198	3.81	158	7.39	198	6.24
39	19.27	79	15.47	119	10.43	159	5.91	199	3.77	159	7.34	199	6.23
40	19.19	80	15.36	120	10.28	160	5.83	200	3.73	160	7.29	200	6.22

*K taken as 1.0 for secondary members.

Note: $C_c = 126.1$

[a]Reproduced from the *Manual of Steel Construction*, 8th ed., courtesy of AISC

METRIC CONVERSION TABLE

Table A.13. U.S. Customary Units to SI Metric Units

Length	1 in. = 25.4 mm	1 ft = 0.3048 m
Area	1 in.^{2} = 645.2 mm^2	1 ft^2 = 0.0929 m^2
Volume or section modulus	1 in.^{3} = 16.39 (10^3) mm^3	1 ft^3 = 0.0283 m^3

Table A.13. (continued)

Moment of inertia	$1 \text{ in.}^4 = 0.4162(10^6) \text{ mm}^4$	$1 \text{ ft}^4 = 0.008631 \text{ m}^4$
Single load, force	$1 \text{ lb} = 4.448 \text{ N}$	$1 \text{ kip} = 4.448 \text{ kN}$
Linear loads	$1 \text{ lb/ft} = 14.59 \text{ N/m}$	$1 \text{ kip/ft} = 14.59 \text{ kN/m}$
Surface loads, stress, or modulus of elasticity	$1 \text{ lb/ft}^2 = 0.0479 \text{ kN/m}^2$	$1 \text{ lb/in}^2 = 0.006895 \text{ MPa}$
Mass	$1 \text{ lb} = 0.454 \text{ kg}$	$1 \text{ ton} = 907.18 \text{ kg}$
Density	$1 \text{ lb/ft}^3 = 16.03 \text{ kg/m}^3$	
Moment	$1 \text{ lb-ft} = 1.356 \text{ Nm}$	$1 \text{ kip-ft} = 1.356 \text{ kNm}$
Velocity	$1 \text{ ft/sec} = 0.3048 \text{ m/sec}$	$1 \text{ mi/hr} = 1.6094 \text{ km/hr}$
Acceleration	$1 \text{ ft/sec}^2 = 0.3048 \text{ m/sec}^2$	
Temperature	$t°C = (t°F - 32)/1.8$	

BEAM DEFLECTION AND ROTATION CASES

Table A.14. Beam Deflection and Rotation Cases

Lateral Deflection	Flexure	Shear
	$\dfrac{Ph^3}{3EI}$	$\dfrac{3Ph}{EA}$
	$\dfrac{Ph^3}{12EI}$	$\dfrac{3Ph}{EA}$
	$\dfrac{wh^4}{8EI}$	$\dfrac{1.5wh^2}{EA}$
	$\dfrac{wh^4}{64EI}$	$\dfrac{1.5wh^2}{EA}$
	$\dfrac{11wh^4}{120EI}$	$\dfrac{wh^2}{EA}$
Moments Due to Joint Translation and Rotation		
	$M = 6EI\Delta/L^2$ $P = 2M/L$	
	$M = 4EI\theta/L$ $P = 1.5M/L$	
	$M = 6EI\theta/L$ $P = 2M/L$	

APPENDIX B: LIST OF BUILDINGS IN FIGURES

Figure 1.1.

(a.) *The Marshall Field Warehouse, Chicago [1887], Henry Hobson Richardson;* (b.) *The Home Life Insurance Co. Building, Chicago [1885], William Le Baron Jenney;* (c.) *Wainwright Building, St. Louis [1891], Adler and Sullivan;* (d.) *Guaranty Building, Buffalo [1895], Adler and Sullivan;* (e.) *Monadnock Building, Chicago [1891], Burnham and Root;* (f.)*The Reliance Building, Chicago [1894], Burnham and Company;* (g.) *Carson Pirie Scott Department Store, Chicago [1904], Louis Sullivan;* (h.) *Metropolitan Life Insurance Tower, New York [1909], Napoleon LeBrun and Sons;* (i.) *Woolworth Building, New York [1913], Cass Gilbert;* (j.) *Chicago Tribune Competition entry by Eliel Saarinen [1922];* (k.) *Chrysler Building, New York [1930], William Van Alen;* (l.) *Empire State Building, New York [1931], Shreve, Lamb, and Harmon;* (m.) *Daily News Building, New York [1930], Howels and Hood;* (n.) *McGraw-Hill Building, New York [1931], Hood and Fouilhoux;* (o.) *RCA Building, New York [1933], Rockefeller Center Associates;* (p.) *Philadelphia Saving Fund Society Building, Philadelphia [1932], Howe and Lescaze.*

Figure 1.2.

Seagram Building, New York [1959], Mies van der Rohe with Philip Johnson.

Figure 1.4.

The buildings (from left to right) are: Top row—John Hancock, Boston (I.M. Pei); United Nations Plaza Hotel, New York (Kevin Roche, John Dinkeloo); a wire frame model (Bayley, 1983); City Center 4, Denver (Metz, Train, Youngren); a wire frame model (Bayley, 1983); Three First National Plaza, Chicago (SOM); Republic Bank Center, Houston (Johnson/Burgee); 701 4th Ave. South, Minneapolis (Murphy/Jahn); Middle row—Saint Louis Place, St. Louis (Peckham, Guyton, Albers and Viets); 333 Wacker Drive, Chicago (Kohn, Pedersen and Fox); Gleisdreieck, Frankfurt (O.M. Ungers); The Palace, Miami (Arquitectonica); ziggurat model; Transamerica Building, San Francisco (W. L. Pereira); 11 Diagonal Street, Johannesburg (Murphy/Jahn); Procter and Gamble Headquarters, Cincinnati (Kohn, Pedersen, Fox); Bottom row—Seagram Building, New York (Mies van der Rohe); E-type building block; Y-type building tower; International River-Center, New Orleans (Hellmuth, Obata and Kassabaum); tower example; Two Dallas Centre, Dallas (Cossutta and Associates); Royal Insurance Group Building, Liverpool (Tripe and Wakeham); Hyatt Regency Hotel, Dallas (W. Becket).

Figure 1.6.

Some of the architects of the buildings are: E.L. Barnes; Welton Becket; Eli Attia; Hellmuth, Obata, Kassabaum; Johnson/Burgee; Kohn, Pedersen, Fox; I. M. Pei; Cesar Pelli; Kevin Roche, John Dinkeloo; Paul Rudolph; Schooley, Cornelius and Schooley; Swanke, Hayden and Connell; E. H. Zeidler; SOM; Murphy/Jahn.

Figure 1.7.

Some of the buildings, from left to right, are: Top row—Deutsche Welle, Cologne (1st case; Hanig, Scheid, Schmidt); Metropolitan Life Insurance Co., Chicago (2nd case; Murphy/Jahn); Civic Center, Chicago (4th case; C. F. Murphy); Apartment building, Saarlouis (5th case; H. Ohl and C. Schnaidt); Gleisdreieck, Frankfurt (7th case; O. M. Ungers); Overseas-Chinese Banking Co., Singapore (8th case; I. M. Pei); Eau Claire Estates Tower, Calgary (9th case; SOM); Bottom row—Banco Bilbao, Madrid (1st case; Fernandez Alba); Brunswick Center, London (4th case; P. Hodgkinson and Leslie Martin).

Figure 1.8.

Conceptual investigation of the Gleisdreieck Office Tower, Frankfurt, FRG. O. M. Ungers, Architect, Baumeister 1/84, TA 1985.

Figure 1.9.

Some of the buildings, from left to right and from top to bottom, are: Documenta Urbana, Kassel, FRG (Hermann Hertzberger); 2440 Boston Road, Bronx, NY (Davis, Brody and Assoc.); Health Science Center, Stony Brook, NY (Bertrand Goldberg); Mercantile Financial Center, Dallas, TX (Johnson/Burgee); Park Centre, Calgary, Canada (Kohn, Pedersen, Fox); The Palace, Miami, FL (Arquitectonica); Student Dormitories, Boulder, CO (Wagener Van der Vorste); Sony Tower, Osaka, Japan (Kisho Kurokawa); N.E.G. Employees' Service Facilities, Ootsu, Japan (Arata Isozaki); Casa Milà, Barcelona, Spain (Antonio Gaudi); apartment building, Caracas, Venezuela (Maricarmen Sánchez).

Figure 1.13.

(a.) *Thyssenhaus (26-story), Düsseldorf, FRG. Hentrich-Petschnigg, Architect,* Bauen and Wohnen, 11/60, ENR, Jan. 1961; **(b.)** *New Haven Savings Bank (18-story), New Haven, CT. Wm. F. Pedersen and Associates,* Connecticut Arch. 3/4, 1975; **(c.)** *Brookhollow Central III (12-story), Houston, TX. 3D/International,* Building Design and Construction 4/82; **(d.)** *Central Beheer, Apeldoorn, Holland. Herman Hertzberger, Architect,* Arch + Sept./Oct. 1974, A + U 8312; **(e.)** *College Life Insurance Co. (11-story), Indianapolis, IN. Kevin Roche–John Dinkeloo,* Sieverts, 1980; **(f.)** *Hessische Landesbank (31-story), Frankfurt, FRG: Novotmy/Mähner, Architects,* Sieverts, 1980; **(g.)** *Concorde Office Building, Sao Paulo, Brazil. Carlos Bratke Architect,* A + U 8311; **(h.)** *IBM 590 Building (43-story), 590 Madison Ave., New York. Edward Larrabee Barnes, Architect, James Ruderman, Structural Engineer,* Building Design and Construction Jan. 1984, A.R. May 1984, P/A 7, 1979; **(i.)** *Valley National Bank of Phoenix (40-story), Phoenix, AZ. Welton Becket and Associates,* Building Design and Construction 3/74.

Figure 1.14.

(a.) *The World Trade Center, New York. Minoru Yamasaki and Associates, Architects,* The World Trade Center, *the Port Authority of NY and NJ, Oct. 1974;* **(b.)** *AT&T Corporate Headquarters, New York. Johnson/Burgee Architects,* A.R. Oct. 1980; **(c.)** *Georgia-Pacific Cen-*

ter, Atlanta, GA. SOM, Architects, Building Design and Construction *Aug. 1983;* (**d.**) *Office building 410, 17th Street, Denver, CO. Muchow Associates, Architects,* USS Building Report *Sept. 1978;* (**e.**) *Commerce Court, Toronto, Canada. I.M. Pei and Associates, Architects,* Canadian Arch. *March 1973;* (**f.**) *Olympia Tower, 5th Ave., New York. SOM, Architects,* P/A *12, 1975;* (**g.**) *Bank of America, San Francisco, CA. Pietro Belluschi Consulting, Architect,* A.R. *July 1970.*

Figure 1.15.

(**a.**) *Ford Foundation Office Building. New York, Kevin Roche, John Dinkeloo,* Press Fact Sheet, P/A *Feb. 1968;* (**b.**) *National Gallery of Art, Washington, D.C. I.M. Pei and Partners,* A.R. *Mid-August 1975,* Building Design and Construction *Oct. 1974,* AIA Journal *Mid-May 1979;* (**c.**) *State of Illinois Center, Chicago, IL. C.F. Murphy/H. Jahn,* ENR *Sept. 22, 1983,* A.R. *Aug. 1980;* (**d.**) *Loews Anatole Dallas Hotel, Dallas, TX. Beran and Shelmire,* Brick in Architecture *Vol. 37, #1;* (**e.**) *Shinuku NS Building. Tokyo. Nikken Sekkei,* JA *Feb. 1983, 12* Kenchiku Bunka *Dec. 1982;* (**f.**) *North Loop Redevelopment Project, Chicago, IL. C.F. Murphy/ Helmuth Jahn,* A.R. *Aug. 1980;* (**g.**) *Law Courts Complex, Vancouver, Canada. Arthur Erickson,* A.R. *Dec. 1980;* (**h.**) *Time Square Hotel, New York. John Portman Associates,* ENR *Feb. 23, 1984;* (**i.**) *Chicago Board of Trade Addition, Chicago, IL. Murphy/Jahn,* A + U *8311;* (**j.**) *The Euram Building, Washington, D.C. Hartman-Fox,* Forum *May 1972;* (**k.**) *Hercules Inc. Headquarters, Wilmington, DE. Kohn Pedersen Fox Associates,* A.R. *June 1981,* AIA *Feb. 1984;* (**l.**) *33 W. Monroe building. Chicago, SOM,* Building Design and Construction *June 1981,* Modern Steel Construction *1st Qtr. 1979;* (**m.**) *The John Fitzgerald Kennedy Library. Boston, I.M. Pei,* A.R. *Feb. 1980.*

Figure 1.17.

(**a.**) *Harbor House, Chicago, IL. Hausner and Macsai Architects [1965] (see also Macsai, 1982);* (**b.**) *Apartment buildings "Tatzelwurm" (9 stories), Roszdorf, FRG. Werkgemeinschaft Roszdorf, Architects;* (**c.**) *Apartment building "Fasan" (20 stories), Stuttgart-Fasanenhof, FRG. Tiedje and Lehmbrock, Architects, Aregger, 1967;* (**d.**) *Apartment building "Triemliplatz" (16 stories), Zürich, Switzerland. R. and E. Guyer, Architects [1966], Aregger, 1967;* (**e.**) *Apartment building "Romeo," Stuttgart, FRG. Hans Scharoun, Architect (see also Sting, 1969);* (**f.**) *Apartment building in Cologne (15 stories), FRG. Schneider Architects (see also Sting, 1969);* (**g.**) *2440 Boston Road apartments (20 stories), Bronx, NY. Davis Brody and Associates, Architects,* AIA Journal *May 1975;* (**h.**) *Twin Parks West-Site 10–12, Bronx, NY. G. Pasanella, Architect [1971] (see also Macsai, 1982);* (**i.**) *Apartment building in London (15 stories), England. Lasdun Architects (see also Sting, 1969);* (**j.**) *Crawford Manor Housing (14 stories), New Haven, CT. Paul Rudolph, Architect (see also Sting, 1969);* (**k.**) *Rokeby Condominiums, Nashville, TN. Barber and McMurray, Architects [1974] (see also Macsai, 1982);* (**l.**) *Apartment building in Bietigheim (21 stories), FRG. Weber and Hoffmann, Architects [1967], Aregger, 1967;* (**m.**) *Eastwood Urban Housing, Roosevelt Island, NY. Sert Jackson and Associates, Architects,* A.R. *Aug. 1976.*

Figure 1.18.

Esslingen-Zollberg/Süd, FRG. D. Raichle and R. Götz, Architects, "Esslingen-Zollberg/Süd— Differenzierung und Typisierung an terrassierten Bauten", Schriftenreihe des Bundesministers für Raumordnung, Bauwesen und Städtebau, 1977.

Figure 1.19.

Barrio Gaudi (1965), Reus (Tarragona), Spain. Ricardo Bofill, Architect, P/A *Sept. 1975,* AD *July 1975, (middle left); Habitat 67, Montreal Canada. Moshe Safdie, Architect, A. E.*

Komendant Structural Engineering Consultant, P/A Oct. 1966, Forum May 1967, AD March 1967, etc., (top right); School for Field Studies, Hatsevah, Israel. I. M. Goodovitch, Architect, AA Nov. 1976 (middle right); Wohnhügel in Marl (1967), FRG. P. Faller, H. Schröder, and R. Frey, Architects, Riccabona, 1974 (2nd case from bottom); Apartment and hotel building, München-Bogenhausen (1969), FRG. Walter Ebert, Architect, Riccabona, 1974. (bottom)

Figure 1.20.

Stufendomino Lyngsberg, Bonn-Bad Godesberg, FRG. A. Wetzel Wohnbau GmbH, Schriften-reihe des Bundesministers für Raumordnung, Bauwesen, und Städtebau, 1975.

Figure 2.1.

Read from left to right: Telecommunication Tower, Donnersberg, FRG (207 m); Empire State Building [1931], New York (381 m); Pirelli Building [1956], Milan, Italy (126 m); Telecom-munication Tower, Hamburg [1965], FRG (272 m); CN Tower [1976], Toronto, Canada (553 m), Stapelhaus (22 stories, project 1964, Wolfgang Döring, FRG); Space Needle [1962], Seattle (188 m); the Great Pyramids [2570–2500, BC] at Giza, Egypt. (Cheops pyramid, originally 147 m high)

Figure 2.7.

(a.) *Tempo Bay Resort Hotel (14 stories), Walt Disney World, Orlando, FL. Welton Becket and Associates, Architects, ENR Dec. 1970, Civil Engineering Jan. 1973, Arch. Franc. Dec. 1974;* **(b.)** *Hyatt Regency San Francisco (17 stories), San Francisco, CA. John Portman and Associates, Architects and Engineers, A.R. Sept. 1973, Arch Franc. Dec. 1974;* **(c.)** *Our Lady of Le Plateau, Abidjan, Ivory Coast. R. Olivieri Architect, Ricardo Morandi Structural Engineer, A + Sept./ Oct. 1974;* **(d.)** *Tempe Municipal Building (3 stories), Tempe, AZ. Michael and Kemper Good-win, Ltd., Architects, Mann and Anderson, Structural Engineers, ENR Dec. 1970, Modern Steel Construction 2nd Qtr. 1972, 1971 AISC Architectural Awards of Excellence;* **(e.)** *GTE Corporation Headquarters (ten stories), Stamford, CT. Victor H. Bisharat, Architect, Jensen and Adams, Structural Engineers, Bethlehem Building Case History No. 33, 1974, ENR Aug. 1972;* **(f.)** *UCSD Library, San Diego, CA. Pereira and Associates, Architects, ENR May 1970, AIA Journal August 1977;* **(g.)** *Hemicycle for the European Parliament, Luxembourg. M. Bohler, Architect, Acier-Stahl-Steel April 1979, Detail June 1981;* **(h.)** *Project of office building, Barn-stone and Keeland, Architects, G. Cunningham, Structural Engineer, ENR March 1963;* **(i.)** *Municipal Administration Center, Dallas, TX. I. M. Pei and Partners, P/A May 1979;* **(j.)** *The Pyramid Condominium, Ocean City, MD. William Morgan, Architects, Sherrer-Bauman and Associates, P/A Jan. 1975 and Sept. 1976, AA Oct. 1977;* **(k.)** *Hotel Inter-Continental (16 stories), Sharjah, United Emirates. The Architects Collaborative, Wayman C. Wing, Structural Engineer, A/R May 1979;* **(l.)** *Dubiner Apartment House [1960], Ramat Gan, Tel Aviv, Israel. Alfred Neumann and Zvi Hecker, Architects, ariel #36, 1974;* **(m.)** *Mar-riott Corporation Headquarters, Bethesda, MD. Mills and Petticord, Architects, Gillum-Colaco, Structural Engineers, USS Building Report March 1979;* **(n.)** *Olympic Village, Montreal, Canada. Roger D'Astous and Luc Durand, Architects.*

Figure 2.10.

(a.) *IBM Office Building, Seattle, WA. Yamasaki and Associates, Architects, Worthington, Skilling, Helle and Jackson Structural Engineers, A.R. Feb. 1965, ENR Oct. 8, 1964;* **(b.)** *Kansas City Bank Tower, Kansas City, MO. Harry Weese and Associates, Architects, Jack D. Gillum and Associates, Structural Engineers, Modern Steel Construction 1st Qtr. 1974, ASCE Jan. 1976, ENR June 6, 1974;* **(c.)** *Brunswick Building, Chicago, IL. SOM, Architects and Structural Engineers. Forum April 1966;* **(d.)** *Century Plaza Towers, Los Angeles, CA. Minoru*

Yamasaki, Architects, Skilling, Helle, Christiansen, Robertson, Structural Engineers, ENR March 28, 1974; (**e.**) *Seattle-First National Bank, Seattle, WA. Naramore, Bain, Brady and Johanson, Architects, Skilling, Helle, Christiansen, Robertson Structural Engineers, A.R. June 1970;* (**f.**) *St. Joseph Hospital, Tacoma, WA. Bertrand Goldberg and Associates, Architects, ABAM, Structural Engineers, ENR July 1974;* (**g.**) *Student Dormitory, City University of Paris, Paris, France. C. Parent and M. Foroughi, E. Ghiai, Architects, Acier-Stahl-Steel June 1968;* (**h.**) *Tennessee State Office Building, Memphis, TN. Gassner/Nathan/Browne, Architects, O. Clarke Mann, Structural Engineers;* (**i.**) *Prentice Women's Hospital, Chicago, IL. B. Goldberg, Architects and Structural Engineers, A.R. July 1976, ENR July 1974;* (**j.**) *Australian Embassy, Paris, France. Harry Seidler and Associates, Architects, Pier Luigi Nervi, Structural Consultant, Domas 588, Nov. 1978;* (**k.**) *IBM Development Engineering Laboratory, La Gaude, France. Marcel Breuer, Architect, P/A Feb. 1967, A.R. Feb. 1964 and May 1979;* (**l.**) *One Corporate Center, Hartford, CT. Irwin Joseph Hirsch and Associates, Architects, Irwin G. Cantor, Structural Engineer, A.R. Sept. 1982;* (**m.**) *Unité d'Habitation, Marseille, France. Le Corbusier, Architect, Arch + Jan./Feb. 1974;* (**n.**) *U.S. Embassy, Tokyo, Japan. Cesar Pelli, Architect, JA March 1977, A.R. April 1977;* (**o.**) *Hirado Resort Hotel Ranpu, Nagasaki Prefecture, Japan. Kuni-ken, Architects and Engineers, JA 7801, 1977.*

Figure 2.17.

(**a.**) *Administration Building, U. of Illinois (28 stories), Chicago, IL. Forum Aug./Sept. 1964, May 1966, A.R. Oct. 1961, Aug. 1963;* (**b.**) *IBM Building (13 stories), Pittsburgh, PA. USS AIA File No. 17A, 1963;* (**c.**) *Houston Lighting and Power Company Electric Tower (27 stories, 1967), Houston, TX. AISI Building Report Vol. 4, No. 1;* (**d.**) *Tokyo Tower (333 m), Tokyo, Japan;* (**e.**) *Water Tower Place (76 stories), Chicago, IL. ENR July 17, 1975;* (**f.**) *The World Trade Center (110 stories), New York. Contemporary Steel Design AISI Vol. 1, No. 4, 1965.*

Figure 2.19.

(**f.**) *Russian Residence (20 stories), New York. SOM, Architect,* Modern Steel Construction 3rd Qtr. 1975, Building Design and Construction *Sept. 1974;* (**h.**) *BMW Tower, Munich, FRG. Karl Schwanzer, Architect, A + Oct. 1973;* (**i.**) *Knights of Columbus Building, New Haven, CT. Kevin Roche and John Dinkeloo Architects, A.R. Aug. 1970;* (**q.**) *USS Corporate Center, Pittsburgh, PA. Harrison and Abramovitz & Abbe, Architects,* The Steel Triangle, *US Steel Publ. July 1969.*

- Crane literature of American Pecco Corporation, Liebherr Crane Corporation, and Emscor Inc.
- Design for Erection Considerations, A. A. Yee and F. R. Masuda, PCI *Nov./Dec. 1974*
- Push-up construction: Civil Engineering-ASCE *March 1979*, AISC Engineering Journal *July 1969.*

Figure 2.21.

(**a.**) *Georgia Power Tower, Atlanta, GA. Heery and Heery, Architects and Engineers, ENR Nov. 1, 1979;* (**b.**) *Nonoalco Tower, Mexico City, Mexico. Mario Pani and Associates, Architects, ENR April 5, 1962 and May 30, 1963;* (**c.**) *BASF Office Building, Ludwigshafen, FRG. ref.: C. Santo, DBV Vorträge Betontag, 1957;* (**d.**) *Sears Tower, Chicago, IL. SOM, Architects, ENR July 8, 1971;* (**e.**) *Embankment Place, London, England. Terry Farrell Partnership, Architects, Ove Arup and Partners, Engineers, A.R. Sept. 1987;* (**f.**) *AMOCO Building, Chicago, IL. Perkins and Will Corp., ENR Nov. 25, 1971,* Engineering Journal AISC 2nd Qtr. 1975; (**g.**) *Footing plan for an eight-story bearing wall structure, Flexicore Co., Inc. Literature;* (**h.**) *Copley Place, Boston, MA. Zaldastani Associates, Structural Engineers,* Civil Engineering/ ASCE Nov. 1984.

Figure 3.6.

The following references have been used: Aynsley, 1972, 1973; Gandemer, 1977, 1979. The cases are read from left to right and top to bottom: Toronto City Hall, Toronto, Canada. Viljo Revell, Architect, P/A March 1963; Earth Sciences Building, MIT, Cambridge, MA. I. M. Pei, Architect, P/A March 1967; Xerox Centre, Chicago, IL. C. F. Murphy and H. Jahn, Architects, P/A 12, 1980; Georgia-Pacific Corp. Tower, Atlanta, GA. SOM, Architects, ENR March 27, 1980; U.S. Steel Building, Pittsburgh, PA. Harrison and Abramovitz and Abbe Architects, Forum Dec. 1971, U.S. Steel: The Steel Triangle, 1969.

Figure 3.15.

Some of the cases are: Second row top from left to right: Xerox Center, Chicago, IL, C. F. Murphy/H. Jahn; Brunswick Building, Chicago, IL, SOM; conceptual study; National Life Building, Nashville, TN, SOM; left column from top to bottom: Xerox Center, Chicago, IL, C. F. Murphy/H. Jahn; 100 William St., New York, Davis, Brody and Assoc.; CBS Building, New York, Eero Saarinen; U.S. Steel Building, Pittsburgh, PA, Harrison and Abramovitz and Abbe; right bottom: John Hancock Center, Chicago, IL, SOM (gravity type vertical air circulation).

Figure 5.3.

(e.) Tower building (18 stories), Little Rock, AR. Harry A. Barry, Architects, AISC, 1965; (f.) 101 California Building (48 stories), San Francisco, CA. Johnson/Burgee Architects, CBM Engineers, ENR Feb. 28, 1980; (g.) Equitable Life Building (25 stories), San Francisco, CA. Paquette and Maurer Struct. Eng., Boris Bresler, 1968; (h.) One Chemung Canal Plaza (7 stories), Elmira, NY. Haskell and Conner and Frost Architects, Bethlehem Steel Building Case History No. 23, Aug. 1972; (i.) Citicorp Center (59 stories), New York. Hugh Stubbins and Associates, Architects, LeMessurier Structural Engineers, A.R. Mid-Aug. 1976; (j.) World Trade Center (110 stories), New York. Minoru Yamasaki and Emery Roth and Sons Architects, Worthington, Skilling, Helle and Jackson Structural Engineers, A.R. May 1964; (k.) IBM Building (13 stories), Pittsburgh, PA. Curtis and Davis Architects, Worthington, Skilling, Helle, and Jackson Structural Engineers, USS Structural Report, Aug. 1963; (l.) Life and Casualty Insurance Co. (30 stories), Nashville, TN. Edwin A. Keeble Associates, ENR Sept. 6, 1956.

Figure 5.4.

(e.) Henry J. Pariseau Apartments (11 stories), Manchester, NH. Isaak Mayer Architect, Bethlehem Steel Building Case History No. 30, July 1973; (f.) Eastern Properties Office Building (4 stories), Lexington, KY. Johnson/Romanowitz Architects, A.R. May 1976; (g.) Bank of America Plaza (20 stories), San Diego, CA. Tucker, Sadler and Associates Architects, USS Building Report, Jan. 1983; (h.) One Shell Plaza (52 stories), Houston, TX. SOM, Architects, Arch. and Eng. News Oct. 1968; (i.) First City Tower (50 stories), Houston, TX. Morris-Aubry, Architect, and Walter B. Moore and Associates, Structural Engineers, ENR Feb. 19, 1981, Structural Engineer Sept. 1982 and Sept. 1983; (j.) Torre Banaven (280 ft, 470 ft, 680 ft), Caracas, Venezuela. Gillum Colaco, Structural Engineer, ENR July 1979; (k.) AMOCO (1136 ft), Chicago, IL. Edward Durell Stone and Associates and The Perkins and Will Corporation, Architects, Engineering Journal AISC, 2nd. Qtr. 1975, ENR Nov. 25, 1971; (l.) 33 West Monroe (28 stories), Chicago, IL. SOM, Architects, A.R. Nov. 1981; (m.) Sears Tower (109 stories), Chicago, IL. SOM, Architects, Engineering Journal AISC, 3rd. Qtr. 1973.

Figure 5.5.

Some of the building cases included in the study are: Center for Creative Studies, Detroit, MI, W. Kessler and Associates; A. N. Richards Medical Research Building, U. of Pennsylvania, Philadelphia, PA, Louis I. Kahn, Architect, Auguste Kommendant Structural Engineer; Office Building Central Beheer, Apeldoorn, Holland, Herman Hertzberger; Administration Building of the Magdeburg Insurance, Hannover, FRG. Walter Henn

Figure 5.6.

(a.) Precast concrete plank with suspended fire-rated ceiling; (b.) Open-web steel joists with 2 1/2-in. concrete slab on permanent steel forming (i.e., noncomposite) and suspended fire-rated ceiling; (c.) 8-in. hollow-core precast concrete plank floor spanning 24 ft supported on steel frame, Bethlehem Steel Building Case History No. 66, 1980; (d.) Dyna-Frame precast concrete framing system, Strescon Industries, Inc., Baltimore, MD, 1984; (e.) 36/24-in. haunch steel girder of 41- and 32-ft spans, Bethlehem Steel Building Case History No. 15, 1971; (f.) 4 1/2-in. concrete slab on permanent steel forming (in noncomposite action) composite with steel beams and suspended fire-rated ceiling, rather than fire protected beams with nonrated ceilings; (g.) Floor system composite stub-girder (40-ft span), with cantilever beams, Bethlehem Steel Building Case History No. 21, 1972.

Figure 5.7.

(a.) Pennzoil Place (37 stories), Houston, TX. Johnson/Burgee, Architects, Ellisor, Structural Engineers, Bethlehem Building Case History No. 47, Aug. 1976, A.R. Nov. 1976, P/A Aug. 1977; (b.) El Paso Tower (70 stories), Houston, TX. I. M. Pei and Partners Architects, Colaco Structural Engineers, ENR Aug. 3, 1978, Oct. 4, 1979, March 5, 1981; (c.) Capital National Bank Plaza (50 stories), Houston, TX. Lloyd Jones Brewer Associates, Architects, Ellisor and Tanner, Structural Engineers, ENR April 12, 1979, Building Design and Construction May 1981, USS Building Report May 1980; (d.) Dravo Tower (54 stories), Pittsburgh, PA. Welton Becket Associates, Architects, Lev Zetlin Associates, Structural Engineers, P/A Dec. 1980, A.R. Mid-Aug. 1981, ENR May 1982; (e.) Marina City Tower (60 stories), Chicago, IL. Bertrand Goldberg, Architect, ENR Feb. 22, 1962, July 18, 1974; (f.) U.S. Steel Headquarters Building (64 stories, 841 ft high), Pittsburgh, PA. Harrison and Abramovitz and Abbe, Architects, Skilling, Helle, Christiansen, Robertson, Structural Engineers, A.R. April 1967, ASCE April 1970, Forum Dec. 1971, The Steel Triangle U.S. Steel, July 1969; (g.) Sixty State Street (38 stories), Boston, MA. SOM, Architects and Structural Engineers, AISC Awards of Excellence 1979; (h.) M.L.C. Centre (68 stories), Sydney, Australia. Harry Seidler and Associates, Architects and Structural Engineers, ENR June 30, 1977; (i.) ICAO Office Building (28 stories), Montreal, Canada. PCI Journal Nov./Dec. 1975; (j.) One Dallas Centre (30 stories), Dallas, TX. I. M. Pei, Architect, Weiskopf and Pickworth, Structural Engineers, AIA Journal Mid-May 1981; (k.) Fourth and Blanchard Building (25 stories), Seattle, WA. Chester L. Lindsey Architects, KPFF Structural Engineers, USS Building Report Jan. 1980; (l.) Georgia Pacific Headquarters Building (52 stories), Atlanta, GA. SOM, Architects, Weidlinger Associates, Structural Engineers, P/A Dec. 1980, ENR March 27, 1980, Oct. 11, 1979, AA 220 April 1982.

Figure 5.12.

(a.) 535 Madison Ave., New York. Edward Larabee Barnes Architect, Irwin G. Cantor, Structural Engineer, A.R. March 1980; (b.) Xerox Centre, Chicago, IL. C. F. Murphy and Helmut Jahn Architects, AIA Journal Mid-May 1981, P/A Dec. 1980; (c.) National Permanent Building, Washington, D.C. Hartman-Cox Architects, P/A Dec. 1977, CRSI Case History Report 7901; (d.) Lake Point Tower, Chicago, IL Schipporeit-Heinrich Architects, A.R. Oct. 1969; (e.) Immeuble d'habitation a Santa Marinella, Rome, Italy, Portoghesi and Gigliotti, Architects. Norberg-Schulz, 1971.

Figure 6.2.

Read from left to right and top to bottom: Toronto-Dominion Centre, Toronto, Canada. Mies van der Rohe, A.R. March, 1971; One Liberty Plaza, New York. SOM, A.R. July 1973; U.N. Plaza Hotel and Office Building, New York. Kevin Roche, John Dinkeloo and Associates, P/A June 1976, G. A. Document 1970–80; PPG Place, Pittsburgh, PA. John Burgee/Philip Johnson, A.R. Oct. 1984, Building Design and Construction Oct. 1984; The 400 Colony Square Office Building, Atlanta, GA. Jova/Daniels/Busby, Schokbeton Publ. March/April 1973; IBM Office Building, Southfield, MI. Gunnar Birkerts and Associates, P/A Sept. 1975, Bethlehem Steel Building Case History No. 70, 1981.

Figure 6.4.

(a.) 88 Pine Street, New York. I. M. Pei and Partners, A.R. April 1975, AIA Journal June 1974 (left case); (b.) Sixty State Street, Boston, MA. SOM Awards of Excellence 1979.

Figure 6.10.

(a.) 8-in. reinforced concrete masonry wall with brick veneer and cast-in-place concrete floor; (b.) Joist floor and: left) 12-in. brick cavity wall with mechanical space; right) 8-in. double-wythe solid stretcher brick wall; (c.) 12-in. solid brick wall and 8-in. concrete pan joist floor construction; (d.) Jespersen-Kay Systems Inc., originally from Denmark; (e.) Left: composite brick and block wall and concrete slab; right: 8-in. concrete block wall and steel joist floor; (f.) Cast-in-place 10-in. concrete wall and slab; (g.) 8-in. reinforced concrete block wall with 4-in. brick veneer and precast concrete slab with bond beams; (h.) 10-in. reinforced brick wall and cast-in-place concrete slab; (i.) 8-in. concrete block with veneer and precast concrete planks; (j.) Recommended practice for the construction of tied large-panel concrete structures according to PCA.

Figure 7.2.

(a.) Apartment building Wienerplatz (18 stories, reinforced concrete), Köln, FRG. Karl Hell, Architect, (Sting, 1969); (b.) Apartment building Märkisches Viertel (18 stories), Berlin, FRG. Mathias Ungers Architect, (Sting, 1969); (c.) Apartment building Schwamendingen (18 stories, brick), Switzerland. P/A Feb. 1966, A.R. Nov. 1963; (d.) Hilliard housing center (16 stories, reinforced concrete), Chicago, IL. B. Goldberg Associates, Architects, Perspecta 13/14, 1971, A.R. Jan. 1966, AA Sept. 1965; (e.) Apartment building "Julia" (12 stories), Stuttgart, FRG. Hans Scharoun, Architect, (Sting, 1969); (f.) Torres Blancas (19 stories), Madrid, Spain. F. J. Saenz de Oiza, Architect, (Sting, 1969); (g.) Sheraton Hotel, (14 stories, brick), Monroeville, PA. Murovich Associates, Architects, BIA Case Study #37, 1974; (i.) La Gradelle apartment building (17 stories), Geneva, Switzerland. J. Hentsch, Architect, 1965, Aregger, 1967; (j.) Neue Vahr apartments (20 stories), Bremen, FRG. Alvar Aalto, Architect, Bauen und Wohnen Nov. 1963, etc.; (k.) Apartment building "Salute" (18 stories, reinforced concrete), Stuttgart, FRG. Hans Scharoun, Architect, (Sting, 1969); (l.) Grünegg A. G. apartment building (11 stories), St. Gall, Switzerland. Graf Architects, 1968, Aregger, 1967; (m.) Edgewater Tower (12 stories, concrete block), New Haven, CT. Kosinski, Architect, Building Design and Construction Nov. 1974; (n.) Apartment tower, University of Delaware (17 stories, precast concrete). Charles Luckman Associates, Building Design and Construction Oct. 1972; (o.) St. Joseph Hospital (10 stories, reinforced concrete), Tacoma, WA. B. Goldberg, Architect, ENR July 18, 1974; (p.) Apartment building (15 stories) in Bagnols-sur-Cèze, France. Caudilis, Josic, Woods Architects, (Sting, 1969); (q.) Apartment building Ile Verte (28 stories, reinforced concrete), Grenoble, Switzerland. Anger and Puccinelli, Architects, 1964, Aregger, 1967; (r.) Unité d'Habitation [1954] (17 stories, reinforced concrete), Marseille, France. Le Corbusier, Architect, Arch + Jan./Feb. 1974, etc.

Figure 7.3.

(**a.**) *Singapore Treasury Building (52 stories), Singapore. The Stubbins Associates, Architects, Le Messurier Associates Structural Engineers, A.R. Feb. 1985;* (**b.**) *Swire Bottlers redevelopment project (31-stories), Hong Kong. Ove Arup Partnership,* The Arup Journal Oct. 1982; (**c.**) *Richards Medical Research Laboratory, University of Pennsylvania, Philadelphia, PA. Louis I. Kahn, Architect, August E. Kommendant, Structural Engineer, A.R. Aug. 1960, etc.;* (**d.**) *Quaker Square Hilton, Akron, OH. Curtis and Rasmussen Inc.* Builder Oct. 1982; (**e.**) *Torrington Towers (14 stories), Torrington, CT. Ulrich Franzen, Architect,* National Concrete Masonry Association (NCMA) Vol. 31, No. 1; (**f.**) *The National Westminster Tower (52 stories), London, England. R. Seifert and Partners, Architects, P. Frishmann and Partners, Engineers,* RIBAJ Dec. 1981, ENR Nov. 25, 1976, Bauingenieur May 1984, Aug. 1984; (**g.**) *Christian Science Center (28 stories), Boston, MA. I. M. Pei, Architect,* P/A Oct. 1966; (**h.**) *King George Tower (34 stories), Sydney, Australia. John Andrews, Architect,* P/A Feb. 1973; (**i.**) *Banco Bilbao, Madrid, Spain. Fernandez Alba, Architect,* AA 159, Dec./Jan. 1973; (**j.**) *Price Tower (19 stories), Bartlesville, OK. Wright, 1956;* (**k.**) *Hong Kong Arts Centre (19 stories), Hong Kong. Tao Ho, Architect, 1977;* (**l.**) *Faculty of Sciences Complex, University of Paris, Paris, France. Seassal/Cassau/Coulon/Albert/de Gortchakoff, Architects,* Acier-Stahl-Steel May 1967; (**m.**) *Savings Bank Hannover, Hannover, FRG. H. Wilke, Architect,* Detail April 1978, Acier-Stahl-Steel Nov. 1975.

Figure 7.4.

(**a.**) *Olivetti Frankfurt, Frankfurt, FRG. Egon Eiermann, Architect, 1972,* Arch + Sept. 1973; (**b.**) *N.C. Mutual Life Insurance Co., Durham, NC. Welton Becket Associates, Architects, Seelye, Stevenson, Value, and Knecht, Structural Engineers,* ENR Aug. 20, 1964, PCI: Post-Tensioning Manual, Chicago, 1972; (**c.**) *Westcoast Building, Vancouver, B.C., Canada. Rhone and Iredale, Architects, Bogue Babicki and Associates, Structural Engineers,* Forum May 1972, P/A Oct. 1969, Arch. and Eng. News Nov. 1969, ENR June 1969; (**d.**) *Office building 410, 17th St., Denver, CO. Muchow Associates, Architects, Ketchum, Konkel, Barrett, Nickel, Austin, Structural Engineers,* USS Building Report, Sept. 1978; (**e.**) *U.S. Steel Headquarters, Pittsburgh, PA. Harrison, Abramowitz and Abbe, Architects, Skilling, Helle, Christiansen, Robertson, Structural Engineers, A.R. April 1967,* USS: The Steel Triangle, ADUSS 88-4434-01, July 1969; (**f.**) *American Center, American Motors Corp. Tower, Southfield, MI. Smith, Hinchman and Grylls Associates, Architects and Engineers,* Bethlehem Steel Building Case History No. 35, Nov. 1974; *top: Laboratory Tower for Johnson Wax Company [1939], Racine, WI. Frank Lloyd Wright, Architect,* (Michaels, 1950)

Figure 7.5.

(**a.**) *Rainier National Bank, Seattle, WA. Minoru Yamasaki and Associates, Architects, SHCR, Structural Engineers,* ENR Oct. 1975; (**b.**) *Shrine of the Missionaries tower, Sault St. Marie, MI. Progressive Design Associates, Architects, Johnston-Sahlman, Structural Engineers,* Archit. and Eng. News Jan. 1968; (**c.**) *Airport Mirabel, Montreal, Canada;* (**d.**) *Restaurant Tower, Berlin-Steglitz, FRG. R. Schüler and U. Witte, Architects, M. F. Manleitner and Caner, Structural Engineers,* Architektur + Bauwelt June 1977; (**e.**) *Russian Residence apartment building, New York. SOM, Architects, Irwin G. Cantor Associates, Structural Engineers,* Modern Steel Construction 3rd Qtr. 1975, Domus 503 Oct. 1971, Building Design and Construction Sept. 1974, Constructioneer Aug. 1974; (**f.**) *OCBC Centre, Singapore. I. M. Pei and Partners, Architects, Ove Arup and Partners, Structural Engineers,* ENR June 1975, A.R. April 1980; (**g.**) *Shizuoka Newspaper Co. Building, Nishi-Ginza, Tokyo, Japan. Kenzo Tange, Architect,* Forum March 1968; (**h.**) *Kibogaoka Youth Castle, Shiga Prefecture, Japan. Urban Science Laboratory T. Nakajima, Architect, Kinki Structural Laboratory, Structural Engineers,* JA Oct. 1972; (**i.**) *Westcoast Building, Vancouver, Canada. Rhone and Iredale, Architects, Bogue B. Babicki Associates, Structural Engineers,* Forum May 1972, P/A Oct. 1969, Archit. and Engin.

News *Nov. 1969,* ENR *June 1969,* Civil Engineering–ASCE *Oct. 1971;* (**j.**) *Knights of Columbus Building, New Haven, CT. Kevin Roche, John Dinkeloo, Architects, Pfisterer Tor and Associates, Structural Engineers,* A.R. *Aug. 1970,* P/A *Sept. 1970;* (**k.**) *Hypo Bank, Munich, FRG. Walter and Bea Betz, Architects,* ENR *June 1979,* A. Rev. *June 1981,* Bauingenieur 56, 1981; (**l.**) *UN City, Vienna, Austria. Johann Staber, Architect,* ENR *April 1976.*

Figure 7.6.

(**a.**) *Yamanashi Communications Center, Kofu, Yamanashi, Japan. Kenzo Tange and The Urtec Team, Architects, Fugaku Yokoyama, Structural Engineer,* Forum *Sept. 1967, etc.;* (**b.**) *Hennepin County Medical Center, Minneapolis, MN. Medical Facilities Associates, The Architects Collaborative (TAC), Bakke Kopp Ballou McFarlin, Structural Engineers,* Interstitial Systems in Healthcare Facilities, *USS, Dec. 1979,* A.R. *Sept. 1973;* (**c.**) *College of Medicine, Technical University Aachen, FRG. Weber Brand and Partners, Architects, Philipp Holzmann AG, Structural Engineers,* Bauingenieur *May 1973,* A. Rev. *Oct. 1986.*

Figure 7.7.

(**a.**) *The Hong Kong Club and Office Building, Hong Kong. Harry Seidler and Associates,* Vision 6, 1983, A + U *Sept. 1986;* (**b.**) *The First National Bank of Chicago, Chicago, IL. C. F. Murphy Associates and the Perkins and Will Partnership,* A.R. *April 1969, Sept. 1970;* (**c.**) *Johnson Art Center, Cornell University, Ithaca, NY. I. M. Pei and Partners,* Arch + *Jan./ Feb. 1974;* (**d.**) *University of Birmingham, England. Arup and Associates, Architects,* Detail 3, *1973;* (**e.**) *Time Square Hotel, New York. John Portman Associates,* ENR *Feb. 23, 1984;* (**f.**) *Bank in Lugano, Switzerland. Guido Tallone, Architect,* Revista Technica della Svizzera Italiana *Nov. 1982;* (**g.**) *Hotel in Nashville, TN. Yearwood and Johnson, Architects,* Civil Engineering/ASCE *June 1985;* (**h.**) *Seattle-First National Bank, Seattle, WA. Naramore, Bain, Brady and Johnson, Architects; Skilling, Helle, Christiansen, Robertson, Structural Engineers,* A.R. *June 1970;* (**i.**) *Marriott Hotel, Cambridge, MA. Moshe Safdie and Associates, Architects, LeMessurier, Structural Engineers,* Building Design and Construction *March 1987;* (**j.**) *Beinecke Rare Book and Manuscript Library, Yale University, New Haven, CT. SOM Architects,* ENR *Feb. 1962, etc.*

Figure 7.8.

Read from left to right; top row: BP Office Tower (12 stories), Antwerp, Belgium. Stijnen, De Meyer and Reussens, Architects, P/A *Sept. 1961; Administration Building, Siemens S.A., Saint-Denis, France. Bernard H. Zehrfuss, Architect, 1971, Joedicke, 1975; Olivetti Building (5 stories), Florence, Italy, Alberto Galardi, Architect,* Arch + *Sept. 1973,* P/A 8, *1973; BMW Administration Building (20 stories), Munich, FRG. Karl Schwanzer, Architect,* Arch. + *Oct. 1973; Standard Bank Centre (35 stories), Johannesburg, South Africa. Hentrich-Petschnigg and Partners,* Arch. Review *Aug. 1971,* ENR *Nov. 1968; second row: AUVA, Vienna, Austria. K. Hlaweniczka, Architect,* Architektur Aktuell *Dec. 1975; Federal Reserve Bank (11 stories), Minneapolis, MN. Gunnar Birkerts and Associates,* A.R. *Oct. 1971,* ENR *Nov. 1971; City Hall (11 stories), Marl, FRG. Van den Broek and Bakema, Architects,* ENR *May 1965,* A&A *Jan. 1964; Kansas City Office Building (32 stories, project), Kansas City, MO. Louis Kahn, Architect, A. E. Komendant, 1975.*

Figure 7.10.

Read from left to right: top row:Velasca Tower [1957], Milan, Italy. Belgiojoso/Peressutti/Rogers; Lloyd's of London, London, England. R. Rogers; Takara Beautillion, Expo 70, Osaka, Japan. Kisho Kurokawa; bottom row: Olympia Center, Chicago, IL. SOM (left and right side); 701 4th Ave. S. Minneapolis, MN. Murphy/Jahn; Project, Houston, TX. Kevin Roche and John Dinkeloo.

Figure 7.11.

(**a.**) *Pan Am Building (59 stories), New York. Emery Roth and Sons and Walter Gropius, ENR June 21, 1962;* (**b.**) *860 Lake Shore Drive (26 stories), Chicago, IL. Mies van der Rohe, 1951; A.R. Oct. 1963;* (**c.**) *Civic Center (648 ft, 38 stories), Chicago, IL. C. F. Murphy Associates and SOM, Forum Oct. 1966, P/A Oct. 1966;* (**d.**) *Married student dormitories (20 stories), Harvard University, Cambridge, MA. Sert, Jackson and Gourley, A.R. Sept. 1963;* (**e.**) *Akasaka Prince Hotel (48 stories), Tokyo, Japan. Kenzo Tange and Urtec, J.A. Aug./Sept. 1976 and Jan. 1984;* (**f.**) *Wulfen Metastadt (7–10 stories), Wulfen-Barkenberg, FRG. Dietrich und Steigerwald, C. Riccabona, 1974;* (**g.**) *Housing for the U.S. Embassy (up to 14 stories), Tokyo, Japan. Harry Weese and Assoc., J.A. Sept. 1983;* (**i.**) *Waterside (40 stories), New York. Davis, Brody and Associates, A.R. April 1976;* (**j.**) *Lake Village East (25 stories), Chicago, IL. Harry Weese and Associates, A.R. April 1975;* (**k.**) *UNESCO Secretariat (7 stories), Paris, France. Marcel Breuer, Bernard Zehrfuss and Pier Luigi Nervi, 1958 (Nervi, 1963).*

Figure 7.15.

Read from left to right: top row: (**a.**) *88 Pine Street, New York. I. M. Pei and Associates, A.R. April 1975, AIA Journal June 1974, May 1975;* (**b.**) *See Domus Sept. 1973;* (**c.**) *Orange County Airport Building, CA. Craig Ellwood Associates, Forum May 1972;* (**d.**) *Central Beheer, Apeldoorn, Holland. Herman Hertzberger, Arch + Sept./Oct. 1974, A + U 8312, Beton Atlas 1980, etc.; Second row:* (**e.**) *Harvard U. Science Center, Cambridge, MA. Sert, Jackson and Associates, A.R. Feb. 1972, March 1974;* (**f.**) *The Boston Five Cents Savings Bank, Boston, MA. Kallman McKinnel, Forum March 1973;* (**g.**) *Dokumenta Urbana 1982, Kassel-Dönche, FRG. Steidle and Partner, Bauen + Wohnen Nr. 1/2, 1981; Third row:* (**h.**) *See Domus 597, August 1979;* (**i.**) *The Yale Rare Book Library, New Haven, CT. SOM, ENR Feb. 1962;* (**j.**) *U.S. Embassy, Dublin, Ireland. John M. Johansen, P/A Feb. 1964;* (**k.**) *Crown Street Parking Garage, New Haven, CT. Carleton Granberry Associates, PCI Vol. 17 No. 1, Jan./Feb. 1972.*

Figure 7.24.

Read from left to right: top: AT&T Building [1980], New York. Johnson/Burgee; Humana Building [1985], Louisville, KY. Michael Graves; Empire State Building [1931], New York. Shreve, Lamb and Harmon; PPG Tower [1982], Pittsburgh, PA. Johnson/Burgee; Chrysler Building [1930], New York. William Van Allen; Equitable Tower West [1985], New York. Edward Larrabee Barnes Associates; middle: Republic Bank Center [1982], Houston, TX. Johnson/Burgee; 525 Vine Building [1985], Cincinnati, OH. Glaser and Myers Associates; bottom: Norstar Bancorp Building [1982], Buffalo, NY. Cannon; Statue of Liberty [1886], New York. F. A. Bartholdi and A. G. Eiffel; timber framed houses [1430 to 1590] in Germany.

Figure 7.25.

(**a.**) *First International office tower (56 stories), Dallas, TX. Hellmuth, Obata and Kassabaum Inc., Architects, Ellisor and Tanner, Structural Engineers, Modern Steel Construction 3rd/4th Qtr. 1976;* (**b.**) *Boston Company Building (42 stories), Boston, MA. Pietro Belluschi, Architects, James Ruderman, Structural Engineer, Bethlehem Steel Building Case History No. 1, 1970; P/A Oct. 1969;* (**c.**) *Torre Banaven office tower (680 ft central tower), proposal, Caracas, Venezuela. Enrique Gomez and Jorge Landi, Architects, Gillum Colaco, Structural Engineer, ENR July 1979;* (**d.**) *Onterie Center (58 stories), Chicago, IL. SOM, Architects, A.R. Mid-Aug. 1981, Civil Engineering Oct. 1986;* (**e.**) *John Hancock Center (100 stories), Chicago, IL. SOM, Architects, A.R. Jan. 1967: ENR Sept. 1965, Forum July/Aug. 1970;* (**f.**) *Whitney Museum Tower (35 stories), proposal, New York. Foster Associates, Architects, Frank Newby, Structural Engineer, A.R. Mid-Aug. 1979;* (**g.**) *Hennepin County Government Center (24 stories), Minneapolis, MN. John Carl Warnecke and Associates, Architects, KKBNA, Structural Engineers, A.R. March 1977;* (**h.**) *The First Wisconsin Center (42 stories), Milwaukee, WI.*

SOM, *Architects*, Modern Steel Construction *4th Qtr. 1973*, Building Design and Construction *Sept. 1975*; (**j.**) *Georgia-Pacific Corp. tower (52 stories), Atlanta, GA.* SOM, *Architects, Weidlinger Associates, Structural Engineers*, ENR *Oct. 1979*; *P/A Dec. 1980*: Building Design and Construction *Aug. 1983*; (**i.**) *Medical Mutual of Cleveland (448 ft), Cleveland, OH. Hugh Stubbins Associates, Architects, LeMessurier, Structural Engineers*, *P/A Dec. 1980*; (**k.**) *Federal Reserve Bank (33 stories), Boston, MA. Hugh Stubbins and Associates, Architects, LeMessurier, Structural Engineers*, ENR *April 1975*, A.R. *Mid-Aug. 1974*; Building Design and Construction *Jan. 1975*; Modern Steel Construction *1978*; (**l.**)*Bush Lan House (10 stories), London, England. Arup Associates, Architects*, The Structural Engineer *Feb. 1977*; ENR *Jan. 20, 1977*; (**m.**) *Alcoa Building (26 stories), San Francisco, CA.* SOM, *Architects*, Contemporary Structures AISI, *1970*, ENR *July 1965*; (**n.**) *IBM Building (13 stories), Pittsburgh, PA. Curtis and Davis, Architects, Worthington, Skilling, Helle and Jackson, Structural Engineers*, USS Structural Report *August 1963*, ENR *Sept. 6, 1962*; (**o.**) *The Century Tower (20-story, 1988), proposal, Tokyo, Japan. Foster Associates, Architects, Ove Arup and Partners, Structural Engineers*, A. Review *Aug. 1988*; (**p.**) *Superskyscraper (142 stories), proposal, Chicago, IL. Kay Vierk Janis, Architect*, Civil Engineering *March 1989*.

Figure 7.27.

Read from left to right starting with the top row: (**a.**) *Arthur Schomburg Plaza (35 stories), New York. Gruzen, Architect*, Building Design and Construction *April 1972*; (**b.**) *Tour Maine-Montparnasse (64 stories), Paris, France. Baudoin, Cassan, de Marien and Saubot, Architects*, AA No. 178, *March/April, 1975*; (**c.**) *1515 Poydras (27 stories), New Orleans, LA.* SOM, *Architects*, Concrete International *Sept. 1984*; (**d.**) *Seagram Building (38 stories), New York. Mies van der Rohe, Architect*; (**e.**) *One Biscayne Tower (40 stories), Miami, FL. Fraga Associates, Architects*, A.R. *Feb. 1974*, ENR *July 20, 1972*, Concrete Today *Vol. 7, No. 2*; (**f.**) *Fiduciary Trust Co. (17 stories), Boston, MA. The Architects Collaborative (TAC)*, Building Design and Construction *Nov. 1976*; (**g.**) *Portland Plaza Condominium (25 stories), Portland, OR. Daniel, Mann, Johnson and Mendenhall, Architects*, A.R. *Oct. 1973*; (**h.**) *Citizens Bank Center (13 stories), Richardson, TX. Omniplan, Architects*, A.R. *July 1975*, Civil Engineering-ASCE *April 1977*; (**i.**) *Inter First Plaza (28 stories), Houston, TX.* SOM, *Architects*, Architecture *Feb. 1985*; (**j.**) *Imperial Chemical Office Building (8 stories), Stamford, CT. W. F. Pedersen, Architect*, P/A *March 1971*; (**k.**) *Manufacturers and Traders Trust Co. (21 stories), Buffalo, NY. Minoru Yamasaki, Architect*, Bethlehem Steel project description; (**l.**) *Harbor Point (54 stories), Chicago, IL. Solomon, Cordwell, Buenz and Associates, Architects*, ENR *Aug. 14, 1975*; (**m.**) *City Administration Building (30 stories), Brussels, Belgium.* ENR *May 3, 1979*; (**n.**) *Westin Hotel (36 stories), Boston, MA. The Architects Collaborative*, CRSI Case History No. 24, Concrete International *Oct. 1983*, Building Design and Construction *Dec. 1983*; (**o.**) *Great Western Savings Center (10 stories), Beverly Hills, CA. W. L. Pereira Associates, Architects*, Building Design and Construction *March 1973*; (**p.**) *IDS Center (57 stories), Minneapolis, MN. Johnson/Burgee, Architects*, AIA Journal *June 1979*, A.R. *Aug. 1970*, ENR *Nov. 1971*; (**q.**) *Australia Square (51 stories), Sydney, Australia. Seidler Associates, Architects*, Art and Arch. *Nov. 1965*; (**r.**) *Gulf Life Tower (27 stories), Jacksonville, FL. Welton Becket and Associates, Architects*, A.R. *Nov. 1966*, *Jan. 1967*, ENR *July 7, 1966*.

Figure 7.28.

(**a.**) *333 West Wacker Building, Chicago, IL. Kohn, Pedersen, Fox Architects, GCE, Structural Engineers*; Building Design and Construction *June 1983*, Modern Steel Construction *1st Qtr. 1984*, P/A *Oct. 1983*; (**b.**) *311 S. Wacker Dr., Chicago, IL. Kohn, Pedersen, Fox Architects, Brackette Davis Drake, Structural Engineers*, ENR *Feb. 4, 1988*, Civil Engineering *March 1988*; (**c.**) *Apartment building, 33 Rue Croulebarbe, Paris, France. Albert-Boileau, Labourdette Architects*, Bouwcentrum, *1963*; (**d.**) *Peachtree Plaza Hotel Tower, Atlanta. John Portman and Associates, Architects and Structural Engineers*, A.R. *June 1976*, ENR *Jan. 22, 1976*; (**e.**) *City Center Office Towers, Fort Worth, TX. Paul Rudolph, Architect, CBM, Structural Engineers*, A.R. *July 1982*, G.A. Document *Oct. 1983*; (**f.**) *Luth Tower, Kuala Lumpur, Malaysia. Hijjas*

Kasturi Associates, Architects, Ranhill Berseketu Sdn, Bhd, Structural Engineers, ENR July 1, 1982, Sept. 20, 1984, Asian Building and Construction March 1984; **(g.)** *Mercantile Tower, St. Louis, MO. Sverdrup and Parcel Associates and Thompson Ventulett and Stainbeck, Architects, Ellisor, Structural Engineers, ENR Oct. 16, 1975, A.R. Aug. 1975;* **(h.)** *Raffles City Tower, Singapore. I. M. Pei and Partners, Architects, Weiskopf and Pickworth, Structural Engineers;* **(i.)** *Gulf Life Tower, Jacksonville, FL. Welton Becket Associates, Architects, R. R. Bradshaw, Inc., Structural Engineers, A.R. Nov. 1966, Jan. 1967; ENR July 7, 1966.*

Figure 7.29.

First Interstate World Center [1989] (73 stories), Los Angeles, CA. I. M. Pei, Architect, CBM, Structural Engineers; Civil Engineering Nov. 1988; ENR Sept. 14, 1989.

Figure 7.30.

(f.) *US Steel Headquarters (64 stories), Pittsburgh, PA, Harrison and Abramovitz and Abbe, Architects, Skilling, Helle, Christiansen, Robertson, Structural Engineers, The Steel Triangle US Steel, July 1969, ASCE April 1970, A.R. April 1967, Forum Dec. 1971, ENR Jan. and June 1970;* **(i.)** *IDS Center (57 stories), Minneapolis, MN. Johnson/Burgee Architects, Severud/ Perrone/Sturm/Conlin/Bandel, Structural Engineers, A.R. Aug. 1970, ENR Nov. 1971, AIA Journal June 1979.*

Figure 7.31.

Trump Tower (68 stories), New York. Swanke Hayden Connell, Architects, I.G. Cantor, Structural Engineer, CRSI Case History No. 23, Concrete International March 1984, ENR April 15, 1982, Building Design and Construction Aug. 1983, P/A Dec. 1980, AA 220, April 1982.

Figure 7.32.

(a.) *Allied Bank Plaza [1983] (71 stories), Houston, TX. SOM, Architects, ENR July 15, 1982; Architecture April 1984; Building Design and Construction Sept. 1980;* **(b.)** *17 State Street building [1988] (42 stories), New York. Emery-Roth and Sons, Architects, DiSimone Caplin and Associates, Structural Engineers, Modern Steel Construction #4, 1988; Civil Engineering June 1988.*

Figure 7.33.

Read from left to right and top to bottom: Allied Bank Tower (60 stories), Dallas, TX. I. M. Pei, Architect, Modern Steel Construction #2, 1987; Architecture Dec. 1986, Building Design and Construction Sept. 1987; AT&T Corporate Headquarters (37 stories), New York. Johnson/ Burgee, Architects, A.R. Oct. 1980; P/A Dec. 1980; Modern Steel Construction AISC, 1st Qtr. 1982; A. Rev. Aug. 1984; Momentum Place (60 stories), Dallas, TX. Burgee/Johnson, Architects, Civil Engineering, June 1986; ENR Nov. 13, 1986; Republic Bank Center (56 stories), Houston, TX. Johnson/Burgee, Architects, Modern Steel Construction 2nd Qtr. 1984; Building Design and Construction June 1983, P/A Feb. 1984, Architecture April 1984; One Magnificent Mile (57 stories), Chicago, IL. SOM, Architects, Building Design and Construction Oct. 1984, Inland Architect May/June 1984, A.R. Mid-Aug. 1980; Sears Tower (110 stories), Chicago, IL. SOM, Architects, Forum Jan./Feb. 1974, Engineering Journal AISC, 3rd Qtr. 1973, Arch. + Aug. 1973, Civil Engineering Nov. 1972; Sunshine 60 (60 stories), Tokyo, Japan. Mitsubishi Estate Co., Architects, Beedle, 1986; Four Allen Center (50 stories), Houston, TX. Lloyd Jones Brewer Associates, Architects, Ellisor and Tanner, Structural Engineers, USS Building Report Dec. 1984, Modern Steel Construction AISC, 2nd Qtr. 1983, Building Design and Construction June 1984; 780 Third Ave. (50 stories), New York. SOM, Architects, ENR June 30, 1983,

Concrete International *Feb. 1985; CBS Building (38 stories), New York. Eero Saarinen, Architect, A.R. July 1965*, ENR *May 21, 1964;* (**a.**) *El Paso Tower (70 stories), Houston, TX. I. M. Pei, Architect;* (**b.**) *Western Pennsylvania National Bank Building (32 stories, proposal), Pittsburgh, PA. SOM, Architects;* (**c.**) *Three Houston Center Gulf Tower Building (52 stories), Houston, TX;* (**d.**) *Southeast Financial Center (53 stories), Miami, FL. SOM, Architects;* (**e.**) *First Canadian Centre (64 and 43 stories), Calgary, Canada. SOM, Architects;* (**f.**) *NCNB Plaza (40 stories), Charlotte, NC. Thompson, Ventulett and Stainback, Architects. (framed tube with interior rigid frame bracing);* (**g.**) *Office building (52 stories), Tokyo, Japan. Nikken Sekkei, Architect (triple tube);* (**h.**) *Pan-American Life Insurance Co. Building (27 stories), New Orleans, LA. (concrete tube).*

Figure 7.36.

Read from left to right: bottom: Mile-high Tower. Frank Lloyd Wright, Architect, A. Forum *Nov. 1956, A.R. March 1957; The Bank of China, Hong Kong. I. M. Pei, Architect, and L. E. Robertson, Structural Engineer, A.R. Sept. 1985,* Civil Engineering *Aug. 1986,* ENR *Oct. 13, 1988; The Bank of the Southwest Tower, Houston, TX. proposal, Murphy/Jahn, Architects, and LeMessurier, Structural Engineers, A.R. Jan 1985,* Arch. Technology *Fall 1983; Trussed megatube, proposal, J. P. Colaco of CBM Engineers,* Civil Engineering *April 1986; Citicorp Center, New York. Hugh Stubbins, Architect, and LeMessurier, Structural Engineers,* ENR *June 24, 1976, A.R. Mid-Aug. 1976 and June 1978; Medical Mutual, Cleveland, OH. Hugh Stubbins, Architect, and LeMessurier, Structural Engineers, P/A Dec. 1980; top: Proposal for a 150-story superframe building in steel. Alfred Swenson (I.I.T., 1969),* Forum *Sept. 1971; Proposal for a 210-story guyed tower, Harry Weese, Architect, and Charles H. Thornton, Structural Engineer of Lev Zetlin Associates,* ENR *Nov. 3, 1983,* High Technology *Jan. 1985; Proposal for a 168-story telescoping superframe, SOM, Architects,* ENR *Nov. 3, 1983,* High Technology *Jan. 1985,* Civil Engineering/ASCE *Jan. 1984; Dearborn Center, Chicago, IL. SOM, Architects,* Civil Engineering-ASCE *March 1985; Lateral sway of megaframe; Proposal for a 200-story linked tower system. DeSimone and Chaplin, Architects,* High Technology *Jan. 1985; Proposal for a 135-story trussed tube. SOM, Architects,* Popular Science, *Dec. 1985; Proposal for a 130-story trussed tube. Eli Attia, Arch.,* Popular Science *Dec. 1985; 383 Madison Ave., New York. KPF and LeMessurier Associates, Architects, A.R. Feb. 1987; InterFirst Plaza, Dallas, TX. Jarvis Putty Jarvis, Architects and LeMessurier Associates, Structural Engineers,* ENR *June 6, 1983, A.R. Jan. 1985; Erewhon Center (proposal). LeMessurier Associates, A.R. Feb. 1985.*

Figure 7.37.

Hongkong Bank, Hong Kong. Foster, Architect, and Ove Arup, Structural Engineer, A. Rev. *May 1981, AA Oct. 1982, A + U Oct. 1983, Vision Jan. 1983, TA Nov. 1983,* Building *July 1984,* ENR *Oct. 4, 1984, P/A March 1986, DBZ Sept. 1986, etc.*

Figure 7.38.

(**a.**) *Tarics A. G.: Concrete-filled steel columns for multistory construction,* Modern Steel Construction *1st Qtr. 1972;* (**b.**) *One Mellon Bank Center (54 stories), Pittsburgh, PA. Welton Becket Associates, Architects, Lev Zetlin Associates, Inc., Structural Engineers, A.R. Mid-Aug. 1981,* Modern Steel Construction *3rd Qtr. 1984, P/A Dec. 1980,* ENR *May 20, 1982;* (**c.**) *Belford, Don: Composite steel-concrete building frame,* Civil Engineering-ASCE *July 1972;* (**d.**) *First City Tower (49 stories), Houston, TX. Morris-Aubry, Architects, Walter P. Moore and Associates, Structural Engineers,* Consulting Eng. *March 1985,* ENR *Feb. 18, 1981, P/A Dec. 1980,* The Structural Engineer *Sept. 1982, Sept. 1983;* (**e.**) *One Shell Square Tower, New Orleans, LA. SOM, Architects,* ENR *June 3, 1971;* Civil Engineering-ASCE *April 1976;* (**f.**) *Tower 49, New York. SOM, Architects,* Civil Engineering-ASCE *March 1985;* (**g.**) *Eau Claire Estate Tower, Calgary, Canada. SOM, Architects;* (**h.**) *3 First National*

Plaza, Chicago, IL. SOM, Architects; (**i.**) *Two Union Square Tower, Seattle, WA. NBBJ Group, Architects, SWMB, Structural Engineers, ENR Feb. 16, 1989;* (**j.**) *Cityplace Tower, Dallas, TX. Cossutta and Associates, Architects, Weiskopf and Pickworth, Structural Engineers, ENR Dec. 10, 1987;* (**k.**) *Pacific First Center, Seattle, WA. Callison Partnership, Architects, SWMB, Structural Engineers, ENR Feb. 16, 1989.*

Figure 7.39.

(**a.**) *One Liberty Place, Philadelphia, PA. Murphy/Jahn Architects, Lev Zetlin Associates, Structural Engineers,* Civil Engineering-ASCE *March 1986,* Modern Steel Construction #2/1987; (**b.**) *383 Madison Ave. (proposal), New York. Kohn Pedersen Fox Associates, Architects, LeMessurier Associates, Structural Engineers, A.R. Feb. 1985 and Feb. 1987;* (**c.**) *Wilshire Finance Building, Los Angeles, CA. Albert C. Martin and Associates, Architects/Engineers, ENR Sept. 19, 1985;* Civil Engineering-ASCE *March 1986;* (**d.**) *900 North Michigan, Chicago, IL. Kohn Pedersen Fox, Architects, Alfred Benesch Co., Structural Engineers,* Building Design and Construction *March 1987.*

Figure 7.40.

(**a.**) *Chicago Mercantile Exchange (44 stories), Chicago, IL. Fujikawa Johnson and Associates, Architects, Alfred Benesch and Co., Structural Engineers,* Concrete International *Dec. 1983, A.R. Nov. 1983,* Modern Steel Construction *2nd Qtr. 1983;* (**b.**) *Ticor Title Insurance Co. Office Building (6 stories), San Diego, CA. Deems/Lewis and Partners, Architects, Atkinson, Johnson and Spurrier, Structural Engineers,* Modern Steel Construction, *3rd Qtr. 1986;* (**c.**) *Centrust Tower (37 stories), Miami, FL. I. M. Pei, Architect, CBM Engineering, Inc., Structural Engineers,* Civil Engineering *Dec. 1986.*

Figure 7.41.

Read from left to right: top row: Dyodon project, pneumatic residential cells. Jean-Paul Jungmann, 1967; Modular housing structures project. J. Francois Gabriel, Housing, Pergamon, 1981; *Office tower project. Carlo Moretti,* Domus 558, 1976; *Nakagin Capsule Tower [1972], Tokyo, Japan. Kisho Kurokawa, A.R. Feb. 1973; Vertical Assembly Building, Cape Kennedy, FL. Urbahn, Roberts, Seelye, Moran, AA March 1975; Centre Pompidou, Paris, France. Piano + Rogers, Architects, and Ove Arup and Associates, Structural Engineers, AA 190 April 1977, P/A May 1977, A. Review May 1977, AD Profiles Feb. 1977, A.R. Feb. 1978, GA 44; bottom row: Montreal Entertainments Tower project. Peter Cook, 1963; Helicoidal skyscraper project. Manfredi Nicoletti A + Nov. 1973,* L'architettura 2172; *Space frame tower projects. Louis I. Kahn and Griswold Tyng, ENR June 5, 1958; Arcosanti. Paoli Soleri, AIA Journal Feb. 1971,* Arts + Architecture V2 #4, 1984.

Figure 8.2.

Read from left to right: Kuwait Water Towers [1980], Kuwait City, Kuwait; octagonal hyperboloid frame for a water tower (Zusse Levinton, 1967); microwave tower [1979] in Key Largo, FL; ski jump tower, Obersdorf, FRG; bell tower [1963], Priory of the Annunciation, Bismarck, ND (Marcel Breuer); Minneapolis/Plymouth Radio Tower [1975]; 233-ft pithead tower [1981], Bergkamen, FRG; bridge on interstate 8, 40 mi east of San Diego; Friedrich Ebert Bridge [1967], Bonn, FRG; Expo Tower [1970], Osaka, Japan (Kisho Kurokawa); gravity type oil drilling platform (1970s); control tower [1974], Dallas/Fort Worth Airport; two transmission towers; olympic ski jump [1979], Lake Placid, NY; 1572-ft Okalhoma City Television Tower; hyperboloid cooling tower; Humber Bridge [1978], England; airshafts for exhaust fumes [1978], Ymuiden, the Netherlands; crane; CN Tower [1976], Toronto, Canada.

Figure 8.3.

Read from left to right and top to bottom: Needle Tower [1968] (Kenneth Snelson); Force System (sculpture by Walter Kaitna, 1963); giant redwood, northern California along the Pacific Ocean; cereal grass; microwave tower; Centrepoint Tower [1972], Sydney, Australia; Spinal Column (kinetic sculpture by Frei Otto, 1963); Eiffel Tower [1889], Paris, France (984 ft, Gustave Eiffel); funicular polygon of revolution tower [1962] (Robert Le Ricolais); Telecommunication Tower [1976], Frankfurt, FRG (Leonardt and Andrae); 377-ft tower [1350] of Freiburg Cathedral, Freiburg, FRG: Leaning Tower of Pisa, Italy (thirteenth century); Banco National de Desenvolvimento Economico [1981], Rio de Janeiro, Brazil (22-story); 15-story laboratory tower [1939], Johnson Wax Company, Racine, WI, (Frank Lloyd Wright).

Figure 8.4.

Read from left to right: top: conceptual layout for a repository in a salt dome, Underground Space (US) 1/4, 1982; radioactive waste disposal in salt dome, Tunnelling and Underground Space Technology, Vol. 1, #2, 1986; portion of underground repository layout for nuclear waste, T&US Vol. 1, #3/4, 1986; subsurface development proposal for the University of Minnesota; underground parking garage, Geneva, Switzerland; section through proposed power installations for a two-step underground pumped hydro storage arrangement, US May/June, 1983; salt dome cavern for crude oil storage, US May/June 1982; bottom: Queen of the Guadalupes, cave in the Guadalupe Mountains, Eddy County, NM; cavern plan for liquefied petroleum gas storage facility, US Vol. 3, #5, 1979; general view of urban underground space; room and pillar mining, Kansas City, MO, Smithonian Feb. 1979.

Figure 8.6.

Read from left to right: top: floating island, Monaco. E. Albert and J. Cousteau; two floating airports, proposals. Lev Zetlin; exhibition pavilions [1971] in Toronto, Canada. Craig-Zeidler-Strong; floating airport, proposal. Lev Zetlin; Aquapolis [1975], Okinawa, Japan. Kiyonori Kikutake; middle: Tecnomare steel gravity platform; Sea Tank Cormorant A (UK) [1978]; Tripod 300, concrete gravity structure, proposal; Doris, Ninian Central Platform (UK) [1978]; Hutton Tension-Leg Platform (UK) [1983]; semisubmersible drilling rig; bottom: Magnus, steel-template-jacket platform (UK) [1982]; jack-up rick; Cognac Platform, Gulf of Mexico (USA) [1978]; Mandrill 400, tripod steel structure, proposal; Lena Platform Guyed Tower, Gulf of Mexico, (USA) [1985]; Condeep, Statfjord B concrete gravity-base platform (Norway) [1981]; articulated loading platform.

APPENDIX C: ANSWERS TO SELECTED PROBLEMS

Problem 2.1	$h = 3600$ ft
Problem 2.2	$t = 5/16$ in.
Problem 2.4	$\Delta = 13.84$ in.
Problem 2.6	Gravity tower.
Problem 2.7	$f_c = 11.18$ ksi, S.F. $= 2.47$
Problem 2.9	$f_c = 114.71$ psi, $f_t = 28.09$ psi
Problem 2.12	$f_c = 6.84$ ksi
Problem 2.14	$f_b = 0.96$ ksi
Problem 2.21e	$I = 1616$ ft^4
Problem 3.3	BM$_{3-4}$: W18×40, BM$_{1-5}$: W21×68
Problem 3.4	W14×370
Problem 3.8a	$T = 1.04$ sec
Problem 3.9	$V = 3605$ k, S.F. $= 5.45$
Problem 3.12	$V = 1718$ k, S.F. $= 2.29$
Problem 3.17d	$V = 473$ k, S.F. $= 3.25$
Problem 3.21	$T = 2745$ lb
Problem 3.25	$f_c = 0.326$ ksi
Problem 3.26	$f_a = 14.25$ ksi, $f_b = 31.97$ ksi
Problem 3.28a	$\Delta = 0.212$ in.
Problem 3.30	5/8 in.
Problem 3.32	3/8 in.
Problem 3.34	$T = -23.30°F$
Problem 4.3	#3 @ 6 in. and #3 @ 12 in.
Problem 4.6	1 5/8 × 20 in. and 3/8 × 68 1/4 in.
Problem 4.8	W14×159
Problem 4.10	W12×120
Problem 4.12	W14×120
Problem 4.16	4 #9, #3 ties @ 12 in.
Problem 4.19a	20 × 20 in.
Problem 4.21	34 × 34 in.
Problem 4.23	10 × 10 in., 4 #5, #3 ties @ 10 in.
Problem 4.26	14 × 14 in.
Problem 4.28	18 × 18 in.

Problem 4.29	28 × 28 in.
Problem 4.31	6 #9
Problem 5.1	$P_A = 490$ k, $P_B = 980$ k, $P_C = 0$
Problem 5.3b	$P_C = P_D = 240$ k, $P_A = P_B = 0$
Problem 5.4	$t = 18$ in.
Problem 5.8	$P = 78.29$ k
Problem 5.9	$P_i = 5.67$ k, $P_o = 9.50$ k
Problem 5.12	$P_y = 46.90$ k, $P_x = 8.21$ k
Problem 5.14	$P_{y1} = 48.91$ k, $P_{y2} = 51.09/2$ k, $P_x = 1.81$ k
Problem 5.18b	5-in. slab, #4 @ 8 1/2 in. at support, #4 @ 13 in. field, #3 @ 11 in. temperature steel.
Problem 5.22b	4 1/2-in. slab, #4 @ 7 in., #3 @ 12 in. temperature steel
Problem 5.25	7-in. slab, #5 @ 5 1/2 in. above columns
Problem 5.28	5-3/4 in. alloy bars
Problem 5.30	6 1/2-in. slab, in each direction strands with a diameter of 0.438 in. @ 12 1/2 in. o.c., $f_c = 517$ psi, $f_t = 91$ psi
Problem 6.1	$t = 2\ 3/4$ in.
Problem 6.4	$f_t = 62.86$ psi NG
Problem 6.6	8-in. brick wall.
Problem 6.7	Additives must be used.
Problem 6.10	$I_s = 3.07$ in.^{4}/ft
Problem 6.12	$I = 9.75b$, $I = 0.56b$
Problem 6.14	#6 @ 22 in. o.c.
Problem 6.16	$f_t = 17.34$ psi ≤ 28 psi OK
Problem 6.19	6000 psi units
Problem 6.21	#5 @ 20 in. o.c.
Problem 6.24	The wall does not work.
Problem 6.26	S.F. $= 1.65 \ge 1.5$ OK
Problem 6.27	2 #6 and 2 #8
Problem 6.30	4 floors
Problem 6.32	$M = 26.64$ ft-k
Problem 6.33	M14×18
Problem 6.34	L 3 1/2×3 1/2×1/4
Problem 6.36a	WT 7×11
Problem 6.37	2 #4
Problem 6.38	2 #4
Problem 7.2	Core is in compression.
Problem 7.3	$A_s = 75$ in.^{2}
Problem 7.4	$k = 1.9$
Problem 7.5	W14×211
Problem 7.7	36 × 36 in.
Problem 7.9	W16×31 beam, W12×53 column.
Problem 7.11d	W30×108 girder, W14×370 column.
Problem 7.16	W14×159
Problem 7.20	W16×36 beam, L 8×6×9/16 bracing, W14×109 column.
Problem 7.23	4.02 ft
Problem 7.24	16 × 16 in., 2 to 3 percent
Problem 7.29	W14×109

REFERENCES

Ambrose, James: *Building Structures*. New York: John Wiley, 1988.

American Institute of Steel Construction: Temperature effects on tall steel framed buildings, *AISC Engineering Journal*, Oct. 1970.

American Institute of Steel Construction: *Manual of Steel Construction*, 8th ed. Chicago: AISC, 1980.

American Institute of Steel Construction: *Save with Steel in Multi-Story Buildings*. New York: AISC, 1965.

American Iron and Steel Institute: *Fire-Resistant Steel-Frame Construction*, 2nd ed. Washington, D.C.: AISI, 1974.

American National Standards Institute: Minimum Design Loads for Buildings and Other Structures (ANSI A58.1-1982). New York: ANSI, 1982.

Architects and Earthquakes. Washington, D.C.: AIA Research Corporation, 1977.

Aregger, Hans and Glaus, Otto: *Highrise Building and Urban Design*, New York: Praeger, 1967.

Arnold, Christopher and Reitherman, Robert: *Building Configuration Seismic Design*. New York: John Wiley, 1982.

Aynsley, Richard M.: Wind effects around buildings. *Arch. Science Review*, March 1972.

Aynsley, Richard M.: Wind effects on high and low rise housing. *Arch. Science Review*, Sept. 1973.

Bakke and Kopp, Inc., Struct. Eng.: Staggered truss framing systems for high-rise buildings. *USS Technical Report*, Dec. 1972.

Banham, Reyner: *Megastructure*. New York: Harper and Row, 1976.

Barnett, Jonathan: *An Introduction to Urban Design*. New York: Harper and Row, 1982.

Bartos, Michael J., Jr.: Building starscrapers in orbit. *Civil Engineering/ASCE*, July 1979.

Bayley, Roger W.: Designing with computers. *The Canadian Architect*, October 1983.

Bednar, Michael J.: *The New Atrium*. New York: McGraw-Hill, 1986.

Beedle, Lynn S., ed.: *Advances in Tall Buildings*. New York: Van Nostrand Reinhold, 1986.

Beedle, Lynn S., ed.: *Second Century of the Skyscraper*. New York: Van Nostrand Reinhold, 1988.

Belford, Don M.: Composite steel-concrete building frame. *Civil Engineering/ASCE*, July 1972.

Bergman, Magnus S.: The development and utilization of subsurface space. *Tunnelling and Underground Space Technology*, Vol. 1, Number 2, 1986.

Bethlehem Steel Corporation: Building Case History, (reports), Bethlehem, PA.

Billington, David P. and Mark, R.: *Structures and Urban Environment*. Department of Civil Engineering, Princeton University, 1983.

Billington, David P., and Goldsmith, Myron, eds.: *Technique and Aesthetics in the Design of Tall Buildings*. Lehigh University, Bethlehem, PA: Institute for the Study of the High-Rise Habitat, 1986.

Bouwcentrum/Rotterdam: *Modern Steel Construction in Europe*. Amsterdam, Netherlands: Elsevier, 1963.

Bradshaw, Vaughn: *Building Control Systems*. New York: John Wiley, 1985.

Bresler, Boris, Lin, T.Y., and Scalzi, John B.: *Design of Steel Structures*, 2nd ed. New York: John Wiley, 1968.

Brick Institute of America: Technical Notes on Brick Construction from BIA, Reston, VA.

Burns, Joseph G.: *The Engineering Aesthetics of Tall Buildings*. New York: ASCE, 1985.

Chang, Fu-Kuei: Human response to motions in tall buildings. *Journal of the Structural Division,* ASCE, Vol. 99, No. ST6, June 1973.

Chopra, Anil K.: *Dynamics of Structures.* Berkeley, CA: Earthquake Engineering Research Institute, 1981.

Colaco, Joseph P.: Partial tube concept for mid-rise structures. *Engineering Journal,* AISC, 4th Qtr, 1974.

Coleman, Robert A.: *Structural Systems Design.* Englewood Cliffs, NJ: Prentice-Hall, 1983.

Composite steel-concrete construction, *Journal of the Structural Division,* ASCE, Vol. 100, No. ST5, Proc. Paper 10561, May 1974.

Concrete Reinforcing Steel Institute Committee on Fire Ratings: *Reinforced Concrete Fire Resistance.* Chicago, IL: CRSI, 1980.

Condit, Carl W.: *American Building Art: The Nineteenth Century.* New York: Oxford U. Press, 1960.

Council on Tall Buildings and Urban Habitat: Monograph on Planning and Design of Tall Buildings (5 volumes): CB *Structural Design of Tall Concrete and Masonry Buildings,* 1978; SB *Structural Design of Tall Steel Buildings,* 1979; SC *Tall Building Systems and Concepts,* 1980; CL *Tall Building Criteria and Loading,* 1980; PC *Planning and Environmental Criteria for Tall Buildings,* 1981. New York: American Society of Civil Engineers.

Council on Tall Buildings and Urban Habitat: *High-Rise Buildings, Recent Progress.* Bethlehem, PA: Lehigh University, 1986.

Cowan, Henry J.: *Science and Building.* New York: John Wiley, 1978.

Coull, A. and Choudhury, J. R.: Stresses and deflections in coupled shear walls. *ACI Journal,* Feb. 1967.

Crump, Ralph W.: Games that buildings play with winds. *AIA Journal,* March 1974.

Dahinden, Justus: *Urban Structures for the Future.* New York: Praeger, 1972.

Dampers blunt the wind's force on tall buildings. *A.R.,* Sept. 1971.

Degenkolb, Henry J.: *Earthquake. Bethlehem Steel Booklet* 2717A., 1978.

Deilman, Harald: *The Dwelling: Use-Types, Plan-Types, Dwelling-Types, Building-Types,* 3rd ed. Stuttgart, FRG: K. Krämer, 1980.

Drechsel, Walther: *Turmbauwerke.* Wiesbaden, FRG: Bauverlag GMBH, 1967.

Drew, Philip: *Third Generation.* New York: Praeger, 1972.

Dym, Clive L. and Klabin, Don: Architectural implications of structural vibration. *A.R.,* Sept. 1975.

Ellers, Fred S.: Advanced offshore oil platforms. *Scientific American,* April 1982.

Evans, Martin: *Housing, Climate and Comfort.* London: Architectural Press, 1980.

Fairweather, Virginia: Building in space. *Civil Engineering/ASCE,* June 1985.

Falconer, Daniel W. and Beedle, Lynn S.: *Classification of Tall Building Systems.* Lehigh University: Fritz Engineering Lab Report 442.3, 1984.

Faller, Peter and Schröder, Hermann: *Terrassierte Bauten in der Ebene, Beispiel Wohnhügel.* Bonn, FRG: Schriftenreihe des Bundesministers für Raumordnung, Bauwesen und Städtebau, 1973.

Fintel, Mark, ed.: *Handbook of Concrete Engineering,* 2nd ed. New York: Van Nostrand Reinhold, Co., 1985.

Fintel, Mark and Ghosh, S.K.: *Economics of Long-Span Concrete Slab Systems for Office Buildings—A Survey.* Skokie, IL: PCA, 1982.

Fintel, Mark and Schultz, Donald M.: A philosophy for structural integrity of large panel buildings. *PCI Journal,* May/June 1976.

Fintel, Mark and Schultz, Donald M.: Structural integrity of large panel buildings. *ACI Journal,* May 1979.

Fitch, James M.: *American Building: The Historical Forces that Shaped It,* 2nd ed. New York: Schocken Books, 1973.

Fling, Russell S.: *Practical Design of Reinforced Concrete.* New York: John Wiley, 1987.

Gandemer, Par J.: *Wind Environment Around Buildings: Aerodynamic concepts.* Proceedings of the 4th International Conference on Wind Effects on Buildings and Structures, Heathrow 1975, Cambridge, UK: Cambridge University Press, 1977.

Gandemer, Par J.: Les effects aerodynamiques du vent dans les ensembles batis. *TA* 325, June/July 1979.

Gaylord, Edwin H., Jr. and Gaylord, Charles N., eds.: *Structural Engineering Handbook,* 2nd ed. New York: McGraw-Hill, 1979.

Gerwick, Ben C., Jr.: *Construction of Offshore Structures.* New York: John Wiley, 1986.

Giedion, Siegfried: *Space, Time, and Architecture*, 5th ed. Cambridge, MA: Harvard U. Press, 1967.

Goldberger, Paul: *The Skyscraper*. New York: Alfred A. Knopf, 1982.

Graff, W. J.: *Introduction to Offshore Structures*. Houston: Gulf Publ. Co., 1981.

Grütter, Jörg Kurt: *Ästhetik der Architektur*. Stuttgart, FRG: Kohlhammer Verlag, 1987.

Guise, David: *Design and Technology in Architecture*. New York: John Wiley, 1985.

Hart, F., Henn W., and Sontag, H.: *Multi-Storey Buildings in Steel*. London: Granada Publ., 1978.

Hooper, Ira: Design of beam-columns. *AISC Engineering Journal*, April 1967.

Innovation in Masonry. *P/A*, Feb. 1979.

International Conference on Planning and Design of Tall Buildings (1972: Lehigh University): *Proceedings* (5 volumes): v.C *Introduction, Conference Record, Indexes;* v.Ia *Tall Building Systems and Concepts;* v.Ib *Tall Building Criteria and Loading;* v.II *Structural Design of Tall Steel Buildings;* v.III *Structural Design of Tall Concrete and Masonry Buildings*. New York: American Society of Civil Engineers, 1972.

International Masonry Institute: *The New and Modern Capabilities of Engineered Loadbearing Walls*. Washington, D.C.: IMI, 1976.

Jahn, Helmut: Genesis of a tower. *Architectural Technology*, Fall 1983.

Jencks, Charles: *Skyscrapers—Skycities*. New York: Rizzoli, 1980.

Joedicke, Jürgen: *Office and Administration Buildings*. Stuttgart, FRG: K. Krämer Verlag, 1975.

Khan, Fazlur R.: The future of highrise structures. *P/A*, Oct. 1972.

Knoll, Franz: Structural design concepts for the Canadian National Tower, Toronto, Canada. *Canadian Journal of Civil Engineering*, v.2, no. 2, 1975.

Komendant, August E.: *18 Years with Architect Louis I. Kahn*. Englewood, NJ: Aloray, 1975.

Kraemer, F. W. and Meyer, Dirk: *Bürohaus-Grundrisse*. Stuttgart, FRG: A. Koch, Publ., 1974.

Leba, Theodore: *The Application of Nonreinforced Concrete Masonry Load-Bearing Walls in Multistory Structures*. McLean, VA: NCMA, 1969.

Leet, Keneth: *Reinforced Concrete Design*. New York: McGraw-Hill, 1982.

Leffler, Robert E.: Efficiency of tubular framing for medium-height buildings. *Engineering Journal*, AISC, 4th Qtr., 1979.

Lehman, Conrad R.: Multi-storey suspension structures. *A.D.*, Nov. 1963.

LeMessurier, William J.: Supertall structures. *A.R.* Jan. 1985 and Feb. 1985.

Leonhardt, Fritz and Mönnig, E.: *Vorlesungen über Massivbau, Dritter Teil*, 3rd ed. Berlin, FRG: Springer Verlag, 1977.

Lerup, Lars, Cronrath, David, Liu, Chiang, and Koh, John: *Learning from Fire: A Fire Protection Primer for Architects*. Washington, D.C.: National Fire Prevention and Control Administration, U.S. Department of Commerce, PB 283 163, 1977.

Lin, T. Y. and Burns, Ned H.: *Design of Prestressed Concrete Structures*, 3rd ed. New York: John Wiley, 1981.

Lin, T. Y. and Stotesbury, Sydney D.: *Structural Concepts and Systems for Architects and Engineers*, 2nd ed. New York: Van Nostrand Reinhold Co., 1988.

Lüchinger, Arnulf: *Structuralism in Architecture and Urban Planning*. Stuttgart, FRG: K. Krämer Verlag, 1980.

Macaulay, David: *Underground*. Boston: Houghton Mifflin, 1976.

MacDonald, Angus J.: *Wind Loading on Buildings*. New York: John Wiley, 1975.

Mackintosh, Albyn: *The Application of Reinforced Concrete Masonry Load-Bearing Walls in Multi-Storied Structures*. McLean, VA: NCMA, 1973.

Macsai, John, Holland, Eugene P., Nachman, Harry S., Anderson, James R., Shlaes, Jared, and Hidvegi, Alfred J.: *Housing*, 2nd ed. New York: John Wiley, 1982.

Malhotra, H. L.: *Design of Fire-Resisting Structures*. Glasgow: Surrey U. Press, 1982.

Martin, Leslie and March, Lionel, eds.: *Urban Space and Structures*. London: Cambridge University Press, 1972.

Melaragno, Michele: *Wind in Architectural and Environmental Design*. New York: Van Nostrand Reinhold, 1982.

Messler, Norbert: *The Art Deco Skyscraper in New York*. New York: Peter Lang, 1986.

Meyer, John: Gophers go underground. *Civil Engineering/ASCE*, Oct. 1978.

Michaels, Leonard: *Contemporary Structure in Architecture*. New York: Reinhold Publ. Co., 1950.

Morris, A. E. J.: *Precast Concrete in Architecture*. London: The Whitney Library of Design, 1978.

National Concrete Masonry Association: An Information Series from NCMA. Herndon, VA: NCMA.

National Concrete Masonry Association: *Tall Buildings with Concrete Masonry Bearing Walls*. Herndon, VA: NCMA, 1978.

Nervi, Pier Luigi: *Buildings, Projects, Structures 1953–1963*. New York: Praeger, 1963.

Neville, Gerald B., ed.: *Simplified Design, Reinforced Concrete Buildings of Moderate Size and Height*. Skokie, IL: PCA, 1984.

Nilson, Arthur H.: *Design of Prestressed Concrete*. New York: John Wiley, 1978.

Norberg-Schulz, Christian: *Existence, Space and Architecture*. New York: Praeger, 1971.

O'Hare, Michael: Wind whistles through MIT tower. *P/A*, March 1967.

Oppenheim, Irving J.: *Structural Systems in Design*. Pittsburgh, PA: Carnegie-Mellon University, 1979.

Optimizing the structure of the skyscraper. *A.R.*, Oct. 1972.

Pelli, Cesar: Skyscrapers. *Perspecta* 18, 1982.

Petroski, Henry: *To Engineer Is Human: The Role of Failure in Successful Design*. New York: St. Martin's, 1985.

Popov, Egor P.: Seismic steel framing systems for tall buildings. *Engineering Journal*, AISC, 3rd Qtr., 1982.

Portland Cement Association: *Building Movement and Joints*. Skokie, IL: PCA, 1982.

Post-Tensioning Institute: *Post-Tensioning Manual*, 2nd ed. Glenview, IL: PTI, 1976.

Prestressed Concrete Institute: *PCI Design Handbook, Precast and Prestressed Concrete*, 3rd ed. Chicago, IL: PCI, 1985.

Prestressed Concrete Institute: *PCI Manual for Structural Design of Architectural Precast Concrete*. PCI Publ. MNL-121,77. Chicago, IL: PCI, 1977.

Prestressed Concrete Institute: *PCI Manual on Design of Connections for Precast Prestressed Concrete*. Chicago, IL: PCI, 1973.

Riccabona, Christof, and Wachberger, Michael: *Terrassenhäuser, e + p 14*. Munich, FRG: Verlag Georg D.W. Callwey, 1974.

Rush, Richard: Mixed framing systems. *Building Design and Construction*. March 1987.

Salvadori, Mario: *Why Buildings Stand Up*. New York: McGraw-Hill, 1982.

Saxon, Richard: *Atrium Buildings*. New York: Van Nostrand Reinhold, 1983.

Scalzi, John B.: The Staggered Truss System—Structural Considerations. *AISC Journal*, Oct. 1971.

Schlaich, Jörg, Schäfer, Kurt, and Jennewein, Mattias: Toward a consistent design of structural concrete. *PCI Journal*, May/June 1987.

Schneider, Robert R. and Dickey, Walter L.: *Reinforced Masonry Design*. Englewood Cliffs, NJ: Prentice-Hall, 1980.

Schriftenreihe des Bundesministers für Raumordnung, Bauwesen und Städtebau: *Aktivierung des Gebäudeinnern von terrassierten Bauten in der Ebene—Beispiel Stufendomino*. Bonn-Bad Godesberg, FRG: publication #03.033, 1975.

Schriftenreihe des Bundesministers für Raumordnung, Bauwesen und Städtebau: *Esslingen-Zollberg/Süd—Differenzierung und Typisierung an terrassierten Bauten*. Bonn-Bad Godesberg, FRG: publication #01.059, 1977.

Schueller, Wolfgang: *Horizontal-Span Building Structures*. New York: John Wiley, 1983.

Schueller, Wolfgang: *High-Rise Building Structures*. New York: John Wiley, 1977.

Schueller, Wolfgang: *Exercise Manual to The Vertical Building Structure*. Virginia Tech, Blacksburg, VA: College of Architecture and Urban Studies, 1990.

Schultz, Donald M.: *Design and Construction of Large Panel Concrete Structures*, Report 6, Design Methodology. Washington, D.C.: U.S. Department of Housing and Urban Development, 1979.

Steel Joist Institute: *Catalogue of Standard Specifications, Load Tables, and Weight Tables for Steel Joists and Joist Girders*. Myrtle Beach, SC: SJI, 1987.

Sieverts, Ernst: *Bürohaus und Verwaltungsbau*. Stuttgart, FRG: Verlag W. Kohlhammer, 1980.

Standard Building Code (SBCCI), 1985 Edition. Birmingham, AL: Southern Building Code Congress International, Inc.

Sting, Hellmuth: *Der Grundriss im Mebrgeschossigen Wohnungsbau*. Stuttgart, FRG: A. Koch Publ., 1969.

Structural Clay Products Institute: *Recommended Practice for Engineered Brick Masonry*. McLean, VA: SCPI, 1969.

Suspended Ceiling Systems: The down side of up. *P/A*, Sept. 1980.

Taranath, Bungale S.: *Structural Analysis and Design of Tall Buildings*. New York: McGraw-Hill, 1988.

Tarics, Alexander G.: Concrete-filled steel columns for multi-story construction. *Modern Steel Construction,* 1st Qtr., 1972.

Teal, Edward J.: Seismic design practice for steel buildings. *AISC Journal,* 4th Qtr., 1975.

Teng, Wayne C.: *Foundation Design*. Englewood Cliffs, N.J.: Prentice-Hall, 1962.

The American Institute of Architects: *Energy in Design*. AIA Catalog #N911 and #N912, Washington, D.C.: AIA, 1982.

Building Officials & Code Administrators International, Inc.: The BOCA Basic Building Code/1981, 8th ed. Homewood, IL: BOCA, 1981.

The quiet revolution in skyscraper design. *Civil Engineering/ASCE,* May 1983.

The sky's the limit. *ENR,* Nov. 3, 1983.

The staggered truss system. *USS Report,* May 1977.

The Structures Group Metropolitan Section—ASCE and The Council of Tall Buildings and Urban Habitat: *Structural Design of Tall Buildings*. New York: SGMA, 1988.

Thinking tall. *P/A,* Dec. 1980.

Thin sheets of air. *P/A,* June 1985.

Thornton, Charles H.: Avoiding wall problems by understanding structural movement. *A.R.,* Dec. 1981.

Timoshenko, Stephen P., and Gere, James M.: *Theory of Elastic Stability,* 2nd ed. New York: McGraw-Hill, 1961.

Tucker, Jonathan B.: Super-skyscrapers. *High Technology,* Jan. 1985.

Uniform Building Code (UBC), 1985 Edition, International Conference of Building Officials, Whittier, CA.

United States Steel: *Interstitial Systems in Healthcare Facilities*. A-DUSS 27-7601-01, Dec. 1979.

Ural, Oktay and Krapfenbauer, Robert, eds.: *Housing: The Impact of Economy and Technology*. New York: Pergamon Press, 1981.

White, Richard N. and Salmon, Charles G., eds.: *Building Structural Design Handbook*. New York: John Wiley, 1987.

Wiesner, Kenneth B.: Taming lively buildings. *Civil Engineering,* June 1986.

Wilson, Forrest: *Emerging Form in Architecture, Conversations with Lev Zetlin*. Boston: Cahners Books, 1975.

Wilson, Forrest: The perils of using thin stone and safeguards against them. *Architecture,* February 1989.

Wind Analysis: Preventive medicine for cladding, structural problems. *ENR,* March 27, 1980.

Wright, Frank Lloyd: *The Story of the Tower*. New York: Horizon Press, 1956.

Zuk, William: Kinetic structures. *Civil Engineering,* Dec. 1968.

INDEX